国外优秀数学著作
原版系列

正交多项式和 q-级数的前沿

Frontiers in Orthogonal Polynomials and q-Series

●［美］M. 祖海尔·纳什德（M. Zuhair Nashed）
●［美］李欣（Xin Li）著

（英文）

HITP
哈爾濱工業大學出版社
HARBIN INSTITUTE OF TECHNOLOGY PRESS

黑版贸审字 08－2020－192 号

图书在版编目(CIP)数据

正交多项式和 q－级数的前沿＝Frontiers in Orthogonal Polynomials and q－Series：英文／(美) M.祖海尔·纳什德(M. Zuhair Nashed),(美)李欣著.—哈尔滨:哈尔滨工业大学出版社,2022.9

ISBN 978－7－5767－0374－0

Ⅰ.①正… Ⅱ.①M… ②李… Ⅲ.①正交多项式－英文②级数－英文 Ⅳ.①O174.21②O173

中国版本图书馆 CIP 数据核字(2022)第 156099 号

ZHENGJIAO DUOXIANGSHI HE Q－JISHU DE QIANYAN

World Scientific

策划编辑　刘培杰　杜莹雪
责任编辑　刘家琳　李　烨
封面设计　孙茵艾
出版发行　哈尔滨工业大学出版社
社　　址　哈尔滨市南岗区复华四道街 10 号　邮编 150006
传　　真　0451－86414749
网　　址　http://hitpress.hit.edu.cn
印　　刷　哈尔滨市颉升高印刷有限公司
开　　本　787 mm×960 mm　1/16　印张 39.25　字数 540 千字
版　　次　2022 年 9 月第 1 版　2022 年 9 月第 1 次印刷
书　　号　ISBN 978－7－5767－0374－0
定　　价　98.00 元

Preface

The idea of preparing this volume was inspired by the invited lectures at the "International Conference on Orthogonal Polynomials and q-Series", held at the University of Central Florida in Orlando, on May 10–12, 2015. The conference was dedicated to Professor Mourad Ismail on his 70th birthday. It was strongly felt that it would be useful and timely to prepare an edited book highlighting important directions of research on orthogonal polynomials, q-series, and related topics in approximation theory, number theory, combinatorics, applied and computational harmonic analysis. Ms. Rochelle Kronzek, Executive Editor of World Scientific, who attended the conference, encouraged the idea, which ultimately led to a publishing agreement.

This volume contains 27 chapters exploring various topics on the frontier of research on orthogonal polynomials, q-series, and related topics mentioned above. We strived for a book that would inspire young researchers and provide pleasant readings for the other generations of researchers in these fields.

We would like to express our appreciation to all the authors who contributed to this book, and to all the referees who helped in the reviews. We like to thank Ms. Rochelle Kronzek and Ms. Lai Fun Kwong, of World Scientific, for their encouragement and support.

M. Zuhair Nashed
Xin Li

Orlando, FL, USA

About the Editors

M. Zuhair Nashed obtained his S.B. and S.M. degrees in Electrical Engineering from MIT and his Ph.D. in Mathematics from the University of Michigan. He has served for many years as a Professor at Georgia Tech and the University of Delaware and has held visiting professor positions at the University of Michigan, University of Wisconsin, AUB, and KFUPM. He also holds distinguished visiting scholar positions at many universities around the world. He is a fellow of the American Mathematical Society (Inaugural Class of 2013) and the recipient of awards such as the Lester Ford Award of the Mathematical Association of America, the Sigma Xi Faculty Research Award and Sustained Research Award in Science from Georgia Tech, Dr. Zakir Husain Award of the Indian Society of Industrial and Applied Mathematics, as well as several other international awards. He has published over 140 papers in mathematics and 30 papers in applied sciences, physics and engineering, has written 20 expository papers and book chapters, and authored and edited 12 books. He is Editor-in-Chief of *Numerical Functional Analysis and Optimization*, Executive Editor of *Sampling Theory in Signal and Image Processing*, founding and past coeditor of *Journal of Integral Equations*, and a member of the editorial board of 30 journals. He is also Editor-in-Chief of the Springer *Handbook of Geomathematics*, Editor-in-Chief of the Birkhäuser book series *Geosystems Mathematics* and of the Birkhäuser *Lecture Notes Geosystems Mathematics and Computing*, and Editor-in-Chief of *Handbook of Mathematical Geodesy.* He was invited to give a lecture at the American Mathematical Society and has also given three plenary lectures at meetings of the Mathematical Association of America and plenary lectures at meetings of the French, Tunisian and Lebanese Mathematical Societies, Indian Society of Industrial and Applied Mathematics, and Japan Society of Mechanical Engineers. He has been invited to give over 400 talks at conferences and colloquia and has organized over 30 conferences and mini-symposia.

Xin Li obtained his B.S. and M.S. degrees in Mathematics from Zhejiang University, China, in 1983 and 1986, respectively, under supervision of Professors Zhu Rui Guo and Zhen Sha, and his Ph.D. in Mathematics from University of South Florida, USA in 1989 under the supervision of Dr. Edward B. Saff. He joined the University of Central Florida as an assistant professor in 1990 and has been serving as professor since 2001. He has published over 70 papers in mathematics, statistics, computer vision, and education.

Contents

1. Mourad Ismail — 1
   Richard Askey

2. Binomial Andrews–Gordon–Bressoud Identities — 7
   Dennis Stanton

3. Symmetric Expansions of Very Well-Poised Basic Hypergeometric Series — 21
   George E. Andrews

4. A Sturm–Liouville Theory for Hahn Difference Operator — 35
   M. H. Annaby, A. E. Hamza and S. D. Makharesh

5. Solvability of the Hankel Determinant Problem for Real Sequences — 85
   Andrew Bakan and Christian Berg

6. Convolution and Product Theorems for the Special Affine Fourier Transform — 119
   Ayush Bhandari and Ahmed I. Zayed

7. A Further Look at Time-and-Band Limiting for Matrix Orthogonal Polynomials — 139
   M. Castro, F. A. Grünbaum, I. Pacharoni and I. Zurrián

8. The Orthogonality of Al-Salam–Carlitz Polynomials for Complex Parameters 155
Howard S. Cohl, Roberto S. Costas-Santos and Wenqing Xu

9. Crouching AGM, Hidden Modularity 169
Shaun Cooper, Jesús Guillera, Armin Straub and Wadim Zudilin

10. Asymptotics of Orthogonal Polynomials and the Painlevé Transcendents 189
Dan Dai

11. From the Gaussian Circle Problem to Multivariate Shannon Sampling 213
Willi Freeden and M. Zuhair Nashed

12. Weighted Partition Identities and Divisor Sums 239
F. G. Garvan

13. On the Ismail–Letessier–Askey Monotonicity Conjecture for Zeros of Ultraspherical Polynomials 251
Walter Gautschi

14. A Discrete Top-Down Markov Problem in Approximation Theory 267
Walter Gautschi

15. Supersymmetry of the Quantum Rotor 291
Vincent X. Genest, Luc Vinet, Guo-Fu Yu and Alexei Zhedanov

16. The Method of Brackets in Experimental Mathematics 307
Ivan Gonzalez, Karen Kohl, Lin Jiu and Victor H. Moll

17. Balanced Modular Parameterizations 319
Tim Huber, Danny Lara and Esteban Melendez

18. Some Smallest Parts Functions from Variations of Bailey's Lemma 343
Chris Jennings-Shaffer

19. Dual Addition Formulas Associated with Dual Product Formulas 373
Tom H. Koornwinder

20. Holonomic Tools for Basic Hypergeometric Functions 393
Christoph Koutschan and Peter Paule

21. A Direct Evaluation of an Integral of Ismail and Valent 417
Alexey Kuznetsov

22. Algebraic Generating Functions for Gegenbauer Polynomials 425
Robert S. Maier

23. q-Analogues of Two Product Formulas of Hypergeometric Functions by Bailey 445
Michael J. Schlosser

24. Summation Formulas for Noncommutative Hypergeometric Series 451
Michael J. Schlosser

25. Asymptotics of Generalized Hypergeometric Functions 497
Y. Lin and R. Wong

26. Mock Theta-Functions of the Third Order of Ramanujan in Terms of Appell–Lerch Series 523
Changgui Zhang

27. Certain Positive Semidefinite Matrices of Special Functions 545
Ruiming Zhang

Index 557

编辑手记 563

Chapter 1

Mourad Ismail

Richard Askey

Department of Mathematics
University of Wisconsin-Madison, USA
askey@math.wisc.edu

Mourad spent a year in Madison as a postdoc at the Mathematics Research Center. During that year, arrangements were made for George Andrews to visit the next year. As a result, I needed to start learning something about q-series, so I gave a number of very short talks on q-series before my afternoon lectures in a course of special functions. Mourad, Dennis Stanton and Jim Wilson heard the q-series talks and I hope that they would be a little more positive about these lectures than my assessment of them. They were terrible, and it is surprising that any of the listeners carried on and eventually learned a lot about these functions. To a large extent, we have George Andrews to thank for that. My first thanks to Mourad is for the work he did on the Pollaczek polynomials, starting with the AMS Memoir we wrote [1]. Legendre polynomials satisfy the three term recurrence relation

$$(2n+1)xP_n(x) = (n+1)P_{n+1}(x) + nP_{n-1}(x), P_0(x) = 1, P_1(x) = x. \tag{1.1}$$

There are many natural ways to modify this recurrence relation. One is to replace the first and last coefficients with $2(n+\lambda)$ and $n+2\lambda-1$ and require the first degree polynomial to be $2\lambda x$. This is the classical way, and the resulting polynomials are the ultraspherical polynomials. A second way is to replace $2n+1$ by $2n+1+a$. This was found by Felix Pollaczek. The changes in the polynomials and their orthogonality are strikingly different. The ultraspherical polynomials are orthogonal on $(-1,1)$ with respect to $(1-x^2)^{\lambda-1/2}$ and the polynomials on $[-1,1]$ grow at most as a power of n. For the Pollaczek polynomials, the weight function on $(-1,1)$ goes to zero much faster than a power of $(1-x^2)$, and the polynomials grow at $x^2=1$ faster than a power of n. There are further generalizations with infinitely many mass points outside of $[-1,1]$ which have 1 and/or -1 as limits. There is also a basic hypergeometric extension. If I had written this paper alone it would have been significantly shorter and contained fewer interesting examples, assuming it would have been finished. Mourad has gone on and written a number of other papers related to this one. There is a third extension of the Legendre polynomial recurrence relation, replace the n in the coefficients by $n+a$. This can be done for the recurrence relation for every set of orthogonal polynomials when the recurrence coefficients can be given explicitly as say a rational function of n, and Mourad has written papers on this type of extension. There are other variants for some of the more general polynomials and Mourad has worked on some of these also.

I want to mention a different type of problem where I owe Mourad a significant debt, but few would be able to guess why. The classical generating function for Legendre polynomials can be written as

$$\sum P_n(\cos\theta)r^n = |(1-re^{i\theta})^{-1/2}|^2, \tag{1.2}$$

where the sum goes from 0 to infinity and $-1<r<1$. In the 1930s, Fejer and Szego considered a more general generating function, with $(1-z)^{-1/2}$ replaced by an analytic function $f(z)$, real for real z, and various conditions on successive differences of the power series coefficients of $f(z)$. They were interested in cases where all the zeros of $p_n(x)$ are real and lie in $-1<x<1$ when $x=\cos(\theta)$ and

$$|f(re^{i\theta})|^2 = \sum p_n(\cos\theta)r^n. \tag{1.3}$$

A little later, two mathematicians asked when these polynomial coefficients were orthogonal. Independently, Feldheim and Lanzewizky showed that this only happens when

$$f(z) = \prod(1-zq^n)/(1-azq^n),$$

where the product goes from $n = 0$ to infinity and $|q| < 1$. This generating function had been found by L. J. Rogers in the middle 1890s, but he was not aware that the polynomials were orthogonal. That is almost the full story, but a related result was obtained by William Allaway in his Ph.D. thesis under Waleed Al-Salam, who was also Mourad Ismail's major professor. Allaway had a different characterization problem, which he solved, and he did two things which had not been done by Feldheim and Lanzewizky. In one important special case, he found the orthogonality explicitly. In the general case, he noticed that there were cases when to move to the next value of n in the recurrence relation, it was necessary to determine what happens in a limiting case where the formal limit is $0/0$. He worked out what happens, and it turns out to be two new limiting cases of the continuous q-ultraspherical polynomials. One of these polynomials has the following recurrence relation:

$$2x(m+\lambda)c^{\lambda}_{mk}(x;k) = (m+2\lambda)c^{\lambda}_{mk+1}(x;k) + mc^{\lambda}_{mk-1}(x;k),$$

and

$$2xc^{\lambda}_{n}(x;k) = c^{\lambda}_{n+1}(x;k) + c\lambda_{n-1}(x;k)$$

when n is not a multiple of k. The weight function for these polynomials is

$$(1-x^2)^{\lambda-1/2}|U_{k-1}(x)|^{2\lambda}$$

when $\lambda > -1/2$, where $U_{k-1}(\cos\theta) = \sin(k\theta)/\sin(\theta)$ are the Chebyshev polynomials of the second kind. Allaway did not have the orthogonality for these polynomials. If he had, I would not be able to mention an error I made. The thesis was written in 1972 and Al-Salam sent me a copy. At that time, I knew nothing about q-series. The recurrence relation above was in Allaway's thesis, and I did not understand it. What I should have done is look very carefully at this set of polynomials and see if I could make sense of them. What I did was put the thesis on a book shelf where many other theses were stored and then forget about it. A few years later, I learned enough about q-series and some of the associated orthogonal polynomials, so I could have figured out what was behind Allaway's new polynomials. Fortunately, a few years later I was invited to give a colloquium talk in Edmonton, where Al-Salam was. I talked on the continuous q-ultraspherical polynomials of Rogers and mentioned the work of Feldheim and Lanzewizky. After the talk, Al-Salam mentioned that the work I discussed seemed similar to work done by Allaway in his thesis. He brought in a copy the next day, and after looking at it, I said that the strange polynomials must come from the polynomials of Rogers by letting q tend to a root of unity. They do, and that led to a couple of joint papers with Al-Salam and Allaway. The interest in these polynomials is that they provide a clue to a relationship between recurrence relation coefficients and weight function zeros interior to the support of the orthogonality measure. The only example before these polynomials was much earlier work of Szego on polynomials orthogonal with respect to $|x|^a(1-x^2)^b$ which was too special to generalize further without having q-cases with an absolutely continuous orthogonality relation. Finally, why did I mention this work as something I want to thank Mourad for? There has been work on basic hypergeometric series (q-series) for more than 200 years. I was afraid that showing that there were results when q is a root of unity might lead to a large number of papers pointing out what happens when q tends to a root of unity. I did not need more papers like that to referee, nor did I want to see the literature cluttered with papers of this type. What Mourad did was write a number of first rate papers about other cases when q tends to a root of unity, so if I got papers to referee which did routine work by letting q tend to a root of unity, all I had to do is refer an author to one or two of Mourad's papers and write that this is the level of work which should be published. Here is a reference to one of these papers, "On sieved orthogonal polynomials, V", a joint paper with Charris [2]. The feared deluge never started.

There are a few of Mourad's other papers I want to mention. His gem of a proof [3] of the summation of Ramanujans ${}_1\psi_1$ is a proof from "The Book", and I do not know how to improve on it. The idea behind it is classical, but no one thought of using it here. It has been used to prove other harder results from easier ones. Another proof which Mourad and Dennis Stanton gave is the use of results of Rogers to give a simple evaluation [4] of the q-beta integral which is the weight function of the polynomials Jim Wilson and I found. When Mourad told me about this in a phone call, my immediate response was: "I hate you". This is a proof I should have found. There is a related paper by Mourad, Dennis and Gerard Viennot titled "The combinatorics of q-Hermite polynomials and the Askey–Wilson integral" [5]. This is a paper I never could have written, and it still amazes me.

The last of the papers mentioned above was published almost 30 years ago. There are many newer results, including a long book on specific sets of orthogonal polynomials. The quality of the more recent work has held up very well, and there are many papers I cannot summarize due to a lack of knowledge. One hopes that graduate students and postdocs will write papers you have trouble reading. Let me thank Mourad for doing this.

References

[1] R. Askey and M. Ismail, *Recurrence Relations, Continued Fractions, and Orthogonal Polynomials*, Memoirs of American Mathematical Society, Vol. **49**(300) (American Mathemathical Society, Providence, RI, 1984), iv + 108 pp.

[2] J. A. Charris and M. E. H. Ismail, Sieved orthogonal polynomials, V: Sieved Pollaczek polynomials, *SIAM J. Math. Anal.* **18** (1987) 1177–1218.

[3] M. E. H. Ismail, A simple proof of Ramanujans ${}_1\psi_1$ sum, *Proc. Amer. Math. Soc.* **63** (1977) 185–186.

[4] M. E. H. Ismail and D. Stanton, On the Askey–Wilson and Rogers polynomials, *Canadian J. Math.* **40** (1988) 1025–1045.

[5] M. E. H. Ismail, D. Stanton and G. Viennot, The combinatorics of q-Hermite polynomials and the Askey–Wilson integral, *European J. Combin.* **8** (1987) 379–392.

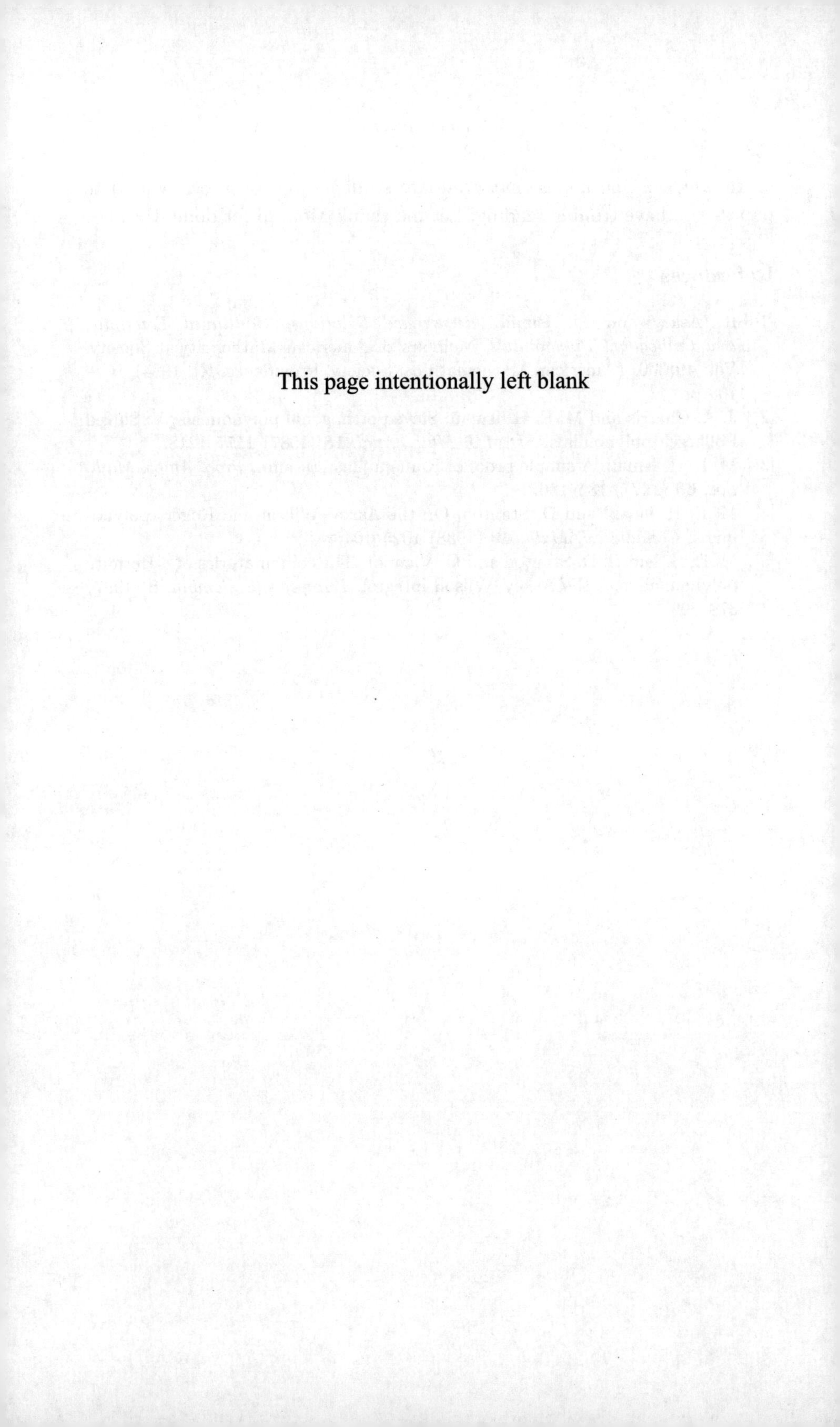

This page intentionally left blank

Chapter 2

Binomial Andrews–Gordon–Bressoud Identities

Dennis Stanton

School of Mathematics, University of Minnesota
Minneapolis, MN 55455, USA
stanton@math.umn.edu

Binomial versions of the Andrews–Gordon–Bressoud identities are given.

Keywords: Rogers–Ramanujan; integer partition.

Mathematics Subject Classification 2010: 05A17, 11P84

1. Introduction

The Rogers–Ramanujan identities

$$\sum_{s=0}^{\infty} \frac{q^{s^2}}{(q;q)_s} = \frac{1}{(q,q^4;q^5)_\infty}, \quad \sum_{s=0}^{\infty} \frac{q^{s^2+s}}{(q;q)_s} = \frac{1}{(q^2,q^3;q^5)_\infty},$$

where

$$(A;q)_n = \prod_{i=0}^{n-1}(1-Aq^i), \quad (A,B;q)_n = (A;q)_n(B;q)_n,$$

were generalized to odd moduli at least five by Andrews [1]. These identities are called the Andrews–Gordon identities

$$\sum_{s_1\geq s_2\geq\cdots\geq s_k\geq 0} \frac{q^{s_1^2+\cdots+s_k^2+s_{k-r+1}+\cdots+s_k}}{(q)_{s_1-s_2}\cdots(q)_{s_{k-1}-s_k}(q)_{s_k}}$$

$$= \frac{(q^{k+1-r},q^{k+2+r},q^{2k+3};q^{2k+3})_\infty}{(q;q)_\infty}, \quad 0\leq r\leq k. \tag{1.1}$$

(If the base q is understood, we sometimes abbreviate $(A;q)_n$ as $(A)_n$. We also assume that $0 < q < 1$.)

Bressoud [7, 9] gave a version of these identities for even moduli

$$\sum_{s_1\geq s_2\geq\cdots\geq s_k\geq 0}\frac{q^{s_1^2+\cdots+s_k^2+s_{k-r+1}+\cdots+s_k}}{(q)_{s_1-s_2}\cdots(q)_{s_{k-1}-s_k}(q^2;q^2)_{s_k}}$$
$$=\frac{(q^{k+1-r},q^{k+1+r},q^{2k+2};q^{2k+2})_\infty}{(q;q)_\infty},\quad 0\leq r\leq k. \tag{1.2}$$

Bressoud's beautiful and efficient proof [8] established both sets of identities when $r = 0$. Moreover, he had other closely related identities, e.g., [9, (3.3), p. 15]

$$\sum_{s_1\geq s_2\geq\cdots\geq s_k\geq 0}\frac{q^{s_1^2+\cdots+s_k^2-(s_1+\cdots+s_j)}}{(q)_{s_1-s_2}\cdots(q)_{s_{k-1}-s_k}(q)_{s_k}}$$
$$=\sum_{s=0}^{j}\frac{(q^{k+1+j-2s},q^{k+2-j+2s},q^{2k+3};q^{2k+3})_\infty}{(q;q)_\infty},\ 0\leq j\leq k. \tag{1.3}$$

The purpose of this paper is to examine Bressoud's proof, and develop new variations and generalizations of these Andrews–Gordon–Bressoud identities. The new results are given in Sections 3–5.

2. The Motivating Question

In [4], which was presented by George Andrews at the *May 2015 UCF Meeting in Honor of Mourad Ismail*, Andrews reconsidered Bressoud's elementary proof [8]. He asked a specific question (see Question 2.6) about Bressoud's proof that we answer in this section.

We shall need a few of the relevant definitions and facts in a recapitulation Bressoud's simple proof [8]. We shall also use these facts in later sections. Bressoud's key idea was to use the following Laurent polynomials, which have arbitrary quadratic exponents.

Definition 2.1. Let

$$H_{2n}(z,a|q)=\sum_{s=-n}^{n}\begin{bmatrix}2n\\ n-s\end{bmatrix}_q q^{as^2}z^s.$$

Bressoud's main lemma [8, Lemma 2], which allowed the quadratic exponent to change, is next.

Lemma 2.2.

$$\frac{H_{2n}(z,a|q)}{(q;q)_{2n}} = \sum_{s=0}^{n} \frac{q^{s^2}}{(q;q)_{n-s}} \frac{H_{2s}(z,a-1|q)}{(q;q)_{2s}}.$$

This lemma may be iterated.

Proposition 2.3.

$$\frac{H_{2n}(z,a+k+1|q)}{(q;q)_{2n}} = \sum_{n \geq s_1 \geq s_2 \geq \cdots \geq s_{k+1} \geq 0} \frac{q^{s_1^2+\cdots+s_{k+1}^2} H_{2s_{k+1}}(z,a|q)}{(q)_{n-s_1}(q)_{s_1-s_2}\cdots(q)_{s_k-s_{k+1}}(q)_{2s_{k+1}}}.$$

The value of $a = 1/2$ is nice because the polynomials $H_{2n}(z,1/2|q)$ factor by the q-binomial theorem. Moreover

$$H_{2n}(-zq^{1/2},1/2|q) = (qz,1/z;q)_n. \tag{2.1}$$

Theorem 2.4 (Bressoud [8, (14)]).

$$\frac{H_{2n}(-zq^{1/2},k+3/2|q)}{(q;q)_{2n}} = \sum_{n \geq s_1 \geq s_2 \geq \cdots \geq s_{k+1} \geq 0} \frac{q^{s_1^2+\cdots+s_{k+1}^2}(qz,1/z;q)_{s_{k+1}}}{(q)_{n-s_1}(q)_{s_1-s_2}\cdots(q)_{s_k-s_{k+1}}(q)_{2s_{k+1}}}. \tag{2.2}$$

We now take the limit $n \to \infty$ of Theorem 2.4. The right side has a clear limit. For the left side we show that Definition 2.1 has a limit as an infinite product when $n \to \infty$.

If $0 < q < 1$ and $-n \leq s \leq n$, we have

$$\begin{bmatrix} 2n \\ n-s \end{bmatrix}_q \leq \frac{1}{(q;q)_\infty},$$

because the q-binomial coefficient $\begin{bmatrix} 2n \\ n-s \end{bmatrix}_q$ is the generating function for partitions inside an $(n-s)\times(n+s)$ rectangle, and right side is the generating function for all partitions. For any fixed integer s, this also shows that

$$\lim_{n\to\infty} \begin{bmatrix} 2n \\ n-s \end{bmatrix}_q = \frac{1}{(q;q)_\infty}.$$

So the limit $n \to \infty$ in Definition 2.1 converges uniformly,

$$\lim_{n\to\infty} H_{2n}(-z,a|q) = \frac{1}{(q;q)_\infty} \lim_{n\to\infty} \sum_{s=-n}^{n} q^{as^2}(-z)^s$$
$$= \frac{(q^{2a}, zq^a, q^a/z; q^{2a})_\infty}{(q)_\infty}, \tag{2.3}$$

using the Jacobi triple product identity.

Corollary 2.5. *For a nonnegative integer k,*

$$\frac{(q^{2k+3}, zq^{k+2}, q^{k+1}/z; q^{2k+3})_\infty}{(q;q)_\infty}$$
$$= \sum_{s_1\geq s_2\geq\cdots\geq s_{k+1}\geq 0} \frac{q^{s_1^2+\cdots+s_{k+1}^2}(qz,1/z;q)_{s_{k+1}}}{(q)_{s_1-s_2}\cdots(q)_{s_k-s_{k+1}}(q)_{2s_{k+1}}}.$$

Note that Corollary 2.5 immediately gives the Andrews–Gordon identities (1.1) for $r=0$. If $z=1$, this choice of z forces $s_{k+1}=0$. The choice of $z=q^r$ does give the right side of the Andrews–Gordon identities (1.1), but not the left side. There is an extra sum over s_{k+1}, and the power of q does not match.

Andrews' Question 2.6. Is there a simple way to understand why the choice of $z=q^r$ eliminates the s_{k+1} sum and replaces it with a power of q?

We now answer Andrews' question, and we will use this answer in subsequent sections. The ingredient we need appeared in a paper of Garrett, Ismail and Stanton [10].

Proposition 2.7. *For any c,*

$$H_{2n}(-q^c, c|q) = q^n H_{2n}(-q^{c-1}, c|q).$$

To answer Andrews' question, start with $H_{2n}(-q^{r+1/2}, k+3/2|q)$, apply Lemma 2.2 $k-r+1$ times to obtain $H_{2s_{k-r+1}}(-q^{r+1/2}, r+1/2|q)$. Next apply Proposition 2.7 once to obtain $q^{s_{k-r+1}} H_{2s_{k-r+1}}(-q^{r-1/2}, r+1/2|q)$. This is the linear exponent in q we need. The remaining exponents arise from again applying Lemma 2.2 followed by Proposition 2.7. The final sum on s_{k+1} is now eliminated, because the final term becomes $H_{2s_{k+1}}(-q^{1/2}, 1/2|q) = (q,1;q)_{s_{k+1}}$, which forces $s_{k+1}=0$.

3. New Andrews–Gordon Identities

In this section we prove two new Andrews–Gordon identities for odd moduli. The first has binomial factors.

Theorem 3.1. *For $0 \le j, r \le k$, and $j + r \le k$,*

$$\sum_{s_1 \ge s_2 \ge \cdots \ge s_k \ge 0} \frac{q^{-s_1-\cdots-s_j}(1+q^{s_1+s_2})(1+q^{s_2+s_3})\cdots(1+q^{s_{j-1}+s_j})}{(q)_{s_1-s_2}\cdots(q)_{s_{k-1}-s_k}(q)_{s_k}}$$
$$\times q^{s_1^2+\cdots+s_k^2+s_{k-r+1}+\cdots+s_k}$$
$$= \sum_{s=0}^{j} \binom{j}{s} \frac{(q^{k+1-r+j-2s}, q^{k+2+r-j+2s}, q^{2k+3}; q^{2k+3})_\infty}{(q;q)_\infty}.$$

Moreover, the j factors of q^{-s_1} and $q^{-s_i}(1+q^{s_{i-1}+s_i})$, $2 \le i \le j$, may be replaced by any j-element subset of $\{q^{-s_1}\} \cup \{q^{-s_i}(1+q^{s_{i-1}+s_i}) : 2 \le i \le k-r\}$.

For example, the binomial factors could occur as the last j of the first $k-r$ summation indices instead of the first j indices, namely

$$\prod_{t=0}^{j-1} q^{-s_{k-r-t}}(1+q^{s_{k-r-1-t}+s_{k-r-t}}).$$

A corollary of Theorem 3.1 is an identity which contains the Andrews–Gordon identities (1.1) when $j=0$ and Bressoud's identities (1.3) when $r=0$.

Theorem 3.2. *For $0 \le j, r \le k$, and $j + r \le k$,*

$$\sum_{s_1 \ge s_2 \ge \cdots \ge s_k \ge 0} \frac{q^{s_1^2+\cdots+s_k^2-(s_1+\cdots+s_j)+(s_{k-r+1}+\cdots+s_k)}}{(q)_{s_1-s_2}\cdots(q)_{s_{k-1}-s_k}(q)_{s_k}}$$
$$= \sum_{s=0}^{j} \frac{(q^{k+1-r+j-2s}, q^{k+2+r-j+2s}, q^{2k+3}; q^{2k+3})_\infty}{(q;q)_\infty}.$$

We need a new fact about the Laurent polynomials $H_{2n}(z, a|q)$.

Proposition 3.3. *For a nonnegative integer n,*

$$\frac{H_{2n}(zq, a+1|q) + H_{2n}(q/z, a+1|q)}{(q;q)_{2n}} = \sum_{s=0}^{n} \frac{q^{s^2-s}(1+q^{n+s})}{(q;q)_{n-s}} \frac{H_{2s}(z,a|q)}{(q;q)_{2s}}.$$

Proof. Note that the left side of Proposition 3.3 is invariant under $z \to 1/z$ so it does have an expansion in terms of $H_{2s}(z, a|q)$. If the coefficient of $z^k q^{ak^2}$ is computed for each side, we must show

$$\begin{bmatrix} 2n \\ n-k \end{bmatrix}_q q^k q^{k^2} + \begin{bmatrix} 2n \\ n-k \end{bmatrix}_q q^{-k} q^{k^2} = \sum_{s=k}^{n} \frac{q^{s^2-s}(1+q^{n+s})}{(q)_{2s}} \frac{(q)_{2n}}{(q)_{n-s}} \begin{bmatrix} 2s \\ s-k \end{bmatrix}_q .$$

The s-sum for the term q^{n+s} is summable as a product by a limiting case of the q-Vandermonde sum, see [11, p. 354, (II.6)]. The s-sum for the term 1 is nearly summable, it is a sum of two products. Putting these terms together yields the two terms on the left side. The details are not given. □

We need some functions which generalize the $H_{2n}(z, a|q)$.

Definition 3.4. For a nonnegative integer j, let

$$F_n^{(0)}(z,a) = H_{2n}(z,a|q), \quad F_n^{(j+1)}(z,a) = F_n^{(j)}(zq,a) + F_n^{(j)}(q/z,a), \quad j \geq 0.$$

Proposition 3.3 can be rewritten using these new functions. The proof is by induction on j.

Proposition 3.5. *For nonnegative integers n and j,*

$$\frac{F_n^{(j+1)}(z,a+1)}{(q;q)_{2n}} = \sum_{s=0}^{n} \frac{q^{s^2-s}(1+q^{n+s})}{(q;q)_{n-s}} \frac{F_s^{(j)}(z,a)}{(q;q)_{2s}}.$$

Iterating Proposition 3.5 is the next proposition.

Proposition 3.6. *For a nonnegative integer n and a positive integer j,*

$$\frac{F_n^{(j)}(z,a)}{(q)_{2n}} = \sum_{n \geq s_1 \geq s_2 \geq \cdots \geq s_j \geq 0} \frac{q^{s_1^2-s_1}(1+q^{n+s_1})}{(q)_{n-s_1}} \times \prod_{t=2}^{j} \frac{q^{s_t^2-s_t}(1+q^{s_{t-1}+s_t})}{(q)_{s_{t-1}-s_t}} \frac{F_{2s_j}^{(0)}(z,a-j)}{(q)_{2s_j}}.$$

Finally the functions F also satisfy Lemma 2.2 because the H functions do.

Proposition 3.7. *For a nonnegative integer n and a positive integer j,*

$$\frac{F_n^{(j)}(z,a)}{(q;q)_{2n}} = \sum_{s=0}^{n} \frac{q^{s^2}}{(q;q)_{n-s}} \frac{F_s^{(j)}(z,a-1|q)}{(q;q)_{2s}}.$$

Any of these functions may be written as a linear combination of $F_n^{(0)}(z,a) = H_{2n}(z,a|q)$.

Proposition 3.8. *For any nonnegative integer j,*

$$F_n^{(j+1)}(z,a) = \sum_{s=0}^{j} \binom{j}{s} (F_n^{(0)}(zq^{j+1-2s},a) + F_n^{(0)}(q^{j+1-2s}/z,a)).$$

Proof. By induction on j we have

$$\begin{aligned}
F_n^{(j+1)}(z,a) &= \sum_{s=0}^{j-1} \binom{j-1}{s} (F_n^{(0)}(zq^{j+1-2s},a) + F_n^{(0)}(q^{j-1-2s}/z,a)) \\
&\quad + \sum_{s=0}^{j-1} \binom{j-1}{s} (F_n^{(0)}(q^{j+1-2s}/z,a) + F_n^{(0)}(zq^{j-1-2s},a)) \\
&= \sum_{s=0}^{j} F_n^{(0)}(zq^{j+1-2s},a) \left(\binom{j-1}{s} + \binom{j-1}{s-1} \right) \\
&\quad + \sum_{s=0}^{j} F_n^{(0)}(q^{j+1-2s}/z,a) \left(\binom{j-1}{s-1} + \binom{j-1}{s} \right) \\
&= \sum_{s=0}^{j} \binom{j}{s} (F_n^{(0)}(zq^{j+1-2s},a) + F_n^{(0)}(q^{j+1-2s}/z,a)). \qquad \square
\end{aligned}$$

We have two expressions for $F_n^{(j)}(z,a)$: Propositions 3.8 and 3.6. The proof of Theorem 3.1 uses these two expressions after taking a limit as $n \to \infty$. We record the appropriate $n \to \infty$ limit of Proposition 3.8.

Proposition 3.9. *If j is a nonnegative integer,*

$$\begin{aligned}
\lim_{n\to\infty} F_n^{(j+1)}(-z,a) &= \frac{1}{(q)_\infty} \sum_{s=0}^{j+1} \binom{j+1}{s} \\
&\quad \times (q^{2a}, zq^{a+j+1-2s}, q^{a-j-1+2s}/z; q^{2a})_\infty.
\end{aligned}$$

Proof. Applying (2.3) and Proposition 3.8 we have

$$\begin{aligned}
\lim_{n\to\infty} F_n^{(j+1)}(-z,a) &= \frac{1}{(q)_\infty} \sum_{s=0}^{j} \binom{j}{s} ((q^{2a}, zq^{a+j+1-2s}, q^{a-j-1+2s}/z; q^{2a})_\infty \\
&\quad + (q^{2a}, q^{a+j+1-2s}/z, zq^{a-j-1+2s}; q^{2a})_\infty)
\end{aligned}$$

$$= \frac{1}{(q)_\infty} \sum_{s=0}^{j+1} \left(\binom{j}{s} + \binom{j}{j+1-s} \right)$$
$$\times (q^{2a}, zq^{a+j+1-2s}, q^{a-j-1+2s}/z; q^{2a})_\infty$$
$$= \frac{1}{(q)_\infty} \sum_{s=0}^{j+1} \binom{j+1}{s}$$
$$\times (q^{2a}, zq^{a+j+1-2s}, q^{a-j-1+2s}/z; q^{2a})_\infty. \qquad \square$$

Proof of Theorem 3.1. We see from Proposition 3.9 that the right side of Theorem 3.1 is

$$\lim_{n\to\infty} F_n^{(j)}(-z, k+3/2), \quad z = q^{-1/2-r}.$$

or at $z = q^{r+1/2}$ since all functions are symmetric under $z \to 1/z$. We apply Proposition 3.7 to obtain j sums and a factor of

$$\frac{F_{2s_j}^{(0)}(-z, k+3/2-j)}{(q)_{2s_j}} = \frac{H_{2s_j}(-z, k+3/2-j|q)}{(q)_{2s_j}}.$$

Now we are in the realm of the Andrews–Gordon proof in Section 2. We finish the proof as before, by applying Lemma 2.2 $k-r$ times, and then inserting the linear factors r times.

Since the functions $F_n^{(j)}(z,a)$ also satisfy Proposition 3.7, we could apply Proposition 3.7 anytime before we use Proposition 3.5 in the first $k-r$ iterates. This gives the arbitrary choice of the binomials. $\square$

Next we derive Theorem 3.2 from Theorem 3.1. The idea is take an appropriate linear combination of Theorem 3.1 to replace the binomial factors in Theorem 3.1 by a single term $q^{-s_1-s_2-\cdots-s_j}$. For example if $j=3$,

$$q^{-s_1-s_2-s_3}(1+q^{s_1+s_2})(1+q^{s_2+s_3}) - q^{-s_3}(1+q^{s_2+s_3}) - q^{-s_1}$$
$$= q^{-s_1-s_2-s_3} \tag{3.1}$$

yields, for the right side of Theorem 3.1,

$$\sum_{s=0}^{3} \binom{3}{s} \frac{(q^{k+1-r+3-2s}, q^{k+2+r-3+2s}, q^{2k+3}; q^{2k+3})_\infty}{(q;q)_\infty}$$
$$-2\sum_{s=0}^{1} \binom{1}{s} \frac{(q^{k+1-r+1-2s}, q^{k+2+r-1+2s}, q^{2k+3}; q^{2k+3})_\infty}{(q;q)_\infty}$$

$$= \sum_{s=0}^{3} \frac{(q^{k+1-r+1-2s}, q^{k+2+r-1+2s}, q^{2k+3}; q^{2k+3})_\infty}{(q;q)_\infty}$$

as predicted by Theorem 3.2.

The version of (3.1) we need for general j uses edges in a graph which is a path from 1 to j: $1-2-3-\cdots-j$. A pair of edges in this graph do not overlap if they do not share a vertex. For a set E of nonoverlapping edges let

$$wt(E) = \prod_{i\notin E, i\geq 2} q^{-s_i}(1+q^{s_{i-1}+s_i}) \times \begin{cases} q^{-s_1} & \text{if } 1\notin E, \\ 1 & \text{if } 1\in E. \end{cases}$$

Here are the three possible sets of nonoverlapping edges E for $j=3$,

$$\begin{aligned} &E=\emptyset, \quad wt(E) = q^{-s_1-s_2-s_3}(1+q^{s_1+s_2})(1+q^{s_2+s_3}), \\ &E = 1-2, \quad wt(E) = q^{-s_3}(1+q^{s_2+s_3}), \\ &E = 2-3, \quad wt(E) = q^{-s_1}. \end{aligned}$$

These are the three terms in (3.1).

Lemma 3.10. *We have*

$$\sum_E (-1)^{|E|} wt(E) = q^{-s_1-s_2-\cdots-s_j},$$

where the sum is over all nonoverlapping edge sets E of $1-2-3-\cdots-j$.

Proof. Again we do an induction on j. Suppose E is a set of nonoverlapping edges for $1-2-3-\cdots-(j+1)$. If the last edge $j-(j+1)$ is in E, the remaining edges are nonoverlapping for $j-1$, so by induction

$$\sum_{E, j-(j+1)\in E} (-1)^{|E|} wt(E) = -q^{-s_1-s_2-\cdots-s_{j-1}}.$$

If the last edge $j-(j+1)$ is not in E, the remaining edges are nonoverlapping for j, so by induction

$$\sum_{E, j-(j+1)\notin E} (-1)^{|E|} wt(E) = q^{-s_{j+1}}(1+q^{s_j+s_{j+1}})q^{-s_1-s_2-\cdots-s_j}.$$

Because

$$\begin{aligned} &-q^{-s_1-s_2-\cdots-s_{j-1}} + q^{-s_{j+1}}(1+q^{s_j+s_{j+1}})q^{-s_1-s_2-\cdots-s_j} \\ &= q^{-s_1-s_2-\cdots-s_{j+1}} \end{aligned}$$

we are done. □

Proof of Theorem 3.2. It remains to show that the linear combination given by Lemma 3.10 gives the correct constants for the infinite products on the right side of Theorem 3.2 (namely 1).

There are $\binom{j-t}{t}$ such nonoverlapping E with t edges, where $2t \leq j$, and therefore $j-2t$ vertices not in E. So the right side becomes

$$\sum_{t=0}^{[j/2]} \binom{j-t}{t} (-1)^t \sum_{s=0}^{j-2t} \binom{j-2t}{s} \frac{\begin{matrix}(q^{k+1-r+j-2t-2s}, \\ q^{k+2+r-j+2t+2s}, q^{2k+3}; q^{2k+3})_\infty\end{matrix}}{(q;q)_\infty}.$$

The coefficient of

$$\frac{(q^{k+1-r+j-2u}, q^{k+2+r-j+2u}, q^{2k+3}; q^{2k+3})_\infty}{(q;q)_\infty}$$

for $0 \leq u \leq j$ is

$$\sum_{s,t,s+t=u} \binom{j-t}{t} (-1)^t \binom{j-2t}{s} = 1$$

by the Chu–Vandermonde evaluation, namely

$$\binom{j}{u} {}_2F_1\left(\begin{matrix} -u, u-j \\ -j \end{matrix} \middle| 1 \right) = 1$$

for $0 \leq u \leq j$. □

4. New Bressoud–Type Identities for Even Moduli

The Bressoud identities for even moduli (1.2) can be proven in the same way. The only change is to replace (2.1) by

$$\frac{H_{2s}(-1,1|q)}{(q)_{2s}} = \frac{1}{(q^2;q^2)_s}.$$

Here we state (without proof) the analogous binomial results for even moduli.

Theorem 4.1. *For $0 \leq j, r \leq k$, and $j + r \leq k$,*

$$\sum_{s_1 \geq s_2 \geq \cdots \geq s_k \geq 0} \frac{q^{-s_1-\cdots-s_j}(1+q^{s_1+s_2})(1+q^{s_2+s_3})\cdots(1+q^{s_{j-1}+s_j})}{(q)_{s_1-s_2}\cdots(q)_{s_{k-1}-s_k}(q^2;q^2)_{s_k}}$$

$$\times q^{s_1^2+\cdots+s_k^2+s_{k-r+1}+\cdots+s_k}$$

$$= \sum_{s=0}^{j} \binom{j}{s} \frac{(q^{k+1-r+j-2s}, q^{k+1+r-j+2s}, q^{2k+2}; q^{2k+2})_\infty}{(q;q)_\infty}.$$

Moreover, the j factors of q^{-s_1} and $q^{-s_i}(1+q^{s_{i-1}+s_i})$, $2 \le i \le j$, may be replaced by any j-element subset of $\{q^{-s_1}\} \cup \{q^{-s_i}(1+q^{s_{i-1}+s_i}) : 2 \le i \le k-r\}$.

Again using Lemma 3.10 we have the version without binomial coefficients.

Theorem 4.2. *For $0 \le j, r \le k$, and $j + r \le k$,*

$$\sum_{s_1 \ge s_2 \ge \cdots \ge s_k \ge 0} \frac{q^{s_1^2+\cdots+s_k^2-(s_1+\cdots+s_j)+(s_{k-r+1}+\cdots+s_k)}}{(q)_{s_1-s_2}\cdots(q)_{s_{k-1}-s_k}(q^2;q^2)_{s_k}}$$
$$= \sum_{s=0}^{j} \frac{(q^{k+1-r+j-2s}, q^{k+1+r-j+2s}, q^{2k+2}; q^{2k+2})_\infty}{(q;q)_\infty}.$$

The case $j = 0$ in Theorem 4.2 is given by Bressoud [9, (3.4), p. 15] while $r = 0$ is [9, (3.5), p. 16].

5. Overpartitions

Finally, for completeness, we give two analogous results for overpartitions, see [3]. Proposition 2.7, which is used to insert linear exponents, requires a special choice of z. But in the two results of this section we have a general z, so we cannot insert the linear factors as before.

The first result is a binomial version of Corollary 2.5.

Theorem 5.1. *For $0 \le j \le k+1$,*

$$\sum_{s_1 \ge s_2 \ge \cdots \ge s_{k+1} \ge 0} \frac{q^{-s_1-\cdots-s_j}(1+q^{s_1+s_2})(1+q^{s_2+s_3})\cdots(1+q^{s_{j-1}+s_j})}{(q)_{s_1-s_2}\cdots(q)_{s_k-s_{k+1}}}$$
$$\times \frac{(-z,-q/z;q)_{s_{k+1}}}{(q)_{2s_{k+1}}} q^{s_1^2+\cdots+s_{k+1}^2}$$
$$= \sum_{s=0}^{j} \binom{j}{s} \frac{(-zq^{k+1+j-2s}, -q^{k+2-j+2s}/z, q^{2k+3}; q^{2k+3})_\infty}{(q;q)_\infty}.$$

Moreover, the j factors of q^{-s_1} and $q^{-s_i}(1+q^{s_{i-1}+s_i})$, $2 \le i \le j$, may be replaced by any j-element subset of $\{q^{-s_1}\} \cup \{q^{-s_i}(1+q^{s_{i-1}+s_i}) : 2 \le i \le k+1\}$.

Theorem 5.2. *For* $0 \leq j \leq k+1$,

$$\sum_{s_1 \geq s_2 \geq \cdots \geq s_{k+1} \geq 0} \frac{q^{s_1^2 - s_1 + \cdots + s_j^2 - s_j + s_{j+1}^2 + \cdots + s_{k+1}^2}}{(q)_{s_1 - s_2} \cdots (q)_{s_k - s_{k+1}}} \frac{(-z, -q/z; q)_{s_{k+1}}}{(q)_{2s_{k+1}}}$$

$$= \sum_{s=0}^{j} \frac{(-zq^{k+1+j-2s}, -q^{k+2-j+2s}/z, q^{2k+3}; q^{2k+3})_\infty}{(q;q)_\infty}.$$

We mention a different expansion for the infinite product in Theorem 5.2 when $j = 0$. This final result comes from a version of the Laurent polynomials $H_{2n}(z, a|q)$ with an odd index. We do not develop the corresponding results here.

Theorem 5.3. *If* k *is a nonnegative integer,*

$$\sum_{s_1 \geq s_2 \geq \cdots \geq s_{k+1} \geq 0} \frac{q^{s_1^2 + \cdots + s_{k+1}^2 + s_1 + \cdots + s_{k+1}}}{(q)_{s_1 - s_2} \cdots (q)_{s_k - s_{k+1}}} \frac{(-q^{k+1}/z)_{s_{k+1}+1} (-zq^{-k})_{s_{k+1}}}{(q)_{2s_{k+1}+1}}$$

$$= \frac{(-zq^{k+1}, -q^{k+2}/z, q^{2k+3}; q^{2k+3})_\infty}{(q;q)_\infty}.$$

If $k = 0$ in Theorems 5.2 and 5.3, we have the curious result (see [3, (5.1)])

$$\frac{(-zq, -q^2/z, q^3; q^3)_\infty}{(q)_\infty} = \sum_{s=0}^{\infty} \frac{(-z, -q/z; q)_s}{(q)_{2s}} q^{s^2}$$

$$= \sum_{s=0}^{\infty} \frac{(-zq)_{s+1} (-1/z)_s}{(q)_{2s+1}} q^{s^2+s}. \tag{5.1}$$

6. Remarks

The Andrews–Gordon identities have combinatorial interpretations for integer partitions, three of which are (see [2]):

(1) those with modular conditions on parts,
(2) those with difference conditions on parts,
(3) those with conditions on iterated Durfee squares.

This paper offers no insightful versions of these results for the binomial versions given here.

Berkovich and Paule [5, 6] have versions of the Andrews–Gordon identities where the linear forms are also modified.

Griffin, Ono, and Warnaar [12] give new infinite families (e.g., [12, Theorem 1.1]) of Rogers–Ramanujan identities. See [12, (2.7)] for the Andrews–Gordon–Bressoud identities in their paper.

Seo and Yee [13] combinatorially study singular overpartitions, whose generating function is given by $j = 0$ in Theorem 5.1 with a special choice of z.

Acknowledgment

The author was supported by NSF grant DMS-1148634.

References

[1] G. Andrews, An analytic generalization of the Rogers–Ramanujan identities for odd moduli, *Proc. Natl. Acad. Sci. USA* **71** (1974) 4082–4085.

[2] G. Andrews, Partitions and Durfee dissection, *Amer. J. Math.* **101**(3) (1979) 735–742.

[3] G. Andrews, Singular overpartitions, *Int. J. Number Theory* **11**(5) (2015) 1523–1533.

[4] G. Andrews, Bressoud polynomials, Rogers–Ramanujan type identities, and applications, *Ramanujan J.* **41** (2016) 287–304.

[5] A. Berkovich and P. Paule, Variants of the Andrews–Gordon identities, *Ramanujan J.* **5**(4) (2001) 391–404.

[6] A. Berkovich and P. Paule, Lattice paths, q-multinomials and two variants of the Andrews–Gordon identities, *Ramanujan J.* **5**(4) (2001) 409–425.

[7] D. Bressoud, An analytic generalization of the Rogers–Ramanujan identities with interpretation, *Quart. J. Math. Oxford Ser.* (2) **31**(124) (1980) 385–399.

[8] D. Bressoud, An easy proof of the Rogers–Ramanujan identities, *J. Number Theory* **16**(2) (1983) 235–241.

[9] D. Bressoud, *Analytic and Combinatorial Generalizations of the Rogers–Ramanujan Identities*, Memoirs of the American Mathematical Society, Vol. 24, No. 227 (American Mathematical Society, Providence, RI, 1980), 54 pp.

[10] K. Garrett, M. E. Ismail and D. Stanton, Variants of the Rogers–Ramanujan identities, *Adv. Appl. Math.* **23**(3) (1999) 274–299.

[11] G. Gasper and M. Rahman, *Basic Hypergeometric Series*, 2nd edn. (Cambridge University Press, Cambridge, 2004).

[12] M. Griffin, K. Ono and S. O. Warnaar, A framework of Rogers–Ramanujan identities and their arithmetic properties, *Duke Math. J.* **165**(8) (2016) 1475–1527.

[13] S. Seo and A. J. Yee, Overpartitions and singular overpartitions, preprint (2016).

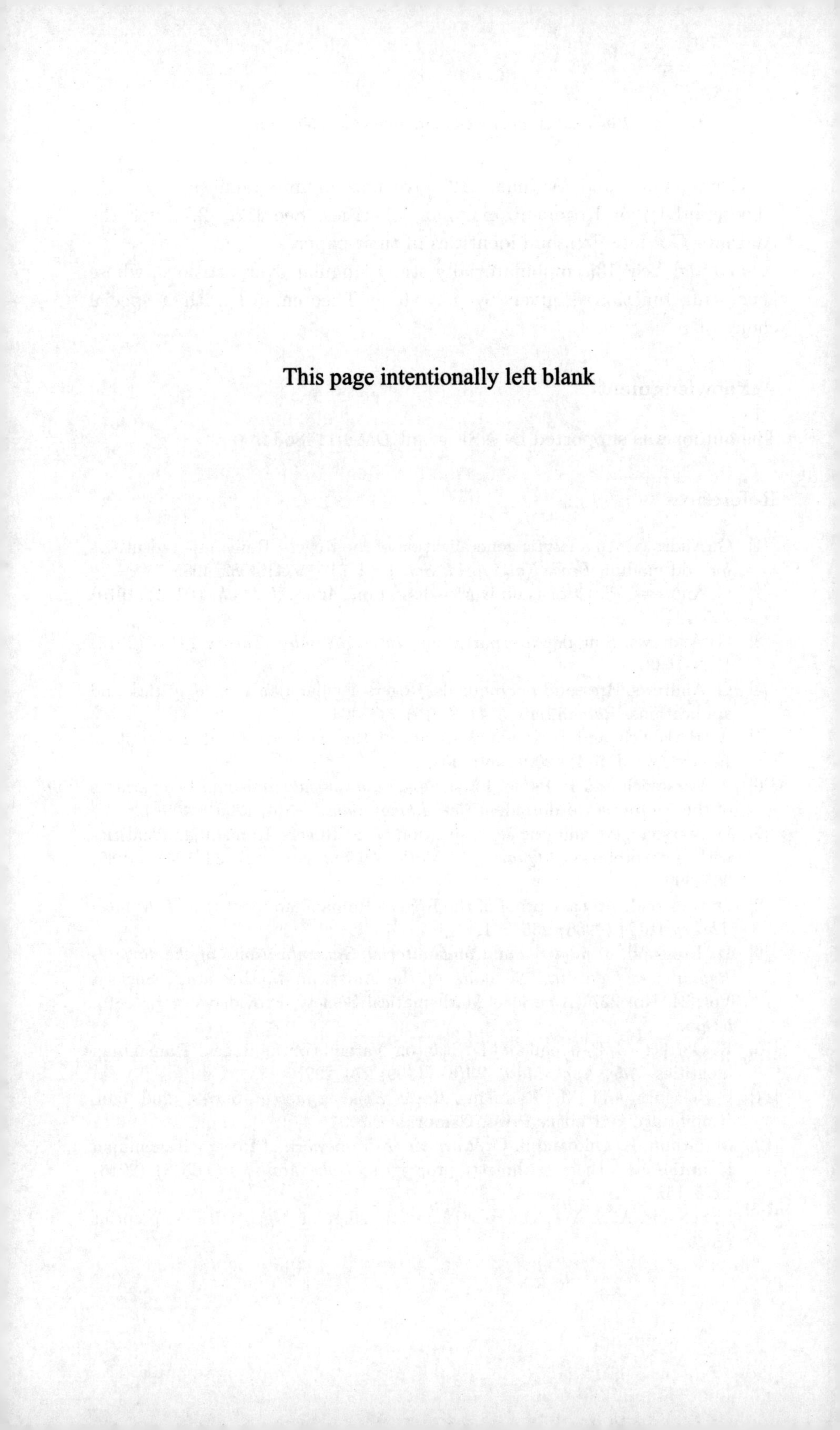

This page intentionally left blank

Chapter 3

Symmetric Expansions of Very Well-Poised Basic Hypergeometric Series

George E. Andrews
The Pennsylvania State University
University Park, PA 16802, USA
gea1@psu.edu

Dedicated to a grand mathematician and a good friend, Mourad Ismail

The classical transformation of the very well-poised ${}_{2k+4}\phi_{2k+3}$ reduces the symmetry of the original series from the full symmetric group, S_{2k}, in the $2k$ parameters to S_2^k symmetry. Thus, the symmetry drops from a group of $(2k)!$ elements to a group of 2^k elements. In this paper, a more symmetric expansion is obtained where the image symmetry group is $S_k \times S_2^k$.

Keywords: Symmetric expansions; q-series; Rogers–Ramanujan identities.

Mathematics Subject Classification 2010: 33D70

1. Introduction

Symmetric expansions have played a vital role in the study of basic or q-hypergeometric functions. Indeed, the road to the Rogers–Ramanujan identities started with Rogers in 1893 [9]. He observed a hidden symmetry in the Heine series

$$\sum_{n\geq 0}\frac{(a)_n(b)_n t^n}{(q)_n(c)_n}, \tag{1.1}$$

where $(a)_n = (a;q)_n = (1-a)(1-aq)\cdots(1-aq^{n-1})$. He set himself the task of finding an expansion of this series that made all the symmetries transparent. A full account of the evolution of Rogers's papers [9–11] into the elaborate expansions of today is given in [4]. It should also be noted that Bowman has greatly extended Rogers's original efforts [7].

The Rogers–Ramanujan identities are two elegant identities:

$$\sum_{n\geq 0}\frac{q^{n^2}}{(q)_n}=\frac{1}{(q;q^5)_\infty(q^4;q^5)_\infty}, \tag{1.2}$$

and

$$\sum_{n\geq 0}\frac{q^{n^2+n}}{(q)_n}=\frac{1}{(q^2;q^5)_\infty(q^3;q^5)_\infty}. \tag{1.3}$$

They first appear on the tenth page of Rogers's paper [11] which was the natural follow-up to [9, 10]. The quintessential q-hypergeometric proof of (1.2) and (1.3) was given by Watson in 1929 in his wonderful identity [12]:

$$\begin{aligned}
&{}_8\phi_7\left(\begin{matrix} a, q\sqrt{a}, -q\sqrt{a}, b_1, c_1, b_2, c_2, q^{-N}; q, \dfrac{a^2q^{2+N}}{b_1c_1b_2c_2} \\ \sqrt{a}, -\sqrt{a}, \dfrac{aq}{b_1}, \dfrac{aq}{c_1}, \dfrac{aq}{b_2}, \dfrac{aq}{c_2}, aq^{N+1}\end{matrix}\right) \\
&= \frac{(aq)_N(\frac{aq}{b_2c_2})_N}{(\frac{aq}{b_2})_N(\frac{aq}{c_2})_N}\,{}_4\phi_3\left(\begin{matrix} \dfrac{aq}{b_1c_1}, b_2, c_2, q^{-N}; q, q \\ \dfrac{aq}{b_1}, \dfrac{aq}{c_1}, \dfrac{b_2c_2q^{-N}}{a}\end{matrix}\right),
\end{aligned} \tag{1.4}$$

where

$${}_{R+1}\phi_R\left(\begin{matrix}\alpha_0, \alpha_1, \ldots, \alpha_R; q, t \\ \beta_1, \ldots, \beta_R\end{matrix}\right) := \sum_{j\geq 0}\frac{(\alpha_0)_j(\alpha_1)_j\cdots(\alpha_R)_j t}{(q)_j(\beta_1)_j\cdots(\beta_R)_j}. \tag{1.5}$$

The left-hand side series in (1.4) is called "well-poised" because the product of every column is the same (in this case, the product is aq) and the adverb "very" is added to describe the special second and third columns. The series on the right-hand side of (1.4) is called "balanced" because the product of the four upper entries times q equals the product of the three lower entries.

Watson deduced (1.1) and (1.2) from (1.4) by letting b_1, c_1, b_2, c_2, and N all tend to ∞ and then setting $a = 1$ to obtain an equivalent result to (1.1) and obtaining (1.2) by $a = q$.

At this point, we note that the left-hand side is a symmetric function on the four parameters b_1, c_1, b_2, c_2 while on the right-hand side, the symmetry has been reduced to $b_1 \leftrightarrow c_1$ and $b_2 \leftrightarrow c_2$.

Around 45 years later, (1.1) and (1.2) were extended to a multiple series generalization [2]:

$$\sum_{n_1,\ldots,n_{k-1}\geq 0} \frac{q^{N_1^2+N_2^2+\cdots+N_{k-1}^2+N_i+\cdots+N_{k-1}}}{(q)_{n_1}(q)_{n_2}\cdots(q)_{n_{k-1}}} = \prod_{\substack{n=1\\ n\neq 0,\pm i \pmod{2x+1}}}^{\infty} \frac{1}{1-q^n}, \tag{1.6}$$

where $N_m = n_m + n_{m+1} + \cdots + n_{k-1}$.

Then, in 1976, the massive generalization of (1.4) was proved [3]: For $k \geq 1$, N a nonnegative integer,

$$
\begin{aligned}
&{}_{2k+4}\phi_{2k+3}\left[\begin{matrix} a, q\sqrt{a}, -q\sqrt{a}, b_1, c_1, b_2, c_2, \ldots, b_k, c_k, q^{-N}; q, \dfrac{a^k q^{k+N}}{b_1\cdots b_k c_1\cdots c_k} \\ \sqrt{a}, -\sqrt{a}, \dfrac{aq}{b_1}, \dfrac{aq}{c_1}, \dfrac{aq}{b_2}, \dfrac{aq}{c_2}, \ldots, \dfrac{aq}{b_k}, \dfrac{aq}{c_k}, aq^{N+1} \end{matrix}\right] \\
&= \frac{(aq)_N(\frac{aq}{b_kc_k})_N}{(\frac{aq}{b_k})_N(\frac{aq}{c_k})_N} \sum_{m_1,\ldots,m_{k-1}\geq 0} \frac{(\frac{aq}{b_1c_1})_{m_1}(\frac{aq}{b_2c_2})_{m_2}\cdots(\frac{aq}{b_{k-1}c_{k-1}})_{m_{k-1}}}{(q)_{m_1}(q)_{m_2}\cdots(q)_{m_{k-1}}} \\
&\times \frac{(b_2)_{m_1}(c_2)_{m_1}(b_3)_{m_1+m_2}(c_3)_{m_1+m_2}\cdots(b_k)_{m_1+\cdots+m_{k-1}}}{(\frac{aq}{b_1})_{m_1}(\frac{aq}{c_1})_{m_1}(\frac{aq}{b_2})_{m_1+m_2}(\frac{aq}{c_2})_{m_1+m_2}\cdots(\frac{aq}{b_{k-1}})_{m_1+\cdots+m_{k-1}}} \\
&\times \frac{(c_k)_{m_1+\cdots+m_{k-1}}}{(\frac{aq}{c_{k-1}})_{m_1+\cdots+m_{k-1}}} \times \frac{(q^{-N})_{m_1+m_2+\cdots+m_{k-1}}}{(b_kc_k\frac{q^{-N}}{a})_{m_1+m_2+\cdots+m_{k-1}}} \\
&\times \frac{(aq)^{m_{k-2}+2m_{k-3}+\cdots+(k-2)m_1} q^{m_1+m_2+\cdots+m_{k-1}}}{(b_2c_2)^{m_1}(b_3c_3)^{m_1+m_2}\cdots(b_{k-1}c_{k-1})^{m_1+m_2+\cdots+m_{k-2}}}.
\end{aligned} \tag{1.7}
$$

Note now that the S_{2k} symmetry of the left-hand side reduces to S_2^k symmetry on the right. Often, this loss of symmetry seems quite significant. In almost all applications, the pairs (b_i, c_i) are naturally kept together, so it would be valuable to have a transformation of the left-hand side of (1.7) that was symmetric in these pairs. To produce such a transformation is the object of this paper.

Theorem 1.1. *For $k \geq 1$, N a nonnegative integer,*

$$
{}_{2k+4}\phi_{2k+3}\left[\begin{matrix} a, q\sqrt{a}, -q\sqrt{a}, b_1, c_1, b_2, c_2, \ldots, b_k, c_k, q^{-N}; q, \dfrac{a^k q^{k+N}}{b_1 \cdots b_k c_1 \cdots c_k} \\ \sqrt{a}, -\sqrt{a}, \dfrac{aq}{b_1}, \dfrac{aq}{c_1}, \dfrac{aq}{b_2}, \dfrac{aq}{c_2}, \ldots, \dfrac{aq}{b_k}, \dfrac{aq}{c_k}, aq^{N+1} \end{matrix}\right]
$$

$$
= \sum_{m_1, \ldots, m_k \geq 0} \prod_{i=1}^{k} \frac{(\frac{aq}{b_i c_i})_{m_i} q^{m_i}}{(q)_{m_i} (\frac{aq}{b_i})_{m_i} (\frac{aq}{c_i})_{m_i}} \times K_k(a, N; m_1, m_2, \ldots, m_k), \quad (1.8)
$$

where K_k is symmetric in $m_1, m_2, \ldots, m_k$ and has the following properties for $k > 1$:

$$
K_k(a, N; m_1, \ldots, m_k) = 0 \quad \text{if } N > m_1 + m_2 + \cdots + m_k. \quad (1.9)
$$

$$
\begin{aligned} &K_k(a, N; m_1, \ldots, m_k) &(1.10) \\ &\quad = q^{-\sigma_2(m_1, \ldots, m_k) - \sigma_1(m_1, \ldots, m_k)} (aq)_N (q)_N \quad \text{if } N = m_1 + \cdots + m_k. \end{aligned}
$$

$$
\begin{aligned} &K_k(a, N; m_1, \ldots, m_k) \\ &\quad = \sum_{j=0}^{N} \frac{(a)_j (1 - aq^{2j}) (q^{-N})_j q^{N_j}}{(q)_j (1 - a)(aq^{N+1})_j} \prod_{r=1}^{k} (q^{-j})_{m_r} (aq^j)_{m_r}, \quad (1.11) \end{aligned}
$$

where $\sigma_s(m_1, \ldots, m_k)$ is the s-th elementary symmetric function in $m_1, \ldots, m_k$.

I would note that neither (1.9) nor (1.10) is an immediate consequence of (1.11) at all. Indeed, one would hope that there might be representations of K_k that would make (1.9) and (1.10) as well as the symmetry clear. To that end, we have the following theorem.

Theorem 1.2.

$$
K_1(a, N; m_1) = \delta_{N, m_1}, \quad (1.12)
$$

$$
\begin{aligned} &K_2(a, N; m_1, m_2) \\ &\quad = \begin{bmatrix} m_1 + m_2 \\ N \end{bmatrix} \frac{(-1)^N q^{\binom{N}{2}} (1 - aq^N)(q^{-N})_{m_1} (q^{-N})_{m_2} (a)_{m_1 + m_2}}{(1 - a)(q)_{m_1 + m_2}}, \quad (1.13) \end{aligned}
$$

$$K_3(a,N;m_1,m_2,m_3)$$

$$= \begin{bmatrix} m_1+m_2+m_3 \\ N \end{bmatrix} \frac{(-1)^N q^{\binom{N}{2}}(1-aq^N)(q^{-N})_{m_1}(q^{-N})_{m_2}(q^{-N})_{m_3} \times (a)_{m_1+m_2+m_3}}{(1-a)(q)_{m_1+m_2+m_3}}$$

$$\times {}_4\phi_3 \left(\begin{matrix} q^{-m_1}, q^{-m_2}, q^{-m_3}, \dfrac{q^{1-N}}{a}; q, q \\ q, q^{-N}, \dfrac{q^{1-m_1-m_2-m_3}}{a}. \end{matrix} \right). \tag{1.14}$$

Section 2 will be devoted to a proof of Theorem 1.1 as well as (1.13). Section 3 will be devoted to the remaining two assertions in Theorem 1.2. Section 4 concludes with possible applications and open problems.

2. Proof of Theorem 1.1

We start with the easiest assertion, namely (1.11). To prove this result, we require the following formulation of the q-Pfaff–Saalschiitz identity [8, equation (2.2.1), p. 32]:

$$\sum_{r=0}^{m} \frac{(q^{-m})_r (aq^m)_r \left(\frac{aq}{bc}\right)_r q^r}{(q)_r \left(\frac{aq}{b}\right)_r \left(\frac{aq}{c}\right)_r} = \frac{a^m q^m (b)_m (c)_m}{b^m c^m \left(\frac{aq}{b}\right)_m \left(\frac{aq}{c}\right)_m}. \tag{2.1}$$

Hence,

$${}_{2k+4}\phi_{2k+3} \begin{bmatrix} a, q\sqrt{a}, -q\sqrt{a}, b_1, c_1, b_2, c_2, \ldots, b_k, c_k, q^{-N}; q, \dfrac{a^k q^{k+N}}{b_1\cdots b_k c_1 \cdots c_k} \\ \sqrt{a}, -\sqrt{a}, \dfrac{aq}{b_1}, \dfrac{aq}{c_1}, \dfrac{aq}{b_2}, \dfrac{aq}{c_2}, \ldots, \dfrac{aq}{b_k}, \dfrac{aq}{c_k}, aq^{N+1} \end{bmatrix}$$

$$= \sum_{j=0}^{N} \frac{(a)_j (1-aq^{2j})(q^{-N})_j q^{Nj}}{(q)_j (1-a)(aq^{N+1})_j} \prod_{i=1}^{k} \frac{a^j q^j (b_i)_j (c_i)_j}{b_i^j c_i^j (\frac{aq}{b_i})_j (\frac{aq}{c_i})_j}$$

$$= \sum_{j=1}^{N} \frac{(a)_j (1-aq^2)(q^{-N})_j q^{Nj}}{(q)_j (1-a)(aq^{N+1})_j} \times \sum_{m_1,\ldots,m_k \geq 0} \frac{(q^{-j})_{m_i} (aq^j)_{m_i} (\frac{aq}{b_i c_i})_{m_i} q^{m_i}}{(q)_{m_i} (\frac{aq}{b_i})_{m_i} (\frac{aq}{c_i})_{m_i}}$$

$$= \sum_{m_1,\ldots,m_k\geq 0} \prod_{i=1}^{k} \frac{(\frac{aq}{b_ic_i})_{m_i} q^{m_i}}{(q)_{m_i}(\frac{aq}{b_i})_{m_i}(\frac{aq}{c_i})_{m_i}}$$

$$\times \sum_{j=0}^{N} \frac{(a)_j(1-aq^{2j})(q^{-N})_j q^{Nj}}{(q)_j(1-a)(aq^{N+1})_j} \prod_{i=1}^{k} (q^{-j})_{m_i}(aq^j)_{m_i}, \tag{2.2}$$

as asserted in (1.11).

In order to treat the other two assertions, we need to rewrite K_k as a very well-poised series. Given the symmetry of K_k in the m's, we shall assume that m_k is at least as large as all the other m_i. Also, we note that

$$(aq^j)_{m_i} = \frac{(a)_{m_i+j}}{(a)_j},$$

and

$$(q^{-j})_{m_i} = (1-q^j)\cdots(1-q^{j-m_i+1}) \cdot q^{-jm_i+\binom{m_i}{2}}(-1)^{m_i}$$

$$= \frac{(q)_j}{(q)_{j-m_i}}(-1)^{m_i} q^{-jm_i+\binom{m_i}{2}}.$$

Hence,

$$K_k(a,N;m_1,\ldots,m_k)$$
$$= \sum_{j=0}^{N} \frac{(a)_j(1-aq^{2j})(q^{-N})_j q^{Nj}}{(q)_j(1-a)(aq^{N+1})_j} \times \prod_{i=1}^{k} \frac{(a)_{m_i+j}(q)_j(-1)^{m_i} q^{\binom{m_i}{2}-jm_i}}{(a)_j(q)_{j-m_i}}.$$

Now, $\frac{1}{(q)_M} = 0$ for $M < 0$, thus if j is less than any m_i, the term is zero. So, we may replace j by $j + m_k$ and no nonzero terms will be deleted, and to make the role of m_k, clear we replace m_k by t. Thus,

$$K_k(a,N;m_1,\ldots,m_{k-1},t)$$

$$= \sum_{j\geq 0} \frac{(a)_{j+t}(1-aq^{2j+2t})(q^{-N})_{j+t} q^{N(j+t)}}{(q)_{j+t}(1-a)(aq^{N+1})_{j+t}}$$

$$\times \prod_{i=1}^{k-1} \frac{(a)_{m_i+j+t}(q)_{j+t}(-1)^{m_i} q^{\binom{m_i}{2}-(j+t)m_i}}{(a)_{j+t}(q)_{j+t-m_i}}$$

$$\times \frac{(a)_{j+2t}(q)_{j+t}(-1)^t q^{\binom{t}{2}-(j+t)t}}{(a)_{j+t}(q)_j}$$

$$= \frac{(a)_{2t}(1-aq^{2t})(q^{-N})_t q^{Nt-\binom{t+1}{2}}(-1)^t}{(aq^{N+1})_t(1-a)}$$

$$\times \prod_{i=1}^{k-1} \frac{(a)_{m_i+t}(q)_t(-1)^{m_i} q^{\binom{m_i}{2}-tm_i}}{(a)_t(q)_{t-m_i}}$$

$$\times {}_{2k+2}\phi_{2k+1}\left(\begin{matrix} aq^{2t}, q\sqrt{aq^{2t}}, -q\sqrt{aq^{2t}}, aq^{m_1+t}, q^{t+1}, \ldots, aq^{m_{k-1}+t}, \\ q^{t+1}, q^{-N+t}; q, q^{N-t-m_1-\cdots-m_{k-1}} \\ \\ \sqrt{aq^{2t}}, -\sqrt{aq^{2t}}, q^{t-m_1+1}, aq^t, \ldots, \\ q^{t-m_{k-1}+1}, aq^t, aq^{N+t+1} \end{matrix}\right). \tag{2.3}$$

Now, (2.3) allows us to obtain a recurrence for K_k by applying (1.8) to the inner series appearing in (2.3). Hence,

$$\begin{aligned} &K_k(a, N; m_1, m_2, \ldots, m_{k-1}, t) \\ &= \frac{(a)_{2t}(1-aq^{2t})(q)_N q^{-t}}{(aq^{N+1})_t(1-a)(q)_{N-t}} \prod_{i=1}^{k-1}(aq^{t+1})_{m_i}(q^{-t})_{m_i} \\ &\times \sum_{\mu_1,\ldots,\mu_{k-1}\geq 0} \prod_{i=1}^{k-1} \frac{(q^{-m_i})_{\mu_i} q^{\mu_i}}{(q)_{\mu_i}(q^{t-m_i+1})_{\mu_i}(aq^{t+1})_{\mu_i}} \\ &\times K_{k-1}(aq^{2t}, N-t; \mu_1, \mu_2, \ldots, \mu_{k-1}). \end{aligned} \tag{2.4}$$

We now proceed to prove (1.9) and (1.10) by mathematical induction on k. The initial case is $k=2$. By (2.3) with $m_1 = m, m_2 = t$,

$$\begin{aligned} &K_2(a, N; m, t) \\ &= \frac{(a)_{2t}(1-aq^{2t})(q)_N q^{-t}(aq^t)_m(q^{-t})_m}{(aq^{N+1})_t(1-a)(q)_{N-t}} \\ &\times {}_6\phi_5\left(\begin{matrix} aq^{2t}, q\sqrt{aq^{2t}}, -q\sqrt{aq^{2t}}, aq^{m+t}, q^{t+1}, q^{-N+t}; q, q^{N-t-m} \\ \sqrt{aq^{2t}}, -\sqrt{aq^{2t}}, q^{t-m+1}, aq^t, aq^{N+t+1} \end{matrix}\right) \end{aligned}$$

$$= \frac{(aq)_{2t}(q)_N(a)_{m+t}(q^{-t})_m(aq^{2t+1})_{N-t}(q^{-m})_{N-t}q^{-t}}{(aq^{N+1})_t(q)_{N-t}(a)_t(q^{t+1-m})_{N-t}(aq^t)_{N-t}}$$

by [8, equation (II.21), p. 238]

$$= \frac{(a)_{m+t}(q)_N(q^{-t})_m(q^{-m})_{N-t}(1-aq^N)q^{-t}}{(1-a)(q)_{N-t}(q^{t+1-m})_{N-t}(aq^{2n+1})_{N-t}}$$

$$= \begin{bmatrix} m+t \\ N \end{bmatrix} \frac{(-1)^N q^{\binom{N}{2}}(1-aq^N)(a)_{m+t}(q^{-N})_m(q^{-N})_t(q)_m(q)_t}{(1-a)(q)_{m+t}}. \tag{2.5}$$

Now, the factor $(q)_{m+t-N}$ in the denominator reveals that K_2 is 0 if $N > m+t$, and if $N = m+t$, then

$$K_2(a, N; m, t)$$

$$= \frac{(-1)^N q^{\binom{N}{2}}(1-aq^N)(a)_N}{(1-a)(q)_N} \times \frac{(q)_N^2(-1)^{m+t}q^{-N(m+t)+\binom{m}{2}+\binom{t}{2}}}{(q)_t(q)_m}$$

$$\times (q)_t(q)_m$$

$$= q^{-\binom{N+1}{2}+\binom{m}{2}+\binom{t}{2}}(aq)_N(q)_N$$

$$= q^{-mt-m-t}(aq)_N(q)_N.$$

Thus, we have established (1.9) and (1.10) in the case $k = 2$.

Now, we must utilize the recurrence to complete the induction proof of (1.9) and (1.10). We assume that (1.9) and (1.10) are valid for each k less than a given k. Suppose $N > m_1 + m_2 + \cdots + m_{k-1} + t$ in (2.4). We see that the terms in the sum on the right-hand side of (2.4) must vanish if any $\mu_i > m_i$ because of the factor $(q^{-m_i})_{\mu_i}$. Given that

$$N > m_1 + m_2 + \cdots + m_{k-1} + t,$$

we see that

$$N - t > m_1 + \cdots + m_{k-1} \geq \mu_1 + \cdots + \mu_{k-1}.$$

Hence, every term of the sum in (2.4) is 0; therefore, (1.9) is valid for K_k.

Next, suppose that

$$N = m_1 + \cdots + m_{k-1} + t.$$

The previous argument shows that now the only nonvanishing term in the inner sum occurs for $\mu_i = m_i$, $1 \leq i \leq k-1$. Hence, in this case, by (2.4) and the induction hypothesis,

$$
\begin{aligned}
&K_k(a, N; m_1, \ldots, m_{k-1}, t) \\
&= \frac{(a)_{2t}(1-aq^{2t})(q)_N q^{-t}}{(aq^{N+1})_t(1-a)(q)_{N-t}} \prod_{i=1}^{k-1} (aq^{t+1})_{m_i}(q^{-t})_{m_i} \\
&\quad \times \prod_{i=1}^{k-1} \frac{(q^{-m_i})_{m_i} q^{m_i}}{(q)_{m_i}(q^{t-m_i+1})_{m_i}(at^{t+1})_{m_i}} \\
&\quad \times q^{-\sigma_2(m_1,\ldots,m_{k-1})-\sigma_1(m_1,\ldots,m_{k-1})}(aq^{2t+1})_{N-t}(q)_{N-t} \\
&= \frac{(aq)_{2t}(q)_N q^{-t}}{(aq^{N+1})_t(q)_{N-t}} \prod_{i=1}^{k-1} \frac{(aq^{t+1})_{m_i}(-1)^{m_i} q^{-tm_i+\binom{m_i}{2}}(q)_t}{(q)_{t-m_i}} \\
&\quad \times \prod_{i=1}^{k-1} \frac{q^{-\binom{m_i}{2}}(q)_{m_i}(q)_{t-m_i}}{(q)_{m_i}(q)_t(aq^{t+1})_{m_i}} \\
&\quad \times q^{-\sigma_2(m_1,\ldots,m_{k-1})-\sigma_1(m_1,\ldots,m_{k-1})}(aq^{2t+1})_{N-t}(q)_{N-t} \\
&= q^{-\sigma_2(m_1,\ldots,m_{k-1},t)-\sigma_1(m_1,\ldots,m_{k-1},t)}(aq)_N(q)_N.
\end{aligned}
$$

Thus, (1.9) and (1.10) have been established by mathematical induction on k. □

3. Proof of Theorem 1.2

As noted previously (1.13) was proved in Section 2. Equation (1.12) follows immediately from the classical summation [8, equation (II.21), p.238]:

$$
{}_6\phi_5\left(\begin{matrix} a, q\sqrt{a}, -q\sqrt{a}, b_1, c_1, q^{-N}; q, \dfrac{aq^{1+N}}{b_1c_1} \\[2ex] \sqrt{a}, -\sqrt{a}, \dfrac{aq}{b_1}, \dfrac{aq}{c_1}, aq^{N+1} \end{matrix}\right) = \frac{(aq)_N(\frac{aq}{b_1c_1})_N}{(\frac{aq}{b_1})_N(\frac{aq}{c_1})_N}. \tag{3.1}
$$

To treat K_3, we must utilize (2.3). So, we shall assume m_3 is not exceeded by m_1 or m_2. Symmetry allows these assumptions without loss of generality.

Hence,

$$K_3(a,N;m_1,m_2,m_3)$$

$$= \frac{(a)_{2m_3}(1-aq^{2m_3})(q^{-N})_{m_3} q^{Nm_3-\binom{m_3+1}{2}}(-1)^{m_3}}{(aq^{N+1})_{m_3}(1-a)}$$

$$\times \frac{(a)_{m_1+m_3}(a)_{m_2+m_3}(q)^2_{m_3}(-1)^{m_1+m_2} q^{\binom{m_1}{2}+\binom{m_2}{2}-m_3(m_1+m_2)}}{(a)^2_{m_3}(q)_{m_3-m_1}(q)_{m_3-m_2}}$$

$$\times {}_8\phi_7\left(\begin{matrix} aq^{2m_3}, q^{m_3+1}\sqrt{a}, -q^{m_3+1}\sqrt{a}, aq^{m_1+m_3}, q^{m_3+1}, aq^{m_2+m_3}, \\ q^{m_3+1}, q^{-N+m_3}; q, q^{N-m_1-m_2-m_3} \\ q^{m_3}\sqrt{a}, -q^{m_3}\sqrt{a}, q^{m_3-m_1+1}, aq^{m_3}, q^{m_3-m_2+1}, \\ aq^{m_3}, aq^{N+m_3+1} \end{matrix}\right) \tag{3.2}$$

$$= C(a,N;m_1,m_2,m_3)$$

$$\times {}_8\phi_7\left(\begin{matrix} aq^{2m_3}, q^{m_3+1}\sqrt{a}, -q^{m_3+1}\sqrt{a}, aq^{m_1+m_3}, q^{m_3+1}, aq^{m_2+m_3}, \\ q^{m_3+1}, q^{-N+m_3}; q, q^{N-m_1-m_2-m_3} \\ q^{m_3}\sqrt{a}, -q^{m_3}\sqrt{a}, q^{m_3-m_1+1}, aq^{m_3}, \\ q^{m_3-m_2+1}, aq^{m_3}, aq^{N+m_3+1} \end{matrix}\right), \tag{3.3}$$

where we have written $C(a,N;m_1,m_2,m_3)$ for the multiplying product.

We now apply Watson's q-analog of Whipple's theorem [8, equation (III.18), p. 242] to the ${}_8\phi_7$. Hence,

$$K_3(a,N;m_1,m_2,m_3) = C(a,N;m_1,m_2,m_3)$$

$$\times \frac{(aq^{2m_3+1})_{N-m_3}(q^{-m_2})_{N-m_3}}{(q^{m_3-m_2+1})_{N-m_3}(aq^{m_3})_{N-m_3}}$$

$$\times {}_4\phi_3\left(\begin{matrix} q^{-m_1}, aq^{m_2+m_3}, q^{m_3+1}, q^{-N+m_3}; q, q \\ q^{m_3-m_1+1}, aq^{m_3}, q^{m_2,+m_3-N+1} \end{matrix}\right)$$

$$= \frac{C(a,N;m_1,m_2,m_3)(aq^{2m_3+1})_{N-m_3}(q^{-m_2})_{N-m_3}}{(q^{m_3-m_2+1})_{N-m_3}(aq^{m_3})_{N-m_3}}$$

$$\times \frac{(q^{-m_1})_{m_1}(q^{m_2-N})_{m_1}}{(q^{m_3-m_1+1})_{m_1}(q^{m_2+m_3-N+1})_{m_1}}$$

$$\times {}_4\phi_3\left(\begin{matrix} q^{-m_1}, q^{m_3+1}, q^{-m_2}, aq^N; q, q \\ aq^{m_3}, q, q^{1-m_1-m_2+N} \end{matrix}\right)$$

(by [8, equation (III.15), $n = m_1, a = q^{m_3+1}$, $b = aq^{m_2+m_3}, c = q^{-N+m_3}, d = aq^{m_3}$, $e = q^{m_3-m_1+1}, f = q^{m_2+m_3-N+1}$, p. 242])

$$= C(a,N;m_1,m_2,m_3)(aq^{2m_3+1})_{N-m_3} \frac{(q^{-m_2})_{N-m_3}(q^{-m_1})_{m_1}(q^{m_2-N})_{m_1}}{\{(q^{m_3-m_2+1})_{N-m_3}(aq^{m_3})_{N-m_3} \times (q^{m_3-m_1+1})_{m_1}(q^{m_2+m_3-N+1})_{m_1}\}}$$

$$\times \frac{(aq^{m_3+m_2})_{m_1}(q^{1-m_1+N})_{m_1}}{(aq^{m_3})_{m_1}(q^{1-m_1-m_2+N})_{m_1}}$$

$$\times {}_4\phi_3\left(\begin{matrix} q^{-m_1}, q^{-m_2}, q^{-m_3}, \dfrac{q^{1-N}}{a}; q; q \\ q, \dfrac{q^{1-m_1-m_2-m_3}}{a}, q^{-N} \end{matrix}\right)$$

(by [8, equation (III.15), $n = m_1, a = q^{-m_2}$, $b = q^{m_3+1}, c = aq^N, d = q$, $e = aq^{m_3}, f = q^{1-m_1-m_2-N}$, p. 242]).

Simplification of the multiplying products yields (1.14).

4. Conclusion

There are many unanswered questions about the pair-symmetric kernel K_k. Here are some of the most important ones:

(1) Are there simplified expansions like (1.12), (1.13) and (1.14) for $k > 3$ that both exhibit symmetry explicitly and yield (1.9) and (1.10) reasonably directly. We should note that (1.14) reduces to (1.10) when $N = m_1 + m_2 + m_3$ because the ${}_4\phi_3$ in (1.14) reduces to a balanced ${}_3\phi_2$ which is summable [8, equation (II.12), p. 237].

(2) What is the relationship of (1.7) to (1.8)? In the case $k = 2$, one can pass easily from (1.8) to (1.7) by noting that both the m_1 and m_2 sums are each balanced (and thus summable ${}_3\phi_2$'s).
(3) The most notable instance of using the paired symmetry of the k pairs of parameters in (1.7) occurs in the study of Durfee symbols (cf. [1, 5]). Indeed, obvious symmetry of partition statistics was very difficult to establish [6]. It would be of interest to pursue this question using an expansion like (1.8) where the symmetry is clearly in evidence.
(4) The proof of Theorem 1.1 is some sort of "multiple Bailey Lemma". It is clear that multiple simultaneous applications of (2.1) can be used instead of the sequential applications in the standard Bailey chain productions [4]. The possibilities here are endless.

References

[1] C. Alfes, K. Bringmann and J. Lovejoy, Automorphic properties of generating functions for generalized odd rank moment and odd Durfee symbols, *Math. Proc. Combridge Philos. Soc.* **151** (2011) 385–406.

[2] G. E. Andrews, An analytic generalization of the Rogers–Ramanujan identities for odd moduli, *Proc. Natl. Acad. Sci. USA* **71** (1974) 4082–4085.

[3] G. E. Andrews, Problems and prospects for basic hypergeometric functions, in *Theory and Applications of Special Functions*, ed. R. Askey (Academic Press, New York, 1975), pp. 191–224.

[4] G. E. Andrews, *q-Series: Their development and Application in Analysis, Number Theory, Combinatorics, Physics, and Computer Algebra*, CBMS Regional Conference Series in Mathematics, No. 66 (American Mathematical Society, Providence, 1986).

[5] G. E. Andrews, Partitions, Durfee-symbols, and the Atkin–Garvan moments of ranks, *Invent. Math.* **169** (2007) 37–73.

[6] C. Boulet and K. Kursungoz, Symmetry of k-marked Durfee symbols, *Int. J. Number Theory* **7** (2011) 215–230.

[7] D. Bowman, *q-Difference Operators, Orthogonal Polynomials, and Symmetric Expansions*, Memoirs of the American Mathematical Society, Vol. 159, No. 757 (American Mathematical Society, Providence, RI, 2002).

[8] G. Gasper and M. Rahman, *Basic Hypergeometric Series*, Encyclopedia of Mathematics and its Applications, Vol. 25 (Cambridge University Press, Cambridge, 1990).

[9] L. J. Rogers, On a three-fold symmetry in the elements of Heine's series, *Proc. London Math. Soc.* **24** (1893) 171–179.

[10] L. J. Rogers, On the expansion of some infinite products, *Proc. London Math. Soc.* **24** (1893) 337–352.
[11] L. J. Rogers, Second memoir on the expansion of certain infinite products, *Proc. London Math. Soc.* **25** (1894) 318–343.
[12] G. N. Watson, A new proof of the Rogers–Ramanujan identities, *J. London Math. Soc.* **4** (1929) 4–9.

This page intentionally left blank

Chapter 4

A Sturm–Liouville Theory for Hahn Difference Operator

M. H. Annaby*, **A. E. Hamza**† **and S. D. Makharesh**‡

Department of Mathematics, Faculty of Science
Cairo University, P.O. Box 12613, Giza, Egypt
**mhannaby@sci.cu.edu.eg*
†hamzaaeg2003@yahoo.com
‡sameeermakarish@yahoo.com

Dedicated to Professor Mourad Ismail
on the occasion of his 70th birthday

This chapter introduces a comprehensive study for Sturm–Liouville theory of the q, ω-Hahn difference operators in the regular setting. We define a Hilbert space of q, ω-square summable functions in terms of Jackson–Nörlund integral. The formulation of the self-adjoint operator and the properties of the eigenvalues and the eigenfunctions are discussed. The construction of Green's function is developed and a study for q, ω-Fredholem integral operator is established. Hence, an eigenfunctions expansion theorem is derived and illustrative examples are exhibited. We also introduce a separate section for numerical simulations and illustrations. We give some comparisons between trigonometric functions and the q and q, ω counterparts. We also test numerically the asymptotic behavior of the zeros of q and q, ω trigonometric functions. The numerical experiments precisely reflect the theoretical results with this respect.

Keywords: Hahn difference operator; Sturm–Liouville theory; q-difference operator; Green's function; eigenfunctions expansion.

Mathematics Subject Classification 2010: 39A70, 39A12, 33D15

1. Introduction

Sturm–Liouville theory (SLT) plays a major role in several applications of a wide variety of physical and engineering problems. The simplest formulation of the problem is that of separate-type conditions, cf. [63],

$$\left.\begin{aligned} -y'' + \nu(t)y = \lambda y, \quad -\infty < a \le t \le b < \infty,\\ a_{11}y(a) + a_{12}y'(b) = 0,\\ a_{21}y(a) + a_{22}y'(b) = 0. \end{aligned}\right\} \tag{1.1}$$

Here, $\nu(\cdot)$ is taken to be continuous and real-valued on $[a, b]$, $\lambda \in \mathbb{C}$ is an eigenvalue parameter, $a_{ij} \in \mathbb{R}$ such that $|a_{i1}| + |a_{i2}| \neq 0$, $i = 1, 2$. It is known, cf. [63], that there is only a countable number of real numbers $\lambda_1 < \lambda_2 < \cdots, \lambda_n \to \infty$ with ∞ as the unique limit point such that problem (1.1) has nontrivial solutions, $\varphi_0, \varphi_1, \ldots$. These numbers are called the eigenvalues of (1.1) and the solutions are called eigenfunctions of the problem. Moreover, for each eigenvalue λ_n, there corresponds one solution $\varphi_n(\cdot)$ up to a nonzero multiplicative constant. Thus, the eigenvalues are geometrically simple. They are also algebraically simple since they are simple zeros of the characteristic function associated with (1.1). In addition, the set $\{\varphi_n\}_{n=0}^{\infty}$ is an orthonormal basis of $L_2(a, b)$, cf. [63]. These properties, i.e., reality and simplicity of eigenvalues, the basis property of the eigenfunctions are the major properties of the Sturm–Liouville problem (1.1). For more details, see the monographs [32, 54, 66].

For practical reasons, many counterparts of Sturm–Liouville problems (SLPs) have been developed, where the differential operator of (1.1) is replaced by other functional operators. Among these counterparts, one finds the discrete SLPs. In [24, 48], SLT is developed when the differential operator of (1.1) is replaced by a couple of forward and backward operators

$$\Delta y(n) = y(n+1) - y(n), \qquad \nabla y(n) = y(n) - y(n-1), \tag{1.2}$$

$n = 1, \ldots, N$. See also [58]. In [4], an SLP is developed in time scales, unifying the discrete and continuous settings mentioned above, see also [7, 31, 51, 55, 57, 67]. Although the derivation of the SLPs on time scales involves many cases and offers a technique to deal with generalized problems, see, e.g., [52], it does not involve the case when the differential operator of (1.1) is replaced by the q-difference operator

$$D_q y(x) := \frac{y(qx) - y(x)}{x(1-q)}, \quad 0 < q < 1. \tag{1.3}$$

In [20], by modifying the primitive work of [34, 35] a q-SLT is developed in the regular setting. It is proved that the properties of eigenvalues and eigenfunctions are pertained, see also [22]. The theory is extended and applied in many directions; see, e.g., [1, 6, 14, 17, 33, 53, 56]. In this chapter, we develop an SLT in the regular setting with separate-type boundary conditions associated with the q, ω-Hahn difference operator

$$D_{q,\omega} f(t) := \frac{f(qt+\omega) - f(t)}{(qt+\omega) - t}. \tag{1.4}$$

For this aim, we define an appropriate Hilbert space in terms of Jackson–Nörlund integral and define a self-adjoint problem. This is done in Section 3. In Section 4, we investigate the existence and the analyticity properties of the solutions of the difference equations. Green's function is constructed in Section 5. Section 6 is devoted to deriving the eigenfunctions expansion theorem of the q, ω-SLP. It also involves two illustrative examples, where the eigenvalues and the eigenfunctions are computed in terms of q, ω-trigonometric functions and their zeros. Section 7 contains several numerical experiments that demonstrate and test the theoretical results of q-functions and q, ω-functions. It is worthwhile to mention that studies in this direction are rare, where authors concentrate only on deriving theoretical results. This numerical study adds a nice feature to the chapter from one side, and on the other side it assures the theoretical results. Nevertheless, serious numerical studies associated with q and q, ω-functions are required to approximate these functions as well as their zeros and to explore the asymptotic studies carried out in the last 50 years. Probably this may lead to have the q and q, ω-functions invloved in the computer algebra systems (CAS).

We would like to mention that theory of different counterparts of SLPs is extended in many aspects to higher order problem as in [9, 11, 13, 42], to singular problems as in [19, 28], generalized boundary conditions and nonself-adjoint problems as in, e.g., [33, 64], to Askey–Wilson operators as in [28, 53, 61]. The theory of generalized SLPs is applied in mathematical and physical problems as in [1, 5, 15, 17, 23, 53]. See also the monographs [26, 27, 45, 61].

2. Preliminaries

This section briefly introduces the calculus of Hahn-difference operators as established in [16]. Let $q \in (0, 1)$, $\omega > 0$ be fixed, $\omega_0 := \omega/(1-q)$, and I be

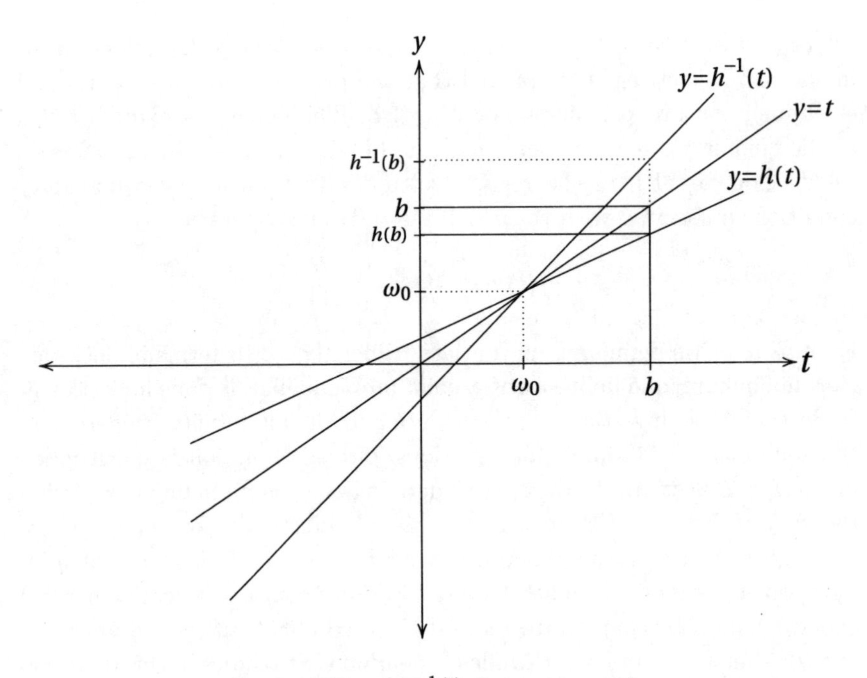

Fig. 1. $h(t)$ and $h^{-1}(t)$, $0 < q < 1$, $\omega > 0$.

an interval of $\mathbb{R}$ containing ω_0. Let $h(t) := qt + \omega,\ t \in I$. Both $h(t)$ and $h^{-1}(t) = (t - \omega)/q$ are illustrated in Fig. 1. The number ω_0 plays the role played by $t = 0$ in the q-setting. In fact

$$\omega_0 = \lim_{k\to\infty} (\underbrace{h \circ h \circ \cdots \circ h}_{k\text{-times}})(t),$$

see Fig. 2. The q, ω-Hahn difference operator introduced by Hahn in [39] can be defined as follows.

Definition 2.1. Let f be a function defined on I. The *Hahn difference operator* is defined by

$$D_{q,\omega} f(t) := \begin{cases} \dfrac{f(qt+\omega) - f(t)}{(qt+\omega) - t}, & t \neq \omega_0, \\ f'(\omega_0), & t = \omega_0, \end{cases} \tag{2.1}$$

provided that f is differentiable at ω_0. In this case, we call $D_{q,\omega} f$, the q, ω-derivative of f. We say that f is q, ω-differentiable, i.e., throughout I, if $D_{q,\omega} f(\omega_0)$ exists.

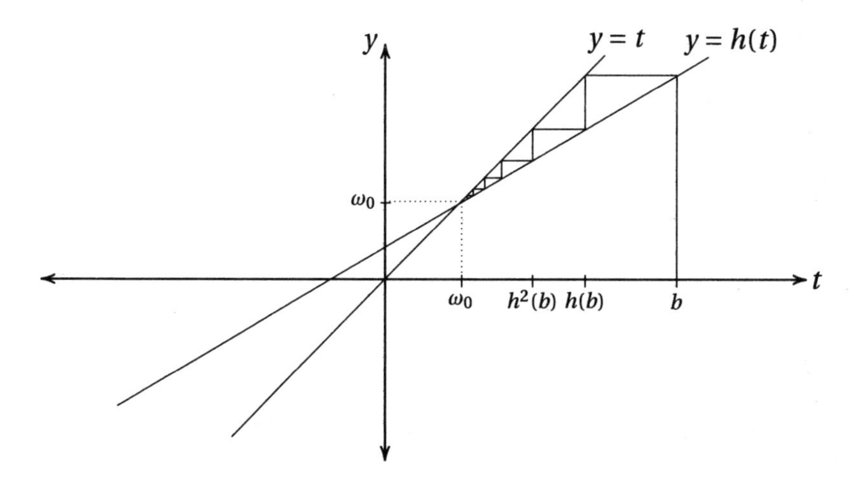

Fig. 2. The iterations of h for $0 < q < 1$, $\omega > 0$.

One can easily check that if f, g are q,ω-differentiable at $t \in I$, then

$$D_{q,\omega}(\alpha f + \beta g)(t) = \alpha D_{q,\omega} f(t) + \beta D_{q,\omega} g(t), \quad \alpha, \beta \in \mathbb{C},$$

$$D_{q,\omega}(fg)(t) = D_{q,\omega}(f(t))g(t) + f(qt+\omega) D_{q,\omega} g(t),$$

$$D_{q,\omega}\left(\frac{f}{g}\right)(t) = \frac{D_{q,\omega}(f(t))g(t) - f(t) D_{q,\omega} g(t)}{g(t)g(qt+\omega)},$$

provided that in the last identity $g(t)g(qt+\omega) \neq 0$ (cf. [16]). The right inverse for $D_{q,\omega}$ is defined in [16] in terms of Jackson–Nörlund sums as follows (cf. [49]). Let $a, b \in I$. The q,ω-integral of f from a to b is defined to be

$$\int_a^b f(t) d_{q,\omega} t := \int_{\omega_0}^b f(t) d_{q,\omega} t - \int_{\omega_0}^a f(t) d_{q,\omega} t, \tag{2.2}$$

$$\int_{\omega_0}^x f(t) d_{q,\omega} t := (x(1-q) - \omega) \sum_{k=0}^{\infty} q^k f(xq^k + \omega [k]_q), \quad x \in I, \tag{2.3}$$

where the series in (2.3) converges at $x = a$ and $x = b$. Here $[k]_q$ is the q-number $[k]_q := \frac{1-q^k}{1-q}, k \in \mathbb{N}_0$. In this case, f is called q,ω-integrable on $[a, b]$ and the sum to the right-hand side of (2.3) will be called the Jackson–Nörlund sum. See [49] for relationship between Nörlund sums and the difference operators. The fundamental theorem of q,ω-calculus given in [16] states that if $f : I \longrightarrow \mathbb{R}$ is continuous at ω_0, and

$$F(t) := \int_{\omega_0}^{t} f(x) d_{q,\omega} x . t \in I,$$

then F is continuous at ω_0. Furthermore, $D_{q,\omega}F(t)$ exists for every $t \in I$ and

$$D_{q,\omega}F(t) = f(t).$$

Conversely,

$$\int_a^b D_{q,\omega} f(t) d_{q,\omega} t = f(b) - f(a) \quad \text{for all } \ a, b \in I.$$

The q, ω-integration by parts for continuous functions f, g is given in [16] by

$$\int_a^b f(t) D_{q,\omega} g(t) d_{q,\omega} t$$

$$= f(t)g(t)|_a^b - \int_a^b D_{q,\omega}(f(t)) g(qt+\omega) d_{q,\omega} t, \quad a, b \in I. \tag{2.4}$$

Also in [16], some simple q, ω-difference equations are studied. As in the q-setting, see, e.g., [36, 47], there are two types of the q, ω-exponential functions, introduced in [16], namely the two parameter q, ω-exponential functions

$$e_{q,\omega}(t, z) = \sum_{k=0}^{\infty} \frac{(z(t(1-q)-\omega)^k}{(q,q)_k}, \quad |t-\omega_0| < \frac{1}{|z(1-q)|}, \tag{2.5}$$

$$E_{q,\omega}(t, z) = \sum_{k=0}^{\infty} \frac{q^{\frac{1}{2}k(k-1)}(z(t(1-q)-\omega)^k}{(q,q)_k}, \quad t \in \mathbb{C}, \tag{2.6}$$

where $(q;q)_k$ is the q-shifted factorial, and $z \in \mathbb{C}$ is a parameter. They satisfy first-order initial value problem (cf. [16])

$$D_{q,\omega} e_{q,\omega}(t, z) = z e_{q,\omega}(t, z), \ e_{q,\omega}(\omega_0, z) = 1, \ |t-\omega_0| < \frac{1}{|z(1-q)|},$$

$$D_{q,\omega} E_{q,\omega}(t, z) = -z E_{q,\omega}(qt+\omega, z), \quad E_{q,\omega}(\omega_0, z) = 1, \quad z, t \in \mathbb{C}.$$

The existence and uniqueness theorems for general first order q, ω-initial value problems are studied in [41]. The rest of this section includes some results from [41], which will be needed in the sequel. Let $(X, \|\cdot\|)$ be a

Banach space, $b > 0$ and

$$S(x_0, b) := \{x \in X : \|x - x_0\| \leq b\}.$$

Let R be the rectangle $R := [\omega_0, \omega_0 + a] \times S(x_0, b)$ where a is positive fixed numbers.

Theorem 2.2. *Assume that $f : R \longrightarrow X$ satisfies the following conditions:*

(i) *$f(t, x)$ is continuous at $t = \omega_0$, for every $x \in S(x_0, b)$.*
(ii) *There exists a positive constant A such that the following Lipschitz condition is fulfilled*

$$\|f(t, x) - f(t, y)\| \leq A\|x - y\|, \quad \forall x, y \in S(x_0, b).$$

Then, there exists $l > 0$ such that the initial value problem

$$D_{q,\omega}x(t) = f(t, x), \qquad x(\omega_0) = x_0.$$

has a unique solution $x(t)$ valid in $[\omega_0, \omega_0 + l]$.

Theorem 2.2 is of a generalized sense, because the Banach space $(X, \|\cdot\|)$ is not characterized. In the next section, we will introduce a specific Hilbert space and we will study the existence and uniqueness theorem for our problem in Section 4. Similar to the study of the q-Wronskian established in [3, 62], the q, ω-Wronskian is defined in [40] for functions $x_1, x_2, x_3, \ldots, x_n$ defined on I, to be

$$W_{q,\omega}(x_1, x_2, \ldots, x_n)(t)$$

$$= \begin{vmatrix} x_1(t) & x_2(t) & \cdots & x_n(t) \\ D_{q,\omega}x_1(t) & D_{q,\omega}x_2(t) & \cdots & D_{q,\omega}x_n(t) \\ \vdots & \vdots & \ddots & \vdots \\ D_{q,\omega}^{n-1}x_1(t) & D_{q,\omega}^{n-1}x_2(t) & \cdots & D_{q,\omega}^{n-1}x_n(t) \end{vmatrix}, \quad t \in I,$$

where $x_1, x_2, x_3, \ldots, x_n$ are q, ω-differentiable functions of $n - 1$ order. It is proved for suitable functions $p_0(\cdot), p_1(\cdot), \ldots, p_n(\cdot)$ that, if $x_1, x_2, \ldots, x_n$ are solutions of the equation,

$$p_0(t)D_{q,\omega}^n x(t) + p_1(t)D_{q,\omega}^{n-1}x(t) + \cdots + p_n(t)x(t) = 0, \tag{2.7}$$

then their q, ω-Wronskian satisfies the first-order Hahn difference equation

$$D_{q,\omega}W_{q,\omega}(t) = -R(t)W_{q.\omega}(t), \quad \forall t \in J \setminus \{\omega_0\}, \tag{2.8}$$

for some interval $J \subset I$ that includes ω_0, where

$$R(t) := \sum_{k=0}^{n-1} (t - h(t))^k p_{k+1}/p_0. \tag{2.9}$$

Moreover, cf. [40], if $(h(t) - t)R(t) \neq 1, t \in J$, then for any set of solutions $\{\psi_i(t)\}_{i=1}^n$, of (2.7) valid in J, we have

$$W_{q,\omega}(t) = \frac{W_{q,\omega}(\omega_0)}{\prod_{k=0}^{\infty}(1 + q^k(t(1-q) - \omega)R(h^k(t)))}, \quad t \in J. \tag{2.10}$$

Consequently, as in the case of linear differential and q-difference equations $W_{q,\omega}(t) \neq 0$ in J if and only if $\{\psi_i\}_{i=1}^n$ is a fundamental set of solutions of (2.7) valid in J.

3. Self-Adjoint Formulations

In the following, we formulate a self-adjoint q,ω-Sturm–Liouville problem (q,ω-SLP) in a suitable Hilbert space. The construction of the Hilbert space is a crucial step in defining a self-adjoint SLP of any type. Similar works in other settings could be found in [10, 12, 19, 20, 38, 57]. Let $b > 0$ be fixed and $L^2_{q,\omega}(\omega_0, b)$ be the set of all complex-valued functions defined on $[\omega_0, b]$ for which

$$\|f(\cdot)\| = \left(\int_{\omega_0}^{b} |f(t)|^2 d_{q,\omega}t\right)^{\frac{1}{2}} < \infty.$$

It is not hard to see that $L^2_{q,\omega}(\omega_0, b)$ is a linear space over $\mathbb{C}$, with the standard addition and scalar multiplication. The space $L^2_{q,\omega}(\omega_0, b)$ is formed from equivalence classes, where f, g belong to the same equivalence class if $f(h^k(b)) = g(h^k(b))$, $k \in \mathbb{N}$. In particular, the zero element is the class of all functions f such that $f(h^k(b)) = 0$, $k \in \mathbb{N}$. As in the classical setting, a class is represented by any of its members. Figures 1 and 2 illustrate the functions $h(t)$, $h^{-1}(t)$ and that $h^k(t) \to \omega_0$ as $k \to \infty$.

Lemma 3.1. *$L^2_{q,\omega}(\omega_0, b)$ is a separable Hilbert space, under the inner product*

$$\langle f, g \rangle := \int_{\omega_0}^{b} f(t)\overline{g(t)} d_{q,\omega}t,$$

where $\overline{z}$ denotes the complex conjugate of $z \in \mathbb{C}$.

Proof. Because of Hölder's inequality, the inner product $\langle \cdot, \cdot \rangle$ is well defined, and the rest of its properties can be easily checked. Thus, $L^2_{q,\omega}(\omega_0, b)$ is pre-Hilbert space. Let $f_n(\cdot)$ be a Cauchy sequence of $L^2_{q,\omega}(\omega_0, b)$, i.e.,

$$\|f_n - f_m\|^2 = (b(1-q)-\omega) \times \sum_{k=0}^{\infty} q^k |(f_n(h^k(b)) - f_m(h^k(b))|^2 \tag{3.1}$$

tends to zero as n, m tend ∞. Hence, for each $k \in \mathbb{N}$, $f_n(h^k(b))$ is a Cauchy sequence of $\mathbb{C}$. Define the function $f(\cdot)$ on (ω_0, b) to be

$$\lim_{n \longrightarrow \infty} f_n(h^k(b)) := f(h^k(b)), \quad k \in \mathbb{N},$$

and f takes any other values on the remaining points of (ω_0, b). We prove that $f \in L^2_{q,\omega}(\omega_0, b)$ and $\|f_n - f\| \longrightarrow 0$ as $n \longrightarrow \infty$. Indeed, for any positive integer j,

$$\begin{aligned}\sum_{k=0}^{j} q^k |f(h^k(b))|^2 &= \lim_{n \longrightarrow \infty} \sum_{k=0}^{j} q^k |(f_n(h^k(b))|^2 \\ &\leqslant M, \quad M := \sup_n \|f_n\| < \infty.\end{aligned} \tag{3.2}$$

Therefore, $f \in L^2_{q,\omega}(\omega_0, b)$. Now, we prove that $\|f_n - f\| \longrightarrow 0$. Let $\varepsilon > 0$ be given. There exists a positive integer N such that if $n, m \geqslant N$, and P is any positive integer, then

$$(b(1-q)-\omega) \sum_{k=0}^{P} q^k |f_n(h^k(b)) - f_m(h^k(b))|^2 \leqslant \|f_n - f_m\| < \varepsilon. \tag{3.3}$$

Fix $n \geqslant N$. From (3.1) and (3.3), we have for all P

$$\begin{aligned}&\sum_{k=0}^{P} q^k |f_n(h^k(b)) - f(h^k(b))|^2 \\ &\quad \leqslant \lim_{m \longrightarrow \infty} \sum_{k=0}^{P} q^k |f_n(h^k(b)) - f_m(h^k(b))|^2 < \frac{\varepsilon}{(b(1-q)-\omega)}.\end{aligned} \tag{3.4}$$

Thus, for $n \geqslant N$,

$$\|f_n - f\| = (b(1-q)-\omega) \sum_{k=0}^{\infty} q^k |f_n(h^k(b)) - f(h^k(b))|^2 < \varepsilon. \tag{3.5}$$

Recall that $b(1-q)-\omega > \omega_0(1-q)-\omega = 0$. To prove separability, it suffices to show that

$$\phi_n(t) = \begin{cases} \dfrac{1}{\sqrt{t-h(t)}}, & t = h^n(b),\ n \in \mathbb{N}_0, \\ 0 & \text{otherwise} \end{cases} \tag{3.6}$$

is an orthonormal basis of $L^2_{q,\omega}(\omega_0, b)$. A simple calculation yields

$$\begin{aligned}\langle \phi_n, \phi_m \rangle &= \int_{\omega_0}^{b} \phi_n(t)\overline{\phi_m(t)}d_{q,\omega}t \\ &= (b(1-q)-\omega)\sum_{k=0}^{\infty} q^k \phi_n(h^k(b))\overline{\phi_m(h^k(b))} = \delta_{nm}.\end{aligned}$$

Let $f \in L^2_{q,\omega}(\omega_0, b)$ be such that $\langle f, \phi_n \rangle = 0, n \in \mathbb{N}_0$. One can see that f is the zero element. Indeed,

$$\begin{aligned}0 = \langle f, \phi_n \rangle &= \int_{\omega_0}^{b} f(t)\overline{\phi_n(t)}d_{q,\omega}t \\ &= \sum_{k=0}^{\infty}(h^k(b) - h^{k+1}(b))f(h^k(b))\overline{\phi_n(h^k(b))} \\ &= \big(h^n(b) - h^{n+1}(b)\big)^{\frac{1}{2}} f(h^n(b)), \quad \forall n \in \mathbb{N}_0.\end{aligned}$$

Clearly $(h^n(b) - h^{n+1}(b))^{\frac{1}{2}} \neq 0$, implying that $f(h^n(b)) = 0$ for all $n \in \mathbb{N}_0$. □

As in the classical setting, see, e.g., [37, 65], the space $L^2_{q,\omega}\big((\omega_0, b) \times (\omega_0, b)\big)$ consists of all complex-valued functions $f(t,\xi)$ defined on $(\omega_0, b) \times (\omega_0, b)$ such that

$$\|f(\cdot,\cdot)\|_2 = \left(\int_{\omega_0}^{b}\int_{\omega_0}^{b} |f(t,\xi)|^2 d_{q,\omega}t d_{q,\omega}\xi\right)^{\frac{1}{2}} < \infty.$$

The zero element is the equivalence class of all functions $f(t,\xi)$ satisfying $f(h^m(b), h^n(b)) = 0$ for all $m, n \in \mathbb{N}_0$. The inner product of $L^2_{q,\omega}\big((\omega_0, b) \times (\omega_0, b)\big)$ is defined by

$$\langle f, g \rangle_2 = \int_{\omega_0}^{b}\int_{\omega_0}^{b} f(t,\xi)\overline{g(t,\xi)}d_{q,\omega}t \ \ d_{q,\omega}\xi.$$

$L^2_{q,\omega}((\omega_0, b) \times (\omega_0, b))$ is also a separable Hilbert space, cf. [14] for a similar treatment when $\omega = 0$. We can easily verify that

$\{\phi_{ij}(t,s) := \phi_i(t) \times \phi_j(s)\}_{i,j\in\mathbb{N}_0}$ is an orthonormal basis of $L^2_{q,\omega}((\omega_0,b) \times (\omega_0,b))$. Before defining the q,ω-Sturm–Lioville problem in $L^2_{q,\omega}(\omega_0,b)$, we derive some preliminary results.

Lemma 3.2. *Let $f(\cdot)$, $g(\cdot) \in L^2_{q,\omega}(\omega_0,b)$ be defined on $[\omega_0, h^{-1}(b)]$ and be continuous at ω_0. Then, for $t \in (\omega_0, b]$, we have*

$$(D_{q,\omega}f)(h^{-1}(t)) = D_{\frac{1}{q},\frac{-\omega}{q}}f(t), \tag{3.7}$$

$$\langle D_{q,\omega}f, g\rangle = f(b)\overline{g(h^{-1}(b))} - f(\omega_0)\overline{g(\omega_0)} + \left\langle f, -\frac{1}{q}D_{\frac{1}{q},\frac{-\omega}{q}}g\right\rangle, \tag{3.8}$$

$$\left\langle -\frac{1}{q}D_{\frac{1}{q},\frac{-\omega}{q}}f, g\right\rangle = -f(h^{-1}(b))\overline{g(\omega_0)} + f(\omega_0)\overline{g(\omega_0)} + \langle f, D_{q,\omega}g\rangle. \tag{3.9}$$

Proof. Equation (3.7) directly follows from the definition $D_{q,\omega}$. Using formula (2.4) of q,ω-integration by parts, the substitution $u = h(t)$ and the fundamental theorem of q,ω-calculus, we obtain

$$\begin{aligned}
\langle D_{q,\omega}f, g\rangle &= \int_{\omega_0}^{b} D_{q,\omega}f(t)\overline{g(t)}d_{q,\omega}t \\
&= f(b)\overline{g(b)} - f(\omega_0)\overline{g(\omega_0)} - \int_{\omega_0}^{b} f(h(t))\overline{D_{q,\omega}g(t)}d_{q,\omega}t \\
&= f(b)\overline{g(b)} - f(\omega_0)\overline{g(\omega_0)} - \int_{\omega_0}^{h(b)} f(u)\frac{1}{q}\overline{D_{\frac{1}{q},\frac{-\omega}{q}}g(u)}d_{q,\omega}u \\
&= f(b)\overline{g(b)} - f(\omega_0)\overline{g(\omega_0)} + \frac{1}{q}(b-h(b))f(b)\overline{D_{\frac{1}{q},\frac{-\omega}{q}}g(b)} \\
&\quad - \int_{\omega_0}^{b} f(u)\frac{-1}{q}\overline{D_{\frac{1}{q},\frac{-\omega}{q}}g(u)}d_{q,\omega}u.
\end{aligned} \tag{3.10}$$

Relation (3.8) results from (3.10) after applying relation (3.7). Also, (3.9) results by the use of (3.8), since

$$\begin{aligned}
\left\langle -\frac{1}{q}D_{\frac{1}{q},\frac{-\omega}{q}}f, g\right\rangle &= \overline{\left\langle g, -\frac{1}{q}D_{\frac{1}{q},\frac{-\omega}{q}}f\right\rangle} \\
&= \overline{-g((b))\overline{f(h^{-1}(b))} + g(\omega_0)\overline{f(\omega_0)} + \langle D_{q,\omega}g, f\rangle} \\
&= -f(h^{-1}(b))\overline{g((b)} + f(\omega_0)\overline{g(\omega_0)} + \langle f, D_{q,w}g\rangle.
\end{aligned}$$

$\square$

Remark 3.3. The previous lemma generalizes and corrects Lemma 4.1 of [20], where the condition of continuity at 0 is dropped.

Now, we are ready to define a self-adjoint q,ω-SLP in $L^2_{q,\omega}(\omega_0, b)$. We define the eigenvalue problem, which consists of the q,ω-difference equation

$$\ell y(t) := -\frac{1}{q} D_{\frac{1}{q}, \frac{-\omega}{q}} D_{q,\omega} y(t) + p(t)y(t) = \lambda y(t), \tag{3.11}$$

where $\omega_0 \leq t \leq b < \infty, \lambda \in \mathbb{C}$, and the boundary conditions

$$U_1(y) := a_1 y(\omega_0) + a_2 D_{\frac{1}{q}, \frac{-\omega}{q}} y(\omega_0) = 0, \tag{3.12}$$

$$U_2(y) := b_1 y(b) + b_2 D_{\frac{1}{q}, \frac{-\omega}{q}} y(b) = 0. \tag{3.13}$$

Here, $p(\cdot)$ is a real-valued continuous function on $[\omega_0, b]$ and $\{a_i, b_i\}_{i=1,2}$ are arbitrary real numbers for which $|a_1| + |a_2| \neq 0 \neq |b_1| + |b_2|$.

Theorem 3.4. *The Sturm–Liouville eigenvalue problem* (3.11)–(3.13) *is self-adjoint in* $L^2_{q,\omega}(\omega_0, b)$.

Proof. For $y(\cdot)$, $z(\cdot)$ in $L^2_{q,\omega}(\omega_0, b)$, we have the following q,ω-Lagrange's identity:

$$\int_{\omega_0}^{b} (\ell y(t)\overline{z(t)} - y(t)\overline{\ell z(t)}) d_{q,\omega} t = [y,\ \overline{z}](b) - [y,\ \overline{z}](\omega_0), \tag{3.14}$$

where

$$[y\ ,\ z](t) = y(t)\overline{D_{\frac{1}{q}, \frac{-\omega}{q}} z(t)} - D_{\frac{1}{q}, \frac{-\omega}{q}} y(t)\overline{z(t)}. \tag{3.15}$$

Indeed, applying (3.9) with $f(t) = D_{q,\omega} y(t)$, and $g(t) = z(t)$, we obtain

$$\begin{aligned}\left\langle -\frac{1}{q} D_{\frac{1}{q}, \frac{-\omega}{q}} D_{q,\omega} y(t), z(t) \right\rangle &= -(D_{q,\omega} y)(h^{-1}(b))\overline{z(b)} + D_{q,\omega} y(\omega_0)\overline{z(\omega_0)} \\ &\quad + \langle D_{q,\omega} y(t), D_{q,w} z(t)\rangle \\ &= -D_{\frac{1}{q}, \frac{-\omega}{q}} y(b)\overline{z(b)} + D_{\frac{1}{q}, \frac{-\omega}{q}} y(\omega_0)\overline{z(\omega_0)} \\ &\quad + \langle D_{q,\omega} y(t), D_{q,\omega} z(t)\rangle. \end{aligned} \tag{3.16}$$

Applying (3.9) again with $f(t) = y(t)$, $g(t) = D_{q,\omega}z(t)$,

$$\begin{aligned}\langle D_{q,\omega}y(t), D_{q,\omega}z(t)\rangle &= y(b)D_{q,\omega}\overline{z(h^{-1}(b))} - y(\omega_0)D_{q,\omega}\overline{z(\omega_0)}\\ &\quad + \left\langle y(t), -\frac{1}{q}D_{\frac{1}{q},\frac{-\omega}{q}}D_{q,\omega}z(t)\right\rangle\\ &= y(b)D_{\frac{1}{q},\frac{-\omega}{q}}\overline{z(b)} - y(\omega_0)D_{\frac{1}{q},\frac{-\omega}{q}}\overline{z(\omega_0)}\\ &\quad + \left\langle y(t), -\frac{1}{q}D_{\frac{1}{q},\frac{-\omega}{q}}D_{q,\omega}z(t)\right\rangle. \end{aligned} \tag{3.17}$$

Combining (3.16) and (3.17) leads to

$$\begin{aligned}\left\langle -\frac{1}{q}D_{\frac{1}{q},\frac{-1}{\omega}}D_{q,\omega}y(t), z(t)\right\rangle &= [y,\ z](b) - [y,\ z](\omega_0)\\ &\quad + \left\langle y(t), -\frac{1}{q}D_{\frac{1}{q},\frac{-\omega}{q}}D_{q,\omega}z(t)\right\rangle. \end{aligned} \tag{3.18}$$

Assuming that y, z satisfy the boundary conditions and since $|a_1| + |a_2| \neq 0 \neq |b_1| + |b_2|$, it is concluded that

$$y(\omega_0)D_{\frac{1}{q},\frac{-\omega}{q}}\overline{z(\omega_0)} - \overline{z(\omega_0)}D_{\frac{1}{q},\frac{-\omega}{q}}y(\omega_0) = 0,$$

$$y(b)D_{\frac{1}{q},\frac{-\omega}{q}}\overline{z(b)} - \overline{z(b)}D_{\frac{1}{q},\frac{-\omega}{q}}y(b) = 0.$$

Since $p(t)$ is real valued, then

$$\begin{aligned}\langle \ell y, z\rangle &= \left\langle -\frac{1}{q}D_{\frac{1}{q},\frac{-\omega}{q}}D_{q,\omega}y(t) + p(t)y(t), z(t)\right\rangle\\ &= \left\langle -\frac{1}{q}D_{\frac{1}{q},\frac{-\omega}{q}}D_{q,\omega}y(t), z(t)\rangle + \langle p(t)y(t), z(t)\right\rangle\\ &= \left\langle y, -\frac{1}{q}D_{\frac{1}{q},\frac{-\omega}{q}}D_{q,\omega}z(t)\right\rangle + \langle p(t)y(t), z(t)\rangle = \langle y, \ell z\rangle. \qquad \square\end{aligned}$$

In the following we investigate the properties of eigenvalues and eigenfunctions of problem (3.11)–(3.13). A number $\lambda \in \mathbb{C}$ is said to be an eigenvalue of problem (3.11)–(3.13) if the problem (3.11)–(3.13) has a nontrivial solution $\phi(t)$, in the sense of the space $L^2_{q,\omega}(\omega_0, b)$, corresponding to the eigenvalue λ. In this case $\phi(t)$ is called an eigenfunction of the problem (3.11)–(3.13) corresponding to the eigenvalue λ. The set of all eigenfunctions of (3.11)–(3.13) corresponding an eigenvalue λ together with the zero function defined above form a subspace of $L^2_{q,\omega}(\omega_0, b)$, which is called the eigenspace of λ. The dimension of the eigenspace of an eigenvalue λ is called its geometric multiplicity. In particular λ is geometrically simple if the dimension of the associated eigenspace is 1. The following lemma contains some expected properties of eigenvalues and eigenfunctions of the self-adjoint problem (3.11)–(3.13).

Lemma 3.5. *All eigenvalues of problem* (3.11)–(3.13) *are real and simple from geometric point of view. Eigenfunctions corresponding different eigenvalues are orthogonal.*

Proof. Let λ_0 be an eigenvalue and $\psi(t)$ be a corresponding eigenfunction. Since

$$\ell\psi(t) = \lambda_0\psi(t), \qquad \overline{\ell\psi(t)} = \overline{\lambda_0}\,\overline{\psi(t)},$$

then

$$\int_{\omega_0}^{b} \ell\psi(t)\overline{\psi(t)}d_{q,w}t = \int_{\omega_0}^{b} \psi(t)\overline{\ell\psi(t)}d_{q,\omega}t.$$

Hence

$$\lambda_0 \int_{\omega_0}^{b} |\psi(t)|^2 d_{q,w}t = \overline{\lambda_0} \int_{\omega_0}^{b} |\psi(t)|^2 d_{q,\omega}t,$$

and consequently

$$(\lambda_0 - \overline{\lambda_0}) \int_{\omega_0}^{b} |\psi(t)|^2 d_{q,\omega}t = 0.$$

Since $\psi(t)$ is an eigenfunction, then $\lambda_0 = \overline{\lambda_0}$. Now, let λ_0 be an eigenvalue with two eigenfunctions ϕ_1 and ϕ_2. We prove that the ϕ_1 and ϕ_2 are linearly dependent by proving that their q,ω-Wronskian vanishes at $t = \omega_0$. Indeed,

$$\begin{aligned} W_{q,\omega}(\phi_1,\phi_2)(\omega_0) &= \phi_1(\omega_0)D_{q,\omega}\phi_2(\omega_0) - \phi_2(\omega_0)D_{q,\omega}\phi_1(\omega_0) \\ &= \phi_1(\omega_0)D_{\frac{1}{q},\frac{-\omega}{q}}\phi_2(\omega_0) - \phi_2(\omega_0)D_{\frac{1}{q},\frac{-\omega}{q}}\phi_2(\omega_0) \\ &= [\phi_1,\phi_2](\omega_0) = 0, \end{aligned}$$

since ϕ_1, ϕ_2 satisfy (3.12). Finally, let $\lambda_1 \neq \lambda_2$ be distinct eigenvalues with eigenfunctions $\psi_1(t)$ and $\psi_2(t)$, respectively. From q,ω-Lagrange's identity

(3.18) we obtain

$$\lambda_1 \int_{\omega_0}^{b} \psi_1(t)\overline{\psi_2(t)}d_{q,\omega}t = \lambda_2 \int_{\omega_0}^{b} \psi_1(t)\overline{\psi_2(t)}d_{q,\omega}t.$$

Since $\lambda_1 \neq \lambda_2$, we obtain the desired orthogonality relation. □

4. Fundamental Solutions

This section is devoted to study existence, uniqueness and the analytic properties of the fundamental solutions of the q,ω-Sturm–Liouville equation (3.11). Let $C^2_{q,\omega}(\omega_0, b)$ be the subspace of $L^2_{q,\omega}(\omega_0, b)$, which consists of all functions $y(\cdot)$ for which $y(\cdot)$, $D_{q,\omega}y(\cdot)$ are continuous at ω_0. By a solution of equation (3.11), we mean a $C^2_{q,\omega}(\omega_0, b)$-function that satisfies (3.11). It is proved in [40] that (3.11) has a fundamental set of solutions which consists of two linearly independent solutions $y_1(\cdot), y_2(\cdot)$. However, the analyticity of solutions of (3.11) as functions of $\lambda \in \mathbb{C}$ has not been discussed. For this reason, we follow the technique of [20] mimicking that of [63], to study existence, uniqueness and analyticity of solutions of (3.11). For this aim, we define q,ω-trigonometric functions to be

$$C_{q,\omega}(t) = \sum_{n=0}^{\infty} \frac{(-1)^n q^{n^2}(t(1-q)-\omega)^{2n}}{(q;q)_{2n}}, \tag{4.1}$$

$$S_{q,\omega}(t) = \sum_{n=0}^{\infty} \frac{(-1)^n q^{n(n+1)}(t(1-q)-\omega)^{2n+1}}{(q;q)_{2n+1}}, \quad t \in \mathbb{C}. \tag{4.2}$$

It is not hard to see that $C_{q,\omega}(\cdot)$ and $S_{q,\omega}(\cdot)$ satisfy the initial value problems

$$\frac{-1}{q} D_{\frac{1}{q},\frac{-\omega}{q}} D_{q,\omega}y(t) = y(t), \quad t \in \mathbb{R}, \tag{4.3}$$

subject to the initial conditions

$$y(\omega_0) = 1, \quad D_{q,\omega}y(\omega_0) = 0; \quad y(\omega_0) = 0, \quad D_{q,\omega}y(\omega_0) = 1,$$

respectively. If we let $t \in \mathbb{C}$, then it can be proved that $C_{q,\omega}(t)$ and $S_{q,\omega}(t)$ are entire functions of order zero. We also prove the following lemma that we will need later on.

Lemma 4.1. *The q,ω-Wronskian of solutions of* (3.11) *is independent of both t and λ.*

Proof. Equation (3.11) is equivalent to

$$\frac{-1}{q} D^2_{q,\omega} y(t) = (\lambda - p(h(t)))y(h(t)), \quad t \in [\omega_0, h(b)], \ \lambda \in \mathbb{C}. \tag{4.4}$$

But $y(h(t)) = y(t) - (t - h(t))D_{q,\omega}y(t)$. Hence

$$D^2_{q,\omega}y(t) + q(\lambda - p(h(t)))(t - h(t))D_{q,\omega}y(t) - q(\lambda - p(h(t)))y(t) = 0. \tag{4.5}$$

Comparing (4.5) with (2.7), we get

$$p_0 = 1, \ \ p_1 = q(\lambda - p(h(t)))(t - h(t)) \quad \text{and} \quad p_2 = -q(\lambda - p(h(t))).$$

Thus $R(t) = 0$; cf. (2.9). Therefore, from (2.10) the q, ω-Wronskian is

$$\begin{aligned} W_{q,\omega}(t) &= \frac{W_{q,\omega}(\omega_0)}{\prod_{k=0}^{\infty}(1 + q^k(t(1-q) - \omega)R(h^k(t)))} \\ &= W_{q,\omega}(\omega_0), \quad t \in (\omega_0, h(b)), \quad \lambda \in \mathbb{C}. \end{aligned}$$

□

Theorem 4.2. *For $\lambda \in \mathbb{C}$, equation (3.11) has a unique solution $\phi(t, \lambda)$ subject to the initial conditions*

$$\phi(\omega_0, \lambda) = c_1, \quad D_{\frac{1}{q}, -\frac{w}{q}}\phi(\omega_0, \lambda) := \lim_{t \longrightarrow \omega_0^+} \frac{\phi(t,\lambda) - \phi(\omega_0, \lambda)}{t - \omega_0} = c_2, \tag{4.6}$$

where $c_1, c_2 \in \mathbb{C}$. Furthermore, for $t \in (\omega_0, b)$, $\phi(t, \lambda)$ is entire in λ.

Proof. Let $\mu := \sqrt{\lambda}$ be the principal branch. Hence

$$\chi_1(t,\lambda) = C_{q,\omega}(t,\mu) \quad \text{and} \quad \chi_2(t,\lambda) = \begin{cases} \dfrac{S_{q,\omega}(t,\mu)}{\mu}, & \mu \neq 0, \\ t, & \mu = 0 \end{cases} \tag{4.7}$$

form a fundamental set of solutions for

$$-\frac{1}{q} D_{\frac{1}{q}, \frac{-\omega}{q}} D_{q,\omega} y(t) - \lambda y(t) = 0. \tag{4.8}$$

Here

$$C_{q,\omega}(t,\mu)=\sum_{n=0}^{\infty}\frac{(-1)^n q^{n^2}(\mu(t(1-q)-\omega))^{2n}}{(q;q)_{2n}},$$

$$S_{q,\omega}(t,\mu)=\sum_{n=0}^{\infty}\frac{(-1)^n q^{n(n+1)}(\mu(t(1-q)-\omega))^{2n+1}}{(q;q)_{2n+1}},\quad t\in\mathbb{C}.$$

From the previous lemma, $W_{q,\omega}(\chi_1(t,\lambda),\chi_2(t,\lambda))\equiv 1$. Define the sequence $\{\varphi_m(\cdot,\lambda)\}_{m=1}^{\infty}$ of successive approximations by

$$\varphi_1(t,\lambda)=c_1\chi_1(t,\lambda)+c_2\chi_2(t,\lambda), \tag{4.9}$$

$$\begin{aligned}\varphi_{m+1}(t,\lambda)&=c_1\chi_1(t,\lambda)+c_2\chi_2(t,\lambda)+q\int_{\omega_0}^{t}[\chi_2(t,\lambda)\chi_1(h(s),\lambda)\\&\quad-\chi_1(t,\lambda)\chi_2(h(s),\lambda)]p(h(s))\varphi_m(h(s))d_{q,\omega}s,\end{aligned} \tag{4.10}$$

$m=1,2,3,\ldots$. For a fixed $\lambda\in\mathbb{C}$, there exist positive numbers $B(\lambda)$ and A independent of t such that

$$|p(t)|\leqslant A,\quad\text{and}\quad|\chi_i(t,\lambda)|\leqslant\sqrt{\frac{B(\lambda)}{2}},\quad i=1,2,\quad t\in(\omega_0,b). \tag{4.11}$$

Let $K(\lambda):=(|c_1|+|c_2|)\sqrt{\frac{B(\lambda)}{2}}$. Then $|\varphi_1(t,\lambda)|\leqslant K(\lambda)$ for all $t\in(\omega_0,b)$. Using mathematical induction, we prove that

$$\begin{aligned}&|\varphi_{m+1}(t,\lambda)-\varphi_m(t,\lambda)|\\&\leqslant K(\lambda)q^{\frac{m(m+1)}{2}}\frac{(AB(\lambda)(t(1-q)-\omega))^m}{(q,q)_m},\quad m=1,2,\ldots.\end{aligned} \tag{4.12}$$

First, triangle inequality implies

$$\begin{aligned}|\varphi_2(t,\lambda)-\varphi_1(t,\lambda)|=q\bigg|\int_{\omega_0}^{t}&[\chi_2(t,\lambda)\chi_1(h(s),\lambda)-\chi_1(t,\lambda)\chi_2(h(s),\lambda)]\\&\times p(h(s))\varphi_1(h(s))d_{q,\omega}s\bigg|\end{aligned}$$

$$\begin{aligned} &\leqslant q\,A\,K(\lambda)\,2\left(\sqrt{\frac{B(\lambda)}{2}}\right)^2 \int_{\omega_0}^{t} d_{q,\omega}s \\ &= q\,A\,K(\lambda)\,B(\lambda)\frac{[t(1-q)-\omega]}{1-q}. \end{aligned} \tag{4.13}$$

Assume the correctness of (4.12) for some m. Now

$$\begin{aligned} &|\varphi_{m+2}(t,\lambda) - \varphi_{m+1}(t,\lambda)| \\ &\leqslant q\left|\int_{\omega_0}^{t} [\chi_2(t,\lambda)\chi_1(h(s),\lambda) - \chi_1(t,\lambda)\chi_2(h(s),\lambda)]\right. \\ &\qquad \left.\times\, p(h(s))(\varphi_{m+1}(h(s),\lambda) - \varphi_m(h(s),\lambda))d_{q,\omega}s\right| \\ &\leqslant qAB(\lambda)\int_{\omega_0}^{t} |\varphi_{m+1}(h^{k+1}(t),\lambda) - \varphi_m(h^{k+1}(t),\lambda)|d_{q,\omega}s \\ &\leqslant qK(\lambda)q^{\frac{m(m+1)}{2}}\frac{(AB(\lambda)(t(1-q)-\omega))^{m+1}}{(q,q)_m}\sum_{k=0}^{\infty} q^{k+mk+m} \\ &\leqslant K(\lambda)q^{\frac{m^2+3m+2}{2}}\frac{(AB(\lambda)(t(1-q)-\omega))^{m+1}}{(q,q)_m}\frac{1}{(1-q^{m+1})} \\ &\leqslant K(\lambda)q^{\frac{(m+1)(m+2)}{2}}\frac{(AB(\lambda)(t(1-q)-\omega))^{m+1}}{(q,q)_{m+1}}. \end{aligned} \tag{4.14}$$

Therefore, inequality (4.12) is true for every $m \in \mathbb{N}$. Inequality (4.12) guarantees the uniform convergence of the series

$$\varphi_1(t,\lambda) + \sum_{m=1}^{\infty} \varphi_{m+1}(t,\lambda) - \varphi_m(t,\lambda) \tag{4.15}$$

on (ω_0, b). Since the mth partial sums of the series is $\varphi_{m+1}(\cdot,\lambda)$, then $\varphi_{m+1}(\cdot,\lambda)$ approaches a function $\phi(\cdot,\lambda)$ uniformly on (ω_0, b). We prove by induction on m that $\varphi_m(t,\lambda)$ and $D_{q,\omega}\varphi_m(t,\lambda)$ are continuous at ω_0, where

$$\begin{aligned} D_{q,\omega}\varphi_{m+1}(t,\lambda) &= c_1 D_{q,\omega}\chi_1(t,\lambda) + c_2 D_{q,\omega}\chi_2(t,\lambda) \\ &\quad + q\int_{\omega_0}^{t} [D_{q,\omega}\chi_2(t,\lambda)\chi_1(h(s),\lambda) \\ &\quad - D_{q,\omega}\chi_1(t,\lambda)\chi_2(h(s),\lambda)]p(h(s))\varphi_m(h(s))d_{q,\omega}s, \end{aligned} \tag{4.16}$$

$m = 1, 2, 3, \ldots$. Therefore, the functions $\phi(\cdot,\lambda)$ and $D_{q,\omega}\phi(\cdot,\lambda)$ are continuous at ω_0, i.e., $\phi(\cdot,\lambda) \in C^2_{q,\omega}(\omega_0, b)$. Let $m \to \infty$ in (4.9). We obtain

$$\begin{aligned}\phi(t,\lambda) = c_1\chi_1(t,\lambda) + c_2\chi_2(t,\lambda) \\ + q\int_{\omega_0}^{t} [\chi_2(t,\lambda)\chi_1(h(s),\lambda) \\ - \chi_1(t,\lambda)\chi_2(h(s),\lambda)]\, p(h(s))\phi(h(s))d_{q,\omega}s. \qquad (4.17)\end{aligned}$$

The interchange of integration and limit is guaranteed because of the uniform convergence. Now we prove that $\phi(\cdot,\lambda)$ solves (3.11) subject to (4.6). First of all, $\phi(\omega_0,\lambda) = c_1$ and

$$\begin{aligned}D_{\frac{1}{q},-\frac{\omega}{q}}\phi(\omega_0,\lambda) &= \lim_{t\longrightarrow\omega_0^+}\frac{\phi(t,\lambda)-\phi(\omega_0,\lambda)}{t-\omega_0} \\ &= c_1 D_{\frac{1}{q},-\frac{\omega}{q}}\phi_1(\omega_0,\lambda) + c_2 D_{\frac{1}{q},-\frac{\omega}{q}}\phi_2(\omega_0,\lambda) = c_2. \qquad (4.18)\end{aligned}$$

To prove that $\phi(\cdot,\lambda)$ satisfies (3.11), we distinguish between two cases. First, let $t \neq \omega_0$. From (4.17), we have

$$\begin{aligned}D_{q,\omega}\phi(t,\lambda) = c_1 D_{q,\omega}\chi_1(t,\lambda) + c_2 D_{q,\omega}\chi_2(t,\lambda) \\ + q\int_{\omega_0}^{t} [D_{q,\omega}\chi_2(t,\lambda)\chi_1(h(s),\lambda) \\ - D_{q,\omega}\chi_1(t,\lambda)\chi_2(h(s),\lambda)]\, p(h(s))\phi(h(s))d_{q,\omega}s. \qquad (4.19)\end{aligned}$$

Therefore,

$$\begin{aligned}&-\frac{1}{q}D_{\frac{1}{q},\frac{-\omega}{q}}D_{q,\omega}\phi(t,\lambda) \\ &= -\frac{1}{q}D_{\frac{1}{q},\frac{-\omega}{q}}D_{q,\omega}\chi_2(t,\lambda)\left(c_1 - q\int_{\omega_0}^{t}\chi_1(h(s),\lambda)p(h(s))\phi(h(s))d_{q,\omega}s\right) \\ &\quad -\frac{1}{q}D_{\frac{1}{q},\frac{-\omega}{q}}D_{q,\omega}\chi_1(t,\lambda) \\ &\quad \times\left(c_2 + q\int_{\omega_0}^{t}\chi_2(h(s),\lambda)p(h(s))\phi(h(s))d_{q,\omega}s\right) - p(t)\phi(t,\lambda) \\ &= \lambda\phi(t,\lambda) - p(t)\phi(t,\lambda). \qquad (4.20)\end{aligned}$$

If $t = \omega_0$, then (3.11) is nothing but

$$D^2 y(\omega_0) - qp(\omega_0)y(\omega_0) = q\lambda y(\omega_0). \tag{4.21}$$

By direct computations, we see that

$$\begin{aligned} D^2_{q,\omega}\phi(\omega_0,\lambda) &= c_1 D^2_{q,\omega}\chi_1(\omega_0,\lambda) + c_2 D^2_{q,\omega}\chi_2(\omega_0,\lambda) + qp(\omega_0)\phi(\omega_0,\lambda) \\ &= -q\lambda c_1\chi_1(\omega_0,\lambda) - q\lambda c_2\chi_2(\omega_0,\lambda) + qp(\omega_0)\phi(\omega_0,\lambda) \\ &= -q\lambda\phi(\omega_0,\lambda) + qp(\omega_0)\phi(\omega_0,\lambda), \end{aligned} \tag{4.22}$$

i.e., $\phi(t,\lambda)$ satisfies (3.11) also at $t = \omega_0$. For the uniqueness assume $\Psi_i(t,\lambda), i = 1,2$, are two solutions of (3.11), (4.6). Let $\Upsilon(t,\lambda) = \Psi_1(t,\lambda) - \Psi_2(t,\lambda), t \in (\omega_0, b)$. Then $\Upsilon(t,\lambda)$ is a solution of (3.11) subject to the initial conditions $\Upsilon(\omega_0,\lambda) = D_{\frac{1}{q},\frac{-\omega}{q}}\Upsilon(\omega_0,\lambda) = 0$. Applying the q,ω-integration by parts after integration twice to (3.11) yields

$$\Upsilon(t,\lambda) = \int_{\omega_0}^{t} (t-s)(\lambda - p(s))\Upsilon(s,\lambda)d_{q,\omega}s. \tag{4.23}$$

Since $\Upsilon(t,\lambda), p(t)$ are continuous at ω_0, then there exist positive numbers $P_{t,\lambda}, R_{t,\lambda}$ such that

$$P_{\lambda,t} = \sup_{n\in\mathbb{N}} |\Upsilon(h^n(t),\lambda)|, \qquad R_{\lambda,t} = \sup_{n\in\mathbb{N}} |\lambda - p(h^n(t)|. \tag{4.24}$$

Again we can prove by mathematical induction on k that

$$|\Upsilon(t,\lambda)| \leqslant P_{\lambda,t}R^k_{\lambda,t}q^{k^2}(1-q)^{2k}\frac{(t-\omega_0)^{2k}}{(q,q)_{2k}}, \quad k \in \mathbb{N}, \quad t \in (\omega_0, b). \tag{4.25}$$

Indeed, if (4.25) holds at $k \in \mathbb{N}$, then from (4.23)

$$\begin{aligned} |\Upsilon(t,\lambda)| &\leqslant P_{\lambda,t}R^{k+1}_{\lambda,t}q^{k^2}\frac{(1-q)^{2k}}{(q,q)_{2k}}\int_{\omega_0}^{t}(t-s)(s-\omega_0)^{2k}d_{q,\omega}s \\ &= P_{\lambda,t}R^{k+1}_{\lambda,t}q^{k^2}\frac{(1-q)^{2k}}{(q,q)_{2k}} \\ &\quad \times \sum_{n=0}^{\infty} q^n(t(1-q)-\omega)(t-h^n(t))(h^n(t)-\omega_0)^{2k} \end{aligned}$$

$$= P_{\lambda,t} R_{\lambda,t}^{k+1} q^{k^2} \frac{(1-q)^{2k}}{(q,q)_{2k}}$$

$$\times \sum_{n=0}^{\infty} q^n (1-q)(t-\omega_0)(1-q^n)(t-\omega_0)(q^n(t-\omega_0))^{2k}$$

$$= P_{\lambda,t} R_{\lambda,t}^{k+1} q^{k^2} \frac{(1-q)^{2k+1}(t-\omega_0)^{2k+2}}{(q,q)_{2k}} \sum_{n=0}^{\infty} q^{(2k+1)n}(1-q^n)$$

$$= P_{\lambda,t} R_{\lambda,t}^{k+1} q^{k^2} \frac{(1-q)^{2k+1}(t-\omega_0)^{2k+2}}{(q,q)_{2k}} \left(\frac{q^{2k+1}(1-q)}{(1-q^{2n+1})(1-q^{2n+2})} \right)$$

$$= P_{\lambda,t} R_{\lambda,t}^{k+1} q^{(k+1)^2} \frac{(1-q)^{2k+2}(t-\omega_0)^{2k+2}}{(q,q)_{2k+2}}.$$

Hence (4.25) holds true at $k+1$. Consequently (4.25) is true for all $k \in \mathbb{N}$ because from (4.23) it is satisfied at $k=0$. Since

$$\lim_{k \longrightarrow \infty} P_{\lambda,t} R_{\lambda,t}^{k} q^{k^2} (1-q)^{2k} \frac{(t-\omega_0)^{2k}}{(q,q)_{2k}} = 0,$$

then $\Upsilon(t,\lambda) = 0$, for all $t \in (\omega_0, b)$. This proves the uniqueness.

It remains to prove that $\phi(t,\lambda), t \in (\omega_0, b)$, is entire in λ. It suffices to prove that $\phi(t,\lambda)$ is analytic in each disc $\mathscr{D}_\rho; \mathscr{D}_\rho := \{\lambda \in \mathbb{C} : |\lambda| \leqslant \rho\}$ for $t \in [\omega_0, b]$, for arbitrary positive number ρ. We first prove that for $t \in [\omega_0, b]$ both $\varphi_m(t,\lambda)$ and $\frac{\partial}{\partial \lambda}\varphi_m(t,\lambda)$ are analytic on $\mathscr{D}_\rho$ and for all $\lambda \in \mathscr{D}_\rho$, $\frac{\partial}{\partial \lambda}\varphi_m(t,\lambda)$ is continuous at (ω_0, λ). As is previously mentioned $\chi_1(t,\lambda), \chi_2(t,\lambda)$ are entire functions of λ for $t \in (\omega_0, b)$. In addition, by direct computation, $\frac{\partial}{\partial \lambda}\chi_i(t,\lambda)$ are continuous at (ω_0, λ) for each $\lambda \in \mathbb{C}$. Hence the statement is true for $m=1$. Assume the correctness of the statement for $m \geqslant 1$. Then for $t_0 \in (\omega_0, b)$, $\lambda_0 \in \mathscr{D}_\rho$ we obtain

$$\left. \frac{\partial \varphi_{m+1}(t_0,\lambda)}{\partial \lambda} \right|_{\lambda=\lambda_0}$$

$$= q \left. \frac{\partial \chi_2(t_0,\lambda)}{\partial \lambda} \right|_{\lambda=\lambda_0} \int_{\omega_0}^{t_0} \chi_1(h(t),\lambda)\varphi_m(h(t),\lambda) d_{q,\omega}t$$

$$+ \left. \frac{\partial \varphi_1(t_0,\lambda)}{\partial \lambda} \right|_{\lambda=\lambda_0} - q \left. \frac{\partial \chi_1(t_0,\lambda)}{\partial \lambda} \right|_{\lambda=\lambda_0} \int_{\omega_0}^{t_0} \chi_2(h(t),\lambda)\varphi_m(h(t),\lambda) d_{q,\omega}t$$

$$+ q\chi_2(t_0,\lambda)\frac{\partial}{\partial\lambda}\left(\int_{\omega_0}^{t_0}\chi_1(h(t),\lambda)\varphi_m(h(t),\lambda)d_{q,\omega}t\right)\Bigg|_{\lambda=\lambda_0}$$

$$- q\chi_1(t_0,\lambda)\frac{\partial}{\partial\lambda}\left(\int_{\omega_0}^{t_0}\chi_2(h(t),\lambda)\varphi_m(h(t),\lambda)d_{q,\omega}t\right)\Bigg|_{\lambda=\lambda_0}. \tag{4.26}$$

From the induction hypothesis, we conclude that $\frac{\partial}{\partial\lambda}(\chi_i(h(t),\lambda)$ $\varphi_m(h(t),\lambda)), i = 1,2$, are continuous at ω_0, λ_0. Therefore, there exist constants $M, \eta > 0$ such that

$$\left|\frac{\partial}{\partial\lambda}(\chi_i(h^n(t_0),\lambda)\varphi_m(h^n(t_0),\lambda))\right| < M, \ n \in \mathbb{N}, \ \lambda \in D_\eta(\lambda_0), \tag{4.27}$$

where $D_\eta(\lambda_0) = \{z \in \mathbb{C} : |z-\lambda_0| < \eta\}$. Hence

$$\begin{aligned}(t_0(1-q)-\omega)q^n&\left|\frac{\partial}{\partial\lambda}(\chi_i(h^{n+1}(t_0),\lambda)\varphi_m(h^{n+1}(t_0),\lambda))\right| \\ &\leqslant (t_0(1-q)-\omega)q^n M, \ n \in \mathbb{N},\end{aligned}$$

for all $\lambda \in D_\eta(\lambda_0)$. Accordingly, the series corresponding to the q,ω-integrals

$$\int_{\omega_0}^{t_0}\frac{\partial}{\partial\lambda}(\chi_i(h(t),\lambda)\varphi_m(h^n(t),\lambda))d_{q,\omega}t, \quad i=1,2 \tag{4.28}$$

is uniformly convergent in a neighborhood of $\lambda = \lambda_0$. This proves that $\varphi_{m+1}(t_0,\lambda)$ is analytic at λ_0. Now, we can interchange the differentiation and integration processes in (4.26) and since t_0, and λ_0 are arbitrary, then

$$\begin{aligned}&\frac{\partial}{\partial\lambda}\varphi_{m+1}(t,\lambda)\\ &\quad= \frac{\partial}{\partial\lambda}\varphi_1(t,\lambda)\\ &\qquad+ q\int_{\omega_0}^{t}\frac{\partial}{\partial\lambda}\left(\chi_2(t,\lambda)\chi_1(h(s),\lambda)\varphi_m(h(s),\lambda)\right)\,p(h(s))\,d_{q,\omega}s\\ &\qquad- q\int_{\omega_0}^{t}\frac{\partial}{\partial\lambda}\left(\chi_1(t,\lambda)\chi_2(h(s),\lambda)\varphi_m(h(s),\lambda)\right)\,p(h(s))\,d_{q,\omega}s\end{aligned} \tag{4.29}$$

for all $t \in (\omega_0, b)$, $\lambda \in \Omega_M$. Again, from the induction hypothesis, the integrals in (4.29) are continuous at (ω_0, λ). Consequently $\frac{\partial}{\partial\lambda}\varphi_{m+1}(t,\lambda)$ is continuous at (ω_0, λ). Let $t_0 \in [\omega_0, b]$ be arbitrary. Then there exist $\gamma(t_0), \beta(t_0) > 0$ such that

$$|\chi_i(t_0,\lambda)| \leqslant \sqrt{\frac{\gamma(t_0)}{2}}, \quad i = 1,2, \quad |\varphi_1(t_0,\lambda)| \leqslant \beta(t_0), \quad \lambda \in \mathscr{D}_\rho. \tag{4.30}$$

Finally, the use of the mathematical induction yields

$$\begin{aligned} &|\varphi_{m+1}(t_0,\lambda) - \varphi_m(t_0,\lambda)| \\ &\leqslant \beta(t_0) q^{\frac{m(m+1)}{2}} \frac{\left(A\gamma(t_0)(t(1-q)-\omega)\right)^m}{(q,q)_m}, \quad m = 1,2,\ldots. \end{aligned} \tag{4.31}$$

Consequently, the series (4.15), with $t = t_0$, converges uniformly in $\mathscr{D}_\rho$ to $\phi(t_0,\lambda)$. Hence $\phi(t_0,\lambda)$ is analytic in $\mathscr{D}_\rho$, i.e., it is entire. □

The technique developed above is mimicking that of [20] and similar to that of Titchmarsh's monograph [63]. We can also apply the technique of Eastham [32] by transforming (3.11) and (4.6) to a first-order system of equations. Next, we indicate how to calculate the eigenvalues of the q,ω-Sturm–Liouville problem. We also prove that the eigenvalues are algebraically simple. Let $\phi_1(\cdot,\lambda)$ and $\phi_2(\cdot,\lambda)$ be the linearly independent solution determined by the initial conditions.

$$\begin{aligned} &\phi_1(\omega_0,\lambda) = 1, \quad D_{q,\omega}\phi_1(\omega_0,\lambda) = 0; \quad \phi_2(\omega_0,\lambda) = 0, \\ &\quad D_{q,\omega}\phi_2(\omega_0,\lambda) = 1, \quad \lambda \in \mathbb{C}. \end{aligned}$$

Every solution of (3.11) can be written as

$$y(t,\lambda) = B_1\phi_1(t,\lambda) + B_2\phi_2(t,\lambda), \tag{4.32}$$

where B_1, B_2 are arbitrary constants. A solution $y(\cdot,\lambda)$ of (3.11) will be an eigenfunction if it satisfies the boundary conditions (3.12)–(3.13), i.e., if we can find a nontrivial solutions for the linear system

$$B_1U_1(\phi_1) + B_2U_1(\phi_2) = 0, \tag{4.33}$$

$$B_1U_2(\phi_1) + B_2U_2(\phi_2) = 0. \tag{4.34}$$

Hence, λ is an eigenvalue if and only if

$$\Delta(\lambda) = \begin{vmatrix} U_1(\phi_1) & U_1(\phi_2) \\ U_2(\phi_1) & U_2(\phi_2) \end{vmatrix} = 0.$$

The function $\Delta(\lambda)$ is called the characteristic determinant associated with the eigenvalue problem (3.11)–(3.13). The zeros of $\Delta(\lambda)$ are exactly the eigenvalues of the problem (3.11)–(3.13). Since $\phi_1(t,\lambda)$ and $\phi_2(t,\lambda)$ are entire in λ for each fixed $t \in [\omega_0, b]$, then $\Delta(\lambda)$ is also entire. Thus the eigenvalues of the problem are at most countable with no finite limit points.

Theorem 4.3. *The eigenvalues of problem* (3.11)–(3.13) *are simple zeros of* $\Delta(\lambda)$.

Proof. For $\lambda \in \mathbb{C}$ we define $\Theta_1(t,\lambda)$ and $\Theta_2(t,\lambda)$ to be

$$\begin{aligned} \Theta_1(t,\lambda) &:= U_1(\phi_2)\phi_1(t,\lambda) - U_1(\phi_1)\phi_2(t,\lambda), \\ \Theta_2(t,\lambda) &:= U_2(\phi_2)\phi_1(t,\lambda) - U_2(\phi_1)\phi_2(t,\lambda). \end{aligned} \tag{4.35}$$

Hence $\Theta_1(t,\lambda)$ and $\Theta_2(t,\lambda)$ are solutions of (3.11) such that

$$\begin{aligned} \Theta_1(\omega_0,\lambda) &= a_2, \quad D_{\frac{1}{q},\frac{-\omega}{q}}\Theta_1(\omega_0,\lambda) = -a_1, \\ \Theta_2(b,\lambda) &= b_2, \quad D_{\frac{1}{q},\frac{-\omega}{q}}\Theta_2(b,\lambda) = -b_1. \end{aligned} \tag{4.36}$$

One can verify that

$$W_{q,\omega}(\Theta_1(\cdot,\lambda),\Theta_2(\cdot,\lambda))(t,\lambda) = \Delta(\lambda)W_{q,\omega}(\phi_1,\phi_2)(t,\lambda) = \Delta(\lambda). \tag{4.37}$$

Since

$$W_{q,\omega}(\phi_1,\phi_2)(t,\lambda) = W_{q,\omega}(\phi_1,\phi_2)(\omega_0,\lambda) = 1,$$

then

$$W_{q,\omega}(\Theta_1(\cdot,\lambda),\Theta_2(\cdot,\lambda))(t,\lambda) = \Delta(\lambda)W_{q,\omega}(\phi_1,\phi_2)(t,\lambda) = \Delta(\lambda). \tag{4.38}$$

Let λ_0 be an eigenvalue of (3.11)–(3.13). Then λ_0 is a real number and consequently $\Theta_i(t,\lambda_0)$ can be taken to be real-valued, $i = 1,2$. Also $\Delta(\lambda_0) = 0$. Form (4.37) we conclude that $W_{q,\omega}(\Theta_1(t,\lambda_0),\Theta_2(t,\lambda_0)) = 0$, i.e.,

$$\Theta_1(t,\lambda_0)D_{q,\omega}\Theta_2(t,\lambda_0) - \Theta_2(t,\lambda_0)D_{q,\omega}\Theta_1(t,\lambda_0) = 0.$$

By geometric simplicity there exists a nonzero constant k_0 such that

$$\Theta_1(t,\lambda_0) = k_0\Theta_2(t,\lambda_0), \tag{4.39}$$

i.e., $\Theta_1(t,\lambda_0), \Theta_2(t,\lambda_0)$ are linearly dependent eigenfunctions. From (4.39) we get

$$\Theta_1(b,\lambda_0) = k_0 a_{22} = k_0 \Theta_2(b,\lambda), \tag{4.40}$$

$$D_{\frac{1}{q},\frac{-\omega}{q}} \Theta_1(b,\lambda_0) = -k_0 a_{21} = D_{\frac{1}{q},\frac{-\omega}{q}} \Theta_1(b,\lambda). \tag{4.41}$$

From q,ω-Lagrange's identity (3.18) and by taking $z(t) = \Theta_1(t,\lambda)$ and $y(t) = \Theta_1(t,\lambda_0)$, we obtain

$$\begin{aligned}
&(\lambda - \lambda_0) \int_{\omega_0}^{b} \Theta_1(t,\lambda_0)\Theta_1(t,\lambda) d_{q,\omega} t \\
&= \Theta_1(b,\lambda) D_{\frac{1}{q},\frac{-\omega}{q}} \Theta_1(b,\lambda_0) - \Theta_1(b,\lambda_0) D_{\frac{1}{q},\frac{-\omega}{q}} \Theta_1(b,\lambda) \\
&= k_0 (\Theta_1(b,\lambda) D_{\frac{1}{q},\frac{-\omega}{q}} \Theta_2(b,\lambda) - \Theta_2(b,\lambda) D_{\frac{1}{q},\frac{-\omega}{q}} \Theta_1(b,\lambda) \\
&= k_0 W_{q,\omega}(\Theta_1(\cdot,\lambda), \Theta_2(\cdot,\lambda))(h^{-1}(b)) \\
&= k_0 \Delta(\lambda).
\end{aligned} \tag{4.42}$$

Since $\Delta(\lambda)$ is entire in λ then

$$\Delta'(\lambda_0) = \lim_{\lambda \longrightarrow \lambda_0} \frac{\Delta(\lambda)}{\lambda - \lambda_0} = \frac{1}{k_0} \int_{\omega_0}^{b} \Theta_1(t,\lambda_0)^2 d_{q,\omega} t \neq 0. \tag{4.43}$$

Therefore λ_0 is a simple zero of $\Delta(\lambda)$. □

Remark 4.4. We can prove the uniqueness in Theorem 4.2 by using q,ω-Gronwall's inequality. Indeed, form (4.23) we obtain the inequality

$$|\Upsilon(t,\lambda)| \leq |b - \omega_0| \int_{\omega_0}^{t} |\lambda - p(s)||\Upsilon(s,\lambda)| d_{q,\omega} s. \tag{4.44}$$

By q,ω-Gronwall's inequality (see [41]), we deduce that $\Upsilon(t,\lambda) = 0$, $t \in (\omega_0, b)$.

5. Green's Function

The construction of Green's function plays a crucial rule in SLT. It transforms the SLP into a Fredholm integral equation of the second kind and gives a concrete solution of the nonhomogeneous problem, cf. [8, 11, 20, 23, 43, 44]. Let $f(t)$ be defined and continuous on $[\omega_0, b]$. Consider the nonhomogeneous boundary-value problem which consists of the Hahn difference equation

$$-\frac{1}{q}D_{\frac{1}{q},\frac{-\omega}{q}}D_{q,\omega}y(t)+\{p(t)-\lambda\}y(t)=f(t),\quad t\in(\omega_0,b),\quad \lambda\in\mathbb{C},\qquad (5.1)$$

and the boundary conditions (3.12)–(3.13). The major theorem of this section, Theorem 5.1, introduces a solution of the nonhomogeneous problem in terms of Green's function provided that $\lambda\in\mathbb{C}$ is not an eigenvalue. The uniqueness of such a solution is guaranteed because if we assume that $\Phi_1(t,\lambda)$, $\Phi_2(t,\lambda)$ are two solutions of (5.1) then $\Phi_1(t,\lambda)-\Phi_2(t,\lambda)$ is a solution of the homogeneous equation corresponding to (5.1) when $f\equiv 0$, where λ is not an eigenvalue.

Theorem 5.1. *Assume that λ is not an eigenvalue of* (3.11)–(3.13), *and f is continuous on $[\omega_0, b]$. Then problem* (5.1), (3.12)–(3.13), *has a unique solution, which can be represented in the form*

$$\Phi(t,\lambda)=\int_{\omega_0}^{b}G(t,\xi,\lambda)f(\xi)d_{q.\omega}\xi,\qquad (5.2)$$

where $G(t,\xi,\lambda)$ is the Green's function of eigenvalue problem (3.11)–(3.13), *defined by*

$$G(t,\xi,\lambda)=-\frac{1}{\Delta(\lambda)}\begin{cases}\Theta_1(\xi,\lambda)\Theta_2(t,\lambda), & \omega_0\leqslant\xi\leqslant t\leqslant b,\\ \Theta_2(\xi,\lambda)\Theta_1(t,\lambda), & \omega_0\leqslant t\leqslant\xi\leqslant b.\end{cases}\qquad (5.3)$$

Here Θ_1,Θ_2 are the solutions defined in (4.35).

Proof. Assume that $\lambda\in\mathbb{C}$ is not an eigenvalue of (3.11)–(3.13). To solve equation (5.1), we use the method of variation of parameters derived in [40]. Assume that a particular solution of (5.1) is given by

$$\Phi(t,\lambda)=p_1(t)\Theta_1(t,\lambda)+p_2(t)\Theta_2(t,\lambda),\qquad (5.4)$$

$p_1(\cdot)$ and $p_2(\cdot)$ are the variable parameters satisfying the first-order q,ω-difference equations

$$D_{q,\omega}p_r(t)=\frac{W_r(\Theta_1,\Theta_2)(h(t),\lambda)}{W_{q,\omega}(\Theta_1,\Theta_2)(t,\lambda)}\frac{f(h(t))}{a_0(t)},\qquad r=1,2.\qquad (5.5)$$

From (4.37), $W_{q,\omega}(\Theta_1, \Theta_2)(t, \lambda) = \Delta(\lambda) \neq 0$. Hence

$$
\begin{aligned}
D_{q,\omega} p_1(t) &= \frac{q}{\Delta(\lambda)} \Theta_2(h(t), \lambda) f(h(t)), \\
D_{q,\omega} p_2(t) &= \frac{-q}{\Delta(\lambda)} \Theta_1(h(t), \lambda) f(h(t)). \qquad (5.6)
\end{aligned}
$$

Therefore, using the fundamental theorem of Hahn-calculus we obtain

$$
\begin{aligned}
p_1(t) &= p_1(\omega_0) + \frac{q}{\Delta(\lambda)} \int_{\omega_0}^{t} \Theta_2(h(\xi), \lambda) f(h(\xi)) d_{q,\omega}\xi, \\
p_2(t) &= p_2(b) + \frac{q}{\Delta(\lambda)} \int_{t}^{b} \Theta_1(h(\xi), \lambda) f(h(\xi)) d_{q,\omega}\xi.
\end{aligned} \qquad (5.7)
$$

Hence the general solution of (5.1) is given by

$$
\begin{aligned}
\Phi(t, \lambda) &= \alpha_1 \Theta_1(t, \lambda) + \alpha_2 \Theta_2(t, \lambda) \\
&\quad + \frac{q}{\Delta(\lambda)} \Theta_1(t, \lambda) \int_{\omega_0}^{t} \Theta_2(h(\xi), \lambda) f(h(\xi)) d_{q,\omega}\xi \\
&\quad + \frac{q}{\Delta(\lambda)} \Theta_2(t, \lambda) \int_{t}^{b} \Theta_1(h(\xi), \lambda) f(h(\xi)) d_{q,\omega}\xi, \qquad (5.8)
\end{aligned}
$$

where $t \in (\omega_0, b)$ and α_1, α_2 are arbitrary constants, which are determined by the boundary conditions. Indeed, simple calculations yield

$$
\begin{aligned}
\Phi(\omega_0, \lambda) &= \alpha_1 \Theta_1(\omega_0, \lambda) + \alpha_2 \Theta_2(\omega_0, \lambda) \\
&\quad + \frac{q}{\Delta(\lambda)} \Theta_2(\omega_0, \lambda) \int_{\omega_0}^{b} \Theta_1(h(\xi)\lambda) f(h(\xi)) d_{q,\omega}\xi, \qquad (5.9)
\end{aligned}
$$

$$
\begin{aligned}
D_{\frac{1}{q}, \frac{-\omega}{q}} \Phi(\omega_0, \lambda) &= \lim_{t \longrightarrow \omega_0^+} \frac{\Phi(t, \lambda) - \Phi(\omega_0, \lambda)}{t - \omega_0} = \alpha_1 D_{\frac{1}{q}, \frac{-\omega}{q}} \Theta_1(\omega_0, \lambda) \\
&\quad + \left(\alpha_2 + \frac{q}{\Delta(\lambda)} \int_{\omega_0}^{b} \Theta_1(h(\xi), \lambda) f(h(\xi)) d_{q,\omega}\xi \right) \\
&\quad \times D_{\frac{1}{q}, \frac{-\omega}{q}} \Theta_2(\omega_0, \lambda). \qquad (5.10)
\end{aligned}
$$

The boundary condition $a_1\Phi(\omega_0,\lambda)+a_2D_{\frac{1}{q},\frac{-\omega}{q}}\Phi(\omega_0,\lambda)=0$ implies that

$$\left(\alpha_2+\frac{q}{\Delta(\lambda)}\int_{\omega_0}^{b}\Theta_1(h(\xi),\lambda)f(h(\xi))d_{q,\omega}\xi\right)W_{q,\omega}(\Theta_1,\Theta_2)=0,$$

i.e.,

$$\alpha_2=-\frac{q}{\Delta(\lambda)}\int_{\omega_0}^{b}\Theta_1(h(\xi),\lambda)f(h(\xi))d_{q,\omega}\xi.$$

Hence,

$$\Phi(t,\lambda)=\alpha_1\Theta_1(t,\lambda)+\frac{q}{\Delta(\lambda)}\int_{\omega_0}^{t}(\Theta_1(t,\lambda)\Theta_2(h(\xi),\lambda)$$

$$-\Theta_2(t,\lambda)\Theta_1(h(\xi),\lambda))f(h(\xi))d_{q,\omega}\xi. \tag{5.11}$$

From (5.8), we obtain

$$\Phi(b,\lambda)=\alpha_1\Theta_1(b,\lambda)+\frac{q}{\Delta(\lambda)}\int_{\omega_0}^{b}(\Theta_1(b,\lambda)\Theta_2(h(\xi),\lambda)$$

$$-\Theta_2(b,\lambda)\Theta_1(h(\xi),\lambda))f(h(\xi))d_{q,\omega}\xi$$

$$=\alpha_1\Theta_1(b,\lambda)+\frac{q}{\Delta(\lambda)}\int_{\omega_0}^{h^{-1}(b)}(\Theta_1(b,\lambda)\Theta_2(h(\xi),\lambda)$$

$$-\Theta_2(b,\lambda)\Theta_1(h(\xi),\lambda))f(h(\xi))d_{q,\omega}\xi,$$

and

$$D_{\frac{1}{q},\frac{-\omega}{q}}\Phi(b,\lambda)=D_{\frac{1}{q},\frac{-\omega}{q}}\Theta_1(b,\lambda)$$

$$\times\left(\alpha_1+\frac{q}{\Delta(\lambda)}\int_{\omega_0}^{h^{-1}(b)}\Theta_2(h(\xi),\lambda)f(h(\xi))d_{q,\omega}\xi\right)$$

$$-\frac{q}{\Delta(\lambda)}D_{\frac{1}{q},\frac{-\omega}{q}}\Theta_2(b,\lambda)\int_{\omega_0}^{h^{-1}(b)}\Theta_1(h(\xi),\lambda))f(h(\xi))d_{q,\omega}\xi.$$

The boundary condition $b_1\Phi(b,\lambda)+b_2D_{\frac{1}{q},\frac{-\omega}{q}}\Phi(b,\lambda)=0$ implies that

$$\left(\alpha_1+\frac{q}{\Delta(\lambda)}\int_{\omega_0}^{h^{-1}(b)}\Theta_2(h(\xi),\lambda)f(h(\xi))d_{q,\omega}\xi\right)$$

$$\times W_{q,\omega}(\Theta_1,\Theta_2)(h^{-1}(b))=0,$$

$$\alpha_1=-\frac{q}{\Delta(\lambda)}\int_{\omega_0}^{h^{-1}(b)}\Theta_2(h(\xi),\lambda)f(h(\xi))d_{q,\omega}\xi.$$

Therefore

$$\Phi(t,\lambda) = -\frac{q}{\Delta(\lambda)}\Theta_2(t,\lambda)\int_{\omega_0}^{t}\Theta_1(h(\xi),\lambda)f(h(\xi))d_{q,\omega}\xi$$

$$-\frac{q}{\Delta(\lambda)}\Theta_1(t,\lambda)\int_{t}^{h^{-1}(b)}\Theta_2(h(\xi),\lambda)f(h(\xi))d_{q,\omega}\xi$$

$$= -\frac{1}{\Delta(\lambda)}\Theta_2(t,\lambda)\int_{\omega_0}^{h(t)}\Theta_1(\xi,\lambda)f(\xi)d_{q,\omega}\xi$$

$$-\frac{1}{\Delta(\lambda)}\Theta_1(t,\lambda)\int_{h(t)}^{b}\Theta_2(\xi,\lambda)f(\xi)d_{q,\omega}\xi$$

$$= -\frac{1}{\Delta(\lambda)}\Theta_2(t,\lambda)\int_{\omega_0}^{t}\Theta_1(\xi,\lambda)f(\xi)d_{q,\omega}\xi$$

$$-\frac{1}{\Delta(\lambda)}\Theta_1(t,\lambda)\int_{t}^{b}\Theta_2(\xi,\lambda)f(\xi)d_{q,\omega}\xi,$$

which is (5.2)–(5.3). To prove the uniqueness of G, suppose that there exists another function, $\widehat{G}(t,\xi,\lambda)$, such that

$$\Psi(t,\lambda) = \int_{\omega_0}^{b}\widehat{G}(t,\xi,\lambda)f(\xi)d_{q,\omega}\xi \tag{5.12}$$

is solution of (5.1), which satisfies (3.12)–(3.13). For convenience let

$$\widehat{G}(t,\xi,\lambda) = \begin{cases} \widehat{G}_1(t,\xi,\lambda), & \omega_0 \leqslant \xi \leqslant t \leqslant b, \\ \widehat{G}_2(t,\xi,\lambda), & \omega_0 \leqslant t \leqslant \xi \leqslant b, \end{cases} \tag{5.13}$$

$$G(t,\xi,\lambda) = \begin{cases} G_1(t,\xi,\lambda), & \omega_0 \leqslant \xi \leqslant t \leqslant b, \\ G_2(t,\xi,\lambda), & \omega_0 \leqslant t \leqslant \xi \leqslant b. \end{cases} \tag{5.14}$$

By subtraction, we obtain

$$\int_{\omega_0}^{b}\{G(t,\xi,\lambda) - \widehat{G}(t,\xi,\lambda)\}f(t)d_{q,\omega}\xi = 0, \tag{5.15}$$

for all functions $f(t) \in L^2_{q,\omega}[\omega_0, b]$. Let $f(t) = \overline{G(t,\xi,\lambda) - \widehat{G}(t,\xi,\lambda)}$ and $t = h^m(b)$, $m \in \mathbb{N}_0$. Then

$$\int_{\omega_0}^{b} \left|G(t,\xi,\lambda) - \widehat{G}(t,\xi,\lambda)\right|^2 d_{q,\omega}\xi$$

$$= \int_{\omega_0}^{h^m(b)} \left|G(h^m(b),\xi,\lambda) - \widehat{G}(h^m(b),\xi,\lambda)\right|^2 d_{q,\omega}\xi$$

$$- \int_{h^m(b)}^{b} \left|G(h^m(b),\xi,\lambda) - \widehat{G}(h^m(b),\xi,\lambda)\right|^2 d_{q,\omega}\xi,$$

which implies that

$$(b((1-q)-\omega)\sum_{k=0}^{\infty} q^k \left|G(h^m(b),h^k(b),\lambda) - \widehat{G}(h^m(b),h^k(b),\lambda)\right|^2 = 0. \tag{5.16}$$

Therefore $G(h^m(b),h^k(b),\lambda) = \widehat{G}(h^m(b),h^k(b),\lambda)$, proving the uniqueness. □

Theorem 5.2. *Let λ be not an eigenvalue of problem* (3.11)–(3.13). *Then we have the following properties*:

(i) *$G(t,\xi,\lambda)$ is continuous at (ω_0,ω_0).*
(ii) *$G(t,\xi,\lambda)$ is symmetric, i.e., $G(t,\xi,\lambda) = \overline{G(\xi,t,\lambda)}$.*
(iii) *For each fixed $\xi \in (\omega_0, b]$ as a function of t, $G(t,\xi,\lambda)$ satisfies the q,ω-difference equation* (3.11) *in the intervals $[\omega_0;\xi)$, $(\xi;b]$ and it satisfies the boundary conditions* (3.12)–(3.13).
(iv) *Let λ_0 be a zero of $\Delta(\lambda)$. Then λ_0 is a simple pole of $G(t;\xi;\lambda)$, and*

$$G(t;\xi;\lambda) = -\frac{\nu_0(\xi)\nu_0(t)}{\lambda-\lambda_0} + G_1(t,\xi,\lambda), \tag{5.17}$$

where $G_1(t,\xi,\lambda)$ is analytic function of λ in a neighborhood of λ_0 and ν_0 is a normalized eigenfunction corresponding to λ_0.

Proof. (i) It follows directly from the continuity of $\Theta_1(t,\lambda)$, $\Theta_2(t,\lambda)$ at (ω_0,ω_0) for each fixed $\lambda \in \mathbb{C}$. (ii) Easy to be checked.

(iii) Let $\xi \in (\omega_0, b]$ be fixed. If $t \in [\omega_0, \xi)$ then

$$G(t,\xi,\lambda) = -\frac{1}{\Delta(\lambda)}\Theta_1(t,\lambda)\Theta_2(\xi,\lambda).$$

So

$$\begin{aligned}\ell G(t,\xi,\lambda) &= -\frac{1}{\Delta(\lambda)}\Theta_2(\xi,\lambda)\ell\Theta_1(t,\lambda)\\ &= -\frac{\lambda}{\Delta(\lambda)}\Theta_2(\xi,\lambda)\Theta_1(t,\lambda) = \lambda G(t,\xi,\lambda).\end{aligned}$$

Similarly for $t \in (\xi, b]$. For the boundary conditions one can see that

$$\begin{aligned}&a_1G(\omega_0,\xi,\lambda) + a_2D_{\frac{1}{q},-\frac{\omega}{q}}G(\omega_0,\xi,\lambda)\\ &= -\frac{\Theta_2(\xi,\lambda)}{\Delta(\lambda)}\{a_1\Theta_1(\omega_0,\lambda) + a_2D_{\frac{1}{q},-\frac{\omega}{q}}\Theta_1(\omega_0,\lambda)\} = 0,\\ &b_1G(b,\xi,\lambda) + b_2D_{\frac{1}{q},-\frac{1}{\omega}}G(b,\xi,\lambda)\\ &= -\frac{\Theta_1(\xi,\lambda)}{\Delta(\lambda)}\{b_1\Theta_2(b,\lambda) + b_2D_{\frac{1}{q},-\frac{\omega}{q}}\Theta_2(b,\lambda)\} = 0.\end{aligned}$$

(iv) Since λ_0 is a simple zero of $\Delta(\lambda)$, then λ_0 is a simple pole of $G(t,\xi,\lambda)$. Indeed, from (4.39) and (4.40) we get

$$\begin{aligned}R(t,\xi) &= \lim_{\lambda\to\lambda_0}(\lambda-\lambda_0)G(t,\xi,\lambda)\\ &= -\lim_{\lambda\to\lambda_0}\frac{\lambda-\lambda_0}{\Delta(\lambda)}\Theta_1(t,\lambda)\Theta_2(\xi,\lambda)\\ &= -\Theta_1(t,\lambda_0)\Theta_1(\xi,\lambda_0)\lim_{\lambda\to\lambda_0}\frac{\lambda-\lambda_0}{\Delta(\lambda)}\\ &= -k_0^{-1}\Theta_1(t,\lambda_0)\Theta_1(\xi,\lambda_0)\frac{k_0}{\int_{\omega_0}^{b}|\Theta_1(s,\lambda_0)|^2d_{q,\omega}s}\\ &= -\frac{\Theta_1(t,\lambda_0)\Theta_1(\xi,\lambda_0)}{\int_{\omega_0}^{b}|\Theta_1(s,\lambda_0)|^2d_{q,\omega}s} = -\nu_0(t,\lambda_0)\nu_0(\xi,\lambda_0).\end{aligned}$$

Therefore $G(t,\xi,\lambda)$ has a simple pole at $\lambda = \lambda_0$ with residue $R(t,\xi)$. □

6. Eigenfunctions Expansion Theorem

In this section, the existence of a denumerable set of eigenvalues of problem (3.11)–(3.13) with no finite limit points will be proved by using the spectral theorem of compact self-adjoint operators in Hilbert spaces; see, e.g., [37, 65]. We also prove that the set of eigenfunctions is an orthonormal basis of $L^2_{q,\omega}(\omega_0, b)$. Define the operator

$$\mathscr{A} : D_{\mathscr{A}} \longrightarrow L^2_{q,\omega}[\omega_0, b], \quad y \longmapsto \ell y,$$

where $D_{\mathscr{A}} := \{y \in C^2_{q,\omega}[\omega_0, b] : U_1(y) = U_2(y) = 0\}$ is a subspace of $L^2_{q,\omega}[\omega_0, b]$. We assume without any loss of generality that $\lambda = 0$ is not an eigenvalue. Thus $\ker \mathscr{A} = \{0\}$. From the previous section, the solution of the problem

$$\mathscr{A}y = f(t), \qquad f \in L^2_{q,\omega}(\omega_0, b), \tag{6.1}$$

is given uniquely in $L^2_{q,\omega}(\omega_0, b)$ by

$$y(t) = \int_{\omega_0}^{b} G(t, \xi) f(\xi) d_{q,\omega}\xi, \tag{6.2}$$

where

$$G(t,\xi) = \begin{cases} c\Theta_1(\xi)\Theta_2(t), & \omega_0 \leqslant \xi \leqslant t \leqslant b, \\ c\Theta_2(\xi)\Theta_1(t), & \omega_0 \leqslant t \leqslant \xi \leqslant b, \end{cases} \qquad c = \frac{-1}{W_{q,\omega}(\Theta_1, \Theta_2)}. \tag{6.3}$$

Replacing $f(t)$ by λy in (6.1). Then eigenvalue problem $(\mathscr{A}y)(t) = \lambda y(t)$ is equivalent to the integral equation

$$y(t) = \lambda \int_{\omega_0}^{b} G(t, \xi) y(\xi) d_{q,\omega}\xi. \tag{6.4}$$

The next theorem proves that the integral operator (6.4) and the Hahn difference operator $\mathscr{A}$ are inverses to each other in the sense prescribed in Theorem 6.1 below. Let $\mathscr{G} : L^2_{q,\omega}(\omega_0, b) \longrightarrow L^2_{q,\omega}(\omega_0, b)$ be the integral operator

$$(\mathscr{G}f)(t) = \int_{\omega_0}^{b} G(t, \xi) f(\xi) d_{q,\omega}\xi. \tag{6.5}$$

Since Green's function $G(t, \xi, \lambda)$ is continuous at (ω_0, ω_0), and $f(t)$ is continuous on $[\omega_0, b]$, then we conclude that the integral operator is bounded and well defined. Indeed

$$\begin{aligned} ||\mathscr{G}f||^2 &= \int_{\omega_0}^{b} \left| \int_{\omega_0}^{b} G(t, \xi) f(\xi) d_{q,\omega}\xi \right|^2 d_{q,\omega}t \\ &= \int_{\omega_0}^{b} \left| (b(1-q) - \omega \sum_{n=0}^{\infty} G(t, h^n(b)) f(h^n(b)) \right|^2 d_{q,\omega}t \end{aligned}$$

$$\leqslant (b(1-q)-\omega)^3 \sum_{m=0}^{\infty} q^m \left(\sum_{n=0}^{\infty} |q^{\frac{n}{2}} G(h^m(b), h^n(b))|^2 \right)$$

$$\times \left(\sum_{n=0}^{\infty} |q^{\frac{n}{2}} f(h^n(b))|^2 \right)$$

$$= \left(\int_{\omega_0}^{b} \int_{\omega_0}^{b} |G(t,\xi)|^2 \, d_{q,\omega}\xi \, d_{q,\omega}t \right) \left(\int_{\omega_0}^{b} |f(t)|^2 d_{q,\omega}t \right)$$

$$= \left(\int_{\omega_0}^{b} \int_{\omega_0}^{b} |G(t,\xi)|^2 \, d_{q,\omega}\xi \, d_{q,\omega}t \right) \|f(\cdot)\|^2. \tag{6.6}$$

Thus, as in the classical Fredholm theory, $\mathscr{G}$ is a bounded linear operator with

$$\|\mathscr{G}\| \leq \left(\int_{\omega_0}^{b} \int_{\omega_0}^{b} |G(t,\xi)|^2 d_{q,\omega}\xi d_{q,\omega}t \right)^{1/2} = \|G(\cdot,\cdot)\|.$$

Equation (6.5) will be called a Hahn–Fredholm integral equation of the second kind.

Theorem 6.1. *The Hahn difference operator $\mathscr{A}$ and the Hahn–Fredholm integral operator $\mathscr{G}$ are inverses to each other in the sense that*

$$(\mathscr{A}\mathscr{G})f = f, \quad f \in L^2_{q,\omega}(\omega_0, b), \quad (\mathscr{G}\mathscr{A})(y) = y, \quad y \in D_{\mathscr{A}}. \tag{6.7}$$

Proof. The proof is given by direct computations. We first show that $y = \mathscr{G}f \in D_{\mathscr{A}}$ for $f \in L^2_{q,\omega}(\omega_0, b)$. Indeed from (6.3), we have

$$y(t) = (\mathscr{G}f)(t) = \Theta_2(t)y_1(t) + \Theta_1(t)y_2(t),$$

where

$$y_1(t) = c\int_{\omega_0}^{t} \Theta_1(\xi)f(\xi)d_{q,\omega}\xi, \quad y_2(t) = c\int_{t}^{b} \Theta_2(\xi)f(\xi)d_{q,\omega}\xi.$$

Hence

$$D_{q,\omega}y(t) = D_{q,\omega}\Theta_2(t)y_1(h(t)) + D_{q,\omega}\Theta_1(t)y_2(h(t)), \tag{6.8}$$

$$D^2_{q,\omega}y(t) = qp(h(t))y(h(t)) - qf(h(t)) \in L^2_{q,\omega}(\omega_0, b), \tag{6.9}$$

$t \in (\omega_0, b)$. Since $D_{q,\omega}\Theta_i(t), y_i(t), i = 1, 2$, are continuous at ω_0, then, so is $D_{q,\omega}y(t)$,

$$D_{\frac{1}{q},\frac{-\omega}{q}}y(\omega_0) = D_{q,\omega}y(\omega_0) = \lim_{t \longrightarrow \omega_0^+} \frac{y(t) - y(\omega_0)}{t - \omega_0} = D_{\frac{1}{q},-\frac{\omega}{q}}\Theta_1(\omega_0)y_2(\omega_0),$$

$y_1(\omega_0) = 0, y_2(b) = 0$. Therefore

$$a_1 y(\omega_0) + a_2 D_{\frac{1}{q},\frac{-\omega}{q}}y(\omega_0) = \big(a_1\Theta_1(\omega_0) + a_2 D_{\frac{1}{q},\frac{-\omega}{q}}\Theta_1(\omega_0)y_2(\omega_0) = 0,$$

$$b_1 y(b) + b_2 D_{\frac{1}{q},\frac{-\omega}{q}}y(b) = \big(b_1\Theta_2(b) + b_2 D_{\frac{1}{q},\frac{-\omega}{q}}\Theta_2(b)\big)y_1(b) = 0.$$

Hence $y \in D_{\mathscr{A}}$. Equation (6.9) implies that $(\mathscr{A}\mathscr{G})f = f$. As for the second equality of (6.7), replacing f in the first equality of (6.7) by $\mathscr{A}y$ leads us directly to $\mathscr{A}\mathscr{G}\mathscr{A}y = y$. Since Ker $\mathscr{A} = \{0\}$, then $y = \mathscr{G}\mathscr{A}y$. □

Consequently as in the case of Sturm–Liouville theory, the Ker $\mathscr{G} = \{0\}$ and Ψ is an eigenfunction of $\mathscr{G}$ corresponding to an eigenvalue μ if and only if Ψ is eigenfunction of $\mathscr{A}$ with eigenvalue $\frac{1}{\mu}$.

Theorem 6.2. *The Fredholm integral operator $\mathscr{G}$ is compact and self-adjoint.*

Proof. We have previously shown that $\mathscr{G}$ is well defined, linear and bounded on $L^2_{q,\omega}(\omega_0, b)$. The self-adjointness arises directly from the symmetry of $G(t, \xi)$. To prove compactness, let

$$\phi_{ij} = \phi_i\phi_j \quad (i, j \in \mathbb{N})$$

be an orthonormal basis of $L^2_{q,\omega}(\omega_0, b) \times (\omega_0, b)$. Hence,

$$G = \sum_{i,j=1}^{\infty} \langle G, \phi_{ij}\rangle \phi_{ij}$$

with convergence in $L^2_{q,\omega}((\omega_0, b) \times (\omega_0, b))$. Let

$$G_n = \sum_{i,j=1}^{n} \langle G, \phi_{ij}\rangle \phi_{ij} \quad (n \in \mathbb{N})$$

and $\mathscr{G}_n$ be the finite rank operator

$$\mathscr{G}_n(f)(t) := \int_{\omega_0}^{b} G_n(t, \xi)f(\xi)d_{q,\omega}\xi, \qquad f \in L^2_{q,\omega}(\omega_0, b).$$

Now

$$\begin{aligned}
\|(\mathscr{G}_n - \mathscr{G})f\| &= \left(\int_{\omega_0}^{b} |(\mathscr{G}_n - \mathscr{G})f(t)|^2 d_{q,\omega}\xi\right)^{\frac{1}{2}} \\
&= \left(\int_{\omega_0}^{b} \left|\int_{\omega_0}^{b} (\mathscr{G}_n - \mathscr{G})(t,\xi)f(\xi)d_{q,\omega}\xi\right|^2 d_{q,\omega}t\right)^{\frac{1}{2}} \\
&\leqslant \left(\int_{\omega_0}^{b}\int_{\omega_0}^{b} |(G_n - G)(t,\xi)|^2 d_{q,\omega}t \;\; d_{q,\omega}\xi\right)^{\frac{1}{2}} \\
&\quad \times \left(\int_{\omega_0}^{b} |f(t)|^2 d_{q,\omega}t\right)^{\frac{1}{2}} \\
&= \|G_n - G\|\|f\|.
\end{aligned}$$

Therefore

$$\|\mathscr{G}_n - \mathscr{G}\| \leqslant \|G_n - G\| \longrightarrow 0 \quad \text{as} \quad n \longrightarrow \infty.$$

The compactness follows directly from the compactness of finite rank operator [37, 65]. □

Corollary 6.3. (i) *The eigenvalues of the operator $\mathscr{A}$ form an infinite sequence $\{\lambda_i\}_{i=1}^{\infty}$ of real numbers which can be ordered so that*

$$|\lambda_1| < |\lambda_2| < |\lambda_2| < \cdots \longrightarrow \infty.$$

(ii) *The set of all eigenfunctions of $\mathscr{A}$ forms an orthonormal basis for $L^2_{q,\omega}(\omega_0, b)$.*

Proof. (i) From the theory of compact self-adjoint operators in Hilbert spaces, cf. [37, 65], the operator $\mathscr{G}$ has an infinite sequence of nonzero real eigenvalues $\{\mu_n\}_{n=1}^{\infty}$, $\mu_n \longrightarrow 0$ as $n \longrightarrow \infty$. Let $\{\phi_n\}_{n=1}^{\infty}$ be the eigenfunctions corresponding to $\{\mu_n\}_{n=1}^{\infty}$. The spectral decomposition theorem implies

$$\mathscr{G}(f) = \sum_{n=1}^{\infty} \mu_n \langle f, \phi_n\rangle \phi_n. \tag{6.10}$$

Since the eigenvalues $\{\lambda_n\}_{n=1}^{\infty}$ of the operator $\mathscr{A}$ are the reciprocal of those of $\mathscr{G}$, we obtain

$$|\lambda_n| = \frac{1}{|\mu_n|} \longrightarrow \infty \quad \text{as } n \longrightarrow \infty. \tag{6.11}$$

(ii) Let $y \in D_{\mathscr{A}}$, then $y = \mathscr{G}(f)$ for some $f \in L^2_{q,\omega}[\omega_0, b]$. Hence

$$\begin{aligned} y &= \sum_{n=1}^{\infty} \mu_n \langle f, \phi_n \rangle \phi_n = \sum_{n=1}^{\infty} \mu_n \langle \ell y, \phi_n \rangle \phi_n \\ &= \sum_{n=1}^{\infty} \mu_n \langle y, \ell \phi_n \rangle \phi_n = \sum_{n=1}^{\infty} \langle y, \phi_n \rangle \phi_n. \end{aligned}$$

Following a technique similar to that of [59], we can see that $\{\phi_i\}_{i=1}^{\infty}$ is an orthonormal basis of eigenfunctions. If zero is an eigenvalue of $\mathscr{A}$, then we can choose $s \in \mathbb{R}$ such that it is not an eigenvalue of $\mathscr{A}$. Applying the above results to $\mathscr{A} - sI$ in place $\mathscr{A}$ yields the corollary. □

Example 6.4. Consider the q, ω-Sturm–Liouville boundary value problem

$$-\frac{1}{q} D_{\frac{1}{q}, \frac{-\omega}{q}} D_{q,\omega} y(t) = \lambda y(t), \quad \omega_0 \le t \le 1, \tag{6.12}$$

where we select q, ω such that $1 > \omega_0$, i.e., $1 - q > \omega > 0$ with q, ω-Dirichlet conditions

$$U_1(y) = y(\omega_0) = 0, \quad U_2(y) = y(1) = 0. \tag{6.13}$$

A fundamental set of solutions of (6.12) is

$$\phi_1(t, \lambda) = C_{q,\omega}(t, \mu), \quad \phi_2(t, \lambda) = \frac{S_{q,\omega}(t, \mu)}{\mu}. \tag{6.14}$$

Now, the eigenvalues of problem (6.12)–(6.13) are the zeros of the determinant

$$\Delta(\lambda) = \begin{vmatrix} U_1(\phi_1) & U_1(\phi_2) \\ U_2(\phi_1) & U_2(\phi_2) \end{vmatrix} = \phi_2(1, \lambda) = \frac{S_{q,\omega}(1, \mu)}{\mu}.$$

Hence, the eigenvalues $\{\lambda_n\}_{n=1}^{n=\infty}$ are the squares of the zeros of $S_{q,\omega}(1, \mu)$. From [18, 21]

$$\lambda_n = \frac{q^{-2n}(1-q)^{-2}(1 + O(q^n))}{1 - \omega_0}, \quad n \geqslant 1. \tag{6.15}$$

The solution $\phi_2(t, \lambda)$ satisfies the first boundary condition of (6.13). Hence the set of eigenfunction is

$$\phi_n(t) = \phi_2(t, \lambda_n) = \frac{S_{q,\omega}(t, \mu_n)}{\mu_n}, \quad n = 1, 2, \ldots.$$

Notice that zero is not eigenvalue because if we take $\lambda = 0$ in (6.12) then the solution that satisfies (6.12) and $U_1(y) = 0$ is $y(t, 0) = t$. Hence

$y(1) = 1 \neq 0$, i.e., $y(t, 0)$ is not an eigenfunction. The result holds true if we take the interval $[\omega_0, b]$ instead of $[\omega_0, 1]$ (cf. [18, 20]).

Example 6.5. Consider equation (6.12) with the q, ω-Neumann boundary condition

$$U_1(y) = D_{\frac{1}{q}, -\frac{\omega}{q}} y(\omega_0) = 0, \qquad U_2(y) = D_{\frac{1}{q}, \frac{\omega}{q}} y(1) = 0. \tag{6.16}$$

Now, the eigenvalues of problem (6.12) with the q, ω-Neumann boundary condition are the zeros of the determinant

$$\Delta(\lambda) = \begin{vmatrix} U_1(\phi_1) & U_1(\phi_2) \\ U_2(\phi_1) & U_2(\phi_2) \end{vmatrix} = \sqrt{q}\mu S_{q,\omega}(1, q^{-\frac{1}{2}}\mu).$$

Hence, the eigenvalues are $\lambda_0 = 0$, and from [18, 21],

$$\lambda_n = \frac{q^{-2n+1}(1-q)^{-2}(1+O(q^n))}{1-\omega_0}, \qquad n \geqslant 1. \tag{6.17}$$

In this case, the eigenfunctions are the functions

$$\phi_1(t) = 1, \qquad \phi_n(t) = C_{q,\omega}(t, \mu_n), \qquad n = 2, 3, \ldots. \tag{6.18}$$

7. Numerical Simulations

This section is devoted to exhibit several results from the theory of q-functions and q, ω-functions. To the best we know, there are few works in this interesting direction. For instance, the only references we know with such studies are those of Bustoz–Cardoso and Suslov [29, 60] for zeros of q-sine functions

$$S_q(z) := \frac{z}{1-q} \sum_{n=0}^{\infty} \frac{(-1)^n q^{n[n+\frac{1}{2}]} z^{2n}}{(q^2; q^3; q^2)_n}, \tag{7.1}$$

$$S_q(\eta, w) := \frac{1}{(-qw^2; q^2)} \sum_{n=0}^{\infty} \frac{(-1)^n q^{n[n+\frac{1}{2}]} w^{2n+1}}{(q^{\frac{1}{2}}; q^{\frac{1}{2}})}, \tag{7.2}$$

respectively; see [30, 60] for more details. However, for testing our results (cf. [18, 21]), we consider the q-trigonometric functions defined via (7.3) and (7.4) below.

7.1. q, ω-Trigonometric functions

In this subsection, we compare both q-trigonometric and q, ω-trigonometric functions with the trigonometric functions. We carry out a comparison in both numerical values and figures. The behavior of these families of functions as $q \to 1^-$ and $\omega \to 0^+$ in comparison with the classical trigonometric functions is exhibited. Figures 3–5 demonstrate that when $q \to 1^-$ and $\omega \to 0^+$ the q-sine and $S_{q,\omega}$ tend to the classical function, as expected and discussed in [36, 45]. Tables 1–3 involve some numerical values that observe this behavior. The basic trigonometric functions $\cos(x; q)$ and $\sin(x; q)$ are defined on $\mathbb{C}$ by

$$\cos(x; q) := \sum_{n=0}^{\infty} (-1)^n \frac{q^{n^2} (x(1-q))^{2n}}{(q; q)_{2n}}, \tag{7.3}$$

$$\sin(x; q) := \sum_{n=0}^{\infty} (-1)^n \frac{q^{n(n+1)} (x(1-q))^{2n+1}}{(q; q)_{2n+1}}. \tag{7.4}$$

We see from the tables and the figures that $\cos(x; q)$, $C_{q,\omega}(x)$ and $\sin(x; q)$, $S_{q,\omega}(x)$ become closer to $\cos x$ and $\sin x$ respectively as $q \to 1^-$ and $\omega \to 0^+$, although we are using a few number of terms of (7.3)–(7.4).

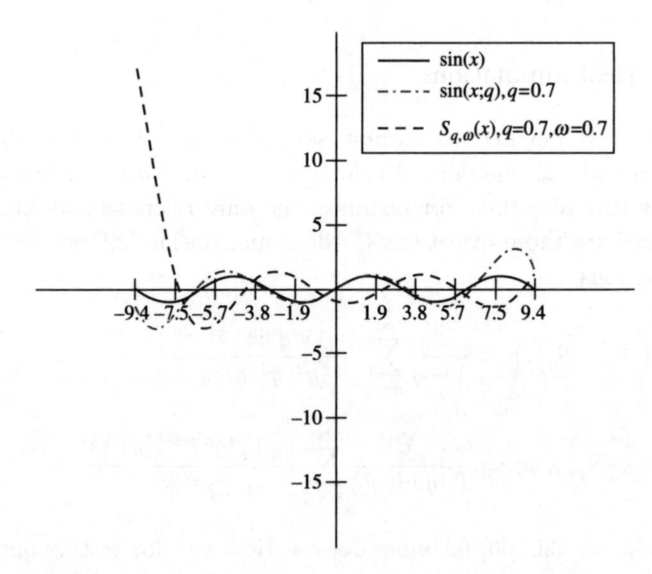

Fig. 3. Curves of $\sin(\cdot)$ and truncated power series for $\sin(\cdot; q)$, $S_{q,\omega}(\cdot)$ on $[-3\pi, 3\pi]$, $q = 0.7, \omega = 0.7$.

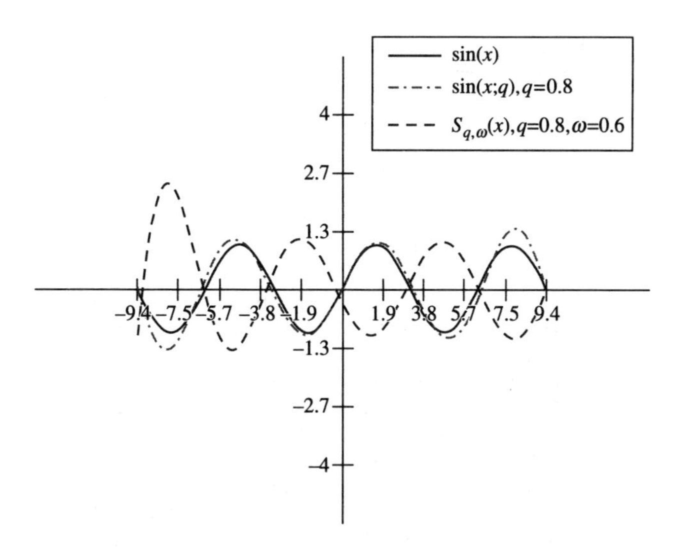

Fig. 4. Curves of $\sin(\cdot)$ and truncated Taylor's series for $\sin(\cdot;q), S_{q,\omega}(\cdot)$ in $[-3\pi, 3\pi], q = 0.8, \omega = 0.6$.

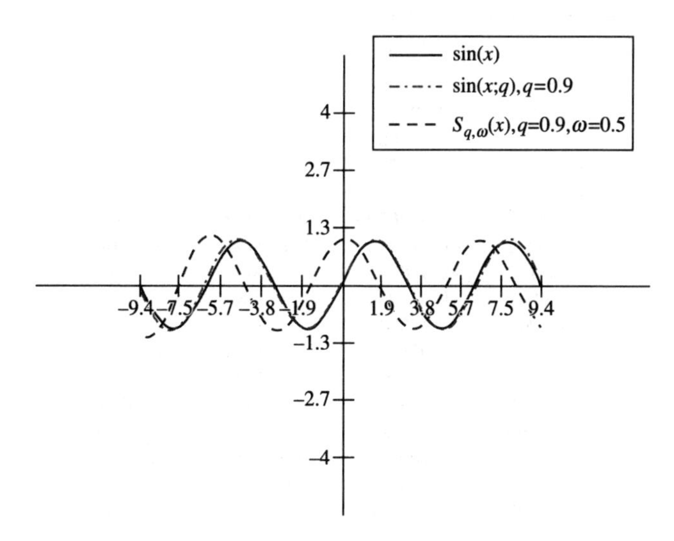

Fig. 5. Curves of $\sin(\cdot)$ and truncated $\sin(\cdot;q), S_{q,\omega}(\cdot)$ series in the interval $[-3\pi, 3\pi], q = 0.9, \omega = 0.5$.

Table 1. Some approximate values at $q = 0.7$, $\omega = 0.7$.

x	$\sin(x)$	$\sin(x;q)$	$S_{q,\omega}(x)$
−9.424778	0.000000	−1.304340	17.555583
−8.432696	−0.837166	−3.118750	8.165112
−6.448532	−0.164595	0.135657	−2.937279
−5.456450	0.735724	1.300975	−2.524767
−4.464369	0.969400	1.122248	−0.540601
−3.472287	0.324699	0.100972	1.044420
−2.480205	−0.614213	−0.861624	1.322244
−1.488123	−0.996584	−1.086285	0.500879
−0.496041	−0.475947	−0.480112	−0.582463
0.496041	0.475947	0.480112	−1.111123
1.488123	0.996584	1.086285	−0.767664
2.480205	0.614213	0.861624	0.146455
3.472287	−0.324699	−0.100972	0.952975
4.464369	−0.969400	−1.122248	1.043591
5.456450	−0.735724	−1.300975	0.289324
6.448532	0.164595	−0.135657	−0.814245
8.432696	0.837166	3.118750	−0.694201
9.424778	0.000000	1.304340	1.158931

Table 2. Selected approximate values at $q = 0.8$, $\omega = 0.6$.

x	$\sin(x)$	$\sin(x;q)$	$S_{q,\omega}(x)$
−9.424778	0.000000	0.059816	−1.207807
−8.432696	−0.837166	−1.278289	2.113101
−6.448532	−0.164595	0.026888	0.105624
−5.456450	0.735724	0.998277	−1.263238
−4.464369	0.969400	1.018192	−1.136624
−3.472287	0.324699	0.181080	−0.003039
−2.480205	−0.614213	−0.762504	0.984651
−1.488123	−0.996584	−1.050092	1.029822
−0.496041	−0.475947	−0.478435	0.205956
0.496041	0.475947	0.478435	−0.745320
1.488123	0.996584	1.050092	−1.053898
2.480205	0.614213	0.762504	−0.499557
3.472287	−0.324699	−0.181080	0.457076
4.464369	−0.969400	−1.018192	1.045757
5.456450	−0.735724	−0.998277	0.779282
6.448532	0.164595	−0.026888	−0.156142
8.432696	0.837166	1.278289	−1.011310
9.424778	0.000000	−0.059816	−0.056695

Table 3. Approximate values for $q = 0.9$ and $\omega = 0.5$.

x	$\sin(x)$	$\sin_q(x)$	$S_{q,\omega}(x)$
−9.424778	0.000000	−0.130903	−1.060978
−8.432696	−0.837166	−0.974381	−0.456269
−6.448532	−0.164595	−0.038576	1.028667
−5.456450	0.735724	0.849422	0.597574
−4.464369	0.969400	0.977009	−0.348872
−3.472287	0.324699	0.252411	−0.977648
−2.480205	−0.614213	−0.682620	−0.752415
−1.488123	−0.996584	−1.020686	0.120777
−0.496041	−0.475947	−0.477059	0.885673
0.496041	0.475947	0.477059	0.885673
1.488123	0.996584	1.020686	0.120777
2.480205	0.614213	0.682620	−0.752415
3.472287	−0.324699	−0.252411	−0.977648
4.464369	−0.969400	−0.977009	−0.348872
5.456450	−0.735724	−0.849422	0.597574
6.448532	0.164595	0.038576	1.028667
8.432696	0.837166	0.974381	−0.456269
9.424778	0.000000	0.130903	−1.060978

7.2. Zeros of q-trigonometric and q,ω-trigonometric functions

There are many ways to investigate the zeros of q-trigonometric and q,ω-trigonometric functions, see, e.g., [2, 21, 25, 29, 46, 50]. Here we explore the results of [21]. It is known that $\sin(\cdot,q)$, $\cos(\cdot,q)$ have only real and simple zeros $\{0,\pm x^m\}_{m=1}^{\infty}$ and $\{\pm y_m\}_{m=1}^{\infty}$ respectively, where $x_m, y_m > 0, m \geqslant 1$ and

$$x_m = q^{-m}(1-q)^{-1}(1+O(q^m)), \tag{7.5}$$
$$y_m = q^{-m+1/2}(1-q)^{-1}(1+O(q^m)), \quad m \geq 1. \tag{7.6}$$

Consequently q,ω-sine and cosine functions defined by (4.2), (4.2) have real and simple zeros $\{\pm g^m\}_{m=1}^{\infty}$, $\{\pm j_m\}_{m=1}^{\infty}$

$$g_m = \omega_0 + q^{-m}(1-q)^{-1}(1+O(q^m)), \tag{7.7}$$
$$j_m = \omega_0 + q^{-m+1/2}(1-q)^{-1}(1+O(q^m)), \quad m \geq 1. \tag{7.8}$$

It is also known that zeros of $\sin(\cdot,q)$ and $\cos(\cdot,q)$ interlace with each other. Figures 6 and 7 exhibit this behavior and Tables 4 and 5 numerically assure this behavior. Moreover, the asymptomatic behavior (7.5)–(7.6) is verified. It should be noted that due to scaling of the

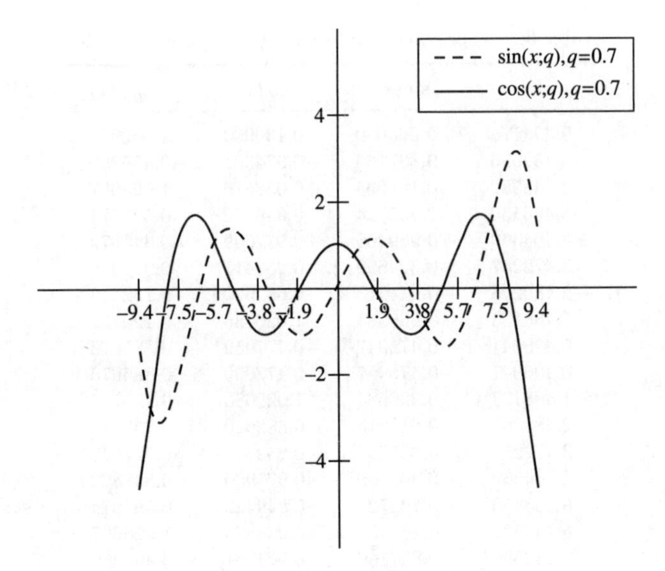

Fig. 6. Approximate curves of $\sin(\cdot;q)$, $\cos(\cdot;q)$ in $[-3\pi, 3\pi], q = 0.7$. The approximate zeros interlace with each other.

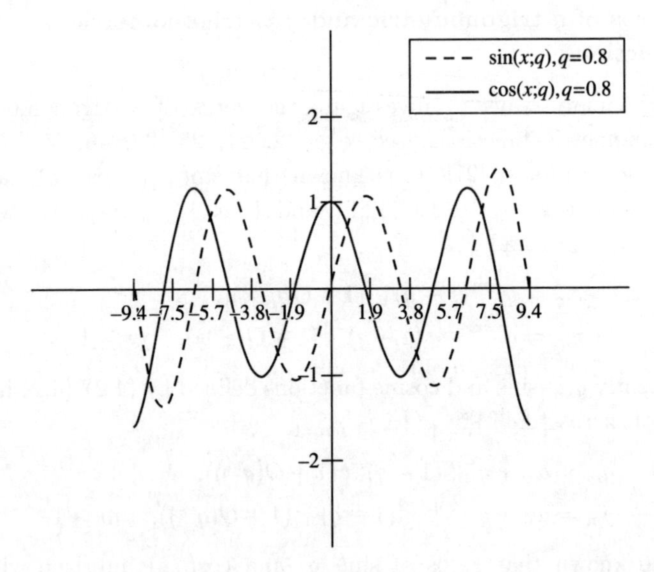

Fig. 7. Truncated curves of $\sin(\cdot;q)$, $\cos(\cdot;q)$ in $[-3\pi, 3\pi]$, $q = 0.8$. The interlace property of zeros is illustrated.

Table 4. Approximate zeros of $\sin(\cdot; q)$, $\cos(\cdot; q)$, $q = 0.7$.

m	$\frac{q^{-m-1/2}}{1-q}$	y_m	$\frac{q^{-m}}{1-q}$	x_m
1	5.691565	4.988727	4.761904	3.384173
2	8.130807	8.051016	6.802721	6.522808
3	11.615438	11.613597	9.718172	9.703067
4	16.593483	16.593476	13.883104	13.882958
5	23.704976	23.704976	19.833006	19.833006
6	33.864252	33.864252	28.332865	28.332865
7	48.377503	48.377503	40.475522	40.475522
8	69.110718	69.110718	57.822175	57.822175
9	98.729597	98.729597	82.603107	82.603107
10	141.042282	141.042282	118.004439	118.004439
11	201.488974	201.488974	168.577770	168.577770
12	287.841391	287.841391	240.825386	240.825386
13	411.201987	411.201987	344.036266	344.036266
14	587.431411	587.431411	491.480379	491.480379
15	839.187729	839.187729	702.114828	702.114828
16	1198.839614	1198.839614	1003.021183	1003.021183

Table 5. Approximate zeros for $\sin(\cdot; q)$, $\cos(\cdot; q)$ when $q = 0.8$.

m	$\frac{q^{-m-1/2}}{1-q}$	y_m	$\frac{q^{-m}}{1-q}$	x_m
1	6.987712	4.909814	6.250000	3.300352
2	8.734641	7.965883	7.812500	6.469878
3	10.918300	10.774155	9.765625	9.393455
4	13.647876	13.638682	12.207031	12.164612
5	17.059845	17.059670	15.258789	15.257325
6	21.324806	21.324804	19.073486	19.073470
7	26.656007	26.656007	23.841858	23.841858
8	33.320009	33.320009	29.802322	29.802322
9	41.650012	41.650012	37.252903	37.252903
10	52.062515	52.062515	46.566129	46.566129
11	65.078143	65.078143	58.207661	58.207661
12	81.347679	81.347679	72.759576	72.759576
13	101.684599	101.684599	90.949470	90.949470
14	127.105749	127.105749	113.686838	113.686837
15	158.882186	158.882186	142.108547	142.108547
16	198.602732	198.602732	177.635684	177.635684

zeros we start counting the zeros of the q-cosine and q, ω-cosine functions together with the behaviors $\frac{q^{-m}}{1-q}$ and $\omega_0 + \frac{q^{-m}}{1-q}$, respectively.

Figures 8 and 9 together with Tables 6 and 7 are devoted to verify (7.7)–(7.8) numerically.

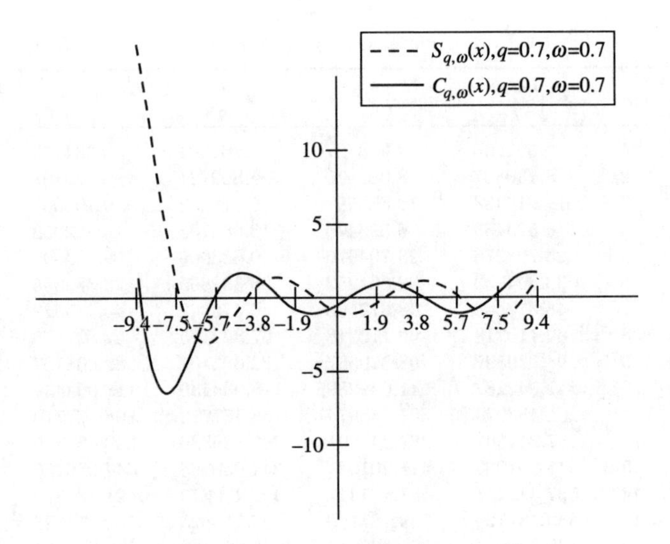

Fig. 8. Curves for truncated series of $S_{q,\omega}(\cdot)$, $C_{q,\omega}(\cdot)$ on $[-3\pi, 3\pi]$, $q = 0.7$, $\omega = 0.7$.

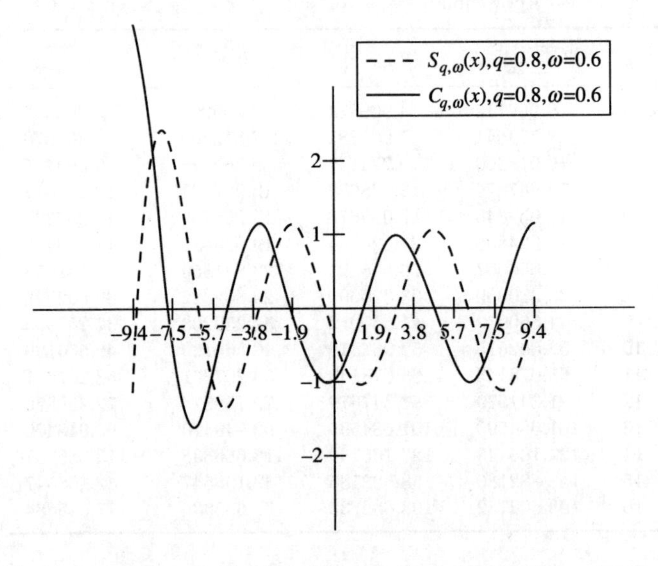

Fig. 9. Approximate curves for $S_{q,\omega}(\cdot)$, $C_{q,\omega}(\cdot)$ in $[-3\pi, 3\pi]$, $q = 0.8$, $\omega = 0.6$.

Table 6. Approximate values of zeros of $S_{q,\omega}(\cdot)$ and $C_{q,\omega}(\cdot)$, $q = 0.7$, $\omega = 0.7$.

m	$\frac{q^{-m-1/2}}{1-q}$	y_m	$\frac{q^{-m}}{1-q}$	x_m
1	8.024898	7.322061	7.095237	5.717507
2	10.464139	10.383496	9.136054	8.856141
3	13.948771	13.946930	12.051506	12.036400
4	18.926816	18.926809	16.216437	16.216291
5	26.038309	26.038309	22.2166339	22.166339
6	36.197585	36.197585	30.666199	30.666199
7	50.710836	50.710836	42.808856	42.808856
8	71.444051	71.444051	60.155508	60.155508
9	101.062930	101.062931	84.936441	84.936441
10	143.375615	143.375615	120.337772	120.337772
11	203.822307	203.822307	170.911103	170.911103
12	290.174724	290.174725	243.158719	243.158719
13	413.535321	413.535321	346.369599	346.369599
14	589.764743	589.764744	493.813713	493.813713
15	841.521063	841.521063	704.448161	704.448161
16	1201.172947	1201.172947	1005.354516	1005.354516

Table 7. Approximations of zeros of $S_{q,\omega}(\cdot)$, $C_{q,\omega}(\cdot)$, $q = 0.8$, $\omega = 0.6$.

m	$\omega_0 + \frac{q^{-m-1/2}}{1-q}$	j_m	$\omega_0 + \frac{q^{-m}}{1-q}$	g_m
1	9.987712	7.909814	9.250000	6.300352
2	11.734641	10.965883	10.812500	9.469878
3	13.918300	13.774155	12.765625	12.393455
4	16.647876	16.6386823	15.207031	15.164611
5	20.059845	20.059670	18.258789	18.257325
6	24.324806	24.324804	22.073486	22.073471
7	29.656007	29.656007	26.841858	26.841858
8	36.320009	36.320009	32.802322	32.802322
9	44.650012	44.650012	40.252903	40.252903
10	55.062515	55.062515	49.566129	49.566129
11	68.078143	68.078143	61.207661	61.207660
12	84.347679	84.347679	75.759576	75.759576
13	104.684599	104.684599	93.949470	93.949470
14	130.105749	130.105749	116.686837	116.686837
15	161.882186	161.882186	145.108547	145.108547
16	201.602732	201.602732	180.635684	180.635684

References

[1] L. D. Abreu, Sampling theory associated with q-difference equations of the Sturm–Liouville type, *J. Phys. A: Math. Gen.* **38** (2005) 10311.

[2] L. D. Aberu, J. Bustoz and J. L. Caradoso, The roots of the third Jackson q-Bessel functions, *Int. J. Math. Math. Sci.* **67** (2003) 4241–4248.

[3] M. H. Abu Risha, M. H. Annaby, M. E. H. Ismail and Z. S. Mansour, Linear q-difference equations, *Z. Anal. Anwendungen* **26** (2007) 481–494.

[4] R. Agarwal, M. Bohner and P. Wong, Sturm–Liouville eigenvalue problems on time scales, *Appl. Math. Comput.* **99** (1999) 153–166.

[5] C. Ahlbrandt, M. Bohner and J. Ridenhour, Hamiltonian systems on time scales, *J. Math. Anal. Appl.* **250** (2000) 561–578.

[6] B. P. Allahverdiev, Spectral problems of non-self-adjoint q-Sturm–Liouville operators in limit-point case, *Kodai Math. J.* **39** (2016) 1–15.

[7] P. Amster, P. De Nápoli and J. Pinasco, Eigenvalue distribution of second-order dynamic equations on time scales considered as fractals, *J. Math. Anal. Appl.* **343** (2008) 573–584.

[8] D. R. Anderson, Positivity of Green's function for an n-point right focal boundary value problem on measure chains, *Math. Comput. Modelling* **31** (2002) 29–50.

[9] D. R. Anderson, Eigenvalue intervals for even-order Sturm–Liouville dynamic equations, *Comm. Appl. Nonlinear Anal.* **12** (2005) 1–13.

[10] D. R. Anderson, G. Sh. Guseinov and J. Hoffacker, Higher-order self-adjoint boundary-value problems on time scales, *J. Comput. Appl. Math.* **194** (2006) 309–342.

[11] D. R. Anderson and J. Hoffacker, Green's function for an even order mixed derivative problem on time scales, *Dynam. Systems Appl.* **12** (2003) 9–22.

[12] D. R. Anderson and J. Hoffacker, A stacked delta–nabla self-adjoint problem of even order, *Math. Comput. Modelling* **38** (2003) 481–494.

[13] D. R. Anderson and R. I. Avery, An even-order three-point boundary value problem on time scales, *J. Math. Anal. Appl.* **291** (2004) 514–525.

[14] M. H. Annaby, q-Type sampling theorems, *Result. Math.* **44** (2003) 214–225.

[15] M. H. Annaby, H. A. Hassan and Z. S. Mansour, Sampling theorems associated with singular q-Sturm–Liouville problems, *Results Math.* **62** (2012) 121–136.

[16] M. H. Annaby, A. E. Hamza and K. A. Aldwoah, Hahn difference operator and associated Jackson–Nörlund integrals, *J. Optim. Theory Appl.* **154** (2012) 133–153.

[17] M. H. Annaby, J. Bustoz and M. E. H. Isamil, On sampling theory and basic Sturm–Liouville systems, *J. Comput. Appl. Math.* **206** (2007) 73–85.

[18] M. H. Annaby and Z. S. Mansour, Asymptotic formulae for eigenvalues and eigenfunctions of q-Sturm–Liouville problems, *Math. Nach.* **284** (2011) 443–470.

[19] M. H. Annaby and Z. S. Mansour and I. A. Soliman, q-Titchmarsh–Weyl theory: series expansion, *Nagoya Math. J.* **205** (2012) 67–118.

[20] M. H. Annaby and Z. S. Mansour, Basic Sturm Liouville problems, *J. Phys. A: Math. Gen.* **38** (2005) 3775–3797.

[21] M. H. Annaby and Z. S. Mansour, On the zeros of the second and third Jackson q-Bessel functions and their associated q-Hankel transforms, *Math. Proc. Cambridge Philos. Soc.* **147** (2009) 47–67.
[22] M. H. Annaby and Z. S. Mansour, *q-Fractional Calculus and Equations* (Springer, Berlin, 2012).
[23] G. Sh. Atici and G. Sh. Guseinov, On Green's functions and positive solutions for boundary value problem on time scales, *J. Comput. Appl. Math.* **141** (2002) 75–99.
[24] F. Atkinson, *Discrete and Continuous Boundary Value Problem* (Academic Press, New York, 1964).
[25] W. Bergweiler and W. K. Hayman, Zeros of solutions of a functional equation, *Comput. Methods Funct. Theory* **3** (2003) 55–78.
[26] M. Bohner and A. Peterson, *Dynamic Equations on Time Scales An Introduction with Applications* (Birkhäuser, Basel, 2001).
[27] M. Bohner and A. Peterson, *Advances in Dynamic Equations on Time Scales* (Birkhäuser, Basel, 2003).
[28] B. M. Brown, W. D. Evans and M. E. H. Ismail, The Askey–Wilson polynomials and q-Sturm–Liouville problems, *Math. Proc. Cambridge Philos. Soc.* **119** (1996) 1–16.
[29] J. Bustoz and J. Cardoso, Basic analog of Fourier series on a q-linear grid, *J. Approx. Theory* **112** (2001) 154–157.
[30] J. Bustoz and S. K. Suslov, Basic analog of Fourier series on a q-quadratic grid, *Methods Appl. Anal.* **5** (1998) 1–38.
[31] F. A. Davidson and P. Rynne, Eigenfunction expansions in L^2 spaces for boundary value problems on time-scales, *J. Math. Anal. Appl.* **335** (2007) 1038–1051.
[32] M. S. P. Eastham, *Theory of Ordinary Differential Equations* (Van Nostrand-Reinhold, London, 1970).
[33] A. Erylmaz, Spectral analysis of q-Sturm–Liouville problem with the spectral parameter in the boundary condition, *J. Func. Spaces Appl.* **2012** (2012) Article ID 736437, 17 pp.
[34] H. Exton, *q-Hypergeometric Functions and Applications* (Halsted Press, New York, 1983).
[35] H. Exton Basic Sturm–Liouville theory, *Rev. Tecn. Fac. Ingr. Univ. Zulia* **11** (1992) 85–100.
[36] G. Gasper and M. Rahaman, *Basic Hypergeometric Series* (Cambridge University Press, New York, 1990).
[37] I. Gohberg and S. Goldberg, *Basic Operator Theory* (Birkhäuser, 1981).
[38] G. S. Guseinov, Self-adjoint boundary value problems on time scales and symmetric Green's functions, *Turkish J. Math.* **29** (2005) 365–380.
[39] W. Hahn, Ein beitrag zur theorie der orthogonalpolynome, *Monatsh. Math.* **95** (1983) 19–24.
[40] A. E. Hamza and S. A. Ahmed, Theory of linear Hahn difference equations, *J. Adv. Math.* **4**(2) (2013) 440–460.
[41] A. E. Hamza and S. A. Ahmed, Existence and uniqueness of solutions of Hahn difference equations, *Adv. Difference Equations* **316** (2013) 1–15.

[42] J. Henderson, Multiple solutions for $2m$th-order Sturm–Liouville boundary value problems on a measure chain, *J. Difference Equations Appl.* **6** (2000) 427–429.
[43] J. Hoffacker, Green's functions for higher order equations on time scales, *Panamer. Math. J.* **11**(4) (2001) 1–28.
[44] J. Hoffacker, Green's functions and eigenvalue comparisons for a focal problem on time scales, *Comput. Math. Appl.* **45** (2003) 1339–1368.
[45] M. E. H. Ismail, *Classical and Quantum Orthogonal Polynomials in One Variable* (Cambridge University Press, Cambridge, 2005).
[46] M. E. H. Ismail, The zeros of basic Bessel functions, the functions $J_{(v+ax)}(x)$ and associated orthogonal polynomials, *J. Math. Anal. Appl.* **86** (1982) 11–19.
[47] F. H. Jackson, q-Difference equations, *Amer. J. Math.* **32** (1910) 14–125.
[48] A. Jirari, *Second Order Sturm–Liouville Difference Equations and Orthogonal Polynomials*, Memories of the American Mathematical Society, Vol. 113, No. 542 (American Mathematical Society, Providence, RI, 1995).
[49] C. Jordan, *Calculus of Finite Differences* (Chelsea, New York, 1965).
[50] H. T. Koelink and R. F. Swarttouw, On the zeros of the Hahn–Exton q-Bessel function and associated q-Lommel polynomials, *J. Math. Anal. Appl.* **186** (1994) 690–710.
[51] Q. Kong, Sturm–Liouville problems on time scales with separated boundary conditions, *Results Math.* **52** (2008) 111–121.
[52] Q. Kong and Q.-R. Wang, Using time scales to study multi-interval Sturm–Liouville problems with interface conditions, *Results Math.* **63** (2011) 451–465.
[53] A. Lavagno, Basic-deformed quantum mechanics, *Reports Math. Phys.* **64** (2009) 79–91.
[54] B. M. Levitan and I. S. Sargsian, *Sturm–Liouville and Dirac Operators* (Kluwer, Academic Publishers, Dordrecht, 1991).
[55] W. T. Li and X. L. Liu, Eigenvalue problems for second-order nonlinear dynamic equations on time scales, *J. Math. Anal. Appl.* **318** (2005) 578–592.
[56] A. Nemri A and A. Fitouhi, Polynomial expansions for solution of wave equation in quantum calculus, *Le Matematiche.* **65** (2010) 73–82.
[57] B. P. Rynne, L^2 spaces and boundary value problems on time-scales, *J. Math. Anal. Appl.* **328** (2007) 1217–1236.
[58] G. Shi and H. Wu, Spectral theory of Sturm–Liouville difference operators, *Linear Algebra Appl.* **430** (2009) 830–846.
[59] I. Stakgold and J. Michael, *Green's Functions and Boundary Value Problems* (Wiley, New Jersey, 2011).
[60] S. K. Suslov, Asymptotics of zeros of basic sine and cosine functions, *J. Approx. Theory* **121** (2003) 292–335.
[61] S. K. Suslov, *An Introduction to Basic Fourier Series* (Kluwer Academic Publishers, London, 2003).

[62] R. F. Swarttouw and H. G. Meijer, A q-analogue of the Wronskian and a second solution of the Hahn–Exton q-Bessel difference equation, *Proc. Amer. Math. Soc.* **120** (1994) 855–864.
[63] E. C. Titchmarsh, *Eigenfunction Expansions Associated with Second Order Differential Equations, Part I*, 2nd edn. (The Clarendon Press, Oxford, 1962).
[64] H. Tuna, Completeness of the root vectors of a dissipative Sturm–Liouville operators on time scales, *Appl. Math. Comput.* **228** (2014) 108–115.
[65] J. Weidmann, *Linear Operators in Hilbert Space* (Springer, Berlin, 1980).
[66] A. Zettl, *Sturm–Liouville Theory* (American Mathematical Society, Providence, RI, 2005).
[67] Y. Zhang and L. Ma, Solvability of SturmLiouville problems on time scales at resonance, *J. Comput. Appl. Math.* **233** (2010) 1785–1797.

Chapter 5

Solvability of the Hankel Determinant Problem for Real Sequences

Andrew Bakan[*,‡] **and Christian Berg**[†,§]

[*] *Institute of Mathematics, National Academy of Sciences of Ukraine Tereschenkivska Street 3, Kyiv 01601, Ukraine*

[†] *Department of Mathematical Sciences, University of Copenhagen Universitetsparken 5, DK-2100 Copenhagen, Denmark*

[‡] *andrew@bakan.kiev.ua*

[§] *berg@math.ku.dk*

To each nonzero sequence $\mathrm{s} := \{s_n\}_{n\geq 0}$ of real numbers, we associate the Hankel determinants $D_n = \det \mathcal{H}_n$ of the Hankel matrices $\mathcal{H}_n := (s_{i+j})_{i,j=0}^n$, $n \geq 0$, and the nonempty set $\mathbb{N}_\mathrm{s} := \{n \geq 1 \,|\, D_{n-1} \neq 0\}$. We also define the Hankel determinant polynomials $P_0 := 1$, and P_n, $n \geq 1$ as the determinant of the Hankel matrix $\mathcal{H}_n$ modified by replacing the last row by the monomials $1, x, \ldots, x^n$. Clearly P_n is a polynomial of degree at most n and of degree n if and only if $n \in \mathbb{N}_s$. Kronecker established in 1881 that if $\mathbb{N}_\mathrm{s}$ is finite then rank $\mathcal{H}_n = r$ for each $n \geq r - 1$, where $r := \max \mathbb{N}_\mathrm{s}$. By using an approach suggested by Iohvidov in 1969 we give a short proof of this result and a transparent proof of the conditions on a real sequence $\{t_n\}_{n\geq 0}$ to be of the form $t_n = D_n$, $n \geq 0$ for a real sequence $\{s_n\}_{n\geq 0}$. This is the Hankel determinant problem. We derive from the Kronecker identities that each Hankel determinant polynomial P_n satisfying $\deg P_n = n \geq 1$ is preceded by a nonzero polynomial P_{n-1} whose degree can be strictly less than $n - 1$ and which has no common zeros with P_n. As an application of our results we obtain a new proof of a recent theorem by Berg and Szwarc about positive semidefiniteness of all Hankel matrices provided that $D_0 > 0, \ldots, D_{r-1} > 0$ and $D_n = 0$ for all $n \geq r$.

Keywords: Hankel matrices; Frobenius rule; Kronecker theorem; orthogonal polynomials.

Mathematics Subject Classification 2010: 44A60, 47B36, 15A15, 15A63

1. Introduction

We use the notation $\mathbb{N} := \{1, 2, \ldots\}$ and $\mathbb{N}_0 := \mathbb{N} \cup \{0\}$. To a sequence $\mathsf{s} := \{s_n\}_{n\geq 0}$ of real numbers we associate the Hankel matrices $\mathcal{H}_n := (s_{i+j})_{i,j=0}^{n}$, $n \geq 0$ and the determinants $D_n = D_n(\mathsf{s}) := \det \mathcal{H}_n$, $n \geq 0$. In this way we get a mapping $D : \{s_n\}_{n\geq 0} \mapsto \{D_n(\mathsf{s})\}_{n\geq 0}$ in the space $\mathbb{R}^{\mathbb{N}_0}$ of sequences of real numbers. We call this mapping the Hankel determinant transform. It was introduced and studied by Layman in [13] who emphasized that such a transform is far from being injective by proving that a sequence s and its binomial transform $\beta(\mathsf{s})$ defined by

$$\beta(\mathsf{s})_n := \sum_{k=0}^{n} \binom{n}{k} s_k, \quad n \geq 0,$$

have the same image under this mapping. Concerning the missing injectivity let us here just point out that the Hankel determinant transform of all the sequences $\{a^n\}_{n\geq 0}$, $a \in \mathbb{R}$, is $\{1, 0, 0, \ldots\}$.

Several authors have been concerned with the sign pattern of the sequence $D(\mathsf{s})$ in order to use this for the determination of the rank and signature of the Hankel matrices. This is given in rules of, e.g., Jacobi, Gundelfinger and Frobenius. See [8, 10] for a treatment of these questions, which become quite technical when zeros occur in the sequence $D(\mathsf{s})$.

The Hankel determinant problem for real sequences is to characterize the image $D(\mathbb{R}^{\mathbb{N}_0})$ in $\mathbb{R}^{\mathbb{N}_0}$, i.e., to find a necessary and sufficient condition for a sequence $\mathsf{t} \in \mathbb{R}^{\mathbb{N}_0}$ to be of the form

$$\begin{vmatrix} s_0 & s_1 & \cdots & s_n \\ s_1 & s_2 & \cdots & s_{n+1} \\ \cdots & \cdots & \cdots & \cdots \\ s_n & s_{n+1} & \cdots & s_{2n} \end{vmatrix} = t_n, \quad n \geq 0. \tag{1.1}$$

with some sequence s of real numbers. It turns out that such conditions are similar to those that were obtained by Frobenius [6, p. 207] in 1894 for all possible signs of the numbers $\{t_n\}_{n\geq 0}$. His arguments were simplified by Gantmacher [8, p. 348] in 1959 and by Iohvidov [10, (12.8), p. 83] in

1982 in an essential way. The purpose of the present chapter is to obtain a further simplification of the Frobenius reasoning by giving in Theorem 4.1 a new setting of the approach suggested by Iohvidov in [10, Chapter II]. This allows to give in Section 7 a self-contained proof of the following theorem.

Theorem 1.1. *Let* $\mathrm{t} := \{t_n\}_{n\geq 0}$ *be a sequence of real numbers and*

$$Z_{\mathrm{t}} := \{n \geq 0 \mid t_n \neq 0\}.$$

If $Z_{\mathrm{t}} = \emptyset$ *then equation* (1.1) *is satisfied if and only if* $s_n = 0$ *for all* $n \geq 0$. *If* $Z_{\mathrm{t}} \neq \emptyset$ *consists of* $1 \leq m \leq \infty$ *distinct elements* $\{n_k\}_{0\leq k<m}$ *arranged in increasing order, then equation* (1.1) *is solvable if and only if the following Frobenius conditions* (*see* [8, p. 348]) *hold:*

$$\begin{aligned} (-1)^{\frac{n_0+1}{2}} t_{n_0} &> 0 \quad \text{if } n_0+1 \in 2\mathbb{N}, \\ (-1)^{\frac{n_{k+1}-n_k}{2}} t_{n_{k+1}} t_{n_k} &> 0 \quad \text{if } n_{k+1}-n_k \in 2\mathbb{N}, \\ & \qquad 0 \leq k < m-1,\ 2 \leq m \leq \infty. \end{aligned}$$

It follows from Theorem 1.1 that (1.1) is solvable if $t_n \neq 0$ for all $n \geq 0$, and not solvable if $\mathrm{t} = \{0,1,0,0,\ldots\}$. Furthermore, the condition $(-1)^{\frac{n(n+1)}{2}} t_n \geq 0$ for all $n \geq 0$ is sufficient for the existence of at least one solution of (1.1).

Let us formulate an elementary result about existence and uniqueness of solutions to (1.1) and which is independent of Theorem 1.1. For this we need the following notation. For a $n \times n$ determinant A, we denote by $A^{k,m}$, $1 \leq k, m \leq n$, the $(n-1)\times(n-1)$ determinant obtained by deleting the kth row and mth column of A. For Hankel determinants we follow Frobenius [6, p. 212] in writing $D'_{n+1} = D^{n+2,n+1}_{n+1}$, $n \geq 0$, i.e.,

$$D'_1 = s_1, \quad D'_2 = \begin{vmatrix} s_0 & s_2 \\ s_1 & s_3 \end{vmatrix},$$

$$D'_{n+1} = \begin{vmatrix} s_0 & s_1 & \cdots & s_{n-2} & s_{n-1} & s_{n+1} \\ s_1 & s_2 & \cdots & s_{n-1} & s_n & s_{n+2} \\ \cdots & \cdots & \cdots & \cdots & \cdots & \cdots \\ s_{n-1} & s_n & \cdots & s_{2n-3} & s_{2n-2} & s_{2n} \\ s_n & s_{n+1} & \cdots & s_{2n-2} & s_{2n-1} & s_{2n+1} \end{vmatrix}, \quad n \geq 2. \qquad (1.2)$$

Proposition 1.2. *Given two sequences* t, t' *of real numbers such that* $t_n \neq 0$ *for all* $n \geq 0$, *there exists a unique sequence* s *of real numbers such that*

$$D_n = t_n \in \mathbb{R} \setminus \{0\}, \quad D'_{n+1} = t'_n \in \mathbb{R}, \quad n \geq 0.$$

To see this we use the Laplace expansion of D_n and D'_{n+1} along the last column and note that $(D'_{n+1})^{n-k,n+1} = D_n^{n-k,n+1}$. This gives the following recurrence formulas:

$$s_0 = D_0, \; s_1 = D'_1; \; s_2 D_0 = D_1 + s_1^2, \; s_3 D_0 = D'_2 + s_1 s_2;$$

$$s_{2n} D_{n-1} = D_n + \sum_{k=0}^{n-1} (-1)^k s_{2n-1-k} \, D_n^{n-k,n+1} =: D_n + F_n(s_0, \ldots, s_{2n-1}),$$

$$s_{2n+1} D_{n-1} = D'_{n+1} + \sum_{k=0}^{n-1} (-1)^k s_{2n-k} \, D_n^{n-k,n+1} =: D'_{n+1}$$

$$+ G_n(s_0, \ldots, s_{2n}), \; n \geq 1,$$

where $D_n^{n-k,n+1}$, $0 \leq k \leq n-1$, depend only on s_j, $0 \leq j \leq 2n-1$, F_n is a function of $s_0, \ldots, s_{2n-1}$ and G_n is a function of $s_0, \ldots, s_{2n}$. If $t'_n = D'_{n+1}, t_n = D_n \neq 0$, $n \geq 0$ are assumed to be given, these relations determine the sequence s uniquely, and the assertion follows.

A complete description of all solutions of (1.1), when some of the numbers t_n vanish, can be derived from the Frobenius results in [6], but this is of no relevance in the present context.

Let $\mathcal{P}[\mathbb{R}]$ denote the set of all algebraic polynomials with real coefficients. Given a sequence $\{s_n\}_{n \geq 0}$ of real numbers, we introduce two sequences of polynomials in $\mathcal{P}[\mathbb{R}]$:

$$P_0(x) := 1, \quad P_1(x) := \begin{vmatrix} s_0 & s_1 \\ 1 & x \end{vmatrix},$$

$$P_n(x) := \begin{vmatrix} s_0 & s_1 & s_2 & \ldots & s_n \\ s_1 & s_2 & s_3 & \ldots & s_{n+1} \\ \ldots & \ldots & \ldots & \ldots & \ldots \\ s_{n-1} & s_n & s_{n+1} & \ldots & s_{2n-1} \\ 1 & x & x^2 & \ldots & x^n \end{vmatrix};$$

$$Q_0(x) := 0, \quad Q_1(x) := s_0^2,$$

$$Q_n(x) := \begin{vmatrix} s_0 & s_1 & s_2 & \dots & s_n \\ s_1 & s_2 & s_3 & \dots & s_{n+1} \\ \dots & \dots & \dots & \dots & \dots \\ s_{n-1} & s_n & s_{n+1} & \dots & s_{2n-1} \\ 0 & s_0 & s_0x+s_1 & \dots & \sum_{k=0}^{n-1} s_k x^{n-1-k} \end{vmatrix}, \quad n \geq 1. \tag{1.3}$$

Note that $D_n^{n+1,n+1-k} = D_n^{n+1-k,n+1}$, $0 \leq k \leq n$, $n \geq 1$ and

$$P_1(x) = D_0 x - D_1',$$

$$P_n(x) = D_{n-1}x^n - D_n' x^{n-1} + \sum_{k=1}^{n-1} (-1)^{k-1} x^{n-1-k} D_n^{n-k,n+1}, \quad n \geq 2. \tag{1.4}$$

The polynomials $\{P_n\}_{n\geq 0}$ are called *Hankel determinant polynomials* with respect to the sequence $\{s_n\}_{n\geq 0}$. Let $L : \mathcal{P}[\mathbb{R}] \to \mathbb{R}$ denote the linear functional determined by

$$L(x^n) = s_n, \ n \geq 0. \tag{1.5}$$

Then

$$L(x^k P_n(x)) = 0, \ 0 \leq k \leq n-1, \ n \geq 1, \tag{1.6}$$

and also

$$L(P_n(x)^2) = D_n D_{n-1}, \ n \geq 0, \ D_{-1} := 1. \tag{1.7}$$

Already Stieltjes considered this kind of functional, see [18, p. 25]. It is also used in [3, Definition 2.1, p. 6]).

In the classical case where all the Hankel determinants $D_n > 0$, these polynomials are proportional to the classical orthonormal polynomials (see [3, Exercise 3.1(a), p. 17; pp. 10, 15])

$$\mathrm{p}_n(x) := \frac{P_n(x)}{\sqrt{D_n D_{n-1}}}, \ n \geq 0, \ D_{-1} := 1, \tag{1.8}$$

and those of the second kind.

In the general case of an arbitrary sequence $\{s_n\}_{n\geq 0}$ of real numbers Frobenius [6, (5), p. 212] obtained in 1894 a recurrence relation for the polynomials $\{P_n\}_{n\geq 0}$ in the following determinant form:

$$D_{n-1}D_n \, xP_n(x) = D_{n-1}^2 \, P_{n+1}(x) + (D_{n-1}D_{n+1}' - D_n D_n')P_n(x) + D_n^2 \, P_{n-1}(x), \tag{1.9}$$

where $n \geq 0$, $P_{-1}(x) := 0$, $P_0(x) = 1$ and $D_{-1} := 1$, $D_0' := 0$. If $D_n \neq 0$ for all $n \geq 0$, then the functional L is called *quasi-definite* (see [3, Definition 3.2, p. 16]) and the monic polynomials

$$p_n(x) := P_n(x)/D_{n-1}, \; n \geq 0, \tag{1.10}$$

are usually considered for which the recurrence (1.9) is written in the Jacobi form (see [3, Theorem 4.1, p. 18])

$$p_{n+1}(x) = (x - a_n)p_n(x) - b_n p_{n-1}(x), \; n \geq 0, \; p_0(x) = 1, \; p_{-1}(x) = 0, \tag{1.11}$$

$$a_n = \frac{D'_{n+1}}{D_n} - \frac{D'_n}{D_{n-1}}, \; n \geq 0, \; b_n = \frac{D_n D_{n-2}}{D^2_{n-1}}, \; n \geq 1, \; a_0 = \frac{D'_1}{D_0}, \; b_0 = D_0, \tag{1.12}$$

where the relations (1.12) are invertible (cf. [3, Theorem 4.2, p. 19])

$$D_n = \prod_{k=0}^{n} b_k^{n+1-k}, \; D'_{n+1} = \left(\sum_{k=0}^{n} a_k\right) \prod_{k=0}^{n} b_k^{n+1-k}, \; n \geq 0, \tag{1.13}$$

and $b_n \neq 0$ for all $n \geq 0$. Conversely, given the recurrence formula (1.11) for monic polynomials $\{p_n\}_{n\geq 0}$ with two arbitrary real sequences $\{a_n\}_{n\geq 0}$ and $\{b_n\}_{n\geq 0}$ satisfying $b_n \neq 0$ for all $n \geq 0$, we determine by (1.13) and Proposition 1.2 a quasi-definite functional L such that $L(p_n(t)p_m(t)) = 0$ and $L(p_n(t)^2) \neq 0$ for all $n, m \geq 0$, $n \neq m$, by virtue of (1.5), (1.10), (1.6) and (1.7). This fact is known as the generalized Favard theorem for quasi-definite functionals (see [3, Theorem 4.4, p. 21]).

By Theorem 1.1, if $\{s_n\}_{n\geq 0}$ is assumed nonzero, i.e., $s_n \neq 0$ for at least one $n \geq 0$, there exists $r \geq 1$ such that $D_{r-1} \neq 0$, and then P_r is a polynomial of degree r. In Theorem 2.4 we derive from the Kronecker identities (2.7) a simple result about zeros of the polynomials P_r and Q_r.

In 1881 Kronecker [12] also characterized all those nonzero sequences $\{s_n\}_{n\geq 0}$ of real numbers whose Hankel matrices $(s_{i+j})_{i,j=0}^{\infty}$ are of finite rank, see Theorem 2.1 of Section 2.

In Corollary 5.1 we provide a new interpretation of this result based on Theorem 4.1.

The results obtained by Frobenius [6] in 1894 are formulated in Theorem 3.1 of Section 3.

In Section 5.3 we use Theorem 4.1 to derive a recent theorem of Berg and Szwarc [1], see Theorem 5.2.

2. Kronecker's Results from 1881

Let $1 \le m \le n$ and $A_n = \{a_{i,j}\}_{i,j=1}^{n}$ be a nonzero square matrix of order n, where A_n being nonzero means that it has at least one nonzero element. For arbitrary $1 \le i_1 < i_2 < \cdots < i_m \le n$ and $1 \le j_1 < j_2 < \cdots < j_m \le n$ the determinant $\det\{a_{i_{k_1},j_{k_2}}\}_{k_1,k_2=1}^{m}$ is called a minor of A_n of order m. The largest order of the nonzero minors of A_n is called the rank of the matrix A_n and is denoted by $\operatorname{rank} A_n$ (see [8, p. 2]). The rank of an infinite matrix $A_\infty = \{a_{i,j}\}_{i,j=1}^{\infty}$ is defined by $\operatorname{rank} A_\infty := \sup_{n\ge 1} \operatorname{rank} A_n \in \mathbb{N} \cup \{\infty\}$, where $\mathbb{N} := \{1, 2, \ldots\}$ (see [9, p. 205]).

Theorem 2.1 (Kronecker [12]). *Let $\{s_n\}_{n\ge 0}$ be a nonzero sequence of real numbers, $\mathcal{H}_n := (s_{i+j})_{i,j=0}^{n}$, $D_n := \det \mathcal{H}_n$, $n \ge 0$, and $\mathcal{H}_\infty := (s_{i+j})_{i,j=0}^{\infty}$. Then a necessary and sufficient condition for $\mathcal{H}_\infty$ to have a finite rank $r \in \mathbb{N}$ is that*

$$D_{r-1} \ne 0, \quad D_n = 0, \ n \ge r. \tag{2.1}$$

The necessity of the condition is formulated in [12, p. 560; 6, p. 204; 9, p. 206; 10, p. 74], while the sufficiency, proved by Kronecker [12, p. 563], is less known and can be found in [10, item 11, p. 79].

It has been proved by Kronecker in [12, $(G^{(m)})$, (G'), p. 567] that if $(s_{i+j})_{i,j=0}^{\infty}$ is of finite rank r then

$$Q_r(x)/P_r(x) = \sum_{k\ge 0} s_k x^{-k-1}. \tag{2.2}$$

Since (1.3) yields $x^r P_r(1/x) \restriction_{x=0} = D_{r-1} \ne 0$, the change of variable $x \to 1/z$ in (2.2) shows that it is equivalent to the Taylor expansion at the origin

$$\psi_r^{\mathrm{s}}(z) := \frac{z^{r-1} Q_r(1/z)}{z^r P_r(1/z)} = \sum_{k\ge 0} s_k z^k$$

of the analytic function ψ_r^{s} on the open disk $|z| < 1/\rho_r$ where $\rho_r := \max\{|z| \mid P_r(z) = 0\}$. Therefore, the series in the right-hand side of (2.2) converges absolutely for every $|x| > \rho_r$, and (2.3) below holds by the Cauchy–Hadamard formula (see [16, (2), p. 200]).

Conversely, Kronecker proved in [12, p. 568] that if the numbers $\{s_n\}_{n\ge 0}$ are the coefficients in the expansion (2.5) of q/p for $p, q \in \mathcal{P}[\mathbb{R}]$, $\deg p = r \in \mathbb{N}$ and $\deg q < r$ then $\mathcal{H}_\infty$ has the rank r, provided $D_{r-1} \ne 0$ (see [14, Section 45, p. 198; 9, Theorem 8, p. 207]). Thus, the following characterization of the Hankel matrices of finite rank holds.

Theorem 2.2. *Let $\{s_n\}_{n\ge 0}$ be a nonzero sequence of real numbers and $\mathcal{H}_\infty := (s_{i+j})_{i,j=0}^{\infty}$.*

(a) *If $\mathcal{H}_\infty$ has a finite rank $r \in \mathbb{N}$ then* $\deg P_r = r$,

$$\overline{\lim_{k\to\infty}} \sqrt[k]{|s_k|} = \max\{|z| \mid z \in \mathbb{C},\ P_r(z) = 0\}, \tag{2.3}$$

and

$$\sum_{k\geq 0} \frac{s_k}{x^{k+1}} = \frac{Q_r(x)}{P_r(x)}, \tag{2.4}$$

where the series is absolutely convergent for every $|x| > \max\{|z||z \in \mathbb{C},\ P_r(z) = 0\}$.

(b) *If* $R := \overline{\lim}_{k\to\infty} \sqrt[k]{|s_k|} < +\infty$ *and there exist* $p, q \in \mathcal{P}[\mathbb{R}]$, *$p$ of degree $r \in \mathbb{N}$ and q of degree at most $r-1$ such that*

$$\sum_{k\geq 0} \frac{s_k}{z^{k+1}} = \frac{q(z)}{p(z)}, \quad |z| > R, \tag{2.5}$$

then rank $\mathcal{H}_\infty \leq r$, *where the equality is attained if p and q have no common roots.*

The following theorem of Kronecker [12, pp. 560, 561 and 571] clarifies the structure of the sequences satisfying rank $\mathcal{H}_\infty < \infty$ (see also [9, Theorem 7, pp. 205 and 234]).

Theorem 2.3. *Let $\{s_n\}_{n\geq 0}$ be a nonzero sequence of real numbers and* $\mathcal{H}_\infty := (s_{i+j})_{i,j=0}^{\infty}$.

(a) *$\mathcal{H}_\infty$ has a finite rank $r \in \mathbb{N}$ if and only if $D_{r-1} \neq 0$ and there exist r numbers $d_0, d_1, \ldots, d_{r-1}$ such that*

$$\sum_{k=0}^{r-1} d_k\ s_{k+m} = s_{r+m}, \quad m \geq 0. \tag{2.6}$$

(b) *If $\mathcal{H}_\infty$ has a finite rank $r \in \mathbb{N}$ then for every $n \geq 0$ there exist r numbers $d_{n,0}, d_{n,1}, \ldots, d_{n,r-1}$ such that*

$$\sum_{k=0}^{r-1} d_{n,k}\ s_{k+m} = s_{r+n+m}, \quad m \geq 0,$$

where $d_{0,k}$ is equal to d_k from (2.6) for each $0 \leq k \leq r-1$.

(c) *If $\mathcal{H}_\infty$ has a finite rank $r \in \mathbb{N}$ then the sequence $\{s_n\}_{n\geq 0}$ is uniquely determined by the values of $s_0, s_1, \ldots, s_{2r-1}$.*

Finally, we note that the equality (2.4) proved by Kronecker in [12, $(G^{(m)})$, (G'), p. 567]) asserts implicitly that the polynomials P_r and Q_r

have no common roots provided that $D_{r-1} \neq 0$. Furthermore, this fact also follows from the identity

$$P_{r-1}(x)\, Q_r(x) - P_r(x) Q_{r-1}(x) = D_{r-1}^2, \tag{2.7}$$

written by Kronecker in [12, (F), p. 564] for arbitrary $r \geq 1$ (see also [6, (14), p. 220]).

Observe that (2.7) can easily be proved when $D_n \neq 0$ for all $n \geq 0$ (see [7, III.15, p. 48]; [2, Theorem 2.12, p. 54]). These restrictions can be removed by the so-called perturbation technique. More precisely, by using the Hilbert matrix $\mathcal{M}_n^1 := ((i+j+1)^{-1})_{i,j=0}^n$, for every $\varepsilon > 0$ we introduce the perturbed sequence

$$\{s_n^\varepsilon\}_{n\geq 0}, \quad s_n^\varepsilon := s_n + \frac{\varepsilon^{n+1}}{n+1}, \; \mathcal{M}_n^\varepsilon := \left(\frac{\varepsilon^{i+j+1}}{i+j+1}\right)_{i,j=0}^n, \; n \geq 0,$$

whose Hankel determinant $D_n^\varepsilon = \det(\mathcal{H}_n + \mathcal{M}_n^\varepsilon) = \mathfrak{m}_n \varepsilon^{(n+1)^2} + \ldots$ for every $n \geq 0$ is a polynomial of degree $(n+1)^2$ in the variable ε with positive leading coefficient $\mathfrak{m}_n := \det \mathcal{M}_n^1 > 0$ (see [15, 3, p. 92]) . Since the zeros of all polynomials D_n^ε, $n \geq 0$, form an at most countable set, there exists a sequence $\{\varepsilon_k\}_{k\geq 0}$ of positive numbers ε_k tending to zero as $k \to \infty$ such that

$$\det \left\{s_{i+j}^{\varepsilon_k}\right\}_{i,j=0}^n \neq 0, \; n, k \geq 0.$$

With (2.7) in hand for $\{s_n^{\varepsilon_k}\}_{n\geq 0}$, $k \geq 0$, we conclude by the continuous dependence of (2.7) on $s_m^{\varepsilon_k}$, $0 \leq m \leq 2r-1$, that (2.7) holds for $\{s_n\}_{n\geq 0}$.

It also follows from (2.7) that P_r and P_{r-1} have no common roots, provided that $D_{r-1} \neq 0$. We have therefore proved the following property (cf. [2, Theorem 2.14, p. 57]).

Theorem 2.4. *Let $\{s_n\}_{n\geq 0}$ be an arbitrary nonzero sequence of real numbers and r be a positive integer satisfying $D_{r-1} \neq 0$. Then $\deg P_r = r$, $P_{r-1} \not\equiv 0$, $Q_r \not\equiv 0$ and the polynomial P_r has no common zeros with the polynomials P_{r-1} and Q_r.*

Observe that Theorem 2.4 can also be easily deduced from [5, Theorems 1.3(ii) and 1.9, pp. 44 and 80]. We will in the sequel use the following notion.

Definition 2.5. Let $\{s_n\}_{n\geq 0}$ be a nonzero sequence of real numbers. The rank of the infinite Hankel matrix $(s_{i+j})_{i,j=0}^{\infty}$ is called the Hankel rank of $\{s_n\}_{n\geq 0}$.

Since $\operatorname{rank}(s_{i+j})_{i,j=0}^{\infty} \in \mathbb{N}\cup\{\infty\}$, the Hankel rank of a real nonzero sequence can be equal to any positive integer or infinity.

3. Frobenius' Theorem from 1894

Let $\mathrm{s} := \{s_n\}_{n\geq 0}$ be an arbitrary nonzero sequence of real numbers and

$$\mathbb{N}_{\mathrm{s}} := \{r \in \mathbb{N} \mid D_{r-1} \neq 0\}. \tag{3.1}$$

Theorem 1.1 yields $\mathbb{N}_{\mathrm{s}} \neq \emptyset$. Suppose that $\mathbb{N}_{\mathrm{s}}$ consists of m $(1 \leq m \leq \infty)$ distinct elements $\{n_k\}_{1\leq k<m+1}$ arranged in increasing order and $n_0 := 0$, where it is assumed that $a + \infty = \infty$ for arbitrary $a \in \mathbb{R}$. Then

$$\{0\} \cup \mathbb{N}_{\mathrm{s}} = \{n_k\}_{0\leq k<m+1}, \;\; 1 \leq m \leq \infty, \;\; 0 = n_0 < n_1 < \cdots. \tag{3.2}$$

We say that the Hankel determinant polynomial P_n defined by (1.3) is of *full degree* if $\deg P_n = n$. It follows from (1.3), (1.4), (3.1) and (3.2) that P_n is of full degree if and only if $n = n_k$ for some $0 \leq k < m+1$, i.e.,

$$\begin{aligned} \{P_n \mid \deg P_n = n, \; n \geq 0\} &= \{P_{n_k}\}_{0\leq k<m+1} \\ &= \{P_{n_0} \equiv 1, P_{n_1}, \ldots, P_{n_k}, P_{n_{k+1}}, \ldots\}, \\ \deg P_{n_k} = n_k, \;\; &0 \leq k < m+1. \end{aligned} \tag{3.3}$$

Theorem 2.4 states that the identities (2.7) proved by Kronecker in 1881 imply that for each $0 \leq k < m$ the polynomial $P_{n_{k+1}}$ is preceded by a nonzero polynomial $P_{n_{k+1}-1}$ which has no common zeros with $P_{n_{k+1}}$ and whose degree can be strictly less than $n_{k+1} - 1$.

In 1894 Frobenius established [6, (10), p. 210] that $P_{n_{k+1}-1}$ for such k is proportional with a nonzero real constant of proportionality to the previous polynomial P_{n_k} of full degree provided that $\deg P_{n_{k+1}-1} < n_{k+1} - 1$, i.e., there exists $\gamma_k \in \mathbb{R} \setminus \{0\}$ such that

$$P_{n_{k+1}-1}(x) = \gamma_k P_{n_k}(x),$$

if $n_{k+1} - n_k \geq 2$ and $0 \leq k < m$ (see also [5, Theorem 1.3(ii), p. 44]). Furthermore, he proved in [6, (8), p. 214] that for $m \geq 2$ the recurrence relations

$$p_{n_{k+1}}(x) = a_k(x)p_{n_k}(x) - \beta_k p_{n_{k-1}}(x), \quad 1 \leq k < m,$$

hold between the monic polynomials

$$p_{n_k}(x) := P_{n_k}(x)/D_{n_k-1}, \quad 0 \le k < m+1, \quad D_{-1} := 1,$$

corresponding to the polynomials P_n of full degree, where $\{\beta_k\}_{1\le k<m}$ are nonzero real numbers and $a_k(x) \in \mathcal{P}[\mathbb{R}]$ is a monic polynomial of degree $n_{k+1} - n_k$ for every $1 \le k < m$ (see also [5, Remark 1.2, p. 71]). It is also proved in [6, (9), p. 210] that

$$P_{n_k+1}(x) \equiv \cdots \equiv P_{n_{k+1}-2}(x) \equiv 0$$

provided that $n_{k+1} - n_k \ge 3$ and $0 \le k < m$ (see also [5, Theorem 1.3, p. 44]). Thus, the following theorem was proved by Frobenius [6] in 1894.

Theorem 3.1. *Let $\{s_n\}_{n\ge 0}$ be an arbitrary nonzero sequence of real numbers and $\mathbb{N}_s$, m and $\{n_k\}_{0\le k<m+1}$ be defined as in (3.1) and (3.2).*

For the Hankel determinant polynomials $\{P_n\}_{n\ge 0}$ defined by (1.3) the following assertions hold.

(a) *If $n_1 \ge 2$, then there exists $\gamma_0 \in \mathbb{R}\setminus\{0\}$ such that*

$$P_0 \equiv 1,\ P_1 \equiv \gamma_0,\ \deg P_{n_1} = n_1 = 2,$$

when $n_1 = 2$ and

$$P_0 \equiv 1,\ P_1 \equiv 0, \ldots, P_{n_1-2} \equiv 0,\ P_{n_1-1} \equiv \gamma_0,\ \deg P_{n_1} = n_1,$$

when $n_1 \ge 3$.

(b) *If $m \ge 2$, $1 \le k < m$ and $n_{k+1} - n_k \ge 2$, then there exists $\gamma_k \ne 0$ such that*

$$\deg P_{n_k} = n_k,\ P_{n_{k+1}-1} = \gamma_k P_{n_k},\ \deg P_{n_{k+1}} = n_{k+1},$$

when $n_{k+1} - n_k = 2$ and

$$\deg P_{n_k} = n_k,\ P_{n_k+1} \equiv 0, \ldots, P_{n_{k+1}-2} \equiv 0,$$
$$P_{n_{k+1}-1} = \gamma_k P_{n_k},\ \deg P_{n_{k+1}} = n_{k+1},$$

when $n_{k+1} - n_k \ge 3$.

(c) *If $m \ge 2$, then, for the monic polynomials*

$$p_0(x) = 1,\ p_{n_k}(x) := \frac{P_{n_k}(x)}{D_{n_k-1}},\ 0 \le k < m+1,$$

there exist monic polynomials in $\mathcal{P}[\mathbb{R}]$

$$a_k(x),\ \deg a_k(x) = n_{k+1} - n_k \ge 1,\ 0 \le k < m,$$

and nonzero real numbers $\{\beta_k\}_{0\le k<m}$ *such that*

$$p_{n_{k+1}}(x) = a_k(x)p_{n_k}(x) - \beta_k p_{n_{k-1}}(x),\ 0 \le k < m,\ p_{n_{-1}} := 0. \quad (3.4)$$

(d) *If* $m < \infty$, *then* $n_m = \max \mathbb{N}_s$ *and* $P_n \equiv 0$ *for all* $n \ge n_m + 1$.

It should be noted that Theorem 3.1(d) follows directly from Theorem 2.1 and Theorem 2.3(a). Indeed, the conditions of Theorem 3.1(d) imply the validity of (2.1) for $r = n_m$ and in view of Theorem 2.1 we obtain that $\mathcal{H}_\infty$ has a finite rank n_m. But for arbitrary $n \ge n_m + 1$ the $(n_m + 1)$th row of the determinant for P_n in (1.3) is the linear combination of the first n_m rows by virtue of (2.6). Hence, $P_n \equiv 0$ and the desired result is proved.

Theorem 3.1 shows that except of polynomials of full degree and proportional to them the sequence $\{P_n\}_{n\ge 0}$ defined in (1.3) contains no other nonzero polynomials. Furthermore, if $n \ge 1$ then it follows from $P_n \equiv 0$ that $\deg P_{n+1} < n + 1$ while $P_n \not\equiv 0$ and $\deg P_n < n$ imply $\deg P_{n+1} = n+1$. Observe that Theorem 3.1(c) was essentially generalized by Draux [5, Theorem 6.2, p. 477] in 1983.

4. Iohvidov's Approach from 1969

Throughout this section we fix an arbitrary nonzero sequence $\mathrm{s} := \{s_n\}_{n\ge 0}$ of real numbers and use the set $\mathbb{N}_s$ defined as in (3.1). The analysis below will not use the statements from Sections 2 and 3.

In 1969 Iohvidov [11] (see also [10]) suggested a new technique for dealing with Hankel matrices. For every $r \in \mathbb{N}_s$ he proposed to use the approximating sequence $\mathrm{s}^{(r)}$ defined as follows.

We first put

$$s_n^{(r)} = s_n, \quad 0 \le n \le 2r - 1. \quad (4.1)$$

Since the first $2r - 1$ numbers $s_0, s_1, s_2, \ldots, s_{2r-2}$ of the sequence s satisfy

$$D_{r-1} = \begin{vmatrix} s_0 & s_1 & \ldots & s_{r-2} & s_{r-1} \\ s_1 & s_2 & \ldots & s_{r-1} & s_r \\ \ldots & \ldots & \ldots & \ldots & \ldots \\ s_{r-2} & s_{r-1} & \ldots & s_{2r-4} & s_{2r-3} \\ s_{r-1} & s_r & \ldots & s_{2r-3} & s_{2r-2} \end{vmatrix} \ne 0, \quad (4.2)$$

it is possible to determine uniquely all r numbers $d_0^{(r)}, d_1^{(r)}, \ldots, d_{r-1}^{(r)}$ from the system

$$\begin{pmatrix} s_0 & s_1 & \ldots & s_{r-2} & s_{r-1} \\ s_1 & s_2 & \ldots & s_{r-1} & s_r \\ \ldots & \ldots & \ldots & \ldots & \ldots \\ s_{r-2} & s_{r-1} & \ldots & s_{2r-4} & s_{2r-3} \\ s_{r-1} & s_r & \ldots & s_{2r-3} & s_{2r-2} \end{pmatrix} \begin{pmatrix} d_0^{(r)} \\ d_1^{(r)} \\ \ldots \\ d_{r-2}^{(r)} \\ d_{r-1}^{(r)} \end{pmatrix} = \begin{pmatrix} s_r \\ s_{r+1} \\ \ldots \\ s_{2r-2} \\ s_{2r-1} \end{pmatrix}, \tag{4.3}$$

and then to define recursively the numbers $s_{2r}^{(r)}, s_{2r+1}^{(r)}, \ldots$ by the formulas

$$s_{2r}^{(r)} = s_r d_0^{(r)} + s_{r+1} d_1^{(r)} + \cdots + s_{2r-2} d_{r-2}^{(r)} + s_{2r-1} d_{r-1}^{(r)},$$

$$s_{2r+m}^{(r)} = \sum_{k=0}^{r-1} s_{r+m+k}^{(r)} d_k^{(r)}, \ m \geq 1. \tag{4.4}$$

We obtain the sequence

$$\mathrm{s}^{(r)} := \{s_n^{(r)}\}_{n\geq 0}, \ \mathrm{s}^{(r)} = \{s_0, \ s_1, \ s_2, \ldots, s_{2r-1}, \ s_{2r}^{(r)}, \ s_{2r+1}^{(r)}, \ldots\}, \tag{4.5}$$

whose terms satisfy a homogeneous linear recurrence relation with constant coefficients of the form

$$\sum_{k=0}^{r-1} d_k^{(r)} s_{k+m}^{(r)} = s_{r+m}^{(r)}, \ m \geq 0, \tag{4.6}$$

which is equivalent to the simultaneous validity of (4.4) and (4.3) provided that (4.1) holds.

The statements of Theorem 4.1 (d) and (e) below follow easily from the main results of Iohvidov in [10, Chapter II], while their proofs given in Sections 6.3 and 6.4 are based on a somewhat different approach than the one used in [10].

Theorem 4.1. *Let* $\mathrm{s} := \{s_n\}_{n\geq 0}$ *be a nonzero sequence of real numbers and* $\mathcal{H}_n := (s_{i+j})_{i,j=0}^n$, $D_n := \det \mathcal{H}_n$, $n \geq 0$.

(a) *For arbitrary* $n \in \mathbb{N}$ *the property*

$$s_0 = s_1 = s_2 = \cdots = s_{n-1} = 0, \ s_n \neq 0, \tag{4.7}$$

is equivalent to

$$D_0 = D_1 = \cdots = D_{n-1} = 0, \ D_n \neq 0.$$

If $n \in \mathbb{N}$ and $D_0 = D_1 = \cdots = D_{n-1} = 0$ then

$$D_n = (-1)^{\frac{n(n+1)}{2}} s_n^{n+1}.$$

(b) *The set*

$$\mathbb{N}_s := \{r \geq 1 \mid D_{r-1} \neq 0\} \tag{4.8}$$

is nonempty and for every $r \in \mathbb{N}_s$ the formulas (4.1), (4.3) *and* (4.4) *produce the sequence*

$$\mathrm{s}^{(r)} := \{s_n^{(r)}\}_{n\geq 0},\ \mathrm{s}^{(r)} = \{s_0,\ s_1,\ s_2, \ldots, s_{2r-1},\ s_{2r}^{(r)},\ s_{2r+1}^{(r)}, \ldots\},$$

such that

$$\operatorname{rank}\big(s_{i+j}^{(r)}\big)_{i,j=0}^{\infty} = r. \tag{4.9}$$

(c) *For every $r \in \mathbb{N}_s$ we have*

$$\overline{\lim_{k\to\infty}} \sqrt[k]{|s_k^{(r)}|} = \max\{|z| \mid z \in \mathbb{C},\ P_r(z) = 0\}, \tag{4.10}$$

and

$$\frac{s_0}{x} + \frac{s_1}{x^2} + \cdots + \frac{s_{2r-1}}{x^{2r}} + \sum_{k\geq 2r} \frac{s_k^{(r)}}{x^{k+1}} = \frac{Q_r(x)}{P_r(x)}, \tag{4.11}$$

where the series is absolutely convergent for every $|x| > \max\{|z| \mid z \in \mathbb{C},\ P_r(z) = 0\}$.

(d) *For arbitrary $d \in \mathbb{N}$ and $r \in \mathbb{N}_s$, the following statements hold:*

$$D_r = (s_{2r} - s_{2r}^{(r)})D_{r-1};$$

$$D'_{r+1} = (s_{2r+1} - s_{2r+1}^{(r)})D_{r-1} - (s_{2r} - s_{2r}^{(r)})D'_r; \tag{4.12}$$

$$D_r = \cdots = D_{r+d-1} = 0 \Leftrightarrow s_{2r} = s_{2r}^{(r)},$$

$$s_{2r+1} = s_{2r+1}^{(r)}, \ldots s_{2r+d-1} = s_{2r+d-1}^{(r)}; \tag{4.13}$$

$$D_r = \cdots = D_{r+d-1} = 0 \Rightarrow$$

$$D_{r+d} = (-1)^{\frac{d(d+1)}{2}} (s_{2r+d} - s_{2r+d}^{(r)})^{d+1} D_{r-1}. \tag{4.14}$$

(e) *For every $r \in \mathbb{N}_s$, the Iohvidov characteristic function*

$$d_r := \inf\{m \geq 0 \mid s_{2r+m} \neq s_{2r+m}^{(r)}\} \in \{0, 1, 2, \ldots\} \cup \{\infty\}, \tag{4.15}$$

where it is assumed that $\inf \emptyset := \infty$, *possesses the following property*

$$d_r = \inf\{m \geq 0 \,|\, D_{r+m} \neq 0\}. \tag{4.16}$$

In particular, for arbitrary $r \in \mathbb{N}_{\mathrm{s}}$ *we have*

$$d_r = \infty \Leftrightarrow \mathrm{s} = \mathrm{s}^{(r)} \Leftrightarrow D_{r-1} \neq 0,\ D_r = D_{r+1} = \cdots = 0. \tag{4.17}$$

The case of Theorem 4.1(a) was first considered by Frobenius [6, p. 206] in 1894. In 1969 Iohvidov [11, (5), p. 244] introduced the characteristic function d_r and established in [11, (7), p. 246] the equalities (4.14) (see also [10, (10.5), p. 62; (11.2), p. 70]). A formula similar to (4.14) has been established in [1, Lemma 2.3]. However, the setting of Theorem 4.1 (d) and (e) differs from that of [10, Chapter 2] because only the finite Hankel matrices $(s_{i+j})_{i,j=0}^{n}$ are considered there.

5. Consequences of Theorem 4.1

5.1. Approximating sequence

The first immediate consequence of Theorem 4.1 is that the sequence $\mathrm{s}^{(r)}$ for every $r \in \mathbb{N}_{\mathrm{s}}$ can equivalently be defined by the expansion (4.11) which in view of $x^r P_r(1/x) \restriction_{x=0} = D_{r-1} \neq 0$ can be considered as the Taylor expansion at the origin of the rational function

$$\frac{z^{r-1}Q_r(1/z)}{z^r P_r(1/z)} = \sum\nolimits_{k\geq 0} s_k^{(r)} z^k,\ |z| < \min\{|\zeta| \mid \zeta \in \mathbb{C} \setminus \{0\},\ P_r(1/\zeta) = 0\}.$$

5.2. Finiteness of rank

Assume now that for a given nonzero sequence $\{s_n\}_{n\geq 0}$ of real numbers the infinite Hankel matrix $(s_{i+j})_{i,j=0}^{\infty}$ has a finite rank. Then the set $\mathbb{N}_{\mathrm{s}}$ defined as in (4.8) is finite because $D_{r-1} \neq 0$ means that the first r rows and the first r columns of the matrix $(s_{i+j})_{i,j=0}^{\infty}$ are linearly independent. There exists therefore a maximal element $r_* := \max \mathbb{N}_{\mathrm{s}} \geq 1$ of $\mathbb{N}_{\mathrm{s}}$ for which we have $d_{r_*} = \infty$ by virtue of (4.16). Then (4.15) yields $\mathrm{s} = \mathrm{s}^{(r_*)}$ and (4.9) implies $\operatorname{rank}(s_{i+j})_{i,j=0}^{\infty} = r_*$. Conversely, if $\mathrm{s} = \mathrm{s}^{(r)}$ for a certain $r \in \mathbb{N}_{\mathrm{s}}$, then we have the validity of (4.9) in view of Theorem 4.1(b). Thus, $\operatorname{rank}(s_{i+j})_{i,j=0}^{\infty} = r$ if and only if $D_{r-1} \neq 0$ and $\mathrm{s} = \mathrm{s}^{(r)}$. Combining this assertion with (4.11), (4.17) and with (4.1), (4.6), (6.15) we obtain the validity of the following corollary which contains the statements of

Theorems 2.1, 2.2 and 2.3(a), while Theorem 2.3(b) follows from (6.3) below.

Corollary 5.1. *Let* $\mathrm{s} := \{s_n\}_{n\geq 0}$ *be a nonzero sequence of real numbers,* $D_n := \det(s_{i+j})_{i,j=0}^n$, $n \geq 0$, *and let the sequence* $\mathrm{s}^{(m)} := \{s_n^{(m)}\}_{n\geq 0}$ *for every* $m \geq 1$ *satisfying* $D_{m-1} \neq 0$ *be defined by the expansion*

$$\frac{Q_m(x)}{P_m(x)} = \sum_{k\geq 0} \frac{s_k^{(m)}}{x^{k+1}}, \quad |x| > \max\{|z| \mid P_m(z) = 0\}.$$

Then $s_n^{(m)} = s_n$, $0 \leq n \leq 2m-1$ *for every such* m, *and the infinite Hankel matrix* $\mathcal{H}_\infty := (s_{i+j})_{i,j=0}^\infty$ *has a finite rank* $r \geq 1$ if *and only if* $D_{r-1} \neq 0$ *and one of the following equivalent condition holds*:

$$\text{(a)} \quad \mathrm{s} = \mathrm{s}^{(r)}; \tag{5.1}$$

$$\text{(b)} \quad D_n = 0, \ n \geq r; \tag{5.2}$$

$$\text{(c)} \quad \sum_{k\geq 0} \frac{s_k}{x^{k+1}} = \frac{Q_r(x)}{P_r(x)}, \ |x| > \max\{|z| \mid z \in \mathbb{C}, \ P_r(z) = 0\}; \tag{5.3}$$

$$\text{(d)} \quad D_{r-1}s_{r+m} + \sum_{k=0}^{r-1} p_{r,k}s_{k+m} = 0, \ m \geq r, \ \textit{where}$$

$$P_r(x) = D_{r-1}x^r + \sum_{k=0}^{r-1} p_{r,k}\, x^k. \tag{5.4}$$

In other words, for arbitrary $r \geq 1$ *satisfying* $D_{r-1} \neq 0$ *we have*

$$\text{(a)} \Leftrightarrow \text{(b)} \Leftrightarrow \text{(c)} \Leftrightarrow \text{(d)} \Leftrightarrow \operatorname{rank} \mathcal{H}_\infty = r, \tag{5.5}$$

while rank $\mathcal{H}_\infty = r$ *implies* $D_{r-1} \neq 0$ *for arbitrary positive integer* r.

5.3. Positive semidefiniteness

The method used by Darboux in deriving formula [4, (68), p. 413], now called the Christoffel–Darboux summation formula, can be applied to the polynomials P_n. This leads to the following formula (see [3, Theorem 4.5, p. 23; 2, Theorem 2.6, p. 50])

$$D_{r-1}^2 \sum_{k=0}^{r-1} \frac{P_k(x)P_k(y)}{D_k D_{k-1}} = \frac{P_r(x)P_{r-1}(y) - P_r(y)P_{r-1}(x)}{x-y}, \ D_{-1} := 1, \ x \neq y,$$

provided that $D_k \neq 0$ for all $0 \leq k \leq r-1$ and r is a positive integer. Thus,

$$D_{r-1}^2 \sum_{k=0}^{r-1} \frac{P_k(x)^2}{D_k D_{k-1}} = P_r'(x)P_{r-1}(x) - P_r(x)P_{r-1}'(x), \tag{5.6}$$

and

$$D_{r-1}^2 \sum_{k=0}^{r-1} \frac{|P_k(z)|^2}{D_k D_{k-1}} = \frac{\operatorname{Im} P_r(z) P_{r-1}(\overline{z})}{\operatorname{Im} z}, \quad z \in \mathbb{C} \setminus \mathbb{R}, \tag{5.7}$$

where $\overline{z}$ is a complex conjugate of z.

Under the conditions

$$D_0 > 0, D_1 > 0, \dots, D_{r-1} > 0, \tag{5.8}$$

we then get

$$\begin{aligned} \frac{\operatorname{Im} P_r(z) P_{r-1}(\overline{z})}{\operatorname{Im} z} &= \frac{\operatorname{Im} P_r(z) \overline{P_{r-1}(z)}}{\operatorname{Im} z} \\ &= D_{r-1}^2 \sum_{k=0}^{r-1} \frac{|P_k(z)|^2}{D_k D_{k-1}} \geq \frac{D_{r-1}}{D_{r-2}} |P_{r-1}(z)|^2, \ \operatorname{Im} z \neq 0, \end{aligned}$$

so if $P_r(z) = 0$ for a z with $\operatorname{Im} z \neq 0$, the last inequality above implies $P_{r-1}(z) = 0$, which is contradicting that P_r and P_{r-1} have no common zeros according to (2.7). Therefore all zeros of P_r are real, and if λ is a real zero of the polynomial P_r we have from (5.6)

$$P_r'(\lambda) P_{r-1}(\lambda) = D_{r-1}^2 \sum_{k=0}^{r-1} \frac{P_k(\lambda)^2}{D_k D_{k-1}}, \tag{5.9}$$

while (2.7) yields

$$P_{r-1}(\lambda) Q_r(\lambda) = D_{r-1}^2. \tag{5.10}$$

Since (5.10) means that $P_{r-1}(\lambda) \neq 0$, (5.9) implies $P_r'(\lambda) \neq 0$ because

$$\sum_{k=0}^{r-1} \frac{P_k(\lambda)^2}{D_k D_{k-1}} \geq \frac{P_{r-1}(\lambda)^2}{D_{r-1} D_{r-2}} > 0.$$

Thus, all zeros $\{\lambda_n\}_{n=1}^r$ of P_r are simple and by virtue of (5.9) and (5.10) we have

$$\mu_n := \frac{Q_r(\lambda_n)}{P_r'(\lambda_n)} = \left(\sum_{k=0}^{r-1} \frac{P_k(\lambda_n)^2}{D_k D_{k-1}} \right)^{-1} \in (0, +\infty), \ 1 \leq n \leq r,$$

which gives the following form of the partial fraction decomposition of Q_r/P_r:

$$\frac{Q_r(x)}{P_r(x)} = \sum_{n=1}^r \frac{Q_r(\lambda_n)}{P_r'(\lambda_n)(x - \lambda_n)} = \sum_{m=0}^{\infty} \frac{1}{x^{m+1}} \sum_{n=1}^r \mu_n \lambda_n^m, \ |x| > \max_{1 \leq n \leq r} |\lambda_n|. \tag{5.11}$$

Assume now that (2.1) and (5.8) hold. Then $r \in \mathbb{N}_s$ and (5.5) implies the validity of (5.3) which in view of (5.11) yields

$$s_m = \sum_{n=1}^{r} \mu_n \lambda_n^m,\ m \geq 0,\ \mu_n = \left(\sum_{k=0}^{r-1} \frac{P_k(\lambda_n)^2}{D_k D_{k-1}} \right)^{-1} > 0,$$

$$P_r(\lambda_n) = 0,\ 1 \leq n \leq r.$$

We have completely proved the following assertion.[a]

Theorem 5.2 (2015, [1, Theorem 1.1, p. 1569]). *Let $\{s_n\}_{n\geq 0}$ be an arbitrary sequence of real numbers and $\mathcal{H}_n := (s_{i+j})_{i,j=0}^{n}$, $n \geq 0$. Assume that there exists a positive integer n_0 such that*

$$D_n := \det \mathcal{H}_n > 0,\ 0 \leq n \leq n_0 - 1,\ \det \mathcal{H}_n = 0,\ n \geq n_0.$$

Then there exist n_0 distinct real numbers $\{x_k\}_{k=1}^{n_0}$ and n_0 positive numbers $\{\mu_k\}_{k=1}^{n_0}$ such that

$$s_n = \int_{-\infty}^{+\infty} x^n d\mu(x),\ n \geq 0,\ \mu := \sum_{k=1}^{n_0} \mu_k \delta_{x_k},$$

where δ_y denotes the Dirac measure placed at the point $y \in \mathbb{R}$.

6. Proof of Theorem 4.1

6.1. Proof of Theorem 4.1(a)

If (4.7) holds then the first column in the matrices $\mathcal{H}_0, \ldots, \mathcal{H}_{n-1}$ is zero, and therefore $D_0 = D_1 = \cdots = D_{n-1} = 0$, while

$$D_n = \begin{vmatrix} 0 & 0 & \ldots & 0 & s_n \\ 0 & 0 & \ldots & s_n & s_{n+1} \\ \ldots & \ldots & \ldots & \ldots & \ldots \\ 0 & s_n & \ldots & s_{2n-2} & s_{2n-1} \\ s_n & s_{n+1} & \ldots & s_{2n-1} & s_{2n} \end{vmatrix} = (-1)^{\frac{n(n+1)}{2}} s_n^{n+1}. \tag{6.1}$$

Conversely, the identity $D_0 = s_0$ together with the condition $D_0 = 0$ imply $s_0 = 0$. Then (6.1) gives $D_1 = -s_1^2$, which by virtue of the

[a] During the preparation of the present chapter the second author learned that the result is formulated in [17, Theorem 1.2, p. 5].

condition $D_1 = 0$ yields $s_1 = 0$. Pursuing a finite number of repetitions of this fact, we arrive at $s_0 = s_1 = \cdots = s_{n-1} = 0$. In view of (6.1), $D_n = (-1)^{\frac{n(n+1)}{2}} s_n^{n+1} \neq 0$ and therefore $s_n \neq 0$. This concludes the proof of Theorem 4.1(a).

6.2. Proof of Theorem 4.1(b)

Since s is a nonzero sequence, Theorem 4.1 (a) implies that $\mathbb{N}_{\mathrm{s}}$ is nonempty. To prove that the Hankel rank of $\mathrm{s}^{(r)}$ is equal to r, we observe that the relations (4.6) and (4.3) can also be written as follows:

$$\begin{pmatrix} s_{m+r}^{(r)} \\ s_{m+r-1}^{(r)} \\ s_{m+r-2}^{(r)} \\ \vdots \\ s_{m+2}^{(r)} \\ s_{m+1}^{(r)} \end{pmatrix} = \begin{pmatrix} d_{r-1}^{(r)} & d_{r-2}^{(r)} & \ldots & d_2^{(r)} & d_1^{(r)} & d_0^{(r)} \\ 1 & 0 & \ldots & 0 & 0 & 0 \\ 0 & 1 & \ldots & 0 & 0 & 0 \\ \vdots & \vdots & \ldots & \vdots & \vdots & \vdots \\ 0 & 0 & \ldots & 1 & 0 & 0 \\ 0 & 0 & \ldots & 0 & 1 & 0 \end{pmatrix} \begin{pmatrix} s_{m+r-1}^{(r)} \\ s_{m+r-2}^{(r)} \\ \vdots \\ s_{m+2}^{(r)} \\ s_{m+1}^{(r)} \\ s_m^{(r)} \end{pmatrix}, \quad m \geq 0.$$

Therefore

$$\begin{pmatrix} s_{m+n+r}^{(r)} \\ s_{m+n+r-1}^{(r)} \\ s_{m+n+r-2}^{(r)} \\ \vdots \\ s_{m+n+2}^{(r)} \\ s_{m+n+1}^{(r)} \end{pmatrix} = \begin{pmatrix} d_{r-1}^{(r)} & d_{r-2}^{(r)} & \ldots & d_2^{(r)} & d_1^{(r)} & d_0^{(r)} \\ 1 & 0 & \ldots & 0 & 0 & 0 \\ 0 & 1 & \ldots & 0 & 0 & 0 \\ \vdots & \vdots & \ldots & \vdots & \vdots & \vdots \\ 0 & 0 & \ldots & 1 & 0 & 0 \\ 0 & 0 & \ldots & 0 & 1 & 0 \end{pmatrix}^{n+1} \begin{pmatrix} s_{m+r-1}^{(r)} \\ s_{m+r-2}^{(r)} \\ \vdots \\ s_{m+2}^{(r)} \\ s_{m+1}^{(r)} \\ s_m^{(r)} \end{pmatrix}, \quad m, n \geq 0, \tag{6.2}$$

and the first row of the matrix in the right-hand side of (6.2) gives the existence of r numbers $d_{n,0}^{(r)}, d_{n,1}^{(r)}, \ldots, d_{n,r-1}^{(r)}$ satisfying

$$\sum_{k=0}^{r-1} d_{n,k}^{(r)} s_{k+m}^{(r)} = s_{r+n+m}^{(r)}, \ m \geq 0, \tag{6.3}$$

where $d_{0,k}^{(r)}$ is equal to $d_k^{(r)}$ from (4.6) for each $0 \le k \le r-1$. This means that

$$d_{n,0}^{(r)} \begin{pmatrix} s_0 \\ s_1 \\ \vdots \\ s_r \\ s_{r+1} \\ s_{r+2} \\ \vdots \\ s_{2r-1} \\ s_{2r}^{(r)} \\ \vdots \end{pmatrix} + d_{n,1}^{(r)} \begin{pmatrix} s_1 \\ s_2 \\ \vdots \\ s_{r+1} \\ s_{r+2} \\ s_{r+3} \\ \vdots \\ s_{2r}^{(r)} \\ s_{2r+1}^{(r)} \\ \vdots \end{pmatrix} + \cdots + d_{n,r-2}^{(r)} \begin{pmatrix} s_{r-2} \\ s_{r-1} \\ \vdots \\ s_{2r-2} \\ s_{2r-1} \\ s_{2r}^{(r)} \\ \vdots \\ s_{3r-3}^{(r)} \\ s_{3r-2}^{(r)} \\ \vdots \end{pmatrix}$$

$$+ d_{n,r-1}^{(r)} \begin{pmatrix} s_{r-1} \\ s_r \\ \vdots \\ s_{2r-1} \\ s_{2r}^{(r)} \\ s_{2r+1}^{(r)} \\ \vdots \\ s_{3r-2}^{(r)} \\ s_{3r-1}^{(r)} \\ \vdots \end{pmatrix} = \begin{pmatrix} s_{r+n}^{(r)} \\ s_{1+r+n}^{(r)} \\ \vdots \\ s_{r+r+n}^{(r)} \\ s_{r+1+r+n}^{(r)} \\ s_{r+2+r+n}^{(r)} \\ \vdots \\ s_{2r-1+r+n}^{(r)} \\ s_{2r+r+n}^{(r)} \\ \vdots \end{pmatrix}, \tag{6.4}$$

i.e., for arbitrary $n \ge 0$ the $(r+n+1)$th column $(s_{n+r+j}^{(r)})_{j=0}^{\infty}$ of the infinite matrix $(s_{i+j}^{(r)})_{i,j=0}^{\infty}$ is a linear combination of the first r columns which are linearly independent by virtue of (4.2). Thus,

$$\operatorname{rank}\big(s_{i+j}^{(r)}\big)_{i,j=0}^{\infty} = r,$$

and we conclude that the Hankel rank of $\mathrm{s}^{(r)}$ is equal to r. Theorem 4.1(b) is proved.

6.3. Proof of Theorem 4.1(d)

For the sequence $\widehat{\mathrm{s}}^{(r)} = \{\widehat{s}_n^{(r)}\}_{n\geq 0}$ defined by

$$\widehat{s}_n^{(r)} := s_n - s_n^{(r)},\ n \geq 0,\ \widehat{\mathrm{s}}^{(r)} = \mathrm{s} - \mathrm{s}^{(r)},$$

we have, by virtue of (4.5),

$$\widehat{\mathrm{s}}^{(r)} = \{s_n - s_n^{(r)}\}_{n\geq 0} = \{0, 0, 0, \ldots, 0, \widehat{s}_{2r}^{(r)}, \widehat{s}_{2r+1}^{(r)}, \ldots\}.$$

It is appropriate at this point to recall (see [8, Definition 8, p. 61]) that two square matrices A and B are called equivalent if there exist two square matrices P and Q with nonzero determinants such that $B = PAQ$. If $\det P = \det Q = 1$ we say that A and B are 1-*equivalent* and write

$$A \overset{1}{\sim} B.$$

It is evident that for two 1-equivalent square matrices A and B we have $\det A = \det B$, and also $\operatorname{rank} A = \operatorname{rank} B$ in view of [8, Theorem 2, p. 62].

For arbitrary $r \in \mathbb{N}_{\mathrm{s}}$ consider the matrix

$$\mathcal{H}_r = \begin{pmatrix} s_0 & s_1 & \cdots & s_{r-2} & s_{r-1} & s_r \\ s_1 & s_2 & \cdots & s_{r-1} & s_r & s_{r+1} \\ \cdots & \cdots & \cdots & \cdots & \cdots & \cdots \\ s_{r-1} & s_r & \cdots & s_{2r-3} & s_{2r-2} & s_{2r-1} \\ s_r & s_{r+1} & \cdots & s_{2r-2} & s_{2r-1} & s_{2r} \end{pmatrix}$$

$$= \begin{pmatrix} s_0^{(r)} & s_1^{(r)} & \cdots & s_{r-2}^{(r)} & s_{r-1}^{(r)} & s_r^{(r)} \\ s_1^{(r)} & s_2^{(r)} & \cdots & s_{r-1}^{(r)} & s_r^{(r)} & s_{r+1}^{(r)} \\ \cdots & \cdots & \cdots & \cdots & \cdots & \cdots \\ s_{r-1}^{(r)} & s_r^{(r)} & \cdots & s_{2r-3}^{(r)} & s_{2r-2}^{(r)} & s_{2r-1}^{(r)} \\ s_r^{(r)} & s_{r+1}^{(r)} & \cdots & s_{2r-2}^{(r)} & s_{2r-1}^{(r)} & s_{2r} \end{pmatrix}.$$

Subtracting from the last column a linear combination of the first r columns with the coefficients from (6.4) with $n = 0$, we conclude that

$$\mathcal{H}_r \overset{1}{\sim} \begin{pmatrix} s_0 & s_1 & \dots & s_{r-1} & 0 \\ s_1 & s_2 & \dots & s_r & 0 \\ \dots & \dots & \dots & \dots & \dots \\ s_{r-1} & s_r & \dots & s_{2r-2} & 0 \\ s_r & s_{r+1} & \dots & s_{2r-1} & \widehat{s}_{2r}^{(r)} \end{pmatrix}. \tag{6.5}$$

The Laplace expansion of the determinant of the right-hand side of (6.5) by minors along column $r+1$ leads to the validity of the left-hand equality in (4.12),

$$D_r = D_{r-1} \cdot (s_{2r} - s_{2r}^{(r)}), \; r \in \mathbb{N}_{\mathbf{s}}. \tag{6.6}$$

Thus, $D_r = 0$ if and only if $s_{2r} = s_{2r}^{(r)}$, which proves (4.13) for $d = 1$.
To prove (4.13) for $d > 1$ assume that

$$s_{2r} = s_{2r}^{(r)}, \; s_{2r+1} = s_{2r+1}^{(r)}, \dots, s_{2r+d-1} = s_{2r+d-1}^{(r)}. \tag{6.7}$$

Consider the matrix

$$\mathcal{H}_{r+d}$$

$$= \begin{pmatrix} s_0^{(r)} & s_1^{(r)} & \dots & s_{r-1}^{(r)} & s_r^{(r)} & s_{r+1}^{(r)} & \dots & s_{r+d-1}^{(r)} & s_{r+d}^{(r)} \\ s_1^{(r)} & s_2^{(r)} & \dots & s_r^{(r)} & s_{r+1}^{(r)} & s_{r+2}^{(r)} & \dots & s_{r+d}^{(r)} & s_{r+d+1}^{(r)} \\ \dots & \dots & \dots & \dots & \dots & \dots & \dots & \dots & \dots \\ s_{r-1}^{(r)} & s_r^{(r)} & \dots & s_{2r-2}^{(r)} & s_{2r-1}^{(r)} & s_{2r}^{(r)} & \dots & s_{2r+d-2}^{(r)} & s_{2r+d-1}^{(r)} \\ s_r^{(r)} & s_{r+1}^{(r)} & \dots & s_{2r-1}^{(r)} & s_{2r}^{(r)} & s_{2r+1}^{(r)} & \dots & s_{2r+d-1}^{(r)} & s_{2r+d} \\ s_{r+1}^{(r)} & s_{r+2}^{(r)} & \dots & s_{2r}^{(r)} & s_{2r+1}^{(r)} & s_{2r+2}^{(r)} & \dots & s_{2r+d} & s_{2r+d+1} \\ \dots & \dots & \dots & \dots & \dots & \dots & \dots & \dots & \dots \\ s_{r+d-1}^{(r)} & s_{r+d}^{(r)} & \dots & s_{2r+d-2}^{(r)} & s_{2r+d-1}^{(r)} & s_{2r+d} & \dots & s_{2r+2d-2} & s_{2r+2d-1} \\ s_{r+d}^{(r)} & s_{r+d+1}^{(r)} & \dots & s_{2r+d-1}^{(r)} & s_{2r+d} & s_{2r+d+1} & \dots & s_{2r+2d-1} & s_{2r+2d} \end{pmatrix}.$$

For every $0 \le n \le d$ we subtract from the $(r+n+1)$th column a linear combination of the first r columns with the coefficients from the equality

(6.4), and we obtain

$$\mathcal{H}_{r+d}$$

$$\overset{1}{\sim}\begin{pmatrix} s_0^{(r)} & s_1^{(r)} & \dots & s_{r-1}^{(r)} & 0 & 0 & \dots & 0 & 0 \\ s_1^{(r)} & s_2^{(r)} & \dots & s_r^{(r)} & 0 & 0 & \dots & 0 & 0 \\ \dots & \dots & \dots & \dots & \dots & \dots & \dots & \dots & \dots \\ s_{r-1}^{(r)} & s_r^{(r)} & \dots & s_{2r-2}^{(r)} & 0 & 0 & \dots & 0 & 0 \\ s_r^{(r)} & s_{r+1}^{(r)} & \dots & s_{2r-1}^{(r)} & 0 & 0 & \dots & 0 & \widehat{s}_{2r+d}^{(r)} \\ s_{r+1}^{(r)} & s_{r+2}^{(r)} & \dots & s_{2r}^{(r)} & 0 & 0 & \dots & \widehat{s}_{2r+d}^{(r)} & \widehat{s}_{2r+d+1}^{(r)} \\ \dots & \dots & \dots & \dots & \dots & \dots & \dots & \dots & \dots \\ s_{r+d-1}^{(r)} & s_{r+d}^{(r)} & \dots & s_{2r+d-2}^{(r)} & 0 & \widehat{s}_{2r+d}^{(r)} & \dots & \widehat{s}_{2r+2d-2}^{(r)} & \widehat{s}_{2r+2d-1}^{(r)} \\ s_{r+d}^{(r)} & s_{r+d+1}^{(r)} & \dots & s_{2r+d-1}^{(r)} & \widehat{s}_{2r+d}^{(r)} & \widehat{s}_{2r+d+1}^{(r)} & \dots & \widehat{s}_{2r+2d-1}^{(r)} & \widehat{s}_{2r+2d}^{(r)} \end{pmatrix}. \tag{6.8}$$

Expanding the determinant of the matrix in the right-hand side of (6.8) after the last row $d+1$ times or by using that the matrix is quasi-triangular (see [8, p. 43]) and using the formulas [8, (67), p. 43], (6.7) and (4.1), we get

$$D_{r+d} = (-1)^{\frac{d(d+1)}{2}}(s_{2r+d} - s_{2r+d}^{(r)})^{d+1} D_{r-1}. \tag{6.9}$$

Furthermore, it follows from (6.8) that for all $0 \le n \le d-1$ the matrix $\mathcal{H}_{r+n}$ is 1-equivalent to the matrix with zero $(r+1)$th column. Therefore

$$D_r = \cdots = D_{r+d-1} = 0, \tag{6.10}$$

which proves the implication $\Leftarrow$ in (4.13) for $d \ge 1$.

To prove the inverse implication in (4.13) for such d assume that (6.10) holds for some $1 \le d < \infty$. Then $D_r = 0$ implies $s_{2r} = s_{2r}^{(r)}$ by virtue of (6.6). Therefore (6.7) holds for $d = 1$, and we can use the expression (6.9) for D_{r+1} to give $s_{2r+1} = s_{2r+1}^{(r)}$ if $D_{r+1} = 0$. Pursuing a finite number of repetitions of this trick, we get at last $s_{2r+d-1} = s_{2r+d-1}^{(r)}$ which completes the proof of (4.13).

Finally, the equivalence (4.13) and the implication (6.7) $\Rightarrow$ (6.9) just deduced give the validity of (4.14).

To prove the right-hand equality in (4.12) we take $r \in \mathbb{N}_s$ and consider

$$D'_{r+1} := \begin{vmatrix} s_0 & s_1 & \dots & s_{r-2} & s_{r-1} & s_{r+1} \\ s_1 & s_2 & \dots & s_{r-1} & s_r & s_{r+2} \\ \dots & \dots & \dots & \dots & \dots & \dots \\ s_{r-2} & s_{r-1} & \dots & s_{2r-4} & s_{2r-3} & s_{2r-1} \\ s_{r-1} & s_r & \dots & s_{2r-3} & s_{2r-2} & s_{2r} \\ s_r & s_{r+1} & \dots & s_{2r-2} & s_{2r-1} & s_{2r+1} \end{vmatrix}$$

$$= \begin{vmatrix} s_0^{(r)} & s_1^{(r)} & \dots & s_{r-2}^{(r)} & s_{r-1}^{(r)} & s_{r+1}^{(r)} \\ s_1^{(r)} & s_2^{(r)} & \dots & s_{r-1}^{(r)} & s_r^{(r)} & s_{r+2}^{(r)} \\ \dots & \dots & \dots & \dots & \dots & \dots \\ s_{r-2}^{(r)} & s_{r-1}^{(r)} & \dots & s_{2r-4}^{(r)} & s_{2r-3}^{(r)} & s_{2r-1}^{(r)} \\ s_{r-1}^{(r)} & s_r^{(r)} & \dots & s_{2r-3}^{(r)} & s_{2r-2}^{(r)} & s_{2r} \\ s_r^{(r)} & s_{r+1}^{(r)} & \dots & s_{2r-2}^{(r)} & s_{2r-1}^{(r)} & s_{2r+1} \end{vmatrix}.$$

Subtracting from the last column a linear combination of the first r columns with the coefficients from (6.4) with $n = 0$, we obtain

$$D'_{r+1} = \begin{vmatrix} s_0^{(r)} & s_1^{(r)} & \dots & s_{r-2}^{(r)} & s_{r-1}^{(r)} & 0 \\ s_1^{(r)} & s_2^{(r)} & \dots & s_{r-1}^{(r)} & s_r^{(r)} & 0 \\ \dots & \dots & \dots & \dots & \dots & \dots \\ s_{r-2}^{(r)} & s_{r-1}^{(r)} & \dots & s_{2r-4}^{(r)} & s_{2r-3}^{(r)} & 0 \\ s_{r-1}^{(r)} & s_r^{(r)} & \dots & s_{2r-3}^{(r)} & s_{2r-2}^{(r)} & s_{2r} - s_{2r}^{(r)} \\ s_r^{(r)} & s_{r+1}^{(r)} & \dots & s_{2r-2}^{(r)} & s_{2r-1}^{(r)} & s_{2r+1} - s_{2r+1}^{(r)} \end{vmatrix}. \tag{6.11}$$

The Laplace expansion of the determinant in the right-hand side of (6.11) by minors along the column $r + 1$ and (4.1) lead to the validity of the right-hand equality in (4.12). This finishes the proof of Theorem 4.1(d).

6.4. Proof of Theorem 4.1(e)

The formula (6.6) proves that $D_r \neq 0$ if and only if $s_{2r} \neq s_{2r}^{(r)}$. Thus, the definitions of d_r given in (4.15) and in (4.16) are the same in the case where $d_r = 0$.

Assume now that the number d_r defined by (4.15) is finite and $1 \le d_r < \infty$. Then (6.7) holds for $d = d_r$ and $s_{2r+d_r} \ne s^{(r)}_{2r+d_r}$ which by (6.9) means that $D_{r+d_r} \ne 0$. Thus, d_r coincides with the infimum in the right-hand side of (4.16).

Conversely, if d_r is defined by (4.16) and $1 \le d_r < \infty$ then $D_{r+d_r} \ne 0$ and (6.10) holds for $d = d_r$ as well as (6.7) by virtue of (4.13). We can therefore apply the formula (6.9) for $d = d_r$ to conclude that $D_{r+d_r} \ne 0$ yields $s_{2r+d_r} \ne s^{(r)}_{2r+d_r}$. This means that d_r coincides with the infimum in the right-hand side of (4.15). Thus, two definitions of d_r given in (4.15) and in (4.16) also coincide when $1 \le d_r < \infty$.

Finally, (4.17) follows directly from (4.13) applied for every positive integer d. This completes the proof of Theorem 4.1(e).

6.5. Proof of Theorem 4.1(c)

If $p(x,t) = \sum_{k=0}^{n}\sum_{j=0}^{m} p_{k,j}\, x^k t^j$ is an algebraic polynomial with real coefficients of two variables x and t, we use the linear functional L from (1.5) with respect to the t-variable to get

$$L_t(p(x,t)) = \sum_{k=0}^{n}\sum_{j=0}^{m} s_j p_{k,j}\, x^k = \sum_{k=0}^{n}\left(\sum_{j=0}^{m} s_j p_{k,j}\right) x^k \in \mathcal{P}[\mathbb{R}].$$

For example,

$$L_t\left(\frac{1-1}{x-t}\right) = 0,\ L_t\left(\frac{x-t}{x-t}\right) = s_0,\ L_t\left(\frac{x^2-t^2}{x-t}\right) = s_0 x + s_1,$$
$$L_t\left(\frac{x^n-t^n}{x-t}\right) = \sum_{j=0}^{n-1} s_j x^{n-1-j},\ n \ge 1.$$

Comparing these equalities with (1.3) we see that

$$Q_n(x) := L_t\left(\frac{P_n(x)-P_n(t)}{x-t}\right),\ n \ge 0,$$

and if $P_n(x) =: \sum_{k=0}^{n} p_{n,k}x^k$, $Q_n(x) =: \sum_{k=0}^{n-1} q_{n,k}x^k$, $n \ge 1$, we obtain

$$P_0(x) = 1,\ P_1(x) = s_0 x - s_1,\ P_n(x) = D_{n-1}x^n + \sum_{k=0}^{n-1} p_{n,k}\, x^k,\ p_{n,n} = D_{n-1},$$

$$\frac{P_n(x)-P_n(t)}{x-t} = \sum_{m=1}^{n} p_{n,m}\frac{x^m-t^m}{x-t} = \sum_{m=0}^{n-1} p_{n,m+1}\sum_{k=0}^{m} x^k t^{m-k}$$
$$= \sum_{k=0}^{n-1}\left[\sum_{m=k}^{n-1} p_{n,m+1}t^{m-k}\right]x^k,$$

$$Q_1(x) := s_0^2,\ Q_n(x) = L_t\left(\frac{P_n(x)-P_n(t)}{x-t}\right)$$
$$= \sum_{k=0}^{n-1}\left(\sum_{m=k}^{n-1} p_{n,m+1}s_{m-k}\right)x^k,$$

$$q_{n,m} = \sum_{k=m}^{n-1} p_{n,k+1}s_{k-m} = \sum_{k=0}^{n-m-1} p_{n,k+m+1}s_k,$$
$$0 \le m \le n-1,\ n \ge 1,$$

where the latter equalities can be written in the following form:

$$q_{n,n-1-m} = \sum_{k=0}^{m} p_{n,n-(m-k)}s_k, 0 \le m \le n-1,\ n \ge 1. \tag{6.12}$$

Let $r \in \mathbb{N}_s$, i.e., $D_{r-1} \neq 0$. According to (1.3) we have $L(t^m P_r(t)) = 0$ for every $0 \le m \le r-1$ and therefore

$$\sum_{k=0}^{r} p_{r,k}s_{k+m} = 0,\ 0 \le m \le r-1, \tag{6.13}$$

which gives

$$\sum_{k=0}^{r-1}(-p_{r,k})s_{k+m} = D_{r-1}s_{r+m},\ 0 \le m \le r-1,$$

or,

$$\begin{pmatrix} s_0 & s_1 & \dots & s_{r-2} & s_{r-1} \\ s_1 & s_2 & \dots & s_{r-1} & s_r \\ \dots & \dots & \dots & \dots & \dots \\ s_{r-2} & s_{r-1} & \dots & s_{2r-4} & s_{2r-3} \\ s_{r-1} & s_r & \dots & s_{2r-3} & s_{2r-2} \end{pmatrix} \begin{pmatrix} -p_{r,0}/D_{r-1} \\ -p_{r,1}/D_{r-1} \\ \dots \\ -p_{r,r-2}/D_{r-1} \\ -p_{r,r-1}/D_{r-1} \end{pmatrix} = \begin{pmatrix} s_r \\ s_{r+1} \\ \dots \\ s_{2r-2} \\ s_{2r-1} \end{pmatrix}. \tag{6.14}$$

Since $D_{r-1} \neq 0$ it follows from (4.3) and (6.14) that

$$d_k^{(r)} = -\frac{p_{r,k}}{D_{r-1}}, \ 0 \le k \le r-1, \ P_r(x) = D_{r-1}x^r + \sum_{k=0}^{r-1} p_{r,k}\, x^k, \ r \in \mathbb{N}_s, \tag{6.15}$$

and therefore the recursive formulas (4.4) can be written in the following manner:

$$D_{r-1}s_{2r+m}^{(r)} + \sum_{k=0}^{r-1} p_{r,k}s_{r+m+k}^{(r)} = 0, \ m \ge 0,$$

which together with (6.13) gives

$$\sum_{k=0}^{r} p_{r,k}s_{k+m}^{(r)} = 0, \ m \ge 0,$$

or

$$D_{r-1}s_{r+m}^{(r)} + \sum_{k=0}^{r-1} p_{r,k}s_{k+m}^{(r)} = 0, \ m \ge 0. \tag{6.16}$$

Denote $P_r^\star(x) := x^r P_r(1/x)$ and $Q_r^\star(x) := x^{r-1}Q_r(1/x)$. Then

$$P_r^\star(x) = D_{r-1} + \sum_{k=0}^{r-1} p_{r,k}\, x^{r-k}, \ P_r^\star(0) = D_{r-1} \neq 0, \ Q_r^\star(x) = \sum_{k=0}^{r-1} q_{r,k}x^{r-1-k},$$

and the function $\psi_r(z) := Q_r^\star(z)/P_r^\star(z)$ is analytic on the open disk $|z| < 1/\rho_r$ where $\rho_r := \max\{|z| \mid z \in \mathbb{C}, \ P_r(z) = 0\}$. Let a_n, $n \ge 0$, be the coefficients of the Taylor expansion of $\psi_r(z)$ at the origin,

$$\psi_r(z) = \sum_{n\ge 0} a_n z^n, \ |z| < 1/\rho_r, \tag{6.17}$$

which is obviously equivalent to the expansion of the form

$$\frac{Q_r(x)}{P_r(x)} = \sum_{m\ge 0} \frac{a_m}{x^{m+1}}, \ |x| > \rho_r. \tag{6.18}$$

The identity

$$\left(\sum_{m=0}^{r} p_{r,m}x^m\right) \sum_{m\ge 0} \frac{a_m}{x^{m+1}} = \sum_{m\ge 0} \frac{\sum_{k=0}^{r} p_{r,k}a_{m+k}}{x^{m+1}} + \sum_{m=0}^{r-1} x^m \sum_{k=m}^{r-1} p_{r,k+1}a_{k-m},$$

and (6.18) imply

$$\sum_{k=m}^{r-1} p_{r,k+1}a_{k-m} = q_{r,m}, \quad 0 \le m \le r-1,$$

$$\sum_{k=0}^{r} p_{r,k}a_{m+k} = 0, \quad m \ge 0,$$

which can be written as

$$\begin{aligned} &\sum_{k=0}^{m} p_{r,r-(m-k)}a_k = q_{r,r-1-m}, \quad 0 \le m \le r-1, \\ &D_{r-1}a_{m+r} + \sum\nolimits_{k=0}^{r-1} p_{r,k}a_{m+k} = 0, \quad m \ge 0. \end{aligned} \tag{6.19}$$

Observe that (4.1), (6.12) for $n = r$ and (6.16) mean that

$$\sum_{k=0}^{m} p_{r,r-(m-k)}s_k^{(r)} = q_{r,r-1-m}, \quad 0 \le m \le r-1,$$

$$D_{r-1}s_{m+r}^{(r)} + \sum_{k=0}^{r-1} p_{r,k}s_{m+k}^{(r)} = 0, \quad m \ge 0.$$

and by subtracting from these equalities the corresponding equalities in (6.19) we obtain

$$D_{r-1}[s_m^{(r)} - a_m] + \sum_{k=0}^{m-1} p_{r,r-(m-k)}[s_k^{(r)} - a_k] = 0, \quad 0 \le m \le r-1,$$

$$D_{r-1}[s_{m+r}^{(r)} - a_{m+r}] + \sum_{k=0}^{r-1} p_{r,k}[s_{m+k}^{(r)} - a_{m+k}] = 0, \quad m \ge 0,$$

where it is assumed that $\sum_0^{-1} := 0$. From these recurrence relations we obtain $s_m^{(r)} = a_m$ for all $m \ge 0$ as a consequence of $D_{r-1} \ne 0$. Together with (4.1) this proves (4.11). Since the radius of convergence of the Taylor series (6.17) is known we get the validity of (4.10) by virtue of the Cauchy–Hadamard formula (see [16, (2), p. 200]). Theorem 4.1(c) is proved.

7. Proof of Theorem 1.1

If $t_n = 0$ for all $n \ge 0$ then Theorem 4.1 (a) yields $s_n = 0$ for every $n \ge 0$ and therefore in this case (1.1) has only one solution as stated in the theorem.

To examine the case $Z_t \neq \emptyset$ we introduce the notation

$$\Delta_0 := (-1)^{\frac{n_0+1}{2}} t_{n_0},\ \Delta_{k+1} := (-1)^{\frac{n_{k+1}-n_k}{2}} t_{n_{k+1}} t_{n_k},$$
$$0 \leq k < m-1,\ 2 \leq m \leq \infty. \tag{7.1}$$

7.1. Necessity of Theorem 1.1

To prove necessity, we assume that (1.1) has at least one solution $s := \{s_n\}_{n\geq 0}$ for a given $\{t_n\}_{n\geq 0}$. Then by Theorem 4.1(a),

$$t_{n_0} = (-1)^{\frac{n_0(n_0+1)}{2}} s_{n_0}^{n_0+1}. \tag{7.2}$$

Furthermore, in (4.8) we have

$$\mathbb{N}_s = \{n_k + 1\}_{0\leq k<m},$$

and for every $0 \leq k < m$ the formulas (4.1), (4.3) and (4.4) for $r = n_k + 1$ and the numbers $s_0, s_1, \ldots, s_{2n_k+1}$ produce the sequence

$$s^{(n_k+1)} = \{s_0, s_1, s_2, \ldots, s_{2n_k+1}, s_{2n_k+2}^{(n_k+1)}, s_{2n_k+3}^{(n_k+1)}, \ldots\},\ 0 \leq k < m. \tag{7.3}$$

For arbitrary $0 \leq k < m-1$, $2 \leq m \leq \infty$, it follows from Theorem 4.1(d) with $r = n_k + 1$ and $d = n_{k+1} - n_k - 1$ that

$$t_{n_{k+1}} = (-1)^{\frac{(n_{k+1}-n_k-1)(n_{k+1}-n_k)}{2}} (s_{n_{k+1}+n_k+1} - s_{n_{k+1}+n_k+1}^{(n_k+1)})^{n_{k+1}-n_k} t_{n_k}. \tag{7.4}$$

But in view of (7.1), (7.2) and (7.4) we have

$$\Delta_0 = (-1)^{\frac{(n_0+1)^2}{2}} s_{n_0}^{n_0+1} > 0,$$
$$\Delta_{k+1} = (-1)^{\frac{(n_{k+1}-n_k)^2}{2}} (s_{n_{k+1}+n_k+1} - s_{n_{k+1}+n_k+1}^{(n_k+1)})^{n_{k+1}-n_k} t_{n_k}^2 > 0,$$
$$0 \leq k < m-1,$$

provided that $n_0 + 1 \in 2\mathbb{N}$ and $n_{k+1} - n_k \in 2\mathbb{N}$ for $0 \leq k < m-1$, $2 \leq m \leq \infty$. This proves the necessity part of Theorem 1.1.

7.2. Sufficiency of Theorem 1.1

The proof of sufficiency proceeds by induction on k. Given a sequence $\{t_n\}_{n\geq 0}$ satisfying the conditions of Theorem 1.1 we will determine the terms of the sequence $s := \{s_n\}_{n\geq 0}$ such that (1.1) holds. Let $D_n := \det(s_{i+j})_{i,j=0}^n$, $n \geq 0$.

If $n_0 = 0$ we put $s_0 = t_0$ to obtain $D_0 = s_0 = t_0$. If $n_0 \geq 1$ and $n_0 + 1 \in 2\mathbb{N} + 1$, we set

$$s_0 = s_1 = \cdots = s_{n_0-1} = 0, \; s_{n_0} = (-1)^{\frac{n_0}{2}} t_{n_0}^{\frac{1}{n_0+1}}.$$

According to Theorem 4.1(a) we have

$$D_0 = D_1 = \cdots = D_{n_0-1} = 0, \; D_{n_0} = (-1)^{\frac{n_0(n_0+1)}{2}}$$
$$\times s_{n_0}^{n_0+1} = (-1)^{n_0(n_0+1)} t_{n_0} = t_{n_0}.$$

Assume now that $n_0 + 1 \in 2\mathbb{N}$. Then, in view of (7.1),

$$t_{n_0} = (-1)^{\frac{n_0+1}{2}} \Delta_0, \; \Delta_0 > 0,$$

and if we put

$$s_0 = s_1 = \cdots = s_{n_0-1} = 0, \; s_{n_0} = \Delta_0^{\frac{1}{n_0+1}},$$

then by Theorem 4.1(a) we obtain

$$D_0 = D_1 = \cdots = D_{n_0-1} = 0, \; D_{n_0} = (-1)^{\frac{n_0(n_0+1)}{2}}$$
$$\times s_{n_0}^{n_0+1} = (-1)^{\frac{n_0(n_0+1)}{2}} \Delta_0 = t_{n_0}.$$

Therefore in both cases the numbers $s_0, \ldots, s_{2n_0}$ with $s_{n_0+1}, \ldots, s_{2n_0}$ chosen arbitrarily satisfy (1.1) for $0 \leq n \leq n_0$.

Suppose that for a certain k satisfying $0 \leq k < m-1$ where $2 \leq m \leq \infty$, the numbers $s_0, \ldots, s_{2n_k}$ satisfy (1.1) for $0 \leq n \leq n_k$. We prove that it is possible to determine the numbers $s_{2n_k+1}, s_{2n_k+2}, \ldots, s_{2n_{k+1}}$ such that (1.1) holds for $0 \leq n \leq n_{k+1}$.

Choosing arbitrarily the number s_{2n_k+1} we construct the sequence $s^{(n_k+1)}$ as in (7.3).

Assume first that $n_{k+1} = n_k + 1$. In view of (4.12) for $r = n_k + 1$,

$$D_{n_{k+1}} = (s_{2n_{k+1}} - s_{2n_{k+1}}^{(n_k+1)}) t_{n_k},$$

and by putting

$$s_{2n_{k+1}} = s_{2n_{k+1}}^{(n_k+1)} + \frac{t_{n_{k+1}}}{t_{n_k}},$$

we obtain the desired equality $D_{n_{k+1}} = t_{n_{k+1}}$.

Assume now that $n_{k+1} - n_k \geq 2$. Then

$$t_{n_k+1} = \cdots = t_{n_{k+1}-1} = 0$$

and if we set

$$s_{2n_k+2} = s^{(n_k+1)}_{2n_k+2},\ s_{2n_k+3} = s^{(n_k+1)}_{2n_k+3}, \ldots, s_{n_{k+1}+n_k} = s^{(n_k+1)}_{n_{k+1}+n_k},$$

we obtain, by virtue of (4.13) with $r = n_k + 1$ and $d = n_{k+1} - n_k - 1$,

$$D_{n_k+1} = 0 = t_{n_k+1}, \ldots, D_{n_{k+1}-1} = 0 = t_{n_{k+1}-1},$$

and in view of (4.14),

$$D_{n_{k+1}} = (-1)^{\frac{(n_{k+1}-n_k-1)(n_{k+1}-n_k)}{2}} (s_{n_{k+1}+n_k+1} - s^{(n_k+1)}_{n_{k+1}+n_k+1})^{n_{k+1}-n_k} t_{n_k}. \tag{7.5}$$

If $n_{k+1} - n_k \in 2\mathbb{N} + 1$ we can choose

$$s_{n_{k+1}+n_k+1} = s^{(n_k+1)}_{n_{k+1}+n_k+1} + (-1)^{\frac{(n_{k+1}-n_k-1)}{2}} \left(\frac{t_{n_{k+1}}}{t_{n_k}}\right)^{\frac{1}{n_{k+1}-n_k}},$$

to have from (7.5), $D_{n_{k+1}} = t_{n_{k+1}}$.

But if $n_{k+1} - n_k \in 2\mathbb{N}$ we set

$$s_{n_{k+1}+n_k+1} = s^{(n_k+1)}_{n_{k+1}+n_k+1} + \left(\frac{\Delta_{k+1}}{t^2_{n_k}}\right)^{\frac{1}{n_{k+1}-n_k}},$$

where according to the conditions of the theorem

$$\Delta_{k+1} = (-1)^{\frac{n_{k+1}-n_k}{2}} t_{n_{k+1}} t_{n_k},\ \Delta_{k+1} > 0.$$

Then (7.5) gives

$$\begin{aligned} D_{n_{k+1}} &= (-1)^{\frac{(n_{k+1}-n_k-1)(n_{k+1}-n_k)}{2}} (s_{n_{k+1}+n_k+1} - s^{(n_k+1)}_{n_{k+1}+n_k+1})^{n_{k+1}-n_k} t_{n_k} \\ &= (-1)^{\frac{(n_{k+1}-n_k-1)(n_{k+1}-n_k)}{2}} \frac{\Delta_{k+1}}{t_{n_k}} \\ &= (-1)^{\frac{(n_{k+1}-n_k-1)(n_{k+1}-n_k)}{2}} (-1)^{\frac{n_{k+1}-n_k}{2}} t_{n_{k+1}} \\ &= (-1)^{\frac{(n_{k+1}-n_k)^2}{2}} t_{n_{k+1}} = t_{n_{k+1}}. \end{aligned}$$

Therefore in both cases the numbers $s_0, \ldots, s_{n_{k+1}+n_k+1}$ satisfy (1.1) for $0 \leq n \leq n_{k+1}$ independently on the choice of $s_{n_{k+1}+n_k+2}, \ldots, s_{2n_{k+1}}$. Choosing arbitrarily the latter numbers we obtain the desired result. This finishes the proof of Theorem 1.1.

References

[1] C. Berg and R. Szwarc, A determinant characterization of moment sequences with finitely many mass points, *Linear Multilinear Algebra* **63** (2015) 1568–1576.

[2] C. Brezinski, *Padé-Type Approximation and General Orthogonal Polynomials*, International Series of Numerical Mathematics, Vol. 50 (Birkhäuser, Basel, 1980).

[3] T. S. Chihara, *An Introduction to Orthogonal Polynomials*, Mathematics and its Applications, Vol. 13 (Gordon and Breach Science Publishers, New York, 1978).

[4] G. Darboux, Mémoire sur l'approximation des fonctions de très-grands nombres, et sur une classe étendue de développements en série, *J. Math. Pures Appl.* **4** (1878) 5–56 377–416 (in French).

[5] A. Draux, *Polynômes Orthogonaux Formels—Applications*, Lecture Notes in Mathematics, Vol. 974 (Springer, Berlin, 1983) (in French).

[6] G. Frobenius, Über das Trägheitsgesetz der quadratischen Formen, *J. Reine Angew. Math.* **114** (1895) 187–230. Reprinted from: *Sitzungsber. Königl. Preuss. Akad. Wiss.* (1894) 241–256, 407–431.

[7] Ya. L. Geronimus, Orthogonal polynomials, translation of the appendix to the Russian translation of Szegő's book. in *Two Papers on Special Functions*, American Mathematical Society Translation , Series (2), Vol. 108 (American Mathematical Society, Providence, RL, 1977), 37–130 (English).

[8] F. R. Gantmacher, *The Theory of Matrices*, Vol. 1 (AMS Chelsea Publishing, Providence, RI, 1998), Translated from the Russian by K. A. Hirsch. Reprint of the 1959 translation.

[9] F. R. Gantmacher, *The Theory of Matrices*, Vol. 2. (AMS Chelsea Publishing, Providence, RI, 2000), Translated from the Russian by K. A. Hirsch. Reprint of the 1959 translation.

[10] I. S. Iohvidov, *Hankel and Toeplitz Matrices and Forms: Algebraic Theory*, (Birkhäuser, Boston, MA 1982) Translated from the Russian by G. Philip A. Thijsse. With an introduction by I. Gohberg.

[11] I. S. Iohvidov, Hankel matrices and forms, *Mat. Sb.* (*N.S.*) **80**(122) (1969) 241–252 (In Russian).

[12] L. Kronecker, Zur Theorie der Elimination einer Variabeln aus zwei algebraischen Gleichungen, *Monatsber. Königl. Preuss. Akad. Wiss. Berlin, Sitzung der Phys.-Math. Klasse* **16** (1881) 535–600.

[13] J. W. Layman, The Hankel transform and some of its properties, *J. Integer Seq.* **4** (2001) Article 01.1.5, 11 pp.

[14] V. V. Prasolov, *Problems and Theorems in Linear Algebra*, Translations of Mathematical Monographs, Vol. 134 (American Mathematical Society, Providence, RI, 1994), Translated from the Russian manuscript by D. A. Leites.

[15] G. Polya and G. Szegő, *Problems and Theorems in Analysis. II: Theory of Functions, Zeros, Polynomials, Determinants, Number Theory, Geometry*, Classics in Mathematics (Springer-Verlag, Berlin, 1998).

[16] W. Rudin, *Real and Complex Analysis*, 2nd edn. McGraw-Hill Series in Higher Mathematics (McGraw-Hill Book Co., New York, 1974).
[17] J. A. Shohat and J. D. Tamarkin, *The Problem of Moments*, American Mathematical Society Surveys, Vol. 1 (American Mathematical Society, New York, 1943).
[18] T. J. Stieltjes, Recherches sur les fractions continues, *Ann. Fac. Sci. Toulouse Sci. Math. Sci. Phys.* **8**(4) (1894) J1–J122 (in French).

This page intentionally left blank

Chapter 6

Convolution and Product Theorems for the Special Affine Fourier Transform

Ayush Bhandari[*,‡] **and Ahmed I. Zayed**[†,§]

[*]*Massachusetts Institute of Technology, Cambridge, MA 02139, USA*
[†]*Department of Mathematical Sciences*
DePaul University, Chicago, IL 60614, USA
[‡]*ayush@MIT.edu*
[§]*azayed@depaul.edu*

The Special Affine Fourier Transform (SAFT) generalizes a number of well-known unitary transformations as well as signal processing and optics related mathematical operations. Unlike the Fourier transform, the SAFT does not work well with the standard convolution operation. Recently, Q. Xiang and K. Y. Qin introduced a new convolution operation that is more suitable for the SAFT and by which the SAFT of the convolution of two functions is the product of their SAFTs and a phase factor. However, their convolution structure does not work well with the inverse transform insofar as the transform of the product of two functions is not equal to the convolution of the transforms. In this chapter we introduce a new convolution operation that works well with both the SAFT and its inverse leading to an analogue of the convolution and product formulas for the Fourier transform. Furthermore, we introduce a second convolution operation that leads to the elimination of the phase factor in the convolution formula obtained by Q. Xiang and K. Y. Qin. We conclude the chapter by introducing a convolution operation associated with the SAFT that will enable one to convolve a function and a sequence of numbers. Such an operation is needed in the study of shift-invariant spaces associated with the SAFT.

Keywords: Convolution; Special Affine Fourier Transform; fractional Fourier transform; linear canonical transform; phase-space; and time-frequency representation.

Mathematics Subject Classification 2010: 42B10, 44A35, 44A15

1. Introduction

The Fourier transform is a very valuable tool in many branches of applied science such as physics, electrical engineering, and optics. If we denote the Fourier transformation of a function f by

$$\mathcal{F}[f](\omega) = \widehat{f}(\omega) = \frac{1}{\sqrt{2\pi}} \int_{\mathbb{R}} f(t)\, e^{\jmath \omega t} dt, \tag{1.1}$$

then for an appropriate function f the inversion formula takes on the form

$$\mathcal{F}^{-1}[\widehat{f}](t) = f(t) = \frac{1}{\sqrt{2\pi}} \int_{\mathbb{R}} \widehat{f}(\omega)\, e^{-\jmath \omega t} d\omega, \tag{1.2}$$

where throughout this article we will adopt the engineering notation $\jmath = \sqrt{-1}$. The above two equations are called the Fourier transform pair.

If we denote by $h_n(x)$ the Hermite function $h_n(x) = e^{-x^2/2} H_n(x)$, where $H_n(x)$ is the Hermite polynomial of degree n, it is known that the Hermite functions are the eigenfunctions of the Fourier transformation with eigenvalues $e^{in\pi/2}$, that is

$$\mathcal{F}[h_n](x) = e^{in\pi/2} h_n(x).$$

In [14], Namais was searching for an operator $\mathcal{F}_\alpha$ that satisfies the relation

$$\mathcal{F}_\alpha[h_n](x) = e^{in\alpha} h_n(x),$$

so that when $\alpha = \pi/2$, one has $\mathcal{F}_{\pi/2} h_n(x) = e^{in\pi/2} h_n(x)$, that is $\mathcal{F}_{\pi/2}$ is the standard Fourier transformation. This generalized operator $\mathcal{F}_\alpha$ may be represented in the form $e^{i\alpha A}$ so that

$$e^{i\alpha A} h_n(x) = e^{in\alpha} h_n(x).$$

Differentiating both sides with respect to α and setting $\alpha = 0$ yields

$$A h_n(x) = n h_n(x).$$

In view of the relation

$$H_n''(x) - 2x H_n'(x) + 2n H_n(x) = 0,$$

one can easily show that

$$h_n''(x) + (1 - x^2)h_n(x) = -2nh_n(x),$$

and hence

$$A = -\frac{1}{2}\frac{d^2}{dx^2} + \frac{1}{2}(x^2 - 1).$$

The Hermite functions satisfy the orthogonality relation

$$\int_{\mathbb{R}} h_m(x)h_n(x)dx = 2^n n!\sqrt{\pi}\delta_{m,n},$$

and, in fact, they form an orthogonal basis of $L^2(\mathbb{R})$. Therefore, it follows that for any $f \in L^2(\mathbb{R})$, we have the following orthogonal expansion in terms of the Hermite functions

$$f(x) = \sum_{n=0}^{\infty} \hat{f}_n h_n(x), \tag{1.3}$$

with

$$\hat{f}_n = c_n \int_{\mathbb{R}} f(x)h_n(x)dx, \tag{1.4}$$

where

$$c_n = \frac{1}{2^n n!\sqrt{\pi}}.$$

Therefore, if we apply the operator $\mathcal{F}_\alpha$ to both sides of equation (1.3), we obtain

$$\mathcal{F}_\alpha f(x) = \sum_{n=0}^{\infty} e^{in\alpha}\hat{f}_n h_n(x),$$

which, when combined with (1.4), yields

$$\mathcal{F}_\alpha[f](x) = \sum_{n=0}^{\infty} c_n e^{in\alpha} h_n(x) \int_{\mathbb{R}} f(t)h_n(t)dt \tag{1.5}$$

$$= \int_{\mathbb{R}} f(t)\left(\sum_{n=0}^{\infty} c_n e^{in\alpha} h_n(x)h_n(t)\right) dt. \tag{1.6}$$

In view of Mehler's formula

$$\sum_{n=0}^{\infty} \frac{z^n}{2^n n!\sqrt{\pi}} h_n(x)h_n(t)$$

$$= \frac{1}{\sqrt{\pi}\sqrt{1-z^2}} \exp\left\{-(x^2+t^2)\left(\frac{1+z^2}{2(1-z^2)}\right) + \frac{2xtz}{1-z^2}\right\}$$

and some easy calculations, we obtain

$$\mathcal{F}_\alpha[f](x) = \frac{c(\alpha)}{\sqrt{2\pi}} \int_{\mathbb{R}} f(t) \exp\left\{-j\left[a(\alpha)(x^2+t^2) - b(\alpha)xt\right]\right\} dt, \tag{1.7}$$

where

$$a(\alpha) = \frac{\cot\alpha}{2},\ b(\alpha) = \csc\alpha,\ c(\alpha) = \sqrt{1 + j\cot\alpha}.$$

The integral transform (1.7) is known as the *Fractional Fourier Transform* (FrFT) and sometimes it is also called the *angular Fourier transform* because it depends on an angle α. It can be shown that the operators $\{\mathcal{F}_\alpha\}$ satisfy the semi-group property

$$\mathcal{F}_\alpha \mathcal{F}_\beta = \mathcal{F}_{\alpha+\beta}, \quad -\pi \le \alpha, \beta \le \pi.$$

Moreover, it is easy to see that $\mathcal{F}_0 = I$, the identity operator, and consequently we have the inversion formula $\mathcal{F}_\alpha^{-1} = \mathcal{F}_{-\alpha}$.

The fractional Fourier transform is not only mathematically interesting, but also very useful in solving some problems in quantum physics, optics, and signal processing; see [9, 12, 15, 16, 23].

Because of the importance of the fractional Fourier transform, it is not surprising that a number of its generalizations have recently been introduced. In [30] the fractional Fourier transform was extended to a large class of functions and generalized functions and in [28] a systematic and unified approach to fractional integral transforms was presented. A relationship between the fractional Fourier transform on the one hand and the Wigner distribution and the ambiguity function on the other hand was presented in [11, 15, 16, 18].

In [8, 17], discrete versions of the FrFT were introduced, which included fractional Fourier series and discrete-time fractional Fourier transforms. A unified approach to the discrete and continuous fractional Fourier transforms was proposed in [10] which accommodates continuous-time, periodic continuous-time, discrete-time, and periodic discrete-time signals.

One of the earliest generalizations of the fractional Fourier transform was the *Linear Canonical Transform* (LCT) [13], which depends on four parameters

$$\begin{pmatrix} a & b \\ c & d \end{pmatrix}, \quad ab - cd = 1.$$

The linear canonical transform, which includes the Fourier, fractional Fourier, Laplace, Gauss–Weierstrass, and Bargmann transforms as special cases, appeared in problems in physics and quantum mechanics; see [25, Chapter 9]. The linear canonical transform reduces to the fractional

Fourier transform when

$$a = \cos\alpha,\ b = \sin\alpha,\ c = -\sin\alpha,\ d = \cos\alpha.$$

For more on the LCT, see [19–22].

2. Phase-Space Transformations

Phase-space transformations, such as fractional Fourier Transform (FrFT) [15] and the Linear Canonical Transform (LCT) [13], are becoming increasingly popular in the areas of signal processing, optics, and communications. A remarkable feature of some of these *phase-space* transformations is that they generalize the Fourier transformation and at the same time they retain properties compatible with Fourier analysis.

In recent years, a number of fundamental mathematical results for phase-space and time-frequency representations have been developed. Some examples include convolution theorems [3, 4, 6, 24, 27, 29], sampling theorems [6, 19, 20, 22, 23, 26, 31], time-frequency transformations [9, 21], shift-invariant signal approximation [6], sparse sampling [7] and super-resolution [5].

The Special Affine Fourier Transformation (SAFT), which was introduced in [1, 2], is an integral transformation associated with a general inhomogeneous lossless linear mapping in phase-space that depends on six parameters independent of the phase-space coordinates. It maps the position x and the wave number k into

$$\begin{pmatrix} x' \\ k' \end{pmatrix} = \begin{pmatrix} a & b \\ c & d \end{pmatrix} \begin{pmatrix} x \\ k \end{pmatrix} + \begin{pmatrix} p \\ q \end{pmatrix}, \tag{2.1}$$

with

$$ad - bc = 1. \tag{2.2}$$

This transformation, which can model many general optical systems [1, 2, 25], maps any convex body into another convex body and equation (2.2) guarantees that the area of the body is preserved by the transformation. Such transformations form the inhomogeneous special linear group ISL$(2, \in \mathbb{R})$.

The integral representation of the wavefunction transformation associated with the transformation (2.1) and (2.2) is given by

$$F(\omega) = \frac{1}{\sqrt{2\pi|b|}} \times \int_{\mathbb{R}} \exp\left\{\frac{j}{2b}\left(at^2 + d\omega^2 + 2t(p-\omega) + 2\omega(bq - dp)\right)\right\} f(t)dt. \tag{2.3}$$

Evidently, the FrFT is a special case of the SAFT. The SAFT offers a unified viewpoint of known optical operations on light waves. For example,

$$g_1(\theta) = \begin{pmatrix} \cos\theta & \sin\theta \\ -\sin\theta & \cos\theta \end{pmatrix} \quad \text{(rotation)},$$

$$g_2(\theta) = \begin{pmatrix} 1 & 0 \\ \theta & 1 \end{pmatrix} \quad \text{(lens transformation)},$$

$$g_3(\theta) = \begin{pmatrix} 1 & \theta \\ 0 & 1 \end{pmatrix} \quad \text{(free space propagation)},$$

$$g_4(\theta) = \begin{pmatrix} e^{\theta} & 0 \\ 0 & e^{-\theta} \end{pmatrix} \quad \text{(magnification)},$$

$$g_5(\theta) = \begin{pmatrix} \cosh\theta & \sinh\theta \\ \sinh\theta & \cosh\theta \end{pmatrix} \quad \text{(hyperbolic transformation)}.$$

The inversion formula for the SAFT is easily shown to be

$$f(t) = \frac{1}{\sqrt{2\pi|b|}} \times \int_{\mathbb{R}} F(\omega) \exp\left\{ \frac{-j}{2b} \left(at^2 + d\omega^2 + 2t(p-\omega) + 2\omega(bq - dp) \right) \right\} d\omega. \tag{2.4}$$

Some of the interesting transformations and signal/optical operations that can be obtained from the SAFT as special cases are listed in Table 1. When $p = 0 = q$, one obtains the homogeneous special group $\mathrm{SL}(2, \in \mathbb{R})$, and the associated integral transform is the *Linear Canonical Transform*, which is associated with the unimodular matrix

$$M = \begin{pmatrix} a & b \\ c & d \end{pmatrix}.$$

Let $\mathscr{T}$ be a linear integral transformation from a function space X into another function space Y. Let us denote the transform of $f, g \in X$ by $F = \mathscr{T}[f]$, $G = \mathscr{T}[g]$, $F, G \in Y$. We define the convolution operation $\star$ in X as the operation, if it exists, that renders $\mathscr{T}[f \star g] = \mathscr{T}[f]\mathscr{T}[g] = FG$. For example, if $\mathscr{T} = \mathcal{F}$ is the Fourier transform operator, then $\star$ is the standard convolution operation $*$ defined by

$$(f * g)(t) = \frac{1}{\sqrt{2\pi}} \int_{\mathbb{R}} f(x) g(t-x) dx,$$

which yields

$$\mathcal{F}[f * g](\omega) = \mathcal{F}[f](\omega)\mathcal{F}[g](\omega). \tag{2.5}$$

Table 1. SAFT, Unitary transformations and operations.

SAFT parameters ($\mathbf{\Lambda}_{\mathrm{S}}$)	Corresponding unitary transform
$\begin{bmatrix} a & b\,| & 0 \\ c & d\,| & 0 \end{bmatrix} = \mathbf{\Lambda}_{\mathrm{L}}$	Linear Canonical Transform
$\begin{bmatrix} \cos\theta & \sin\theta\,| & p \\ -\sin\theta & \cos\theta\,| & q \end{bmatrix} = \mathbf{\Lambda}_{\theta}^{\mathrm{O}}$	Offset Fractional Fourier Transform
$\begin{bmatrix} \cos\theta & \sin\theta\,| & 0 \\ -\sin\theta & \cos\theta\,| & 0 \end{bmatrix} = \mathbf{\Lambda}_{\theta}$	Fractional Fourier Transform
$\begin{bmatrix} 0 & 1\,| & p \\ -1 & 0\,| & q \end{bmatrix} = \mathbf{\Lambda}_{\mathrm{FT}}^{\mathrm{O}}$	Offset Fourier Transform (FT)
$\begin{bmatrix} 0 & 1\,| & 0 \\ -1 & 0\,| & 0 \end{bmatrix} = \mathbf{\Lambda}_{\mathrm{FT}}$	Fourier Transform (FT)
$\begin{bmatrix} 0 & J\,| & 0 \\ J & 0\,| & 0 \end{bmatrix} = \mathbf{\Lambda}_{\mathrm{LT}}$	Laplace Transform (LT)
$\begin{bmatrix} J\cos\theta & J\sin\theta\,| & 0 \\ J\sin\theta & -J\cos\theta\,| & 0 \end{bmatrix}$	Fractional Laplace Transform
$\begin{bmatrix} 1 & b\,| & 0 \\ 0 & 1\,| & 0 \end{bmatrix}$	Fresnel Transform
$\begin{bmatrix} 1 & J^b\,| & 0 \\ J & 1\,| & 0 \end{bmatrix}$	Bilateral Laplace Transform
$\begin{bmatrix} 1 & -J^b\,| & 0 \\ 0 & 1\,| & 0 \end{bmatrix}, b \geq 0$	Gauss–Weierstrass Transform
$\frac{1}{\sqrt{2}}\begin{bmatrix} 0 & e^{-J\pi/2}\,| & 0 \\ e^{-J\pi/2} & 1\,| & 0 \end{bmatrix}$	Bargmann Transform
SAFT parameters ($\mathbf{\Lambda}_{\mathrm{S}}$)	**Corresponding single operation**
$\begin{bmatrix} 1/\alpha & 0\,| & 0 \\ 0 & \alpha\,| & 0 \end{bmatrix} = \mathbf{\Lambda}_{\alpha}$	Time Scaling
$\begin{bmatrix} 1 & 0\,| & \tau \\ 0 & 1\,| & 0 \end{bmatrix} = \mathbf{\Lambda}_{\tau}$	Time Shift
$\begin{bmatrix} 1 & 0\,| & 0 \\ 0 & 1\,| & \xi \end{bmatrix} = \mathbf{\Lambda}_{\xi}$	Frequency Shift

(*Continued*)

Table 1. (*Continued*)

SAFT parameters ($\mathbf{\Lambda}_{\mathrm{S}}$)	Corresponding optical operation
$\begin{bmatrix} \cos\theta & \sin\theta & \vert & 0 \\ -\sin\theta & \cos\theta & \vert & 0 \end{bmatrix} = \mathbf{\Lambda}_\theta$	Rotation
$\begin{bmatrix} 1 & 0 & \vert & 0 \\ \tau & 1 & \vert & 0 \end{bmatrix} = \mathbf{\Lambda}_\tau$	Lens Transformation
$\begin{bmatrix} 1 & \eta & \vert & 0 \\ 0 & 1 & \vert & 0 \end{bmatrix} = \mathbf{\Lambda}_\eta$	Free Space Propogation
$\begin{bmatrix} e^{\beta} & 0 & \vert & 0 \\ 0 & e^{-\beta} & \vert & 0 \end{bmatrix} = \mathbf{\Lambda}_\beta$	Magnification
$\begin{bmatrix} \cosh\alpha & \sinh\alpha & \vert & 0 \\ \sinh\alpha & \cosh\alpha & \vert & 0 \end{bmatrix} = \mathbf{\Lambda}_\eta$	Hyperbolic Transformation

This result is known as the Fourier *convolution theorem.* We also have, in view of the Fourier inversion formula,

$$\mathcal{F}[f \cdot g] = \mathcal{F}[f] * \mathcal{F}[g] = F * G,$$

which is known as the product formula.

The purpose of this chapter is to extend the above results and establish a convolution and a product formulas for the **Special Affine Fourier Transform** (SAFT).

3. Convolution and Product Theorems

In this section we discuss convolution and product theorems for different phase-space transformations and then introduce our main results in Section 3.2.

3.1. Convolution theorems for the FrFT and the LCT

In context of signal processing theory, Almeida, to the best of our knowledge, was the first to derive convolution and product theorems for the fractional Fourier Transform (FrFT) [4]. Unfortunately, Almeida's formulation did not conform with the classical Fourier convolution-multiplication property. That is, the convolution of functions in time domain did not result in multiplication of their respective FrFT spectrums. As a follow up, Zayed formulated the convolution operation for the FrFT which resulted in an elegant convolution-multiplication property in FrFT domain [29].

In [24] Wei, Ran, Li, Ma and Tan derived convolution theorems for the linear canonical transform.

Recently, Xiang and Qin [27] introduced a new convolution operation that is more suitable for the SAFT and by which the SAFT of the convolution of two functions is the product of their SAFTs and a phase factor. However, their convolution structure does not work well with the inverse transform insofar as the transform of the product of two functions is not equal to the convolution of the transforms.

3.2. Convolution and product theorems for the SAFT

In this section we introduce a new convolution operation that works well with both the SAFT and its inverse leading to an analogue of the convolution and product formulas for the Fourier transform. Furthermore, we introduce a second convolution operation that leads to the elimination of the phase factor in the convolution formula obtained in [27]; see Table 1.

Let f^* denote the complex-conjugate of f and $\langle f, g\rangle = \int f(t)\, g^*(t)\, dt$ be the standard L_2 inner-product. The SAFT operation, that is, $\mathscr{T}_{\mathsf{SAFT}} : f \to \widehat{f}_{\mathbf{\Lambda}_{\mathsf{S}}}$, is defined as

$$\widehat{f}_{\mathbf{\Lambda}_{\mathsf{S}}}(\omega) = \begin{cases} \langle f, \kappa_{\mathbf{\Lambda}_{\mathsf{S}}}(\cdot, \omega)\rangle, & b \neq 0, \\ \sqrt{d} e^{\jmath \frac{cd}{2}(\omega - p)^2 + \jmath \omega q} x\left(d(\omega - p)\right), & b = 0, \end{cases} \tag{3.1}$$

where

$$\kappa_{\mathbf{\Lambda}_{\mathsf{S}}}(t, \omega) = K_b^* \exp\left(-\frac{\jmath}{2b}\left(at^2 + d\omega^2 + 2t(p - \omega) - 2\omega(dp - bq)\right)\right) \tag{3.2}$$

and $K_b = \frac{1}{\sqrt{2\pi b}}$,

- $\mathbf{\Lambda}_{\mathsf{S}}^{(2\times 3)}$ is the augmented SAFT parameter matrix of the form

$$\mathbf{\Lambda}_{\mathsf{S}} = \left[\mathbf{\Lambda} \,|\, \underline{\lambda}\right], \tag{3.3}$$

  which is in turn parameterized by the LCT matrix $\mathbf{\Lambda}_{\mathsf{L}}$ [13] (see Table 1) and an offset vector $\underline{\lambda}$ such that

$$\mathbf{\Lambda} = \begin{bmatrix} a & b \\ c & d \end{bmatrix} \text{ with } ad - bc = 1 \quad \text{and} \quad \underline{\lambda} = \begin{bmatrix} p \\ q \end{bmatrix}.$$

  This is the reason that the SAFT is sometimes referred to as the *Offset Linear Canonical Transform* or the OLCT.
- $\kappa_{\mathbf{\Lambda}_{\mathsf{S}}}(t, \omega)$ in (3.2) is the SAFT kernel parameterized by SAFT matrix $\mathbf{\Lambda}_{\mathsf{S}}$.

Thanks to the additive property of the SAFT [17], the inverse-SAFT (or the iSAFT) is simply the SAFT evaluated using matrix $\mathbf{\Lambda}_{\mathsf{S}}^{\text{inv}}$ with parameters

$$\mathbf{\Lambda}_{\mathsf{S}}^{\text{inv}} \stackrel{\text{def}}{=} \left[\begin{array}{cc|c} +d & -b & bq-dp \\ -c & +a & cp-aq \end{array}\right] = \left[\begin{array}{cc|c} +d & -b & p_0 \\ -c & +a & q_0 \end{array}\right]. \tag{3.4}$$

As a result, we are able to define the inverse transform iSAFT,

$$f(t) = C_{\mathbf{\Lambda}_{\mathsf{S}}^{\text{inv}}} \langle \widehat{f}_{\mathbf{\Lambda}_{\mathsf{S}}}, \kappa_{\mathbf{\Lambda}_{\mathsf{S}}^{\text{inv}}}(\cdot, t) \rangle \tag{3.5}$$

with some transform-dependent phase constant

$$C_{\mathbf{\Lambda}_{\mathsf{S}}^{\text{inv}}} = \exp\left(\frac{\jmath}{2}\left(cdp^2 + abq^2 - 2adpq\right)\right).$$

Next, we develop a convolution structure for the SAFT denoted by $*_{\mathbf{\Lambda}_{\mathsf{S}}}$ so that we can obtain a representation of form,

$$\mathscr{T}_{\mathsf{SAFT}}[f *_{\mathbf{\Lambda}_{\mathsf{S}}} g] \propto \mathscr{T}_{\mathsf{SAFT}}[f]\, \mathscr{T}_{\mathsf{SAFT}}[g]$$

which is consistent with the Fourier convolution theorem (2.5).

Before we define the convolution operation in the SAFT domain, let us introduce the chirp modulation operation.

Definition 3.1 (Chirp Modulation). Let $\mathbf{A} = [a_{j,k}]$ be a 2×2 matrix. We define the modulation function,

$$m_{\mathbf{A}}(t) \stackrel{\text{def}}{=} \exp\left(\jmath \frac{a_{11}}{2a_{12}} t^2\right). \tag{3.6}$$

Furthermore, for a given function f, we define its chirp modulated functions associated with the matrix $\mathbf{A}$ as,

$$\overrightarrow{f}(t) \stackrel{\text{def}}{=} m_{\mathbf{A}}(t) f(t) \quad \text{and} \quad \overleftarrow{f}(t) \stackrel{\text{def}}{=} m_{\mathbf{A}}^*(t) f(t). \tag{3.7}$$

For example, when $\mathbf{A} = \mathbf{\Lambda}_{\mathsf{S}}$, then we have $\overrightarrow{f}(t) = m_{\mathbf{\Lambda}_{\mathsf{S}}} f(t) = e^{\jmath \frac{at^2}{2b}} f(t)$. For the case when $\mathbf{A} = \mathbf{\Lambda}_{\mathsf{S}}^{\text{inv}}$, we get $\overrightarrow{f}(t) = m_{\mathbf{\Lambda}_{\mathsf{S}}^{\text{inv}}}(t) f(t) = e^{-\jmath \frac{dt^2}{2b}} f(t)$.

Next we define the SAFT convolution operator.

Definition 3.2 (SAFT Convolution/Filtering). Let f and g be two given functions and $*$ denote the usual convolution operation (see (2.5)). The SAFT convolution is defined by

$$h(t) = (f *_{\mathbf{\Lambda}_{\mathsf{S}}} g)(t) \stackrel{\text{def}}{=} K_b m_{\mathbf{\Lambda}_{\mathsf{S}}}^*(t) \left(\overrightarrow{f}(t) * \overrightarrow{g}(t)\right). \tag{3.8}$$

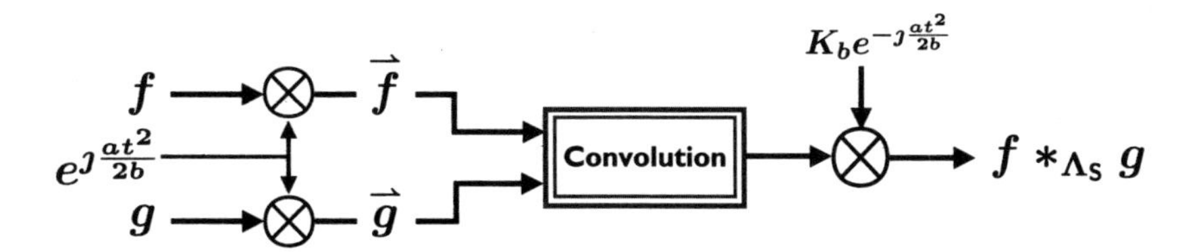

Fig. 1. Conceptual definition of SAFT convolution.

Figure 1 illustrates the block diagram for SAFT domain convolution defined in Definition 3.2. Next, we state the convolution and product theorem for the SAFT domain.

Theorem 3.3 (SAFT Convolution and Product Theorem). *Let f and g be any two given functions for which the convolution $*_{\mathbf{\Lambda}_\mathsf{S}}$ exists and set*

$$h(t) = (f *_{\mathbf{\Lambda}_\mathsf{S}} g)(t).$$

Furthermore, let $\widehat{f}_{\mathbf{\Lambda}_\mathsf{S}}(\omega), \widehat{g}_{\mathbf{\Lambda}_\mathsf{S}}(\omega)$ and $\widehat{h}_{\mathbf{\Lambda}_\mathsf{S}}(\omega)$ be the SAFT of f, g and h, respectively. Then we have

$$h(t) = (f *_{\mathbf{\Lambda}_\mathsf{S}} g)(t) \xrightarrow{\mathsf{SAFT}} \widehat{h}_{\mathbf{\Lambda}_\mathsf{S}}(\omega) = \Phi_{\mathbf{\Lambda}_\mathsf{S}}(\omega)\, \widehat{f}_{\mathbf{\Lambda}_\mathsf{S}}(\omega)\, \widehat{g}_{\mathbf{\Lambda}_\mathsf{S}}(\omega),$$

where $\Phi_{\mathbf{\Lambda}_\mathsf{S}}(\omega) = e^{\jmath\frac{\omega}{b}(dp-bq)} e^{-\jmath\frac{d\omega^2}{2b}}$. Moreover, let

$$h(t) = \Phi_{\mathbf{\Lambda}_\mathsf{S}^{\mathrm{inv}}}(t)\, f(t)\, g(t) \quad \text{with } \Phi_{\mathbf{\Lambda}_\mathsf{S}^{\mathrm{inv}}}(t) = e^{\jmath\frac{at^2}{2b}} e^{-\jmath\frac{t}{b}(ap_0+bq_0)};$$

then we have $\widehat{h}_{\mathbf{\Lambda}_\mathsf{S}}(\omega) = C_{\mathbf{\Lambda}_\mathsf{S}^{\mathrm{inv}}}(\widehat{f} *_{\mathbf{\Lambda}_\mathsf{S}^{\mathrm{inv}}} \widehat{g})(\omega)$.

Proof. We begin with computing the SAFT of h

$$\begin{aligned}
\widehat{h}_{\mathbf{\Lambda}_\mathsf{S}}(\omega) &\overset{(3.1)}{=} \mathscr{T}_{\mathsf{SAFT}}[h](\omega) = \langle h(t), \kappa_{\mathbf{\Lambda}_\mathsf{S}}(t,\omega)\rangle \\
&= \int_{\mathbb{R}} h(t)\, \kappa^*(t,\omega)\, dt \\
&\overset{(3.8)}{=} \int_{\mathbb{R}} \left(K_b m^*_{\mathbf{\Lambda}_\mathsf{S}}(t) \int_{\mathbb{R}} \vec{f}(z)\, \vec{g}(t-z)\, dz \right) \kappa^*(t,\omega)\, dt \\
&= \underbrace{K_b^2 e^{\jmath\frac{d\omega^2}{2b}} e^{-\jmath\frac{\omega}{b}(dp-bq)}}_{C(\omega)} \int_{\mathbb{R}} e^{\jmath\frac{t}{b}(p-\omega)} \boxed{m_{\mathbf{\Lambda}_\mathsf{S}}(t)} \\
&\quad \times \left(\boxed{m^*_{\mathbf{\Lambda}_\mathsf{S}}(t)} \int_{\mathbb{R}} \vec{f}(z)\, \vec{g}(t-z)\, dz \right) dt.
\end{aligned}$$

In the above development, note that the items in the box cancel one another because $m^*_{\mathbf{\Lambda}_\mathsf{S}} m_{\mathbf{\Lambda}_\mathsf{S}} = 1$ (see Definition 3.1). Setting $t - z = v$ and using (3.7), we obtain an integral of separable form, that is, $\widehat{h}_{\mathbf{\Lambda}_\mathsf{S}}(\omega) = I_f(\omega) I_g(\omega)$ because

$$\underbrace{K_b^2 \Phi^*_{\mathbf{\Lambda}_\mathsf{S}}(\omega) \iint_{\mathbb{R}} f(x) m_{\mathbf{\Lambda}_\mathsf{S}}(x) g(v) m_{\mathbf{\Lambda}_\mathsf{S}}(v) e^{\jmath \frac{v+x}{b}(p-\omega)} dx dv}_{\widehat{h}_{\mathbf{\Lambda}_\mathsf{S}}(\omega)}$$

$$= I_f(\omega) I_g(\omega) = \widehat{h}_{\mathbf{\Lambda}_\mathsf{S}}(\omega), \tag{3.9}$$

where, for a given function f, we define

$$I_f(\omega) \stackrel{\text{def}}{=} K_b \sqrt{\Phi^*_{\mathbf{\Lambda}_\mathsf{S}}(\omega)} \int_{\mathbb{R}} f(z) m_{\mathbf{\Lambda}_\mathsf{S}}(z) e^{\jmath \frac{z}{b}(p-\omega)} dz. \tag{3.10}$$

Indeed, using (3.10) and (3.1), it is easy to see that

$$I_f(\omega) = \sqrt{\Phi^*_{\mathbf{\Lambda}_\mathsf{S}}(\omega)} \Phi_{\mathbf{\Lambda}_\mathsf{S}}(\omega) \widehat{f}_{\mathbf{\Lambda}_\mathsf{S}}(\omega), \tag{3.11}$$

and this result extends to $I_g(\omega)$ by symmetry.

We conclude

$$\begin{aligned}
\widehat{h}_{\mathbf{\Lambda}_\mathsf{S}}(\omega) &\stackrel{(3.9)}{=} I_f(\omega) I_g(\omega) \\
&\stackrel{(3.11)}{=} \sqrt{\Phi^*_{\mathbf{\Lambda}_\mathsf{S}}(\omega)} \Phi_{\mathbf{\Lambda}_\mathsf{S}}(\omega) \widehat{f}_{\mathbf{\Lambda}_\mathsf{S}}(\omega) \cdot \sqrt{\Phi^*_{\mathbf{\Lambda}_\mathsf{S}}(\omega)} \Phi_{\mathbf{\Lambda}_\mathsf{S}}(\omega) \widehat{g}_{\mathbf{\Lambda}_\mathsf{S}}(\omega) \\
&= \Phi_{\mathbf{\Lambda}_\mathsf{S}}(\omega) \widehat{f}_{\mathbf{\Lambda}_\mathsf{S}}(\omega) \widehat{g}_{\mathbf{\Lambda}_\mathsf{S}}(\omega),
\end{aligned}$$

which is the first conclusion of the Theorem.

Now we establish the product theorem for the SAFT

$$\Phi_{\mathbf{\Lambda}_\mathsf{S}^{\text{inv}}}(t) f(t) g(t) \xrightarrow{\mathsf{SAFT}} C_{\mathbf{\Lambda}_\mathsf{S}^{\text{inv}}} (\widehat{f} *_{\mathbf{\Lambda}_\mathsf{S}^{\text{inv}}} \widehat{g})(\omega).$$

Since the inverse-SAFT is the SAFT of a function with $\mathbf{\Lambda}_\mathsf{S} = \mathbf{\Lambda}_\mathsf{S}^{\text{inv}}$ in (3.4), we have

$$\begin{aligned}
h(t) &= C_{\mathbf{\Lambda}_\mathsf{S}^{\text{inv}}} K_b^* \int_{\mathbb{R}} \widehat{h}_{\mathbf{\Lambda}_\mathsf{S}}(\omega) e^{-\jmath \frac{(at^2 + d\omega^2)}{2b}} e^{-\jmath \frac{\omega}{b}(p_0 - t)} e^{\jmath \frac{t}{b}(ap_0 + bq_0)} d\omega \\
&= C_{\mathbf{\Lambda}_\mathsf{S}^{\text{inv}}} K_b^* \Phi^*_{\mathbf{\Lambda}_\mathsf{S}^{\text{inv}}}(t) \int_{\mathbb{R}} \widehat{h}_{\mathbf{\Lambda}_\mathsf{S}}(\omega) e^{-\jmath \frac{d\omega^2}{2b}} e^{-\jmath \frac{\omega(p_0 - t)}{b}} d\omega.
\end{aligned}$$

By setting,

$$\begin{aligned}
\widehat{h}_{\mathbf{\Lambda}_\mathsf{S}}(\omega) &= C_{\mathbf{\Lambda}_\mathsf{S}^{\text{inv}}} (\widehat{f} *_{\mathbf{\Lambda}_\mathsf{S}^{\text{inv}}} \widehat{g})(\omega) \\
&= C_{\mathbf{\Lambda}_\mathsf{S}^{\text{inv}}} K_b^* m^*_{\mathbf{\Lambda}_\mathsf{S}^{\text{inv}}}(\omega) (\overrightarrow{\widehat{f}}(\omega) * \overrightarrow{\widehat{g}}(\omega)),
\end{aligned}$$

where, for $\mathbf{\Lambda}_{\mathsf{S}} = \mathbf{\Lambda}_{\mathsf{S}}^{\mathrm{inv}}$ (see (3.4)), we have

$$m_{\mathbf{\Lambda}_{\mathsf{S}}^{\mathrm{inv}}}(\cdot) = e^{-\jmath\frac{d(\cdot)^2}{2b}} \text{ and } \vec{f}(\omega) = m_{\mathbf{\Lambda}_{\mathsf{S}}^{\mathrm{inv}}}(\omega)\, f(\omega).$$

Upon simplification, we obtain the separable integrals

$$\begin{aligned}
h(t) &= (C_{\mathbf{\Lambda}_{\mathsf{S}}^{\mathrm{inv}}} K_b^*)^2 \Phi^*_{\mathbf{\Lambda}_{\mathsf{S}}^{\mathrm{inv}}}(t) \\
&\quad \times \iint \widehat{f}(\nu)\, m_{\mathbf{\Lambda}_{\mathsf{S}}^{\mathrm{inv}}}(\nu)\, \widehat{g}(\omega-\nu)\, m_{\mathbf{\Lambda}_{\mathsf{S}}^{\mathrm{inv}}}(\omega-\nu)\, d\nu e^{-\jmath\frac{\omega(p_0-t)}{b}}\, d\omega \\
&= \int \widehat{\vec{f}}(\nu)\, e^{-\jmath\frac{\nu(p_0-t)}{b}}\, d\nu \int \widehat{\vec{g}}(\omega)\, e^{-\jmath\frac{\omega(p_0-t)}{b}}\, d\omega \\
&= I_f(t)\, I_g(t),
\end{aligned}$$

where

$$\begin{aligned}
I_f(t) &= C_{\mathbf{\Lambda}_{\mathsf{S}}^{\mathrm{inv}}} K_b^* \sqrt{\Phi^*_{\mathbf{\Lambda}_{\mathsf{S}}^{\mathrm{inv}}}(t)} \int_{\mathbb{R}} \vec{f}(\omega) e^{-\jmath\frac{\omega(p_0-t)}{b}}\, d\omega. \\
&= \sqrt{\Phi^*_{\mathbf{\Lambda}_{\mathsf{S}}^{\mathrm{inv}}}(t)\Phi_{\mathbf{\Lambda}_{\mathsf{S}}^{\mathrm{inv}}}(t)}\, f(t).
\end{aligned}$$

As a result, we have

$$\begin{aligned}
h(t) &= I_f(t)\, I_g(t) \\
&= \sqrt{\Phi^*_{\mathbf{\Lambda}_{\mathsf{S}}^{\mathrm{inv}}}(t)\Phi_{\mathbf{\Lambda}_{\mathsf{S}}^{\mathrm{inv}}}(t)}\, f(t) \cdot \sqrt{\Phi^*_{\mathbf{\Lambda}_{\mathsf{S}}^{\mathrm{inv}}}(t)\Phi_{\mathbf{\Lambda}_{\mathsf{S}}^{\mathrm{inv}}}(t)}\, g(t) \\
&= \Phi_{\mathbf{\Lambda}_{\mathsf{S}}^{\mathrm{inv}}}(t)\, f(t)\, g(t)
\end{aligned}$$

which is the desired result. □

4. Comparison and Alternative Results

In this section we compare our results with those of [27]. We show that our approach is not only easier to derive, but also provides more symmetric formulas to implement. Our convolution formula is the same as the one given in [27], both assert that the SAFT of the convolution of two functions is the product of their SAFT and a phase factor given by $\Phi_{\mathbf{\Lambda}_{\mathsf{S}}}(\omega)$. But our product formula is different from that in [27] which states that the SAFT of the product of two functions f and g is

$$K_b^2 \Phi^*_{\mathbf{\Lambda}_{\mathsf{S}}}(\omega) \left(\widehat{f}_{\mathbf{\Lambda}_{\mathsf{S}}}(\omega)\, \Phi_{\mathbf{\Lambda}_{\mathsf{S}}}(\omega) * \widehat{g}_{\mathbf{\Lambda}_{\mathsf{FT}}}\left(\frac{\omega}{b}\right)\right). \tag{4.1}$$

The reason our convolution and product formulas are more symmetric and simpler goes back to our definition of the chirp modulation, see Definition 3.1, which uses the adaptive matrix $\mathbf{\Lambda}_{\mathsf{S}}$ that accommodates both the forward and backward SAFT.

Furthermore, we will now derive another convolution for SAFT which eliminates the phase factor $\Phi_{\mathbf{\Lambda}_\mathsf{S}}$ from the convolution formula.

Definition 4.1 (Phase-free SAFT Convolution). Let f and g be two given functions and $*$ denote the usual convolution operation. The second SAFT convolution $\star$ is defined by

$$h(t) = (f \star g)(t) \stackrel{\text{def}}{=} \sqrt{2}K_b m^*_{\mathbf{\Lambda}_\mathsf{S}}(t)\,(\vec{f}\,(.) * \vec{g}\,(.))(\sqrt{2}t).$$

In view of this SAFT-convolution, we have the following theorem.

Theorem 4.2. *Let $h(t) = (f \star g)(t)$. Then, we have*

$$\widehat{h}_{\mathbf{\Lambda}_\mathsf{S}}(\omega) = \widehat{f}_{\mathbf{\Lambda}_1}(\omega/\sqrt{2})\widehat{g}_{\mathbf{\Lambda}_1}(\omega/\sqrt{2}),$$

where $\widehat{f}_{\mathbf{\Lambda}_1}$ denotes the SAFT of f with respect to the matrix $\mathbf{\Lambda}_1 = [\mathbf{\Lambda} \mid \underline{\lambda}/\sqrt{2}]$ (cf. (3.3)).

Proof. Let $\Omega = bq - dp$. We have

$$\widehat{h}_{\mathbf{\Lambda}_\mathsf{S}}(\omega) = \sqrt{2}K_b^2 e^{-\frac{\jmath a t^2}{2b}} \int_{\mathbb{R}} e^{\frac{\jmath}{2b}(at^2+d\omega^2+2tp-2t\omega+2\omega\Omega)}dt$$
$$\times \int_{\mathbb{R}} f(\tau)e^{\frac{\jmath a t^2}{2b}} g(\sqrt{2}t-\tau)e^{\frac{\jmath a}{2b}(\sqrt{2}t-\tau)^2}d\tau.$$

Setting $x = \sqrt{2}t - \tau$ and simplifying the integrals, we obtain

$$\begin{aligned}\widehat{h}_{\mathbf{\Lambda}_\mathsf{S}}(\omega) &= K_b^2\Phi^*_{\mathbf{\Lambda}_\mathsf{S}}(\omega)\int_{\mathbb{R}} e^{\frac{\jmath a\tau^2}{2b}}f(\tau)d\tau \\ &\quad\times \int_{\mathbb{R}} g(x)e^{\frac{\jmath}{2b}(ax^2+2(p-\omega)(x+\tau)/\sqrt{2})}dx \\ &= K_b^2\Phi^*_{\mathbf{\Lambda}_\mathsf{S}}(\omega)\int_{\mathbb{R}} e^{\frac{\jmath}{2b}(a\tau^2+2(p-\omega)\tau/\sqrt{2})}f(\tau)d\tau \\ &\quad\times \int_{\mathbb{R}} g(x)e^{\frac{\jmath}{2b}(ax^2+2(p-\omega)x/\sqrt{2})}dx \\ &= K_b^2\Phi^*_{\mathbf{\Lambda}_\mathsf{S}}(\omega)\int_{\mathbb{R}} e^{\frac{\jmath}{2b}(a\tau^2+\sqrt{2}p\tau-\sqrt{2}\tau\omega)}f(\tau)d\tau \\ &\quad\times \int_{\mathbb{R}} e^{\frac{\jmath}{2b}(ax^2+\sqrt{2}px-\sqrt{2}x\omega)}g(x)dx.\end{aligned}$$

But since

$$\Phi^*_{\mathbf{\Lambda}_\mathsf{S}}(\omega) = e^{\jmath\frac{1}{2b}(2d(\frac{\omega}{\sqrt{2}})^2+2\sqrt{2}\Omega\frac{\omega}{\sqrt{2}})},$$

it follows that $\widehat{h}_{\mathbf{\Lambda}_\mathsf{S}}(\omega) = I_f(\omega)I_g(\omega)$, where

$$I_f(\omega) = K_b \int_{\mathbb{R}} e^{\frac{j}{2b}(a\tau^2 + d(\frac{\omega}{\sqrt{2}})^2 + \sqrt{2}p\tau - \sqrt{2}\tau\omega + \sqrt{2}\Omega(\omega/\sqrt{2}))} f(\tau)d\tau$$

and similar expression for $I_g(\omega)$. But it is easy to see that

$$I_f(\omega) = \widehat{f}_{\mathbf{\Lambda}_1}(\omega/\sqrt{2}),$$

and this completes the proof. □

Relation to convolution theory of LCTs: In the special case where $p = q = 0 \Leftrightarrow \mathbf{\Lambda}_\mathsf{S} = \mathbf{\Lambda}_\mathsf{LCT}$, the SAFT reduces to the LCT and the last convolution theorem takes the simple form

$$\mathscr{T}_\mathsf{LCT}[f \star g](\omega) = \mathscr{T}_\mathsf{LCT}[f](\omega/\sqrt{2})\mathscr{T}_\mathsf{LCT}[g](\omega/\sqrt{2}).$$

5. Another Convolution Structure

In this section we introduce a convolution operation associated with the SAFT that will enable us to convolve a function and a sequence of numbers. This operation is needed for the study of shift-invariant spaces associated with SAFT.

Definition 5.1 (Discrete Time SAFT (DT-SAFT)). Let $P = \{p(k)\}$ be a sequence in ℓ^2, that is, $\sum_k |p(k)|^2 < \infty$. We define the discrete time SAFT of P as

$$\widehat{P}_{\mathbf{A}}(\omega) = \frac{1}{\sqrt{2\pi|b|}} \sum_k p(k) \exp\left\{\frac{j}{2b}(ak^2 + d\omega^2 - 2\omega k + \Omega\omega + 2pk)\right\}, \tag{5.1}$$

and define the convolution of a sequence P and a function $\phi \in L^2(\mathbb{R})$ as

$$h(t) = (P *_{\mathbf{\Lambda}_\mathsf{S}} \phi)(t) = \frac{1}{\sqrt{2\pi|b|}} e^{-jat^2/2b} \sum_k e^{jak^2/2b} p(k) e^{ja(t-k)^2/b} \phi(t-k).$$

Theorem 5.2. *Let P and ϕ be as above and $h(t) = (P *_A \phi)(t)$. Then*

$$H(\omega) = \widehat{h}_{\mathbf{A}}(\omega) = \overline{\eta}_A(\omega)\widehat{P}_{\mathbf{A}}(\omega)\,\Phi_A(\omega),$$

where Φ_A is the SAFT of ϕ. Moreover, $|\widehat{P}_{\mathbf{A}}(\omega)|$ is periodic with period $\Delta = 2\pi b$.

Proof. From the definition of SAFT, we have

$$
\begin{aligned}
H(\omega) &= \int_{\in\mathbb{R}} h(t)k(t,\omega)dt \\
&= \frac{1}{2\pi|b|}\sum_k e^{jak^2/2b}p(k)\int_{\in\mathbb{R}} e^{-jat^2/2b}\phi(t-k)e^{ja(t-k)^2/2b} \\
&\quad \times \exp\left\{\frac{j}{2b}(at^2+d\omega^2-2\omega t+\Omega\omega+2pt)\right\}dt \\
&= \frac{1}{2\pi|b|}\sum_k e^{jak^2/2b}p(k)\int_{\in\mathbb{R}} \phi(u) \\
&\quad \times \exp\left\{\frac{j}{2b}(au^2+d\omega^2-2\omega(u+k)+\Omega\omega+2p(u+k))\right\}du \\
&= \frac{\overline{\eta}_A(\omega)}{2\pi|b|}\sum_k \exp\left\{\frac{j}{2b}(ak^2-2k\omega+2pk)\right\}p(k) \\
&\quad \times \int_{\in\mathbb{R}} \phi(u)\exp\left\{\frac{j}{2b}(au^2+d\omega^2-2\omega u+\Omega\omega+2pu)\right\}du \\
&= \frac{\overline{\eta}_A(\omega)}{\sqrt{2\pi|b|}}\left[\sum_k \exp\left\{\frac{j}{2b}(ak^2+d\omega^2-2k\omega+\Omega\omega+2pk)\right\}p(k)\right]\Phi_A(\omega) \\
&= \overline{\eta}_A(\omega)\hat{P}_A(\omega)\Phi_A(\omega),
\end{aligned}
$$

where $\eta_A(\omega) = \exp\left\{\frac{j}{2b}\left(d\omega^2+\Omega\omega\right)\right\}$ and $\Omega = 2(bq-dp)$. Furthermore, since $e^{-jk\Delta/b} = e^{-2jk\pi} = 1$, we have

$$
\begin{aligned}
\hat{P}_A(\omega+\Delta)(\omega) &= \frac{1}{\sqrt{2\pi|b|}}\sum_k p(k)\exp\left\{\frac{j}{2b}(ak^2+d(\omega+\Delta)^2-2k(\omega+\Delta)\right. \\
&\qquad \left. +\Omega(\omega+\Delta)+2pk)\right\} \\
&= \frac{1}{\sqrt{2\pi|b|}}\sum_k p(k)\exp\left\{\frac{j}{2b}(ak^2+d\omega^2-2k\omega+\Omega\omega+2pk)\right\} \\
&\quad \times \exp\left\{\frac{j}{2b}(d\Delta^2+2d\omega\Delta-2k\Delta+\Omega\Delta)\right\} \\
&= \hat{P}_A(\omega)\exp\left\{\frac{j}{2b}(d\Delta^2+2d(\omega)\Delta+\Omega\Delta)\right\}.
\end{aligned}
$$

Thus,

$$|\hat{P}_A(\omega + \Delta)| = |\hat{P}_A(\omega)|. \qquad \square$$

6. Conclusion

In this chapter, we introduced two definitions of convolution operation that establish the convolution-product theorem for the Special Affine Fourier Transform (SAFT) introduced by Abe and Sheridan [1, 2]. Our result is quite general in that our convolution-product theorem is applicable to all the listed unitary transformations in Table 1.

Furthermore, we also presented a product theorem for the SAFT which establishes the fact that the product of functions in the time domain amounts to convolution in SAFT domain. We conclude that our construction of the convolution structure for the SAFT domain establishes the SAFT duality principle, that is, convolution in one domain amounts to multiplication in the transform domain and vice versa. Our results can be used to develop the semi-discrete convolution structure [6] for sampling and approximation theory linked with the Special Affine Fourier Transform, but this will be done in a separate project.

We concluded the chapter by introducing a convolution operation associated with the SAFT that will enable us to convolve a function and a sequence of numbers. Such an operation is useful in the study of shift-invariant spaces associated with the SAFT which will be presented in a future work.

References

[1] S. Abe and J. T. Sheridan, Optical operations on wave functions as the abelian subgroups of the special affine Fourier transformations, *Opt. Lett.* **19**(22) (1994) 1801.

[2] S. Abe and J. Sheridan, Generalization of the fractional Fourier transformation to an arbitrary linear lossless transformation an operator approach, *J. Phys. A: Math. Gen.* **27**(12) (1994) 4179.

[3] O. Akay and G. Boudreaux-Bartels, Fractional convolution and correlation via operator methods and an application to detection of linear FM signals, *IEEE Trans. Signal Process.* **49**(5) (2001) 979–993.

[4] L. Almeida, Product and convolution theorems for the fractional Fourier transform, *IEEE Signal Process. Lett.* **4**(1) (1997) 15–17.

[5] A. Bhandari, Y. Eldar and R. Raskar, Super-resolution in phase space, in *IEEE International Conference on Acoustics, Speech and Signal Processing* (*ICASSP*), (2015), pp. 4155–4159.

[6] A. Bhandari and A. I. Zayed, Shift-invariant and sampling spaces associated with the fractional Fourier transform domain, *IEEE Trans. Signal Process.* **60**(4) (2012) 1627–1637.
[7] A. Bhandari and P. Marziliano, Sampling and reconstruction of sparse signals in fractional Fourier domain, *IEEE Signal Process. Lett.* **17**(3) (2010) 221–224.
[8] C. Candan, M. A. Kutay and H. M. Ozakdas, The discrete fractional Fourier transform, *IEEE Trans. Signal Process.* **48**(5) (2000) 1329–1337.
[9] C. Capus and K. Brown, Short-time fractional Fourier methods for the time-frequency representation of chirp signals, *J. Acoustic. Soc. Amer.* **113**(6) (2003) 3253–3263.
[10] T. Erseghe, Kraniauskas and G. Carioraro, Unified fractional Fourier transform and sampling theorem, *IEEE Trans. Signal Process.* **47**(12) (1999) 3419–3423.
[11] A. W. Lohmann, Image rotation, Wigner rotation and the fractional Fourier transform, *J. Opt. Soc. Amer. A* **10** (1993) 2181–2186.
[12] D. Mendlovich and H. M. Ozaktas, Fractional Fourier transforms and their optical implementation 1, *J. Opt. Soc. Amer. A* **10** (1993) 1875–1881.
[13] M. Moshinsky and C. Quesne, Linear canonical transformations and their unitary representations, *J. Math. Phys.* **12**(8) (1971) 1772–1780.
[14] V. Namias, The fractional order Fourier transforms and its application to quantum mechanics, *J. Inst. Math. Appl.* **25** (1980) 241–265.
[15] H. M. Ozaktas, M. A. Kutay and Z. Zalevsky, *The Fractional Fourier Transform with Applications in Optics and Signal Processing* (Wiley, New York, 2001).
[16] H. M. Ozaktas, M. A. Kutay and D. Mendlovic, Introduction to the fractional Fourier transform and its applications, in *Advances in Imaging Electronics and Physics* (Academic Press, New York, 1999), Chapter 4.
[17] S.-C. Pei and J.-J. Ding, Eigenfunctions of the offset Fourier, fractional Fourier, and linear canonical transforms, *J. Opt. Soc. Amer. Opt. Image Sci. Vis.* **20**(3) (2003) 522–532.
[18] V. B. Shakhmurov and A. I. Zayed, Fractional Wigner distribution and ambiguity functions, *J. Frac. Calc. and Appl. Anal.* **6**(4) (2003) 473–490.
[19] J. Shi, X. Liu, X. Sha and N. Zhang, Sampling and reconstruction of signals in function spaces associated with the linear canonical transform, *IEEE Trans. Signal Process.* **60**(11) (2012) 6041–6047.
[20] A. Stern, Sampling of compact signals in offset linear canonical transform domains, *Signal, Image Video Process.* **1**(4) (2007) 359–367.
[21] R. Tao, Y.-L. Li and Y. Wang, Short-time fractional Fourier transform and its applications, *IEEE Trans. Signal Process.* **58**(5) (2010) 2568–2580.
[22] R. Tao, B.-Z. Li, Y. Wang and G. K. Aggrey, On sampling of band-limited signals associated with the linear canonical transform, *IEEE Trans. Signal Process.* **56**(11) (2008) 5454–5464.
[23] R. Tao, B. Deng, W.-Q. Zhang and Y. Wang, Sampling and sampling rate conversion of band limited signals in the fractional Fourier transform domain, *IEEE Trans. Signal Process.* **56**(1) (2008) 158–171.

[24] D. Wei, Q. Ran, Y. Li, J. Ma and L. Tan, A convolution and product theorem for the linear canonical transform, *IEEE Signal Process. Lett.* **16**(10) (2009) 853–856.

[25] K. B. Wolf, *Integral Transforms in Science and Engineering* (Plenum Press, New York, 1979).

[26] X.-G. Xia, On bandlimited signals with fractional Fourier transform, *IEEE Signal Process. Lett.* **3**(3) (1996) 72–74.

[27] Q. Xiang and K. Qin, Convolution, correlation, and sampling theorems for the offset linear canonical transform, *Signal, Image Video Process.* **8**(3) (2014) 433–442.

[28] A. I. Zayed, A class of fractional integral transforms: A generalization of the fractional Fourier transform, *IEEE Trans. Signal Process.* **50** (2002) 619–627.

[29] A. I. Zayed, A convolution and product theorem for the fractional Fourier transform, *IEEE Signal Process. Lett.* **5**(4) (1998) 101–103.

[30] A. I. Zayed, Fractional Fourier transform of generalized functions, *Integral Transforms Spec. Funct.* **7**(4) (1998) 299–312.

[31] H. Zhao, Q.-W. Ran, J. Ma and L.-Y. Tan, On bandlimited signals associated with linear canonical transform, *IEEE Signal Process. Lett.* **16**(5) (2009) 343–345.

This page intentionally left blank

Chapter 7

A Further Look at Time-and-Band Limiting for Matrix Orthogonal Polynomials

M. Castro[*,¶], **F. A. Grünbaum**[†,||], **I. Pacharoni**[‡,**] **and I. Zurrián**[§,††]

[*] *Departamento de Matemática Aplicada II and IMUS*
Universidad de Sevilla
EPS c/ Virgen de Africa 7, 41011 Sevilla, Spain
[†] *Department of Mathematics, University of California*
Berkeley, CA 94720, USA
[‡] *CIEM-FaMAF, Universidad Nacional de Córdoba*
Córdoba 5000, Argentina
[§] *Facultad de Matemáticas, Pontificia Universidad Católica*
Santiago 7820436, Chile
[¶] *mirta@us.es*
[||] *grunbaum@math.berkeley.edu*
[**] *pacharon@famaf.unc.edu.ar*
[††] *zurrian@famaf.unc.edu.ar*

We extend to a situation involving matrix-valued orthogonal polynomials, a scalar result that originates in work of Shannon and a ground-breaking series of papers by Slepian, Landau and Pollak at Bell Labs in the 1960s. While these papers feature integral and differential operators acting on scalar-valued functions, we are dealing here with integral and differential operators acting on matrix-valued functions.

Keywords: Time-band limiting; matrix-valued orthogonal polynomials.

Mathematics Subject Classification 2010: 33C45, 22E45, 33C47

1. Introduction

Shannon [28] posed the question of how to best use the values of the Fourier transform $\mathcal{F}f(k)$ of f for values of k in the band $[-W, W]$ when $f(t)$ is a time-limited signal.

A detailed account of how this led to the series of papers by three workers at Bell Labs in the 1960s: Slepian, Landau and Pollak, see [20, 21, 31–35] is given in [5, 14, 15] and need not be repeated here. The readers unfamiliar with these Bell Lab papers may want to look at these last references.

With this motivation at hand, we can give an account of what we do in this paper: we start with a (matrix-valued) version of a second-order differential operator. Here Shannon would have started with the (scalar-valued) second derivative.

We then build the analog of the "time-and-band limiting" integral operator, which we will denote by S. We then show that the same "lucky accident" that the workers at Bell Labs found holds here too: we can exhibit a second-order differential operator, denoted by $\widetilde{D}$, such that

$$S\widetilde{D} = \widetilde{D}S.$$

This has, as in the original case of Shannon, very important numerical consequences: it gives a reliable way to compute the eigenvectors of S, something that cannot be done otherwise.

The eigenfunctions of S and $\widetilde{D}$ are the same, but using the differential operator instead of the integral one, we have a manageable numerical problem: while the integral operator has a spectrum with eigenvalues that are extremely close together, the differential one has a very spread out spectrum, resulting in a stable numerical computation.

Previous explorations of the commutativity property above in the matrix-valued case can be seen in [5, 14], dealing with a full matrix and a narrow-banded one, and in [15], dealing with an integral operator. In the present paper, we extend the work started in the previous references.

For more details on computational issues, see [2, 18, 22]. For applications involving (sometimes) vector-valued quantities on the sphere, see [17, 26, 29, 30].

2. Preliminaries

Let $W = W(x)$ be a weight matrix of size R in the open interval (a, b). By this we mean a complex $R \times R$-matrix-valued integrable function W on the interval (a,b) such that $W(x)$ is positive definitive almost everywhere and with finite moments of all orders. Let $Q_w(x), w = 0,1,2,\ldots$, be a sequence of real-valued matrix orthonormal polynomials with respect to the weight $W(x)$. Consider the following two Hilbert spaces: the space $L^2((a,b),W(t)dt)$, denoted here by $L^2(W)$, of all measurable matrix-valued functions $f(x)$, $x \in (a,b)$, satisfying $\int_a^b \operatorname{tr}(f(x)W(x)f^*(x))\,dx < \infty$ and the space $\ell^2(M_R,\mathbb{N}_0)$ of all real-valued $R \times R$ matrix sequences $(C_w)_{w\in\mathbb{N}_0}$ such that $\sum_{w=0}^{\infty} \operatorname{tr}(C_w C_w^*) < \infty$.

The map $\mathcal{F} : \ell^2(M_R,\mathbb{N}_0) \longrightarrow L^2(W)$ given by

$$(A_w)_{w=0}^{\infty} \longmapsto \sum_{w=0}^{\infty} A_w Q_w(x)$$

is an isometry. If the polynomials are dense in $L^2(W)$, this map is unitary with the inverse $\mathcal{F}^{-1} : L^2(W) \longrightarrow \ell^2(M_R,\mathbb{N}_0)$ given by

$$f \longmapsto A_w = \int_a^b f(x)\,W(x)\,Q_w^*(x)dx.$$

We denote our map by $\mathcal{F}$ to remind ourselves of the usual Fourier transform. Here $\mathbb{N}_0$ takes up the role of "physical space" and the interval (a,b) the role of "frequency space". This is, clearly, a noncommutative extension of the problem raised by Shannon since he was concerned with scalar-valued functions and we are dealing with matrix-valued ones.

The *time limiting operator*, at level N, acts on $\ell^2(M_R,\mathbb{N}_0)$ by simply setting equal to zero all the components with index larger than N. We denote it by χ_N. The *band-limiting operator*, at level Ω, acts on $L^2(W)$ by multiplication by the characteristic function of the interval (a,Ω), $\Omega \le b$. This operator will be denoted by χ_Ω. One could consider restricting the band to an arbitrary subinterval (a_1,b_1). However, the algebraic properties exhibited here, see Section 4 and beyond, hold only with this restriction. A similar situation arises in the classical case going all the way back to Shannon.

Consider the problem of determining a function f, from the following data: f has support on the finite set $\{0,\ldots,N\}$ and its Fourier transform $\mathcal{F}f$ is known on the compact set $[a,\Omega]$. This can be formalized as follows:

$$\chi_\Omega \mathcal{F} f = g = \text{known}, \quad \chi_N f = f.$$

We can combine the two equations into

$$Ef = \chi_\Omega \mathcal{F} \chi_N f = g.$$

To analyze this problem, we need to compute the singular vectors (and values) of the operator $E : \ell^2(M_R, \mathbb{N}_0) \longrightarrow L^2(W)$. These are given by the eigenvectors of the operators

$$E^*E = \chi_N \mathcal{F}^{-1} \chi_\Omega \mathcal{F} \chi_N \quad \text{and} \quad S_2 = EE^* = \chi_\Omega \mathcal{F} \chi_N \mathcal{F}^{-1} \chi_\Omega.$$

The operator E^*E acting in $\ell^2(M_R, \mathbb{N}_0)$ is just a finite-dimensional block-matrix M, and each block is given by

$$(M)_{m,n} = (E^*E)_{m,n} = \int_a^\Omega Q_m(x) W(x) Q^*{}_n(x) dx, \quad 0 \le m, n \le N.$$

The second operator $S = EE^*$ acts in $L^2((a, \Omega), W(t)dt)$ by means of the integral kernel

$$k(x,y) = \sum_{n=0}^{N} Q_n^*(x) Q_n(y).$$

Consider now the problem of finding the eigenfunctions of E^*E and EE^*. For arbitrary N and Ω there is no hope of doing this analytically, and one has to resort to numerical methods and this is not an easy problem. Of all the strategies one can dream of for solving this problem, none sounds so appealing as that of finding an operator with simple spectrum which would have the same eigenfunctions as the original operators. This is exactly what Slepian, Landau and Pollak did in the scalar case, when dealing with the real line and the actual Fourier transform. They discovered (the analog of) the following properties:

- For each N, Ω, there exists a symmetric tridiagonal matrix L, with simple spectrum, commuting with M.
- For each N, Ω, there exists a self-adjoint differential operator D, with simple spectrum, commuting with the integral operator $S = EE^*$.

To this day nobody has a simple explanation for these miracles, and this paper displays more instances where this holds. Indeed, there has been a systematic effort to see if the "bispectral property" first considered in [7], guarantees the commutativity of these two operators, a global and a local one. A few papers where this question has been taken up, include [8–12, 16, 24, 25].

We recall that while the papers [5, 14] deal with the full matrix E^*E alluded to above, here, as well as in [15], we deal with the integral operator $S = EE^*$ mentioned above.

3. An Example of Matrix-Valued Orthogonal Polynomials

For $\alpha, \beta > -1$, the scalar Jacobi weight is given by

$$w_{\alpha,\beta}(x) = (1-x)^{\alpha}(1+x)^{\beta}, \tag{3.1}$$

supported in the interval $[-1,1]$. In this paper, we consider a Jacobi-type weight matrix of dimension two (see also [13])

$$W(x) = W_{(\alpha,\beta)} = \frac{1}{2}\begin{pmatrix} w_{\alpha,\beta}+w_{\beta,\alpha} & -w_{\alpha,\beta}+w_{\beta,\alpha} \\ -w_{\alpha,\beta}+w_{\beta,\alpha} & w_{\alpha,\beta}+w_{\beta,\alpha} \end{pmatrix}, \quad x \in [-1,1]. \tag{3.2}$$

Let us observe that a particular case of these weight matrices has been studied in [5, 14, 15]. In fact, the weight matrix considered in [14, 15, 23] is

$$W_{p,m}(x) = (1-x^2)^{\frac{m}{2}-1}\begin{pmatrix} px^2+m-p & -mx \\ -mx & (m-p)x^2+p \end{pmatrix}, \quad x \in [-1,1]. \tag{3.3}$$

Taking $\alpha = p-1$, $\beta = p+1$ in (3.2), we obtain a multiple of the previous weight for the special case $m = 2p$.

On the other hand, taking $\beta = \alpha - 1$ in (3.2) we obtain a linear translation $x \longrightarrow 1+x$ of the weight considered in [5],

$$W_{\lambda}(x) = [x(2-x)]^{\lambda-3/2}\begin{pmatrix} 1 & x-1 \\ x-1 & 1 \end{pmatrix}, \quad x \in [0,2], \tag{3.4}$$

with $\lambda = \alpha + \frac{1}{2}$.

A sequence of matrix orthogonal polynomials with respect to the matrix-valued inner product going with (3.2) is given by

$$P_n(x) = P_n^{(\alpha,\beta)}(x) = \frac{1}{2}\begin{pmatrix} p_n^{(\alpha,\beta)}(x)+p_n^{(\beta,\alpha)}(x) & -p_n^{(\alpha,\beta)}(x)+p_n^{(\beta,\alpha)}(x) \\ -p_n^{(\alpha,\beta)}(x)+p_n^{(\beta,\alpha)}(x) & p_n^{(\alpha,\beta)}(x)+p_n^{(\beta,\alpha)}(x) \end{pmatrix}, \tag{3.5}$$

where $p_n^{(\alpha,\beta)}$ are the classical Jacobi polynomials

$$p_n^{(\alpha,\beta)} = \frac{(\alpha+1)_n}{n!}\,{}_2F_1\left(\begin{matrix} -n,\, n+\alpha+\beta \\ \alpha+1 \end{matrix}; \frac{1-x}{2}\right),$$

which are orthogonal with respect to the weight $w_{\alpha,\beta}$ (see, for instance, [36, Chapter VI]).

The norm of the polynomials $P_n(x)$ is given by

$$\langle P_n, P_n \rangle = ||P_n(x)||^2 = \int_{-1}^{1} P_n(x) W(x) P_n(x)^* \, dx = h_n \mathrm{Id},$$

where Id is the matrix identity of size 2×2 and

$$h_n = \frac{2^{\alpha+\beta+1}\,\Gamma(\alpha+n+1)\,\Gamma(\beta+n+1)}{(\alpha+\beta+2n+1)\,n!\,\Gamma(\alpha+\beta+n+1)}. \tag{3.6}$$

The sequence of orthogonal polynomials $(P_n)_n$ satisfies the three term recurrence relation

$$xP_n(x) = A_n P_{n+1}(x) + B_n P_n(x) + C_n P_{n-1}(x), \quad n \geq 1,$$

where

$$A_n = \frac{2(n+1)(n+\alpha+\beta+1)}{(2n+\alpha+\beta+1)(2n+\alpha+\beta+2)} \mathrm{Id},$$

$$B_n = \frac{\alpha^2 - \beta^2}{(2n+\alpha+\beta)(2n+\alpha+\beta+2)} T,$$

$$C_n = \frac{2(\alpha+n)(\beta+n)}{(2n+\alpha+\beta)(2n+\alpha+\beta+1)} \mathrm{Id},$$

with

$$T = \begin{pmatrix} 0 & 1 \\ 1 & 0 \end{pmatrix}. \tag{3.7}$$

The sequence of matrix orthogonal polynomials $(P_n)_n$ satisfies the following differentiation formula:

$$(1-x^2)\frac{d}{dx}P_n(x) = -nxP_n(x) - \frac{n(\alpha-\beta)}{\alpha+\beta+2n} T\, P_n(x) + \gamma_{n-1} P_{n-1}(x), \tag{3.8}$$

where

$$\gamma_{n-1} = \frac{2(n+\alpha)(n+\beta)}{\alpha+\beta+2n}. \tag{3.9}$$

We also have the following Christoffel–Darboux formula for the sequence of orthogonal polynomials $(P_n(x))_n$, introduced for a general sequence of matrix orthogonal polynomials in [6].

$$\frac{\kappa_{n-1}}{\kappa_n\, h_{n-1}}\left(P_{n-1}^*(y)P_n(x)-P_n^*(y)P_{n-1}(x)\right)=(x-y)\sum_{k=0}^{n-1}\frac{P_k^*(y)P_k(x)}{h_k}, \tag{3.10}$$

with

$$\kappa_n=\frac{\Gamma(\alpha+\beta+2n+1)}{2^n\, n!\,\Gamma(\alpha+\beta+n+1)}.$$

Observe that $\kappa_n\mathrm{Id}$ is the leading coefficient of the matrix polynomial $P_n(x)$ and we also have

$$\frac{\kappa_{n-1}}{\kappa_n}=\frac{2n(n+\alpha+\beta)}{(2n+\alpha+\beta)(2n+\alpha+\beta-1)}.$$

The matrix polynomial P_n, for each $n\geq 0$, is an eigenfunction of the second-order differential operator

$$D=\frac{d^2}{dx^2}(1-x^2)+\frac{d}{dx}\left(-x(\alpha+\beta+2)+(\alpha-\beta)T\right), \tag{3.11}$$

with scalar eigenvalues $\Lambda_n=-n(n+\alpha+\beta+1)$. We have that the differential operator D can be factorized as

$$D=\frac{d}{dx}\left(\frac{d}{dx}(1-x^2)W(x)\right)W(x)^{-1},$$

and therefore the sequence of matrix orthogonal polynomials $(P_n(x))_n$ satisfies

$$\frac{d}{dx}\left(\frac{dP_n}{dx}(x)\,(1-x^2)W(x)\right)W(x)^{-1}=\Lambda_n P_n(x). \tag{3.12}$$

4. Time and Band Limiting: Integral and Differential Operators

Given a sequence of matrix orthonormal polynomials $\{Q_w\}_{w\geq 0}$ with respect to the weight W, we fix a natural number N and $\Omega\in(-1,1)$ and we consider the integral kernel

$$k(x,y)=\sum_{w=0}^{N}Q_w^*(x)Q_w(y). \tag{4.1}$$

It defines the integral operator S acting on $L^2((-1,\Omega),W)$ "from the right-hand side":

$$(fS)(x)=\int_{-1}^{\Omega}f(y)W(y)\big(k(x,y)\big)^*dy. \tag{4.2}$$

The restriction to the interval $[1, \Omega]$ implements "band-limiting" while the restriction to the range $0, 1, \ldots, N$ takes care of "time-limiting". In the language of [12], where the authors were dealing with scalar-valued functions defined on spheres, the first restriction gives a "spherical cap" while the second one amounts to truncating the expansion in spherical harmonics.

We search for a self-adjoint differential operator $\widetilde{D}$, defined in $[-1, \Omega]$, commuting with the integral operator S.

The main result of this section is the following theorem.

Theorem 4.1. *Let* $p(x) = (1-x^2)W(x)$. *The symmetric second-order differential operator*

$$\widetilde{D} = \frac{d}{dx}\left((x-\Omega)\frac{d}{dx}p(x)\right)W(x)^{-1} + x\,A, \tag{4.3}$$

with $A = N(N+\alpha+\beta+2)\mathrm{Id}$, *commutes with the integral operator* S *given in* (4.2).

Explicitly, we have

$$\widetilde{D} = \frac{d^2}{dx^2}\widetilde{E}_2(x) + \frac{d}{dx}\widetilde{E}_1(x) + \widetilde{E}_0(x), \tag{4.4}$$

where the coefficients $\widetilde{E}_j$, $j = 0, 1, 2$, are given by

$$\begin{aligned}
\widetilde{E}_2 &= (x-\Omega)(1-x^2)\mathrm{Id},\\
\widetilde{E}_1 &= (-(3+\alpha+\beta)x^2 + x\,\Omega(2+\alpha+\beta)+1)\mathrm{Id} + (\alpha-\beta)(x-\Omega)T,\\
\widetilde{E}_0 &= x\,N(N+\alpha+\beta+2)\mathrm{Id},
\end{aligned}$$

and T is the permutation matrix given in (3.7).

Let us observe that the differential operator $\widetilde{D}$ is somehow related to the differential operator D, given in (3.11) and (3.12). Explicitly, we have

$$\widetilde{D} = (x-\Omega)D + (1-x^2)\frac{d}{dx} + xA.$$

We first show that the operator in (4.3) is indeed symmetric with respect to the (matrix valued) inner product defined by (3.2).

Proposition 4.2. *The differential operator* $\widetilde{D}$ *is a symmetric operator with respect to*

$$\langle f, g\rangle_\Omega = \int_{-1}^{\Omega} f(x)W(x)g^*(x)\,dx. \tag{4.5}$$

Proof. For an appropriate dense set of functions f, g, we have

$$\langle f\widetilde{D}, g\rangle_\Omega = \int_{-1}^{\Omega} \frac{d}{dx}\left((x-\Omega)\frac{df}{dx}(x)\,p(x)\right) g^*(x)\,dx$$

$$+\int_{-1}^{\Omega} xf(x)AW(x)g^*(x)\,dx$$

$$= -\int_{-1}^{\Omega} \frac{df}{dx}(x)\,(x-\Omega)p(x)\frac{dg^*}{dx}(x)\,dx$$

$$+(x-\Omega)\frac{df}{dx}(x)\,p(x)g^*(x)\Big|_{-1}^{\Omega}$$

$$+\int_{-1}^{\Omega} xf(x)AW(x)g^*(x)\,dx$$

$$= \int_{-1}^{\Omega} f(x)\,\frac{d}{dx}\left((x-\Omega)p(x)\frac{dg^*}{dx}(x)\right)\,dx$$

$$-f(x)(x-\Omega)p(x)\frac{dg^*}{dx}(x)\Big|_{-1}^{\Omega}$$

$$+(x-\Omega)\frac{df}{dx}(x)\,p(x)\frac{dg^*}{dx}(x)\Big|_{-1}^{\Omega} + \int_{-1}^{\Omega} xf(x)AW(x)g^*(x)\,dx.$$

Since the factor $(x-\Omega)p(x)$ vanishes at $x=-1$ and $x=\Omega$, we get

$$\langle f\widetilde{D}, g\rangle_\Omega = \int_{-1}^{\Omega} f(x)W(x)\left(\frac{d}{dx}\left[(x-\Omega)\frac{dg}{dx}(x)p(x)\right]W^{-1}\right)^*(x)\,dx$$

$$+\int_{-1}^{\Omega} xf(x)W(x)\,(g(x)A)^*\,dx$$

Therefore

$$\langle f\widetilde{D}, g\rangle_\Omega = \langle f, g\widetilde{D}\rangle_\Omega$$

completing the proof of the proposition. □

Proof of Theorem 4.1. From [15, Proposition 3.1] we have that a symmetric differential operator $\widetilde{D}$ commutes with an integral operator S with kernel k if and only if

$$(k(x,y)^*)\,\widetilde{D}_x = (k(x,y)\widetilde{D}_y)^*.$$

(Here we use D_x to stress that D acts on the variable x.)

Let D be the differential operator introduced in (3.11) and let $(Q_n)_n$ be the orthonormal sequence of matrix-valued polynomials given by

$$Q_n(x) = h_n^{-1/2}\,P_n(x),$$

where $h_n = \langle P_n, P_n\rangle$ is explicitly given in (3.6).

These polynomials are eigenfunctions of the differential operator D, i.e., $Q_nD = \Lambda_nQ_n$, with $\Lambda_n = -n(n+\alpha+\beta+1)$ and they satisfy

$$(1-x^2)\frac{d}{dx}Q_n(x) = -nxQ_n(x) - \frac{n(\alpha-\beta)}{\alpha+\beta+2n}T\,Q_n(x) + \tilde{\gamma}_{n-1}Q_{n-1}(x),$$

with

$$\tilde{\gamma}_{n-1} = \frac{2(n+\alpha)(n+\beta)}{\alpha+\beta+2n}\frac{h_{n-1}^{1/2}}{h_n^{1/2}}.$$

Then

$$\begin{aligned}
(k(x,y)\widetilde{D}_y)^* &= (y-\Omega)\sum_{n=0}^{N} Q_n^*(y)\Lambda_nQ_n(x) + (1-y^2)\sum_{n=0}^{N}\frac{d}{dy}Q_n^*(y)Q_n(x)\\
&\quad + y\sum_{n=0}^{N} AQ_n^*(y)Q_n(x)\\
&= (y-\Omega)\sum_{n=0}^{N}\Lambda_nQ_n^*(y)Q_n(x) + y\sum_{n=0}^{N} AQ_n^*(y)Q_n(x)\\
&\quad + \sum_{n=0}^{N}\left(-ynQ_n^*(y)Q_n(x) - \frac{n(\alpha-\beta)}{(\alpha+\beta+2n)}Q_n^*(y)TQ_n(x)\right.\\
&\quad \left. + \gamma_{n-1}Q_{n-1}^*(y)Q_n(x)\right)
\end{aligned}$$

and similarly

$$(k(x,y)^*)\,\widetilde{D}_x = (x-\Omega)\sum_{n=0}^{N}\Lambda_n Q_n^*(y)Q_n(x) + x\sum_{n=0}^{N} AQ_n^*(y)Q_n(x)$$

$$+\sum_{n=0}^{N}\Bigg(-xnQ_n^*(y)Q_n(x) - \frac{n(\alpha-\beta)}{(\alpha+\beta+2n)}Q_n^*(y)TQ_n(x)$$

$$+\tilde{\gamma}_{n-1}Q_n^*(y)Q_{n-1}(x)\Bigg).$$

Thus

$$(k^*\widetilde{D}_x) - (k\widetilde{D}_y)^* = (x-y)\sum_{n=0}^{N}((\Lambda_n - n)\mathrm{Id} + A)Q_n^*(y)Q_n(x)$$

$$+\sum_{n=0}^{N}\tilde{\gamma}_{n-1}(Q_n^*(y)Q_{n-1}(x) - Q_{n-1}^*(y)Q_n(x)). \quad (4.6)$$

From the Cristoffel–Darboux formula, given in (3.10), we obtain that

$$\tilde{\gamma}_{n-1}(Q_{n-1}^*(y)Q_n(x) - Q_n^*(y)Q_{n-1}(x))$$

$$= (x-y)\,\tilde{\gamma}_{n-1}\frac{\kappa_n\,h_{n-1}^{1/2}}{\kappa_{n-1}h_n^{1/2}}\sum_{k=0}^{n-1}Q_k^*(y)Q_k(x).$$

It is easy to verify that $\tilde{\gamma}_{n-1}\frac{\kappa_n\,h_{n-1}^{1/2}}{\kappa_{n-1}h_n^{1/2}} = \alpha+\beta+2n+1$. Therefore, by exchanging the order of summation we get

$$\sum_{n=0}^{N}\tilde{\gamma}_{n-1}(Q_{n-1}^*(y)Q_n(x) - Q_n^*(y)Q_{n-1}(x))$$

$$= (x-y)\sum_{n=0}^{N}(\alpha+\beta+2n+1)\sum_{k=0}^{n-1}Q_k^*(y)Q_k(x)$$

$$= (x-y)\sum_{n=0}^{N-1}\sum_{j=n+1}^{N}(\alpha+\beta+2j+1)Q_n^*(y)Q_n(x)$$

$$= (x-y)\sum_{n=0}^{N-1}\big(N(N+\alpha+\beta+2) - n(n+\alpha+\beta+2)\big)Q_n^*(y)Q_n(x).$$

Now from (4.6), and using that $(\Lambda_n - n)\mathrm{Id} + A = -n(n+\alpha+\beta+2) + N(N+\alpha+\beta+2)$ we get

$$(k(x,y)\widetilde{D}_y)^* = (k(x,y)^*)\,\widetilde{D}_x.$$

Hence the operators $\widetilde{D}$ and S commute. □

Remark 4.3. It is worth noticing that if one takes the new operator $\widetilde{D}^{(1)} = \widetilde{D}T$, one obtains an operator commuting with the integral operator S in (4.2), linearly independent with the operator $\widetilde{D}$. The space of the differential operators of order two commuting with S is generated by T, $\widetilde{D}T$ and $\widetilde{D}$.

5. A Chebyshev-Type Example

As a particular example, if we put $\alpha = \frac{1}{2}$, $\beta = -\frac{1}{2}$ in (3.2), we have the Chebyshev-type weight

$$W_{(\frac{1}{2},-\frac{1}{2})}(x) = \frac{1}{\sqrt{1-x^2}}\begin{pmatrix} 1 & x \\ x & 1 \end{pmatrix}, \quad x \in [-1,1], \tag{5.1}$$

which was introduced in [1, p. 586] for a different purpose.

The monic family of polynomials orthogonal with respect to this weight matrix is given explicitly in terms of the Chebyshev polynomials of the second kind $U_n(x)$,

$$\widetilde{P}_n(x) = \frac{1}{2^n}\begin{pmatrix} U_n(x) & -U_{n-1}(x) \\ -U_{n-1}(x) & U_n(x) \end{pmatrix},$$

and it satisfies a first-order differential equation as pointed out in [3, 4].

Moreover, the polynomials $P_n(x) = P_n^{(\alpha,\alpha-1)}(x)$ in (3.5) satisfy, for any $\alpha > 0$, the first-order differential equation

$$P_n'(x)\begin{pmatrix} -x & 1 \\ -1 & -x \end{pmatrix} + P_n(x)\begin{pmatrix} -2\alpha & 0 \\ 0 & 0 \end{pmatrix} = \begin{pmatrix} -2\alpha - n & 0 \\ 0 & n \end{pmatrix} P_n(x),$$

as shown in [5].

One considers here the integral operator S in (4.2) defined by the integral kernel

$$k(x,y) = \sum_{n=0}^{N} 4^n \widetilde{P}_n(x)^* \widetilde{P}_n(y).$$

Particularly, the norm of the polynomials $\widetilde{P}_n$ is given by $||\widetilde{P}_n|| = \frac{\sqrt{\pi}}{2^n}$.

Hence, for this particular example, the commuting operator $\widetilde{D}$ is given by

$$\widetilde{D} = \frac{d^2}{dx^2}(1-x^2)(x-\Omega) + \frac{d}{dx}((-3x^2+2\Omega\, x+1)\mathrm{Id} + (x-\Omega)T) + N(N+2)x,$$

where T is the permutation matrix in (3.7).

6. Conclusion and Outlook

The main result derived in the previous sections is the existence of an explicit differential operator $\widetilde{D}$, which, as we proved, commutes with S.

If one compares this result with the one in the celebrated series of papers by Slepian, Landau and Pollak one may say that we are at the stage of their first paper. What is needed now is an argument to conclude that the eigenfunctions of $\widetilde{D}$ will automatically be eigenfunctions of the integral operator S.

In the series of papers mentioned above, the simplicity of the spectrum of $\widetilde{D}$ follows from classical Sturm–Liouville theory and this guarantees that they have found a good way to compute the eigenvectors of S. In our situation, things could eventually be reduced to that case, but in principle $\widetilde{D}$, as well as S, have "matrix-valued eigenvalues", and the appropriate notion of "simple spectrum" requires careful handling.

To be quite explicit: there are two technical points that we are not addressing in this paper. The first one is the issue of "simple spectrum" in the matrix-valued context. A good place to look for this sort of question is [27]. A second issue is that of making precise the correct self-adjoint extension (i.e., boundary conditions) for our second-order differential operator $\widetilde{D}$. This issue has been implicit starting with the first papers of Slepian, Landau and Pollak, and has been considered quite explicitly in the very recent paper [19].

Acknowledgments

The work of the first author was partially supported by MTM2015-6588-C4-1-P (Ministerio de Economía y Competitividad), FQM-262, FQM-7276 (Junta de Andalucía), Feder Funds (European Union) and the program Campus de Excelencia Internacional of the Ministerio de Educación, Cultura y Deporte. The work of the second author was partially supported by AFOSR, through FA9550-16-1-0175. The third and fourth authors were partially supported by SeCyT-UNC and by CONICET grant PIP 112-200801-01533. The work of the fourth author was also supported by FONDECYT 3160646.

References

[1] Ju. M. Berezanskii, *Expansions in Eigenfunctions of Selfadjoint Operators*, Translations of Mathematical Monographs, Vol. 17 (American Mathematical Society, Providence, RI, 1968).

[2] A. Bonami and A. Karoui, Uniform approximation and explicit estimates for the prolate spheroidal wave functions, *Constructive Approx.* **43**(1) (2016) 15–45; See also arXiv:1405.3676.

[3] M. Castro and F. A. Grünbaum, Orthogonal matrix polynomials satisfying first order differential equations: A collection of instructive examples, *J. Nonlinear Math. Phys.* **12**(Suppl. 2) (2005) 63–76.

[4] M. Castro and F. A. Grünbaum, The algebra of differential operators associated to a family of matrix valued orthogonal polynomials: five instructive examples, *Internat. Math. Res. Notices* **2006** (2006) AA.ID 47602, 1–33.

[5] M. Castro and F. A. Grünbaum, The Darboux process and time-and-band limiting for matrix orthogonal polynomials. *Linear Algebra Appl.* **487** (2015) 328–341.

[6] A. J. Durán, Markov's theorem for orthogonal matrix polynomials, *Canad. J. Math.* **48** (1996) 1180–1195.

[7] J. J. Duistermaat and F. A. Grünbaum, Differential equations in the spectral parameter, *Comm. Math. Phys.* **103** (1986) 177–240.

[8] F. A. Grünbaum, A new property of reproducing kernels of classical orthogonal polynomials, *J. Math. Anal. Appl.* **95** (1983) 491–500.

[9] F. A. Grünbaum, Band-time-band limiting integral operators and commuting differential operators, *Algebra Anal.* **8** (1996) 122–126.

[10] F. A. Grünbaum, Some explorations into the mystery of band and time limiting, *Adv. Appl. Math.* **13** (1992) 328–349.

[11] F. A. Grünbaum, The bispectral problem: an overview, in *Special Functions*, eds. J. Bustoz *et al.* (Kluwer Academic Publishers, 2000), pp. 129–140.

[12] F. A. Grünbaum, L. Longhi and M. Perlstadt, Differential operators commuting with finite convolution integral operators: some nonabelian examples, *SIAM J. Appl. Math.* **42** (1982) 941–955.

[13] F. A. Grünbaum, I. Pacharoni and J. A. Tirao, Matrix valued orthogonal polynomials of the Jacobi type, *Indag. Math.* **14**(3,4) (2003) 353–366.

[14] F. A. Grünbaum, I. Pacharoni and I. Zurrián, Time and band limiting for matrix valued functions: An example SIGMA **11** (2015) 044, 14 pp.

[15] F. A. Grünbaum, I. Pacharoni and I. Zurrián, Time and band limiting for matrix valued functions: An integral and a commuting differential operator, *Inverse Problems* **33**(2) (2017) 025005.

[16] F. A. Grünbaum and M. Yakimov, The prolate spheroidal phenomenon as a consequence of bispectrality, in *Superintegrability in Classical and Quantum Systems*, CRM Proceedings & Lecture Notes, Vol. 37 (American Mathematical Society, Providence, RI, 2004), pp. 301–312.

[17] K. Jahn and N. Bokor, Revisiting the concentration problem of vector fields within a spherical cap: A commuting differential operator solution, *J. Fourier Anal. Appl.* **20** (2014) 421–451.

[18] P. Jamming, A. Karoui and S. Spektor, The approximation of almost time and band limited functions by their expansion in some orthogonal polynomial bases, preprint (2015); arXiv:1501.03655.

[19] V. Katsnelson, Self-adjoint boundary conditions for the prolate spheroidal differential operator, preprint (2016); arXiv:1603.07542.

[20] H. J. Landau and H. O. Pollak, Prolate spheroidal Wave Functions: Fourier analysis and uncertainty, II, *Bell Sys. Tech. J.* **40**(1) (1961) 65–84.

[21] H. J. Landau and H. O. Pollak, Prolate spheroical wave Functions: Fourier analysis and uncertainty, III, *Bell Sys. Tech. J.* **41**(4) (1962) 1295–1336.

[22] A. Osipov, V. Rokhlin and H. Xiao, *Prolate Spheroidal Wave Functions of Order Zero: Mathematical Tools for Bandlimited Approximation* (Springer, 2014).

[23] I. Pacharoni and I. Zurrián, Matrix ultraspherical polynomials: the 2×2 fundamental cases, *Constructive Approx.* **43**(2) (2016) 253–271; arXiv:1309.6902 [math.RT].

[24] M. Perlstadt, Chopped orthogonal polynomial expansions—some discrete cases, *SIAM J. Discrete Math.* **4** (1983) 94–100.

[25] M. Perlstadt, A property of orthogonal polynomial families with polynomial duals, *SIAM J. Math. Anal.* **15** (1984) 1043–1054.

[26] A. Plattner and F. Simons, Spatiospectral concentration of vector fields on a sphere, *Appl. Comput. Harmon. Anal.* **36**(1) (2014) 1–22.

[27] F. Rofe-Beketov and A. Kholkin, *Spectral Analysis of Differential Operators*, World Scientific Monograph Series in Mathematics, Vol. 7 (World Scientific, 2005).

[28] C. Shannon, A mathematical theory of communication, *Bell. syst. Tech. J.* **27** (1948), 379–423 (July) and 623–656 (October).

[29] F. J. Simons and F. A. Dahlen, Spherical Slepian functions on the polar gap in geodesy, *Geophys. J. Int.* **166** (2006) 1039–1061.

[30] F. J. Simons, F. A. Dahlen and M. A. Wieczorek, Spatiospectral concentration on a sphere, *SIAM Rev.* **48**(3) (2006) 504–536.

[31] D. Slepian, Prolate spheroidal wave functions: Fourier analysis and uncertainty, IV, *Bell Sys. Tech. J.* **43**(6) (1964) 3009–3058.

[32] D. Slepian, On bandwidth, *Proc. IEEE* **64**(3) (1976) 292–300.

[33] D. Slepian, Prolate spheroidal wave functions: Fourier analysis and uncertainty, V, *Bell Sys. Tech. J.* **57**(5) (1978) 1371–1430.

[34] D. Slepian, Some comments on Fourier analysis, uncertainty and modeling, *SIAM Rev.* **25**(3) (1983) 379–393.

[35] D. Slepian and H. O. Pollak, Prolate spheroidal wave functions: Fourier analysis and uncertainty, I, *Bell Sys. Tech. J.* **40**(1) (1961) 43–64.

[36] G. Szegö, *Orthogonal Polynomials*, Colloquium Publications, Vol. 23 (American Mathematical Society, Providence, RI, 1975).

This page intentionally left blank

Chapter 8

The Orthogonality of Al-Salam–Carlitz Polynomials for Complex Parameters*

Howard S. Cohl[†,¶], **Roberto S. Costas-Santos**[‡,‖] **and Wenqing Xu**[§,**]

[†]*Applied and Computational Mathematics Division*
National Institute of Standards and Technology
Mission Viejo, CA 92694, USA
[‡]*Departamento de Física y Matemáticas, Facultad de Ciencias*
Universidad de Alcalá, 28871 Alcalá de Henares, Madrid, Spain
[§]*Department of Mathematics and Statistics*
California Institute of Technology, CA 91125, USA
[¶]*howard.cohl@nist.gov*
[‖]*rscosa@gmail.com*
[**]*williamxuxu@yahoo.com*

In this chapter, we study the orthogonality conditions satisfied by Al-Salam–Carlitz polynomials $U_n^{(a)}(x;q)$ when the parameters a and q are not necessarily real nor "classical", i.e., the linear functional $\mathbf{u}$ with respect to such a polynomial sequence is quasi-definite and not positive definite. We establish orthogonality on a simple contour in the complex plane which depends on the parameters. In all cases we show that the orthogonality conditions characterize the Al-Salam–Carlitz polynomials $U_n^{(a)}(x;q)$ of degree n up to a constant factor. We

*Portions of this article were prepared by a U.S. Government employee as part of his official duties. Official contribution of the National Institute of Standards and Technology; not subject to copyright in the United States.

also obtain a generalization of the unique generating function for these polynomials.

Keywords: q-orthogonal polynomials; q-difference operator; q-integral representation; discrete measure.

Mathematics Subject Classification 2010: 33C45, 42C05

1. Introduction

The Al-Salam–Carlitz polynomials $U_n^{(a)}(x;q)$ were introduced by Al-Salam and Carlitz in [1] as follows:

$$U_n^{(a)}(x;q) := (-a)^n q^{\binom{n}{2}} \sum_{k=0}^{n} \frac{(q^{-n};q)_k (x^{-1};q)_k}{(q;q)_k} \frac{q^k x^k}{a^k}. \tag{1.1}$$

In fact, these polynomials have a Rodrigues-type formula [2, (3.24.10)]

$$U_n^{(a)}(x;q) = \frac{a^n q^{\binom{n}{2}} (1-q)^n}{q^n w(x;a;q)} \mathscr{D}_{q^{-1}}^n (w(x;a;q)),$$

where

$$w(x;a;q) := (qx;q)_\infty (qx/a;q)_\infty,$$

the q-Pochhammer symbol (q-shifted factorial) is defined as

$$(z;q)_0 := 1, \quad (z;q)_n := \prod_{k=0}^{n-1} (1 - zq^k),$$

$$(z;q)_\infty := \prod_{k=0}^{\infty} (1 - zq^k), \quad |z| < 1,$$

and the q-derivative operator is defined by

$$\mathscr{D}_q f(z) := \begin{cases} \dfrac{f(qz) - f(z)}{(q-1)z} & \text{if } q \neq 1 \text{ and } z \neq 0, \\ f'(z) & \text{if } q = 1 \text{ or } z = 0. \end{cases}$$

Remark 1.1. Observe that by the definition of the q-derivative

$$\mathscr{D}_{q^{-1}} f(z) = \mathscr{D}_q f(qz), \quad \text{and} \quad \mathscr{D}_{q^{-1}}^n f(z) := \mathscr{D}_{q^{-1}}^{n-1}(\mathscr{D}_{q^{-1}} f(z)),\ n = 2, 3, \ldots.$$

The expression (1.1) shows us that $U_n^{(a)}(x;q)$ is an analytic function for any complex-valued parameters a and q, and thus can be considered for general $a, q \in \mathbb{C} \setminus \{0\}$.

The classical Al-Salam–Carlitz polynomials correspond to parameters $a < 0$ and $0 < q < 1$. For these parameters, the Al-Salam–Carlitz polynomials are orthogonal on $[a, 1]$ with respect to the weight function w. More specifically, for $a < 0$ and $0 < q < 1$ [2, (14.24.2)],

$$\int_a^1 U_n^{(a)}(x;q) U_m^{(a)}(x;q)(qx, qx/a; q)_\infty d_q x = d_n^2 \delta_{n,m},$$

where

$$d_n^2 := (-a)^n (1-q)(q;q)_n (q;q)_\infty (a;q)_\infty (q/a;q)_\infty q^{\binom{n}{2}},$$

and the q-Jackson integral [2, (1.15.7)] is defined as

$$\int_a^b f(x) d_q x := \int_0^b f(x) d_q x - \int_0^a f(x) d_q x,$$

where

$$\int_0^a f(x) d_q x := a(1-q) \sum_{n=0}^{\infty} f(aq^n) q^n.$$

Taking into account the previous orthogonality relation, it is a direct result that if a and q are classical, i.e., a, $q \in \mathbb{R}$, with $a \neq 1$, $0 < q < 1$, all the zeros of $U_n^{(a)}(x;q)$ are simple and belong to the interval $[a, 1]$. This is no longer valid for general a and q complex. In this paper, we show that for general a, q complex numbers, but excluding some special cases, the Al-Salam–Carlitz polynomials $U_n^{(a)}(x;q)$ may still be characterized by orthogonality relations. The case $a < 0$ and $0 < q < 1$ or $0 < aq < 1$ and $q > 1$ is classical, i.e., the linear functional $\mathbf{u}$ with respect to such a polynomial sequence is orthogonal, which is positive definite and in such a case there exists a weight function $\omega(x)$ so that

$$\langle \mathbf{u}, p \rangle = \int_a^1 p(x)\, \omega(x)\, dx, \quad p \in \mathbb{P}[x].$$

Note that this is the key for the study of many properties of Al-Salam–Carlitz polynomials I and II. Thus, our goal is to establish orthogonality conditions for most of the remaining cases for which the linear form $\mathbf{u}$ is quasi-definite, i.e., for all $n, m \in \mathbb{N}_0$

$$\langle \mathbf{u}, p_n p_m \rangle = k_n \delta_{n,m}, \quad k_n \neq 0.$$

We believe that these new orthogonality conditions can be useful in the study of the zeros of Al-Salam–Carlitz polynomials. For general

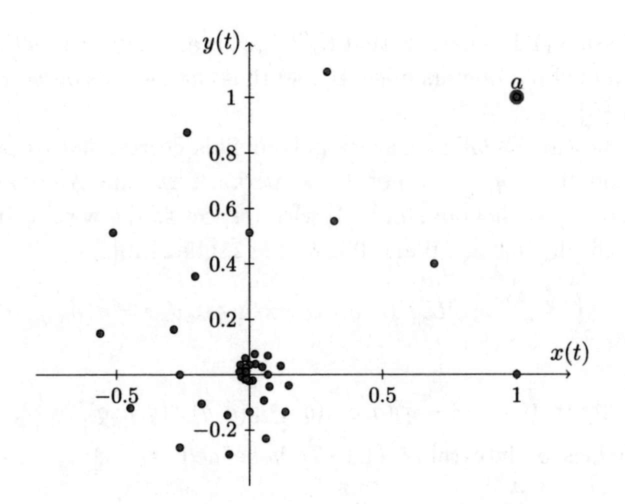

Fig. 1. Zeros of $U_{30}^{(1+i)}(x; \frac{4}{5}\exp(\pi i/6))$.

$a, q \in \mathbb{C} \setminus \{0\}$, the zeros are not confined to a real interval, but they distribute themselves in the complex plane as we can see in Fig. 1. Throughout this paper denote $p := q^{-1}$.

2. Orthogonality in the Complex Plane

Theorem 2.1. *Let $a, q \in \mathbb{C}$, $a \neq 0, 1$, $0 < |q| < 1$. The Al-Salam–Carlitz polynomials are the unique polynomials (up to a multiplicative constant) satisfying the property of orthogonality*

$$\int_a^1 U_n^{(a)}(x;q) U_m^{(a)}(x;q) w(x;a;q) d_q x = d_n^2 \delta_{n,m}. \tag{2.1}$$

Remark 2.2. If $0 < |q| < 1$, the lattice $\{q^k : k \in \mathbb{N}_0\} \cup \{aq^k : k \in \mathbb{N}_0\}$ is a set of points which are located inside on a single contour that goes from 1 to 0, and then from 0 to a, through the spirals

$$S_1 : z(t) = |q|^t \exp(it \arg q), \quad S_2 : z(t) = |a||q|^t \exp(it \arg q + i \arg a),$$

where $0 < |q| < 1$, $t \in [0, \infty)$, which we can see in Fig. 2. Taking into account (2.1), we need to avoid the $a = 1$ case. For the $a = 0$ case, we cannot apply Favard's result [3], because in such a case this polynomial sequence fulfills the recurrence relation (see [2])

$$U_{n+1}^{(0)}(x;q) = (x - q^n) U_n^{(0)}(x;q), \quad U_0^{(0)}(x;q) = 1.$$

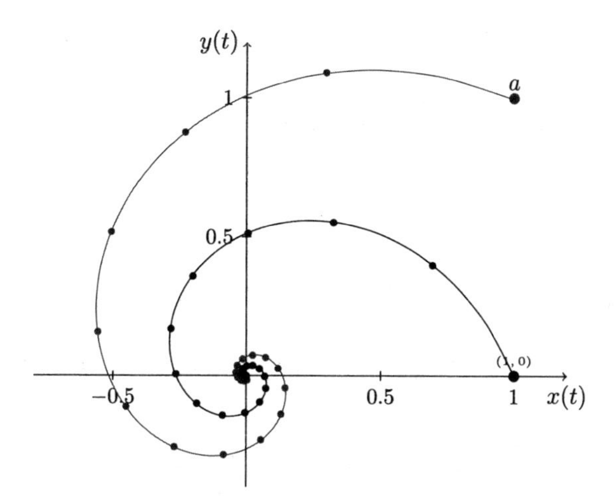

Fig. 2. The lattice $\{q^k : k \in \mathbb{N}_0\} \cup \{(1+i)q^k : k \in \mathbb{N}_0\}$ with $q = 4/5\exp(\pi i/6)$.

Proof of Theorem 2.1. Let $0 < |q| < 1$, and $a \in \mathbb{C}$, $a \neq 0, 1$. We are going to express the q-Jackson integral (2.1) as the difference of the two infinite sums and apply the identity

$$\sum_{k=0}^{M} f(q^k)\mathscr{D}_{q^{-1}} g(q^k) q^k = \frac{f(q^M)g(q^M) - f(q^{-1})g(q^{-1})}{q^{-1}-1} - \sum_{k=0}^{M} g(q^{k-1})\mathscr{D}_{q^{-1}} f(q^k) q^k. \tag{2.2}$$

Let $n \geq m$. Then, for one side, since $w(q^{-1}; a; q) = 0$, and using the identities [2, (14.24.7) and (14.24.9)], one has

$$\begin{aligned}
&\sum_{k=0}^{\infty} U_m^{(a)}(q^k;q)U_n^{(a)}(q^k;q)w(q^k;a;q)q^k \\
&= \frac{a(1-q)}{q^{2-n}} \lim_{M\to\infty} \sum_{k=0}^{M} \mathscr{D}_{q^{-1}}[w(q^k;a;q)U_{n-1}^{(a)}(q^k;q)]U_m^{(a)}(q^k;q)q^k \\
&= aq^{n-1} \lim_{M\to\infty} U_m^{(a)}(q^M;q)U_{n-1}^{(a)}(q^M;q)w(q^M;a;q) \\
&\quad + aq^{n-1}(q^m-1) \lim_{M\to\infty} \sum_{k=0}^{M-1} w(q^k;a;q)U_{n-1}^{(a)}(q^k;q)U_{m-1}^{(a)}(q^k;q)q^k.
\end{aligned}$$

Following an analogous process as before, and since $w(aq^{-1};a;q)=0$, we have

$$\sum_{k=0}^{\infty} U_m^{(a)}(aq^k;q)U_n^{(a)}(aq^k;q)w(aq^k;a;q)aq^k$$

$$= aq^{n-1}\lim_{M\to\infty} U_m^{(a)}(aq^M;q)U_{n-1}^{(a)}(aq^M;q)w(aq^M;a;q)$$

$$+ aq^{n-1}(q^m-1)\lim_{M\to\infty}\sum_{k=0}^{M-1} w(aq^k;a;q)U_{n-1}^{(a)}(aq^k;q)U_{m-1}^{(a)}(aq^k;q)aq^k.$$

Therefore, if $m<n$, and since m is finite, one can first repeat the previous process $m+1$ times obtaining

$$\sum_{k=0}^{\infty} U_m^{(a)}(q^k;q)U_n^{(a)}(q^k;q)w(q^k;a;q)q^k$$

$$= \lim_{M\to\infty}\sum_{\nu=1}^{m+1}(-aq^n)^\nu q^{-\nu(\nu+1)/2}(q^{-m+\nu-1};q)_\nu$$

$$\times U_{m-\nu+1}^{(a)}(q^M;q)U_{n-\nu}^{(a)}(q^M;q)w(q^M;a;q),$$

and

$$\sum_{k=0}^{\infty} U_m^{(a)}(aq^k;q)U_n^{(a)}(aq^k;q)w(aq^k;a;q)aq^k$$

$$= \lim_{M\to\infty}\sum_{\nu=1}^{m+1}(-aq^n)^\nu q^{-\nu(\nu+1)/2}(q^{-m+\nu-1};q)_\nu$$

$$\times U_{m-\nu+1}^{(a)}(aq^M;q)U_{n-\nu}^{(a)}(aq^M;q)w(aq^M;a;q).$$

Hence, since the difference of both limits, term by term, goes to 0 since $|q|<1$, then

$$\int_a^1 U_n^{(a)}(x;q)U_m^{(a)}(x;q)(qx,qx/a;q)_\infty d_qx = 0.$$

For $n=m$, following the same idea, we have

$$\int_a^1 U_n^{(a)}(x;q)U_n^{(a)}(x;q)w(x;a;q)d_qx$$

$$= \frac{a(q^n-1)}{q^{1-n}}\sum_{k=0}^{\infty}(w(q^k;a;q)(U_{n-1}^{(a)}(q^k;q))^2q^k$$

$$- aw(aq^k;a;q)(U_{n-1}^{(a)}(aq^k;q))^2q^k)$$

$$= (-a)^n (q;q)_n q^{\binom{n}{2}} \sum_{k=0}^{\infty} (w(q^k;a;q)q^k - a\, w(aq^k;a;q)q^k)$$

$$= (-a)^n (q;q)_n (q;q)_\infty\, q^{\binom{n}{2}} \sum_{k=0}^{\infty} ((q^{k+1}/a;q)_\infty - a(aq^{k+1};q)_\infty) \frac{q^k}{(q;q)_k},$$

since it is known that in this case [2, (14.24.2)]

$$\int_a^1 U_n^{(a)}(x;q) U_n^{(a)}(x;q) w(x;a;q) d_q x$$
$$= (-a)^n (q;q)_n (q;q)_\infty (a;q)_\infty (q/a;q)_\infty q^{\binom{n}{2}}.$$

Due to the normality of this polynomial sequence, i.e., $\deg U_n^{(a)}(x;q) = n$ for all $n \in \mathbb{N}_0$, the uniqueness is straightforward, thus the result holds. □

From this result, and taking into account that the squared norm for the Al-Salam–Carlitz polynomials is known, we obtained the following consequence for which we could not find any reference.

Corollary 2.3. *Let $a, q \in \mathbb{C} \setminus \{0\}$, $|q| < 1$. Then*

$$\sum_{k=0}^{\infty} ((q^{k+1}/a;q)_\infty - a(aq^{k+1};q)_\infty) \frac{q^k}{(q;q)_k} = (a;q)_\infty (q/a;q)_\infty.$$

The following case, which is just the Al-Salam–Carlitz polynomials for the $|q| > 1$ case, is commonly called the Al-Salam–Carlitz II polynomials.

Theorem 2.4. *Let $a, q \in \mathbb{C}$, $a \neq 0, 1$, $|q| > 1$. Then, the Al-Salam–Carlitz polynomials are unique (up to a multiplicative constant) satisfying the property of orthogonality given by*

$$\int_a^1 U_n^{(a)}(x;q^{-1}) U_m^{(a)}(x;q^{-1}) (q^{-1}x;q^{-1})_\infty (q^{-1}x/a;q^{-1})_\infty d_{q^{-1}} x$$
$$= (-a)^n (1-q^{-1}) (q^{-1};q^{-1})_n (q^{-1};q^{-1})_\infty$$
$$\times (a;q^{-1})_\infty (q^{-1}/a;q^{-1})_\infty\, q^{-\binom{n}{2}} \delta_{m,n}. \tag{2.3}$$

Proof. Let us denote q^{-1} by p; then $0 < |p| < 1$. For $a \in \mathbb{C}$, $a \neq 0, 1$. Then, by using the identity (2.2) replacing $q \mapsto p$, and taking into account that $w(aq;a;p) = w(q;a;p) = 0$ and [2, (14.24.9)], for $m < n$ one has

$$\sum_{k=0}^{\infty} aw(ap^k;a;p)U_m^{(a)}(ap^k;p)U_n^{(a)}(ap^k;p)p^k$$
$$= ap^{n-1} \lim_{M\to\infty} U_m^{(a)}(ap^M;p)U_{n-1}^{(a)}(ap^M;p)w(ap^M;a;p)$$
$$+ ap^{n-1}(1-p^m) \lim_{M\to\infty} \sum_{k=0}^{M-1} aw(ap^k;a;p)U_{n-1}^{(a)}(ap^k;p)U_{m-1}^{(a)}(ap^k;p)p^k.$$

Following the same idea from the previous result, we have

$$\sum_{k=0}^{\infty} w(p^k;a;p)U_m^{(a)}(p^k;p)U_n^{(a)}(p^k;p)p^k$$
$$= ap^{n-1} \lim_{M\to\infty} U_m^{(a)}(p^M;p)U_{n-1}^{(a)}(p^M;p)w(p^M;a;p)$$
$$+ ap^{n-1}(1-p^m) \lim_{M\to\infty} \sum_{k=0}^{M-1} w(p^k;a;p)U_{n-1}^{(a)}(p^k;p)U_{m-1}^{(a)}(p^k;p)p^k.$$

Therefore, the property of orthogonality holds for $m < n$. Next, if $n = m$, we have

$$\int_a^1 U_n^{(a)}(x;p)U_n^{(a)}(x;p)w(x;a;p)\, d_px$$
$$= \frac{a(p^n-1)}{p^{1-n}}\sum_{k=0}^{\infty}(aw(ap^k;a;p)(U_{n-1}^{(a)}(ap^k;p))^2p^k$$
$$- w(p^k;a;p)(U_{n-1}^{(a)}(p^k;p))^2p^k)$$
$$= (-a)^n(p;p)_n p^{\binom{n}{2}} \left(\sum_{k=0}^{\infty} aw(ap^k;a;p)p^k - w(p^k;a;p)p^k\right)$$
$$= (-a)^n\ (q^{-1};q^{-1})_n(p;p)_\infty p^{\binom{n}{2}} \sum_{k=0}^{\infty} \frac{q^k(a(p^{k+1}a;p)_\infty - (p^{k+1}/a;p)_\infty)}{(p;p)_k}$$
$$= (-a)^n(q^{-1};q^{-1})_n(p;p)_\infty(a;p)_\infty(p/a;p)_\infty p^{\binom{n}{2}}.$$

Using the same argument as in Theorem 2.1, the uniqueness holds, so the claim follows. □

Remark 2.5. Observe that in the previous theorems if $a = q^m$, with $m \in \mathbb{Z}$, $a \neq 0$, after some logical cancellations, the set of points where we need to calculate the q-integral is easy to compute. For example, if $0 < aq < 1$ and $0 < q < 1$, one obtains the sum [2, (14.25.2), p. 537].

Remark 2.6. The $a = 1$ case is special because it is not considered in the literature. In fact, the linear form associated with the Al-Salam–Carlitz polynomials $\mathbf{u}$ is quasi-definite and fulfills the Pearson-type distributional equations

$$\mathscr{D}_q[(x-1)^2\mathbf{u}] = \frac{x-2}{1-q}\mathbf{u} \quad \text{and} \quad \mathscr{D}_{q^{-1}}[q^{-1}\mathbf{u}] = \frac{x-2}{1-q}\mathbf{u}.$$

Moreover, the Al-Salam–Carlitz polynomials fulfill the three-term recurrence relation [2, (14.24.3)]

$$xU_n^{(a)}(x;q) = U_{n+1}^{(a)}(x;q) + (a+1)q^n U_n^{(a)}(x;q) - aq^{n-1}(1-q^n)U_{n-1}^{(a)}(x;q), \tag{2.4}$$

where $n = 0, 1, \ldots,$ with initial conditions $U_0^{(a)}(x;q) = 1$, $U_1^{(a)}(x;q) = x - a - 1$.

Therefore, we believe that it will be interesting to study such a case for its peculiarity because the coefficient $q^{n-1}(1-q^n) \neq 0$ for all n, so one can apply Favard's result.

2.1. The $|q| = 1$ case

In this section, we only consider the case where q is a root of unity. Let N be a positive integer such that $q^N = 1$; then, due to the recurrence relation (2.4) and following the same idea that the authors did in [4, Section 4.2], we apply the following process:

(1) The sequence $(U_n^{(a)}(x;q))_{n=0}^{N-1}$ is orthogonal with respect to the Gaussian quadrature

$$\langle \mathbf{v}, p\rangle := \sum_{s=1}^{N} \gamma_1^{(a)} \cdots \gamma_{N-1}^{(a)} \frac{p(x_s)}{(U_{N-1}^{(a)}(x_s))^2},$$

where $\{x_1, x_2, \ldots, x_N\}$ are the zeros of $U_N^{(a)}(x;q)$ for such value of q.

(2) Since $\langle \mathbf{v}, U_n^{(a)}(x;q)U_n^{(a)}(x;q)\rangle = 0$, we need to modify such a linear form. Next, we can prove that the sequence $(U_n^{(a)}(x;q))_{n=0}^{2N-1}$ is orthogonal with respect to the bilinear form

$$\langle p, r\rangle_2 = \langle \mathbf{v}, pq\rangle + \langle \mathbf{v}, \mathscr{D}_q^N p \mathscr{D}_q^N r\rangle,$$

since $\mathscr{D}_q U_n^{(a)}(x;q) = (q^n - 1)/(q-1)U_{n-1}^{(a)}(x;q)$.

(3) Since $\langle U_{2N}^{(a)}(x;q), U_{2N}^{(a)}(x;q)\rangle_2 = 0$, and taking into account the above results, we consider the linear form

$$\langle p,r\rangle_3 = \langle \mathbf{v}, pq\rangle + \langle \mathbf{v}, \mathscr{D}_q^N p \mathscr{D}_q^N r\rangle + \langle \mathbf{v}, \mathscr{D}_q^{2N} p \mathscr{D}_q^{2N} r\rangle.$$

(4) Therefore one can obtain a sequence of bilinear forms such that the Al-Salam–Carlitz polynomials are orthogonal with respect to them.

3. A Generalized Generating Function for Al-Salam–Carlitz Polynomials

For this section, we are going to assume $|q| > 1$, or $0 < |p| < 1$. Indeed, by starting with the generating functions for Al-Salam–Carlitz polynomials [2, (14.25.11) and (14.25.12)], we derive generalizations using the connection relation for these polynomials.

Theorem 3.1. *Let $a, b, p \in \mathbb{C} \setminus \{0\}$, $|p| < 1$, $a, b \neq 1$. Then*

$$U_n^{(a)}(x;p) = (-1)^n (p;p)_n p^{-\binom{n}{2}} \sum_{k=0}^{n} \frac{(-1)^k a^{n-k} (b/a;p)_{n-k} p^{\binom{k}{2}}}{(p;p)_{n-k}(p;p)_k} U_k^{(b)}(x;p). \tag{3.1}$$

Proof. If we consider the generating function for Al-Salam–Carlitz polynomials [2, (14.25.11)]

$$\frac{(xt;p)_\infty}{(t,at;p)_\infty} = \sum_{n=0}^{\infty} \frac{(-1)^n p^{\binom{n}{2}}}{(p;p)_n} U_n^{(a)}(x;p) t^n,$$

and multiply both sides by $(bt;p)_\infty/(bt;p)_\infty$, we obtain

$$\sum_{n=0}^{\infty} \frac{(-1)^n p^{\binom{n}{2}}}{(p;p)_n} U_n^{(a)}(x;p) t^n = \frac{(bt;p)_\infty}{(at;p)_\infty} \sum_{n=0}^{\infty} \frac{(-1)^n p^{\binom{n}{2}}}{(p;p)_n} U_n^{(b)}(x;p) t^n. \tag{3.2}$$

If we now apply the q-binomial theorem [2, (1.11.1)]

$$\frac{(az;p)_\infty}{(z;p)_\infty} = \sum_{k=0}^{\infty} \frac{(ap;p)_n}{(p;p)_n} z^n, \quad 0 < |p| < 1, \quad |z| < 1,$$

to (3.2), and then collect powers of t, we obtain

$$\sum_{k=0}^{\infty} t^k \sum_{m=0}^{k} \frac{(-1)^m a^{k-m} (b/a;p)_{k-m} p^{\binom{m}{2}}}{(p;p)_{k-m}(p;p)_m} U_m^{(b)}(x;p)$$

$$= \sum_{n=0}^{\infty} \frac{(-1)^n p^{\binom{n}{2}}}{(p;p)_n} U_n^{(a)}(x;p) t^n.$$

Taking into account this expression, the result follows. □

Theorem 3.2. *Let $a, b, p \in \mathbb{C} \setminus \{0\}$, $|p| < 1$, $a, b \neq 1$, $t \in \mathbb{C}$, $|at| < 1$. Then*

$$(at;p)_\infty \, {}_1\phi_1 \left(\begin{matrix} x \\ at \end{matrix} ; p, t \right) = \sum_{k=0}^{\infty} \frac{p^{k(k-1)}}{(p;p)_k} \, {}_1\phi_1 \left(\begin{matrix} b/a \\ 0 \end{matrix} ; p, atp^k \right) U_k^{(b)}(x;p) t^k, \tag{3.3}$$

where

$$ {}_r\phi_s \left(\begin{matrix} a_1, a_2, \ldots, a_r \\ b_1, b_2, \ldots, b_s \end{matrix} ; p, z \right)$$

$$= \sum_{k=0}^{\infty} \frac{(a_1;p)_k (a_2;p)_k \cdots (a_r;p)_k}{(b_1;p)_k (b_2;p)_k \cdots (b_s;p)_k} \frac{z^k}{(p;p)_k} (-1)^{(1+s-r)k} p^{(1+s-r)\binom{k}{2}},$$

is the unilateral basic hypergeometric series.

Proof. We start with a generating function for Al-Salam–Carlitz polynomials [2, (14.25.12)]

$$(at;q)_\infty \, {}_1\phi_1 \left(\begin{matrix} x \\ at \end{matrix} ; q, t \right) = \sum_{k=0}^{\infty} \frac{q^{n(n-1)}}{(q;q)_n} V_n^{(a)}(x;q) t^n$$

and (3.1) to obtain

$$(at;p)_\infty \, {}_1\phi_1 \left(\begin{matrix} x \\ at \end{matrix} ; p, t \right)$$

$$= \sum_{n=0}^{\infty} t^n (-1)^n p^{\binom{n}{2}} \sum_{k=0}^{n} \frac{(-1)^k a^{n-k} (b/a;p)_{n-k} p^{\binom{k}{2}}}{(p;p)_{n-k}(p;p)_k} U_k^{(b)}(x;p).$$

We reverse the order of summations, shift the n variable by a factor of k, and use the basic properties of the q-Pochhammer symbol, and [2, (1.10.1)]. Observe that we can reverse the order of summation since our sum is of the form

$$\sum_{n=0}^{\infty} a_n \sum_{k=0}^{n} c_{n,k} U_k^{(a)}(x;p),$$

where

$$a_n = t^n, \quad c_{n,k} = \frac{(-1)^k a^{n-k} (b/a;p)_{n-k} p^{\binom{k}{2}}}{(p;p)_{n-k}(p;p)_k}.$$

In this case, one has

$$|a_n| \leq |t|^n, \quad |c_{n,k}| \leq K(1+n)^{\sigma_1} |a|^n,$$

and $|U_n^{(a)}(x;p)| \leq (1+n)^{\sigma_2}$, where K_1, σ_1, and σ_2 are positive constants independent of n. Therefore, if $|at| < 1$, then

$$\left| \sum_{n=0}^{\infty} a_n \sum_{k=0}^{n} c_{n,k} U_k^{(a)}(x;p) \right| < \infty,$$

and this completes the proof. □

As we saw in Section 2, the orthogonality relation for Al-Salam–Carlitz polynomials for $|q| > 1$, $|p| < 1$, and $a \neq 0, 1$ is

$$\int_{\Gamma} U_n^{(a)}(x;p) U_m^{(a)}(x;p) w(x;a;p) d_p x = d_n^2 \delta_{n,m}.$$

Taking this result in mind, the following result follows.

Theorem 3.3. *Let* $a, b, p \in \mathbb{C} \setminus \{0\}$, $t \in \mathbb{C}$, $|at| < 1$, $|p| < 1$, $m \in \mathbb{N}_0$. *Then*

$$\int_a^1 {}_1\phi_1 \left(\begin{matrix} q^{-x} \\ at \end{matrix} ; q, t \right) U_m^{(b)}(q^{-x};p)(q^{-1}x;q^{-1})_\infty (q^{-1}x/a;q^{-1})_\infty dq^{-1}$$

$$= (-bt)^m q^{3\binom{m}{2}} (b;p)_\infty (p/b;p)_\infty \, {}_1\phi_1 \left(\begin{matrix} b/a \\ 0 \end{matrix} ; q, atq^m \right).$$

Proof. From (3.3), we have $x \mapsto p^x$ and multiply both sides by $U_m^{(b)}(x;p)w(x;a;p)$, and by using the orthogonality relation (2.3), the desired result holds. □

Note that the applications of connection relations to the rest of the known generating functions for Al-Salam–Carlitz polynomials [2, (14.24.11) and (14.25.12)] leave these generating functions invariant.

Acknowledgments

The author R. S. Costas-Santos acknowledges financial support from the National Institute of Standards and Technology. The authors thank the anonymous referee for her/his valuable comments and suggestions. They contributed to improve the presentation of this chapter.

References

[1] W. A. Al-Salam and L. Carlitz, Some orthogonal q-polynomials, *Math. Nach.* **30** (1965) 47–61.

[2] R. Koekoek, P. A. Lesky and R. F. Swarttouw, *Hypergeometric Orthogonal Polynomials and Their q-Analogues.* Springer Monographs in Mathematics (Springer-Verlag, Berlin, 2010). With a foreword by Tom H. Koornwinder.

[3] T. S. Chihara, *An Introduction to Orthogonal Polynomials*, Mathematics and its Applications, Vol. 13 (Gordon and Breach Science Publishers, New York, 1978).

[4] R. S. Costas-Santos and J. F. Sánchez-Lara, Orthogonality of q-polynomials for non-standard parameters, *J. Approx. Theory* **163**(9) (2011) 1246–1268.

This page intentionally left blank

Chapter 9

Crouching AGM, Hidden Modularity

Shaun Cooper[*,||], **Jesús Guillera**[†,**], **Armin Straub**[‡,††] and **Wadim Zudilin**[§,¶,‡‡,§§]

[*]*Institute of Natural and Mathematical Sciences*
Massey University — Albany
Private Bag 102904, North Shore Mail Centre
Auckland 0745, New Zealand
[†]*Department of Mathematics, University of Zaragoza*
50009 Zaragoza, Spain
[‡]*Department of Mathematics and Statistics*
University of South Alabama, 411 University Blvd N
MSPB 325, Mobile, AL 36688, USA
[§]*Institute for Mathematics, Astrophysics and Particle Physics*
Radboud Universiteit, P.O. Box 9010
6500 GL Nijmegen, The Netherlands
[¶]*School of Mathematical and Physical Sciences*
The University of Newcastle, Callaghan, NSW 2308, Australia
[||]*s.cooper@massey.ac.nz*
[**]*jguillera@gmail.com*
[††]*straub@southalabama.edu*
[‡‡]*w.zudilin@math.ru.nl*
[§§]*wadim.zudilin@newcastle.edu.au*

Dedicated to Mourad Ismail, with warm wishes from four continents

Special arithmetic series $f(x) = \sum_{n=0}^{\infty} c_n x^n$, whose coefficients c_n are normally given as certain binomial sums, satisfy "self-replicating" functional identities. For example, the equation

$$\frac{1}{(1+4z)^2} f\left(\frac{z}{(1+4z)^3}\right) = \frac{1}{(1+2z)^2} f\left(\frac{z^2}{(1+2z)^3}\right)$$

generates a modular form $f(x)$ of weight 2 and level 7, when a related modular parameterization $x = x(\tau)$ is properly chosen. In this chapter we investigate the potential of describing modular forms by such self-replicating equations as well as applications of the equations that do not make use of the modularity. In particular, we outline a new recipe of generating AGM-type algorithms for computing π and other related constants. Finally, we indicate some possibilities to extend the functional equations to a two-variable setting.

Keywords: Modular form; arithmetic hypergeometric series; super-congruence; identity for $1/\pi$; AGM iteration.

Mathematics Subject Classification 2010: 11F11, 11B65, 11F33, 11Y55, 11Y60, 33C20, 33F05, 65B10, 65D20

1. Introduction

Modular forms and functions form a unique enterprise in the world of special functions. They are innocent looking q-series, at the same time highly structured and possess numerous links to other parts of mathematics. In our exposition below we will try to hide the modularity of objects under consideration as much as possible, just pointing out related references to the literature where the modular origin is discussed in detail.

One way of thinking of modular forms is through Picard–Fuchs linear differential equations [24, Section 5.4]. The concept can be illustrated on the example

$$f_7(x) = \sum_{n=0}^{\infty} u_n x^n, \tag{1.1}$$

with coefficients

$$u_n = \sum_{k=0}^{n} \binom{n}{k}^2 \binom{n+k}{n} \binom{2k}{n} = \sum_{k=0}^{n} (-1)^{n-k} \binom{3n+1}{n-k} \binom{n+k}{n}^3. \tag{1.2}$$

The function $f_7(x)$ represents a level 7 and weight 2 modular form, e.g., [10, 28] (see also Section 3 below), and satisfies a linear differential equation, which we record in the form of a recurrence relation for the coefficients

$$(n+1)^3 u_{n+1} = (2n+1)(13n^2+13n+4)u_n + 3n(3n-1)(3n+1)u_{n-1} \tag{1.3}$$

where $n = 0, 1, 2, \dots$. The single initial condition $u_0 = 1$ is used to start the sequence. This is sequence A183204 in the *OEIS* [22]. Recursions such as (1.3) are famously known as Apéry-like recurrence equations [2, 4, 10, 25].

The series defining $f_7(x)$ converges in the disk $|x| < 1/27$ as $x = 1/27$ is a singular point of the underlying linear differential equation produced by the recursion.

In this chapter, we will investigate another point of view on the modularity, using functional equations of a different nature. The function $f(x) = f_7(x)$ defined above will be our principal example, while the special functional equation it satisfies will be [12, Theorem 4.4]

$$\frac{1}{(1+4z)^2} f\left(\frac{z}{(1+4z)^3}\right) = \frac{1}{(1+2z)^2} f\left(\frac{z^2}{(1+2z)^3}\right), \tag{1.4}$$

valid in a suitable neighborhood of $z = 0$. Note that this equation, together with the initial condition $f(0) = 1$, uniquely determines the coefficients in the Taylor series expansion $f(x) = \sum_{n=0}^{\infty} u_n x^n$. In Section 2, we will address this fact and discuss some variations of equation (1.4). In particular, we observe that, in arithmetically interesting instances—those with connections to modular forms—the corresponding sequences appear to satisfy (and possibly are characterized by) strong divisibility properties. Without pretending to fully cover the topic of such functional equations, we list several instances for other modular forms in Section 3 together with some details of the underlying true modular forms and functions. An important feature of this way of thinking of modular forms is applications. One such application—to establishing Ramanujan-type formulas for $1/\pi$—has already received attention in the literature, e.g., see [14, 27, 28]. In Section 4 we give some related examples for $f_7(x)$ and equation (1.4), though concentrating mostly on another application of the special functional equations—to AGM-type algorithms. Such algorithms for computing π were first discovered independently by Brent [6] and Salamin [20] in 1976 (see also [3]). A heavy use of the modularity made later, faster versions by other authors somehow sophisticated. Our exposition in Section 4 follows the ideas in [13] and streamlines the production of AGM-type iterations by freeing them from the modularity argument.

Three-term recurrence relations, which are more general than the one given in (1.3), naturally appear in connection with orthogonal polynomials [15]—one of the topics mastered by Mourad Ismail. The polynomials have long-existing ties with arithmetic, and the recent work [9, 19, 23, 28] indicates some remarkable links between generating functions of classical Legendre polynomials and arithmetic hypergeometric functions. Interestingly enough, these links allow one to produce a two-variable functional equation that may be thought of as a generalization of (1.4). This is the subject of Section 5.

2. Self-Replication vs. Modularity

The first natural question to ask is how sensitive the dependence of $f(x)$ on the parameters of equation (1.4) is. In other words, what can be said about a solution $f(x) = f(\lambda, \mu; x) = \sum_{n=0}^{\infty} c_n x^n$, with $c_0 = 1$, to the equation

$$\frac{1}{(1+\mu z)^2} f\left(\frac{z}{(1+\mu z)^3}\right) = \frac{1}{(1+\lambda z)^2} f\left(\frac{z^2}{(1+\lambda z)^3}\right) \tag{2.1}$$

when λ and μ are integers? Writing the equation as

$$\sum_{n=0}^{\infty} \frac{c_n z^n}{(1+\mu z)^{3n+2}} = \sum_{n=0}^{\infty} \frac{c_n z^{2n}}{(1+\lambda z)^{3n+2}}$$

and applying the binomial theorem to expand both sides into a power series in z, we obtain the recursion

$$c_n = \sum_{k=0}^{\lfloor n/2 \rfloor} \binom{n+k+1}{3k+1} (-\lambda)^{n-2k} c_k - \sum_{k=0}^{n-1} \binom{n+2k+1}{3k+1} (-\mu)^{n-k} c_k, \tag{2.2}$$

for $n = 1, 2, \ldots$, which together with $c_0 = 1$ determines the sequence $c_n = c_n(\lambda, \mu)$. Clearly, $c_n(\lambda, \mu) \in \mathbb{Z}$ whenever $\lambda, \mu \in \mathbb{Z}$. Also, the recursion (2.2) implies that the functional equation (2.1) completely recovers the analytic solution $f(x)$. The latter fact allows us to think of (1.4) and, more generally, of (2.1) as *self-replicating* functional equations that define the series and function $f(x)$. In spite of the property $c_n \in \mathbb{Z}$ it remains unclear whether these sequences have some arithmetic significance in general. In the sequel, we will refer to instances as *arithmetic* if the generating function $f(x)$ is of modular origin.

The fact that the sequences arising from self-replicating equations such as (1.4) are integral is useful since this integrality is rather inaccessible from associated differential equations or recurrences such as (1.3). (It remains an open problem to classify all integer solutions to recurrences of the shape (1.3), though it is conjectured that the known Apéry-like sequences form a complete list.) For Apéry-like sequences, the integrality usually follows from binomial sum representations such as (1.2). Integrality can also be deduced by exhibiting a modular parameterization of the differential equation; however, we do not know *a priori* whether such a parameterization should exist [25].

In the arithmetic instances of modular origin, the sequence c_n necessarily satisfies a linear differential equation [24]. On the other hand, for any sequence produced by self-replication, we can (try to) guess a linear differential equation for its generating function by computing a few initial terms. Once guessed, we can then use closure properties of holonomic functions to prove that the solution to the differential equation does indeed satisfy the self-replicating functional equation. If the generating function satisfies a linear differential equation then this somewhat brute experimental approach is guaranteed to find this equation and prove its correctness in finite time. Showing that the sequence is not holonomic is more painful, as the time required by this brute force approach is infinite.

On the other hand, all known arithmetic instances lead to sequences which satisfy strong divisibility properties. For instance, for primes p, these sequences $c(n)$ satisfy the *Lucas congruences*

$$c(n) \equiv c(n_0)c(n_1)\cdots c(n_r) \pmod{p}, \tag{2.3}$$

where $n = n_0 + n_1 p + \cdots + n_r p^r$ is the expansion of n in base p. In the case of Apéry-like numbers, these congruences were shown in [17] and further general results were recently obtained in [1]. In addition, these sequences satisfy, or are conjectured to satisfy, the $p^{\ell r}$-congruences

$$c(mp^r) \equiv c(mp^{r-1}) \pmod{p^{\ell r}} \tag{2.4}$$

with $\ell \in \{1, 2, 3\}$, which in interesting cases hold for $\ell > 1$ and are referred to as *supercongruences*, e.g., see [18] and the references therein.

In the following examples, we therefore consider self-replicating functional equations of basic shapes, and search for values of the involved parameters such that the corresponding sequences satisfy such congruences. In each case, in which we observe the Lucas congruences and p^r-congruences, we are able to identify these sequences and relate them to modular forms.

The recursion (2.2) together with the original self-replicating functional equation (1.4) suggest that the choice $\lambda^2 = \mu$ in (2.1) is of special arithmetic significance. This is substantiated in the next example.

Example 2.1. We consider the functional relation (2.1) with integer parameters λ and μ. The terms $c_n(\lambda, \mu)$ of the corresponding sequence are polynomials in λ and μ. The first few are

$$1,\ 2(\mu - \lambda),\ 7\mu^3 - 10\lambda\mu + 3\lambda^2 + 2\mu - 2\lambda,\ \ldots .$$

Using these initial terms we then determine congruence conditions on the parameters λ, μ, which need to be satisfied in order that the p^r-congruences (2.4), with $\ell = 1$, hold for all primes p. In the case $\lambda = \mu$ we have $c_n(\lambda, \lambda) = 0$ for $n \geq 1$, and so we assume $\lambda \neq \mu$ in the sequel. Remarkably, the empirical data suggests that these congruences only hold in a finite number of cases, all of modular origin, as well as an infinite family of unclear origin. Specifically, the p^r-congruences appear to hold modulo p^r for all primes p only in the cases $\lambda^2 = \mu$ with

$$\lambda \in \{-2, -1, 2, 4, 16\}$$

and the infinite family of cases $\lambda = -2\mu$ where μ is any even integer. If there exist further cases, then $\min\{|\mu|, |\lambda|\} > 10{,}000$ (and $|\lambda| > 100{,}000$ in the particularly relevant case $\lambda^2 = \mu$).

Here is the analysis of the cases when $\lambda^2 = \mu$:

- The case $\lambda = 2$ is the level 7 sequence from the introduction (and Section 3 below).
- The case $\lambda = -2$ is the level 3 sequence $c_n = c_n(-2, 4)$ with

$$\sum_{n=0}^{\infty} c_n z^n = \left(\sum_{n=0}^{\infty} \binom{2n}{n} \binom{3n}{n} z^n \right)^2$$

(see Example 3.1).
- The case $\lambda = -1$ produces the sequence $c_n(-1, 1) = 2^n \binom{2n}{n}$, which clearly satisfies p^r-congruences (but does not satisfy p^{2r}-congruences). The corresponding generating function $\sum_{n=0}^{\infty} c_n z^n$ is algebraic.
- The case $\lambda = 4$ produces the sequence $c_n = c_n(4, 16)$ with

$$\sum_{n=0}^{\infty} c_n z^n = \left(\sum_{n=0}^{\infty} \binom{2n}{n}^2 \binom{3n}{n} z^n \right)^2,$$

which appears to satisfy p^{3r}-supercongruences. The corresponding modular form has level 3 and weight 4.
- The case $\lambda = 16$ produces the sequence $c_n = c_n(16, 256)$ with (note the exponent)

$$\sum_{n=0}^{\infty} c_n z^n = \left(\sum_{n=0}^{\infty} \binom{2n}{n} \binom{3n}{n} \binom{6n}{3n} z^n \right)^4,$$

which appears to satisfy p^{2r}-supercongruences. The modular form has level 1 and weight 8.

Apparently, the sequences corresponding to $(\lambda, \mu) = (-4\alpha, 2\alpha)$, that is,

$$\frac{1}{(1 + 2\alpha z)^2} f\left(\frac{z}{(1 + 2\alpha z)^3} \right) = \frac{1}{(1 - 4\alpha z)^2} f\left(\frac{z^2}{(1 - 4\alpha z)^3} \right),$$

satisfy p^r congruences for all integers α but do not appear to be holonomic. However, these sequences fail to satisfy Lucas congruences (except modulo 2 and 3, for trivial reasons). In the case $\alpha = 1$ the first few terms of the sequence $c_n = c_n(-4, 2)$ are

$$1,\ 12,\ 168,\ 2496,\ 38328,\ 600672,\ 9539808,\ 152891520,\ \ldots.$$

The first 250 terms do not reveal a linear recurrence with polynomial coefficients, suggesting that the sequence is not holonomic, and we were unable to otherwise identify the sequence. It does take the slightly simplified form

$$c_n = \sum_{k=0}^{n} d_k d_{n-k},$$

where

$$d_n = 1,\ 6,\ 66,\ 852,\ 11874,\ 172860,\ 2586108,\ \ldots.$$

Similar comments apply to the case $\alpha = -1$, which results in an alternating sequence. □

There are many natural shapes of self-replicating functional equations that one could similarly analyze. Here, we restrict ourselves to a variation of (2.1).

Example 2.2. In the spirit of the previous example, consider the functional relations

$$\frac{1}{1+\mu z} f\left(\frac{z}{(1+\mu z)^2}\right) = \frac{1}{1+\lambda z} f\left(\frac{z^2}{(1+\lambda z)^2}\right).$$

We find supercongruences modulo p^{3r}, that is (2.4) with $\ell = 3$, in the cases $(\lambda, \mu) = (-2, 0)$, $(\lambda, \mu) = (0, -2)$ and $(\lambda, \mu) = (0, 4)$, which correspond to the sequences $\binom{2n}{n}$, $(-1)^n\binom{2n}{n}$ and $\binom{2n}{n}^2$, respectively.

The case $(\lambda, \mu) = (-8, 16)$ produces the sequence a_n with

$$\sum_{n=0}^{\infty} a_n z^n = \left(\sum_{n=0}^{\infty} \binom{2n}{n}\binom{4n}{2n} z^n\right)^2,$$

which appears to satisfy p^{2r}-supercongruences; this is a modular form of level 2 given in Example 3.1 below.

The cases $(\lambda, \mu) = (-2, 0)$ and $(\lambda, \mu) = (0, -2)$ generalize to the case $(\lambda, \mu) = (-\alpha - 2, \alpha)$, which corresponds to the sequence $(\alpha+1)^n\binom{2n}{n}$. For obvious reasons we have p^r-congruences for all of these, as well as Lucas congruences and recurrences.

In the cases $(\lambda, \mu) = (2\alpha, 0)$, we appear to have p^r-congruences but no Lucas congruences and no recurrences.

We did not find additional instances of congruences modulo p^r. If there exist any, then $\min\{|\mu|, |\lambda|\} > 10{,}000$. □

The arithmetic sequences observed in this section all satisfy strong congruences. It would be interesting to understand to which degree such congruences characterize these arithmetic instances of the self-replication process.

3. Self-Replicating Functional Equations

For levels $\ell \in \{2, 3, 4, 5, 7\}$ we use the modular functions

$$z_\ell(\tau) = q\left(\prod_{j=1}^{\infty}\frac{1-q^{\ell j}}{1-q^j}\right)^{24/(\ell-1)}, \quad \text{where } q = e^{2\pi i\tau},$$

and the corresponding weight two modular forms

$$P_\ell(\tau) = q\frac{\mathrm{d}}{\mathrm{d}q}\log z_\ell$$

to define implicitly the analytic functions $f_\ell(x)$ by

$$P_2(\tau) = f_2\left(\frac{z_2(\tau)}{1+64z_2(\tau)}\right), \quad P_3(\tau) = f_3\left(\frac{z_3(\tau)}{1+27z_3(\tau)}\right),$$

$$P_4(\tau) = f_4\left(\frac{z_4(\tau)}{1+16z_4(\tau)}\right), \quad P_5(\tau) = f_5\left(\frac{z_5(\tau)}{1+22z_5(\tau)+125z_5(\tau)^2}\right)$$

$$\text{and} \quad P_7(\tau) = f_7\left(\frac{z_7(\tau)}{1+13z_7(\tau)+49z_7(\tau)^2}\right).$$

Explicitly we have

$$f_2(x) = \left(\sum_{n=0}^{\infty}\binom{2n}{n}\binom{4n}{2n}x^n\right)^2 = {}_2F_1\left(\begin{matrix}\frac{1}{4}, \frac{3}{4}\\ 1\end{matrix}\,\middle|\, 64x\right)^2,$$

$$f_3(x) = \left(\sum_{n=0}^{\infty}\binom{2n}{n}\binom{3n}{n}x^n\right)^2 = {}_2F_1\left(\begin{matrix}\frac{1}{3}, \frac{2}{3}\\ 1\end{matrix}\,\middle|\, 27x\right)^2,$$

$$f_4(x) = \left(\sum_{n=0}^{\infty}\binom{2n}{n}^2 x^n\right)^2 = {}_2F_1\left(\begin{matrix}\frac{1}{2}, \frac{1}{2}\\ 1\end{matrix}\,\middle|\, 16x\right)^2,$$

where the standard hypergeometric notation is used, and also

$$f_5(x) = \sum_{n=0}^{\infty}\left\{\binom{2n}{n}\sum_{k=0}^{n}\binom{n}{k}^2\binom{n+k}{k}\right\}x^n,$$

while binomial expressions for the coefficients of $f_7(x)$ are given in (1.2).

Example 3.1. Each of these can be characterized by a self-replicating functional equation:

$$\begin{aligned}
\frac{1}{1+16z} f_2\left(\frac{z}{(1+16z)^2}\right) &= \frac{1}{1-8z} f_2\left(\frac{z^2}{(1-8z)^2}\right), \\
\frac{1}{(1+4z)^2} f_3\left(\frac{z}{(1+4z)^3}\right) &= \frac{1}{(1-2z)^2} f_3\left(\frac{z^2}{(1-2z)^3}\right), \\
\frac{1}{(1+4z)^2} f_4\left(\frac{z}{(1+4z)^2}\right) &= f_4(z^2), \\
\frac{1}{1+8z} f_5\left(\frac{z}{(1+4z)(1+8z)^2}\right) &= \frac{1}{1+2z} f_5\left(\frac{z^2}{(1+4z)(1+2z)^2}\right), \\
\frac{1}{(1+4z)^2} f_7\left(\frac{z}{(1+4z)^3}\right) &= \frac{1}{(1+2z)^2} f_7\left(\frac{z^2}{(1+2z)^3}\right).
\end{aligned}$$

The equations for f_3 and f_7 are discussed in Example 2.1, and the one for f_2 shows up in Example 2.2. These instances demonstrate that the example (1.4) for level 7 may be viewed as part of a bigger family.

We expect similar self-replicating equations for levels 9, 13 and 25; see [11, Section 8]. □

In order to give some more weight 2 examples, define

$$\begin{aligned}
\hat{f}_2(x) &= \sum_{n=0}^{\infty} \binom{2n}{n}^2 \binom{4n}{2n} x^n = {}_3F_2\left(\left.\begin{matrix}\frac14, \frac12, \frac34 \\ 1, 1\end{matrix}\right| 256x\right), \\
\hat{f}_3(x) &= \sum_{n=0}^{\infty} \binom{2n}{n}^2 \binom{3n}{n} x^n = {}_3F_2\left(\left.\begin{matrix}\frac13, \frac12, \frac23 \\ 1, 1\end{matrix}\right| 108x\right), \\
\hat{f}_4(x) &= \sum_{n=0}^{\infty} \binom{2n}{n}^3 x^n = {}_3F_2\left(\left.\begin{matrix}\frac12, \frac12, \frac12 \\ 1, 1\end{matrix}\right| 64x\right)
\end{aligned}$$

and

$$\hat{f}_5(x) = \sum_{n=0}^{\infty} \left\{ \sum_{k=0}^{n} (-1)^{n-k} \binom{n}{k}^3 \binom{4n-5k}{3n} \right\} x^n.$$

Example 3.2. A functional equation for $\hat{f}_2$ is given by

$$\frac{1}{1+27z} \hat{f}_2\left(\frac{z}{(1+27z)^4}\right) = \frac{1}{1+3z} \hat{f}_2\left(\frac{z^3}{(1+3z)^4}\right).$$

This may be regarded as a cubic version of the functional equations studied in Examples 2.1 and 2.2.

A functional equation for $\hat{f}_3$ is given by the case $\lambda = 4$ in Example 2.1. Functional equations for $\hat{f}_4$ and $\hat{f}_5$ are of a different type:

$$\frac{1}{1+8z}\hat{f}_4\left(z\left(\frac{1-z}{1+8z}\right)^3\right) = \hat{f}_4\left(z^3\left(\frac{1-z}{1+8z}\right)\right),$$

$$\frac{1}{(1+4z)^2}\hat{f}_4\left(z\left(\frac{1-z}{1+4z}\right)^5\right) = \hat{f}_4\left(z^5\left(\frac{1-z}{1+4z}\right)\right),$$

$$\frac{1}{1-5z}\hat{f}_5\left(z\left(\frac{1-z}{1-5z}\right)^2\right) = \hat{f}_5\left(z^2\left(\frac{1-z}{1-5z}\right)\right). \qquad \square$$

Example 3.2 suggests considering functional equations of the type

$$t(z)\, f\big(r(z)\, s^n(z)\big) = f\big(r^n(z)\, s(z)\big),$$

where $r(z)$, $s(z)$ and $t(z)$ are rational functions with

$$r(z) = z + O(z^2), \quad s(z) = 1 + O(z), \quad t(z) = 1 + O(z),$$

and $n \geq 2$ is an integer.

Example 3.3. We list three more instances that may be compared with the functional equation for f_4 in Example 3.1:

$$g_b(z^2) = \frac{1}{1+3z}\, g_b\left(z\,\frac{1-z}{1+3z}\right),$$

$$g_c(z^3) = \frac{1}{1+2z+4z^2}\, g_c\left(z\,\frac{1-z+z^2}{1+2z+4z^2}\right)$$

and

$$g_5(z^5) = \frac{1}{1+3z+4z^2+2z^3+z^4}\, g_5\left(z\,\frac{1-2z+4z^2-3z^3+z^4}{1+3z+4z^2+2z^3+z^4}\right).$$

The solutions to these functional equations that satisfy the initial condition $g(0) = 1$ are given by

$$g_b(x) = \sum_{n=0}^{\infty}\left\{\sum_{k=0}^{\infty}\binom{n}{k}^2\binom{2k}{k}\right\}x^n, \qquad g_c(x) = \sum_{n=0}^{\infty}\left\{\sum_{k=0}^{\infty}\binom{n}{k}^3\right\}x^n$$

and

$$g_5(x) = \sum_{n=0}^{\infty}\left\{\sum_{k=0}^{\infty}\binom{n}{k}^2\binom{n+k}{k}\right\}x^n.$$

The functions g_b and g_c may be parameterized by level 6 modular forms, while g_5 may be parameterized by level 5 modular forms, e.g., see [4] and [8, Theorem 3.4 and Tables 1 and 2]. The functions g_b, g_c and g_5 are denoted by (c), (a) and (b), respectively, in [2, equation (28)]. They are generating functions for the sequences C, A and D, respectively, in [25]. □

4. Formulas for $1/\pi$ and AGM-Type Iterations

Example 4.1. We can use the transformation (1.4) as $z \to (1/8)^-$, so that $z/(1+4z)^3 \to (1/27)^-$ and $z/(1+4z)^3 \to 1/125$. Applying the asymptotics and techniques from [14, Example 5] we obtain the formula

$$\sum_{n=0}^{\infty} u_n(4+21n) \times \frac{1}{5^{3n+3}} = \frac{1}{8\pi}. \tag{4.1}$$

This can also be obtained using modular forms, e.g., see [10, equation (37) and Table 1].

Consider the limiting case $h \to h_0^+$, where $h_0 = -(5+\sqrt{21})/2$, of the transformation [7, Lemmas 4.1 and 4.3]

$$\frac{1}{\sqrt{1+13h+49h^2}} \sum_{n=0}^{\infty} u_n \left(\frac{h}{1+13h+49h^2}\right)^n$$
$$= \frac{1}{\sqrt{1+5h+h^2}}\, {}_3F_2\left(\begin{matrix}\frac{1}{6}, \frac{1}{2}, \frac{5}{6} \\ 1, 1\end{matrix} \,\middle|\, \frac{1728h^7}{(1+13h+49h^2)(1+5h+h^2)^3}\right),$$

so that the argument of the ${}_3F_2$ series on the right-hand side tends to $-\infty$, and follow [14, Theorem 2 and Example 2] to obtain

$$\sum_{n=0}^{\infty} u_n \left(\frac{3\sqrt{21}-14}{56}\right)^n = \left(\frac{128(\sqrt{7}-\sqrt{3})}{49(5-\sqrt{21})^2}\right)^{1/3} \frac{\sqrt{\pi}}{3\Gamma(\frac{5}{6})^3} \tag{4.2}$$

and

$$\sum_{n=0}^{\infty} u_n(6\sqrt{21}-20+15(\sqrt{21}-2)n) \left(\frac{3\sqrt{21}-14}{56}\right)^n = \frac{8\sqrt{7}}{\pi}. \tag{4.3}$$

The series (4.3) corresponds to the case $N = 21$ of the identity (39) in [10]. □

A principal feature of an AGM-type algorithm is a sequence of algebraic approximations a_k, where $k = 0, 1, 2, \ldots$, that converge to a given number ξ at a certain rate $r > 1$, so that $|\xi - a_{k+1}| < C|\xi - a_k|^r$ for some absolute constant $C > 0$. If $r = 2$, as for the classical AGM algorithm of Brent [6] and

Salamin [20], then it is said to be a quadratic iteration. The next example uses a functional equation to produce quadratic AGM-type iterations to numbers ξ of the form

$$\sum_{n=0}^{\infty} u_n\,(a_0+b_0n)x_0^n=\xi, \tag{4.4}$$

where a_0, b_0 and x_0 are fixed real algebraic numbers, where the sequence u_n is given in (1.2). The iterations are based on the self-replicating functional equation (1.4).

Example 4.2. Let a_0, b_0 and x_0 be chosen in accordance with (4.4), while z_0 is the real solution to

$$\frac{z}{(1+4z)^3}=x_0$$

of the smallest absolute value. Define the sequences a_k, b_k and z_k for all $k\ge 1$ recursively as follows:

$$a_{k+1}=a_k\,\frac{(1+4z_k)^2}{(1+2z_k)^2}+b_k\,\frac{4z_k(1+4z_k)^2}{(1+2z_k)^3(1-8z_k)},$$

$$b_{k+1}=2b_k\,\frac{(1+4z_k)^3(1-z_k)}{(1+2z_k)^3(1-8z_k)},$$

$$z_{k+1}=\frac{2z_k^2}{1+6z_k+(1+2z_k)\sqrt{1+8z_k}}.$$

Then a_k converges to ξ in (4.4) quadratically.

To justify the above iteration, we couple the self-replicating equation (1.4) with the one obtained from it by differentiation:

$$\sum_{n=0}^{\infty} u_n\left(\left(A-\frac{8Bz}{1+4z}\right)+nB\left(1-\frac{12z}{1+4z}\right)\right)\frac{z^k}{(1+4z)^{3k+2}}$$
$$=\sum_{n=0}^{\infty} u_n\left(\left(A-\frac{4Bz}{1+2z}\right)+nB\left(2-\frac{6z}{1+2z}\right)\right)\frac{z^{2k}}{(1+2z)^{3k+2}}. \tag{4.5}$$

In addition to the above three sequences we also consider

$$x_k=\frac{z_k}{(1+4z_k)^3}; \tag{4.6}$$

note that the definition of z_{k+1} implies that

$$x_{k+1}=\frac{z_{k+1}}{(1+4z_{k+1})^3}=\frac{z_k^2}{(1+2z_k)^3}.$$

Then induction on k then shows that

$$\sum_{n=0}^{\infty} u_n (a_k + b_k n) x_k^n = \xi. \tag{4.7}$$

Indeed, on the kth step we first define the numbers A, B and $z = z_k$ such that

$$\frac{z}{(1+4z)^3} = x_k, \quad \frac{B}{(1+4z)^2}\left(1 - \frac{12z}{1+4z}\right) = b_k,$$

$$\frac{1}{(1+4z)^2}\left(A - \frac{8Bz}{1+4z}\right) = a_k$$

and assign to the data

$$x_{k+1} = \frac{z^2}{(1+2z)^3}, \quad b_{k+1} = \frac{B}{(1+2z)^2}\left(2 - \frac{6z}{1+2z}\right) \tag{4.8}$$

and

$$a_{k+1} = \frac{1}{(1+2z)^2}\left(A - \frac{4Bz}{1+2z}\right). \tag{4.9}$$

Application of (4.5) implies the validity of (4.7) for k replaced with $k+1$; taking $z = z_k$ and eliminating A, B lead to the formulas for a_{k+1}, b_{k+1} and x_{k+1} by means of a_k, b_k and z_k only.

In this construction for x_k sufficiently close to 0 we have $x_{k+1} \approx z_k^2 \approx x_k^2$, so that x_k tends to 0 quadratically. Performing the limit in (4.7) as $k \to \infty$ we obtain

$$\xi = \lim_{k\to\infty} \sum_{n=0}^{\infty} u_n (a_k + b_k n) x_k^n = \lim_{k\to\infty} a_k.$$

Furthermore, $|\xi - a_k|$ is roughly of magnitude x_k, so that a_k converges to ξ quadratically.

To execute the algorithm one can use (4.1) to produce the initial conditions

$$a_0 = \frac{4}{125}, \quad b_0 = \frac{21}{125}, \quad x_0 = \frac{1}{125}, \quad z_0 = \left(\frac{\sqrt{2}-1}{2}\right)^3 = 0.00888347\ldots. \tag{4.10}$$

Then for the iteration scheme given by (4.8) and (4.9) we have

$$\lim_{n\to\infty} a_n = \frac{1}{2\pi}$$

and the convergence is quadratic.

Moreover, the formula (4.2) leads to the initial conditions

$$a_0 = 3\left(\frac{128(\sqrt{7}-\sqrt{3})}{49(5-\sqrt{21})^2}\right)^{-1/3}, \quad b_0 = 0, \quad x_0 = \frac{3\sqrt{21}-14}{56},$$

while z_0 is a real solution of the equation $x_0(1+4z)^3 - z = 0$, for which the iteration scheme given by (4.8) and (4.9) has the property that

$$\lim_{n\to\infty} a_n = \frac{\pi^{1/2}}{\Gamma(5/6)^3} = \frac{\Gamma(1/6)^3}{8\pi^{5/2}}$$

with quadratic convergence. In a similar way, the identity (4.3) leads to a different set of initial conditions such that the iteration scheme given by (4.8) and (4.9) generates another sequence that converges to $1/(2\pi)$. □

Our next example makes use of the identities from Example 3.2 to produce an iteration scheme that converges quintically to $1/\pi$.

Example 4.3. Define sequences a_k, b_k, x_k and z_k as follows. Let

$$a_0 = \frac{1}{4}, \quad b_0 = 1, \quad x_0 = -\frac{1}{64}.$$

Recursively, let z_k be the real solution to

$$z^{1/5}\frac{1-z}{1+4z} = x_k^{1/5}$$

that is closest to the origin, and then let

$$a_{k+1} = a_k\,(1+4z_k)^2 + 8\,b_k\,\frac{z_k(1-z_k)(1+4z_k)^2}{1-22z_k-4z_k^2},$$

$$b_{k+1} = 5\,b_k\,(1+4z_k)^2\frac{1+2z_k-4z_k^2}{1-22z_k-4z_k^2}$$

and

$$x_{k+1} = z_k^5\frac{1-z_k}{1+4z_k}.$$

Then

$$\lim_{k\to\infty} a_k = \frac{1}{2\pi}$$

and the rate of convergence is of order 5.

To justify the above algorithm, we differentiate the last formula of Example 3.2 to get

$$\sum_{n=0}^{\infty}\binom{2n}{n}^3\left(\left(A-\frac{8Bx}{1+4x}\right)+Bn\,\frac{1-22x-4x^2}{(1-x)(1+4x)}\right)\frac{x^n(1-5x)^{5n}}{(1+4x)^{5n+2}}$$
$$=\sum_{n=0}^{\infty}\binom{2n}{n}^3\left(A+5Bn\,\frac{1+2x-4x^2}{(1-x)(1+4x)}\right)\frac{x^{5n}(1-x)^n}{(1+4x)^n}.$$

Consider an iteration scheme that goes as follows. Suppose x_k, a_k and b_k are given, for some k. Compute z, B and A from the formulas

$$z\left(\frac{1-z}{1+4z}\right)^5=x_k,\quad B\,\frac{1-22x-4x^2}{(1+4x)^3(1-x)}=b_k$$

and

$$\frac{1}{(1+4x)^2}\left(A-\frac{8Bx}{1+4x}\right)=a_k.$$

Then, let x_{k+1}, b_{k+1} and a_{k+1} be defined by

$$x_{k+1}=\frac{x^5(1-x)}{(1+4x)},\quad b_{k+1}=\frac{5B(1+2z-4z^2)}{(1-z)(1+4z)},\quad a_{k+1}=A.$$

These iterations are equivalent to the ones stated in the theorem.

To get initial conditions, consider Bauer's series [8, Table 6, $N=2$; 26],

$$\sum_{n=0}^{\infty}\binom{2n}{n}^3\left(\frac{1}{4}+n\right)\left(-\frac{1}{64}\right)^n=\frac{1}{2\pi}.$$

This gives $a_0=1/4$, $b_0=1$ and $x_0=-1/64$, as well as the limit

$$\lim_{n\to\infty}a_n=\frac{1}{2\pi}.$$

The data

$$\left|a_0-\frac{1}{2\pi}\right|\approx 9.08\times10^{-2},\qquad \left|a_1-\frac{1}{2\pi}\right|\approx 6.65\times10^{-9},$$
$$\left|a_2-\frac{1}{2\pi}\right|\approx 8.25\times10^{-47},\qquad \left|a_3-\frac{1}{2\pi}\right|\approx 4.57\times10^{-239}$$

are good supporting evidence.

One can also use the series [8, Table 6, $N = 3$]

$$\sum_{n=0}^{\infty} \binom{2n}{n}^3 \left(\frac{1}{6}+n\right)\left(\frac{1}{256}\right)^n = \frac{2}{3\pi}$$

to produce initial conditions $a_0 = 1/6$, $b_0 = 1$, $x_0 = 1/256$ and limiting value

$$\lim_{n\to\infty} a_n = \frac{2}{3\pi}.$$

Another option is to use the series [8, Table 6, $N = 7$]

$$\sum_{n=0}^{\infty} \binom{2n}{n}^3 \left(\frac{5}{42}+n\right)\left(\frac{1}{4096}\right)^n = \frac{8}{21\pi},$$

to give $a_0 = 5/42$, $b_0 = 1$, $x_0 = 1/4096$ and the limit

$$\lim_{n\to\infty} a_n = \frac{8}{21\pi}.$$

These have all been tested numerically. There is no practical advantage to starting with different initial conditions, as the speed of convergence is determined by the quintic nature of the algorithm. □

Observe that the sequence b_k in Example 4.2 with initial conditions given by (4.10) has the property

$$\frac{b_k}{b_0} = 2^k \times \sqrt{\frac{1-26x_k-27x_k^2}{1-26x_0-27x_0^2}},$$

where x_k is given by (4.6). And, for the sequences b_k in Example 4.3 we have

$$\frac{b_k}{b_0} = 5^k \times \sqrt{\frac{1-64x_k}{1-64x_0}}.$$

The factors 2^k and 5^k reflect the quadratic and quintic rates of convergence of the respective iterations, and the factors $1-26x-27x^2$ and $1-64x$ are the leading coefficients in the respective differential equations for the functions $f(x)$, e.g., see [10, p. 176].

5. Self-Replication of Legendre Polynomials

Several generating functions of the Legendre polynomials

$$P_n(x) = {}_2F_1\left(\left.\begin{matrix}-n,\, n+1\\ 1\end{matrix}\,\right|\,\frac{1-x}{2}\right) = \frac{1}{2^n}\sum_{k=0}^{n}\binom{n}{k}^2 (x-1)^k(x+1)^{n-k}$$

are known but the one we are going to treat in this section is given by

$$F(x;z)=\sum_{n=0}^{\infty}\binom{2n}{n}^2 P_n(x)z^n.$$

The Bailey–Brafman identity [9] allows us to recast the two-variable function as a product of two hypergeometric functions,

$$F\left(\frac{U+V-2UV}{U-V};\frac{U-V}{16}\right)={}_2F_1\left(\begin{matrix}\frac12,\frac12\\1\end{matrix}\,\middle|\,U\right)\cdot{}_2F_1\left(\begin{matrix}\frac12,\frac12\\1\end{matrix}\,\middle|\,V\right),$$

and with the help of the quadratic transformation

$${}_2F_1\left(\begin{matrix}\frac12,\frac12\\1\end{matrix}\,\middle|\,u^2\right)=\frac{1}{1+u}\cdot{}_2F_1\left(\begin{matrix}\frac12,\frac12\\1\end{matrix}\,\middle|\,\frac{4u}{(1+u)^2}\right)$$

(essentially, the equation for f_4 in Example 3.1) we get the following self-replicating identity:

$$\begin{aligned}&F\left(\frac{u^2+v^2-2u^2v^2}{u^2-v^2};\frac{u^2-v^2}{16}\right)\\&\quad=\frac{1}{(1+u)(1+v)}F\left(\frac{(1+uv)(u+v)-4uv}{(1-uv)(u-v)};\frac{(1-uv)(u-v)}{4(1+u)^2(1+v)^2}\right).\end{aligned}\tag{5.1}$$

Because of the way the two parameters u and v are tangled in the relation (5.1), it is not easy to perform an analysis of the equation analogous to the one we have had for (1.4). Notice, however, that $F(x;z)$ is a particular case of Appell's F_4 function and that there is a similar-looking identity [16] for Appell's F_1 function (see also [21, Chapter 4]). As the latter identity is known to be linked to a three-term generalization of the arithmetic–geometric mean, we believe that (5.1) and the like equations may find some interesting number-theoretic applications.

Acknowledgments

The work of W. Zudilin was supported by the Max Planck Institute for Mathematics (Bonn).

References

[1] B. Adamczewski, J. P. Bell and E. Delaygue, Algebraic independence of G-functions and congruences "á la Lucas", preprint (2016); arXiv:1603.04187 [math.NT].

[2] G. Almkvist, D. van Straten and W. Zudilin, Generalizations of Clausen's formula and algebraic transformations of Calabi–Yau differential equations, *Proc. Edinburgh Math. Soc.* (2) **54**(2) (2011) 273–295.
[3] D. H. Bailey and J. M. Borwein, *Pi, the Next Generation: A Sourcebook on the Recent History of Pi and its Computation* (Springer, New York 2016).
[4] F. Beukers, Irrationality of π^2, periods of an elliptic curve and $\Gamma_1(5)$, in *Diophantine Approximations and Transcendental Numbers (Luminy, 1982)*, Progress in Mathematics, Vol. 31 (Birkhäuser, Boston, MA, 1983), pp. 47–66.
[5] J. M. Borwein and P. B. Borwein, *Pi and the AGM: A Study in Analytic Number Theory and Computational Complexity*, *Canadian Mathematical Society Series of Monographs and Advanced Texts*, Vol. 4 (Wiley, New York, 1987).
[6] R. P. Brent, Fast multiple-precision evaluation of elementary functions, *J. Assoc. Comput. Mach.* **23**(2) (1976) 242–251.
[7] H. H. Chan and S. Cooper, Eisenstein series and theta functions to the septic base, *J. Number Theory* **128**(3) (2008) 680–699.
[8] H. H. Chan and S. Cooper, Rational analogues of Ramanujan's series for $1/\pi$, *Math. Proc. Cambridge Philos. Soc.* **153**(3) (2012) 361–383.
[9] H. H. Chan, J. Wan and W. Zudilin, Legendre polynomials and Ramanujan-type series for $1/\pi$, *Israel J. Math.* **194**(1) (2013) 183–207.
[10] S. Cooper, Sporadic sequences, modular forms and new series for $1/\pi$, *Ramanujan J.* **29** (2012) 163–183.
[11] S. Cooper and D. Ye, The Rogers–Ramanujan continued fraction and its level 13 analogue, *J. Approx. Theory* **193** (2015) 99–127.
[12] S. Cooper and D. Ye, Level 14 and 15 analogues of Ramanujan's elliptic functions to alternative bases, *Trans. Amer. Math. Soc.* **368**(11) (2016) 7883–7910.
[13] J. Guillera, New proofs of Borwein-type algorithms for Pi, *Integral Transforms Spec. Funct.* **27**(10) (2016) 775–782.
[14] J. Guillera and W. Zudilin, Ramanujan-type formulae for $1/\pi$: The art of translation, in *The Legacy of Srinivasa Ramanujan*, eds. B. C. Berndt and D. Prasad, Ramanujan Mathematical Society, Lecture Notes Series, Vol. 20 (Ramanujan Mathematical Society, 2013) pp. 181–195.
[15] M. E. H. Ismail, *Classical and Quantum Orthogonal Polynomials in One Variable*, Encyclopedia of Mathematics and its Application, Vol. 98 (Cambridge University Press, Cambridge 2005), With two chapters by W. Van Assche and a foreword by R. A. Askey.
[16] K. Koike and H. Shiga, Isogeny formulas for the Picard modular form and a three terms arithmetic geometric mean, *J. Number Theory* **124**(1) (2007) 123–141.
[17] A. Malik and A. Straub, Divisibility properties of sporadic Apéry-like numbers, *Res. Number Theory* **2**(1) (2016) 1–26.
[18] R. Osburn, B. Sahu and A. Straub, Supercongruences for sporadic sequences, *Proc. Edinburgh. Math. Soc.* (2) **59**(2) (2016) 503–518.

[19] M. D. Rogers and A. Straub, A solution of Sun's $520 challenge concerning $520/\pi$, *Intern at J. Number Theory* **9**(5) (2013) 1273–1288.
[20] E. Salamin, Computation of π using arithmetic–geometric mean, *Math. Comp.* **30** (1976) no. 135, 565–570.
[21] D. Schultz, Cubic theta functions and identities for Appell's F_1 function, Ph.D. dissertation, University of Illinois at Urbana-Champaign, Urbana, IL (2014).
[22] N. J. A. Sloane (ed.), The on-line encyclopedia of integer sequences, (2016); `https://oeis.org`.
[23] J. Wan and W. Zudilin, Generating functions of Legendre polynomials: a tribute to Fred Brafman, *J. Approx. Theory* **164** (2012) 488–503.
[24] D. Zagier, Elliptic modular forms and their applications, in *The 1–2–3 of Modular Forms*, Universitext (Springer, Berlin, 2008), pp. 1–103.
[25] D. Zagier, Integral solutions of Apéry-like recurrence equations, *in Groups and Symmetries*, CRM Proceedings Lecture Notes, Vol. 47 (American Mathematical Society, Providence, RI, 2009), pp. 349–366.
[26] W. Zudilin, Ramanujan-type formulae for $1/\pi$: A second wind? in *Modular Forms and String Duality*, eds. N. Yui, H. Verrill and C. F. Doran, Fields Institute Communications, Vol. 54 (American Mathematical Society, Providence, RI, 2008), pp. 179–188.
[27] W. Zudilin, Lost in translation, in *Advances in Combinatorics, Waterloo Workshop in Computer Algebra, W80*, May 26–29, 2011, eds., I. Kotsireas and E. V. Zima (Springer, New York, 2013), pp. 287–293.
[28] W. Zudilin, A generating function of the squares of Legendre polynomials, *Bull. Austral. Math. Soc.* **89**(1) (2014) 125–131.

This page intentionally left blank

Chapter 10

Asymptotics of Orthogonal Polynomials and the Painlevé Transcendents

Dan Dai

Department of Mathematics
City University of Hong Kong, Hong Kong
dandai@cityu.edu.hk

Dedicated to Professor Mourad Ismail on the occasion of his 70th birthday

In this survey, we review asymptotic results of some orthogonal polynomials which are relate to the Painlevé transcendents. They are obtained by applying Deift–Zhou's nonlinear steepest descent method for Riemann–Hilbert problems. In the last part of this chapter, we list several open problems.

Keywords: Orthogonal polynomials; asymptotics; Painlevé transcendents; Riemann–Hilbert problems.

Mathematics Subject Classification 2010: 33E17, 34M55, 41A60

1. Introduction

1.1. Orthogonal polynomials

Let μ be a positive Borel measure on $\mathbb{R}$ and assume that the moments $\int x^n d\mu(x)$ exist for all $n = 0, 1, 2, \ldots$. Then, there exists a unique sequence of monic orthogonal polynomials $\{\pi_n(x)\}_{n=0}^{\infty}$,

$$\pi_n(x) = x^n + \cdots, \tag{1.1}$$

such that

$$\int \pi_m(x)\pi_n(x)d\mu(x) = h_n\delta_{m,n}. \tag{1.2}$$

When $d\mu(x) = w(x)dx$ for some continuous function $w(x)$ on an interval, the corresponding polynomials are called *continuous orthogonal polynomials* and the function $w(x)$ is referred to as the weight function. If μ is a discrete measure, the corresponding polynomials are called *discrete orthogonal polynomials.* For example, when μ is supported on the integers $k \in \mathbb{Z}$ with masses w_k, then instead of (1.2), the polynomials satisfy the following orthogonality relation:

$$\sum_{k\in\mathbb{Z}} \pi_m(k)\pi_n(k)w_k = h_n\delta_{m,n}. \tag{1.3}$$

It is well known that orthogonal polynomials play a remarkable role in many areas of mathematics and physics, such as continued fractions, operator theory (Jacobi operators), approximation theory, numerical analysis, quadrature, combinatorics, random matrices, Radon transform, computer tomography, etc. Of course, the list presented here is far from complete.

Besides their important applications, orthogonal polynomials also satisfy a lot of fascinating properties. Let us mention a few of them below. The classical ones, namely *the Jacobi, Laguerre and Hermite polynomials*, satisfy a second-order differential equation of the following form:

$$c_2(x)y''(x) + c_1(x)y'(x) + c_0(x)y(x) = 0, \tag{1.4}$$

where $c_0(x)$, $c_1(x)$ and $c_2(x)$ are polynomials of degree at most 0, 1 and 2, respectively. From the orthogonality property (1.2), it can be verified that all orthogonal polynomials satisfy a three-term recurrence relation

$$x\pi_n(x) = \pi_{n+1}(x) + a_n\pi_n(x) + b_n\pi_{n-1}(x), \quad n = 1, 2, \ldots, \tag{1.5}$$

with

$$\pi_0(x) = 1, \quad \pi_1(x) = x - a_0, \tag{1.6}$$

where a_n is real and $b_n > 0$ for $n > 0$. In addition, the continuous orthogonal polynomials satisfy the well-known Rodrigues formula

$$\pi_n(x) = \frac{1}{c_n w(x)} \frac{d^n}{dx^n}(w(x)[g(x)]^n), \tag{1.7}$$

where c_n is a constant and $g(x)$ is a polynomial independent of n. Both continuous and discrete orthogonal polynomials have generating functions

$$F(x,z) = \sum_{n=0}^{\infty} d_n \pi_n(x) z^n. \tag{1.8}$$

Here, for the same polynomials $\pi_n(x)$, the function $F(x,z)$ may be different by choosing different coefficients d_n. From the generating function and Cauchy's integral formula, one can easily get an integral representation of the form

$$\pi_n(x) = \frac{1}{2\pi i d_n} \int_C \frac{F(x,z)}{z^{n+1}} dz, \tag{1.9}$$

where C is a closed contour surrounding the origin in the positive direction. For more properties of orthogonal polynomials, we refer to [1, 11, 32, 33, 42, 45, 53], etc.

1.2. Asymptotic methods

In the study of orthogonal polynomials, there is a lot of interest in their asymptotic behaviors as the polynomial degree n is large. In the literature, there are several different methods to derive their asymptotics. For the classical orthogonal polynomials, since they satisfy differential equations of the form (1.4), one can apply the powerful asymptotic methods in the differential equation theory, such as the Liouville–Green approximation (also called the WKB approximation) and the turning point theory. For more details about the differential equation method, we refer to the definitive work of Olver [48]. However, since not all orthogonal polynomials satisfy differential equations, this approach becomes inapplicable and people turn to the integral approach. From the generating function and after a suitable re-scaling of the variable x, one often transforms the integral (1.9) into the form

$$\frac{1}{2\pi i} \int_C g(x,z) e^{nf(x,z)} dz. \tag{1.10}$$

Then, based on the properties of the phase function $f(x,z)$ and its saddle points, one may apply the method of steepest descents to derive the asymptotic expansion of the orthogonal polynomials. For more information about the integral approach, we refer to [59].

Besides the differential equation method and the integral approach mentioned above, researchers have made significant progress on two novel methods in the past twenty years, namely the difference equation method and the Riemann–Hilbert (RH) approach. Starting from the three-term recurrence relation (1.5), Wong and his colleagues developed a turning point theory for second-order difference equations in a series of papers [10, 56, 57]. Their ideas and asymptotic results are similar to those in the differential equation theory. But the asymptotic analysis is far more complicated than the differential cases due to the discrete feature. And the turning point theory for difference equations is completely nontrivial. Nowadays, the difference equation theory is viewed as an analogue of the classical asymptotic theory for linear second-order differential equations; see the survey article by Wong [60].

1.3. The Riemann–Hilbert approach

The RH approach is based on the key observation that orthogonal polynomials are connected with a RH problem as follows (cf. [30]).

The RH problem for $Y(z)$: Assume μ is supported on the real line and $d\mu(x) = w(x)dx$ in (1.2). The problem is to determine a 2×2 matrix-valued function $Y : \mathbb{C} \to \mathbb{C}^{2\times 2}$ such that the following conditions hold:

(Y1) $Y(z)$ is analytic in $\mathbb{C}\backslash\mathbb{R}$.

(Y2) Let $Y_+(x)$ and $Y_-(x)$ denote the limiting values of $Y(z)$ as z approaches $x \in \mathbb{R}$ from the upper and lower half-plane, respectively. Then, $Y(z)$ satisfies the jump condition

$$Y_+(x) = Y_-(x)\begin{pmatrix} 1 & w(x) \\ 0 & 1 \end{pmatrix}, \quad x \in \mathbb{R}. \tag{1.11}$$

(Y3) The asymptotic behavior of $Y(z)$ at infinity is

$$Y(z) = (I + O(1/z))\begin{pmatrix} z^n & 0 \\ 0 & z^{-n} \end{pmatrix}, \quad \text{as } z \to \infty. \tag{1.12}$$

By virtue of the Plemelj formula and Liouville's theorem, one has the following results.

Theorem 1.1 ([30]). *The unique solution to the above RH problem is given by*

$$Y(z) = \begin{pmatrix} \pi_n(z) & \dfrac{1}{2\pi i}\displaystyle\int_{\mathbb{R}} \frac{\pi_n(s)w(s)}{s-z}ds \\ -2\pi i\gamma_{n-1}^2\,\pi_{n-1}(z) & -\gamma_{n-1}^2\displaystyle\int_{\mathbb{R}} \frac{\pi_{n-1}(s)w(s)}{s-z}ds \end{pmatrix}, \tag{1.13}$$

where $\pi_n(z)$ is the monic orthogonal polynomial with respect to the weight $w(x)$, and γ_n is the leading coefficient of the orthonormal polynomial.

Remark 1.2. If the weight function $w(x)$ is supported on certain interval (a, b) instead of $\mathbb{R}$, similar results also hold. The only difference is that one needs to add extra conditions at the endpoints a, b to ensure the uniqueness of the RH problem. For example, see [44] where $(a, b) = (-1, 1)$.

Remark 1.3. One can also formulate an RH problem for the discrete orthogonal polynomials. In this case, the above RH problem becomes an interpolation one and the jump condition (Y2) is replaced by residue conditions; see [4].

On the basis of Theorem 1.1, Deift and Zhou *et al.* [26, 27] developed a very powerful nonlinear steepest descent method to derive the asymptotics of $\pi_n(z)$ from the RH problem for Y; see also [22]. The idea is to obtain, via a series of invertible transformations $Y \to T \to S \to R$, the RH problem for R whose jump is close to the identity matrix.

- $Y \to T$ is to rescale the variable z, and normalize large-z condition (Y3). As a result, $T(z)$ solves an RH problem with oscillatory jumps, normalized at infinity.
- $T \to S$ is to deform the contour. In the meantime, the oscillatory jumps are factorized and $S(z)$ solves an RH problem without oscillation.
- $S \to R$, the final transformation, includes global parametrix construction as well as local parametrix constructions near critical points (usually they are also the self-intersection points of the contour). Then, R satisfies an RH problem with all jumps close to I for large polynomial degree n.

The asymptotic expansion for R can be derived on the whole complex plane. Tracing back, the uniform asymptotics of the orthogonal polynomials is obtained as $n \to \infty$. Since the establishment of Deift–Zhou nonlinear steepest descent method, the RH approach has been successfully applied to study asymptotics of various orthogonal polynomials, as well as their application in proving universality results in random matrix theory, see [3, 4, 17, 23, 26, 44, 55] for a very incomplete list from the literature.

It is well known that, the asymptotic expansions of classical orthogonal polynomials usually involve Airy functions, Bessel functions and trigonometric functions, see [49, Section 18.15]. All of these functions satisfy linear ordinary differential equations (ODEs). In the past a few years, people realize that, when a parameter tends to certain critical values in weight functions, some new functions will appear in the asymptotic expansions. These functions are related to the Painlevé equations, which are nonlinear second-order ODEs. The RH approach plays a crucial role in deriving these Painlevé asymptotics. Indeed, such kind of results have not been obtained by any other methods so far. In this chapter, we will review asymptotic results of some orthogonal polynomials which are related to the Painlevé equations. However, due to the space limitation, it is impossible for us to cover all of the results. For example, we donot mention asymptotics of multiple orthogonal polynomials, which have very important applications in random matrix theory. As a consequence, if any related results are not covered in this survey, the omission certainly does not reflect on the importance of the omitted works.

The rest of the article is arranged as follows. In Section 2, we first briefly introduce the definitions of the Painlevé equations as well as some of their properties. Then, we focus on the RH problems associated with the Painlevé equations. In Section 3, after providing readers with some ideas about why and how the Painlevé asymptotics will appear, we review the asymptotic results of orthogonal polynomials which are related to PI–PV in the literature. In the last section, we list several problems in this research area.

2. Painlevé Equations

In the literature, people are interested in nonlinear ordinary differential equations of the following form

$$w_{xx} = F(x, w, w_x), \tag{2.1}$$

where F is a function meromorphic in x and rational in $w(x)$ and $w'(x)$. Usually the solutions of these equations have movable singularities, which means that the positions of possible singularities depend on the initial conditions of the equations. At the turn of the twentieth century, Painlevé, Gambier, Fuchs *et al.* (1893–1906) classified the equations (2.1) whose solutions are free from movable branch points and essential singularities. This property is called *the Painlevé property* now. It turns out that the equations satisfying the Painlevé property can be reduced to six canonical forms, which are now known as the Painlevé differential equations listed below:

PI: $\quad w_{xx} = 6w^2 + x,$

PII: $\quad w_{xx} = 2w^3 + xw - \alpha,$

$$\text{PIII:}\quad w_{xx} = \frac{1}{w}w_x^2 - \frac{1}{x}w_x + \frac{1}{x}(\alpha w^2+\beta) + \gamma w^3 + \frac{\delta}{w},$$

$$\text{PIV:}\quad w_{xx} = \frac{1}{2w}w_x^2 + \frac{3}{2}w^3 + 4xw^2 + 2(x^2-\alpha)w + \frac{\beta}{w},$$

$$\text{PV:}\quad w_{xx} = \left(\frac{1}{2w}+\frac{1}{w-1}\right)w_x^2 - \frac{1}{x}w_x + \frac{(w-1)^2}{x^2}\left(\alpha w+\frac{\beta}{w}\right) + \frac{\gamma w}{x} + \frac{\delta w(w+1)}{w-1},$$

$$\text{PVI:}\quad w_{xx} = \frac{1}{2}\left(\frac{1}{w}+\frac{1}{w-1}+\frac{1}{w-x}\right)w_x^2 - \left(\frac{1}{x}+\frac{1}{x-1}+\frac{1}{w-x}\right)w_x + \frac{w(w-1)(w-x)}{x^2(x-1)^2}\left(\alpha+\frac{\beta x}{w^2}+\frac{\gamma(x-1)}{(w-1)^2}+\frac{\delta x(x-1)}{(w-x)^2}\right),$$

where α, β, γ and δ are constants. The solutions of PI–PVI are called the *Painlevé transcendents.*

One can see that the study of the Painlevé equations originates from a purely mathematical point of view. However, it turns out that the Painlevé equations satisfy a lot of nice properties and have applications in various areas of physics. Let us briefly mention several properties first. For example, the Painlevé equations appear as similarity reductions of some integrable nonlinear PDEs. For some special parameters, they possess exact solutions which are given in terms of rational functions, algebraic functions or classical special functions. For PII–PVI, the Painlevé transcendents satisfy *Bäcklund transforms*, which can be viewed as nonlinear recurrence relations. Moreover, similar to the role that the classical special functions play in linear physics, the Painlevé transcendents appear in many areas of nonlinear physics, such as statistical mechanics, random matrix theory, quantum gravity and quantum field theory, nonlinear optics and fiber optics, etc. Nowadays, the Painlevé transcendents are viewed as nonlinear analogues of the classical special functions and included in the latest version of mathematical handbook [49, Chapter 32]; also see [19] and references therein for more information.

2.1. Lax pair and Riemann–Hilbert problems

Although the Painlevé equations are nonlinear, each of them can be expressed as the compatibility condition of a system of *linear* differential equations (Lax pair) in the following form:

$$\frac{\partial \Psi}{\partial \lambda} = A(\lambda, x)\Psi, \quad \frac{\partial \Psi}{\partial x} = B(\lambda, x)\Psi. \tag{2.2}$$

Here, $\Psi(\lambda, x)$, $A(\lambda, x)$ and $B(\lambda, x)$ are 2×2 matrix functions, $A(\lambda, x)$ and $B(\lambda, x)$ are rational functions in λ. The compatibility condition is $\frac{\partial^2 \Psi}{\partial \lambda \partial x} = \frac{\partial^2 \Psi}{\partial x \partial \lambda}$, which means

$$A_x - B_\lambda + AB - BA = 0. \tag{2.3}$$

The Lax pair (2.2) is greatly helpful in studying properties of the Painlevé transcendents. One of the direct outcomes is that, from the ordinary differential equation theory and the rational property of $A(\lambda, x)$, it is possible for us to formulate an RH problem for the sectionally holomorphic function $\Psi(\lambda, x)$ in the complex λ-plane, with x appearing as a parameter. The solution of this RH problem is associated with some special solutions of the Painlevé equations if the jump matrices are specified; see [31] for comprehensive information about this method.

Let us take PI as a concrete example to illustrate the relation between an RH problem and the Painlevé transcendents. In [41], Kapaev formulates a model RH problem for $\Psi(\lambda) = \Psi(\lambda; x)$ as follows:

(a) $\Psi(\lambda; x)$ is analytic for $\lambda \in \mathbb{C} \setminus \Gamma_\Psi$, where

$$\Gamma_\Psi = \gamma_{-2} \cup \gamma_{-1} \cup \gamma_1 \cup \gamma_2 \cup \gamma^* \tag{2.4}$$

is illustrated in Fig. 1.

(b) Let $\Psi_\pm(\lambda; x)$ denote the limiting values of $\Psi(\lambda; s)$ as λ tends to the contour Γ_Ψ from the left and right sides, respectively. Then, $\Psi(\lambda; s)$ satisfies the following jump conditions:

$$\Psi_+(\lambda; x) = \Psi_-(\lambda; x) \begin{cases} \begin{pmatrix} 1 & s_k \\ 0 & 1 \end{pmatrix}, & z \in \gamma_k, \ k = \pm 1, \\ \begin{pmatrix} 1 & 0 \\ s_k & 1 \end{pmatrix}, & z \in \gamma_k, \ k = \pm 2, \\ \begin{pmatrix} 0 & -i \\ -i & 0 \end{pmatrix}, & z \in \gamma^*, \end{cases} \tag{2.5}$$

where the complex constants s_k's are the so-called *Stokes multipliers*. They satisfy the following relation:

$$1 + s_k s_{k+1} = -i s_{k+3}, \quad s_{k+5} = s_k, \quad k \in \mathbb{Z}. \tag{2.6}$$

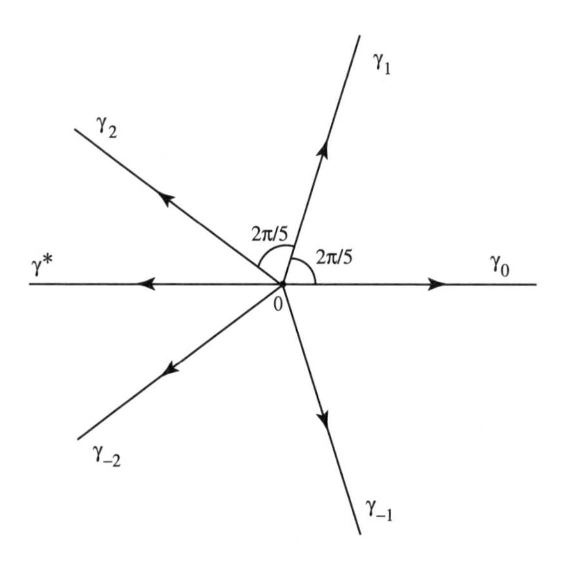

Fig. 1. The contour Γ_Ψ associated with the Painlevé I equation.

(c) As $\lambda \to \infty$, $\Psi(\lambda; x)$ satisfies the asymptotic condition

$$\Psi(\lambda; x) = \lambda^{\frac{1}{4}\sigma_3} \frac{\sigma_3 + \sigma_1}{\sqrt{2}} \times \left(I + \frac{\Psi_{-1}(x)}{\sqrt{\lambda}} + \frac{\Psi_{-2}(x)}{\lambda} + O(\lambda^{-\frac{3}{2}}) \right) e^{\theta(\lambda,x)\sigma_3} \tag{2.7}$$

for $\arg\lambda \in (-\pi, \pi)$, where

$$\theta(\lambda, x) = \frac{4}{5}\lambda^{\frac{5}{2}} + x\lambda^{\frac{1}{2}}, \tag{2.8}$$

σ_1 and σ_3 are the Pauli matrices

$$\sigma_1 = \begin{pmatrix} 0 & 1 \\ 1 & 0 \end{pmatrix}, \quad \sigma_3 = \begin{pmatrix} 1 & 0 \\ 0 & -1 \end{pmatrix}.$$

From the above RH problem, one can prove that

$$w(x) = 2(\Psi_{-2}(x))_{12} \tag{2.9}$$

is a solution of PI, where $\Psi_{-2}(x)$ is given in (2.7). Moreover, it is known that, for every set of Stokes multipliers s_k satisfying (2.6), there exists a unique solution of PI and vice versa. In particular, by choosing

$$s_0 = 0, \quad s_1 = (1-\alpha)i, \quad s_{-1} = \alpha i, \quad s_{\pm 2} = i \tag{2.10}$$

with α being a complex constant, we get a special solution $w_\alpha(x)$ of PI. One can even derive the asymptotic expansions of $w_\alpha(z)$ from the above RH problem when the independent variable z is complex. More precisely, Kapaev [41] showed that $w_\alpha(z)$ is the so-called *tronquée* solution of PI whose asymptotic behavior is given by

$$w_\alpha(z) = w_0(z) + \frac{\alpha i}{\sqrt{\pi}} 2^{-\frac{11}{8}} (-3z)^{-\frac{1}{8}} \exp\left[-\frac{1}{5} 2^{\frac{11}{4}} 3^{\frac{1}{4}} (-z)^{\frac{5}{4}}\right] (1 + O(z^{-\frac{3}{8}})) \tag{2.11}$$

as $z \to \infty$ and $\arg z = \arg(-z) + \pi \in [\frac{3}{5}\pi, \pi]$. Here $w_0(z)$ is the *tritronquée* solution satisfying

$$w_0(z) \sim \sqrt{-z/6}\left[1 + \sum_{k=1}^{\infty} a_k (-z)^{-5k/2}\right] \quad \text{as } z \to \infty,\ -\frac{\pi}{5} < \arg z < \frac{7\pi}{5}, \tag{2.12}$$

where the coefficients a_k can be determined recursively. In general, the PI transcendents are meromorphic functions and possess infinitely many poles in the complex plane. The tronquée and tritronquée solutions mentioned above satisfy the property that the solution $w_\alpha(z)$ is pole-free for z in some sectors of the complex z-plane. For more information about the tritronquée solutions of PI, we refer to [20, 40] as well as references therein.

Remark 2.1. The RH problems for PII–PV can be found in [31, Chapter 5]. About PVI, its Lax pair possesses four regular singular points, which are more than the numbers of singular points of any other Painlevé equations; see [39, Appendix C]. As a consequence, the RH problem for PVI is a little more complicated. For example, from the associated Lax pair, Its, Lisovyy and Prokhorov constructed a RH problem for PVI in [37, Section 3]. Then, they use this problem to study asymptotic behaviors of the corresponding tau functions.

3. The Painlevé Asymptotics

From Section 1.3 and Section 2.1, one can see that both orthogonal polynomials and the Painlevé transcendents satisfy 2×2 RH problems. Then, the natural question is that whether there exist any relations between these two types of RH problems. The answer is positive. Indeed, this is also the reason why the Painlevé type asymptotics appear.

Let us recall the Deift–Zhou nonlinear steepest descent method. The final transform $S \to R$ involves parametrix constructions, which are essentially constructions of approximate solutions to the RH problem S when the parameter n is large. Under certain special situations, the RH problems for orthogonal polynomials near some critical points are very similar to those for the Painlevé equations. Therefore, the RH problems for the Painlevé equations (for example the one in Section 2.1) is adopted to construct the local parametrix in the transform $S \to R$. As a consequence, the Painlevé functions appear in the final asymptotic expansions of orthogonal polynomials.

To make a clearer explanation about how the Painlevé type asymptotics appear, we need the limiting zero distribution of orthogonal polynomials. Given a well-behaved weight function, this distribution can be obtained explicitly from the potential theory; see [50]. Let us rescale the variable and put the weight function $w(x)$ in the form of $e^{-nV(x)}$. The function $V(x)$ is also called *potential*. For simplicity, we first assume the weight function is supported on the whole real axis. It is well known that the limiting zero distribution is the so-called *equilibrium measure* ν^*, which is the unique minimizer of following energy functional among all Borel probability measures ν on $\mathbb{R}$:

$$E_V(\nu) := \int\int \log \frac{1}{|x-y|} d\nu(x)d\nu(y) + \int V(x)d\nu(x). \tag{3.1}$$

If $V(x)$ is real analytic, then

$$\operatorname{supp} \nu^* = \bigcup_{i=1}^{k} [a_i, b_i], \quad a_1 < b_1 < \cdots < a_k < b_k, \tag{3.2}$$

where k is a finite number. Moreover, the density of ν^* is given in the following form:

$$\rho(x) = \frac{d\nu^*}{dx} = \begin{cases} h(x)\sqrt{(x-a_1)(b_1-x)} & \text{if } k=1, \\ h(x)\sqrt{(x-a_1)(x-b_1)\cdots(x-a_k)(b_k-x)} & \text{if } k \geq 2 \end{cases} \tag{3.3}$$

for $x \in \operatorname{supp} \nu^*$, where $h(x)$ is a real analytic function and strictly positive for $x \in \operatorname{supp} \nu^*$; see [24]. Note that the density function $\rho(x)$ is strictly positive in the support and vanishes like a square root at the endpoints. Under this situation, the corresponding asymptotic expansions of orthogonal polynomials are given in terms of trigonometric functions in the compact subset of $\operatorname{supp} \nu^*$ and Airy functions in the neighborhood of a_i and b_i, respectively. If the weight function is supported not on the whole real axis but on an interval (for example $[a, \infty)$), the density function $\rho(x)$ may blow up with an exponent $-1/2$ near the bounded endpoint a, namely,

$$\rho(x) \sim \frac{1}{\sqrt{x-a}} \quad \text{as } x \to a+\,. \tag{3.4}$$

Then, the asymptotic expansions of orthogonal polynomials involve Bessel functions near the endpoint a.

The equilibrium measure depends on the weight function of orthogonal polynomials. For some special weights, the corresponding equilibrium measures are no longer in the form of (3.3) or (3.4). Consequently, some new functions, such as the Painlevé functions, will emerge in the asymptotic expansions of orthogonal polynomials.

3.1. PI asymptotics

When the density of an associated equilibrium measure vanishes with an exponent $3/2$ at an endpoint of its support, the local parametrix construction will involve RH problems of PI. Note that this type of vanishing is impossible in the case of usual orthogonality with respect to exponential weights on the real line. One needs to consider the *non-Hermitian orthogonality* which we will explain below with a concrete example.

In [29], Duits and Kuijlaars consider a varying quartic weight

$$w_N(x) := e^{-NV_t(x)} = e^{-N(tx^4/4+x^2/2)} \quad \text{for } t < 0. \tag{3.5}$$

Obviously, when $t < 0$, all of the moments

$$\mu_{k,N} = \int_{\mathbb{R}} x^k e^{-N(tx^4/4+x^2/2)} dx \tag{3.6}$$

do not exist. As a consequence, the corresponding orthogonal polynomials are not well defined as well. To ensure the convergence of the above integral, one needs to change the original integration contour $\mathbb{R}$ to a contour Γ in the complex plane, such that $\operatorname{Re} V_t(z) \to +\infty$ as $z \to \infty$ along the contour Γ. By choosing Γ to be $\{re^{\pi i/4}, r \in \mathbb{R}\}$, we turn to study the polynomials satisfying the following non-Hermitian orthogonality:

$$\int_\Gamma \pi_{n,N}(z) z^k e^{-N(tz^4/4+z^2/2)} dz = 0, \quad \text{for } k = 0, 1, \ldots, n-1. \tag{3.7}$$

It is interesting to note that the first RH problem for orthogonal polynomials formulated by Fokas, Its and Kitaev [30] is related to the non-Hermitian orthogonality instead of the usual one. However, because the bilinear form

$$\langle p, q\rangle = \int_\Gamma p(z)q(z)e^{-N(tz^4/4+z^2/2)}dz \tag{3.8}$$

is not positive definite, the polynomials $\pi_{n,N}(z)$ may not exist for some n. One of the results in [29] is that they proved the existence of $\pi_{n,n}(z)$ and $\pi_{n\pm1,n}(z)$ when n is large enough.

When $t < 0$, it still makes sense to consider the equilibrium problem in (3.1). Indeed, similar to the case $t \geq 0$, the density of the equilibrium measure can also calculated explicitly:

$$\rho(x) = \frac{t}{2\pi}(x^2 - d_t^2)\sqrt{c_t^2 - x^2} \quad \text{for } x \in [-c_t, c_t], \tag{3.9}$$

where $t_{cr} < t < 0$ with $t_{cr} = -1/12$. Here c_t and d_t are two constants depending on t and $0 < c_t < d_t$. Note that the above density function is similar to that in (3.3). Therefore, like what we have achieved in the classical cases, trigonometric and Airy type asymptotics will also appear. New phenomenon occurs as $t \to t_{cr}+$, where both c_t and d_t tend to $\sqrt{8}$. Then, the density becomes

$$\rho_{cr}(x) = \frac{1}{24\pi}(8 - x^2)^{3/2}, \quad x \in [-\sqrt{8}, \sqrt{8}], \tag{3.10}$$

which vanishes with an exponent 3/2 at endpoints. Taking a *double scaling limit* as $n \to \infty$ and $t \to t_{cr}$, the PI asymptotics appear. To give readers some ideas about the Painlevé type asymptotics, we quote Duits and Kuijlaars' main results below. However, the details of other results will not be included in this chapter due to the space limitation.

Theorem 3.1 ([29, Theorem 1.2]). *Let t vary with n such that*

$$n^{4/5}(t + 1/12) = -c_1 x, \quad c_1 = 2^{-9/5}3^{-6/5}, \tag{3.11}$$

remains fixed, where x is not a pole of $w_1(x)$ in (2.11). Then, for large enough n, the recurrence coefficients $b_{n,n}(t)$ associated with the orthogonal polynomials exist and satisfy

$$b_{n,n}(t) = 2 - 2c_2 w_1(x)n^{-2/5} + O(n^{-3/5}), \quad c_2 = 2^{3/5}3^{2/5}, \tag{3.12}$$

as $n \to \infty$. The above expansion holds uniformly for x in compact subsets of $\mathbb{R}$ not containing any of the poles of $w_1(x)$.

Remark 3.2. To simplify the formulas and make them easier to understand, we choose the parameters $\alpha = \beta = 1$ in [29, Theorem 1.2]. In this case, the other recurrence coefficient $a_{n,n}(t)$ in (1.5) is equal to 0 for all n.

It is well known that the Painlevé transcendents possess infinitely many poles in the complex plane in general. In the above theorem, to ensure the validity of the results, the variable x is required to be bounded away from the poles of the tronquée $w_\alpha(x)$ in (2.11). Recently, through a more delicate *triple scaling limit*, Bertola and Tovbis [5] successfully obtain the asymptotics near the poles of $w_\alpha(x)$.

Similar situation like that in (3.10) also appears for the cubic potentials

$$V_t(z) = tz^3 + z^2/2 \quad \text{for } t < 0. \tag{3.13}$$

In [6, 7], Bleher and Deaño derived the PI asymptotics of the corresponding orthogonal polynomials as well as their applications in random matrix theory. Recently, inspired by the study of Wigner time-delay in mathematical physics (cf. [54, 58]), Xu, Dai and Zhao [63] considered a singular potential as follows:

$$V_t(z) = z - \log z + t/z \quad \text{for } t < 0. \tag{3.14}$$

Note that the above potential has a pole at $z = 0$, which is different from the polynomial potentials considered before. Applying the Deift–Zhou steepest descent analysis, we also obtained the PI asymptotics for the corresponding orthogonal polynomials.

Remark 3.3. PI asymptotics appear only when the non-Hermitian orthogonality is considered. The reason is that, for the usual orthogonality, the density of the equilibrium measure can only vanish at an endpoint with an exponent $(4k+1)/2$ with $k \in \mathbb{N} \cup \{0\}$; see [25, 43] for more details. The generic case $k = 0$ gives the Airy-type asymptotics. The first critical case $k = 1$ yields asymptotic expansion in terms of the second member of the Painlevé I hierarchy; see [18]. For $k \geq 2$, it is expected that the asymptotic expansions will involve higher members of the Painlevé I hierarchy.

3.2. PII asymptotics

The first PII asymptotics for orthogonal polynomials appear in the paper [3] by Baik, Deift and Johansson, where they studied the length of the longest increasing subsequence of a random permutation of $\{1, 2, \ldots, n\}$. In their paper, they need to know the asymptotic behavior of certain *orthogonal polynomials on the unit circle* (OPUC). However, the following quartic potential may be a simpler example to understand. We will come back to the PII asymptotics about OPUC at the end of this section.

In [8], Bleher and Its considered the following varying quartic weight

$$w_N(x) := e^{-NV_t(x)} = e^{-N(x^4/4+tx^2/2)}, \quad x \in \mathbb{R}. \tag{3.15}$$

Since the leading coefficient of $V_t(x)$ is positive, all of the moments $\mu_{k,N}$ given in (3.6) exist and corresponding orthogonal polynomials satisfy the usual orthogonality. One may compare the difference between the two quartic weights in (3.5) and (3.15). For the quartic potential in (3.15), the corresponding equilibrium measure can also be computed explicitly. Depending on the parameter t in (3.15), we have

$$\rho(x) = \frac{x^2+b_t}{2\pi}\sqrt{a_t^2-x^2}, \quad x \in [-a_t, a_t], \quad \text{when } t > -2, \tag{3.16}$$

with $a_t, b_t > 0$;

$$\rho_{cr}(x) = \frac{x^2}{2\pi}\sqrt{4-x^2}, \quad x \in [-2,2], \quad \text{when } t = t_{cr} = -2, \tag{3.17}$$

and

$$\rho(x) = \frac{|x|}{2\pi}\sqrt{(a_t^2-x^2)(x^2-b_t^2)}, \quad 0 < b_t \le |x| \le a_t, \quad \text{when } t < -2, \tag{3.18}$$

with $a_t = \sqrt{2-t}$ and $b_t = \sqrt{-2-t}$; see Fig. 2.

When $t > -2$ and $t < -2$, the density functions in (3.16) and (3.18) are similar to that in (3.3). But when $t = -2$, the density function $\rho_{cr}(x)$ in (3.17) vanishes quadratically at an interior point of its support. Moreover, the case $t = -2$ indicates a transition where the support of $\rho(x)$ splits from one cut (when $t > -2$) to two cuts (when $t < -2$). Consequently, in the neighborhood of 0, a PII RH problem is needed to construct the local parametrix and the PII asymptotics follow. Note that the special PII transcendent involved here is the famous Hastings–McLeod solution for homogeneous PII equation (the constant term $\alpha = 0$ in PII); see [8]. For more properties about the Hastings–McLeod solution, as well as other special solutions of homogeneous PII, we refer to [28] and references therein.

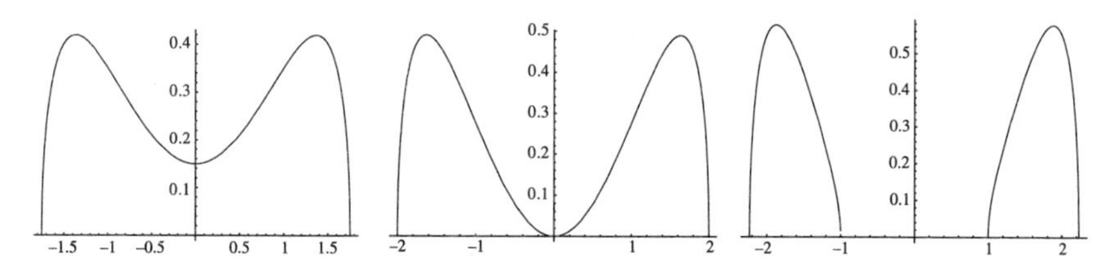

Fig. 2. From the left to right, the density $\rho(x)$ when $t > -2$, $t = -2$ and $t < -2$, respectively.

Soon, the work of Bleher and Its [8] was generalized by Claeys and Kuijlaars [15] and Claeys, Kuijlaars and Vanlessen [17]. Instead of the quartic potential in (3.15), it was shown in [15, 17] that similar results also hold for more general potential $V(x)$, as long as the corresponding equilibrium measure satisfies the similar properties illustrated in Fig. 2. In addition, if an algebraic singularity is added at the point where that cut-split takes place (for example, one may include an extra factor $|x|^\alpha$ in the quartic weight (3.15)), then we will obtain asymptotics related to the inhomogeneous PII equation; see [16, 17].

On the other hand, when studying polynomials orthogonal with respect to the perturbed Hermite weight, people found out that their asymptotics are related to the Painlevé XXXIV equation; see [38] for an algebraic perturbation and Xu and Zhao [64] for a jump perturbation. Note that Painlevé XXXIV can be reduced to PII through a simple transform.

Finally, let us go back to OPUC mentioned at the beginning of this section. Let $p_n(z) = \kappa_n z^n + \cdots$, $\kappa_n > 0$, be a polynomial orthonormal with respect to the weight $w(z)$ on the unit circle, they satisfy the following orthogonality relation

$$\frac{1}{2\pi i}\int_{|z|=1} p_n(z)\,\overline{p_m(z)}\,w(z)\frac{dz}{z} = \delta_{m,n}. \tag{3.19}$$

Sometimes, the above formula is rewritten as

$$\frac{1}{2\pi}\int_{-\pi}^{\pi} p_n(e^{i\theta})\,\overline{p_m(e^{i\theta})}\,w(e^{i\theta})d\theta = \delta_{m,n}. \tag{3.20}$$

Instead of (1.5) in the real case, OPUCs satisfy a recurrence relation as follows

$$\kappa_n p_{n+1}(z) = \kappa_{n+1} z\, p_n(z) + p_{n+1}(0) p_n^*(z), \tag{3.21}$$

where $p_n^*(z) = z^n\,\overline{p_n(1/\bar{z})}$. From the orthogonality condition, it is also possible to formulate a RH problem for OPUC as follows:

(Y1) $Y(z)$ is analytic in $\mathbb{C}\backslash\Sigma$, where Σ is the unit circle oriented counter-clockwise.

(Y2) $Y(z)$ satisfies the jump condition

$$Y_+(z) = Y_-(z)\begin{pmatrix} 1 & w(z)/z^n \\ 0 & 1 \end{pmatrix}, \quad z \in \Sigma. \tag{3.22}$$

(Y3) The asymptotic behavior of $Y(z)$ at infinity is

$$Y(z) = (I + O(1/z)) \begin{pmatrix} z^n & 0 \\ 0 & z^{-n} \end{pmatrix}, \quad \text{as } z \to \infty. \tag{3.23}$$

For more information and properties about OPUC, we refer to Simon's treatise [51, 52].

For the OPUC appearing in [3], they are orthogonal about the varying weight

$$w_N(z) := e^{-NV_t(z)} = e^{\frac{tN}{2}(z+z^{-1})}, \quad z \in \Sigma. \tag{3.24}$$

The associated equilibrium measure is supported on the whole circle or part of it, depending on the parameter t. When t achieves the critical value $t_{\text{cr}} = 1$, the measure becomes

$$d\mu_{\text{cr}}(\theta) = \frac{1}{2\pi}(1 + \cos\theta)d\theta, \quad \theta \in [0, 2\pi]. \tag{3.25}$$

One can see that the above density vanishes quadratically at $\theta = \pi$, which is similar to the real quartic case (3.17). Based on the above properties of the equilibrium measure, PII asymptotics appear under a double scaling limit as $n \to \infty$ and $t \to t_{\text{cr}}$. It should be emphasized that, this is the first time that the Painlevé RH problems emerge in the local parametrix construction of the RH analysis. Since the work of Baik, Deift and Johansson [3], more and more Painlevé asymptotics are discovered while studying orthogonal polynomials with different weight functions.

3.3. PIII asymptotics

Regarding PIII, because both 0 and ∞ are irregular singular points in the first equation of its Lax pair (2.2), then the corresponding RH problem possesses an *essential singularity* at 0. While perturbing the classical Hermite and Laguerre weight with some essential singularities,

$$w(x;t) = e^{-\frac{t}{2x^2} - \frac{x^2}{2}}, \quad t > 0,\ x \in \mathbb{R}, \tag{3.26}$$

$$w(x;t) = x^\alpha e^{-x-\frac{t}{x}}, \quad x \in (0,\infty), \quad t > 0, \quad \alpha > 0, \tag{3.27}$$

Brightmore, Mezzadri and Mo [9] and Xu, Dai and Zhao [61, 62] obtained the PIII asymptotics for polynomials orthogonal with respect to the above weight functions, respectively. Later, Atkin, Claeys and Mezzadri [2] generalized (3.27) and studied the weight

$$w(x;t) = x^\alpha e^{-x-(\frac{t}{x})^k}, \quad k \in \mathbb{N}. \tag{3.28}$$

In the end, the asymptotic expansions in [2] are given in terms of solutions to a hierarchy of the PIII equations. On the other hand, Xu and Zhao [66] and Zeng, Xu and Zhao [67] considered a modified Jacobi weight of the form

$$w(x;t) = (1-x^2)^{\beta}(t^2-x^2)^{\alpha}, \quad x \in (-1,1) \tag{3.29}$$

with $\beta > -1$, $\alpha+\beta > -1$ and $t > 1$. Their asymptotics is essentially related to the generalized PV equation. After a Möbius transformation, this equation can be transformed to a PIII.

3.4. PIV asymptotics

To obtain the PIV asymptotics, like Section 3.1 for the PI cases, we need to change the orthogonality interval on the real line to a contour in the complex plane and consider non-Hermitian orthogonality. In [21], Dai and Kuijlaars considered the following weight

$$z^{-N+\nu}e^{-Nz}(z-1)^{2b}. \tag{3.30}$$

The equilibrium measure is supported on the so-called Szegő curve $\mathcal{S}$

$$\mathcal{S} := \{z \in \mathbb{C} : |ze^{1-z}| = 1 \text{ and } |z| \le 1\}. \tag{3.31}$$

Moreover, the density of the equilibrium measure vanishes linearly at the point $z=1$. Note that such kind of vanishing behavior is impossible for the usual orthogonality. For the usual case, if the density vanishes at an interior point z_0 of its support, it must behaves like $|z-z_0|^{2k}$ for $k \in \mathbb{N}$. When $k=1$, we have a density similar to that in (3.17) and obtain the PII asymptotics. For the current case, we achieved the PIV asymptotics instead.

3.5. PV asymptotics

Note that the original PV equation possesses three singularities: 0, 1 and ∞. The corresponding RH problem also includes three singularities, which are more than those in the RH problems for PI–PIV. Consequently, the contours and the jump conditions in the RH problem for PV become more complicated. On the other hand, if the asymptotics of orthogonal polynomials involve PV, their weight functions are supposed to have some singularities. Moreover, these singularities should influence each other when the parameters in the weight function vary. In [13], Claeys, Its and Krasovsky considered OPUC whose weight function possesses a Fisher–Hartwig singularity, which combines a jump-type and a root-type singularity. Taking a double scaling limits, they derived the PV asymptotics for the related OPUC. See also Claeys and Krasovsky [14] for two merging Fisher–Hartwig singularities for OPUC. Similar situations also exist for polynomials orthogonal on real intervals. For example, Claeys and Fahs [12] studied weight functions with merging algebraic singularities; Xu and Zhao [65, 66] considered modified Jacobi weights where the algebraic singularity tends to the endpoint of the orthogonal interval. For all these cases, the PV asymptotics come into play.

4. Further Problems

In this section, we list some unsolved problems in this area.

1. The PVI asymptotics: To the best of our knowledge, the PVI asymptotics for orthogonal polynomials have never appeared in the literature. Using the isomonodromy method introduced in [31], one can derive RH problems for PVI from its Lax pair; see Remark 3.3. Note that, the RH problems for Painlevé equations may not be unique. If one finds a nice RH problem for PVI, this problem can be employed to construct local parametrices for orthogonal polynomials with certain special weight functions. Then, the PVI asymptotics will be obtained accordingly.

2. Discrete orthogonal polynomials: Although these polynomials satisfy a discrete orthogonality in (1.3), it is still possible to formulate a discrete RH problem (also called interpolation problem) for them; see Remark 1.3. Starting from this problem, Baik *et al.* [4] developed Deift and Zhou's method to derive the asymptotic expansions for discrete orthogonal polynomials. Note that these polynomials possess some special features. For example, there exist the so-called *saturated regions*, where the density of the limiting zero distribution reaches the maximum (the density of the orthogonality nodes distribution) in some intervals. This leads to an *upper constraint* for the equilibrium measure in the energy minimization problem (3.1). However, despite these differences, similar asymptotic expansions are also obtained, which involve trigonometric functions and the Airy functions; for example, see [4]. Recently, while studying nonintersecting Brownian motions on the half-line, Liechty [46] considered discrete orthogonal polynomials with respect to the following weight

$$w(x_k) = \exp\left(-\frac{n\pi^2 t}{2}x^2\right), \quad x_k = \frac{k-\alpha}{n}, \text{ for } k \in \mathbb{Z} \text{ and } \alpha \in [-1/2, 1/2]. \tag{4.1}$$

Note that there exists a critical value $t_{\rm cr} = 1$ when the upper constraint of the equilibrium measure is just active. In a double scaling limit as $n \to \infty$ and $t \to 1$, the PII asymptotics for the orthogonal polynomials were derived in [46]. See also a subsequent work by Liechty and Wang [47] where a class of similar discrete orthogonal polynomials is studied. Now the question is: will there appear any other Painlevé asymptotics for discrete orthogonal polynomials?

3. q-Orthogonal Polynomials: In Ismail's monograph [33], there is a large part related to q-series and q-orthogonal polynomials. With his colleagues, Ismail has also studied their asymptotics, which usually involve the q-Airy functions and Jacobi theta functions; for example see [34–36], etc. It is well known that both q-orthogonal polynomials and q-Airy functions are q-generalizations of the corresponding classical ones. In the literature, there is also a considerable amount of work about q-Painlevé equations. It would be very interesting if some connections between q-orthogonal polynomials and q-Painlevé equations can be established.

Acknowledgments

The author is grateful for Professors Shuai-Xia Xu and Yu-Qiu Zhao for their helpful comments and discussions. The present work was partially supported by grants from the Research Grants Council of the Hong Kong Special Administrative Region, China (Project No. CityU 11300814, CityU 11300115).

References

[1] G. E. Andrews, R. Askey and R. Roy, *Special Functions*, Encyclopedia of Mathematics and its Applications (Cambridge University Press, Cambridge, 1999).

[2] M. R. Atkin, T. Claeys and F. Mezzadri, Random matrix ensembles with singularities and a hierarchy of Painlevé III equations, *Int. Math. Res. Notices* (2016) **2016**(8) 2320–2375.

[3] J. Baik, P. Deift and K. Johansson, On the distribution of the length of the longest increasing subsequence of random permutations, *J. Amer. Math. Soc.* **12**(4) (1999) 1119–1178.

[4] J. Baik, T. Kriecherbauer, K. T.-R. McLaughlin and P. D. Miller, *Discrete Orthogonal Polynomials: Asymptotics and Applications*, Annals of Mathematics Studies, Vol. 164 (Princeton University Press, Princeton, NJ, 2007).

[5] M. Bertola and A. Tovbis, Asymptotics of orthogonal polynomials with complex varying quartic weight: global structure, critical point behaviour and the first Painlevé equation, *Constr. Approx.* **41** (2015) 529–587.

[6] P. M. Bleher and A. Deaño, Topological expansion in the cubic random matrix model, *Int. Math. Res. Notices* **2013** (2013) 2699–2755.

[7] P. M. Bleher and A. Deaño, Painlevé I double scaling limit in the cubic matrix model, *Random Matrices Theory Appl.* **5**(2) (2016) 1650004, 58 pp.

[8] P. Bleher and A. Its, Double scaling limit in the random matrix model: the Riemann–Hilbert approach, *Comm. Pure Appl. Math.* **56**(4) (2003) 433–516.

[9] L. Brightmore, F. Mezzadri and M. Y. Mo, A matrix model with a singular weight and Painlevé III, *Comm. Math. Phys.* **333** (2015) 1317–1364.

[10] L.-H. Cao and Y.-T. Li, Linear difference equations with a transition point at the origin, *Anal. Appl.* (*Singap.*) **12**(1) (2014) 75–106.

[11] T. S. Chihara, *An Introduction to Orthogonal Polynomials* (Gordon and Breach Science Publishers, New York, 1978).

[12] T. Claeys and B. Fahs, Random matrices with merging singularities and the Painlevé V equation, *SIGMA* **12** (2016) Paper 031, 44 pp.

[13] T. Claeys, A. Its and I. Krasovsky, Emergence of a singularity for Toeplitz determinants and Painlevé V, *Duke Math. J.* **160**(2) (2011) 207–262.

[14] T. Claeys and I. Krasovsky, Toeplitz determinants with merging singularities, *Duke Math. J.* **164**(15) (2015) 2897–2987.

[15] T. Claeys and A. B. J. Kuijlaars, Universality of the double scaling limit in random matrix models, *Comm. Pure Appl. Math.* **59** (2006) 1573–1603.

[16] T. Claeys and A. B. J. Kuijlaars, Universality in unitary random matrix ensembles when the soft edge meets the hard edge, in *Integrable Systems and Random Matrices*, Contemporary Mathematics, Vol. 458 (American Mathematical Society, Providence, RI, 2008), pp. 265–279.

[17] T. Claeys, A. B. J. Kuijlaars and M. Vanlessen, Multi-critical unitary random matrix ensembles and the general Painlevé II equation, *Ann. of Math.* **168** (2008) 601–641.

[18] T. Claeys and M. Vanlessen, Universality of a double scaling limit near singular edge points in random matrix models, *Comm. Math. Phys.* **273**(2) (2007) 499–532.

[19] P. A. Clarkson, Painlevé equations-nonlinear special functions, in *Orthogonal Polynomials and Special Functions*, 331–411, Lecture Notes in Mathematics, Vol. 1883 (Springer, Berlin, 2006), pp. 331–411.

[20] O. Costin, M. Huang and S. Tanveer, Proof of the Dubrovin conjecture and analysis of the tritronquée solutions of PI, *Duke Math. J.* **163**(4) (2014) 665–704.

[21] D. Dai and A. B. J. Kuijlaars, Painlevé IV asymptotics for orthogonal polynomials with respect to a modified Laguerre weight, *Stud. Appl. Math.* **122**(1) (2009) 29–83.

[22] P. Deift, *Orthogonal Polynomials and Random Matrices: A Riemann–Hilbert Approach*, Courant Lecture Notes, Vol. 3 (New York University, 1999).

[23] P. Deift, A. Its and I. Krasovsky, Asymptotics of Toeplitz, Hankel, and Toeplitz+Hankel determinants with Fisher–Hartwig singularities, *Ann. of Math.* **174**(2) (2011) 1243–1299.

[24] P. Deift, T. Kriecherbauer and K. T.-R. McLaughlin, New results for the asymptotics of orthogonal polynomials and related problems via the Lax–Levermore method, in *Recent Advances in Partial Differential Equations*, Proceedings of Symposia in Applied Mathematics, Vol. 54 (American Mathematical Society, Providence, RI, 1998), pp. 87–104.

[25] P. Deift, T. Kriecherbauer and K. T.-R. McLaughlin, New results on the equilibrium measure for logarithmic potentials in the presence of an external field, *J. Approx. Theory* **95**(3) (1998) 388–475.

[26] P. Deift, T. Kriecherbauer, K. T.-R. McLaughlin, S. Venakides and X. Zhou, Uniform asymptotics for polynomials orthogonal with respect to varying exponential weights and applications to universality questions in random matrix theory, *Comm. Pure Appl. Math.* **52** (1999) 1335–1425.

[27] P. Deift, T. Kriecherbauer, K. T.-R. McLaughlin, S. Venakides and X. Zhou, Strong asymptotics of orthogonal polynomials with respect to exponential weights, *Comm. Pure Appl. Math.* **52** (1999) 1491–1552.

[28] P. Deift and X. Zhou, Asymptotics for the Painlevé II equation, *Comm. Pure Appl. Math.* **48**(3) (1995) 277–337.

[29] M. Duits and A. B. J. Kuijlaars, Painlevé I asymptotics for orthogonal polynomials with respect to a varying quartic weight, *Nonlinearity* **19** (2006) 2211–2245.

[30] A. S. Fokas, A. R. Its and A. V. Kitaev, The isomonodromy approach to matrix models in 2D quantum gravity, *Comm. Math. Phys.* **147** (1992) 395–430.

[31] A. S. Fokas, A. R. Its, A. A. Kapaev and V. Yu. Novokshenov, in *Painlevé Transcendents: The Riemann–Hilbert Approach*, AMS Mathematical Surveys and Monographs, Vol. 128 (American Mathematical Society, Providence RI, 2006).

[32] W. Gautschi, *Orthogonal Polynomials: Computation and Approximation*, Numerical Mathematics and Scientific Computation (Oxford University Press, New York, 2004).

[33] M. E. H. Ismail, *Classical and Quantum Orthogonal Polynomials in one Variable*, Encyclopedia in Mathematics (Cambridge University Press, Cambridge, 2005).

[34] M. E. H. Ismail, Asymptotics of q-orthogonal polynomials and a q-Airy function, *Int. Math. Res. Notices* 2005(18) 1063–1088.

[35] M. E. H. Ismail and X. Li, Plancherel–Rotach asymptotics for q-orthogonal polynomials, *Constr. Approx.* **37**(3) (2013) 341–356.

[36] M. E. H. Ismail and R. Zhang, Scaled asymptotics for q-orthogonal polynomials, *C. R. Math. Acad. Sci. Paris* **344**(2) (2007) 71–75.

[37] A. Its, O. Lisovyy and A. Prokhorov, Monodromy dependence and connection formulae for isomonodromic tau functions, preprint (2016); arXiv:1604.03082.

[38] A. R. Its, A. B. J. Kuijlaars and J. Östensson, Critical edge behavior in unitary random matrix ensembles and the thirty-fourth Painlevé transcendent, *Int. Math. Res. Notices* **2008** (9) 2008 67 pp.
[39] M. Jimbo and T. Miwa, Monodromy preserving deformation of linear ordinary differential equations with rational coefficients, II, *Physica D* **2**(3) (1981) 407–448.
[40] N. Joshi and A. V. Kitaev, On Boutroux's tritronquée solutions of the first Painlevé equation, *Stud. Appl. Math.* **107** (2001) 253–291.
[41] A. A. Kapaev, Quasi-linear Stokes phenomenon for the Painlevé first equation, *J. Phys. A* **37** (2004) 11149–11167.
[42] R. Koekoek, P. A. Lesky and R. F. Swarttouw, *Hypergeometric Orthogonal Polynomials and Their q-Analogues*, Springer Monographs in Mathematics (Springer, Berlin, 2010).
[43] A. B. J. Kuijlaars and K. T.-R. McLaughlin, Generic behavior of the density of states in random matrix theory and equilibrium problems in the presence of real analytic external fields, *Comm. Pure Appl. Math.* **53**(6) (2000) 736–785.
[44] A. B. J. Kuijlaars and K. T.-R. McLaughlin, W. Van Assche and M. Vanlessen, The Riemann–Hilbert approach to strong asymptotics for orthogonal polynomials on $[-1,1]$, *Adv. Math.* **188**(2) (2004) 337–398.
[45] A. L. Levin and D. S. Lubinsky, *Orthogonal Polynomials for Exponential Weights* (Springer, New York, 2001).
[46] K. Liechty, Nonintersecting Brownian motions on the half-line and discrete Gaussian orthogonal polynomials, *J. Stat. Phys.* **147**(3) (2012) 582–622.
[47] K. Liechty and D. Wang, Nonintersecting Brownian motions on the unit circle, *Ann. Probab.* **44**(2) (2016) 1134–1211.
[48] F. W. J. Olver, *Asymptotics and Special Functions* (A K Peters, Ltd., Wellesley, MA, 1997).
[49] F. Olver, D. Lozier, R. Boisvert and C. Clark, *NIST Handbook of Mathematical Functions* (Cambridge University Press, Cambridge, 2010).
[50] E. B. Saff and V. Totik, *Logarithmic Potentials with External Fields*, Grundlehren der Mathematischen Wissenschaften, Vol. 316 (Springer, Berlin, 1997).
[51] B. Simon, *Orthogonal Polynomials on the Unit Circle, V.1: Classical Theory*, AMS Colloquium Series (American Mathematical Society, Providence, RI, 2005).
[52] B. Simon, *Orthogonal Polynomials on the Unit Circle, V.2: Spectral Theory*, AMS Colloquium Series (American Mathematical Society, Providence, RI, 2005).
[53] G. Szegő, *Orthogonal Polynomials*, 4th edn., Colloquium Publications, Vol. 23 (American Mathematical Society, Providence, RI, 1975).
[54] C. Texier and S. N. Majumdar, Wigner time-delay distribution in chaotic cavities and freezing transition, *Phys. Rev. Lett.* **110** (2013) 250602.

[55] M. Vanlessen, Strong asymptotics of Laguerre-type orthogonal polynomials and applications in random matrix theory, *Constr. Approx.* **25** (2007) 125–175.

[56] Z. Wang and R. Wong, Asymptotic expansions for second-order linear difference equations with a turning point, *Numer. Math.* **94**(1) (2003) 147–194.

[57] Z. Wang and R. Wong, Linear difference equations with transition points, *Math. Comp.* **74**(250) (2005) 629–653.

[58] E. P. Wigner, Lower limit for the energy derivative of the scattering phase shift, *Phys. Rev.* **98** (1955) 145–147.

[59] R. Wong, *Asymptotic Approximations of Integrals* (Academic Press, Boston, 1989). Reprinted by SIAM, Philadelphia, PA, 2001.

[60] R. Wong, Asymptotics of linear recurrences, *Anal. Appl.* (*Singap.*) **12**(4) (2014) 463–484.

[61] S.-X. Xu, D. Dai and Y.-Q. Zhao, Critical edge behavior and the Bessel to Airy transition in the singularly perturbed Laguerre unitary ensemble, *Comm. Math. Phys.* **332** (2014) 1257–1296.

[62] S.-X. Xu, D. Dai and Y.-Q. Zhao, Painlevé III asymptotics of Hankel determinants for a singularly perturbed Laguerre weight, *J. Approx. Theory* **192** (2015) 1–18.

[63] S.-X. Xu, D. Dai and Y.-Q. Zhao, Hankel determinants for a singular complex weight and the first and third Painlevé transcendents, *J. Approx. Theory* **205** (2016) 64–92.

[64] S.-X. Xu and Y.-Q. Zhao, Painlevé XXXIV asymptotics of orthogonal polynomials for the Gaussian weight with a jump at the edge, *Stud. Appl. Math.* **127**(1) (2011) 67–105.

[65] S.-X. Xu and Y.-Q. Zhao, Critical edge behavior in the modified Jacobi ensemble and the Painlevé V transcendents, *J. Math. Phys.* **54**(8) (2013) 083304, 29 pp.

[66] S.-X. Xu and Y.-Q. Zhao, Critical edge behavior in the modified Jacobi ensemble and Painlevé equations, *Nonlinearity* **28**(6) (2015) 1633–1674.

[67] Z.-Y. Zeng, S.-X. Xu and Y.-Q. Zhao, Painlevé III asymptotics of Hankel determinants for a perturbed Jacobi weight, *Stud. Appl. Math.* **135**(4) (2015) 347–376.

Chapter 11

From the Gaussian Circle Problem to Multivariate Shannon Sampling

Willi Freeden[*,‡] **and M. Zuhair Nashed**[†,§]

[*]Geomathematics Group, University of Kaiserslautern
MPI-Building G 26, 67663 Kaiserslautern, Germany
[†]Department of Mathematics, University of Central Florida
Orlando, FL 32816, USA
[‡]freeden@rhrk.uni-kl.de
[§]zuhair.nashed@ucf.edu

Dedicated to Professor Mourad Ismail on the occasion of his 70th birthday

A historical overview leading to present generalizations of Shannon's sampling theorem (1949) and starting from the Gaussian circle problem (1801) is sketched. It is shown that the bridge between Gauss's and Shannon's work is constituted by certain extensions of the famous Hardy–Landau identities in geometric lattice point theory. Particular interest is laid on the matter dealing with bandlimited functions corresponding to, e.g., geoscientifically relevant regions. Emphasis is put on the study of the convergence of the resulting multivariate cardinal series, as well as on the delicate explicit specification of under- and oversampling in multivariate Shannon sampling. Finally, the routes to sampling expansions are exhibited in Paley–Wiener spaces, leading to multivariate sinc-type reproducing kernels.

Keywords: Gaussian circle problem; Hardy-type lattice point summation; Shannon-type sampling; Gaussian summability of cardinal series; explicit over- and undersampling involving regular regions; Paley–Wiener spaces; sinc-type reproducing kernels.

Mathematics Subject Classification 2010: 11P21, 40A25, 62D05, 65B15

1. Introduction

In a much quoted dictum, the "Princeps Mathematicorum" Carl Friedrich Gauss (1777–1855) asserted that

> "*Mathematics is the Queen of the Sciences and the Theory of Numbers is the Queen of Mathematics.*"

Moreover, in the introduction to "Eisenstein's Mathematische Abhandlungen" (1847) C.F. Gauss wrote:

> "*The Higher Arithmetic presents us with an inexhaustible storehouse of interesting truths-of truths, too, which are not isolated but stand in the closest relation to one another, and between which, with each successive advance of the science, we continually discover new and sometimes wholly unexpected points of contact. A great part of the theories of Arithmetic derive an additional charm from the peculiarity that we easily arrive by induction at important propositions which have the stamp of simplicity upon them but the demonstration of which lies so deep as not to be discovered until after many fruitless efforts; and even then it is obtained by some tedious and artificial process while the simpler methods of proof long remain hidden from us.*"

All this is well illustrated by what is perhaps Gauss's most profound publication, namely his "Disquisitiones arithmeticae", original Latin edition by Gerhard Fleischer, Lipsiae (Leipzig) 1801. It has been described, quite justifiably, as the "Magna Carta of Number Theory", and the depth and originality of thoughts to be manifested in this work are particularly remarkable considering that they were written when Gauss was only about 18 years of age. In view of the great impact Gauss had on large areas of modern number theory, anything even approaching a comprehensive representation of their influence seems untenable. It is not surprising that there is a huge amount of literature concerned with Gauss's number theoretical results, and his influence on modern mathematics (see also [15]) is enormous.

The purpose of this contribution is to address the question if settings of the "Queen of Mathematics" indeed admit a strong link to modern sampling theory and settings. Actually, our contribution aims at showing that there are certain important historical stages which form the basis of today's multivariate Shannon sampling theorem, starting from the circle problem (see [17]), leading over to Hardy–Landau identities and Hardy's conjecture (see [18, 19, 30, 31, 35, 36]) about the asymptotic behavior of the number of lattice points inside circles, continuing with multivariate weighted lattice point sums over regular regions (cf. [10, 11]), and resulting in Shannon-type sampling theory (as recently presented by Freeden and Nashed [14]).

2. Lattice Points Inside Circles

We start our bridge from Gaussian concepts of number theory to modern sampling with a recapitulation of some results on the number of lattice points inside circles $\mathbb{S}_N^1 = \{x \in \mathbb{R}^2 : |x| = N\}$ of radii $N > \frac{\sqrt{2}}{2}$ around the origin 0 (Fig. 1); (for more background material and deeper number theoretical concepts, the readers are referred, e.g., to the monographs [11, 15, 16, 32]).

The problem of determining the total number of lattice points of $\mathbb{Z}^2$ inside and on the boundary of disks $\overline{\mathbb{B}_N^2} = \{x \in \mathbb{R}^2 : |x| \leq N\}$, i.e.,

$$\#_{\mathbb{Z}^2}(\overline{\mathbb{B}_N^2}) = \#\{(n_1, n_2)^T \in \mathbb{Z}^2 : n_1^2 + n_2^2 \leq N^2\} \tag{2.1}$$

reaches back to Euler [8]. In today's nomenclature, it can be equivalently expressed as a sum in the form

$$\#_{\mathbb{Z}^2}(\overline{\mathbb{B}_N^2}) = \sum_{\substack{n_1^2+n_2^2 \leq N^2 \\ (n_1,n_2)^T \in \mathbb{Z}^2}} 1. \tag{2.2}$$

Gauss [17] found a simple, but efficient method for its estimation: associate to every square the Northwest edge as lattice point. The union of all squares

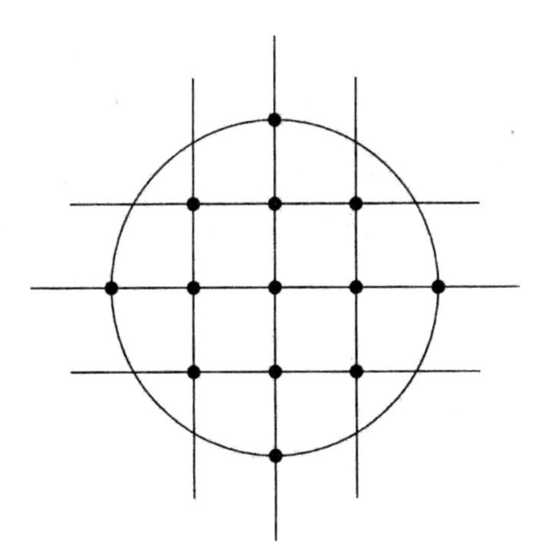

Fig. 1. Lattice points inside a circle.

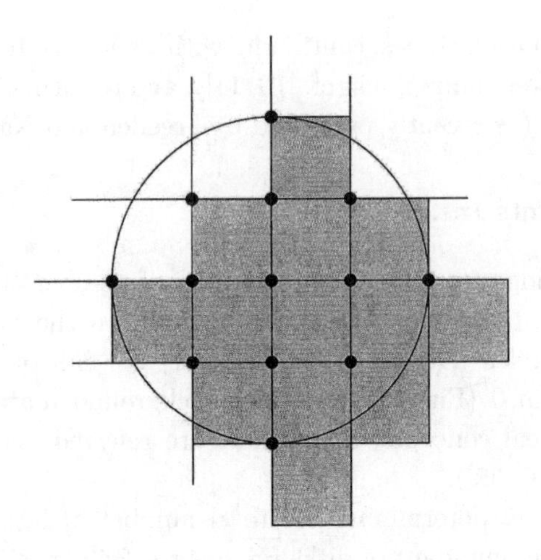

Fig. 2. The polyhedral set $\mathbb{P}_N^2$.

with lattice points inside $\overline{\mathbb{B}_N^2}$ defines a polyhedral set $\mathbb{P}_N^2$ with area

$$\|\mathbb{P}_N^2\| = \#_{\mathbb{Z}^2}(\overline{\mathbb{B}_N^2}) \tag{2.3}$$

(cf. Fig. 2). Since the diagonal of each square is $\sqrt{2}$, the geometry of Fig. 2 tells us that

$$\pi\left(N - \frac{\sqrt{2}}{2}\right)^2 \leq \#_{\mathbb{Z}^2}(\overline{\mathbb{B}_N^2}) \leq \pi\left(N + \frac{\sqrt{2}}{2}\right)^2. \tag{2.4}$$

Therefore, $\#_{\mathbb{Z}^2}(\overline{\mathbb{B}_N^2}) - \pi N^2$ after division by N is bounded for $N \to \infty$, which is usually written with Landau's O-symbol as

$$\#_{\mathbb{Z}^2}(\overline{\mathbb{B}_N^2}) = \pi N^2 + O(N). \tag{2.5}$$

In other words, the number of lattice points in $\overline{\mathbb{B}_N^2}$ is equal to the area of that circle plus a remainder of the order of the boundary. In particular,

$$\#_{\mathbb{Z}^2}(\overline{\mathbb{B}_N^2}) \sim \pi N^2 \tag{2.6}$$

so that a method of determining the irrational, transcendent number π becomes obvious (for alternative approaches to π within the history of analysis the readers are referred to [46]):

$$\lim_{N\to\infty} \frac{\#_{\mathbb{Z}^2}(\overline{\mathbb{B}_N^2})}{N^2} = \pi. \tag{2.7}$$

Gauss [17] illustrated his result by taking $N^2 = 100\,000$. In this case he calculated

$$\sum_{\substack{|g|^2 \leq 100\,000 \\ g \in \mathbb{Z}^2}} 1 = 314\ 197. \tag{2.8}$$

This calculation determines the number π up to three decimals after the comma.

3. Circle Problem and Hardy's Conjecture

In the nomenclature of Landau's O-symbols, the formula (2.5) due to Gauss [17] allows the representation

$$\#_{\mathbb{Z}^2}(\overline{\mathbb{B}^2_N}) = \pi N^2 + O(N). \tag{3.1}$$

The so-called *circle problem* is concerned with the question of determining the bound

$$\alpha_2 = \inf\{\gamma : \#_{\mathbb{Z}^2}(\overline{\mathbb{B}^2_{\sqrt{N}}}) = \pi N + O(N^\gamma)\}. \tag{3.2}$$

Until now, we knew from (3.1) that $\alpha_2 \leq \frac{1}{2}$. An improvement of the Gaussian result, however, turned out to be very laborious, in fact, requiring a great effort. A first remarkable result is due to Sierpinski [45], who proved by use of a method of his teacher Voronoi [49] that

$$\#_{\mathbb{Z}^2}(\overline{\mathbb{B}^2_{\sqrt{N}}}) = \pi N + O(N^{\frac{1}{3}}), \tag{3.3}$$

i.e., $\alpha_2 \leq \frac{1}{3}$. The proof of Sierpinski is elementary (see, e.g., [16] for more details); it is a link between geometry and number theory. Today, his proof can be shortened substantially (see, e.g., [16]).

By use of advanced methods on exponential sums (based on the work by, e.g., Weyl [50], Chen [4], and others) the estimate $1/3$ could be strengthened to some extent. It culminated in the publication by Kolesnik [29], who had as his sharpest result with these techniques

$$\#_{\mathbb{Z}^2}(\overline{\mathbb{B}^2_{\sqrt{N}}}) - \pi N = O(N^{\frac{139}{429}}). \tag{3.4}$$

Huxley [24] devised a substantially new approach (not discussed here); his strongest result was the estimate

$$\#_{\mathbb{Z}^2}(\overline{\mathbb{B}^2_{\sqrt{N}}}) - \pi N = O(N^{\frac{131}{416}}). \tag{3.5}$$

(note that $\frac{139}{429} = 0.324009\ldots$, while $\frac{131}{416} = 0.315068\ldots$)

Table 1. Incremental improvements for the values ε_2 in the estimate (3.7).

0.250000	Gauss [17]
0.083333...	Voronoi [48], Sierpinski [44]
0.080357...	Littlewood, A. Walfisz (1924)
0.079268...	van der Corput [47]
0.074324...	Chen [4]
0.074009...	Kolesnik [29]
0.064903...	Huxley [24]

Hardy's conjecture claims

$$\#_{\mathbb{Z}^2}(\overline{\mathbb{B}^2_{\sqrt{N}}}) - \pi N = O(N^{\frac{1}{4}+\varepsilon}) \tag{3.6}$$

for *every* $\varepsilon > 0$. This conjecture seems to be still a challenge for future work. However, in the year 2007, S. Cappell and J. Shaneson deposited a paper entitled "Some Problems in Number Theory I: The Circle Problem" in the arXiv:math/0702613 claiming to prove the bound of $O(N^{\frac{1}{4}+\varepsilon})$ for $\varepsilon > 0$.

Table 1 lists incomplete incremental improvements for the quantities ε_2 of the upper limit for the circle problem

$$\#_{\mathbb{Z}^2}(\overline{\mathbb{B}^2_{\sqrt{N}}}) - \pi N = O(N^{\frac{1}{4}+\varepsilon_2}). \tag{3.7}$$

For all recent improvements, the proofs became rather long and made use of some heavier machinery in hard analysis.

Summarizing our results about lattice points inside circles (cf. [11]) we are confronted with the following situation:

$$\tfrac{1}{4} \le \alpha_2 \le \tfrac{1}{4} + \varepsilon_2 \tag{3.8}$$

and

$$\#_{\mathbb{Z}^2}(\overline{\mathbb{B}^2_{\sqrt{N}}}) - \pi N \neq O(N^{\frac{1}{4}}), \tag{3.9}$$

$$\#_{\mathbb{Z}^2}(\overline{\mathbb{B}^2_{\sqrt{N}}}) - \pi N = O(N^{\frac{1}{4}+\varepsilon_2}), \tag{3.10}$$

where

$$0 < \varepsilon_2 \le \tfrac{1}{4}. \tag{3.11}$$

4. Variants of the Circle Problem

As already mentioned, there are many perspectives to formulate variants of the Gaussian lattice point problem for the circle. It already was the merit of Landau [31] to point out some particularly interesting areas, such as the first two items of the following list:

- *General two-dimensional lattices*

$$\Lambda = \{g = ng_1 + mg_2 \; : \; n, m \in \mathbb{Z}\} \tag{4.1}$$

  with $g_1, g_2 \in \mathbb{R}^2$ linearly independent (see Fig. 3) can be used instead of the unit lattice $\mathbb{Z}^2$,
- The *remainder term* can be represented as an alternating series, called *Hardy–Landau series* in terms of the Bessel function J_1 of order 1 (for the different facets of the proof, see [18, 19, 30, 35]):

$$\sum_{\substack{|g|\le N\\ g\in\Lambda}}{}' 1 = \frac{\pi N^2}{\|\mathcal{F}_\Lambda\|} + \lim_{R\to\infty} \frac{\pi N^2}{\|\mathcal{F}_\Lambda\|} \sum_{\substack{0<|h|\le R\\ h\in\Lambda^{-1}}} \frac{J_1(2\pi|h|N)}{\pi|h|N}, \tag{4.2}$$

  where Λ is an arbitrary lattice in $\mathbb{R}^2$ and

$$\mathcal{F}_\Lambda = \{x = x_1 g_1 + x_2 g_2 \in \mathbb{R}^2 : -\tfrac{1}{2} \le x_i < \tfrac{1}{2}, i = 1,2\} \tag{4.3}$$

  is the *fundamental cell of* $\Lambda \subset \mathbb{R}^2$ with

$$\|\mathcal{F}_\Lambda\| = \sqrt{\det\left((g_i \cdot g_j)_{i,j=1,2}\right)} \tag{4.4}$$

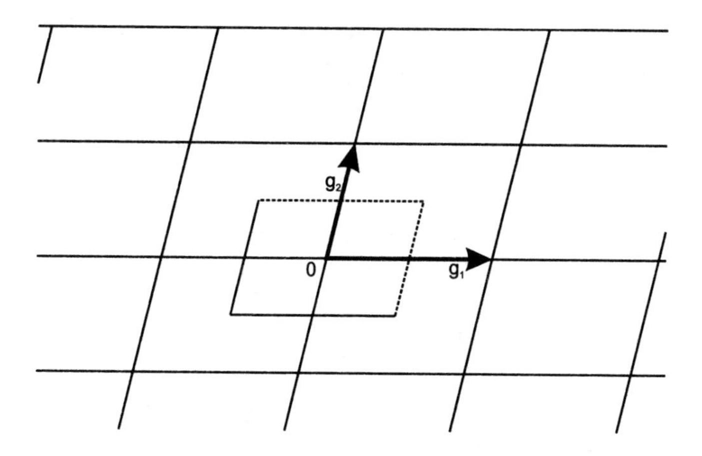

Fig. 3. Two-dimensional lattice Λ generated by $g_1, g_2 \in \mathbb{R}^2$.

as the area $\|\mathcal{F}_\Lambda\|$ *of* $\mathcal{F}_\Lambda$. Moreover, we follow the convention

$$\sum_{\substack{|g|\le N\\ g\in\Lambda}}{}' 1 = \sum_{\substack{|g|<N\\ g\in\Lambda}} 1 + \frac{1}{2}\sum_{\substack{|g|=N\\ g\in\Lambda}} 1 \tag{4.5}$$

used in lattice point theory (note that the last sum only occurs if there is a lattice point $g\in\Lambda$ with $|g|=N$).

- Lattice points can be affected by nonconstant weights (see [10])

$$\sum_{\substack{|a+g|\le N\\ g\in\Lambda}}{}' e^{2\pi i y\cdot(a+g)} F(a+g)$$
$$= \lim_{R\to\infty} \frac{1}{\|\mathcal{F}_\Lambda\|} \sum_{\substack{|h-y|\le R\\ h\in\Lambda^{-1}}} e^{2\pi i a\cdot h} \int_{\substack{|x|\le N\\ x\in\mathbb{R}^2}} F(x)\, e^{-2\pi i x\cdot(h-y)}\, dy, \tag{4.6}$$

where dy is the volume element, $a,y\in\mathbb{R}^2$, F is twice continuously differentiable in $\overline{\mathbb{B}_N^2}$, $N>0$, and the following convention has been used analogously to (4.5)

$$\sum_{\substack{|a+g|\le N\\ g\in\Lambda}}{}' \cdots = \sum_{\substack{|a+g|<N\\ g\in\Lambda}} \cdots + \frac{1}{2}\sum_{\substack{|a+g|=N\\ g\in\Lambda}} \cdots. \tag{4.7}$$

Note that, for $F=1$, this formula leads back to

$$e^{2\pi i a\cdot y} \sum_{\substack{|g+a|\le N\\ g\in\Lambda}}{}' e^{2\pi i g\cdot y}$$
$$= \lim_{R\to\infty} \frac{\pi N^2}{\|\mathcal{F}_\Lambda\|} \sum_{\substack{|h-y|\le R\\ h\in\Lambda^{-1}}} e^{2\pi i a\cdot h} \frac{J_1(2\pi|h-y|N)}{\pi|h-y|N}. \tag{4.8}$$

For $a=y=0$ we obtain the *classical Hardy–Landau identity*, i.e., the identity

$$\sum_{\substack{|g|\le N\\ g\in\Lambda}}{}' 1 = \lim_{R\to\infty} \frac{\pi N^2}{\|\mathcal{F}_\Lambda\|} \sum_{\substack{|h|\le R\\ h\in\Lambda^{-1}}} \frac{J_1(2\pi|h|N)}{\pi|h|N}. \tag{4.9}$$

holds true. Observe that J_1 satisfies the asymptotic relation $J_1(r) = \frac{r}{2}+\cdots$, so that

$$\sum_{\substack{|g|\le N\\ g\in\Lambda}}{}' 1 = \frac{\pi N^2}{\|\mathcal{F}_\Lambda\|} + \lim_{R\to\infty} \frac{\pi N^2}{\|\mathcal{F}_\Lambda\|} \sum_{\substack{0<|h|\le R\\ h\in\Lambda^{-1}}} \frac{J_1(2\pi|h|N)}{\pi|h|N}. \tag{4.10}$$

- Lattice points can be extended over particular regions in $\mathbb{R}^2$ (see [25] for the case of constant weight). More explicitly (see [9]), let Λ be an arbitrary lattice in $\mathbb{R}^2$. Let $\mathcal{G} \subset \mathbb{R}^2$ be a convex region containing the origin and possessing a boundary curve ∂G such that its normal field ν is continuously differentiable and its curvature is nonvanishing. Suppose that F is of class $\mathrm{C}^{(2)}(\overline{\mathcal{G}})$. Then, for all $a, y \in \mathbb{R}^2$, the series

$$\sum_{g\in\Lambda} e^{2\pi i a\cdot g} \underbrace{\int_{\mathcal{G}} F(x)\ e^{-2\pi i x\cdot(g-y)}\ dx}_{=F^{\wedge}_{\mathcal{G}}(g-y)} \tag{4.11}$$

converges in the spherical sense, and we have

$$\frac{1}{\|\mathcal{F}_\Lambda\|} \sideset{}{'}\sum_{\substack{a+h\in\overline{\mathcal{G}}\\ h\in\Lambda^{-1}}} e^{2\pi i y\cdot(a+h)} F(a+h)$$
$$= \lim_{N\to\infty} \sum_{\substack{|g-y|\le N\\ g\in\Lambda}} e^{2\pi i a\cdot g} \int_{\mathcal{G}} F(x) e^{-2\pi i x\cdot(g-y)} dx. \tag{4.12}$$

- Generalizations to lattices $\Lambda \subset \mathbb{R}^q$ and regular regions $\mathcal{G} \subset \mathbb{R}^q, q \geq 2$, and continuous functions on $\overline{\mathcal{G}} = \mathcal{G} \cup \partial\mathcal{G}$ can be formulated in Gaussian summability (see [11] for a more detailed study)

$$\sideset{}{'}\sum_{\substack{a+g\in\overline{\mathcal{G}}\\ g\in\Lambda}} e^{2\pi i y\cdot(a+g)} F(a+g)$$
$$= \lim_{\substack{\tau\to0\\ \tau>0}} \frac{1}{\|\mathcal{F}_\Lambda\|} \sum_{h\in\Lambda^{-1}} e^{-\tau\pi^2 h^2} e^{2\pi i h\cdot a}$$
$$\times \int_{\mathcal{G}} F(x) e^{-2\pi i x\cdot(h-y)}\ dx,\ \ a, y \in \mathbb{R}^q, \tag{4.13}$$

where a *regular region* $\mathcal{G}$ *in* $\mathbb{R}^q$ is understood to be an open and connected set $\mathcal{G} \subset \mathbb{R}^q$, $q \geq 2$, for which (i) its boundary $\partial\mathcal{G}$ constitutes an orientable, piecewise smooth Lipschitzian manifold of dimension $q-1$, (ii) the origin is contained in $\mathcal{G}$, and (iii) $\mathcal{G}$ divides $\mathbb{R}^q$ uniquely into the "inner space" $\mathcal{G}$ and the "outer space" $\mathbb{R}^q\backslash\overline{\mathcal{G}}$, $\overline{\mathcal{G}} = \mathcal{G} \cup \partial\mathcal{G}$.

Clearly, we have

$$\sideset{}{'}\sum_{\substack{a+g\in\overline{\mathcal{G}}\\ g\in\Lambda}} e^{2\pi i y\cdot(a+g)} F(a+g)$$
$$= \frac{1}{\|\mathcal{F}_\Lambda\|} \int_{\mathcal{G}} F(x) e^{2\pi i x\cdot y}\ dx$$

$$+ \lim_{\substack{\tau\to 0\\ \tau>0}} \frac{1}{\|\mathcal{F}_\Lambda\|} \sum_{\substack{0<|h|\le R\\ h\in\Lambda^{-1}}} e^{-\tau\pi^2 h^2} e^{2\pi i h\cdot a}$$

$$\times \int_{\mathcal{G}} F(x) e^{-2\pi i x\cdot(h-y)}\, dx, \; a, y\in\mathbb{R}^q, \tag{4.14}$$

where the following abbreviation has been used consistently

$$\sum_{\substack{a+g\in\overline{\mathcal{G}}\\ g\in\Lambda}}{}' \cdots = \sum_{\substack{a+g\in\mathcal{G}\\ g\in\Lambda}} \cdots + \sum_{\substack{a+g\in\partial\mathcal{G}\\ g\in\Lambda}} \alpha_{\mathcal{G}}(a+g)\ldots \tag{4.15}$$

with $\alpha_{\mathcal{G}}(a+g)$ denoting the *solid angle* subtended by $\partial\mathcal{G}$ at $a+g$ (note that, as geoscientifically relevant regular regions, we may choose the interior of the (actual) Earth's body or parts of it, the interior of geoscientifically relevant surfaces such as the geoid, telluroid, etc., but also ball, ellipsoid, cube, polyhedral bodies, etc., are included in accordance with the above definition) (see Figs. 4 and 5).

5. Multivariate Shannon-Type Sampling

Let us continue with the observation that, for every $y\in\mathbb{R}^q$, the series

$$a \mapsto \lim_{\substack{\tau\to 0\\ \tau>0}} \lim_{N\to\infty} \frac{1}{\|\mathcal{F}_\Lambda\|} \sum_{\substack{|h-y|\le N\\ h\in\Lambda^{-1}}} e^{-\tau\pi^2 h^2}\, e^{2\pi i a\cdot h} \underbrace{\int_{\mathcal{G}} F(x) e^{-2\pi i x\cdot(h-y)}\, dx}_{F^\wedge_{\mathcal{G}}(h-y)} \tag{5.1}$$

as well as the finite sum

$$a \mapsto \sum_{\substack{a+g\in\overline{\mathcal{G}}\\ g\in\Lambda}}{}' e^{2\pi i y\cdot(a+g)} F(a+g), \; a\in\mathbb{R}^q \tag{5.2}$$

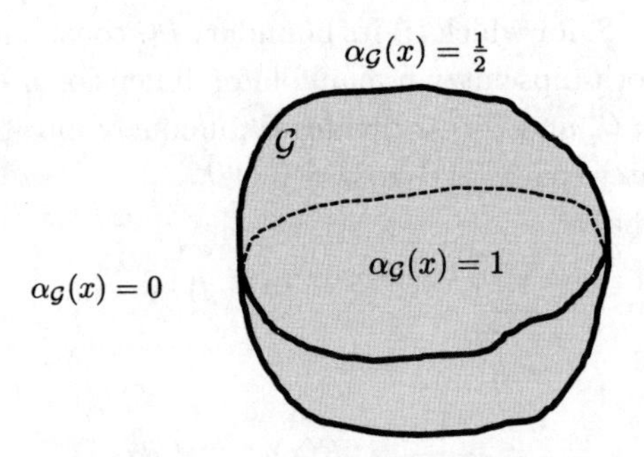

Fig. 4. Solid angle subtended at $x\in\mathbb{R}^3$ by the surface $\partial\mathcal{G}$ of a regular region $\mathcal{G}$ with "smooth boundary".

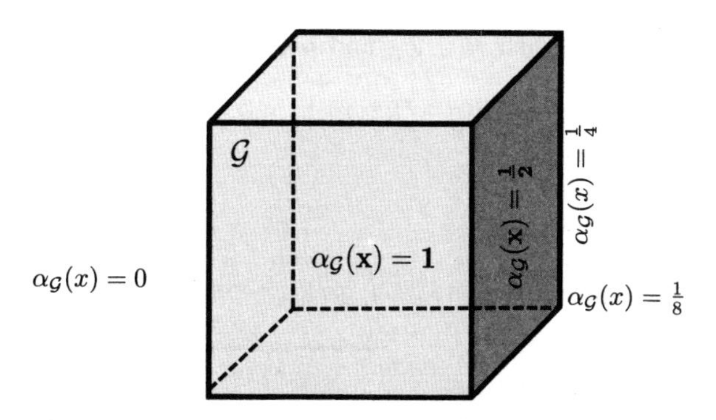

Fig. 5. Solid angle subtended at $x \in \mathbb{R}^3$ by the surface $\partial\mathcal{G}$ of the "nonsmooth" cube $\mathcal{G} = (-1, 1)^3$.

show Λ-periodicity, i.e., as functions of the variable $a \in \mathbb{R}^q$ they are periodic with respect to the lattice $\Lambda \subset \mathbb{R}^q$. As a consequence (see [14] for the details), Shannon-type sampling procedures can be obtained by formal integration of the lattice point identity (4.13) over a regular region $\mathcal{H}$ that is not-necessarily equal to the regular region $\mathcal{G}$

$$\int_{\mathcal{H}} \sum_{\substack{a+g\in\overline{\mathcal{G}} \\ g\in\Lambda}}{}' e^{-2\pi i y\cdot(a+g)} F(a+g)\, da$$

$$= \sum_{\substack{(\mathcal{F}_\Lambda+\{g'\})\cap\overline{\mathcal{H}}\neq\emptyset \\ g'\in\Lambda}} \underbrace{\int_{\mathcal{G}\cap\bigcup_{g\in\Lambda}(((\overline{\mathcal{H}}\cap(\mathcal{F}_\Lambda+\{g'\}))-\{g'\})+\{g\})} F(x) e^{-2\pi i y\cdot x}\, dx}_{=F^\wedge_{\mathcal{G}\cap\bigcup_{g\in\Lambda}(((\overline{\mathcal{H}}\cap(\mathcal{F}_\Lambda+\{g'\}))-\{g'\})+\{g\})}(y)}$$

$$= \lim_{\substack{\tau\to 0 \\ \tau>0}} \frac{1}{\|\mathcal{F}_\Lambda\|} \sum_{h\in\Lambda^{-1}} e^{-\tau\pi^2 h^2} \underbrace{\int_{\mathcal{G}} F(x) e^{-2\pi i h\cdot x}\, dx}_{=F^\wedge_{\mathcal{G}}(h)} \underbrace{\int_{\mathcal{H}} e^{2\pi i a\cdot(h-y)}\, da}_{=K_{\mathcal{H}}(h-y)}. \tag{5.3}$$

The identity (5.3) has many interesting properties. For example, by virtue of the Gaussian summability, the convergence of the cardinal-type series on the right-hand side of (5.3) is exponentially accelerated. Furthermore, all manifestations of over- and undersampling can be explicitly analyzed by the finite sum of Fourier transforms on the left side of the identity, dependent on the geometric configurations of the chosen regular regions $\mathcal{G}, \mathcal{H}$.

Some particular configurations, i.e., special choices of regular regions $\mathcal{G}, \mathcal{H}$ should be studied in more detail, where we almost literally follow [14]: We begin with $\overline{\mathcal{H}} \subset \overline{\mathcal{F}_\Lambda}$ and $\mathcal{G}$ arbitrary. In this case we have

$$\sum_{\substack{(\mathcal{F}_\Lambda+\{g'\})\cap\mathcal{H}\neq\emptyset \\ g'\in\Lambda}} F^{\wedge}_{\mathcal{G}\cap\bigcup_{g\in\Lambda}(((\mathcal{H}\cap(\mathcal{F}_\Lambda+\{g'\}))-\{g'\})+\{g\})}(y)$$
$$= F^{\wedge}_{\mathcal{G}\cap\bigcup_{g\in\Lambda}(\mathcal{H}+\{g\})}(y)$$
$$= \frac{1}{\|\mathcal{F}_\Lambda\|} \lim_{\substack{\tau\to 0\\ \tau>0}} \sum_{h\in\Lambda^{-1}} e^{-\tau\pi^2h^2} \underbrace{\int_{\mathcal{G}} F(x)e^{-2\pi i y\cdot x}\, dx}_{=F^{\wedge}_{\mathcal{G}}(h)} \underbrace{\int_{\mathcal{H}} e^{2\pi i a\cdot(h-y)}\, da.}_{=K_{\mathcal{H}}(h-y)} \tag{5.4}$$

As a consequence, for $\overline{\mathcal{H}} = \overline{\mathcal{F}_\Lambda}$ and $\mathcal{G}$ arbitrary, we obtain

$$\sum_{\substack{(\mathcal{F}_\Lambda+\{g'\})\cap\mathcal{H}\neq\emptyset \\ g'\in\Lambda}} F^{\wedge}_{\mathcal{G}\cap\bigcup_{g\in\Lambda}(((\mathcal{H}\cap(\mathcal{F}_\Lambda+\{g'\}))-\{g'\})+\{g\})}(y)$$
$$= F^{\wedge}_{\mathcal{G}}(y)$$
$$= \frac{1}{\|\mathcal{F}_\Lambda\|} \lim_{\substack{\tau\to 0\\ \tau>0}} \sum_{h\in\Lambda^{-1}} e^{-\tau\pi^2h^2} \underbrace{\int_{\mathcal{G}} F(x)e^{-2\pi i h\cdot x}\, dx}_{=F^{\wedge}_{\mathcal{G}}(h)} \underbrace{\int_{\mathcal{F}_\Lambda} e^{2\pi i a\cdot(h-y)}\, da.}_{=K_{\mathcal{F}_\Lambda}(h-y)} \tag{5.5}$$

We continue with $\overline{\mathcal{G}} \subset \overline{\mathcal{F}_\Lambda}$ and $\mathcal{H}$ arbitrary. This yields the identity

$$\sum_{\substack{(\mathcal{F}_\Lambda+\{g'\})\cap\mathcal{H}\neq\emptyset \\ g'\in\Lambda}} F^{\wedge}_{\mathcal{G}\cap\bigcup_{g\in\Lambda}(((\mathcal{H}\cap(\mathcal{F}_\Lambda+\{g'\}))-\{g'\})+\{g\})}(y)$$
$$= \sum_{\substack{(\mathcal{F}_\Lambda+\{g'\})\cap\mathcal{H}\neq\emptyset \\ g'\in\Lambda}} F^{\wedge}_{\mathcal{G}\cap((\mathcal{H}\cap(\mathcal{F}_\Lambda+\{g'\}))-\{g'\})}(y)$$
$$= \frac{1}{\|\mathcal{F}_\Lambda\|} \lim_{\substack{\tau\to 0\\ \tau>0}} \sum_{h\in\Lambda^{-1}} e^{-\tau\pi^2h^2} \underbrace{\int_{\mathcal{G}} F(x)e^{-2\pi i h\cdot x}\, dx}_{=F^{\wedge}_{\mathcal{G}}(h)} \underbrace{\int_{\mathcal{H}} e^{2\pi i a\cdot(h-y)}\, da.}_{=K_{\mathcal{H}}(h-y)} \tag{5.6}$$

For $\overline{\mathcal{G}}, \overline{\mathcal{H}} \subset \overline{\mathcal{F}_\Lambda}$ we have

$$\sum_{\substack{(\mathcal{F}_\Lambda+\{g'\})\cap\mathcal{H}\neq\emptyset \\ g'\in\Lambda}} F^{\wedge}_{\mathcal{G}\cap\bigcup_{g\in\Lambda}(((\mathcal{H}\cap(\mathcal{F}_\Lambda+\{g'\}))-\{g'\})+\{g\})}(y)$$
$$= F^{\wedge}_{\mathcal{G}\cap\mathcal{H}}(y)$$
$$= \frac{1}{\|\mathcal{F}_\Lambda\|} \lim_{\substack{\tau\to 0\\ \tau>0}} \sum_{h\in\Lambda^{-1}} e^{-\tau\pi^2h^2} F^{\wedge}_{\mathcal{G}}(h)\ K_{\mathcal{H}}(h-y). \tag{5.7}$$

For $\overline{\mathcal{G}} \subset \overline{\mathcal{H}} \subset \overline{\mathcal{F}_\Lambda}$ we have

$$F_{\mathcal{G}}^{\wedge}(y) = \frac{1}{\|\mathcal{F}_\Lambda\|} \lim_{\substack{\tau\to 0\\ \tau>0}} \sum_{h\in\Lambda^{-1}} e^{-\tau\pi^2 h^2} F_{\mathcal{G}}^{\wedge}(h) \; K_{\mathcal{H}}(h-y), \tag{5.8}$$

whereas, for $\overline{\mathcal{H}} \subset \overline{\mathcal{G}} \subset \overline{\mathcal{F}}_\Lambda$,

$$F_{\mathcal{H}}^{\wedge}(y) = \frac{1}{\|\mathcal{F}_\Lambda\|} \lim_{\substack{\tau\to 0\\ \tau>0}} \sum_{h\in\Lambda^{-1}} e^{-\tau\pi^2 h^2} F_{\mathcal{G}}^{\wedge}(h) \; K_{\mathcal{H}}(h-y). \tag{5.9}$$

In particular, for $\overline{\mathcal{G}} = \overline{\mathcal{H}} \subset \overline{\mathcal{F}_\Lambda}$, we are able to formulate the following identity:

$$F_{\mathcal{G}}^{\wedge}(y) = \frac{1}{\|\mathcal{F}_\Lambda\|} \lim_{\substack{\tau\to 0\\ \tau>0}} \sum_{h\in\Lambda^{-1}} e^{-\tau\pi^2 h^2} F_{\mathcal{G}}^{\wedge}(h) \; K_{\mathcal{G}}(h-y), \quad y \in \mathbb{R}^q. \tag{5.10}$$

Obviously, by choosing a small sampling density such that $\overline{\mathcal{F}_\Lambda}$ covers the compact support $\overline{\mathcal{G}}$ of the original signal $F_{\mathcal{G}}^{\wedge}$, the number of samples $F_{\mathcal{G}}^{\wedge}(h), h \in \Lambda^{-1}$, for reconstruction is high, and vice versa. In practice, we are therefore required to find a compromise between sampling density and total number of samples. This can be achieved by a choice of Λ such that $\overline{\mathcal{F}_\Lambda}$ covers $\overline{\mathcal{G}}$ tightly.

For $\overline{\mathcal{G}} \subset \overline{\mathcal{F}}_\Lambda = \overline{\mathcal{H}}$ and the lattice Λ generated by the vectors $g_1, \ldots, g_q \in \mathbb{R}^q$ we have in explicitly written form

$$\begin{aligned} F_{\mathcal{G}}^{\wedge}(y) = \lim_{\substack{\tau\to 0\\ \tau>0}} & \sum_{h\in\Lambda^{-1}} e^{-\tau\pi^2 h^2} F_{\mathcal{G}}^{\wedge}(h) \\ & \times \frac{\sin(\pi(g_1\cdot(h-y)))}{\pi(g_1\cdot(h-y))} \cdot \ldots \cdot \frac{\sin(\pi g_q\cdot(h-y))}{\pi(g_q\cdot(h-y))}. \end{aligned} \tag{5.11}$$

In other words, for sufficiently small $\tau > 0$, $F_{\mathcal{G}}^{\wedge}$ can be expressed by the series on the right-hand side of (5.11) in exponential convergence, i.e.,

$$\begin{aligned} (F^{(\tau)})_{\mathcal{G}}^{\wedge}(y) \simeq & \sum_{h\in\Lambda^{-1}} e^{-\tau\pi^2 h^2} F_{\mathcal{G}}^{\wedge}(h) \\ & \times \frac{\sin(\pi(g_1\cdot(h-y)))}{\pi(g_1\cdot(h-y))} \cdot \ldots \cdot \frac{\sin(\pi g_q\cdot(h-y))}{\pi(g_q\cdot(h-y))}. \end{aligned} \tag{5.12}$$

Replacing the lattice Λ by its dilated lattice $\sigma\Lambda$, $\sigma > 1$, we get

$$\begin{aligned} F_{\mathcal{G}}^{\wedge}(y) = \lim_{\substack{\tau\to 0\\ \tau>0}} \sigma^q & \sum_{h\in\Lambda^{-1}} e^{-\tau\pi^2 \left(\frac{h}{\sigma}\right)^2} F_{\mathcal{G}}^{\wedge}\left(\frac{h}{\sigma}\right) \\ & \times \frac{\sin(\pi(\sigma g_1\cdot(\frac{h}{\sigma}-y)))}{\pi(\sigma g_1\cdot(\frac{h}{\sigma}-y))} \cdot \ldots \cdot \frac{\sin(\pi(\sigma g_q\cdot(\frac{h}{\sigma}-y)))}{\pi(\sigma g_q\cdot(\frac{h}{\sigma}-y))}. \end{aligned} \tag{5.13}$$

The standard form of sampling in Gauss–Weierstrass summability is provided by taking $\overline{\mathcal{G}} = \overline{\mathcal{H}} = \overline{\mathcal{F}_\Lambda}$, i.e.,

$$F^{\wedge}_{\mathcal{F}_\Lambda}(y) = \frac{1}{\|\mathcal{F}_\Lambda\|} \lim_{\substack{\tau\to 0\\ \tau>0}} \sum_{h\in\Lambda^{-1}} e^{-\tau\pi^2 h^2} F^{\wedge}_{\mathcal{F}_\Lambda}(h)\, K_{\mathcal{F}_\Lambda}(h-y),\; y\in\mathbb{R}^q. \quad (5.14)$$

Remark 5.1. Given an arbitrary lattice $\Lambda \subset \mathbb{R}^q$ and an arbitrary regular region $\mathcal{H} \subset \mathbb{R}^q$, we are able to find a constant $\sigma \in \mathbb{R}$ such that $\overline{\mathcal{H}} \subset \overline{\mathcal{F}_{\sigma\Lambda}}$ as tightly as possible. Under these circumstances we obtain

$$\begin{aligned} &F^{\wedge}_{\mathcal{G}\cap\bigcup_{g\in\sigma\Lambda}(\mathcal{H}+\{g\})}(y) \\ &= \frac{1}{\|\mathcal{F}_{\sigma\Lambda}\|} \lim_{\substack{\tau\to 0\\ \tau>0}} \sum_{h\in(\sigma\Lambda)^{-1}} e^{-2\tau\pi^2 h^2} F^{\wedge}_{\mathcal{G}}(h) K_{\mathcal{H}}(h-y) \\ &= \frac{1}{\|\mathcal{F}_\Lambda\|} \lim_{\substack{\tau\to 0\\ \tau>0}} \frac{1}{\sigma^q} \sum_{h\in\Lambda^{-1}} e^{-\tau\pi^2(\frac{h}{\sigma})^2} F^{\wedge}_{\mathcal{G}}\left(\frac{h}{\sigma}\right) K_{\mathcal{H}}\left(\frac{h}{\sigma}-y\right). \end{aligned} \quad (5.15)$$

In comparison to sampling with respect to the lattice Λ the identity (5.15) related to $\sigma\Lambda$ provides *up-sampling* for values $\sigma > 1$ or *down-sampling* for values $\sigma < 1$, respectively. Of particular significance are choices of σ, such that Λ is a sublattice of $\sigma\Lambda, \sigma > 1$, or $\sigma\Lambda, \sigma < 1$, is a sublattice of Λ.

Let $\mathcal{G}, \mathcal{H}$ be regular regions. Suppose that $F = 1$ on $\overline{\mathcal{G}}$. Then, for all $y \in \mathbb{R}^q$, we have

$$\begin{aligned} &\int_{\mathcal{H}} \sum_{\substack{a+g\in\overline{\mathcal{G}}\\ g\in\Lambda}}{}' e^{-2\pi i y\cdot(a+g)}\, da \\ &= \sum_{\substack{(\mathcal{F}_\Lambda+\{g'\})\cap\mathcal{H}\neq\emptyset\\ g\in\Lambda}} \int_{\mathcal{G}\cap\bigcup_{g\in\Lambda}(((\mathcal{H}\cap(\mathcal{F}_\Lambda+\{\gamma'\})-\{g'\})+\{g\}))} e^{-2\pi i y\cdot x}\, dx \\ &= \frac{1}{\|\mathcal{F}_\Lambda\|} \lim_{\substack{\tau\to 0\\ \tau>0}} \sum_{h\in\Lambda^{-1}} e^{-\tau\pi^2 h^2} \int_{\mathcal{G}} e^{-2\pi i h\cdot x}\, dx \int_{\mathcal{H}} e^{2\pi i x\cdot(h-y)}\, dx \end{aligned} \quad (5.16)$$

such that, especially for $\mathcal{H} = \mathcal{G}$,

$$\begin{aligned} &\int_{\mathcal{G}} \sum_{\substack{a+g\in\overline{\mathcal{G}}\\ g\in\Lambda}}{}' e^{-2\pi i y\cdot(a+g)}\, da \\ &= \sum_{\substack{(\mathcal{F}_\Lambda+\{g'\})\cap\mathcal{G}\neq\emptyset\\ g\in\Lambda}} \int_{\mathcal{G}\cap\bigcup_{g\in\Lambda}(((\mathcal{G}\cap(\mathcal{F}_\Lambda+\{\gamma'\})-\{g'\})+\{g\}))} e^{-2\pi i y\cdot x}\, dx \\ &= \frac{1}{\|\mathcal{F}_\Lambda\|} \lim_{\substack{\tau\to 0\\ \tau>0}} \sum_{h\in\Lambda^{-1}} e^{-\tau\pi^2 h^2} \int_{\mathcal{G}} e^{-2\pi i h\cdot x}\, dx \int_{\mathcal{G}} e^{2\pi i x\cdot(h-y)}\, dx. \end{aligned} \quad (5.17)$$

In particular, for $y = 0$, we obtain

$$\int_{\mathcal{H}} \sum_{\substack{a+g\in\overline{\mathcal{G}}\\ g\in\Lambda}}{}' 1 \, da$$

$$= \sum_{\substack{(\mathcal{F}_\Lambda+\{g'\})\cap\mathcal{H}\neq\emptyset\\ g\in\Lambda}} \int_{\mathcal{G}\cap\bigcup_{g\in\Lambda}(((\mathcal{H}\cap(\mathcal{F}_\Lambda+\{g'\})-\{g'\})+\{g\}))} 1 \, dx$$

$$= \frac{1}{\|\mathcal{F}_\Lambda\|} \lim_{\substack{\tau\to 0\\ \tau>0}} \sum_{h\in\Lambda^{-1}} e^{-\tau\pi^2 h^2} \int_{\mathcal{G}} e^{-2\pi i h\cdot x} \, dx \int_{\mathcal{H}} e^{2\pi i x\cdot h} \, dx \tag{5.18}$$

and, especially for $\mathcal{H} = \mathcal{G}$,

$$\int_{\mathcal{G}} \sum_{\substack{a+g\in\overline{\mathcal{G}}\\ g\in\Lambda}}{}' 1 \, da = \sum_{\substack{(\mathcal{F}_\Lambda+\{g'\})\cap\mathcal{G}\neq\emptyset\\ g'\in\Lambda}} \int_{\mathcal{G}\cap\bigcup_{g\in\Lambda}(((\mathcal{G}\cap(\mathcal{F}_\Lambda+\{g'\})-\{g'\})+\{g\}))} 1 \, dx$$

$$= \lim_{\substack{\tau\to 0\\ \tau>0}} \frac{1}{\|\mathcal{F}_\Lambda\|} \sum_{h\in\Lambda^{-1}} e^{-\tau\pi^2 h^2} \left| \int_{\mathcal{G}} e^{-2\pi i h\cdot x} \, dx \right|^2 . \tag{5.19}$$

The identities (5.18) and (5.19) are the canonical preparations to turn over to the Parseval identity involving regular regions $\mathcal{G}, \mathcal{H} \subset \mathbb{R}^q$ within the Gauss–Weierstrass framework.

Remark 5.2. The formula (5.19) was already used in two-dimensional Euclidean space $\mathbb{R}^2$. In fact, Müller [35, 36] noticed that the Parseval-type identity

$$\int_{\mathcal{G}} \sum_{\substack{a+g\in\overline{\mathcal{G}}\\ g\in\mathbb{Z}^2}}{}' 1 \, da = \sum_{h\in\mathbb{Z}^2} \left| \int_{\mathcal{G}} e^{-2\pi i h\cdot x} \, dx \right|^2 \tag{5.20}$$

holds true for all symmetric (with respect to the origin) and convex regions $\mathcal{G} \subset \mathbb{R}^2$. An easy consequence of (5.20) is the inequality

$$\int_{\mathcal{G}} \sum_{\substack{a+g\in\overline{\mathcal{G}}\\ g\in\mathbb{Z}^2}}{}' 1 \, da \geq \|\mathcal{G}\|^2 = \left(\int_{\mathcal{G}} dx \right)^2 , \tag{5.21}$$

such that under the assumption $\|\mathcal{G}\| > 4$, an essential step towards *Minkowski's Theorem* can be made by considering simultaneously the lattice $2\mathbb{Z}^2$ (see [34]). Indeed, the inequality

$$\int_{\mathcal{G}} \sum_{\substack{a+g\in\overline{\mathcal{G}}\\ g\in 2\mathbb{Z}^2}}{}' 1 \, da \geq \frac{\|\mathcal{G}\|}{4} \|\mathcal{G}\| > \|\mathcal{G}\| \tag{5.22}$$

is the key to guarantee that a symmetric (with respect to the origin) and convex region $\mathcal{G} \subset \mathbb{R}^2$ with $\|\mathcal{G}\| > 4$ contains lattice points of $\mathbb{Z}^2$ different from the origin (for further details see, e.g., [23]).

It should be noted that Müller's two-dimensional approach was generalized by Freeden and Nashed [14] in various ways. For example, the Gauss–Weierstrass summability can be avoided independently of the dimension of the Euclidean space $\mathbb{R}^q, q \geq 1$. The Parseval identity of type (5.20) can be extended to arbitrary lattices $\Lambda \subset \mathbb{R}^q$ and regular regions $\mathcal{G} \subset \mathbb{R}^q$. Moreover, arbitrary continuous weight functions can be included instead of constant weights.

Theorem 5.3 (Extended Parseval Identity in Gauss–Weierstrass Summability). *Let Λ be a lattice in $\mathbb{R}^q$. Let $\mathcal{G}, \mathcal{H} \subset \mathbb{R}^q$ be arbitrary regular regions. Suppose that F is of class $\mathrm{C}^{(0)}(\overline{\mathcal{G}})$ and G is of class $\mathrm{C}^{(0)}(\overline{\mathcal{H}})$, respectively. Then we have*

$$\begin{aligned}
\lim_{\substack{\tau\to 0\\ \tau>0}} \frac{1}{\|\mathcal{F}_\Lambda\|} \sum_{h\in\Lambda^{-1}} e^{-\tau\pi^2 h^2} \int_{\mathcal{G}} F(x)e^{-2\pi i x\cdot h} dx \overline{\int_{\mathcal{H}} G(x)e^{-2\pi i x\cdot h}\, dx} \\
= \lim_{\substack{\tau\to 0\\ \tau>0}} \frac{1}{\|\mathcal{F}_\Lambda\|} \sum_{h\in\Lambda^{-1}} e^{-\tau\pi^2 h^2} F_{\mathcal{G}}^{\wedge}(h)\, \overline{G_{\mathcal{H}}^{\wedge}(h)} \\
= \int_{\mathcal{H}} \sum_{\substack{a+g\in\overline{\mathcal{G}}\\ g\in\Lambda}}{}' F(a+g)\, \overline{G(a)}\, da. \qquad (5.23)
\end{aligned}$$

Now, if $\overline{\mathcal{G}}, \overline{\mathcal{H}}$ are subsets of $\overline{\mathcal{F}_\Lambda}$, then we obtain from the extended Parseval identity (cf. Theorem 5.3)

$$\lim_{\substack{\tau\to 0\\ \tau>0}} \frac{1}{\|\mathcal{F}_\Lambda\|} \sum_{g\in\Lambda^{-1}} e^{-\tau\pi^2 h^2} F_{\mathcal{G}}^{\wedge}(h)\, \overline{G_{\mathcal{H}}^{\wedge}(h)} = \int_{\mathcal{G}\cap\mathcal{H}} F(a)\, \overline{G(a)}\, da. \qquad (5.24)$$

If $\overline{\mathcal{G}} \subset \overline{\mathcal{H}} \subset \overline{\mathcal{F}_\Lambda}$, then

$$\lim_{\substack{\tau\to 0\\ \tau>0}} \frac{1}{\|\mathcal{F}_\Lambda\|} \sum_{g\in\Lambda^{-1}} e^{-\tau\pi^2 h^2} F_{\mathcal{G}}^{\wedge}(h)\, \overline{G_{\mathcal{H}}^{\wedge}(h)} = \int_{\mathcal{G}} F(a)\, \overline{G(a)}\, da. \qquad (5.25)$$

Thus, the special configuration $\overline{\mathcal{G}} = \overline{\mathcal{H}} = \overline{\mathcal{F}_\Lambda}$ (see, e.g., [47]) leads back to the conventional Parseval identity of Fourier theory (however, in Gauss–Weierstrass summability).

By virtue of Theorem 5.3 (with $\mathcal{H} = \mathcal{G}$) we arrive at following corollary in Gauss–Weierstrass summability.

Corollary 5.4 (Parseval-Type Identity in Gauss–Weierstrass Summability). *Let Λ be a lattice in $\mathbb{R}^q$. Let $\mathcal{G} \subset \mathbb{R}^q$ be a regular region.*

Suppose that F, G are of class $\mathrm{C}^{(0)}(\overline{\mathcal{G}})$. Then the following variant of the Parseval identity holds true:

$$\lim_{\substack{\tau\to 0\\ \tau>0}} \frac{1}{\|\mathcal{F}_\Lambda\|} \sum_{h\in\Lambda^{-1}} e^{-\tau\pi^2 h^2} F_{\mathcal{G}}^{\wedge}(h)\ \overline{G_{\mathcal{G}}^{\wedge}(h)} = \int_{\mathcal{G}} \sum_{\substack{a+g\in\overline{\mathcal{G}}\\ g\in\Lambda}}{}' F(a+g)\ \overline{G(a)}\ da. \tag{5.26}$$

In particular, we have

$$\lim_{\substack{\tau\to 0\\ \tau>0}} \frac{1}{\|\mathcal{F}_\Lambda\|} \sum_{h\in\Lambda^{-1}} e^{-\tau\pi^2 h^2} \left|F_{\mathcal{G}}^{\wedge}(h)\right|^2 = \int_{\mathcal{G}} \sum_{\substack{a+g\in\overline{\mathcal{G}}\\ g\in\Lambda}}{}' F(a+g)\overline{F(a)}\ da. \tag{5.27}$$

The Gaussian summability of the cardinal series on the right-hand side of (5.3) is of great importance seen from numerical point of view; it enables a fast computation of the series. Nonetheless, Freeden and Nashed [14]) show that the identity (5.3) additionally holds true in ordinary sense.

Corollary 5.5 (Parseval-Type Identity for Regular Regions). *Under the assumptions of Theorem* 5.3, *we have*

$$\frac{1}{\|\mathcal{F}_\Lambda\|} \sum_{h\in\Lambda^{-1}} F_{\mathcal{G}}^{\wedge}(h)\ \overline{G_{\mathcal{H}}^{\wedge}(h)} = \int_{\mathcal{H}} \sum_{\substack{a+g\in\overline{\mathcal{G}}\\ g\in\Lambda}}{}' F(a+g)\ \overline{G(a)}\ da. \tag{5.28}$$

In particular, if $\overline{\mathcal{G}} = \overline{\mathcal{H}}$ and $F = G$, then

$$\frac{1}{\|\mathcal{F}_\Lambda\|} \sum_{h\in\Lambda^{-1}} \left|F_{\mathcal{G}}^{\wedge}(h)\right|^2 = \int_{\mathcal{G}} \sum_{\substack{a+g\in\overline{\mathcal{G}}\\ g\in\Lambda}}{}' F(a+g)\ \overline{F(a)}\ da. \tag{5.29}$$

Under the choice $\overline{\mathcal{G}} = \overline{\mathcal{H}} \subset \overline{\mathcal{F}_\Lambda}$ and $F = G$ we find that

$$\frac{1}{\|\mathcal{F}_\Lambda\|} \sum_{h\in\Lambda^{-1}} |F_{\mathcal{G}}^{\wedge}(h)|^2 = \int_{\mathcal{G}} |F(a)|^2\ da. \tag{5.30}$$

It is remarkable that Corollary 5.5 also enables us to formulate the Shannon sampling theorem in ordinary sense avoiding Gauss–Weierstrass nomenclature (by letting $G(x) = 1$ for $x \in \mathcal{H}$ and replacing $F(x)$ by $F(x)e^{-2\pi i y\cdot x}$ for $x \in \mathcal{G}$).

Theorem 5.6 (Shannon-Type Sampling Theorem). *Let $\mathcal{G}, \mathcal{H}$ be regular regions in $\mathbb{R}^q$. Suppose that F is a member of the class $\mathrm{C}^{(0)}(\overline{\mathcal{G}})$. Then*

$$\int_{\mathcal{H}} \sum_{\substack{a+g\in\overline{G} \\ g\in\Lambda}}{}' e^{-2\pi i y\cdot(a+g)} F(a+g)\, da \tag{5.31}$$

$$= \sum_{\substack{(\mathcal{F}_\Lambda+\{g'\})\cap\mathcal{H}\neq\emptyset \\ g'\in\Lambda}} F^{\wedge}_{\mathcal{G}\cap\bigcup_{g\in\Lambda}(((\mathcal{H}\cap(\mathcal{F}_\Lambda+\{g'\}))-\{g'\})+\{g\})}(y)$$

$$= \frac{1}{\|\mathcal{F}_\Lambda\|} \sum_{h\in\Lambda^{-1}} F^{\wedge}_{\mathcal{G}}(h)\, K_{\mathcal{H}}(h-y) \tag{5.32}$$

is valid for all $y \in \mathbb{R}^q$. For $\overline{\mathcal{G}} \subset \overline{\mathcal{H}} \subset \overline{\mathcal{F}_\Lambda}$, we have

$$F^{\wedge}_{\mathcal{G}}(y) = \frac{1}{\|\mathcal{F}_\Lambda\|} \sum_{h\in\Lambda^{-1}} F^{\wedge}_{\mathcal{G}}(h)\ \ K_{\mathcal{H}}(h-y), \quad y\in\mathbb{R}^q. \tag{5.33}$$

Finally, under the assumption $\overline{\mathcal{G}} = \overline{\mathcal{H}} \subset \overline{\mathcal{F}}_\Lambda$, the identity (5.33) implies that

$$F^{\wedge}_{\mathcal{G}}(y) = \frac{1}{\|\mathcal{F}_\Lambda\|} \sum_{h\in\Lambda^{-1}} F^{\wedge}_{\mathcal{G}}(h)\ \ K_{\mathcal{G}}(h-y), \quad y\in\mathbb{R}^q. \tag{5.34}$$

Remark 5.7. Furthermore, for $\overline{\mathcal{G}} \subset \overline{\mathcal{H}} = \overline{\mathcal{F}}_\Lambda$, we obtain

$$F^{\wedge}_{\mathcal{G}}(y) = \frac{1}{\|\mathcal{F}_\Lambda\|} \sum_{h\in\Lambda^{-1}} F^{\wedge}_{\mathcal{G}}(h)\, K_{\mathcal{F}_\Lambda}(h-y), \quad y\in\mathbb{R}^q. \tag{5.35}$$

Explicitly written out (cf. [43]) we obtain for the lattice Λ generated by the vectors $g_1,\ldots,g_q \in \mathbb{R}^q$

$$F^{\wedge}_{\mathcal{G}}(y) = \sum_{h\in\Lambda^{-1}} F^{\wedge}_{\mathcal{G}}(h) \frac{\sin(\pi(g_1\cdot(h-y)))}{\pi(g_1\cdot(h-y))} \cdot \ldots \cdot \frac{\sin(\pi g_q\cdot(h-y))}{\pi(g_q\cdot(h-y))}, \ y\in\mathbb{R}^q. \tag{5.36}$$

$$F^{\wedge}_{\mathcal{G}}(y) = \frac{1}{\|\mathcal{F}_\Lambda\|} \sum_{h\in\Lambda^{-1}} F^{\wedge}_{\mathcal{G}}(h)\ \ K_{\mathcal{G}}(h-y). \tag{5.37}$$

In fact, the identity (5.37) is a multivariate variant of the Shannon sampling theorem (cf. [44]), but now for multivariate regular regions $\mathcal{G}$. The principal impact of Shannon sampling on information theory is that it allows the replacement of a bandlimited signal $F^{\wedge}_{\mathcal{G}}$ related to $\mathcal{G}$ by a discrete sequence of its samples without loss of any information. Also it specifies the lowest rate, i.e., the Nyquist rate (cf. [14]), that it enables to reproduce the original signal. In other words, Shannon sampling provides the bridge between continuous and discrete versions of a bandlimited function.

The extensions of the Shannon sampling theorem as presented here has many applications in engineering and physics, for example, in signal processing, data transmission, cryptography, constructive approximation, and inverse problems such as, e.g., the antenna problem (see [37] for the univariate case).

6. Paley–Wiener Spaces

Many extensions of our sampling theorems can be studied in more detail. Some of the aspects leading to Paley–Wiener spaces will be finally explained (see also [14]).
Restricting ourselves to regular regions $\mathcal{G} \subset \mathbb{R}^q$ with $\overline{\mathcal{G}} \subset \overline{\mathcal{F}_\Lambda}$, so that the continuous signal

$$F_{\mathcal{G}}^{\wedge}(y) = \int_{\mathcal{G}} F(a) e^{-2\pi i a \cdot y} \, da, \quad y \in \mathbb{R}^q \tag{6.1}$$

is recovered from the sampled signal over lattice points of the inverse lattice $F_{\mathcal{G}}^{\wedge}(h), h \in \Lambda^{-1}$, i.e.,

$$F_{\mathcal{G}}^{\wedge}(y) = \frac{1}{\|\mathcal{F}_\Lambda\|} \sum_{h \in \Lambda^{-1}} F_{\mathcal{G}}^{\wedge}(h) \; K_{\mathcal{G}}(h - y), \quad y \in \mathbb{R}^q \tag{6.2}$$

with

$$K_{\mathcal{G}}(x - y) = \int_{\mathcal{G}} e^{-2\pi i a \cdot (x - y)} \, da, \quad x, y \in \mathbb{R}^q, \tag{6.3}$$

we are able to deduce some interesting results in the area of *approximate integration*: As a matter of fact, the bandlimited function $F_{\mathcal{G}}^{\wedge}$ allows to express its integral over the Euclidean space $\mathbb{R}^q$ by the product of the lattice density and the sum over all samples in points of the inverse lattice

$$\lim_{N \to \infty} \int_{\substack{|x| \leq N \\ x \in \mathbb{R}^q}} F_{\mathcal{G}}^{\wedge}(x) \, dx = \frac{1}{\|\mathcal{F}_\Lambda\|} \sum_{h \in \Lambda^{-1}} F_{\mathcal{G}}^{\wedge}(h). \tag{6.4}$$

Furthermore, the Parseval identity is valid:

$$\frac{1}{\|\mathcal{F}_\Lambda\|} \sum_{h \in \Lambda^{-1}} |F_{\mathcal{G}}^{\wedge}(h)|^2 = \int_{\mathcal{G}} |F(a)|^2 \, da. \tag{6.5}$$

From Fourier theory it follows that

$$\int_{\mathcal{G}} |F(a)|^2 \, da = \int_{\mathbb{R}^q} \left| F_{\mathcal{G}}^{\wedge}(y) \right|^2 \, dx. \tag{6.6}$$

In other words, if $F^{\wedge}_{\mathcal{G}}$, $\overline{\mathcal{G}} \subset \overline{\mathcal{F}_\Lambda}$, belongs to the inner product space

$$\mathrm{C}^{(0)}_{\overline{\mathcal{G}}} = \left\{ y \mapsto \underbrace{\int_{\mathcal{G}} e^{-2\pi i a \cdot y} F(a)\, da}_{=F^{\wedge}_{\mathcal{G}}(y)} : F \in \mathrm{C}^{(0)}(\overline{\mathcal{G}}) \right\}, \tag{6.7}$$

then

$$\int_{\mathbb{R}^q} |F^{\wedge}_{\mathcal{G}}(y)|^2 \, dy = \frac{1}{\|\mathcal{F}_\Lambda\|} \sum_{h \in \Lambda^{-1}} |F^{\wedge}_{\mathcal{G}}(h)|^2. \tag{6.8}$$

Replacing Λ by its inverse lattice Λ^{-1} we find

$$\int_{\mathbb{R}^q} |F^{\wedge}_{\mathcal{G}}(y)|^2 \, dy = \|\mathcal{F}_\Lambda\| \sum_{g \in \Lambda} |F^{\wedge}_{\mathcal{G}}(g)|^2. \tag{6.9}$$

Looking at our approach critically we notice that Shannon sampling is formulated on the reference set $\mathrm{C}^{(0)}_{\overline{\mathcal{G}}}$, that is a strict subset of the associated Paley–Wiener space

$$\mathrm{B}_{\overline{\mathcal{G}}} = \left\{ y \mapsto \int_{\mathcal{G}} e^{-2\pi i a \cdot y} F(a)\, da,\ y \in \mathbb{R}^q : F \in \mathrm{L}^2(\mathcal{G}) \right\}. \tag{6.10}$$

This observation, however, does not bother us very much, since every $F \in \mathrm{L}^2(\mathcal{G})$ can be approximated (in $\mathrm{L}^2(\mathcal{G})$-sense) by a function $F_\varepsilon \in \mathrm{C}^{(0)}(\overline{\mathcal{G}})$ in ε-accuracy such that

$$\begin{aligned}
&\sup_{y \in \mathbb{R}^q} \left| F^{\wedge}_{\mathcal{G}}(y) - (F_\varepsilon)^{\wedge}_{\mathcal{G}}(y) \right| \\
&\leq \sup_{y \in \mathbb{R}^q} \left| \int_{\mathcal{G}} e^{-2\pi i a \cdot y} (F(a) - F_\varepsilon(a))\, da \right| \\
&\leq \left(\int_{\mathcal{G}} |F(a) - F_\varepsilon(a)|^2 \, da \right)^{1/2} \left(\int_{\mathcal{G}} |e^{-2\pi i a \cdot y}|^2 \, da \right)^{1/2} \\
&\leq \left(\int_{\mathbb{R}^q} \left| F^{\wedge}_{\mathcal{G}}(a) - (F_\varepsilon)^{\wedge}_{\mathcal{G}}(a) \right|^2 da \right)^{1/2} \left(\int_{\mathcal{G}} |e^{-2\pi i a \cdot y}|^2 \, da \right)^{1/2} \\
&= \sqrt{\|\mathcal{G}\|}\ \varepsilon.
\end{aligned} \tag{6.11}$$

All in all, if $\mathcal{G} \subset \mathbb{R}^q$ is a regular region with $\overline{\mathcal{G}} \subset \overline{\mathcal{F}_\Lambda}$, then the *Paley–Wiener space* $\mathrm{B}_{\overline{\mathcal{G}}}$ is the completion of the space $\mathrm{C}^{(0)}_{\overline{\mathcal{G}}}$ under the $\mathrm{L}^2(\mathbb{R}^q)$-topology:

$$\mathrm{B}_{\overline{\mathcal{G}}} = \overline{\mathrm{C}^{(0)}_{\overline{\mathcal{G}}}}^{\|\cdot\|_{\mathrm{L}^2(\mathbb{R}^q)}}. \tag{6.12}$$

Under the aforementioned assumption that $\mathcal{G}$ is a regular region with $\overline{\mathcal{G}} \subset \overline{\mathcal{F}_\Lambda}$, the set $\mathrm{B}_{\overline{\mathcal{G}}}$ forms a *reproducing kernel space* with kernel

$$K_{\mathcal{G}}(x-y) = \int_{\mathcal{G}} e^{2\pi i a\cdot(x-y)}\, da. \tag{6.13}$$

In fact, by virtue of (6.6), we see that

$$|F_{\mathcal{G}}^{\wedge}(y)| \leq \sqrt{\|\mathcal{G}\|}\sqrt{\int_{\mathcal{G}} |F(a)|^2\, da} = \sqrt{\|\mathcal{G}\|}\sqrt{\int_{\mathbb{R}^q} |F_{\mathcal{G}}^{\wedge}(w)|^2\, dw}. \tag{6.14}$$

Moreover, standard Fourier inversion (see, e.g., [47]) yields

$$\begin{aligned} F_{\mathcal{G}}^{\wedge}(y) &= \int_{\mathbb{R}^q} F_{\mathcal{G}}^{\wedge}(x)\left(\int_{\mathcal{G}} e^{2\pi i a\cdot(x-y)}\, da\right)\, dx \\ &= \int_{\mathbb{R}^q} F_{\mathcal{G}}^{\wedge}(x)\; K_{\mathcal{G}}(x-y)\, dx \end{aligned} \tag{6.15}$$

for all $y \in \mathbb{R}^q$, where

$$\int_{\mathbb{R}^q} \cdots = \lim_{N\to\infty} \int_{\substack{|x|\leq N \\ x\in\mathbb{R}^q}} \cdots. \tag{6.16}$$

Hence, $B_{\overline{\mathcal{G}}}$ is a reproducing kernel Hilbert space with kernel (6.13).

Remark 6.1. Going over to the Paley–Wiener space $\mathrm{B}_{\overline{\mathcal{F}_\Lambda}}$ we are able to guarantee, in addition, that the reproducing kernels form an orthonormal system with discrete othogonality property (see [39–42]).

7. All Routes to Sampling Expansions Lead to Reproducing Kernels

The classical proof of the Shannon sampling theorem is based on the inverse of the Fourier integral and complex Fourier series. Suppose that the signal F defined on the real line is square integrable and bandlimited, say to $\mathcal{F}_{\mathbb{Z}}$. Then F can be expressed as the inverse of the Fourier transform $F_{\mathbb{Z}}^{\wedge}$, where now the integration is over $\mathcal{F}_{\mathbb{Z}}$. We extend $F_{\mathbb{Z}}^{\wedge}$ periodically to the real line and expand the resulting periodic function by complex Fourier series. Finally we interchange the integration and summation, and the sampling theorem pops up. While this proof is very simple, it is not very revealing in two respects:

(i) One does not get any insight from the interchange of integration and summation; in particular, one does not hear or feel the heart beat of the proof.

(ii) The sufficient condition on the signal F, namely bandlimitedness, is an indirect condition imposed on the Fourier transform, rather than F itself.

In classical Shannon theory it has been recognized that the space of bandlimited functions is the same as the Paley–Weiner space of entire functions whose restriction to the real line is of exponential growth. And this space is a reproducing kernel Hilbert space with the sinc function as a reproducing kernel. However, the reproducing kernel property does not enter the classical derivations of various approaches to the cardinal series. For various perspectives and surveys on sampling expansions, see [1–3, 20–22, 26–28, 51] and others. Nashed and Walter [41] introduced a new approach to general sampling theorems for functions in reproducing kernel Hilbert spaces and showed how many of the sampling results in the earlier literature are special cases of their approach. In another paper Nashed and Walter [42] showed how to construct a reproducing kernel Hilbert space from a function space that admits sampling expansions.

In view of this affinity between reproducing kernel spaces and sampling expansions, it is not surprising that in the past 25 years, reproducing kernel Hilbert and Banach spaces have played major roles in signal analysis and applications in inverse problems and imaging. For some perspectives on the role of various function spaces in sampling expansions, see [38, 40] and also the introduction to Freeden and Nashed [14].

Seminal ideas are often triggered by amazing insights. Many of such ideas are drastically generalized in new directions and applied in diverse fields, far beyond what anyone could have envisioned. The remarkable developments in sampling theory represent such an example.

All in all, the goal of this contribution was to show that significant roots of today's generalizations of Shannon's sampling theorem are basic number theoretical results starting from contributions of Gauss [17]. In fact, our excursion going out from the Gaussian circle problem of the early 19th century via the extensions of the Hardy–Landau lattice point identities of the late 20th century resulted in new multivariate Shannon sampling procedures of high practical applicability, thereby providing general reproducing kernel space structure of the associated Paley–Wiener spaces.

References

[1] P. L. Butzer, A survey of the Whittaker Shannon sampling theorem and some of its extensions, *J. Math. Res. Exposition* **3** (1983) 185–212.

[2] P. L. Butzer, W. Splettstösser and R. L. Stens, The sampling theory and linear prediction in signal analysis, *Jahresber. Deutsch. Math. Verein.* **90** (1988) 1–60.

[3] P. L. Butzer, R. L. Stens, Sampling theory for not necessarily band-limited functions: a historical overview, *SIAM Rev.* **34** (1992) 40–53.

[4] J.-R. Chen, The lattice points in a circle, Sci. Sinica **12** (1963) 633–649.

[5] P. J. Davis, *Interpolation and Approximation* (Blaisdell, New York, 1963).

[6] P. J. Davis and P. Rabinowitz *Numerical Integration* (Blaisdell, Toronto, 1967).

[7] C. Eisenhart, Biography on Carl Friedrich Gauss, in *International Encyclopedia of Social Sciences*, Vol. 6 (The Free Press, New York, 1986), pp. 74–81.

[8] L. Euler, Methodus universalis serierum convergentium summas quam proxime inveniendi, *Comment. Acad. Sci. Petropolitanae.* **8**(3–9), (1736), Opera Omnia (XIV), 101–107.

[9] W. Freeden, Eine Verallgemeinerung der Hardyschen Identität, Ph.D. thesis, RWTH Aachen (1975).

[10] W. Freeden, "Uber eine Verallgemeinerung der Hardy–Landauschen Identität, *Manuscr. Math.* **24** (1978) 205–216.

[11] W. Freeden, *Metaharmonic Lattice Point Theory* (CRC Press, Taylor & Francis Group, Boca Raton, 2011).

[12] W. Freeden, Geomathematics: its role, its aim, and its potential, in *Handbook of Geomathematics* Vol. 1, 2nd edn., eds. W. Freeden, M. Z. Nashed and T. Sonar (Springer, New York, 2015), pp. 3–79.

[13] W. Freeden and M. Gutting, *Special Functions of Mathematical (Geo)Physics* (Birkhäuser, Basel, 2013).

[14] W. Freeden and M. Z. Nashed, Multi-variate Hardy-type lattice point summation and Shannon-type sampling, *Int. J. Geomath.* **6** (2015) 163–249.

[15] W. Freeden, T. Sonar and B. Witte, *Gauss as Scientific Mediator between Mathematics and Geodesy from the Past to the Present. Handbook of Mathematical Geodesy*, eds. W. Freeden and M. Z. Nashed (Birkhäuser, Basel, 2017).

[16] F. Fricker, *Einführung in die Gitterpunktlehre* (Birkhäuser, Basel, 1981).

[17] C. F. Gauss, Untersuchungen über Höhere Arithmetik, in *Disquisitiones Arithmeticae* (Gerhard Fleischer jun., Leipzig, auch: Gauss Werke, Band 1, 1801) (in Latin).

[18] G. H. Hardy, On the expression of a number as the sum of two squares, *Quart. J. Math. (Oxford)* **46** (1915) 263–283.

[19] G. H. Hardy and E. Landau, The lattice points of a circle, *Proc. Roy. Soc. A*, **105** (1924) 244–258.

[20] J. R. Higgins, Five short stories about the cardinal series, *Bull. Amer. Math. Soc.* **12** (1985) 45–89.

[21] J. R. Higgins, *Sampling Theory in Fourier and Signal Analysis. Volume 1: Foundations* (Oxford University Press, Oxford, 1996).

[22] J. R. Higgins and R. L. Stens, *Sampling Theory in Fourier and Signal Analysis: Volume 2: Advanced Topics* (Oxford University Press, Oxford, 2000).

[23] E. Hlawka, J. Schoißengeier and R. Taschner, *Geometric and Analytic Number Theory* (Springer, Berlin, 1991).

[24] M. N. Huxley, Exponential sums and lattice points III, *Proc. London Math.* **87** (2003) 591–609.

[25] V. K. Ivanow, A generalization of the Voronoi–Hardy identity, *Sibirsk. Math. Z.* **3** (1962) 195–212.

[26] J. A. Jerri, On the application of some interpolating functions in physics, *J. Res. Nat. Bur. Standards, Sec. B* **73** (1969) 241–245.

[27] J. A. Jerri, Sampling expansion for Laguerre — L^2 transforms, *J. Res. Nat. Bur. Standards, Sect. B* **80** (1976) 415–418.

[28] J. A. Jerri, The Shannon sampling theorem — its various extensions and applications: A tutorial review, *Proc. IEEE* **65** (1977) 1565–1596.

[29] G. Kolesnik, On the method of exponent pairs, *Acta Arith.* **45** (1985) 115–143.

[30] E. Landau, Über die Gitterpunkte in einem Kreis (Erste Mitteilung), *Gött. Nachr.* (1915) 148–160.

[31] E. Landau, Über die Gitterpunkte in einem Kreis IV, *Gött. Nachr.* (1924) 58–65.

[32] E. Landau, *Vorlesungen über Zahlentheorie* (Chelsea Publishing Compagny, New York, 1969). Reprint from the orignal version published by S. Hirzel, Leipzig, 1927.

[33] J. E. Littlewood and A. Walfisz, The Lattice Points of a Circle. *Proc. Royal Soc. London Ser. (A)*, **106** (1924) 478–487.

[34] H. Minkowski, *Geometrie der Zahlen* (Teubner, Leipzig, 1896).

[35] C. Müller, Eine Verallgemeinerung der Eulerschen Summenformel und ihre Anwendung auf Fragen der analytischen Zahlentheorie, *Abh. Math. Sem. Univ. Hamburg* **19** (1954) 41–61.

[36] C. Müller, Eine Erweiterung der Hardyschen Identität, *Abh. Math. Sem. Univ. Hamburg* **19** (1954) 66–76.

[37] M. Z. Nashed, Operator-theoretic and computational approaches to ill-posed problems with applications to antenna theory, *IEEE Trans. Antennas Propagation* **29** (1981) 220–231.

[38] M. Z. Nashed, Inverse problems, moment problems, signal processing: in Menage a Trois, in *Mathematics of Science and Technology*, eds. A. H. Sidiqqi, R. C. Singh, and P. Manchanda (World Scientific, 2011), pp. 1–19.

[39] M. Z. Nashed and Q. Sun, Sampling and reconstruction of signals in a reproducing kernel subspace of $L^\rho(R^d)$, *J. Funct. Anal.* **258** (2010) 2422–2452.

[40] M. Z. Nashed and Q. Sun, Function spaces for sampling expansions, in Multiscale Signal Analysis and Modeling, eds. X. Shen and A. I. Zayed (Springer, 2013), pp. 81–104.

[41] M. Z. Nashed and G. G. Walter, General sampling theorems for functions in reproducing kernel Hilbert spaces, *Math. Control Signals Syst.* **4** (1991) 363–390.
[42] M. Z. Nashed and G. G. Walter, Reproducing kernel hilbert space from sampling expansions, *Contemporary Math.* **190** (1995) 221–226.
[43] E. Parzen, A simple proof and some extensions of the sampling theorem, Technical Report No. 7, Department of Statistics, Stanford University (1956), pp. 1–10.
[44] C. E. Shannon, Communication in the presence of noise, *Proc. Institute of Radio Engineers* **37** (1949) 10–21.
[45] W. Sierpinski, O pewnem zagadnieniu z rachunku funckcyj asmptotycznych (Über ein Problem des Kalküls der asymptotischen Funktion (polnisch)), *Prace Math.-Fiz.* **17** (1906) 77–118.
[46] T. Sonar, *3000 Jahre Analysis* (Springer, Berlin, 2011).
[47] E. M. Stein and G. Weiss, *Introduction to Fourier Analysis on Euclidean Spaces*, (Princeton University Press, Princeton, NJ, 1971).
[48] J. G. van der Corput, Zum Teilerproblem, *Math. Ann.* **98** (1928) 697–716.
[49] G. Voronoi, Sur un probleme du calcul des fonctions asymptotiques, *JRAM* **126** (1903) 241–282.
[50] H. Weyl, Über die Gleichverteilung von Zahlen mod, *Eins. Math. Ann.* **77** (1916) 313–352.
[51] A. Zayed, *Advances in Shannon's Sampling Theory* (CRC Press, Boca Raton, 1993).

This page intentionally left blank

Chapter 12

Weighted Partition Identities and Divisor Sums

F. G. Garvan

Department of Mathematics, University of Florida
Gainesville, FL 32611-8105, USA
fgarvan@ufl.edu

Dedicated to Mourad Ismail on the occasion of his 70th birthday

We prove two one-parameter q-series identities which specialize to four weighted partition identities. Three of the weighted partition identities have coefficients which are divisor sums. The other is an identity originally due to Alladi whose right side is Gauss's triangular number series. Some of the proofs depend on q-series results related to the arithmetic of $\mathbb{Q}[\sqrt{2}]$ due to Lovejoy and due to Corson, Favero, Liesinger and Zubairy.

Keywords: Partitions; weighted partition identities; divisor sums; basic hypergeometric series.

Mathematics Subject Classification 2010: 05A19, 11P82, 11P84, 33D15

1. Introduction

Many authors have studied weighted counts of partitions. Alladi [1, 2] was the first to undertake a systemic study of weighted partition identities. Fokkink, Fokkink and Wang [10] proved the following theorem.

Theorem 1.1. *Let $\mathcal{D}$ denote the set of partitions π into distinct parts. Then*

$$\sum_{\substack{\pi \in \mathcal{D} \\ |\pi|=n}} (-1)^{\#(\pi)+1} s(\pi) = d(n), \tag{1.1}$$

where $\#(\pi)$ is the number of parts of π, $s(\pi)$ is the smallest part of π, $|\pi|$ is the sum of parts of π, and $d(n)$ is the number of divisors of n.

Andrews [5] gave a short proof of this theorem using the q-analog of Gauss's theorem [4, Corollary 2.4, p. 20] by showing that

$$\sum_{n=1}^{\infty} \frac{(-1)^{n-1} q^{n(n+1)/2}}{(q;q)_n (1-q^n)} = \sum_{n=1}^{\infty} \frac{q^n}{1-q^n}. \tag{1.2}$$

Here we are using the standard q-notation

$$(a;q)_n = (1-a)(1-aq)\cdots(1-aq^{n-1}).$$

It is straightforward to see that the left and right sides of (1.2) correspond combinatorially to the left and right sides of (1.1), respectively. In this chapter we obtain some identities similar to (1.2) and interpret them as weighted partition identities.

Theorem 1.2.

$$\sum_{n=1}^{\infty} \frac{(-1)^{n-1} z^n q^{n^2}}{(zq;q^2)_n (1-zq^{2n})} = \sum_{n=1}^{\infty} \frac{z^n q^{n(n+1)/2} (q;q)_{n-1}}{(zq;q)_n}, \tag{1.3}$$

$$\sum_{n=0}^{\infty} \frac{(-1)^n z^n q^{n^2+n}}{(zq^2;q^2)_n (1-zq^{2n+1})} = \sum_{n=0}^{\infty} \frac{z^n q^{n(n+1)/2} (q;q)_n}{(zq;q)_n}. \tag{1.4}$$

The case $z = 1$ of (1.4) was observed earlier by Alladi [3, equation (6.8), p. 352]. The other cases $z = \pm 1$ of our theorem also lead to some interesting weighted partition theorems. Let $\mathcal{P}_o$ be the set of partitions in which all parts except possibly the largest part are odd and all odd positive integers less than or equal to the largest part occur as parts. Similarly let $\mathcal{P}_e$ be the set of partitions in which all parts except possibly the largest part are even and all even positive integers less than or equal to the largest part occur as parts. For a partition π we let $\ell_O(\pi)$ be the largest odd part and $\ell_E(\pi)$ be the largest even part with the convention that $\ell_E(\pi) = 0$ if π has no even parts. We define four partition weights:

$$\omega_{1a}(\pi) = (-1)^{(\ell_O(\pi)-1)/2}, \tag{1.5}$$

$$\omega_{1b}(\pi) = (-1)^{(\ell_O(\pi)-1)/2+\#(\pi)+1}, \tag{1.6}$$

$$\omega_{2a}(\pi) = (-1)^{\ell_E(\pi)/2}, \tag{1.7}$$

$$\omega_{2b}(\pi) = (-1)^{\ell_E(\pi)/2+\#(\pi)}. \tag{1.8}$$

Corollary 1.3. *Let n be a positive integer. Then*

(i)

$$\sum_{\substack{\pi \in \mathcal{P}_o \\ |\pi|=n}} \omega_{1a}(\pi) = d_1(n),$$

where $d_1(n)$ is the number of odd divisors of n.

(ii)

$$\sum_{\substack{\pi \in \mathcal{P}_o \\ |\pi|=n}} \omega_{1b}(\pi) = (-1)^{n(n-1)/2}\, d_{8,1}(n),$$

where $d_{8,1}(n)$ is the number of divisors of n congruent to $\pm 1 \pmod 8$ minus the number of divisors of n congruent to $\pm 3 \pmod 8$.

(iii)

$$\sum_{\substack{\pi \in \mathcal{P}_e \\ |\pi|=n}} \omega_{2a}(\pi) = \begin{cases} 1 & \text{if } n = m(m+1)/2, \\ 0 & \text{otherwise.} \end{cases}$$

(iv)

$$\sum_{\substack{\pi \in \mathcal{P}_e \\ |\pi|=n}} \omega_{2b}(\pi) = (-1)^n\, d_{8,1}(8n+1).$$

Alladi [3, Theorem 5, p. 352] obtained Corollary 1.3(iii) in a slightly different form. We also note that a different combinatorial interpretation of (iv) was found by Andrews [6, Theorem 2.3, p. 33] in terms of the DE-rank of partitions with no repeated even parts.

We illustrate Corollary 1.3 with some examples.

Example 1.4. We illustrate (i) and (ii) when $n = 9$.

π	$\ell_O(\pi)$	$\#(\pi)$	$\omega_{1a}(\pi)$	$\omega_{1b}(\pi)$
1^9	1	9	1	1
$1^7 2^1$	1	8	1	-1
$1^5 2^2$	1	7	1	1
$1^3 2^3$	1	6	1	-1
$1 2^4$	1	5	1	1
$1^6 3^1$	3	7	-1	-1
$1^3 3^2$	3	5	-1	-1
$1^2 3^1 4^1$	3	4	-1	1
$1^1 3^1 5^1$	5	3	1	1

In the table we are using the frequency notation [4, p. 1] for a partition. We see that

$$\sum_{\substack{\pi \in \mathcal{P}_o \\ |\pi|=9}} \omega_{1a}(\pi) = 3 = d_1(9),$$

and

$$\sum_{\substack{\pi \in \mathcal{P}_o \\ |\pi|=9}} \omega_{1b}(\pi) = 1 = (-1)^{36} d_{8,1}(9).$$

Example 1.5. We illustrate (iii) and (iv) when $n = 15$.

π	$\ell_E(\pi)$	$\#(\pi)$	$\omega_{2a}(\pi)$	$\omega_{2b}(\pi)$
1^{15}	0	15	1	-1
$2^6 3^1$	2	7	-1	1
$2^3 3^3$	2	6	-1	-1
$2^3 4^1 5^1$	4	5	1	-1
$2^1 4^2 5^1$	4	4	1	1

We see that

$$\sum_{\substack{\pi \in \mathcal{P}_e \\ |\pi|=15}} \omega_{2a}(\pi) = 1,$$

as expected since 15 is a triangular number. Also we have

$$\sum_{\substack{\pi \in \mathcal{P}_e \\ |\pi|=15}} \omega_{2b}(\pi) = -1 = (-1)^{15} d_{8,1}(11^2).$$

2. Proof of Theorem 1.2

2.1. Proof of (1.3)

By [7, Entry 1.7.2, p. 29] we have

$$\sum_{n=1}^{\infty} \frac{z^n q^{n(n+1)/2}(q;q)_{n-1}}{(zq;q)_n} = zq \sum_{n=0}^{\infty} \frac{z^n q^n (q^2;q^2)_n}{(zq^2;q^2)_{n+1}}$$

$$= \sum_{n=1}^{\infty} \frac{z^n q^n (q^2;q^2)_{n-1}}{(zq^2;q^2)_n}, \tag{2.1}$$

$$1 + (1-z)\sum_{n=1}^{\infty} \frac{(-1)^n z^n q^{n^2}}{(zq;q^2)_n (1-zq^{2n})} = \sum_{n=0}^{\infty} \frac{(-1)^n z^n q^{n^2} (z;q^2)_n}{(zq;q^2)_n (zq^2;q^2)_n}$$

$$= \lim_{a\to 0} {}_3\phi_2 \left(\begin{matrix} q^2, & a^{-1}, & z \\ & zq, & zq^2 \end{matrix} ; q^2, azq \right)$$

$$= \lim_{a\to 0} \frac{(z, azq^3; q^2)_\infty}{(zq^2, azq; q^2)_\infty} \times {}_3\phi_2 \left(\begin{matrix} q^2, & azq, & q \\ & zq, & azq^3 \end{matrix} ; q^2, z \right)$$

(by [11, equation (III.9), p. 241])

$$= (1-z)\sum_{n=0}^{\infty} \frac{(q;q^2)_n z^n}{(zq;q^2)_n}, \tag{2.2}$$

and

$$\sum_{n=1}^{\infty} \frac{(-1)^n z^n q^{n^2}}{(zq;q^2)_n (1-zq^{2n})} = -\frac{1}{(1-z)} + \sum_{n=0}^{\infty} \frac{(q;q^2)_n z^n}{(zq;q^2)_n}. \tag{2.3}$$

From (2.1) and (2.3) we see that (1.3) is equivalent to

$$\sum_{n=1}^{\infty} \frac{z^n q^n (q^2;q^2)_{n-1}}{(zq^2;q^2)_n} = \frac{z}{(1-z)} - \sum_{n=1}^{\infty} \frac{(q;q^2)_n z^n}{(zq;q^2)_n}. \tag{2.4}$$

We start with the left-hand side of (2.4)

$$\sum_{n=1}^{\infty} \frac{z^n q^n (q^2;q^2)_{n-1}}{(zq^2;q^2)_n} = \sum_{n=1}^{\infty} z^n q^n (q^2;q^2)_{n-1} \sum_{j=0}^{\infty} \begin{bmatrix} n+j-1 \\ j \end{bmatrix}_{q^2} z^j q^{2j} \tag{2.5}$$

(by [4, equation (3.3.7), p. 36])

$$= \sum_{m=1}^{\infty} \sum_{n=1}^{m} \frac{(q^2;q^2)_{m-1} q^{2m-n} z^m}{(q^2;q^2)_{m-n}} = \sum_{m=1}^{\infty} \sum_{k=0}^{m-1} \frac{(q^2;q^2)_{m-1} q^{m+k} z^m}{(q^2;q^2)_k}. \tag{2.6}$$

Next we consider the right-hand side of (2.4)

$$\sum_{n=1}^{\infty} \frac{(q;q^2)_n z^n}{(zq;q^2)_n} = \sum_{n=1}^{\infty} z^n (q;q^2)_n \sum_{j=0}^{\infty} \begin{bmatrix} n+j-1 \\ j \end{bmatrix}_{q^2} z^j q^j$$

$$= \sum_{m=1}^{\infty} \sum_{n=1}^{m} \frac{(q;q^2)_n (q^2;q^2)_{m-1} z^m q^{m-n}}{(q^2;q^2)_{m-n} (q^2;q^2)_{n-1}}. \tag{2.7}$$

By equating coefficients of z^m, dividing by $(q^2;q^2)_{m-1}$, and using (2.5) and (2.7) we see that (2.4) is equivalent to showing that

$$\sum_{k=0}^{m-1} \frac{q^{m+k}}{(q^2;q^2)_k} = \frac{1}{(q^2;q^2)_{m-1}} - \sum_{n=1}^{m} \frac{(q;q^2)_n q^{m-n}}{(q^2;q^2)_{m-n} (q^2;q^2)_{n-1}}, \tag{2.8}$$

for $m \geq 1$. To prove this we multiply by z^m and sum

$$\sum_{m=1}^{\infty} z^m q^m \sum_{k=0}^{m-1} \frac{q^k}{(q^2;q^2)_k} = \sum_{k=0}^{\infty} \frac{q^k}{(q^2;q^2)_k} \sum_{m=k+1}^{\infty} z^m q^m$$

$$= \frac{zq}{(1-zq)} \sum_{k=0}^{\infty} \frac{q^{2k} z^k}{(q^2;q^2)_k} = \frac{zq}{(1-zq)(zq^2;q^2)_\infty}, \tag{2.9}$$

by [11, equation (II.1), p. 236]. Similarly we have

$$\sum_{m=1}^{\infty} \frac{z^m}{(q^2;q^2)_{m-1}} = \sum_{m=0}^{\infty} \frac{z^{m+1}}{(q^2;q^2)_m} = \frac{z}{(z;q^2)_\infty}. \tag{2.10}$$

We have

$$\sum_{m=1}^{\infty} z^m \sum_{n=1}^{m} \frac{(q;q^2)_n q^{m-n}}{(q^2;q^2)_{m-n} (q^2;q^2)_{n-1}} = \sum_{n=1}^{\infty} \sum_{m=n}^{\infty} \frac{z^m q^{m-n} (q;q^2)_n}{(q^2;q^2)_{m-n} (q^2;q^2)_{n-1}},$$

$$= \sum_{n=1}^{\infty} \frac{z^n (q;q^2)_n}{(q^2;q^2)_{n-1}} \sum_{m=0}^{\infty} \frac{z^m q^m}{(q^2;q^2)_m}$$

$$= \frac{1}{(zq;q^2)_\infty} \sum_{n=1}^{\infty} \frac{z^n (q;q^2)_n}{(q^2;q^2)_{n-1}}$$

$$= \frac{z(1-q)}{(zq;q^2)_\infty} \sum_{n=0}^{\infty} \frac{z^n (q^3;q^2)_n}{(q^2;q^2)_n}$$

$$= \frac{z(1-q)(zq^3;q^2)_\infty}{(zq;q^2)_\infty (z;q^2)_\infty} = \frac{z(1-q)}{(1-zq)(z;q^2)_\infty} \tag{2.11}$$

by the q-binomial theorem [11, equation (II.3), p. 236]. Now

$$\frac{zq}{(1-zq)(zq^2;q^2)_\infty} = \frac{zq(1-z)}{(1-zq)(z;q^2)_\infty}$$

$$= \frac{z((1-zq)-(1-q))}{(1-zq)(z;q^2)_\infty} = \frac{z}{(z;q^2)_\infty} - \frac{z(1-q))}{(1-zq)(z;q^2)_\infty}.$$

Therefore by (2.9)–(2.11)

$$\sum_{m=1}^{\infty} z^m \sum_{k=0}^{m-1} \frac{q^{m+k}}{(q^2;q^2)_k} = \sum_{m=1}^{\infty} \frac{z^m}{(q^2;q^2)_{m-1}}$$

$$- \sum_{m=1}^{\infty} z^m \sum_{n=1}^{m} \frac{(q;q^2)_n q^{m-n}}{(q^2;q^2)_{m-n}(q^2;q^2)_{n-1}},$$

and we have (2.8) for $m \geq 1$. This completes the proof of (1.3).

2.2. Proof of (1.4)

By replacing z by zq^{-1} we see that (1.4) is equivalent to

$$\sum_{n=0}^{\infty} \frac{(-1)^n z^n q^{n^2}}{(zq;q^2)_n (1-zq^{2n})} = \sum_{n=0}^{\infty} \frac{z^n q^{n(n-1)/2} (q;q)_n}{(z;q)_n}. \tag{2.12}$$

Now

$$\frac{z}{1-z} = \sum_{n=1}^{\infty} \frac{z^n q^{n(n-1)/2} (q;q)_{n-1}}{(z;q)_n}$$

$$- \sum_{n=1}^{\infty} \frac{z^{n+1} q^{n(n+1)/2} (q;q)_n}{(z;q)_{n+1}}$$

$$= \sum_{n=1}^{\infty} \frac{z^n q^{n(n-1)/2} (q;q)_{n-1}}{(z;q)_{n+1}}$$

$$\times ((1-zq^n) - zq^n(1-q^n))$$

$$
\begin{aligned}
&= \sum_{n=1}^{\infty} \frac{z^n q^{n(n-1)/2}(q;q)_{n-1}}{(z;q)_{n+1}} \\
&\quad \times ((1-zq^n)(1-q^n)+q^n(1-z)) \\
&= \sum_{n=1}^{\infty} \frac{z^n q^{n(n-1)/2}(q;q)_n}{(z;q)_n} \\
&\quad + \sum_{n=1}^{\infty} \frac{z^n q^{n(n+1)/2}(q;q)_{n-1}}{(zq;q)_n}. \qquad (2.13)
\end{aligned}
$$

Then

$$
\begin{aligned}
\sum_{n=0}^{\infty} \frac{(-1)^n z^n q^{n^2}}{(zq;q^2)_n(1-zq^{2n})} &= 1 + \frac{z}{1-z} + \sum_{n=1}^{\infty} \frac{(-1)^n z^n q^{n^2}}{(zq;q^2)_n(1-zq^{2n})} \\
&= 1 + \frac{z}{1-z} - \sum_{n=1}^{\infty} \frac{z^n q^{n(n+1)/2}(q;q)_{n-1}}{(zq;q)_n} \quad \text{(by (1.3))} \\
&= 1 + \sum_{n=1}^{\infty} \frac{z^n q^{n(n-1)/2}(q;q)_n}{(z;q)_n} \\
&= \sum_{n=0}^{\infty} \frac{z^n q^{n(n-1)/2}(q;q)_n}{(z;q)_n} \quad \text{(by (2.13))},
\end{aligned}
$$

which gives (2.12) and completes the proof of (1.4).

3. Proof of Corollary 1.3

We examine the left-hand side of (1.3):

$$
\begin{aligned}
&\sum_{n=1}^{\infty} \frac{(-1)^{n-1} z^n q^{n^2}}{(zq;q^2)_n(1-zq^{2n})} \\
&= \sum_{k=1}^{\infty} (-1)^{k-1} z^k q^{1+3+\cdots+(2k-1)} \times \frac{1}{(1-zq)} \times \frac{1}{(1-zq^3)} \\
&\quad \times \cdots \times \frac{1}{(1-zq^{2k-1})} \times \frac{1}{(1-zq^{2k})} \\
&= \sum_{n=1}^{\infty} \left(\sum_{\substack{\pi \in \mathcal{P}_o \\ |\pi|=n}} z^{\#(\pi)} (-1)^{\frac{1}{2}(\ell_O(\pi)-1)} \right) q^n. \qquad (3.1)
\end{aligned}
$$

In the sum observe that $2k-1=\ell_O(\pi)$ and $k-1=\frac{1}{2}(\ell_O(\pi)-1)$. Thus letting $z=1$ in (1.3) we have

$$\sum_{n=1}^{\infty}\left(\sum_{\substack{\pi\in\mathcal{P}_o\\|\pi|=n}}(-1)^{\frac{1}{2}(\ell_O(\pi)-1)}\right)q^n=\sum_{n=1}^{\infty}\frac{(-1)^{n-1}q^{n^2}}{(q;q^2)_n(1-q^{2n})}$$

$$=\sum_{n=1}^{\infty}\frac{q^{n(n+1)/2}(q;q)_{n-1}}{(q;q)_n}$$

$$=\sum_{n=1}^{\infty}\frac{q^{n(n+1)/2}}{(1-q^n)}=\sum_{n=1}^{\infty}\frac{q^n}{(1-q^{2n})}$$

$$=\sum_{n=1}^{\infty}d_1(n)q^n,$$

by (2.1) or [8, equation (47.1), p. 487], and we have Corollary 1.3(i). Similarly letting $z=-1$ in (1.3) we have

$$\sum_{n=1}^{\infty}\left(\sum_{\substack{\pi\in\mathcal{P}_o\\|\pi|=n}}(-1)^{\frac{1}{2}(\ell_O(\pi)-1)+\#(\pi)+1}\right)q^n=\sum_{n=1}^{\infty}\frac{(-1)^{n-1}q^{n(n+1)/2}(q;q)_{n-1}}{(-q;q)_n}. \tag{3.2}$$

By [12, equation (2.11)]

$$\sum_{n=1}^{\infty}\frac{(-1)^{n-1}q^{n(n+1)/2}(q;q)_{n-1}}{(-q;q)_n}=\sum_{n=1}^{\infty}\sum_{j=-n+1}^{n}(-1)^{n+j}q^{2n^2-j^2}. \tag{3.3}$$

Combining this result with [12, Theorem 1.1, p. 179] we find that

$$\sum_{n=1}^{\infty}\frac{(-1)^{n-1}q^{n(n+1)/2}(q;q)_{n-1}}{(-q;q)_n}=\sum_{n=1}^{\infty}(-1)^{n(n-1)/2}d_{8,1}(n)q^n, \tag{3.4}$$

and

$$\sum_{n=1}^{\infty}\left(\sum_{\substack{\pi\in\mathcal{P}_o\\|\pi|=n}}(-1)^{\frac{1}{2}(\ell_O(\pi)-1)+\#(\pi)+1}\right)q^n=\sum_{n=1}^{\infty}(-1)^{n(n-1)/2}d_{8,1}(n)q^n,$$

which gives Corollary 1.3(ii).

Proceeding as in (3.1) we find that

$$\sum_{n=0}^{\infty} \frac{(-1)^n z^n q^{n^2+n}}{(zq^2;q^2)_n(1-zq^{2n+1})} = 1 + \sum_{n=1}^{\infty} \left(\sum_{\substack{\pi \in \mathcal{P}_e \\ |\pi|=n}} z^{\#(\pi)} (-1)^{\frac{1}{2}\ell_E(\pi)} \right) q^n.$$

This time we note that any partition π, into ones is an element of $\mathcal{P}_e$ and satisfies $\ell_E(\pi) = 0$. This explains the $n = 0$ term in the sum on the left side of (3.1). So letting $z = 1$ in (1.4) we have

$$1 + \sum_{n=1}^{\infty} \left(\sum_{\substack{\pi \in \mathcal{P}_e \\ |\pi|=n}} (-1)^{\frac{1}{2}\ell_E(\pi)} \right) q^n = \sum_{n=0}^{\infty} \frac{(-1)^n q^{n^2+n}}{(q^2;q^2)_n(1-q^{2n+1})} = \sum_{n=0}^{\infty} q^{n(n+1)/2},$$

and we have Corollary 1.3(iii). Finally letting $z = -1$ in (1.4) we have

$$1 + \sum_{n=1}^{\infty} \left(\sum_{\substack{\pi \in \mathcal{P}_e \\ |\pi|=n}} (-1)^{\frac{1}{2}\ell_E(\pi)+\#(\pi)} \right) q^n = \sum_{n=0}^{\infty} \frac{(-1)^n q^{n(n+1)/2}(q;q)_n}{(-q;q)_n}. \quad (3.5)$$

By [9, Theorem 3.2, p. 398]

$$\sum_{n=0}^{\infty} \frac{(-1)^n q^{n(n+1)/2}(q;q)_n}{(-q;q)_n} = \sum_{n=0}^{\infty} (-1)^n \Delta(n) q^n, \quad (3.6)$$

where $\Delta(n)$ = number of inequivalent elements in the ring of integers of $\mathbb{Q}(\sqrt{2})$ with norm $8n+1$. As noted by Andrews [6, p. 34], Dean Hickerson has shown that

$$\Delta(n) = d_{8,1}(8n+1).$$

Thus

$$\sum_{n=1}^{\infty} \left(\sum_{\substack{\pi \in \mathcal{P}_e \\ |\pi|=n}} (-1)^{\frac{1}{2}\ell_E(\pi)+\#(\pi)} \right) q^n = \sum_{n=0}^{\infty} (-1)^n d_{8,1}(8n+1) q^n, \quad (3.7)$$

and we have Corollary 1.3(iv).

Acknowledgments

I would like to thank Krishna Alladi for his comments and suggestions. The author was supported in part by a grant from the Simon's Foundation (#318714).

References

[1] K. Alladi, Partition identities involving gaps and weights, *Trans. Amer Math. Soc.* **349** (1997) 5001–5019.
[2] K. Alladi, Partition identities involving gaps and weights, II, *Ramanujan J.* **2** (1998) 21–37.
[3] K. Alladi, Analysis of a generalized Lebesgue identity in Ramanujan's Lost Notebook, *Ramanujan J.* **29** (2012) 339–358.
[4] G. E. Andrews, *The Theory of Partitions*, Encyclopedia of Mathematics and its Applications, Vol. 2 (Addison-Wesley Reading, MA, 1976). (Reissued: Cambridge University Press, Cambridge, 1985).
[5] G. E. Andrews, The number of smallest parts in the partitions of n, *J. Reine Angew. Math.* **624** (2008) 133–142.
[6] G. E. Andrews, Partitions with distinct evens, in *Advances in Combinatorial Mathematics* (Springer, Berlin, 2009), pp. 31–37.
[7] G. E. Andrews and B. C. Berndt, *Ramanujan's Lost Notebook. Part II* (Springer, New York, 2009).
[8] B. C. Berndt, *Ramanujan's Notebooks, Part V* (Springer, New York, 1998).
[9] D. Corson, D. Favero, K. Liesinger and S. Zubairy, Characters and q-series in $\mathbb{Q}(\sqrt{2})$, *J. Number Theory* **107** (2004) 392–405.
[10] R. Fokkink, W Fokkink and Z. B. Wang, A relation between partitions and the number of divisors, *Amer. Math. Monthly* **102** (1995) 345–346.
[11] G. Gasper and M. Rahman, *Basic Hypergeometric Series*, Encyclopedia of Mathematics and its Applications (Cambridge University Press, Cambridge, 2004).
[12] J. Lovejoy, Overpartitions and real quadratic fields, *J. Number Theory* **106** (2004) 178–186.

Chapter 13

On the Ismail–Letessier–Askey Monotonicity Conjecture for Zeros of Ultraspherical Polynomials

Walter Gautschi

Department of Computer Science, Purdue University
West Lafayette, IN 47907-2107, USA
wgautschi@purdue.edu

Dedicated to Mourad E. H. Ismail on his 70th birthday

A number of conjectured monotonicity properties for zeros of ultraspherical polynomials are reviewed, leading up to the Ismail–Letessier–Askey (ILA) conjecture of the title, which has been proved in 1999 by Elbert and Siafarikas. It is shown that two of the earlier conjectures are consequences of the ILA conjecture. Computational support is provided for strengthening several of these conjectures, including the ILA conjecture, from monotonicity to complete monotonicity.

Keywords: Ultraspherical polynomials; zeros; monotonicity.

Mathematics Subject Classification 2010: 33C45, 65D20

1. Introduction

Inequalities and monotonicity properties for zeros of orthogonal polynomials depending on a parameter is a classical subject; see, e.g., [24, Chapter 6]. The last three or four decades, however, have seen a considerable increase of activity in this area. Several approaches have been pursued concurrently. One is via differential equations, specifically the Sturm comparison theorem. Surveys and tutorials on this are given by L. Lorch, A. Laforgia, and

M. E. Muldoon in, respectively, [22, 20], and [21], and further applications to generalized Laguerre polynomials in [3], to Jacobi polynomials in [5, Section 3], and to ultraspherical polynomials in [1, 2]. Another approach uses Markov's theorem [24, Section 6.12], which has recently been applied in [11] in connection with Freud and sub-range Freud polynomials, and a slight extension thereof in [15, Section 3] with applications to Laguerre, Jacobi, and Meixner polynomials. An approach originating in physics, and promoted primarily by Mourad Ismail, makes use of the Hellmann–Feynman theorem, which looks at the zeros of orthogonal polynomials as eigenvalues of an operator depending on a parameter and states formulas for the derivatives of the eigenvalues with respect to that parameter. This is surveyed in the paper [19] and applied there, and in a number of other papers [14, 17, 18], to zeros of a variety of orthogonal polynomials, including birth and death process polynomials.

Finally, there is an entirely empirical approach based on numerical computation, which is an important vehicle to test conjectured inequalities and monotonicity properties, thereby providing stimuli for further analytical work. Examples of this are a series of papers, [6–8, 13] on the zeros of Jacobi polynomials and also the paper [11] already cited on the zeros of Freud polynomials. This is the approach used here in Section 5 to test conjectured higher monotonicity properties for zeros of ultraspherical polynomials. Ordinary monotonicity properties are surveyed in Section 2. Software tools used in this paper are briefly described in Section 4.

2. Ultraspherical Polynomials and the ILA Conjecture

Ultraspherical polynomials arise in the solution of Laplace's equation in high-dimensional spaces, when written in terms of hyperspherical coordinates and solved by the method of separation of variables. The polynomials, also named after the Austrian mathematician Leopold Gegenbauer (1849–1903), who introduced them in his doctoral thesis of 1875 and studied them in subsequent papers, are commonly denoted by $C_n^{(\lambda)}$ [23, equation 18.7.1] (but see also [24, Section 4.7], where the notation $P_n^{(\lambda)}$ is used), n being the degree and $\lambda > -1/2$ a parameter. They are orthogonal on the interval $(-1, 1)$ relative to the weight function $w_\lambda(t) = (1-t^2)^{\lambda-1/2}$, that is

$$\int_{-1}^{1} C_k^{(\lambda)}(t)C_\ell^{(\lambda)}(t)w_\lambda(t)\mathrm{d}t = 0 \quad \text{if } k \neq \ell. \tag{2.1}$$

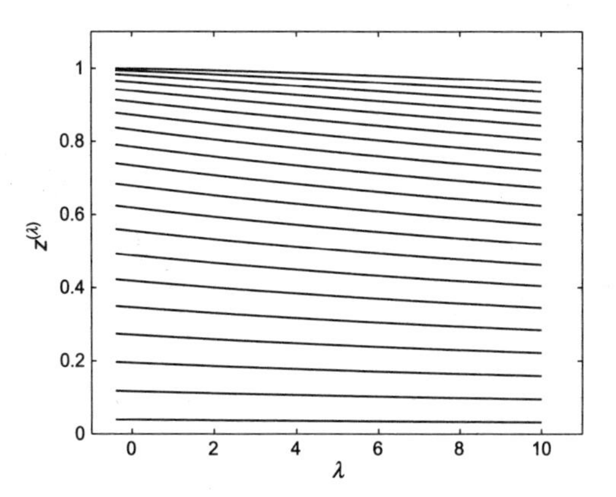

Fig. 1. The positive zeros of $C_n^{(\lambda)}$ for $n = 40$ and $-1/2 < \lambda \leq 10$.

Important special cases include Chebyshev polynomials of the first and second kind, corresponding to $\lambda = 0$ and $\lambda = 1$, and Legendre polynomials, corresponding to $\lambda = 1/2$.

Since the weight function w_λ is even, the polynomial $C_n^{(\lambda)}$ is even or odd, depending on whether n is even or odd, and the zeros therefore are symmetric, or antisymmetric, with respect to the origin. To study them, it thus suffices to look at the positive zeros $z_{n,k}^{(\lambda)}$ of $C_n^{(\lambda)}$. In Fig. 1 we show plots of them as functions of λ for $-0.49 \leq \lambda \leq 10$ and $n = 40$. It appears from the graphs, and has been verified computationally, that the zeros are all monotonically decreasing. This was already known to Stieltjes (see [24, p. 121]) and a proof using Markov's theorem in combination with quadratic transformation of hypergeometric functions is mentioned in [15, p. 188]. Laforgia [20] conjectured that the zeros multiplied by λ become monotonically increasing for all $\lambda > -1/2$ and all $n \geq 2$, and proved this for $0 < \lambda < 1$ using one of Szegö's formulations of the Sturm comparison theorem [24, Theorem 1.82.1]. We verified this computationally for $-1/2 < \lambda \leq 10$ (in steps of of $1/100$) and for all n with $2 \leq n \leq 40$. Graphs for $n = 40$ are shown on the left of Fig. 2. Ahmed, Muldoon, and Spigler [1], also using Sturmian methods, proved that the zeros multiplied by $[\lambda + (2n^2 + 1)/(4n + 2)]^{1/2}$ are increasing when $-1/2 < \lambda \leq 3/2$, $n \geq 2$, which we verified computationally for the same values of λ and n as above and show graphs for $n = 40$ on the left of Fig. 3. This actually implies the validity of Laforgia's conjecture for $-1/2 < \lambda \leq 3/2$ by straightforward differentiation of $\lambda^{-1}[\lambda+(2n^2+1)/(4n+2)]^{1/2}[\lambda z_{n,k}^{(\lambda)}]$. Ismail and Letessier [16] conjectured

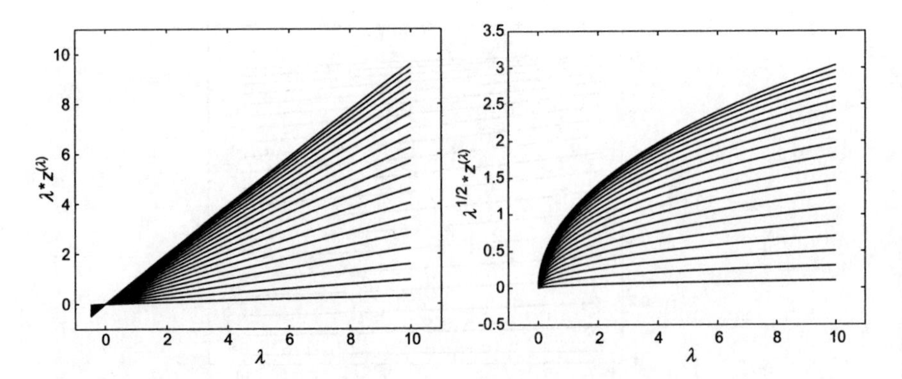

Fig. 2. The positive zeros of $C_n^{(\lambda)}$ multiplied by λ (left) for $n = 40$, $-1/2 < \lambda \le 10$, and multiplied by $\sqrt{\lambda}$ (right) for $n = 40$, $0 \le \lambda \le 10$.

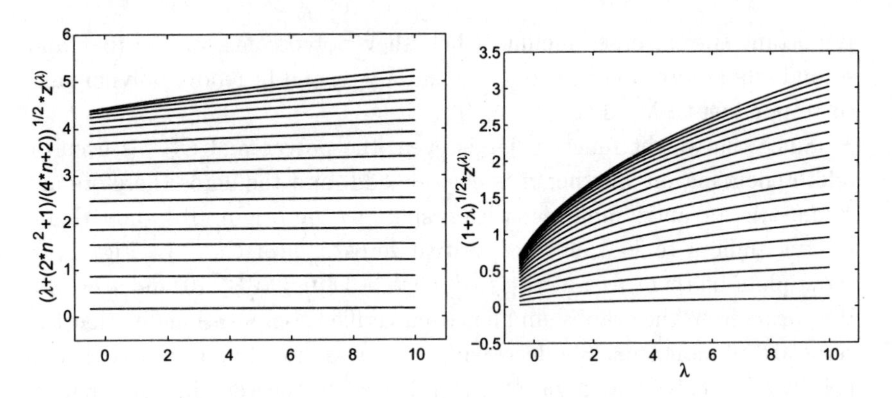

Fig. 3. The positive zeros of $C_n^{(\lambda)}$ multiplied by $[\lambda + (2n^2+1)/(4n+2)]^{1/2}$ (left) and multiplied by $(1+\lambda)^{1/2}$ (right) for $n = 40$, $-1/2 < \lambda \le 10$.

that the zeros multiplied by $\sqrt{\lambda}$ are monotonically increasing for $\lambda \ge 0$ and proved this for the largest zero. We verified the conjecture for all $2 \le n \le 40$ and all positive zeros, for the same values of λ as before, but with $\lambda \ge 0$. Respective graphs are shown on the right of Fig. 2 for $n = 40$. Askey suggested that monotonic growth may also hold when the zeros are multiplied by $(1+\lambda)^{1/2}$, which has become known as the ILA conjecture, named after Ismail, Letessier, and Askey. We verified this conjecture for all $3 \le n \le 40$ (when $n = 2$ the product in question is constant equal to $1/\sqrt{2}$) and for $-1/2 < \lambda \le 10$ in steps of $1/100$; see the graphs on the right of Fig. 3. The conjecture, in fact, has been proven by Elbert and Siafarikas [4] by showing that the Ahmed–Muldoon–Spigler result holds for all $\lambda > -1/2$, which in turn, by straightforward differentiation, implies the ILA conjecture.

3. Some Implications of the ILA Conjecture

With $z(\lambda)$ denoting any positive zero of $C_n^{(\lambda)}$, writing

$$f(\lambda) = \left(\frac{1+\lambda}{\lambda}\right)^{1/2} [\lambda^{1/2} z(\lambda)] = (1+\lambda^{-1})^{1/2} [\lambda^{1/2} z(\lambda)]$$

and using the ILA conjecture $f'(\lambda) > 0$ (proved by Elbert and Siafarikas), differentiation of the right-hand side yields

$$(1+\lambda^{-1})^{1/2}[\lambda^{1/2} z(\lambda)]' = f'(\lambda) + \tfrac{1}{2}\,(1+\lambda^{-1})^{-1/2}\lambda^{-3/2} z(\lambda) > 0 \quad \text{for } \lambda > 0,$$

that is, the validity of the Ismail–Letessier conjecture. Likewise, writing

$$f(\lambda) = \left(\frac{1+\lambda}{\lambda^2}\right)^{1/2} [\lambda z(\lambda)]$$

and differentiating yields

$$\left(\frac{1+\lambda}{\lambda^2}\right)^{1/2} [\lambda z(\lambda)]' = f'(\lambda) + \tfrac{1}{2}\,(1+\lambda)^{-1/2}(1+2/\lambda) z(\lambda) > 0 \quad \text{if } \lambda > 0,$$

proving Laforgia's conjecture for $\lambda > 0$. For $\lambda \le 0$, the conjecture is trivially true since $[\lambda z(\lambda)]' = z(\lambda) + \lambda z'(\lambda) > 0$ because of $z'(\lambda) < 0$.

4. Matlab Software for Orthogonal Polynomials

All computational work in this chapter was done by using the Matlab software package *Orthogonal Polynomials and Quadrature* (`OPQ`), located at

`http : //dx.doi.org/10.4231/R7959FHP`,

and its symbolic (variable-precision) counterpart (`SOPQ`), located at

`http : //dx.doi.org/10.4231/R7ZG6Q6T`.

We describe and illustrate here the routines most relevant for our purposes. Many other applications can be found in [10].

A system $\{\pi_k\}_{k=0}^{\infty}$ of monic polynomials $\pi_k(t) = t^k + \cdots$ is called *orthogonal* on the interval (a, b), $-\infty \le a < b \le \infty$, with respect to a positive weight function w, if (cf. also (2.1))

$$\int_a^b \pi_k(t)\pi_\ell(t)w(t)\mathrm{d}t = 0 \quad \text{when } k \neq \ell. \tag{4.1}$$

It is known that every such system satisfies a three-term recurrence relation

$$\begin{gathered}\pi_{k+1}(t) = (t - \alpha_k)\pi_k(t) - \beta_k\pi_{k-1}(t), \quad k = 0, 1, 2, \ldots, \\ \pi_{-1}(t) = 0, \quad \pi_0(t) = 1,\end{gathered} \tag{4.2}$$

with real $\alpha_k = \alpha_k(w)$ and positive $\beta_k = \beta_k(w)$ that depend on the weight function w. Thus, to obtain the polynomial π_n of degree n from (4.2) requires knowledge of the first n of these coefficients, $\alpha_0, \alpha_1, \ldots \alpha_{n-1}$ and $\beta_0, \beta_1, \ldots, \beta_{n-1}$, where β_0 can be arbitrary, but is conveniently defined to be $\beta_0 = \int_a^b w(t)\mathrm{d}t$. These are provided by the OPQ routine with syntax

$$\texttt{ab} = \texttt{r\_}\mathit{name}(\texttt{n}, \ldots), \tag{4.3}$$

where *name* specifies the name of the orthogonal polynomial, n the degree, and ab is the $n \times 2$ array of recurrence coefficients

$$\texttt{ab} = \begin{bmatrix} \alpha_0 & \beta_0 \\ \alpha_1 & \beta_1 \\ \vdots & \vdots \\ \alpha_{n-1} & \beta_{n-1} \end{bmatrix}.$$

The dots on the right of (4.3) stand for possible parameters defining the orthogonal polynomial.

The polynomials that interest us here are exclusively *Jacobi polynomials*, which are orthogonal on the interval $(-1, 1)$ relative to the weight function $w(t) = (1 - t)^\alpha(1 + t)^\beta$, $\alpha > -1$, $\beta > -1$. More specifically, we are interested in *ultraspherical* polynomials, which are Jacobi polynomials with parameters $\alpha = \beta = \lambda - 1/2$, where $\lambda > -1/2$. Their first n recurrence coefficients are produced by the OPQ routine

```
ab=r_jacobi(n,lambda-1/2)
```

in Matlab double precision, and by the SOPQ routine

```
sab=sr_jacobi(dig,n,lambda-1/2)
```

in `dig`-digit precision. Thus, `ab=r_jacobi(5,1)` and `sab=sr_jacobi (42,5,1)`, which are the same as `ab=r_jacobi(5,1,1)` and `sab=sr_jacobi (42,5,1,1)`, produce

```
ab =

     0     1.333333333333333e+00
     0     2.000000000000000e-01
     0     2.285714285714286e-01
     0     2.380952380952381e-01
     0     2.424242424242424e-01
```

and

```
sab =

[ 0,  1.33333333333333333333333333333333333333333]
[ 0,                                          0.2]
[ 0, 0.228571428571428571428571428571428571428571]
[ 0, 0.238095238095238095238095238095238095238095]
[ 0, 0.242424242424242424242424242424242424242424]
```

that is, the first five recurrence coefficients for the ultraspherical polynomial $C_5^{(3/2)}$ in double and 42-digit precision.

There is no `OPQ` routine that specifically generates the zeros of orthogonal polynomials, but they can be generated by the Gauss quadrature routine

$$\mathtt{xw} = \mathtt{gauss(n, ab)}, \tag{4.4}$$

which obtains in the first column of the n × 2 array `xw` the n nodes (in increasing order), and in the second column the corresponding weights, of the `n`-point Gaussian quadrature rule associated with the weight function w specified by the $n \times 2$ array `ab` of its first n recurrence coefficients. The n Gaussian nodes are nothing but the n zeros of the orthogonal polynomial of degree n relative to the weight function w. Thus, the n zeros of $C_n^{(\lambda)}$ in increasing order are obtained by the following Matlab routine:

```
function z=zeros_us(n,lambda)
%ZEROS_US Zeros of ultraspherical polynomials.
%  Z=ZEROS_US(N,LAMBDA) generates the N zeros in
%  increasing order of the Nth-degree ultraspherical
%  polynomial C_N^{(LAMBDA)}.
```

```
ab=r_jacobi(n,lambda-1/2); xw=gauss(n,ab);
z=xw(:,1);
```

The Matlab script

```
%PLOT_ZEROS_US

n=40;
n0=floor((n+1)/2)+1;
for k=n0:n
  si=(-.49:.01:10)';
%  si=(0:.01:10)';
  s=size(si); y=zeros(s);
  i=0;
  for lam=-.49:.01:10
%  for lam=0:.01:10
    i=i+1;
    z=zeros_us(n,lam);
    y(i,k)=z(k);
%     y(i,k)=lam*z(k);
%     y(i,k)=sqrt(lam)*z(k);
%     y(i,k)=(2*n^2+1+2*lam*(2*n+1))^(1/2)*z(k);
%     y(i,k)=(1+lam)^(1/2)*z(k);
  end
  for i=2:s(1)
    if y(i,k)>=y(i-1,k)
  %    if y(i,k)<=y(i-1,k)
      [n i k]
      [y(i,k) y(i-1,k)]
      error('wrong sign')
    end
  end
  plot(si,y(:,k),'LineWidth',1.5);set(gca,'FontSize',14)
  hold on
  axis([-1 11 0 1.1])
%  axis([-1 11 -1 11])
%  axis([-1 11 -.5 3.5])
%  axis([-1 11 -5 70])
%  axis([-1 11 -.5 3.5])
  xlabel('\lambda'); ylabel('z(\lambda)');
%  xlabel('\lambda'); ylabel('\lambda*z(\lambda)')
%  xlabel('\lambda'); ylabel('\lambda^{1/2}*z(\lambda)')
%  xlabel('\lambda');
%  ylabel('(2*n^2+1+2*\lambda*(2*n+1))^{1/2}*z(\lambda)')
%  xlabel('\lambda'); ylabel('(1+\lambda)^{1/2}*z(\lambda)')
end
```

not only produces all plots in the figures of Section 2 but also checks the validity of the relevant monotonicity properties by selecting the appropriate commands from those commented out or currently active.

5. Conjectured Higher Monotonicity Properties for Ultraspherical Polynomials

Let again $z(\lambda)$ be any positive zero of the nth-degree ultraspherical polynomial $C_n^{(\lambda)}$, $\lambda > -1/2$. (For simplicity of notation, we do not show the dependence of z on n.) The ILA conjecture, proved by Elbert and Siafarikas [4], states that

$$f(\lambda) = (1+\lambda)^{1/2} z(\lambda), \quad \lambda > -1/2, \tag{5.1}$$

is an increasing function of λ for $\lambda > -1/2$ and $n \geq 3$, even though $z(\lambda)$ is decreasing. Likewise, the Ahmed–Muldoon–Spigler result, as extended and proved by Elbert and Siafarikas (see end of Section 2), states that

$$g(\lambda) = \left(\lambda + \frac{2n^2+1}{4n+2}\right)^{1/2} z(\lambda), \quad \lambda > -1/2, \tag{5.2}$$

is an increasing function of λ for $\lambda > -1/2$ and $n \geq 2$. The Ismail–Letessier conjecture, as proved in Section 3, states that

$$h(\lambda) = \sqrt{\lambda}\, z(\lambda), \quad \lambda \geq 0, \tag{5.3}$$

is monotonically increasing for $\lambda \geq 0$ and $n \geq 2$, while Laforgia's conjecture, also proved in Section 3, states that

$$k(\lambda) = \lambda z(\lambda), \quad \lambda > -1/2, \tag{5.4}$$

is monotonically increasing for $\lambda > -1/2$ and $n \geq 2$.

The graphs of f and h on the right of Figs. 3 and 2, showing not only monotonicity, but also concavity of all positive zeros, suggest that these zeros satisfy also higher monotonicity properties. The same may be true for the function g and some of the larger zeros in the graph on the left of Fig. 3, and perhaps even for the function $k(\lambda)$.

5.1. The function $f(\lambda)$

Our conjecture for the function f is as follows.

Conjecture 5.1. *For all $n \geq 3$ the first derivative f' of the function f in (5.1) is completely monotone on $(-1/2, \infty)$, i.e.,*

$$(-1)^m f^{(m+1)}(\lambda) > 0, \quad m = 0, 1, 2, \ldots, \quad \lambda > -1/2. \tag{5.5}$$

The evidence we provide for this and the subsequent conjectures is for $n \leq 15$. It is based on divided differences. Thus, let $s(h) = \{s_j(h)\}_{j=1}^{\infty}$, $h > 0$, be the infinite sequence with

$$s_j(h) = -1/2 + jh, \quad j = 1, 2, 3, \ldots,$$

and $d_{j,m}(h;f) = [s_j(h), s_{j+1}(h), \ldots, s_{j+m}(h)]f$ the mth-order divided difference of f relative to $m+1$ consecutive members of $s(h)$ starting with $s_j(h)$. Our objective is to verify computationally that for given integers $J > 1$ and $M > 0$

$$(-1)^m d_{j,m+1}(h;f) > 0, \quad j = 1, 2, \ldots, J; \quad m = 0, 1, \ldots, M. \tag{5.6}$$

Since the $(m+1)$th divided difference of f is equal to a positive constant times the $(m+1)$th derivative of f evaluated at some intermediate point (see, e.g., [9, equation (2.117)]), it is plausible, especially if h is small, that (5.6) implies (5.5), at least for the λ-range and the m-values indicated by (5.6), but very likely for all $\lambda > -1/2$ and all $m \geq 0$.

In principle, the computational validation of (5.6) is straightforward, but requires caution when m is large, because of numerical instability. By virtue of (see, e.g., [9, Chapter 2, Exercise 54])

$$\Delta^m f_j = m! h^m d_{j,m}(h;f),$$

where $\Delta^m f_j$ is the mth difference of f at s_j, the computation of divided differences and differences is equally stable. Because of cancellation errors, however, computing differences becomes increasingly unstable as m increases, and may even yield the wrong sign. Therefore, high-precision arithmetic is imperative, though time-consuming, when m is large.

We select $h = 0.02$, $M = 14$, and $J = 350$, which are constants that remain fixed for the remainder of this section. The choice of these constants covers divided differences of orders 1–15 and a λ-interval $-1/2 < \lambda \leq 13/2$.

In Matlab double-precision arithmetic, we were able to confirm (5.6) for all $3 \leq n \leq 15$ and $0 \leq m \leq 4$ in a matter of a few seconds runtime. The differences $\Delta^m f_j$ involved were never smaller in absolute value than 1.894×10^{-13}. For the values (n, m) not covered by these computations, that is, for $3 \leq n \leq 15$ and $5 \leq m \leq 14$, it took 36-digit arithmetic in symbolic Matlab and some 70 h of runtime to confirm (5.6). The absolutely smallest difference $\Delta^m f_j$ observed was 7.774×10^{-30}.

The computations described are implemented in the Matlab script[a] `conj_geg.m` for double precision arithmetic, and in the script `sconj_geg.m` for variable-precision arithmetic using the Matlab symbolic toolbox. The former script also uses the `OPQ` routines `r_jacobi.m` and `gauss.m`, computing respectively the recurrence coefficients of the relevant ultraspherical polynomials and the related Gaussian quadrature rules (hence, in particular, the positive zeros of the ultraspherical polynomials), and the latter

[a] All Matlab scripts referenced in this chapter are collected in [12].

script uses the corresponding variable-precision routines `sr_jacobi.m` and `sgauss.m`; cf. Section 4. An additional pair of routines is used, `dd.m` and `sdd.m`, generating (from the bottom up) the appropriate divided differences in double or variable precision; see, e.g., Machine Assignment 7(a) in [9, Chapter 2] and its solution on p. 153. The script `conj_geg.m` and the routine `dd.m` are shown below. Their variable-precision versions are straightforward translations from double-precision arithmetic to variable-precision arithmetic.

```
%CONJ_GEG

f0='%12.4e\n';
h=1/50; J=350; M=15;
for n=3:15
  fprintf('n=%4.0f\n',n)
  nh=floor(n/2); z=zeros(J,1);
  for k=1:nh
    for j=1:J
      b=-1+j*h;
%      b=-1/2-h+j*h;
      ab=r_jacobi(n,b);
      xw=gauss(n,ab);
      z(j)=sqrt(b+3/2)*xw(n+1-k,1);
%      z(j)=sqrt(b+1/2)*xw(n+1-k,1);
%      z(j)=-xw(n+1-k,1);
    end
    for m=1:M
      dmin=10^20; x=zeros(m+1,1); zm=zeros(m+1,1);
      for jm=1:J-m
        for mu=1:m+1
          x(mu)=-1+(jm+mu-1)*h;
          zm(mu)=z(jm+mu-1);
        end
        dm=dd(m,x,zm);
        if sign((-1)^m*dm)>0
          disp('wrong sign')
          fprintf(f0,h^m*factorial(m)*dm)
          error('quit')
        end
```

```
        if abs(dm)<dmin, dmin=abs(dm); end
      end
      fprintf(f0,h^m*factorial(m)*dmin)
    end
    fprintf('\n')
  end
end


function y=dd(n,x,f)
%DD Divided difference.
%   Y=DD(N,X,F) evaluates the Nth divided difference
%   Y of F, where X=[X_0,X_1,...,X_N]^T,
%   F=[F(X_0),F(X_1),...,F(X_N)]^T.

d=zeros(n+1,1);
d=f;
if n==0
  y=d(1);
  return
end
for j=1:n
  for i=n:-1:j
    d(i+1)=(d(i+1)-d(i))/(x(i+1)-x(i+1-j));
  end
end
y=d(n+1);
```

The script `conj_geg.m`, as listed, produces an error message "`wrong sign`" already for $n = 3$ and $m = 5$ and displays the delinquent difference. The latter is close in absolute value to the machine precision, in fact equal to 3.684×10^{-15}, and therefore unreliable. To produce the double-precision result mentioned above, the m-loop in the script has to be run only up to $m = 5$. The 36-digit confirmation of (5.6) is produced by the script `sconj_geg.m`.

5.2. The function $h(\lambda)$

The conjecture for the function h is as follows.

Conjecture 5.2. *For all $n \geq 2$ the first derivative h' of the function h in* (5.3) *is completely monotone on* $[0, \infty)$, *i.e.*,

$$(-1)^m h^{(m+1)}(\lambda) > 0, \quad m = 0, 1, 2, \ldots, \ \lambda \geq 0. \tag{5.7}$$

The numerical validation of Conjecture 5.2 is done by the same scripts `conj_geg.m` and `sconj_geg.m`, slightly modified to deal with the function h in place of f and the interval $\lambda \geq 0$ instead of $\lambda > -1/2$. Here, the double-precision routine does the job quickly for $2 \leq n \leq 15$ and $0 \leq m \leq 6$, and the 36-digit routine for the remaining cases in about 79 h runtime.

With regard to complete monotonicity, the functions g and k in (5.2), (5.4) do not quite measure up to the functions f and h in (5.1) and (5.3). It is true that for the single positive zero $z(\lambda)$, when $n = 2$ or $n = 3$, the functions g' and k' are indeed completely monotone, but for $n = 4$, computation suggests that for g this is true only for the larger of the two positive zeros and not for the other, and for the function k only for the smaller of the two zeros and not for the other.

5.3. The function $g(\lambda)$

Conjecture 5.3. *For all $n \geq 4$, the first derivative g' of the function g in* (5.2) *is completely monotone when $z(\lambda)$ is the largest positive zero of $C_n^{(\lambda)}$, but not so otherwise.*

This was verified in the same manner as Conjectures 5.1 and 5.2, with the same scripts `conj_geg.m`, `sconj_geg.m`, suitably modified. It took another 15 h of runtime.

For the positive zeros $z(\lambda)$ of g other than the largest, additional computations (in 36-digit arithmetic) suggest an "incomplete monotonicity" property, i.e., the existence of a positive integer m_0 such that

$$(-1)^m g^{(m+1)}(\lambda) > 0 \quad \text{for } 0 \leq m \leq m_0, \ \lambda > -1/2, \tag{5.8}$$

with the opposite inequality holding when $m = m_0 + 1$. If we denote the positive zeros of $C_n^{(\lambda)}$ in decreasing order by $z_{n,k}^{(\lambda)}$, $k = 1, 2, \ldots, \lfloor \mathrm{n}/2 \rfloor$, then it is found for $4 \leq n \leq 30$ that

$$m_0 = 1 \text{ holds}$$

$$\text{for } k = 2 \text{ when } 17 \leq n \leq 30$$

$$m_0 = 2 \text{ holds}$$

$$\text{for } k = 2 \text{ when } 4 \leq n \leq 16$$

$$\text{for } k=3 \text{ when } 6 \le n \le 30$$
$$\text{for } k=4 \text{ when } 8 \le n \le 30$$
$$\text{for } k=5 \text{ when } 17 \le n \le 30$$
$$\text{for } k=6 \text{ when } 24 \le n \le 30$$
$$m_0 = 3 \text{ holds}$$
$$\text{for } k=5 \text{ when } 10 \le n \le 16$$
$$\text{for } k=6 \text{ when } 12 \le n \le 23$$
$$\text{for } 7 \le k \le 15 \text{ when } 2k \le n \le 30$$

5.4. The function $k(\lambda)$

It took another 60 h of runtime to verify the following conjecture for the function k.

Conjecture 5.4. *For* $4 \le n \le 10$, *the first derivative of the function* k *in* (5.4) *is incompletely monotone* (in the sense of (5.8), with $m_0 = 3, 2, 1$ for respectively $n = 4$, $5 \le n \le 6$, $n \ge 7$) *when* $z(\lambda)$ *is the largest positive zero of* $C_n^{(\lambda)}$, *and completely monotone otherwise. For* $11 \le n \le 15$, *both the largest and second-largest zero is incompletely monotone* (the former with $m_0 = 1$, and the latter with $m_0 = 8, 5, 3, 2$ for respectively $n = 11$, $n = 12$, $13 \le n \le 14$, $n = 15$), *while all the other zeros are completely monotone.*

This pattern likely continues for $n > 15$, with the first few positive zeros (in decreasing order) being incompletely monotone, and the remaining ones completely monotone.

5.5. The zeros $z(\lambda)$

It seems natural to ask whether higher monotonicity properties may hold also for the zeros themselves. After all, when $n = 2$, we have $-z'(\lambda) = (1/2\sqrt{2})(\lambda+1)^{-3/2}$, which is clearly completely monotone, and the same is true for $n = 3$, where $-z'(\lambda) = (1/2\sqrt{2/3})(\lambda+2)^{-3/2}$, suggesting that $-z'(\lambda)$ might be completely monotone for all $n \ge 2$,

$$(-1)^{m+1} z^{(m+1)}(\lambda) > 0, \quad m = 0, 1, 2, \ldots, \lambda > -1/2.$$

Computations (even in Matlab double precision), however, confirm this only for $m = 0$, the first counterexample occurring already when $n = 6$, $m = 1$, and $\lambda = -1/2 + h$, and others for $n = 7$, $m = 1$, and $\lambda = 0$, or $n = 8$, $m = 1$, and $\lambda = 1/2$.

Acknowledgments

The author is indebted to Mourad E. H. Ismail for the important reference [4].

References

[1] S. Ahmed, M. E. Muldoon and R. Spigler, Inequalities and numerical bounds for zeros of ultraspherical polynomials, *SIAM J. Math. Anal.* **17** (1986) 1000–1007.

[2] Á. Elbert and A. Laforgia, Some monotonicity properties of the zeros of ultraspherical polynomials, *Acta Math. Hung.* **48** (1986) 155–159.

[3] Á. Elbert and A. Laforgia, Monotonicity results on the zeros of generalized Laguerre polynomials, *J. Approx. Theory* **51** (1987) 168–174.

[4] Á. Elbert and P. Siafarikas, Monotonicity properties of the zeros of ultraspherical polynomials, *J. Approx. Theory* **97** (1999) 31–39.

[5] L. Gatteschi, On some approximations for the zeros of Jacobi polynomials, in *Approximation and computation*, International Series of Numerical Mathematics, Vol. 119 (Birkhäuser, Boston, MA, 1994), pp. 207–218.

[6] W. Gautschi, On a conjectured inequality for the largest zero of Jacobi polynomials, *Numer. Algorithms* **49** (2008) 195–198. [Also in *Selected Works*, Vol. 1, 518–521.]

[7] W. Gautschi, On conjectured inequalities for zeros of Jacobi polynomials, *Numer. Algorithms* **50** (2009) 93–96. [Also in *Selected Works*, Vol. 1, 523–526.]

[8] W. Gautschi, New conjectured inequalities for zeros of Jacobi polynomials, *Numer. Algorithms* **50** (2009) 293–296. [Also in *Selected Works*, Vol. 1, 528–531.]

[9] W. Gautschi, *Numerical Analysis*, 2nd edn. (Birkhäuser, New York, 2012).

[10] W. Gautschi, *Orthogonal Polynomials in Matlab: Exercises and Solutions*, Software, Environments, Tools (SIAM, Philadelphia, PA, 2016).

[11] W. Gautschi, Monotonicity properties of the zeros of Freud and sub-range Freud polynomials: analytic and empirical results, *Math. Comp.* **86** (2017) 855–864.

[12] W. Gautschi, Scripts for the Ismail–Letessier–Askey monotonicity conjecture for zeros of ultraspherical polynomials, *Purdue University Research Repository*, `doi:10.4231/R78G8HNQ`.

[13] W. Gautschi and P. Leopardi, Conjectured inequalities for Jacobi polynomials and their largest zeros, *Numer. Algorithms* **45** (2007) 217–230. [Also in *Selected Works*, Vol. 1, 503–516.]

[14] M. E. H. Ismail, The variation of zeros of certain orthogonal polynomials, *Adv. Appl. Math.* **8** (1987) 111–118.

[15] M. E. H. Ismail, Monotonicity of zeros of orthogonal polynomials, in *q-Series and Partitions*, ed. D. Stanton (Springer, New York, 1989), pp. 177–190.

[16] M. E. H. Ismail and J. Letessier, Monotonicity of zeros of ultraspherical polynomials, in *Orthogonal Polynomials and Their Applications*, Lecture Notes in Mathematics, Vol. 1329 (Springer, Berlin, 1988), Problem Section, pp. 329–330.

[17] M. E. H. Ismail and M. E. Muldoon, A discrete approach to monotonicity of zeros of orthogonal polynomials, *Trans. Amer. Math. Soc.* **323** (1991) 65–78.

[18] M. E. H. Ismail and R. Zhang, On the Hellmann–Feynman theorem and the variation of zeros of certain special functions, *Adv. Appl. Math.* **9** (1988) 439–446.

[19] M. E. H. Ismail and R. Zhang, The Hellmann–Feynman theorem and the variation of zeros of special functions, *Ramanujan International Symposium on Analysis* (Macmillan of India, New Delhi, 1989), pp. 151–183.

[20] A. Laforgia, Monotonicity properties for the zeros of orthogonal polynomials and Bessel functions, in *Orthogonal Polynomials and Applications* Lecture, Notes in Mathematics, Vol. 1171 (Springer, Berlin, 1985), pp. 267–277.

[21] A. Laforgia and M. E. Muldoon, Some consequences of the Sturm comparison theorem, *Amer. Math. Monthly* **93** (1986) 89–94.

[22] L. Lorch, Elementary comparison techniques for certain classes of Sturm–Liouville equations, in *Difference Equations*, Sympos. Univ. Upsaliensis Ann. Quingentesimum Celebrantis, No. 7 (Almqvist & Wiksell, Stockholm, 1977), pp. 12–133.

[23] F. W. J. Olver *et al.*, *NIST Handbook of Mathematical Functions* (US Department of Commerce and Cambridge University Press, 2010).

[24] G. Szegö, *Orthogonal Polynomials*, 4th edn. (American Mathematical Society, Providence, RI, 1975).

Chapter 14

A Discrete Top-Down Markov Problem in Approximation Theory

Walter Gautschi
Department of Computer Science, Purdue University
West Lafayette, IN 47907-2107, USA
wgautschi@purdue.edu

The Markov brothers' inequalities in approximation theory concern polynomials p of degree n and assert bounds for the kth derivatives $|p^{(k)}|$, $1 \leq k \leq n$, on $[-1,1]$, given that $|p| \leq 1$ on $[-1,1]$. Here we go in the other direction, seeking bounds for $|p|$, given a bound for $|p^{(k)}|$. For the problem to be meaningful, additional restrictions on p must be imposed, for example, $p(-1) = p'(-1) = \cdots = p^{(k-1)}(-1) = 0$. The problem then has an easy solution in the continuous case, where the polynomial and their derivatives are considered on the whole interval $[-1,1]$, but is more challenging, and also of more interest, in the discrete case, where one focuses on the values of p and $p^{(k)}$ on a given set of $n-k+1$ distinct points in $[-1,1]$. Analytic solutions are presented and their fine structure analyzed by computation.

Keywords: Markov problem; discrete; top-down.

Mathematics Subject Classification 2010: 41A10, 41A17

1. Introduction

For a polynomial p of degree $n = 2$ such that

$$|p(t)| \leq 1 \quad \text{for } -1 \leq t \leq 1, \tag{1.1}$$

the 19th-century chemist Dmitri I. Mendeleev (the creator of the periodic table of elements) raised the question of what this implies with regard to the magnitude of $|p'(t)|$ on $[-1,1]$. He found the answer to be 4 and told this to one of his colleagues, the mathematician Andrei A. Markov, who in turn generalized and solved the question in 1890 for arbitrary n by showing that

$$|p'(t)| \le n^2 \quad \text{for } -1 \le t \le 1. \tag{1.2}$$

For higher, say kth-order, derivatives, $1 \le k \le n$, the same problem was solved by the younger brother Vladimir A. Markov, who showed, rather remarkably, that

$$|p^{(k)}(t)| \le \frac{n^2(n^2-1^2)\cdots[n^2-(k-1)^2]}{1\cdot 3\cdots(2k-1)} \quad \text{for } -1 \le t \le 1. \tag{1.3}$$

The results of the brothers Markov have generated a great deal of interest in the approximation theory community and led to a considerable body of literature; see, e.g., the historical review of Shadrin [5].

We may call these problems "bottom-up", since they go from the 0th derivative up to the kth derivative. Here we look at *top-down Markov problems*, assuming

$$|p^{(k)}(t)| \le 1 \quad \text{for } -1 \le t \le 1, \tag{1.4}$$

and seeking a bound for $|p(t)|$ on $[-1,1]$. Since we can always add a polynomial of degree $k-1$ to p without affecting (1.4), top-down Markov problems become meaningful only if additional restrictions are imposed on the polynomial p. The simplest ones are to prescribe the values of the polynomial and its first $k-1$ derivatives to be zero at the left endpoint of the interval, $p(-1)=p'(-1)=\cdots=p^{(k-1)}(-1)=0$. Even then, the answer is fairly trivial, since $p(t)$ is the k-times iterated integral from -1 to t of $p^{(k)}$, i.e., because of the initial conditions imposed,

$$p(t) = \int_{-1}^{t} \frac{(t-\tau)^{k-1}}{(k-1)!}\, p^{(k)}(\tau)\mathrm{d}\tau. \tag{1.5}$$

Therefore, by (1.4),

$$|p(t)| \le \int_{-1}^{t} \frac{(t-\tau)^{k-1}}{(k-1)!}\,\mathrm{d}\tau = \frac{(t+1)^k}{k!} \le \frac{2^k}{k!} \quad \text{for } -1 \le t \le 1. \tag{1.6}$$

The bound is sharp, being attained when $p^{(k)}(t) \equiv 1$.

This straightforward answer is probably the reason why top-down Markov problems have never been given any attention. However, if we replace (1.4) by

$$|p^{(k)}(\tau_\nu)| \le 1, \quad \nu = 1, 2, \ldots, n-k+1, \tag{1.7}$$

for a set of $n-k+1$ distinct points $\{\tau_\nu\}$ in $[-1,1]$ and, as before, assume $p(-1) = p'(-1) = \cdots = p^{(k-1)}(-1) = 0$, we can ask, for each $\nu = 1, 2, \ldots, n-k+1$, the question of bounding $|p(\tau_\nu)|$ over all polynomials satisfying the constraints imposed. This problem is meaningful and not without interest; we call it a *discrete top-down Markov problem.* To the best of our knowledge, the problem has never been studied before. The special case $k=1$ of the problem, however, came up recently in work of Hager *et al.* [4] on orthogonal collocation methods in optimal control. The n points $\{\tau_\nu\}$ in this application are the zeros of the Legendre polynomial of degree n, and the authors conjectured for $|p(\tau_n)|$ the bound $\tau_n + 1$ and corresponding extremal polynomial $p^*(t) = t+1$ and the same bound for $|p(\tau_\nu)|$, $1 \le \nu < n$.

Thus, the problem we propose is the following.

Problem 1.1 (Discrete Top-Down Markov Problem). *Given integers $n \ge 1$ and $1 \le k \le n$, and given $n-k+1$ distinct points $\mathbb{T}_n^{(k)} = \{\tau_\nu\}$ in $[-1,1]$, $-1 \le \tau_1 < \tau_2 < \cdots < \tau_{n-k+1} \le 1$, consider the following class of polynomials of degree n,*

$$\begin{aligned} \mathbb{Q}_n^{(k)} = \{p \in \mathbb{P}_n : \ & p(-1) = p'(-1) = \cdots = p^{(k-1)}(-1) = 0, \\ & \textit{and } |p^{(k)}(\tau_\nu)| \le 1 \quad \textit{for } \nu = 1, 2, \ldots, n-k+1\}. \end{aligned} \tag{1.8}$$

For each ν, $\nu = 1, 2, \ldots, n-k+1$, determine the maximum possible value $M_{n,\nu}^{(k)}$ of $|p(\tau_\nu)|$ when $p \in \mathbb{Q}_n^{(k)}$,

$$M_{n,\nu}^{(k)} = \max_{p \in \mathbb{Q}_n^{(k)}} |p(\tau_\nu)|, \quad \nu = 1, 2, \ldots, n-k+1. \tag{1.9}$$

It may be instructive to look at two very special cases, in which not only the continuous, but also the discrete problem has an easy solution. The first case is $n = k = 1$, and $\tau_1 \in (-1, 1]$. Here we are dealing with the class of linear functions $p(t)$ that vanish at $t = -1$. In the continuous problem one adds the condition $|p'(t)| \le 1$ on the interval $[-1,1]$, and in the discrete problem, the condition $|p'(\tau_1)| \le 1$ at some point τ_1 in $(-1,1]$. In both problems the extremal polynomials are $p(t) = t+1$, for which

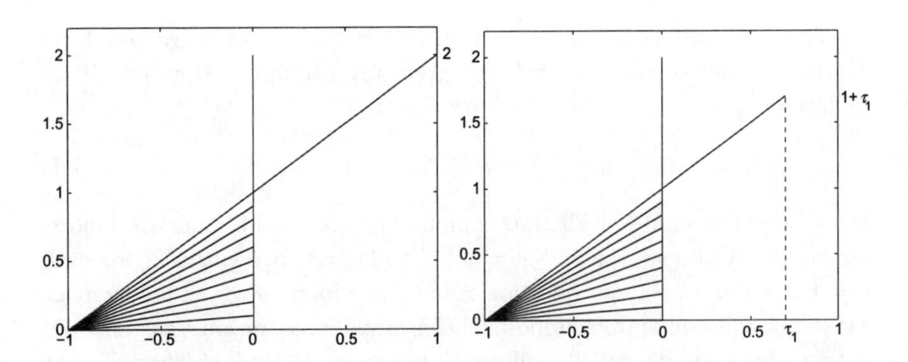

Fig. 1. Continuous vs. discrete top-down Markov problem in the case $n = k = 1$.

$\max_{|t|\le 1}|p(t)| = 2$ in the continuous case, and $M_{1,1}^{(1)} = 1+\tau_1$ in the discrete case; see Fig. 1.

The second special case is $k = n$, $\tau_1 \in (-1, 1]$. Here, the initial conditions in the problem yield $p(t) = p^{(n)}(-1)(t+1)^n/n!$ and the assumption about $p^{(n)}$, which is a constant, is $|p^{(n)}| \le 1$. Therefore, $\max_{|t|\le 1}|p(t)| = 2^n/n!$ in the continuous problem, and

$$M_{n,1}^{(n)} = \frac{(1+\tau_1)^n}{n!} \quad (k = n) \tag{1.10}$$

in the discrete one, the extremal polynomials being $(t+1)^n/n!$ in both cases.

In the general case, explicit answers to the discrete top-down problem can still be given, but seem to require computational work to get detailed information about their properties for concrete choices of the support points. In Section 2 we give a complete solution of our problem and introduce terminology for further study. Computer software to evaluate the solution is provided in Section 3, and specific examples are discussed in Section 4.

2. Solution of the Problem

We present the solution of our problem in the form of a theorem.

Theorem 2.1. *Let ℓ_μ, $\mu = 1, 2, \ldots, n-k+1$, be the elementary Lagrange interpolation polynomials of degree $n-k$ associated with the $n-k+1$ support* *points of* $\mathbb{T}_n^{(k)}$,

$$\ell_\mu(\tau) = \prod_{\substack{\kappa=1\\ \kappa\neq\mu}}^{n-k+1} \frac{\tau-\tau_\kappa}{\tau_\mu-\tau_\kappa}, \quad \mu = 1,2,\ldots,n-k+1 \quad \text{if } k<n;$$

$$\ell_1(\tau) = 1 \quad \text{if } k=n. \tag{2.1}$$

Let $\mathbf{s}_\nu^{(k)} = [s_{\nu,1}^{(k)}, s_{\nu,2}^{(k)}, \ldots, s_{\nu,n-k+1}^{(k)}]$ *be the vector with entries* $s_{\nu,\mu}^{(k)} = 1$ *if the integral*

$$I_{\nu,\mu}^{(k)} = \int_{-1}^{\tau_\nu} (\tau_\nu - \tau)^{k-1} \ell_\mu(\tau)\mathrm{d}\tau \tag{2.2}$$

is positive, $s_{\nu,\mu}^{(k)} = -1$ *if it is negative, and an arbitrary value, for example* $s_{\nu,\mu}^{(k)} = 0$*, otherwise. Define the polynomial* $p_{\nu,k}^*$ *of degree* n *by*

$$p_{\nu,k}^*(t) = \frac{1}{(k-1)!}\int_{-1}^{t} (t-\tau)^{k-1} p_{n-k}(\tau;\mathbf{s}_\nu^{(k)})\mathrm{d}\tau, \tag{2.3}$$

where $p_{n-k}(\,\cdot\,;\mathbf{s}_\nu^{(k)})$ *is the interpolation polynomial of degree* $\leq n-k$ *passing through the* $n-k+1$ *points* $(\tau_\mu, s_{\nu,\mu}^{(k)})$*. Then*

$$M_{n,\nu}^{(k)} = \frac{1}{(k-1)!}\sum_{\mu=1}^{n-k+1}\left|\int_{-1}^{\tau_\nu}(\tau_\nu-\tau)^{k-1}\ell_\mu(\tau)\mathrm{d}\tau\right|,$$

$$\nu = 1,2,\ldots,n-k+1, \tag{2.4}$$

and $p_{\nu,k}^*$ *in* (2.3) *is the associated extremal polynomial, for which*

$$M_{n,\nu}^{(k)} = p_{\nu,k}^*(\tau_\nu), \quad \nu = 1,2,\ldots,n-k+1. \tag{2.5}$$

Proof. Let $p \in \mathbb{Q}_n^{(k)}$, $1 \leq k \leq n$. Because of $p(-1) = p'(-1) = \cdots = p^{(k-1)}(-1) = 0$, one has (cf. (1.5))

$$p(t) = \int_{-1}^{t} \frac{(t-\tau)^{k-1}}{(k-1)!} p^{(k)}(\tau)\mathrm{d}\tau. \tag{2.6}$$

Since the $(n-k+1)$-point Lagrange interpolation formula produces an exact identity when applied to any polynomial of degree $\leq n-k$, we can write

$$p^{(k)}(\tau) = \sum_{\mu=1}^{n-k+1} p^{(k)}(\tau_\mu)\ell_\mu(\tau), \tag{2.7}$$

with ℓ_μ as defined in (2.1). Therefore, by (2.6),

$$p(t) = \int_{-1}^{t} \frac{(t-\tau)^{k-1}}{(k-1)!} \sum_{\mu=1}^{n-k+1} p^{(k)}(\tau_\mu)\ell_\mu(\tau)\mathrm{d}\tau,$$

that is,

$$p(t) = \frac{1}{(k-1)!} \sum_{\mu=1}^{n-k+1} p^{(k)}(\tau_\mu) \int_{-1}^{t} (t-\tau)^{k-1}\ell_\mu(\tau)\mathrm{d}\tau. \tag{2.8}$$

Now consider the case $t = \tau_\nu$. We have

$$|p(\tau_\nu)| = \frac{1}{(k-1)!} \left| \sum_{\mu=1}^{n-k+1} p^{(k)}(\tau_\mu) \int_{-1}^{\tau_\nu} (\tau_\nu-\tau)^{k-1}\ell_\mu(\tau)\mathrm{d}\tau \right|,$$

$$\nu = 1, 2, \ldots, n-k+1. \tag{2.9}$$

The only way we can control the magnitude of this expression is by choosing the values $p^{(k)}(\tau_\mu)$ subject to the constraints of the problem. To maximize $|p(\tau_\nu)|$, we select $p^{(k)}(\tau_\mu) = 1$ if the integral in (2.9), that is, $I_{\nu,\mu}^{(k)}$ in (2.2), is positive, $p^{(k)}(\tau_\mu) = -1$ if it is negative, and an arbitrary value $p^{(k)}(\tau_\mu)$ otherwise. This is always possible, since $p^{(k)}$ is a polynomial of degree $n-k$, and yields (2.4).

The extremal polynomial, by construction and the definition of the sign vector $\mathbf{s}_\nu^{(k)}$ is (2.8) with $p^{(k)}(\tau_\mu) = s_{\nu,\mu}^{(k)}$, $\mu = 1, 2, \ldots, n-k+1$, that is, (2.3). □

We remark that the integral (2.2) may indeed vanish for some μ, in which case $s_{\nu,\mu}^{(k)}$ can be chosen arbitrarily. A simple example is $n = 2$, $k = 1$, and $\tau_1 = -1/3$, $\tau_2 = 1/3$. Then $I_{2,2}^{(1)} = \int_{-1}^{1/3} \ell_2(\tau)\mathrm{d}\tau = \int_{-1}^{1/3}[(3/2)\tau + 1/2]\mathrm{d}\tau = 0$. If we choose $s_{2,2}^{(1)} = \sigma$, with σ arbitrary, one gets from (2.3) and (2.5),

$$M_{2,2}^{(1)} = \int_{-1}^{1/3} p_1(\tau; \mathbf{s}_2^{(1)})\mathrm{d}\tau = \int_{-1}^{1/3} \left[\frac{3}{2}(\sigma-1)\tau + 1 + \frac{1}{2}(\sigma-1)\right]\mathrm{d}\tau$$

$$= \left[\frac{3}{4}(\sigma-1)\tau^2 + (1 + \frac{1}{2}(\sigma-1))\tau\right]_{-1}^{1/3} = \frac{4}{3},$$

which is the correct answer, regardless of the choice of σ.

Note that for $k = n$, since $\ell_1(\tau) = 1$, we get from (2.4)

$$M_{n,1}^{(n)} = \frac{(1+\tau_1)^n}{n!},$$

in agreement with (1.10).

The value of $M_{n,\nu}^{(k)}$ in (2.4) could be further estimated by

$$\begin{aligned}
M_{n,\nu}^{(k)} &< \frac{1}{(k-1)!} \sum_{\mu=1}^{n-k+1} \int_{-1}^{\tau_\nu} (\tau_\nu - \tau)^{k-1} |\ell_\mu(\tau)| \mathrm{d}\tau \\
&< \frac{(\tau_\nu + 1)^{k-1}}{(k-1)!} \int_{-1}^{\tau_\nu} \sum_{\mu=1}^{n-k+1} |\ell_\mu(\tau)| \mathrm{d}\tau \\
&= \frac{(\tau_\nu + 1)^{k-1}}{(k-1)!} \int_{-1}^{\tau_\nu} \lambda_{n-k+1}(\tau) \mathrm{d}\tau \\
&< \frac{(\tau_\nu + 1)^{k}}{(k-1)!} \Lambda_{n-k+1}, \quad \nu = 1, 2, \ldots, n-k+1,
\end{aligned}$$

where $\lambda_{n-k+1}(\tau)$ is the Lebesgue function for $(n-k+1)$-point polynomial interpolation and $\Lambda_{n-k+1} = \max_{-1\le\tau\le1} \lambda_{n-k+1}(\tau)$ the Lebesgue constant. However, when k is fixed and n large, this may be a gross over-estimation, since, even in the best of circumstances, Λ_{n-k+1} grows logarithmically to ∞ as $n \to \infty$ whereas the bounds $M_{n,\nu}^{(k)}$, as will be seen in Section 4.1.2, are often smaller than 2.

The sign pattern for the integrals $I_{\nu,\mu}^{(k)}$ in (2.2) is characterized by the $(n-k+1) \times (n-k+1)$ "sign-pattern matrix"

$$\mathbf{S}_{n-k+1} = [s_{i,j}^{(k)}] = \begin{bmatrix} \mathbf{s}_1^{(k)} \\ \mathbf{s}_2^{(k)} \\ \vdots \\ \mathbf{s}_{n-k+1}^{(k)} \end{bmatrix}, \tag{2.10}$$

where each row vector $\mathbf{s}_\nu^{(k)}$ exhibits the signs of the integrals $I_{\nu,\mu}^{(k)}$ for $\mu = 1, 2, \ldots, n-k+1$. The matrix $\mathbf{S}_{n-k+1}$ is uniquely determined by the $n-k+1$ support points τ_ν if all integrals $I_{\nu,\mu}^{(k)}$ are different from zero. Its features greatly influence the way the bounds $M_{n,\nu}^{(k)}$ behave.

We illustrate this in the case $k = 1$ and τ_ν the zeros of the Legendre polynomial of degree n. In this case (and in many others, cf. the end of

Section 4.1.2), the matrix $\mathbf{S}_n$ is found to have the following characteristic structure: all elements on the diagonal and below are 1, and those in successive upper side diagonals are $-1, +1, -1, \ldots$,

$$\begin{aligned} s_{i,j} &= 1 \quad \text{if } j \le i, \\ s_{i,i+r} &= (-1)^r \quad \text{if } r = 1, 2, \ldots, n-i. \end{aligned} \tag{2.11}$$

We call this the *canonical sign pattern*. Thus, in successive rows of $\mathbf{S}_n$, the number of sign changes diminishes by 1 from one row to the next, being $n-1$ in the first row, and 0 in the last. In particular, for $\nu = n$, by (2.5) and (2.3), since $p_{n-1}(\,\cdot\,; \mathbf{s}_n^{(1)}) \equiv 1$, we get the bound $M_{n,n}^{(1)} = 1 + \tau_n$, which is slightly smaller than the one for $k = 1$ in the continuous case (cf. (1.6)). But is $M_{n,n}^{(1)}$ the largest of all individual bounds $M_{n,\nu}^{(k)}$?

The graphs in Fig. 2, produced by the script[a] `plot_pnmk_leg20.m`, show the polynomials $p_{n-1}(\,\cdot\,; \mathbf{s}_\nu^{(1)})$ in the case $n = 20$, for $\nu = 1, 7, 13$, and 19.

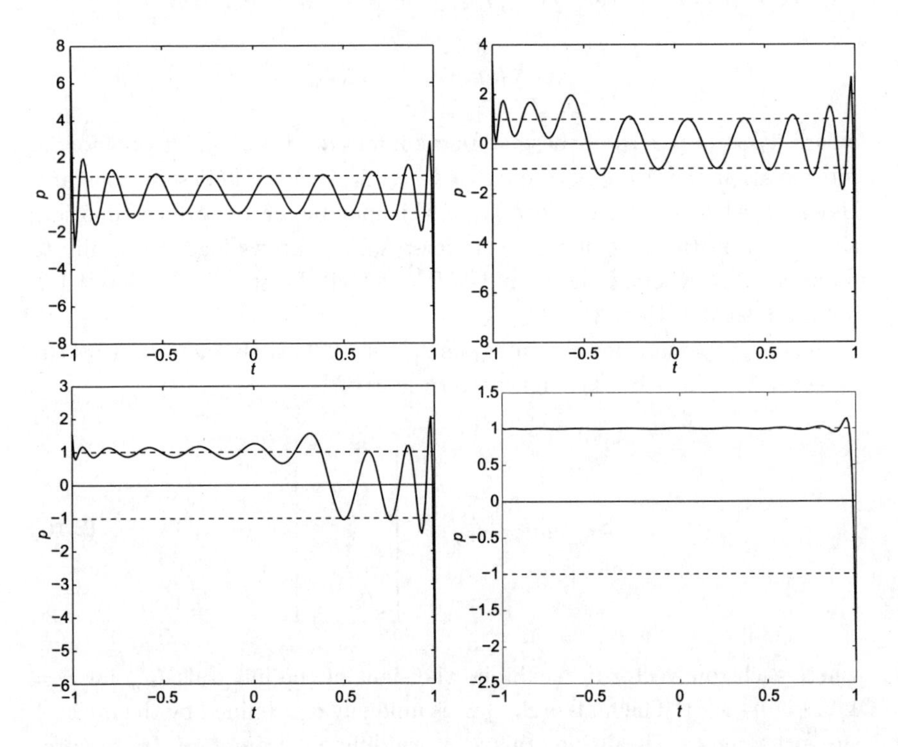

Fig. 2. The polynomial $p_{19}(\,\cdot\,; \mathbf{s}_\nu^{(1)})$ for $\nu = 1, 7, 13, 19$ (from top left to bottom right).

[a] All Matlab scripts referenced in this chapter are collected in [3].

As expected, the oscillations (about 0) diminish with increasing ν and also shift to the right. Combined with the fact that the interval of integration in (2.2) also increases, this contributes to the bounds $M_{n,\nu}^{(k)}$ likely to exhibit monotonic behavior as functions of ν.

Definition 2.2. The set $\mathbb{T}_n^{(k)}$ of $n-k+1$ support points τ_ν is said to have *Property* M_k ("M for "monotone") if

$$0 \le M_{n,1}^{(k)} < M_{n,2}^{(k)} < \cdots < M_{n,n-k+1}^{(k)}. \tag{2.12}$$

A collection of sets $\{\mathbb{T}_n^{(k)}\}$, $k = 1, 2, \ldots, n$, is said to have *Property* M, if each set $\mathbb{T}_n^{(k)}$ has Property M_k.

An explicit criterion for the validity of Property M_1 can be given in the case $n = 2$, $k = 1$, using (2.5) and (2.3). One finds

$$M_{2,1}^{(1)} = (1+\tau_1)\left[1 + \frac{1+\tau_1}{\tau_2 - \tau_1}\right], \quad M_{2,2}^{(1)} = 1 + \tau_2,$$

so that Property M_1 holds precisely if

$$\frac{\tau_2 - \tau_1}{1+\tau_1} > 1,$$

that is, if the interval $[\tau_1, \tau_2]$ is larger than the interval $[-1, \tau_1]$. In particular, it holds when $\tau_1 = -1$. Also note that Property M_n, involving only one bound, $M_{n,1}^{(n)}$, always holds trivially.

In the case where Property M_k holds, we may further distinguish between the largest bound $M_{n,n-k+1}^{(k)}$ to be normal or nonnormal according to the following definition.

Definition 2.3. When Property M_k holds, the largest bound $M_{n,n-k+1}^{(k)}$ is called *normal* if the associated sign-pattern vector $\mathbf{s}_{n-k+1}^{(k)}$ consists of 1s only, so that, by (2.5) and (2.3) with $\nu = n - k + 1$,

$$M_{n,n-k+1}^{(k)} = \frac{(\tau_{n-k+1}+1)^k}{k!} \quad \text{(normal)}. \tag{2.13}$$

It is said to be *supnormal* (respectively, *subnormal*) if $M_{n,n-k+1}^{(k)}$ is larger (respectively smaller) than $(\tau_{n-k+1}+1)^k/k!$.

The fine structure of the solution as defined by these concepts will be examined for concrete support points $\mathbb{T}_n^{(k)}$ by computation in Section 4.

3. Software

A main piece of software, in Matlab double precision, for computing the quantities $M_{n,\nu}^{(k)}$ is the routine Mnu.m, shown below, using (2.4). It first transforms the required integrals to integrals extended over $[-1,1]$ before computing them (exactly) by $\lfloor(n+1)/2\rfloor$-point Gaussian quadrature. We also prepared an analogous symbolic routine sMnu.m that can be run in variable-precision arithmetic to check the accuracy of the double-precision answers. Both routines call on routines ellmu.m resp. sellmu.m evaluating the elementary Lagrange interpolation polynomials. A number of additional Matlab routines and scripts, which are identified in the proper context, have been written, making freely use of routines from the software package OPQ [2, Section 1], for example the routine gauss.m in the program displayed below. Among them is the routine pstar.m evaluating the extremal polynomial $p_{\nu,k}^*$ of (2.3).

Here, we list the principal routines Mnu.m and pstar.m.

```
function [M,S]=Mnu
%MNU The vector of the quantities in Eq. (1.9) and the
%sign-pattern matrix S of Eq. (2.10).
%  Given the (n-k+1)x1 array tau of support points, the
%  function [M,S]=Mnu generates the (n-k+1)x1 array M of
%  the quantities M_{n,1},M_{n,2},...,M_{n,n-k+1} in Eq. (2.4),
%  using floor((n+1)/2)-point Gaussian quadrature to (exactly)
%  compute the integrals in Eq. (2.4). The global variable ab0
%  is the N0x2 array, N0=floor((N+1)/2), of recurrence
%  coefficients for monic Legendre polynomials, useful when
%  the routine is run for many values of n<=N.

global n k tau ab0
M=zeros(n-k+1,1); S=zeros(n-k+1);
n0=floor((n+1)/2);
xw0=gauss(n0,ab0); x0=xw0(:,1); w0=xw0(:,2);
for nu=1:n-k+1
  L=0;
  for mu=1:n-k+1
    if tau(nu)==-1
      s=0;
    else
      t=.5*(tau(nu)-1+(tau(nu)+1)*x0); y=ellmu(mu,t);
      s=sum(w0.*(tau(nu)-t).^(k-1).*y);
    end
    if abs(s)<1e-14
      S(nu,mu)=0;
    elseif s>0
```

```
      S(nu,mu)=1;
    else
      S(nu,mu)=-1;
    end
    L=L+abs(s);
  end
  M(nu)=.5*(tau(nu)+1)*L/factorial(k-1);
end

function y=pstar(nu,t)
%PSTAR Extremal polynomial
%   Y=PSTAR(NU,T) evaluates the extremal polynomial p_{NU,K}^*(T),
%   given the (N-K+1)x1 array TAU of support points and the
%   (N-K+1)x(N-K+1) array S of sign patterns. The variable T has
%   to be a single value.

global n k tau S ab0
xw0=gauss(floor((n+1)/2),ab0); x0=xw0(:,1); w0=xw0(:,2);
sx=size(x0); t1=(t-1+(t+1)*x0)/2;
pnmk=zeros(sx);
for mu=1:n-k+1
  pnmk=pnmk+S(nu,mu)*ellmu(mu,t1);
end
y=sum(w0.*(t-t1).^(k-1).*pnmk);
y=(t+1)*y/(2*factorial(k-1));
```

4. Examples

4.1. Zeros of Jacobi polynomials

In this subsection, $\mathbb{T}_n^{(k)} = \{\tau_\nu\}$ will be the zeros of the Jacobi polynomial $P_{n-k+1}^{(\alpha,\beta)}$ of degree $n-k+1$ with parameters $\alpha > -1$, $\beta > -1$.

4.1.1. *Property M*

We first try to find out in which part of the (α,β)-plane Property M holds for $\{\mathbb{T}_n^{(k)}\}$. We do this by letting β, for fixed α, move upwards from $\beta = -0.9$ (where Property M holds) in steps of 0.1 until Property M fails for the first time. If this happens at $\beta = \beta_0$, we will have two values of β, $\beta_{\text{low}} = \beta_0 - 0.1$ and $\beta_{\text{high}} = \beta_0$, at the first of which Property M holds, and at the second of which it does not. We then apply a bisection type method to determine a more accurate value of β^* such that Property M holds for $\beta < \beta^*$ and fails to hold for $\beta > \beta^*$. If we do this for α = −0.9:0.1:5 and plot β^* vs. α, we obtain a curve in the (α,β)-plane, $\beta = \beta^*(\alpha)$, $-0.9 \le \alpha \le 5$, below of which Property M holds, and above of which it does not. This

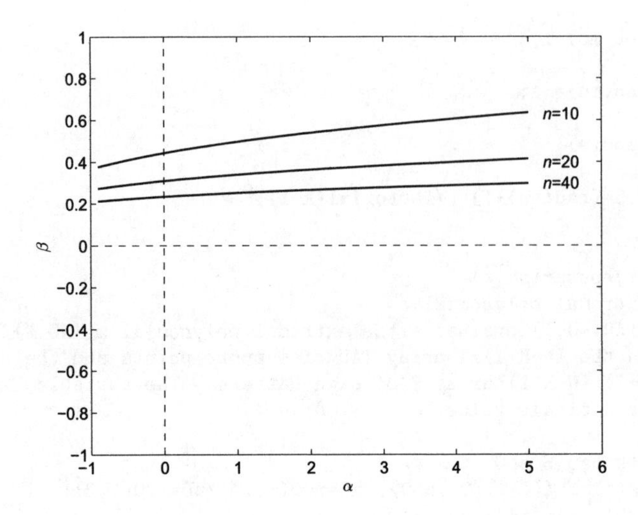

Fig. 3. Domains $\mathcal{D}_M$ for zeros of Jacobi polynomials

is implemented in the script `Mnu_test2.m` and the routine `PropM.m`, for $n = 10$, 20, and 40. The curves produced are shown in Fig. 3, generated by the script `plot_bestar.m`.

The domains $\mathcal{D}_M$ in the (α,β)-plane in which Property M holds are quite large, containing most likely the whole strip $-1<\alpha<\infty$, $-1<\beta\leq 0.15$. The important case $\alpha=\beta=0$ of Legendre polynomials is well within this domain. It looks as if the upper boundary, when $n\to\infty$, would become a horizontal line at the height of about 0.15, an expectation that was reinforced when the script `Mnu_test2.m` was run with $n = 80$.

It may be thought that the discrete Markov problem yields solutions that are similar to those of the continuous problem, that is, producing bounds for $|p|$ nearly equal, or less than 2. The present discussion provides an opportunity to dispel this notion. As can be seen from Fig. 3, Property M holds in a domain that extends far (probably infinitely far) to the right and includes the positive real axis. So, let us look at the following exotic example.

Example 4.1. The case $\alpha = 20$, $\beta = 0$.

For simplicity, we take $k = 1$ and $n = 10$. The bounds $M^{(1)}_{10,\nu}$, produced by the script `Ex4_1.m`, are then as shown in Table 1. We can see that $M^{(1)}_{10,10}$ is anything but "nearly equal, or less than to 2"! (See also Example 4.2.)

Table 1. Bounds for $\alpha = 20$, $\beta = 0$.

ν	$M^{(1)}_{10,\nu}$
1	1.801846843051301e−02
2	5.842152063398374e−02
3	1.268621727178132e−01
4	2.226277805868339e−01
5	3.440754243906574e−01
6	4.890625971135910e−01
7	6.552254810765432e−01
8	1.088942247153418e+00
9	4.231399671654690e+00
10	4.323112886064376e+01

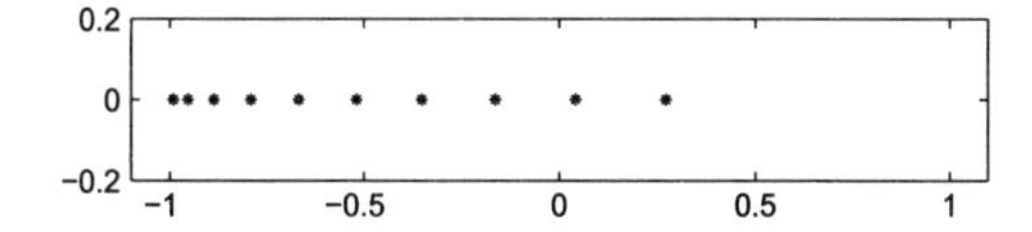

Fig. 4. Zeros of the Jacobi polynomial $P_{10}^{(20,0)}$.

The corresponding sign-pattern matrix is

$$\mathbf{S}_{10}^{(1)} = \begin{bmatrix} 1 & -1 & 1 & -1 & 1 & -1 & 1 & -1 & 1 & -1 \\ 1 & 1 & -1 & 1 & -1 & 1 & -1 & 1 & -1 & 1 \\ 1 & 1 & 1 & -1 & 1 & -1 & 1 & -1 & 1 & -1 \\ 1 & 1 & 1 & 1 & -1 & 1 & -1 & 1 & -1 & 1 \\ 1 & 1 & 1 & 1 & 1 & -1 & 1 & -1 & 1 & -1 \\ 1 & 1 & 1 & 1 & 1 & 1 & -1 & 1 & -1 & 1 \\ 1 & 1 & 1 & 1 & 1 & 1 & 1 & -1 & 1 & -1 \\ 1 & -1 & 1 & -1 & 1 & 1 & 1 & 1 & -1 & 1 \\ -1 & 1 & -1 & 1 & -1 & 1 & -1 & 1 & 1 & -1 \\ 1 & -1 & 1 & -1 & 1 & -1 & 1 & -1 & 1 & 1 \end{bmatrix}.$$

Interestingly, it has the canonical ± 1 Toeplitz pattern in the upper triangular part, but quite a few -1s in the lower triangular part. The support points τ_ν, plotted by the script **plot_tau20.m** and displayed in Fig. 4, are considerably "left-heavy" and therefore allow the interpolation polynomial $p_9(\,\cdot\,;\mathbf{s}_{10}^{(1)})$, plotted by the script **plot_pnmk_jac10.m**, in the

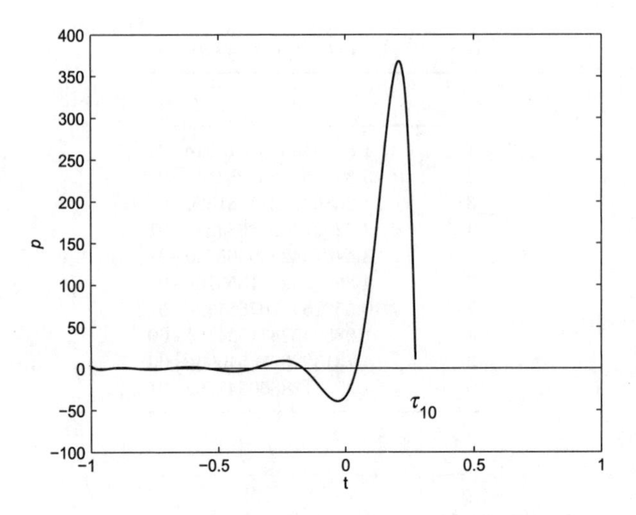

Fig. 5. The interpolation polynomial $p_9(\,\cdot\,; \mathbf{s}_{10}^{(1)})$ on $[-1, \tau_{10}]$.

last quarter of the interval $[-1, 1]$ to unimpededly soar to great heights in absolute value. Just before $t = \tau_{10}$ it has a large hump as shown in Fig. 5, causing the last bound in Table 1 to be unusually large.

Returning to Fig. 3, what can be said about the bounds $M_{n,\nu}^{(k)}$ (for $n = 10, 20, 40$) when β is above the respective curve in Fig. 3? The matter is then more complicated since the zeros of the Jacobi polynomials, when β becomes large, become increasingly "right-heavy" and, as a result, at least when $k = 1$, the interpolation polynomial p_{n-1} in (2.3) is large (possibly very large) at the left end of the interval $[-1, 1]$, regardless of the value of ν. This makes the bounds $M_{n,\nu}^{(1)}$ large for all ν, and the issue of monotonicity becomes blurred. There may no longer prevail any particular pattern with regard to the relative magnitudes of the bounds. This changes, however, when k becomes larger, because of the mitigating influence of the factor $(\tau_\nu - \tau)^{k-1}$ in (2.3). Indeed, already when $k = 2$, the bounds are monotonically increasing for $n = 10, 20, 40$, at least when $\beta \le 10$, but very likely even beyond that.

When $k = 1$, it is meaningful to ask for what values of α, β we have $M_{n,\nu}^{(1)} \le M_{n,n}^{(1)}$ for all $1 \le \nu < n$. This was determined by the script `Mnu_test5.m` and the routine `PropMax.m` similarly as before and led to the curves (in red) of Fig. 6, below of which this maximum property holds. They were produced by the script `plot_bestarMax.m`. The figure, at the bottom, also reproduces the three curves of Fig. 3.

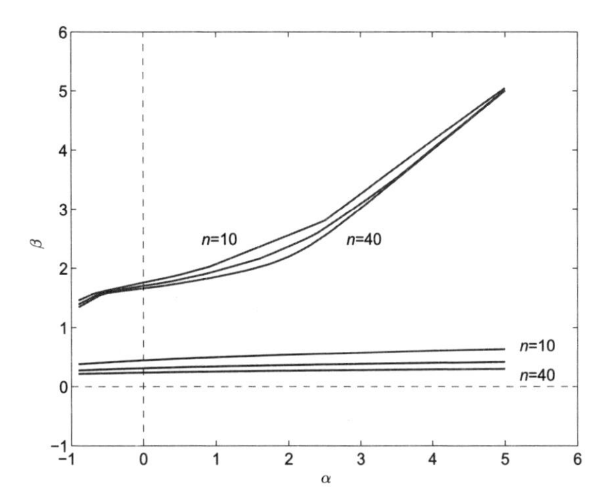

Fig. 6. Domains for zeros of Jacobi polynomials in which $M_{n,n}^{(1)}$ is the largest bound.

Here is another somewhat exotic example.

Example 4.2. The case $\alpha = 0$, $\beta = 5$ (above the red curves in Fig. 6) and $k = 1$.

Taking again $n = 10$, the script `Ex4_2.m` now obtains uniformly large bounds shown in Table 2 and the sign-pattern matrix

$$\mathbf{S}_{10}^{(1)} = \begin{bmatrix} 1 & -1 & 1 & -1 & 1 & -1 & 1 & -1 & 1 & -1 \\ 1 & -1 & 1 & -1 & 1 & -1 & 1 & -1 & 1 & -1 \\ 1 & -1 & 1 & -1 & 1 & -1 & 1 & -1 & 1 & -1 \\ 1 & -1 & 1 & -1 & 1 & -1 & 1 & -1 & 1 & -1 \\ 1 & -1 & 1 & -1 & 1 & -1 & 1 & -1 & 1 & -1 \\ 1 & -1 & 1 & -1 & 1 & -1 & 1 & -1 & 1 & -1 \\ 1 & -1 & 1 & -1 & 1 & -1 & 1 & -1 & 1 & -1 \\ 1 & -1 & 1 & -1 & 1 & -1 & 1 & -1 & 1 & -1 \\ 1 & -1 & 1 & -1 & 1 & -1 & 1 & -1 & 1 & -1 \\ 1 & -1 & 1 & -1 & 1 & -1 & 1 & -1 & 1 & -1 \end{bmatrix}.$$

As can be seen, the columns of $\mathbf{S}_{10}^{(1)}$ are alternately all 1s and all -1s. The support points τ_ν, plotted by the script `plot_tau5.m` and shown in Fig. 7, are now "right-heavy", as already mentioned, causing the interpolation

Table 2. Bounds for $\alpha = 0, \beta = 5$.

ν	$M^{(1)}_{10,\nu}$
1	9.451771476864684e+01
2	9.197235521181210e+01
3	9.282549856814146e+01
4	9.241442728995621e+01
5	9.263482633418995e+01
6	9.252393392561106e+01
7	9.256151952490204e+01
8	9.257534088982851e+01
9	9.252776896050197e+01
10	9.259039977486584e+01

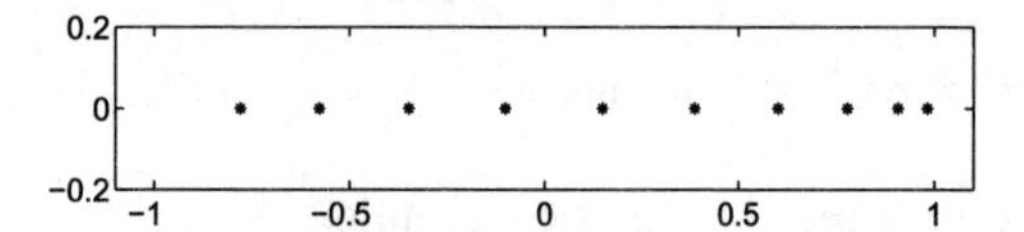

Fig. 7. Zeros of the Jacobi polynomial $P_{10}^{(0,5)}$.

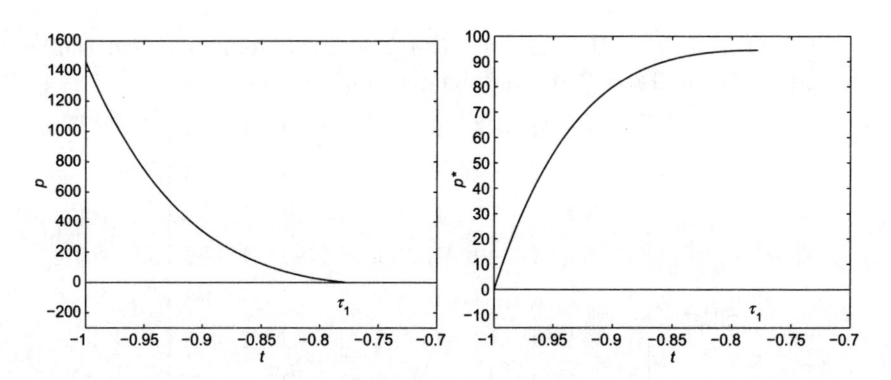

Fig. 8. The interpolation polynomial $p_9(\,\cdot\,;\mathbf{s}_\nu^{(1)})$ and extremal polynomial $p^*_{\nu,1}$ on $[-1, \tau_1]$.

polynomial $p_9(\,\cdot\,;\mathbf{s}_\nu^{(1)})$, which, by the way, is the same for each ν, to be very large at the left end of the interval $[-1, 1]$, as shown on the left of Fig. 8. The extremal polynomial $p^*_{\nu,1}(t)$ (also the same for each ν), accordingly, raises quickly to a large number at $t = \tau_1$, see the graph on the right of Fig. 8, and stays at that level for the rest of the interval $[-1, 1]$, which explains the uniformly large magnitude of all the bounds in Table 2. The graphs of Fig. 8 were produced in the script `Ex4_2.m`.

4.1.2. *Normality of* $M_{n,n-k+1}^{(k)}$ *for* $k = 1$

Now that we know of a fairly large domain $\mathcal{D}_M$ (cf. Fig. 3) of the (α,β)-plane in which Property M holds, i.e., in which the bounds $M_{n,1}^{(k)}, M_{n,2}^{(k)}, \ldots, M_{n,n-k+1}^{(k)}$ are monotonically increasing for each k, it is of interest to determine in which subdomain $\mathcal{D}_N$ the largest of these, $M_{n,n-k+1}^{(k)}$, is normal according to Definition 2.3 in Section 2. Because then, by (2.3) and (2.5), we will have

$$0 \leq M_{n,1}^{(k)} < M_{n,2}^{(k)} < \cdots < M_{n,n-k}^{(k)} < \frac{(1+\tau_{n-k+1})^k}{k!}, \quad 1 \leq k \leq n. \quad (4.1)$$

We analyze this only in the case $k = 1$, since for $k > 1$, normality of the last bound is more the exception than the rule.

Before we proceed, let us digress just a little. Normality of $M_{n,n}^{(1)}$, by definition, means that the sign-pattern vector $\mathbf{s}_n^{(1)}$ consists of all 1s, hence, by (2.2),

$$\int_{-1}^{\tau_n} \ell_\mu(\tau)\mathrm{d}\tau > 0, \quad \mu = 1, 2, \ldots, n. \quad (4.2)$$

Since τ_n, the largest zero of $P_n^{(\alpha,\beta)}$, is usually very close to 1, the inequality (4.2) is almost the same as $\int_{-1}^{1} \ell_\mu(\tau)\mathrm{d}\tau > 0$, which would mean that the Newton–Cotes quadrature formula on $[-1, 1]$ with the nodes being the zeros of the Jacobi polynomial $P_n^{(\alpha,\beta)}$ is positive. This is a property not easy to establish rigorously, but has been studied in the 1970s by R. Askey, and in the 1980s by G. Sottas (for references, see [1]). In particular, a conjecture of Askey, slightly revised in [1, Section 4.1], is positivity of the Newton–Cotes formulas in the (α, β)-domain $\mathcal{D}_{NC}$ shown in Fig. 9.

We have used our script `Mnu_test3.m` and routine `normal.m` to verify (4.2) for $n = 10, 20, 40$ on a grid on $\mathcal{D}_{NC}$ with spacing 0.01. We were successful in all cases except for various values of β near the left boundary $\alpha = -1$ (more precisely, when $\alpha = -0.99$).

Resuming our analysis of normality, we now proceed similarly as in Section 4.1.1, but search for fixed β, $-1 < \beta \leq 0.2$, for the alpha-value α^* for which normality of $M_{n,n}^{(1)}$ holds for all $\alpha < \alpha^*$, but does not hold for $\alpha > \alpha^*$. We can then develop plots of $\alpha = \alpha^*(\beta)$, or by reversing, plots of $\beta = \beta^*(\alpha)$, to delineate the domain $\mathcal{D}_N$. The results for $n = 10, 20, 40$, obtained by the script `Mnu_test3.m` and plotted by `plot_bestar_low.m`,

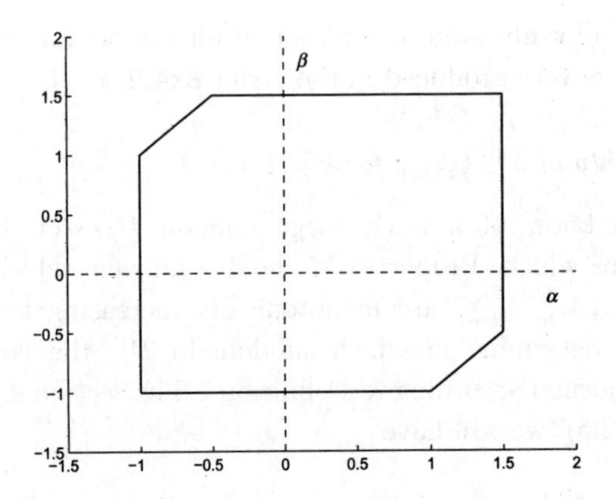

Fig. 9. Positivity domain $\mathcal{D}_{NC}$ for Newton–Cotes formulas with nodes equal to the zeros of the Jacobi polynomial $P_n^{(\alpha,\beta)}$.

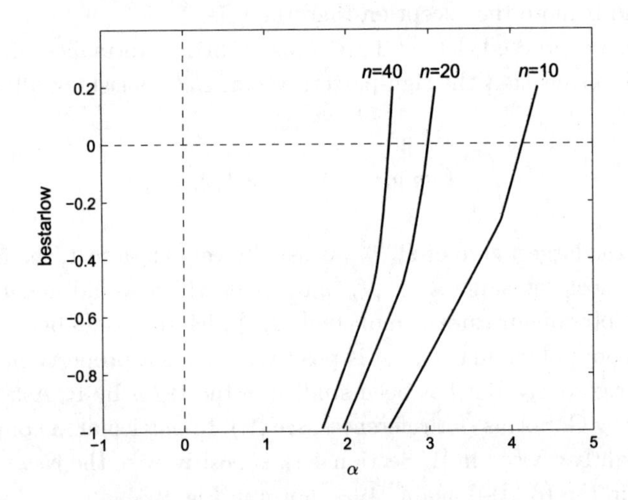

Fig. 10. Domains $\mathcal{D}_N$ for zeros of Jacobi polynomials when $k = 1$.

are shown in Fig. 10. Here, it appears that as $n \to \infty$, there might be a vertical limit boundary at about $\alpha = 1.3$. To the right of the curves, as was determined by the script `Mnu_test4.m`, the bound $M_{n,n}^{(1)}$ is consistently supnormal (cf. Definition 2.3 in Section 2), for α as large as 10, and probably remains so for all $\alpha > 10$.

By comparing the domains $\mathcal{D}_M$ of Fig. 3 with the domains $\mathcal{D}_N$ of Fig. 10, and taking approximate intersections, one can conclude with confidence that, when $k = 1$, the inequalities (4.1) hold in the rectangular domain

$-1 < \alpha \le 1$, $-1 < \beta \le 0.15$, for any value of n. In the rectangle $-1 < \alpha \le 1$, $-1 < \beta \le 0.5$, we used the script `Mnu_test1.m` to verify that (on a grid with spacing 0.05) the sign-pattern matrix $\mathbf{S}_{n-k+1}^{(k)}$, when $k = 1$, rather remarkably, is always canonical (cf. Section 2), at least for $n \le 40$, but very likely for all $n \ge 2$.

4.1.3. *Monotonicity of* $M_n^{(k)}$

The script `Mnu_test6.m` was written to explore in what part of the (α, β)-plane the inequalities

$$M_n^{(1)} > M_n^{(2)} > \cdots > M_n^{(n)} \tag{4.3}$$

hold, where

$$M_n^{(k)} = \max_{1 \le \nu \le n-k+1} M_{n,\nu}^{(k)}.$$

It was found that, for $2 \le n \le 40$, the inequalities (4.3) are valid on a grid with spacing 0.05 within the square $Q = \{\alpha, \beta : |\alpha| < 0.5,\ |\beta| < 0.5\}$, suggesting that they may hold for all $\alpha, \beta \in Q$, perhaps even for all $n \ge 2$.

With regard to monotonicity in the variable n,

$$M_2^{(k)} < M_3^{(k)} < M_4^{(k)} < \cdots, \tag{4.4}$$

the script `Mnu_test7.m`, analogous to `Mnu_test6.m`, suggests that, at least for $n \le 40$, the inequalities are valid on the square $-1 < \alpha, \beta \le 1$ when $k = 1$, on the square $-0.5 \le \alpha, \beta < 0.5$ when $k = 2$, and on $|\alpha|, |\beta| \le 0.25$ when $k = 3$ and $n \ge 3$. They hold for still larger values of k and $n \ge k$, in sufficiently smaller (α, β)-domains, in particular for Legendre polynomials ($\alpha = \beta = 0$), when $4 \le k \le 26$, but not when $k = 27$, where the inequality fails to hold for $n = 32, 33$.

4.2. Gauss–Lobatto quadrature points

In this subsection, $\mathbb{T}_n^{(k)} = \{\tau_\nu\}$ are the $(n - k + 1)$-point Gauss–Lobatto quadrature points relative to the Jacobi weight function when $n - k \ge 2$, the points $\tau_1 = -1$, $\tau_2 = 1$, when $n - k = 1$, and the point $\tau_1 = -1$, when $n - k = 0$. Results analogous to those in Section 4.1 can be obtained by making relatively minor changes to the software used before.

4.2.1. *Property M*

The scripts `Mnu_test2.m`, `plot_bestarlob.m` in combination with the routine `PropM.m`, adjusted to deal with Gauss–Lobatto points, yield Fig. 11,

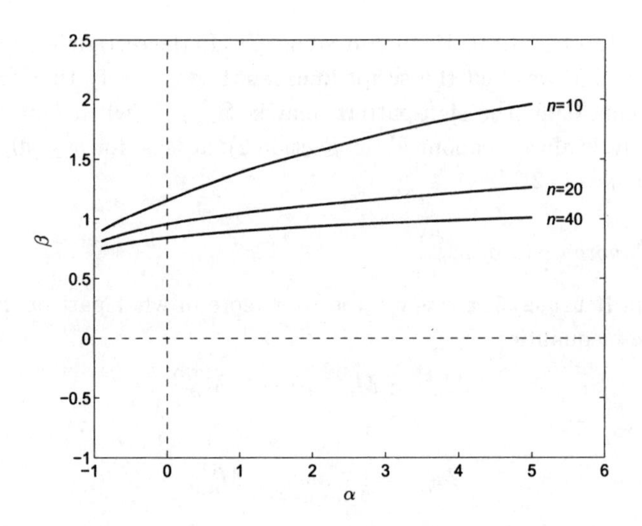

Fig. 11. Domains $\mathcal{D}_M$ for Gauss–Lobatto points.

showing the domains $\mathcal{D}_M$ (below the curves) in which Property M holds. They extend to β-values considerably larger than those in Fig. 3, but otherwise there is not much change.

The analogue of Fig. 6, for Gauss–Lobatto quadrature points, is Fig. 12, produced by the routine `plot_bestarlobMax.m` and the suitably adjusted script `Mnu_test5.m` and routine `PropMax.m`, showing the domains (below the red curves) in which $M_{n,\nu}^{(1)} \leq M_{n,n}^{(1)}$ for $\nu = 1, 2, \ldots, n-1$.

4.2.2. *Normality of* $M_{n,n-k+1}^{(k)}$ *for* $k = 1$

The analogue of Fig. 10 showing the domains $\mathcal{D}_N$ (to the left of the curves) in which $M_{n,n}^{(1)}$ is normal, is Fig. 13, produced by the routine `plot_bestarlob_low.m` and the suitably adjusted script `Mnu_test3.m` and routine `normal.m`.

Comparison of the domains $\mathcal{D}_M$ and $\mathcal{D}_N$ again allows us to conclude with confidence that the inequalities (4.1) with $k = 1$ hold for all $n \geq 2$ in the rectangular domain $-1 < \alpha \leq 1$, $-1 < \beta \leq 0.5$, except, possibly, when β is very close to -1.

The startling property of canonical sign-pattern matrices $\mathbf{S}_n^{(1)}$ that we noted at the end of Section 4.1.2 to hold for zeros of Jacobi polynomials does no longer hold for all n when the support points are Gauss–Lobatto

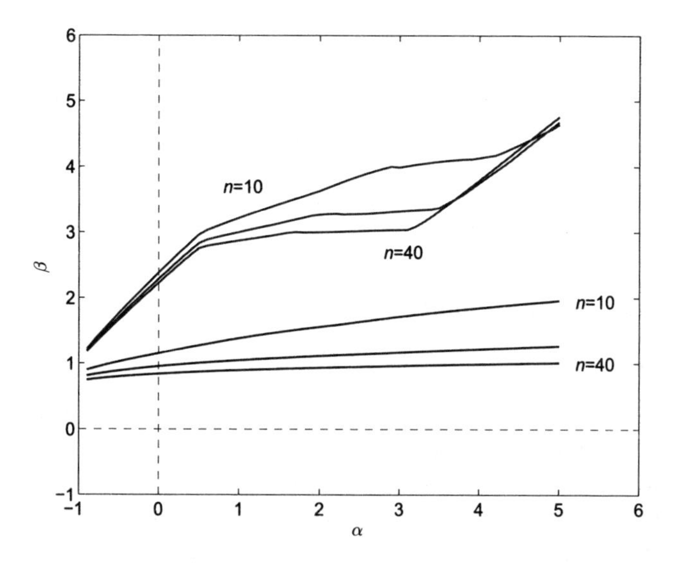

Fig. 12. Domains for Gauss–Lobatto points in which $M_{n,n}^{(1)}$ is the largest bound.

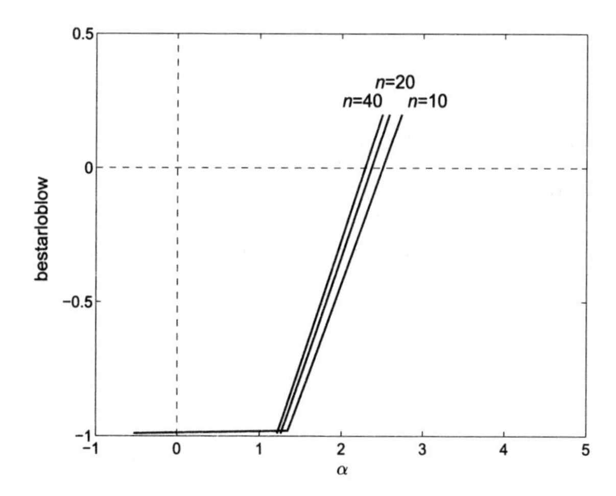

Fig. 13. Domains $\mathcal{D}_N$ for Gauss–Lobatto points when $k = 1$.

points, not even in the special case of Legendre polynomials. In that case, however, it does hold sporadically for selected values of n; those ≤ 100 are shown in Table 3. When it does not hold, the offending elements in the matrix $\mathbf{S}_n^{(1)}$ are consistently located in the first row, either in position 2 through n, or in position $1, 3, 5, \ldots, 2\lfloor(n+1)/2\rfloor - 1$.

Table 3. Values of $n \le 100$ for which $\mathbf{S}_n^{(1)}$ is canonical for Gauss–Lobatto points in the case $\alpha = \beta = 0$.

3	10	17	23	35	39	46	61	70	76	84	92
5	12	19	25	36	40	52	64	73	80	87	94
6	13	20	30	37	41	56	68	74	82	89	
8	16	21	32	38	45	58	69	75	83	91	

4.3. Equally spaced points

4.3.1. *Equally spaced points on* $(-1, 1)$

In this subsection, $\mathbb{T}_n^{(k)}$ will be the points $\tau_\nu = -1 + 2\nu/(n-k+2)$, $\nu = 1, 2, \ldots, n-k+1$, equally spaced in the open interval $(-1, 1)$.

The following properties are found to hold: The canonical sign-pattern matrix $\mathbf{S}_{n-k+1}$ is the $(n-k+1) \times (n-k+1)$ matrix whose columns are alternately all 1s and all -1s (cf. also Example 4.2),

$$s_{i,j} = (-1)^{j-1}, \quad 1 \le i, j \le n-k+1, \tag{4.5}$$

disregarding occasional zero elements. This was verified by the script `Mnu_test1_equal.m` and the routine `Prop_canon_equal.m` for all $n \le 100$.

With regard to the bounds $M_{n,\nu}^{(k)}$, the properties are markedly different depending on whether k is equal to 1 or larger than 1.

In the case $k = 1$, the interpolation polynomial p_{n-1} in (2.3), which by the previous property is the same for each ν, is even if n is odd, and odd if n is even. As a consequence, by (2.5), we have

$$\begin{aligned} &M_{n,\nu}^{(1)} = M_{n,n+1-\nu}^{(1)}, \quad \nu = 1, 2, \ldots, n \quad (n \text{ even}), \\ &M_{n,\nu}^{(1)} - M_{n,(n+1)/2}^{(1)} = -[M_{n,n+1-\nu}^{(1)} - M_{n,(n+1)/2}^{(1)}], \\ &\qquad \nu = 1, 2, \ldots, n \quad (n \text{ odd}), \end{aligned} \tag{4.6}$$

that is symmetry when n is even, and anti-symmetry with respect to the middle bound $M_{n,(n+1)/2}^{(1)}$ when n is odd.

In the case $k > 1$, we consistently have Property M_k,

$$M_{n,1}^{(k)} < M_{n,2}^{(k)} < \cdots < M_{n,n-k+1}^{(k)} \quad (1 < k < n), \tag{4.7}$$

as was verified by the script `Mnu_test2_equal0.m` and routine `Prop_mon_equal.m` for all $n \le 100$.

4.3.2. *Equally spaced points on* $[-1, 1]$

Here, $\mathbb{T}_n^{(k)}$ are the points $\tau_\nu = -1 + 2(\nu - 1)/(n - k)$, $\nu = 1, 2, \ldots, n - k + 1$, where $n \geq 2$, $1 \leq k \leq n - 1$, which are equally spaced in the closed interval $[-1, 1]$.

In this case, no canonical sign-pattern matrix $\mathbf{S}_{n-k+1}$ could be detected. It was found, however, by the script `Mnu_test2_equalC.m` and the same routine `Prop_mon_equal.m` as before, that Property M_k, that is (4.7), holds precisely if

$$(2 \leq n \leq 5 \text{ and } 1 \leq k \leq n - 1) \quad \text{or} \quad (n \geq 6 \text{ and } 2 \leq k \leq n - 1),$$

at least as long as $n \leq 100$, but very likely also for $n > 100$.

Acknowledgments

I am indebted to Mourad E. H. Ismail for bringing the work [4] of W. W. Hager *et al.* to my attention and to W. W. Hager for a critical reading of earlier drafts of the paper.

References

[1] W. Gautschi, Moments in quadrature problems, *Comput. Math. Appl.* **33** (1997) 105–118.

[2] W. Gautschi, *Orthogonal Polynomials in Matlab: Exercises and Solutions*, Software, Environments, and Tools (SIAM, 2016).

[3] W. Gautschi, Scripts for a discrete top-down Markov problem in approximation theory. *Purdue University Research Repository*, `doi:10.4231/R74Q7RX0`.

[4] W. W. Hager, H. Hou and A. V. Rao. Convergence rate for a Gauss collocation method applied to unconstrained optimal control, *J. Optim. Theory Appl.* **169**(3) (2016) 801–824.

[5] A. Shadrin, Twelve proofs of the Markov inequality, in *Approximation Theory: A Volume Dedicated to Borislav Bojanov* (Prof. M. Drinov Academic Publishing House, Sofia, 2004), pp. 233–298.

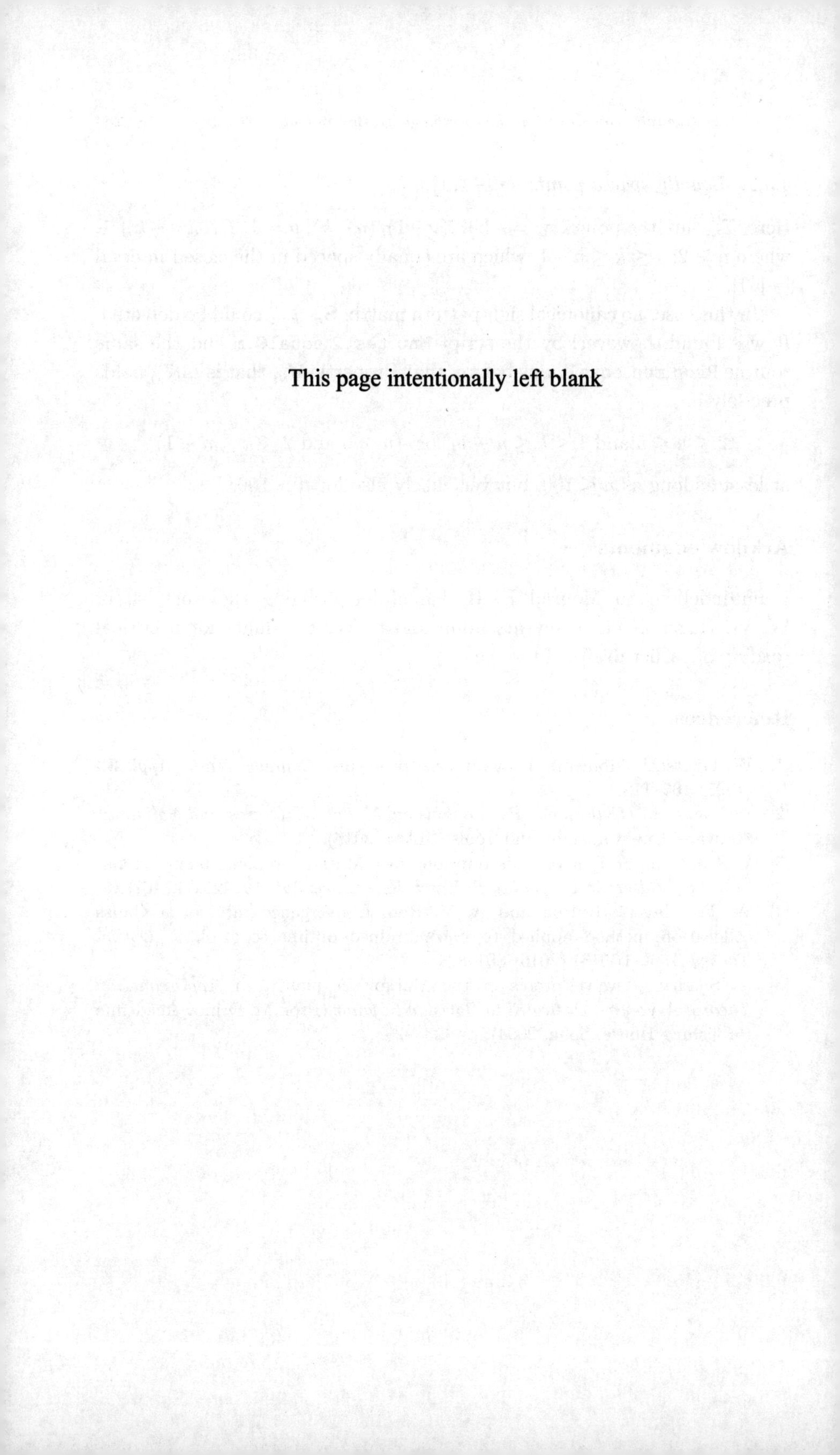

This page intentionally left blank

Chapter 15

Supersymmetry of the Quantum Rotor

Vincent X. Genest[*,¶], Luc Vinet[†,‖], Guo-Fu Yu[‡,] and Alexei Zhedanov[§,††]**

[*]*Department of Mathematics, Massachusetts Institute of Technology Cambridge, MA 02139, USA*

[†]*Centre de Recherches Mathématiques, Université de Montréal Montréal QC, Canada H3C 3J7*

[‡]*Department of Mathematics, Shanghai Jiao Tong University Shanghai 200240, China*

[§]*Department of Applied Mathematics and Physics Graduate School of Informatics, Kyoto University Kyoto 606-8501, Japan*

[¶]*vxgenest@mit.edu*

[‖]*vinet@crm.umontreal.ca*

[**]*gfyu@sjtu.edu.cn*

[††]*zhedanov@fti.dn.ua*

This paper is cordially dedicated to Mourad Ismail on the occasion of his 70th birthday

The quantum rotor is shown to be supersymmetric. The supercharge Q, whose square equals the Hamiltonian, is constructed with reflection operators. The conserved quantities that commute with Q form the algebra $\mathrm{so}(3)_{-1}$, an anticommutator version of $\mathrm{so}(3)$. The subduced representation of $\mathrm{so}(3)_{-1}$ on the space of spherical harmonics with total angular momentum j is constructed and found to decompose into two irreducible components. Two natural bases for the irreducible representation spaces of $\mathrm{so}(3)_{-1}$ are introduced and their overlap coefficients prove expressible in terms of orthogonal polynomials of a discrete variable called anti-Krawtchouk polynomials.

Keywords: Quantum rotor; supersymmetry; special Bannai–Ito algebra; anti-Krawtchouk polynomials.

Mathematics Subject Classification 2010: 33D45, 17B37, 33D80

1. Introduction

It has been appreciated that certain scalar quantum mechanical systems can exhibit supersymmetry with supercharges involving reflection operators or more generally involutions [17, 19]. The simplest example of such an instance is that of the following harmonic oscillator Hamiltonian in one dimension [19]:

$$H = -\frac{1}{2}\frac{\partial^2}{\partial x^2} + \frac{1}{2}x^2 - \frac{1}{2}R,$$

where R is the reflection operator acting as x: $Rf(x) = f(-x)$. Indeed, it is straightforward to see that one then has

$$H = Q^2,$$

with the supercharge Q having the expression

$$Q = \frac{1}{\sqrt{2}}\left(\frac{\partial}{\partial x}R + x\right).$$

The occurrence of involutions in quantum mechanical dynamics is not unfamiliar [2, 9, 15, 18]; it is in particular related to the use of Dunkl operators which are differential-difference operators associated to reflection groups [5]. The purpose of this chapter is to show that another simple system, the quantum linear rigid rotor, is similarly supersymmetric.

When vibrational modes are neglected, a diatomic molecule can be described by the quantum rotor Hamiltonian which is nothing else than the so(3) Casimir operator

$$H = J_1^2 + J_2^2 + J_3^2 + 1/4, \tag{1.1}$$

where the J_1, J_2, J_3 are the angular momentum generators

$$J_1 = \frac{1}{i}(x_2\partial_{x_3} - x_3\partial_{x_2}), \quad J_2 = \frac{1}{i}(x_3\partial_{x_1} - x_1\partial_{x_3}), \quad J_3 = \frac{1}{i}(x_1\partial_{x_2} - x_2\partial_{x_1}).$$

We are choosing units in which $\hbar^2/2I = 1$, with I the moment of inertia, and are shifting the energies by $1/4$ for convenience. It is well known from angular momentum theory that the generators J_i's are symmetries of the quantum rotor, i.e.,

$$[H, J_1] = 0, \quad [H, J_2] = 0, \quad [H, J_3] = 0,$$

and that they satisfy the commutation relations

$$[J_1, J_2] = iJ_3, \quad [J_2, J_3] = iJ_1, \quad [J_3, J_1] = iJ_2. \tag{1.2}$$

The spectrum of H is positive with the energies labeled by a nonnegative integer j and given by

$$E_j = (j + 1/2)^2, \quad j = 0, 1, 2, \ldots.$$

The fact that E_j is a square is an indication that the Hamiltonian might be supersymmetric. Owing to the so(3) symmetry of H, the level E_j has a $(2j+1)$-fold degeneracy. A natural basis for the wavefunctions is obtained by simultaneously diagonalizing J_3 and H. This leads to the spherical harmonics $Y_j^m(\hat{r})$, where $\hat{r} \in S^2$, which satisfy [6]

$$\begin{aligned} H\, Y_j^m(\hat{r}) &= E_j Y_j^m(\hat{r}), \quad j = 0, 1, 2, \ldots, \\ J_3\, Y_j^m(\hat{r}) &= m Y_j^m(\hat{r}), \quad m = -j, -j+1, \ldots, 0, \ldots, j-1, j, \end{aligned} \tag{1.3}$$

and which have the expression

$$Y_j^m(\hat{r}) = \sqrt{\frac{2j+1}{4\pi}\frac{(j-m)!}{(j+m)!}}\, P_j^m\left(\frac{x_3}{\sqrt{x_1^2 + x_2^2 + x_3^2}}\right)\left(\frac{x_1 + ix_2}{\sqrt{x_1^2 + x_2^2}}\right)^m, \tag{1.4}$$

where $P_j^m(z)$ is the associated Legendre polynomials. They satisfy the orthogonality relation

$$\int_{S^2} Y_j^m(\hat{r})\, (Y_{j'}^{m'}(\hat{r}))^*\, \mathrm{d}\Omega = \delta_{jj'}\delta_{mm'},$$

where a^* stands for the complex conjugate of a. For a fixed value of j, the spherical harmonics $Y_j^m(\hat{r})$ span an irreducible representation space of so(3) with the raising and lowering operators $J_\pm = J_1 \pm iJ_2$ acting as follows:

$$J_\pm\, Y_j^m(\hat{r}) = [(j \mp m)(j \pm m + 1)]^{1/2} Y_j^{m\pm 1}(\hat{r}). \tag{1.5}$$

Let R_1, R_2, R_3 be the operators that respectively reflect the coordinates x_1, x_2, x_3. With $\{A, B\} = AB + BA$, it is easily verified that

$$\begin{aligned} \{L_i, R_j\} &= 0, \quad i \neq j, \\ [L_i, R_i] &= 0. \end{aligned} \tag{1.6}$$

It follows that H is also invariant under all reflections R_i, $i = 1, 2, 3$. The constants of the motion of H are thus elements of the universal enveloping algebra of so(3) combined multiplicatively with reflections. This set can be viewed as spanning the full symmetry algebra of the quantum rotor.

Within this ensemble, there are elements Q whose square is equal to H; this explains why we claim that H is supersymmetric.

Given one such Q, we shall consider the subset of conserved quantities that have the property of commuting with Q and shall identify its structure. It will be shown to be a unital algebra freely generated by operators obeying relations expressed as the anticommutator analogs of the so(3) relations (1.2). This algebra will be denoted $\mathrm{so}(3)_{-1}$ since, as shall be explained, it is the $q = -1$ version of $\mathrm{so}(3)_q$, a q-deformation of so(3) that has been introduced and studied in [7, 14]. The algebra $\mathrm{so}(3)_{-1}$ has already been considered per se and its representations have been examined [1, 3, 10, 13, 16]. It is sometimes referred to as the anticommutator spin algebra. We shall here construct its representations on the space of wavefunctions of the rotor, i.e., on the spherical harmonics. These representations are subduced from the $(2j+1)$-dimensional irreducible representations of the full symmetry algebra $U(\mathrm{so}(3)) \otimes \mathbb{Z}_2^3$ of H. For a given value of j, these $\mathrm{so}(3)_{-1}$ representations will be found to have two irreducible components of dimension $j+1$ and j.

The idea of an operator whose square is equal to a Laplacian is familiar in spinorial contexts where γ-matrices or Clifford algebra generators are used. The fact that this can be achieved in a scalar situation using reflections/involutions is much less known. This was observed in the framework of two of the author's analysis of the Laplace–Dunkl equation on S^2 [11]. It is also connected to the tensoring of superalgebra realizations when the grade involution is employed in a manifest fashion [12]. In light of these observations much of Clifford analysis can be translated to the scalar realm. Our goal here is to indicate how this pans out in a simple and physically relevant model.

The remainder of the paper proceeds as follows. In Section 2, we introduce the supercharge Q and exhibit the generators of its commutant. We identify the structure relations they satisfy and indicate why the notation $\mathrm{so}(3)_{-1}$ is appropriate for the algebra they generate. The nonuniqueness of the supercharge will also be commented upon. In Section 3, we construct the representations of $\mathrm{so}(3)_{-1}$ on the space of spherical harmonics $Y_j^m(\hat{r})$ which will be seen to split in two irreducible $\mathrm{so}(3)_{-1}$ modules. In Section 4, we shall indicate how the so-called anti-Krawtchouk polynomials define the transformation matrices between two natural bases for the irreducible representations of $\mathrm{so}(3)_{-1}$.

2. The Supercharge Q and its Symmetry Algebra

In this section, we exhibit the supercharge Q which squares to the Hamiltonian (1.1) of the quantum rigid rotor. We obtain its symmetries, and show that they generate $\mathrm{so}(3)_{-1}$, also known as the anticommutator spin algebra.

Consider the operator Q defined as

$$Q = -iJ_1R_3 + iJ_2R_2R_3 - iJ_3R_2 - 1/2. \tag{2.1}$$

Upon using the relations (1.2) and (1.6), a straightforward calculation shows that one has

$$Q^2 = H, \tag{2.2}$$

where H is the Hamiltonian of the rigid quantum rotor given by (1.1). Since the reflection operators R_i are self-adjoint, that is $R_i^\dagger = R_i$, it can be verified using (1.6) that the supercharge (2.1) is also self-adjoint. Moreover, it follows from (2.2) that on the space of spherical harmonics, the eigenvalues q_j of (2.1) are of the form

$$q_j = \pm(j + 1/2), \quad j = 0, 1, 2, \ldots.$$

Observe that while the angular momentum generators J_1, J_2, J_3 are symmetries of the quantum rotor $H = Q^2$, they are not symmetries of the supercharge Q; indeed, a direct calculation shows that $[J_i, Q] \neq 0$ for $i = 1, 2, 3$. It is also verified that the reflections R_1, R_2, R_3 do not commute with Q either.

2.1. Symmetry Algebra

Let K_1, K_2, K_3 be the self-adjoint operators defined as

$$K_1 = iJ_1R_2 + R_2R_3/2, \quad K_2 = -iJ_2R_1R_2 + R_1R_3/2,$$
$$K_3 = -iJ_3R_1 + R_1R_2/2. \tag{2.3}$$

A straightforward calculation shows that these operators are symmetries of Q, i.e.,

$$[K_1, Q] = 0, \quad [K_2, Q] = 0, \quad [K_3, Q] = 0.$$

In view of (2.2), it follows that K_1, K_2, K_3 are also symmetries of the quantum rotor. These operators are seen to obey the following relations:

$$\{K_1, K_2\} = K_3, \quad \{K_2, K_3\} = K_1, \quad \{K_3, K_1\} = K_2. \tag{2.4}$$

The relations (2.4) define $\mathrm{so}(3)_{-1}$ or the anticommutator spin algebra. It is clear that these relations can be viewed as an anticommutator version of the relations (1.2). The algebra (2.4) has a Casimir operator C defined as

$$C = K_1^2 + K_2^2 + K_3^2, \tag{2.5}$$

which commutes with K_1, K_2, K_3. This fact is easily confirmed using the elementary commutator identity $[A, BC] = \{A, B\}C - B\{A, C\}$. In the realization (2.3), the Casimir operator of the symmetry algebra is related to the supercharge Q through

$$C = Q^2 - Q. \tag{2.6}$$

The notation $\mathrm{so}(3)_{-1}$ for the algebra (2.4) originates from the observation that the relations (2.4) correspond to taking $q \to -1$ in the defining relations of the quantum algebra $\mathrm{so}(3)_q$, which read [7]

$$q^{1/2}I_1I_2 - q^{-1/2}I_2I_1 = I_3, \quad q^{1/2}I_2I_3 - q^{-1/2}I_3I_2 = I_1,$$
$$q^{1/2}I_3I_1 - q^{-1/2}I_1I_3 = I_2.$$

Let us also note that the supercharge Q defined in (2.1) is not unique: there are other operators with similar expressions that square to the rigid rotor. For example, one could take

$$\widetilde{Q} = -iJ_1R_1R_2 + iJ_2R_1 - iJ_3R_1R_3 - R_1R_2R_3/2,$$

which also satisfies $\widetilde{Q}^2 = H$. Other operators, associated to different coproducts of the Lie superalgebra $\mathrm{osp}(1, 2)$ can also be constructed; see [10] for more details.

3. Irreducible Representations of so(3)$_{-1}$

In this section, we construct a basis for the irreducible representations of $\mathrm{so}(3)_{-1}$ in terms of spherical harmonics, and compute the action of the generators on this basis. We proceed in two steps. First, we diagonalize the operator K_3 and obtain its eigenfunctions as linear combination of two spherical harmonics. Second, we diagonalize the supercharge Q and obtain its eigenfunctions. The final expression for the common eigenfunctions of K_3 and Q are expressed as a linear combination of four spherical harmonics.

Let us first observe that the reflection operators have a simple action on the spherical harmonics $Y_j^m(\hat{r})$. Indeed, using the explicit expression (1.4), it is directly verified that

$$R_1Y_j^m(\hat{r}) = Y_j^{-m}(\hat{r}), \quad R_2Y_j^m(\hat{r}) = (-1)^m Y_j^{-m}(\hat{r}),$$
$$R_3Y_j^m(\hat{r}) = (-1)^{j+m}Y_j^m(\hat{r}). \tag{3.1}$$

3.1. Diagonalization of K_3

We now construct the eigenfunctions of the symmetry K_3 in the spherical harmonics basis. Using the above relations, as well as the formulas (1.3) and (2.3), one finds that K_3 has the action

$$K_3 Y_j^m(\hat{r}) = -imY_j^{-m}(\hat{r}) + \frac{(-1)^m}{2} Y_j^m(\hat{r}),$$
$$m = -j, -j+1, \ldots, 0, \ldots, j-1, j.$$

For a given j, the matrix representing K_3 in the spherical harmonics basis is thus block diagonal with j 2×2 blocks and one 1×1 block, corresponding to $Y_j^0(\hat{r})$. This matrix is easily diagonalized. Let $M_j^{m,\epsilon}(\hat{r})$ be the functions defined as

$$M_j^{m,\epsilon}(\hat{r}) = \frac{1}{\sqrt{2}}[Y_j^{-m}(\hat{r}) + i\epsilon Y_j^m(\hat{r})], \quad \epsilon = \pm 1, \quad m = 0, 1, \ldots, j. \tag{3.2}$$

For $m = 0$, one should only consider the value $\epsilon = +1$. With this convention, the space spanned by the basis functions $M_j^{m,\pm}$ is $(j+1)$-dimensional for $\epsilon = +1$ and j-dimensional for $\epsilon = -1$. The functions $M_j^{m,\epsilon}(\hat{r})$ satisfy the eigenvalue equation

$$K_3\, M_j^{m,\epsilon}(\hat{r}) = \left(\epsilon m + \frac{(-1)^m}{2}\right) M_j^{m,\epsilon}(\hat{r}).$$

Since the functions $M_j^{m,\epsilon}(\hat{r})$ are normalized linear combinations of the spherical harmonics, one has

$$\int_{S^2} M_j^{m,\epsilon}(\hat{r})[M_{j'}^{m',\epsilon'}(\hat{r})]^*\, \mathrm{d}\Omega = \delta_{jj'}\delta_{mm'}\delta_{\epsilon\epsilon'}.$$

The inverse relations are

$$Y_j^m(\hat{r}) = -\frac{i}{\sqrt{2}}[M_j^{m,+}(\hat{r}) - M_j^{m,-}(\hat{r})],$$
$$Y_j^{-m}(\hat{r}) = \frac{1}{\sqrt{2}}[M_j^{m,+}(\hat{r}) + M_j^{m,-}(\hat{r})], \quad m = 0, 1, \ldots, j.$$

It can be verified that the functions $M_j^{m,\epsilon}(\hat{r})$ are not eigenfunctions of the supercharge Q. In the next subsection, we construct the eigenfunctions of Q in the $M_j^{m,\epsilon}(\hat{r})$ basis.

3.2. Simultaneous diagonalization of Q and K_3

In order to construct a basis in which the symmetry K_3 and the supercharge Q are both diagonal, we examine the action of Q on the eigenfunctions of K_3. Upon using the explicit expression (2.1) for the supercharge Q, as well as the actions (1.5) of the raising/lowering operators and the explicit decomposition (3.2) of the eigenfunctions $M_j^{m,\epsilon}(\hat{r})$, a direct calculation shows that

$$QM_j^{m,\epsilon}(\hat{r}) = \frac{(-1)^j i[(-1)^{m+1} - \epsilon]}{2} \sqrt{(j-m)(j+m+1)}\, M_j^{m+1,\epsilon}(\hat{r}) - \frac{1+2(-1)^m \epsilon m}{2} M_j^{m,\epsilon}(\hat{r}) + \frac{(-1)^j i[\epsilon - (-1)^m]}{2} \times \sqrt{(j+m)(j-m+1)}\, M_j^{m-1,\epsilon}(\hat{r}). \quad (3.3)$$

As can be seen from the above expression, when m is even and $\epsilon = +1$, or when m is odd and $\epsilon = -1$, the coefficient in front of $M_j^{m-1,\epsilon}(\hat{r})$ vanishes. Similarly, when m is odd and $\epsilon = +1$, or when m is even and $\epsilon = -1$, the coefficient in front of $M_j^{m+1,\epsilon}(\hat{r})$ is zero. As a consequence, the matrix representing Q can easily be diagonalized in the $M_j^{m,\epsilon}(\hat{r})$ basis. Let $\mathscr{F}_j^k(\hat{r})$ be defined as

$$\mathscr{F}_j^k(\hat{r}) = \sqrt{\frac{j-k}{2j+1}} M_j^{k+1,(-1)^k}(\hat{r}) + i(-1)^{j+k+1} \times \sqrt{\frac{j+k+1}{2j+1}} M_j^{k+1,(-1)^k}(\hat{r}), \quad k = 0, 1, \ldots, j. \quad (3.4)$$

The functions $\mathscr{F}_j^k(\hat{r})$ satisfy

$$Q\,\mathscr{F}_j^k(\hat{r}) = -(j+1/2)\,\mathscr{F}_j^k(\hat{r}), \quad K_3\,\mathscr{F}_j^k(\hat{r}) = (-1)^k(k+1/2)\,\mathscr{F}_j^k(\hat{r}), \quad (3.5)$$

and are thus simultaneous eigenfunctions of Q and K_3. It is obvious that $\mathscr{F}_j^k(\hat{r})$ satisfies

$$\int_{S^2} \mathscr{F}_j^k(\hat{r})[\mathscr{F}_{j'}^{k'}(\hat{r})]^*\, d\Omega = \delta_{jj'}\delta_{kk'}.$$

Similarly, let $\mathscr{G}_j^k(\hat{r})$ be another family of functions defined as

$$\mathscr{G}_j^k(\hat{r}) = \sqrt{\frac{j+k+1}{2j+1}} M_j^{k+1,(-1)^k}(\hat{r}) + i(-1)^{k+j} \times \sqrt{\frac{j-k}{2j+1}} M_j^{k,(-1)^k}(\hat{r}), \quad k = 0, \ldots, j-1. \quad (3.6)$$

These functions $\mathscr{G}_j^k(\hat{r})$ satisfy

$$Q\,\mathscr{G}_j^k(\hat{r}) = (j+1/2)\,\mathscr{G}_j^k(\hat{r}), \quad K_3\,\mathscr{G}_j^k(\hat{r}) = (-1)^k(k+1/2)\,\mathscr{G}_j^k(\hat{r}), \tag{3.7}$$

and are again simultaneous eigenfunctions of Q and K_3. They also obey the orthogonality relation

$$\int_{S^2} \mathscr{G}_j^k(\hat{r})[\mathscr{G}_{j'}^{k'}(\hat{r})]^*\,\mathrm{d}\Omega = \delta_{jj'}\delta_{kk'}.$$

We have thus constructed two orthonormal bases of joint eigenfunctions of the supercharge Q and the symmetry operator K_3. For a given value of j, the set of eigenfunctions $\mathscr{F}_j^k(\hat{r})$ is $(j+1)$-dimensional and corresponds to the eigenvalue $-(j+1/2)$ of Q, while the set of eigenfunctions $\mathscr{G}_j^k(\hat{r})$ is j-dimensional and corresponds to the eigenvalue $(j+1/2)$ of Q. Since Q is self-adjoint, the two bases are orthogonal to one another. It is also manifest that taken together they span the $(2j+1)$-dimensional space of spherical harmonics.

3.3. Action of the symmetry algebra

Let us now show that the span of $\mathscr{F}_j^k(\hat{r})$ for $k = 0,\ldots,j$ supports an irreducible representation of the $\mathrm{so}(3)_{-1}$ algebra. To that end, we examine the action of the generator K_1 on the basis elements $\mathscr{F}_j^k(\hat{r})$. By a direct calculation, one finds that K_1 acts in a tridiagonal fashion according to

$$K_1\,\mathscr{F}_j^k(\hat{r}) = U_{k+1}\,\mathscr{F}_j^{k+1}(\hat{r}) + B_k\mathscr{F}_j^k(\hat{r}) + U_k\mathscr{F}_j^{k-1}(\hat{r}), \quad k = 0,\ldots,j, \tag{3.8}$$

where

$$U_k = \begin{cases} 0, & k=0, \\ \sqrt{\dfrac{(j+k+1)(j+1-k)}{4}}, & k\neq 0, \end{cases}$$

$$B_k = \begin{cases} (-1)^j\left(\dfrac{j+1}{2}\right), & k=0, \\ 0, & k\neq 0. \end{cases} \tag{3.9}$$

The action of K_2 is directly obtained from the relation $K_2 = \{K_1, K_3\}$. Since $U_k \neq 0$ for $k = 1,\ldots,j$, it follows that the $\mathrm{so}(3)_{-1}$ algebra acts irreducibly on the space spanned by the functions $\mathscr{F}_j^{k+1}(\hat{r})$. In other words, the functions $\mathscr{F}_j^k(\hat{r})$, for $k = 0,\ldots,j$, support a $(j+1)$-dimensional unitary irreducible representation of $\mathrm{so}(3)_{-1}$. In a similar fashion, the span of $\mathscr{G}_j^k(\hat{r})$ for $k = 0, \ldots, j-1$ also supports an irreducible representation of $\mathrm{so}(3)_{-1}$. The action of K_1 on $\mathscr{G}_j^k(\hat{r})$ is given by

$$K_1 \mathscr{G}_j^k(\hat{r}) = V_{k+1} \mathscr{G}_j^{k+1}(\hat{r}) + C_k \mathscr{G}_j^k(\hat{r}) + V_k \mathscr{G}_j^{k-1}(\hat{r}), \quad k = 0, \ldots, j-1, \tag{3.10}$$

where

$$V_k = \begin{cases} 0, & k = 0, \\ \sqrt{\dfrac{(j+k)(j-k)}{4}}, & k \neq 0, \end{cases} \qquad C_k = \begin{cases} (-1)^{j-1}\left(\dfrac{j}{2}\right), & k = 0, \\ 0, & k \neq 0. \end{cases} \tag{3.11}$$

By the same reasoning, the functions $\mathscr{G}_j^k(\hat{r})$, for $k = 0, \ldots, j-1$, span a unitary irreducible representation of $\mathrm{so}(3)_{-1}$ of dimension j. Upon comparing the formulas (3.9) and (3.11), it is seen that the two representations differ only by their respective dimensions.

3.4. Spherical harmonics and representations of $\mathrm{so}(3)_{-1}$

In light of the above results, we can give the explicit decomposition of the irreducible representations of $\mathrm{so}(3)$ in irreducible representations of $\mathrm{so}(3)_{-1}$. Upon denoting by V_j the $(2j+1)$-dimensional irreducible representations $\mathrm{so}(3)$ defined by the matrix elements (1.3), (1.5), and by S_j the $(j+1)$-dimensional representations of $\mathrm{so}(3)_{-1}$ defined by the matrix elements (3.5), (3.9), one has

$$V_j = S_j \oplus S_{j-1}. \tag{3.12}$$

In other words, each irreducible representation V_j of $\mathrm{so}(3)$ decomposes as a direct sum of two irreducible representations S_j of $\mathrm{so}(3)_{-1}$. A similar result was derived by Brown in [3] using different methods.

4. Anti-Krawtchouk Polynomials

In this section, we show that the anti-Krawtchouk polynomials arise as overlap coefficients between the basis $\mathscr{F}_j^k(\hat{r})$ and an alternative basis $\mathscr{Z}_j^k(\hat{r})$. For the remainder of this section, we shall take $j = N$, where N is a positive integer.

Let us first recall that the basis $\mathscr{F}_N^k(\hat{r})$ with $k = 0, \ldots, N$ supports an $(N+1)$-dimensional irreducible representation of the $\mathrm{so}(3)_{-1}$ algebra and satisfies

$$\begin{aligned} &\Gamma\, \mathscr{F}_N^k(\hat{r}) = -(N+1/2)\, \mathscr{F}_N^k(\hat{r}), \\ &K_3\, \mathscr{F}_N^k(\hat{r}) = (-1)^k (k+1/2)\, \mathscr{F}_N^k(\hat{r}), \quad k = 0, 1, \ldots, N, \\ &K_1\, \mathscr{F}_N^k(\hat{r}) = U_{k+1}\, \mathscr{F}_N^{k+1}(\hat{r}) + B_k \mathscr{F}_N^k(\hat{r}) + U_k \mathscr{F}_N^{k-1}(\hat{r}), \end{aligned} \tag{4.1}$$

with

$$U_k = \begin{cases} 0, & k = 0, \\ \sqrt{\dfrac{(N+k+1)(N+1-k)}{4}}, & k \neq 0, \end{cases}$$

$$B_k = \begin{cases} (-1)^N \left(\dfrac{N+1}{2}\right), & k = 0, \\ 0, & k \neq 0. \end{cases} \tag{4.2}$$

Let $\mathscr{Z}_N^k(\hat{r})$ be the basis functions defined by

$$\mathscr{Z}_N^k(x_1, x_2, x_3) = \mathscr{F}_N^k(x_2, x_3, (-1)^{N+1} x_1), \quad (x_1, x_2, x_3) = \hat{r} \in S^2. \tag{4.3}$$

It is verified that the basis $\mathscr{Z}_N^k(\hat{r})$ also supports an irreducible representation of $\mathrm{so}(3)_{-1}$ in which the action of the generators is

$$\begin{aligned} &\Gamma\, \mathscr{Z}_N^k(\hat{r}) = -(N+1/2)\, \mathscr{Z}_N^k(\hat{r}), \\ &K_1\, \mathscr{Z}_N^k(\hat{r}) = (-1)^k (k+1/2)\, \mathscr{Z}_N^k(\hat{r}), \quad k = 0, 1, \ldots, N, \\ &K_2\, \mathscr{Z}_N^k(\hat{r}) = U_{k+1}\, \mathscr{Z}_N^{k+1}(\hat{r}) + B_k \mathscr{Z}_N^k(\hat{r}) + U_k \mathscr{Z}_N^{k-1}(\hat{r}). \end{aligned} \tag{4.4}$$

Consider the expansion coefficients $W_n(k)$ between the bases $\mathscr{F}_N^k(\hat{r})$ and $\mathscr{Z}_N^k(\hat{r})$. These coefficients appear in the formula

$$\mathscr{Z}_N^k(\hat{r}) = \sum_{n=0}^{N} W_n(k)\, \mathscr{F}_N^n(\hat{r}), \tag{4.5}$$

which holds for any value of the position vector $\hat{r} \in S^2$. The overlap coefficients can be expressed in terms of the integral

$$W_n(k) = \int_{S^2} \mathrm{d}\Omega\, \mathscr{F}_N^n(\hat{r}) [\mathscr{Z}_N^k(\hat{r})]^*.$$

Upon applying K_1 on both sides of the expansion formula (4.5) and using the actions (4.1) and (4.4), it is seen that the expansion coefficients $W_n(k)$

obey the three-term recurrence relation

$$(-1)^k(k+1/2)\,W_n(k) = U_{n+1}W_{n+1}(k) + B_nW_n(k) + U_nW_{n-1}(k), \quad n = 0,\ldots,N.$$

Upon taking $W_n(k) = \omega_k\widehat{P}_n(y_k)$ with $\widehat{P}_0(y_k) = 1$ and

$$w_k = \int_{S^2} \mathrm{d}\Omega\, \mathscr{F}_N^0(\hat{r})[\mathscr{Z}_N^k(\hat{r})]^*, \tag{4.6}$$

one finds that the monic polynomials $\widehat{P}_n(y_k)$ satisfy the recurrence relation

$$y_k\widehat{P}_n(y_k) = U_{n+1}\widehat{P}_{n+1}(y_k) + B_n\widehat{P}_n(y_k) + U_n\widehat{P}_{n-1}(y_k),$$

with $y_k = (-1)^k(k+1/2)$. If one takes $\widehat{P}_n(y_k) = (\frac{2^n}{U_1\cdots U_n})P_n(y_k)$, one finds the relation

$$x_kP_n(x_k) = P_{n+1}(x_k) - (A_n + C_n)P_n(x_k) + A_{n-1}C_nP_{n-1}(x_k), \tag{4.7}$$

where

$$A_n = \frac{(-1)^{n+N+1}(N+1)+n+1}{4}, \quad C_n = \begin{cases} 0, & n = 0, \\ \dfrac{(-1)^{N+n}(N+1)-n}{4}, & n \neq 0, \end{cases} \tag{4.8}$$

and where

$$x_k = (-1)^k(k/2+1/4) - 1/4, \quad k = 0,1,\ldots,N. \tag{4.9}$$

The recurrence relation (4.7) corresponds to that of the Bannai–Ito polynomials $B_n(x_k;\rho_1,\rho_2,r_1,r_2)$, in the notation of [8], with

$$\rho_1 = 0, \quad \rho_2 = \frac{(-1)^N(N+1)}{2}, \quad r_1 = 0, \quad r_2 = \frac{(-1)^N(N+1)}{2}, \tag{4.10}$$

with x_k given by (4.9). As the monic polynomials $P_n(x_k)$ are associated with the anti-commutator spin algebra, we shall refer to them as the anti-Krawtchouk polynomials. The name Krawtchouk is justified by the fact that when calculating the overlap coefficients between two spherical harmonics differing by a cyclic permutation of the coordinates, as in (4.3) (without the change of sign), one naturally finds the Krawtchouk polynomials with parameter $p = 1/2$. The explicit hypergeometric expressions for the polynomials $P_n(x_k)$ involves a ${}_4F_3$ with argument 1 whose explicit expression can be found in [8] using the choice of parameters (4.10). Finally, the polynomials $P_n(x_k)$ satisfy a discrete orthogonality relation of the form

$$\sum_{k=0}^{N} \omega_k P_n(x_k) P_m(x_k) = u_1 \cdots u_n \delta_{nm},$$

where the weight w_k is given by (up to a normalization factor)

$$w_k = (-1)^k \times \begin{cases} \dfrac{(1+\alpha_N)_k}{(1-\alpha_N)_k} & k \text{ even}, \\ \dfrac{(1+\alpha_N)_{k-1}}{(1-\alpha_N)_{k-1}} & k \text{ odd}, \end{cases}$$

where $\alpha_N = (-1)^N(N+1)$; thus result can be found in [10].

5. Conclusion

In this chapter, we have exhibited the supersymmetric nature of the quantum rotor by explicitly constructing a supercharge operator Q whose square is the Hamiltonian of the rotor. We have obtained the symmetries of Q, and shown that these symmetries generate the $\mathrm{so}(3)_{-1}$ algebra, also known as the anticommutator spin algebra. We have constructed a basis for the eigenfunctions of the supercharge in terms of spherical harmonics, and shown that each $(2j+1)$-dimensional irreducible representation of so(3) decomposes in a direct sum of two irreducible representations of $\mathrm{so}(3)_{-1}$ of dimension j and $(j+1)$. Finally, we have explained how a special family of orthogonal polynomials, the anti-Krawtchouk polynomials, arise as overlap coefficients between two eigenbases of the supercharge corresponding to the diagonalization of two different symmetry operators.

These results call for a generalization to the multivariate case. Indeed, it would be interesting to determine a supercharge for the generalized rotor on the n-sphere, whose Hamiltonian is the $\mathrm{so}(n)$ Casimir operator

$$H = \sum_{i<j\leq n} J_{ij}^2,$$

with $J_{ij} = \frac{1}{i}(x_i \partial_{x_j} - x_j \partial_{x_i})$. In this case, one would expect the symmetry algebra of the supercharge to correspond to a special case of the higher rank Bannai–Ito algebra, as determined in [4].

Acknowledgments

VXG holds a postdoctoral fellowship from the Natural Sciences and Engineering Research Council of Canada (NSERC). The research of LV is supported in part by NSERC. GFY is funded by the National Natural Science Foundation of China (Grant #11371251).

References

[1] M. Arik and U. Kayserilioglu, The anticommutator spin algebra, its representations and quantum group invariance, *Int. J. Mod. Phys. A* **18** (2003) 5039–5046.

[2] L. Brink, J. H. Hanssson, S. Konstein and M. A. Vasiliev, The calogero model — anyonic representation, fermionic extension and supersymmety, *Nucl. Phys. B* **401** (1993) 591–612.

[3] G. M. F. Brown, Totally bipartite/abipartite Leonard pairs and Leonard triples of Bannai/Ito type, *Electron J. Linear Algebra* **26** (2013) 258–299.

[4] H. De Bie, V. X. Genest and L. Vinet, The $\mathbb{Z}_2^n$ Dirac–Dunkl operator and a higher rank Bannai–Ito algebra, preprint (2015); arXiv:1511.02177.

[5] C. F. Dunkl, Differential-difference operator associated to reflection groups, *Trans. Amer. Math. Soc.* **311** (1989) 167–183.

[6] A. R. Edmonds, *Angular Momentum in Quantum Mechanics* (Princeton University Press, 1996).

[7] D. B. Fairlie, Quantum deformations of su(2), *J. Phys. A: Math. Gen.* **23** (1990) L183–L187.

[8] V. X. Genest, L. Vinet and A. Zhedanov, Bispectrality of the Complementary Bannai–Ito polynomials, *SIGMA* **9** (2013) 18–37.

[9] V. X. Genest, L. Vinet and A. Zhedanov, The Bannai–Ito algebra and a superintegrable system with reflections on the 2-sphere, *J. Phys. A: Math. Theor.* **47** (2014) 205202.

[10] V. X. Genest, L. Vinet and A. Zhedanov, The Bannai–Ito polynomials as Racah coefficients of the $s\ell_{-1}(2)$ algebra, *Proc. Amer. Math. Soc.* **142** (2014) 1545–1560.

[11] V. X. Genest, L. Vinet and A. Zhedanov, A Laplace–Dunkl equation on S^2 and the Bannai–Ito algebra, *Commun. Math. Phys.* **336** (2015) 243–259.

[12] V. X. Genest, L. Vinet and A. Zhedanov, The quantum superalgebra $osp_q(1|2)$ and a q-generalization of the Bannai–Ito polynomials, *Commun. Math. Phys.* **344** (2016) 465–481.

[13] M. F. Gorodnii and G. B. Podkolzin, Irreducible representations of a graded lie algebra, in *Spectral Theory of Operators and Infinite-Dimensional Analysis*, ed. Yu. M. Berezanski, Vol. 3 (SSR, Inst. Mat., Kiev., 1984) Akad. Nauk Ukrain, pp. 66-77.

[14] M. Havlicek, A. U. Klimyk and S. Posta, Representations of the cyclically symmetric q-deformed algebra $\mathrm{so}_q(3)$, *J. Math. Phys.* **40** (1999) 2135.

[15] L. Lapointe and L. Vinet, Exact operator solution of the Calogero–Sutherland model, *Commun. Math. Phys.* **178** (1996) 425–452.
[16] V. L. Ostrovskyi and S. D. Silvestrov, Representations of the real forms of a graded analogue of the lie algebra $\mathrm{sl}(2,\mathbb{C})$, *Ukrainian Math. J.* **44** (1992) 1395–1401.
[17] M. S. Plyuschay, Deformed Heisenberg algebra, fractional spin fields and supersymmetry without fermions, *Ann. Phys.* **245** (1996) 339–360.
[18] M. P. Polychronakos, Exchange operator formalism for integrable systems of particles, *Phys. Rev. Lett.* **69** (1992) 703–705.
[19] S. Post, L. Vinet and A. Zhedanov, Supersymmetric quantum mechanics with reflections. *J. Phys. A: Math. Theor.* **44** (2011) 435301.

This page intentionally left blank

Chapter 16

The Method of Brackets in Experimental Mathematics

Ivan Gonzalez[*,¶], **Karen Kohl**[†,||], **Lin Jiu**[‡,**]
and Victor H. Moll[§,††]

[*]Instituto de Física y Astronomía, Universidad de Valparaiso
Valparaiso, Chile
[†]Department of Mathematics, University of Southern Mississippi
Long Beach, MS 39560, USA
[‡]Research Institute for Symbolic Computation
Johannes Kepler University, Linz, Austria
[§]Department of Mathematics, Tulane University
New Orleans, LA 70118, USA
[¶]ivan.gonzalez@uv.cl
[||]karen.kohl@usm.edu
[**]ljiu@risc.uni-linz.ac.at
[††]vhm@math.tulane.edu

Dedicated to Professor Mourad Ismail on the occasion of his birthday

The method of brackets is a new method originally developed in the context of definite integrals coming from the evaluation of Feynman diagrams. It consists on a small number of heuristic rules. The work presented here describes these rules and illustrates the method. The example presented here include elementary ones as well as two simple Feynman diagrams.

Keywords: Definite integrals; method of brackets; heuristic rules; Feynman diagrams.

Mathematics Subject Classification 2010: 33C05, 33F10

1. Introduction

The problem of evaluating definite integrals appears in elementary courses. Given a function f and an interval $[a,b] \subset \mathbb{R}$, the task involves expressing

$$\mathcal{I}(f) = \int_a^b f(x)\,dx \tag{1.1}$$

in terms of the (internal) parameters of f. It is surprising that the methods required in solving this problem depend very strongly on subtle forms of the function f. For instance, `Mathematica` gives

$$\int_0^\infty \frac{dx}{e^x+1} = \log 2 \tag{1.2}$$

but it is unable to evaluate

$$\int_0^\infty \frac{dx}{e^x+1+x}. \tag{1.3}$$

The fact that many integrals *cannot be evaluated* is an all-too-familiar experience to both professional mathematicians as well as beginners. Probably the earliest example of such a phenomena comes in the rectification of the ellipse: given $a > b$ and the equation $\frac{x^2}{a^2} + \frac{y^2}{b^2} = 1$, the arc length is given by

$$L(a,b) = 4a\int_0^1 \sqrt{\frac{1-k^2t^2}{1-t^2}}\,dt. \tag{1.4}$$

Here $k = \sqrt{1-b^2/a^2}$ is the eccentricity of the ellipse.

Naturally, the question of whether an integral can be evaluated in *closed form* depends on the type of functions that are allowed in the answer. For example, the integral appearing in (1.4) is the *complete elliptic integral of the second kind*, denoted by $E(k)$. For relevant information, the reader may consult [3, 11].

A source of interesting integrals comes from Feynman diagrams. These are pictorial representations of elementary particle interactions. The reader will find in [10, 13] more information about this topic.

Figure 1 depicts the interaction of three particles corresponding to the three external lines of momenta P_1, P_2 and P_3. In this case, the Schwinger parameterization produces the integral

$$G = \frac{(-1)^{-D/2}}{\Gamma(a_1)\Gamma(a_2)\Gamma(a_3)} \int_0^\infty \int_0^\infty \int_0^\infty \frac{x_1^{a_1-1} x_2^{a_2-1} x_3^{a_3-1}}{(x_1+x_2+x_3)^{D/2}}$$

$$\times \exp\left(\sum_{j=1}^{3} x_j m_j^2 \right) \exp\left(-\frac{C_{11}P_1^2 + 2C_{12}P_1 \cdot P_2 + C_{22}P_2^2}{x_1+x_2+x_3} \right) dx_1 dx_2 dx_3.$$

The algorithms in [7, 8] give the coefficients $C_{i,j}$ as

$$C_{11} = x_1(x_2+x_3), \quad C_{12} = x_1x_3, \quad C_{22} = x_3(x_1+x_2). \tag{1.5}$$

Conservation of momentum implies $P_3 = P_1 + P_2$, and after replacing the coefficients $C_{i,j}$ into the equation for G, we obtain

$$G = \frac{(-1)^{-D/2}}{\prod_{j=1}^{3} \Gamma(a_j)} \int_0^\infty \int_0^\infty \int_0^\infty x^{a_1-1} x^{a_2-1} x^{a_3-1}$$

$$\times \frac{\exp\left(x_1m_1^2 + x_2m_2^2 + x_3m_3^2\right) \exp\left(-\frac{x_1x_2P_1^2 + x_2x_3P_2^2 + x_3x_1P_3^2}{x_1+x_2+x_3}\right)}{(x_1+x_2+x_3)^{D/2}} dx_1 dx_2 dx_3.$$

To solve the Feynman diagram in Fig. 1, one needs to evaluate the integral G as a function of the variables P_1, $P_2 \in \mathbb{R}^4$, the masses m_i, the dimension D and the parameters a_i.

The special massless case $m_1 = m_2 = m_3 = 0$ has been evaluated in [6] by `the method of brackets` described here. A similar problem, the case of the bubble diagram, is discussed in Example 2.3 below.

The main goal of this chapter is to introduce this method to a general audience. This is an algorithm for the evaluation of integrals on the half-line $[0, \infty)$ and it consists of a small number of rules. Some of these have been proven, so the method is partly science, while others have been proposed

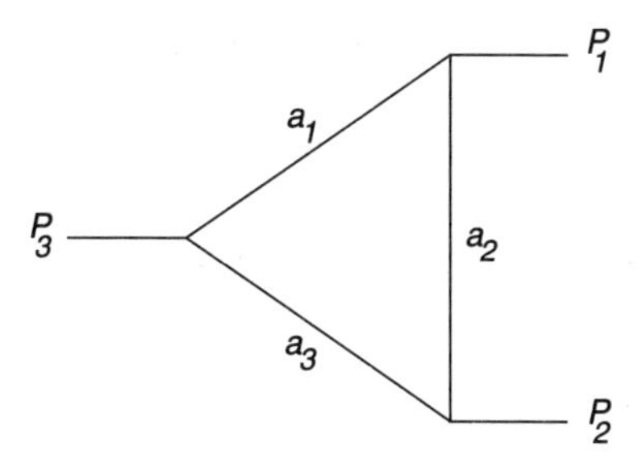

Fig. 1. The triangle.

based on the authors' experience. Thus the method falls in the realm of Experimental Mathematics as described in [2]. Still, some rules of this method are in the process of being created, so the method is partly an art.

The method of brackets has provided a powerful and flexible alternative to classical methods of evaluating a class of definite integrals. The reader will find in [5] a selection of entries of the table by Gradshteyn and Ryzhik [9] checked by this method.

2. The Algorithm

As advertised in the introductory section, we now reveal an algorithm for the evaluation of definite integrals. The starting point is the notion of *bracket*, defined by the divergent integral

$$\langle a \rangle = \int_0^\infty x^{a-1}\, dx, \quad \text{for } a \in \mathbb{C}, \tag{2.1}$$

and a few set of rules described below.

Rule 1. *Assign to the integral* $I(f) = \int_0^\infty f(x)\, dx$ *a bracket series*

$$\sum_n \phi_n a(n) \langle \alpha n + \beta \rangle. \tag{2.2}$$

Here $\phi_n = \frac{(-1)^n}{n!}$ is called the *indicator* and the coefficients $a(n)$ come from an assumed expansion $f(x) = \sum_{n=0}^\infty \phi_n a(n) x^{\alpha n + \beta - 1}$. The extra "$-1$" in the exponent is set for convenience. The coefficients are written as $a(n)$ because these will soon be evaluated at complex numbers n, not necessarily positive integers.

Now we need to state how to convert the bracket series into a number.

Rule 2. *The bracket series* $\sum_n \phi_n a(n) \langle \alpha n + \beta \rangle$ *is assigned the value*

$$\frac{1}{|\alpha|} a(n^*) \Gamma(-n^*). \tag{2.3}$$

Here $\Gamma(x)$ is the Euler's gamma function and n^* is obtained from the vanishing of the brackets; that is, $n^* = -\beta/\alpha$ solves $\alpha n + \beta = 0$. This rule is reminiscent of *Ramanujan's Master Theorem*; for further discussion we refer to [1].

Example 2.1. To compute the integral

$$I_1 = \int_0^\infty e^{-tx}\, dx \tag{2.4}$$

expand the integrand as

$$e^{-tx} = \sum_{n=0}^{\infty} \frac{(-1)^n}{n!} t^n x^n = \sum_{n=0}^{\infty} \phi_n t^n x^n \tag{2.5}$$

so that $\alpha = \beta = 1$ and $a(n) = t^n$. Then the bracket series is $\sum_n \phi_n t^n \langle n+1 \rangle$ and the evaluation of the integral requires to solve the equation $n+1=0$. Therefore $n^* = -1$ and the integral becomes

$$I_1 = \frac{1}{|1|} t^{-1} \Gamma(1) = \frac{1}{t}. \tag{2.6}$$

And that is all.

The first difficulty in this method comes from the prerequisite of having an explicit form of the coefficients in the expansion. The next example illustrates how to proceed.

Example 2.2. To prove the integral evaluation

$$I_2 = \int_0^{\infty} e^{-ax} \sin(bx)\, dx = \frac{b}{a^2+b^2}, \tag{2.7}$$

start with the classical expansions

$$e^{-ax} = \sum_{n_1=0}^{\infty} \phi_{n_1} a^{n_1} x^{n_1} \quad \text{and} \quad \sin bx = \sum_{n_2=0}^{\infty} \phi_{n_2} \frac{n_2!\, b^{2n_2+1}}{(2n_2+1)!} x^{2n_2+1}. \tag{2.8}$$

Replace these in (2.7) to obtain the *two-dimensional bracket series*

$$\begin{aligned} &\sum_{n_1,n_2} \phi_{n_1,n_2} a^{n_1} b^{2n_2+1} \frac{\Gamma(n_2+1)}{\Gamma(2n_2+2)} \langle n_1 + 2n_2 + 2 \rangle \\ &\quad \equiv \sum_{n_1,n_2} \phi_{n_1,n_2} C(n_1,n_2) \langle \alpha_{11} n_1 + \alpha_{12} n_2 + \beta_1 \rangle. \end{aligned} \tag{2.9}$$

Here $\phi_{n_1,n_2} = \phi_{n_1}\phi_{n_2}$. The relation $n! = \Gamma(n+1)$ has been used to convert factorials in terms of gamma function in anticipation of replacing n by a noninteger value.

Rule 3. *Each representation of an integral by a bracket series has associated an* ***index of the representation*** *according to*

$$\textit{index} = \textit{number of sums} - \textit{number of brackets}. \tag{2.10}$$

In the case of a multi-dimensional bracket series of positive index, the system generated by the vanishing of the coefficients has a number of free parameters. The solution is then determined upon computing all the contributions of maximal rank in the system by selecting these free parameters. Any two series expressed in the same variable and converging in a common region are added. Divergent series are discarded.

Thus, to evaluate (2.9) proceed as follows: make the brackets vanish and consider two cases treating n_1 and n_2 as free parameters. In the first case, when n_1 is free, the vanishing of the brackets gives $n_2^* = -\frac{\alpha_{11}}{\alpha_{12}} n_1 - \frac{\beta_1}{\alpha_{12}}$. Then the bracket series generates the classical series

$$\frac{1}{|\alpha_{12}|} \sum_{n_1=0}^{\infty} \phi_{n_1} C(n_1, n_2^*) \Gamma(-n_2^*). \tag{2.11}$$

The case of n_2 is treated in a similar manner.

In the evaluation of (2.7), the case when n_1 is free gives $n_2^* = -\frac{1}{2}n_1 - 1$. Thus one obtains the series

$$T_1 = \frac{1}{2b} \sum_{n_1=0}^{\infty} \phi_{n_1} \left(\frac{a}{b}\right)^{n_1} \frac{\Gamma(-\frac{n_1}{2})}{\Gamma(-n_1)} \Gamma(\tfrac{n_1}{2} + 1). \tag{2.12}$$

To simplify the expression for T_1 observe that the terms with odd n_1 vanish, therefore

$$T_1 = \frac{1}{2b} \sum_{m=0}^{\infty} \frac{1}{(2m)!} \left(\frac{a}{b}\right)^{2m} \frac{\Gamma(-m)\Gamma(m+1)}{\Gamma(-2m)}. \tag{2.13}$$

The duplication formula for the gamma function in the form

$$\frac{\Gamma(-x)}{\Gamma(-2x)} = \frac{\sqrt{\pi}\, 2^{2x+1}}{\Gamma\left(\frac{1}{2} - x\right)}, \tag{2.14}$$

transforms T_1 into

$$T_1 = \frac{\sqrt{\pi}}{b} \sum_{m=0}^{\infty} \frac{m!}{(2m)!} \frac{1}{\Gamma(\frac{1}{2} - m)} \left(\frac{2a}{b}\right)^{2m} = \frac{1}{b} \sum_{m=0}^{\infty} \left(-\frac{a^2}{b^2}\right)^m, \tag{2.15}$$

after using (see [9, 8.339.3])

$$\Gamma\left(\tfrac{1}{2} - m\right) = \frac{(-1)^m 2^{2m} m! \sqrt{\pi}}{(2m)!}. \tag{2.16}$$

Similarly, the case n_2 free gives $n_1^* = -2n_2 - 2$ and it yields

$$T_2 = \frac{b}{a^2} \sum_{m=0}^{\infty} \left(-\frac{b^2}{a^2}\right)^m. \tag{2.17}$$

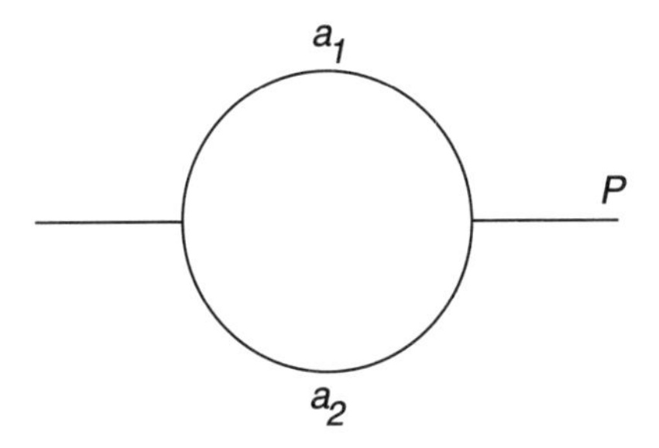

Fig. 2. The bubble.

The conclusion is that T_1 and T_2 are both given by a series (in this case a geometric series) in the parameters $x = -a^2/b^2$ and $1/x$, respectively. Each one of this series represents the value of (2.7) in complementary regions of convergence.

Example 2.3. This is the evaluation of a D-dimensional integral corresponding to the massless bubble Feynman diagram. It is a simpler example than the triangle diagram discussed in Section 1. The result is well known [4]. In momentum space, the corresponding integral is given by

$$G := \int \frac{1}{i\pi^{D/2}} \frac{1}{[q^2]^{a_1}\,[(p-q)^2]^{a_2}}\, d^Dq, \tag{2.18}$$

where the parameters $\{a_i\}$ are arbitrary. The Schwinger representation[a] corresponding to this diagram produces

$$G = \frac{(-1)^{-\frac{D}{2}}}{\Gamma(a_1)\,\Gamma(a_2)} \int_0^\infty \int_0^\infty x^{a_1-1} y^{a_2-1}\, \frac{\exp(-\frac{xy}{x+y}\,p^2)}{(x+y)^{\frac{D}{2}}}\, dx\, dy. \tag{2.19}$$

In order to construct a bracket series for this integral, it is convenient to expand first the exponential function

$$\exp\left(-\frac{xy}{x+y}\,p^2\right) = \sum_{n=0}^{\infty} \frac{(-1)^n}{n!} \left(p^2\right)^n \frac{x^n y^n}{(x+y)^n},$$

[a]There is a canonical procedure associating to each Feynman diagram a multi-dimensional integral. For details, the reader is referred to [12, Chapter 3], under the name *alpha parameters*.

and arrive at

$$G = \frac{(-1)^{-\frac{D}{2}}}{\Gamma(a_1)\Gamma(a_2)} \int_0^\infty \int_0^\infty x^{a_1-1} y^{a_2-1} \times \sum_{n=0}^{\infty} \frac{(-1)^n}{n!} \left(p^2\right)^n \frac{x^n y^n}{(x+y)^{\frac{D}{2}+n}} \, dx \, dy. \tag{2.20}$$

As a next step, apply the binomial expansion to $(x+y)^{-D/2-n}$ so that

$$(x+y)^{-(D/2+n)} = \sum_{k=0}^{\infty} \frac{(-1)^k}{k!} \left(\tfrac{D}{2}+n\right)_k x^{-D/2-n-k} y^k \tag{2.21}$$

and replace in (2.20) to obtain, after the change of variables $x \mapsto 1/x$,

$$\begin{aligned} G &= \frac{(-1)^{-\frac{D}{2}}}{\Gamma(a_1)\Gamma(a_2)} \int_0^\infty \int_0^\infty \sum_{k=0}^{\infty}\sum_{n=0}^{\infty} \\ &\quad \times \frac{(-1)^n}{n!} \frac{(-1)^k}{k!} \left(p^2\right)^n \left(\frac{D}{2}+n\right)_k x^{k-a_1+\frac{D}{2}} y^{k+n+a_2} \frac{dx}{x} \frac{dy}{y} \\ &= \frac{(-1)^{-\frac{D}{2}}}{\Gamma(a_1)\Gamma(a_2)} \sum_{k=0}^{\infty}\sum_{n=0}^{\infty} \phi_{n,k} (p^2)^n \left(\frac{D}{2}+n\right)_k \left\langle k - a_1 + \frac{D}{2} \right\rangle \langle k + a_2 + n \rangle. \end{aligned}$$

The problem has been reduced to the evaluation of a multi-dimensional bracket series where the number of sums is equal to the number of brackets. This is solved by the next rule.

Rule 4. *Assume the matrix $B = (b_{ij})$ is nonsingular. Then the assignment is*

$$\begin{aligned} &\sum_{n_1, n_2, \ldots, n_r} \phi_{n_1 \ldots n_r} a(n_1, \ldots, n_r) \langle b_{11} n_1 + \cdots + b_{1r} n_r + c_1 \rangle \cdots \\ &\qquad \times \langle b_{r1} n_1 + \cdots + b_{rr} n_r + c_r \rangle \\ &\qquad = \frac{1}{|\det(B)|} a(n_1^*, \ldots, n_r^*) \Gamma(-n_1^*) \ldots \Gamma(-n_r^*), \end{aligned}$$

where $\{n_i^*\}$ is the (unique) solution of the linear system obtained from the vanishing of the brackets. There is no assignment if B is singular.

The reader will now easily verify, in view of Rule 4, that

$$G = (-1)^{-\frac{D}{2}} \left(p^2\right)^{\frac{D}{2}-a_1-a_2} \frac{\Gamma(a_1+a_2-\frac{D}{2})\Gamma(\frac{D}{2}-a_1)\Gamma\left(\frac{D}{2}-a_2\right)}{\Gamma(a_1)\Gamma(a_2)\Gamma(D-a_1-a_2)}. \tag{2.22}$$

A similar procedure evaluates the Feynman diagram for the triangle in Fig. 1. The special massless situation, $m_1 = m_2 = m_3 = 0$ and assumption $P_1^2 = P_2^2 = 0$, produces the integral

$$G_1 = \frac{(-1)^{-D/2}}{\Gamma(a_1)\Gamma(a_2)\Gamma(a_3)} \int_{\mathbb{R}_+^3} x_1^{a_1-1} x_2^{a_2-1} x_3^{a_3-1} \times \frac{\exp(-\frac{x_1 x_3}{x_1+x_2+x_3} P_3^2)}{(x_1+x_2+x_3)^{D/2}} \, dx_1 \, dx_2 \, dx_3,$$

and the method of brackets then gives

$$G_1 = \frac{(-1)^{-D/2}}{\Gamma(a_1)\Gamma(a_2)\Gamma(a_3)} (P_3^2)^{D/2-a_1-a_2-a_3} \times \frac{\Gamma(a_1+a_2+a_3-\frac{D}{2})\Gamma(\frac{D}{2}-a_2-a_3)\Gamma(a_2)\Gamma(\frac{D}{2})\Gamma(\frac{D}{2}-a_1-a_2)}{\Gamma(D-a_1-a_2-a_3)}.$$

The final example is elementary and is used to illustrate a new rule.

Example 2.4. Entry 3.725.1 in [9] states that

$$\int_0^\infty \frac{\sin ax}{x(x^2+b^2)} \, dx = \frac{\pi}{2b^2}(1 - e^{-ab}). \tag{2.23}$$

A (possible) classical evaluation begins by differentiation with respect to a to see that the evaluation is equivalent to

$$\int_0^\infty \frac{\cos ax}{x^2+b^2} \, dx = \frac{\pi e^{-ab}}{2b}. \tag{2.24}$$

Rescaling with the change of variables $x = bt$ shows that it suffices to prove

$$\int_0^\infty \frac{\cos \alpha t}{t^2+1} \, dt = \frac{\pi}{2} e^{-\alpha}, \tag{2.25}$$

with $\alpha = ab$. The final step is carried out by contour integration.

As a show case, we propose utilizing the method of brackets to evaluate this integral. The first difficult step is come up with a series expansion for the integrand. This task will be simplified by the following instruction.

Rule 5. *For $\alpha \in \mathbb{C}$, the multinomial power $(u_1 + u_2 + \cdots + u_r)^\alpha$ is assigned to the r-dimension bracket series*

$$\sum_{n_1, n_2, \ldots, n_r} \phi_{n_1 n_2 \cdots n_r} u_1^{n_1} \cdots u_r^{n_r} \frac{\langle -\alpha + n_1 + \cdots + n_r \rangle}{\Gamma(-\alpha)}. \tag{2.26}$$

In the current situation,

$$(x^2+b^2)^{-1} \mapsto \sum_{n_1,n_2} \phi_{n_1,n_2} x^{2n_1} b^{2n_2} \langle 1+n_1+n_2 \rangle. \tag{2.27}$$

The standard expansion

$$\sin ax = \sum_{n_3=0}^{\infty} \frac{(-1)^{n_3}}{(2n_3+1)!} a^{2n_3+1} x^{2n_3+1} = \sum_{n_3=0}^{\infty} \phi_{n_3} \frac{\Gamma(n_3+1)}{\Gamma(2n_3+2)} a^{2n_3+1} x^{2n_3+1}$$

now leads to the bracket series

$$\int_0^\infty \frac{\sin ax \, dx}{x(x^2+b^2)} = \sum_{n_1,n_2,n_3} \phi_{n_1,n_2,n_3} \frac{\Gamma(n_3+1)}{\Gamma(2n_3+2)} a^{2n_3+1} b^{2n_2} \times \langle n_1+n_2+1 \rangle \langle 2n_1+2n_3+1 \rangle.$$

Rule 3 enforces that the solution is obtained by solving the system

$$\begin{aligned} n_1+n_2 &= -1, \\ 2n_1+2n_3 &= -1. \end{aligned} \tag{2.28}$$

Since the system (2.28) is of rank 1, the solution relies on one free parameter.

Case 1: n_1 is free. Then $n_2 = -1-n_1$ and $n_3 = -\frac{1}{2}-n_1$ and hence the bracket series is

$$\sum_{n_1=0}^{\infty} \frac{1}{2} (-1)^{n_1} \Gamma(\tfrac{1}{2}+n_1) \frac{\Gamma(\frac{1}{2}-n_1)}{\Gamma(1-2n_1)} a^{-2n_1} b^{-2n_1-2}. \tag{2.29}$$

This series contains a single nonvanishing term, that for $n_1=0$. It reduces to $\pi/2b^2$. This is the asymptotic value of the integral as $ab \to \infty$. *This is a typical phenomena.* Series that reduce to a finite number of nonzero terms produce asymptotic expansions of the solution.

Case 2: n_2 is free. Then $n_1 = -1-n_2$ and $n_3 = \frac{1}{2}+n_2$ and the series becomes

$$\frac{1}{2} \sum_{n_2=0}^{\infty} (-1)^{n_2} \Gamma(-\tfrac{1}{2}-n_2) \frac{\Gamma(\frac{3}{2}+n_2)}{\Gamma(2n_2+3)} a^{2n_2+2} b^{2n_2}. \tag{2.30}$$

Now use $\Gamma(-\frac{1}{2}-n)\Gamma(\frac{1}{2}+n) = (-1)^{n+1}\pi$ to simplify the previous series to

$$-\frac{\pi a^2}{2} \sum_{n_2=0}^{\infty} \frac{(ab)^{2n_2}}{(2n_2+2)!} = \frac{\pi}{2b^2} \left[1-\cosh(ab)\right]. \tag{2.31}$$

Case 3: n_3 is free. Then $n_1 = -\frac{1}{2} - n_3$ and $n_2 = n_3 - \frac{1}{2}$. Then the series becomes

$$\frac{a}{2b}\sum_{n_3=0}^{\infty}\frac{(-1)^{n_3}}{\Gamma(2n_3+2)}a^{2n_3}b^{2n_3}\Gamma(n_3+\tfrac{1}{2})\Gamma(\tfrac{1}{2}-n_3) = \frac{\pi a}{2b}\sum_{n_3=0}^{\infty}\frac{(ab)^{2n_3}}{(2n_3+1)!}$$
$$= \frac{\pi}{2b^2}\sinh(ab).$$

The process yields the asymptotic behavior of the solution and two convergent series. Rule 3 commands that these two convergent series should be added to produce the result. Indeed,

$$\frac{\pi}{2b^2}\left[1-\cosh(ab)\right] + \frac{\pi}{2b^2}\sinh(ab) = \frac{\pi}{2b^2}\left(1-e^{-ab}\right), \tag{2.32}$$

confirms (2.23).

3. Conclusions

The method of brackets provides a flexible procedure to evaluate a large number of definite integrals on the half-line $[0, \infty)$. It consists of a small number of rules to produce, from the integrand, a bracket series and a second set of rules to evaluate these formal series. Some progress has been made in providing rigorous proofs of these rules, but most of them remain in the experimental stage.

Acknowledgments

The authors wish to thank T. Ambederhan for improvements on an earlier version of this work.

References

[1] T. Amdeberhan, O. Espinosa, I. Gonzalez, M. Harrison, V. Moll and A. Straub, Ramanujan master theorem, *Ramanujan J.* **29** (2012) 103–120.

[2] D. H. Bailey and J. M. Borwein, Experimental mathematics: examples, method and implications, *Notices Amer. Math. Soc.* **52** (2005) 502–514.

[3] J. M. Borwein and P. B. Borwein. *Pi and the AGM — A Study in Analytic Number Theory and Computational Complexity*, 1st edn. (Wiley, New York, 1987).

[4] A. I. Davydychev, Some exact results for n-point massive Feynman integrals, *J. Math. Phys.* **32** (1991) 1052–1060.

[5] I. Gonzalez, K. Kohl, and V. Moll, Evaluation of entries in Gradshteyn and Ryzhik employing the method of brackets, *Scientia* **25** (2014) 65–84.

[6] I. Gonzalez and V. Moll, Definite integrals by the method of brackets. Part 1, *Adv. Appl. Math.* **45** (2010) 50–73.
[7] I. Gonzalez and I. Schmidt, Recursive method to obtain the parametric representation of a generic Feynman diagram, *Phys. Rev. D* **72** (2005) 106006.
[8] I. Gonzalez and I. Schmidt, Optimized negative dimensional integration method (NDIM) and multiloop Feynman diagram calculation, *Nucl. Phys. B* **769** (2007) 124–173.
[9] I. S. Gradshteyn and I. M. Ryzhik, *Table of Integrals, Series, and Products*, 8th edn., eds. D. Zwillinger and V. Moll (Academic Press, New York, 2015).
[10] A. Grozin, *Lectures on QED and QCD: Practical Calculation and Renormalization of One and Multi-loop Feynman Diagrams*, 1st edn. (World Scientific 2007).
[11] D. F. Lawden, *Elliptic Functions and Applications*, Applied Mathematical Sciences, Vol. 80 (Springer, 1989).
[12] V. A. Smirnov, *Feynman Integral Calculus* (Springer, Berlin, 2006).
[13] M. Veltman, *Diagrammatica: The Path to Feynman Diagrams*, Cambridge Lecture Notes in Physics, Vol. 4 (Cambridge University Press, 1994).

Chapter 17

Balanced Modular Parameterizations

Tim Huber*, **Danny Lara**† **and Esteban Melendez**‡

Department of Mathematics, University of Texas
Rio Grande Valley, 1201 West University Avenue
Edinburg, TX 78539, USA
**timothy.huber@utrgv.edu*
†*danny-lara@uiowa.edu*
‡*estebanjms@gmail.com*

For prime levels $5 \leq p \leq 19$, sets of theta quotients are constructed that generate graded rings of modular forms of positive integer weight for $\Gamma_1(p)$. Action by $\Gamma_0(p)$ is shown to cyclically permute the generators. This induces symmetric representations for modular forms. The generators are used to deduce representations for the number of t-core partitions of an integer as convolutions of L-functions. Coupled systems of differential equations of level p are constructed for each basis analogous to Ramanujan's differential equations for Eisenstein series on the full modular group.

Keywords: Modular forms; elliptic functions; Eisenstein series; theta functions; integer partitions; t-cores; differential equations.

Mathematics Subject Classification 2010: 11F03, 11F11

1. Introduction

The graded ring of modular forms for a finite index subgroup of the full modular group is isomorphic to a polynomial ring in two or more generators modulo a finite set of relations [7; 31, p. 249]. Certain classical polynomial representations for modular forms exhibit coefficient symmetry. For example, the *Klein polynomials*, whose roots encode distinguished points of the stereographically projected circumsphere for a regular icosahedron, are symmetric in absolute value about the middle coefficients

$$K_v(\Lambda) = \left(1 - 11\Lambda - \Lambda^2\right)^5,$$

$$K_e(\Lambda) = 1 + 228\Lambda + 494\Lambda^2 - 228\Lambda^3 + \Lambda^4, \tag{1.1}$$

$$K_f(\Lambda) = 1 - 522\Lambda - 10005\Lambda^2 - 10005\Lambda^4 + 522\Lambda^5 + \Lambda^6. \tag{1.2}$$

The polynomials $K_e(\Lambda)$ and $K_f(\Lambda)$, encoding the edge and face points, correspond to representations for Eisenstein series in terms of two modular parameters of level five

$$\begin{aligned} B^{20}K_e(\Lambda) &= 1 + 240\sum_{n=1}^{\infty}\frac{n^3q^n}{1-q^n}, \\ B^{30}K_f(\Lambda) &= 1 - 504\sum_{n=1}^{\infty}\frac{n^5q^n}{1-q^n}, \quad |q| < 1, \end{aligned} \tag{1.3}$$

and

$$A^5(q) = q\frac{(q;q)_\infty^2}{(q^2,q^3;q^5)_\infty^5}, \quad B^5(q) = \frac{(q;q)_\infty^2}{(q,q^4;q^5)_\infty^5}, \quad \Lambda = A^5/B^5, \tag{1.4}$$

where $(a;q)_n = \prod_{k=0}^{n-1} 1 - aq^k$ and $(a_1, a_2, \ldots, a_r; q)_n = \prod_{j=1}^{r}(a_j;q)_n$ for $n \in \mathbb{N} \cup \infty$. These are special cases of more general symmetric parameterizations for modular forms in terms of A^5, B^5 from [5]. In what follows, the symmetry proved in [5] is placed in a more general context and extended to other primes. At level seven, certain modular forms may be expressed in symmetric form in terms of the parameters

$$\begin{aligned} x &= q\frac{(q^2,q^5,q^7,q^7;q^7)_\infty}{(q^3,q^4;q^7)_\infty^2}, \quad y = -q\frac{(q,q^6,q^7,q^7;q^7)_\infty}{(q^2,q^5;q^7)_\infty^2}, \\ z &= \frac{(q^3,q^4,q^7,q^7;q^7)_\infty}{(q,q^6;q^7)_\infty^2}. \end{aligned} \tag{1.5}$$

For example, the Hecke Eisenstein series twisted, respectively by the Jacobi symbol and trivial character $\chi_{1,7}$ modulo seven have the symmetric representations [6, 16]

$$\begin{aligned} x + y + z &= 1 + 2\sum_{n=1}^{\infty}\left(\frac{n}{7}\right)\frac{q^n}{1-q^n}, \quad x^2 + y^2 + z^2 \\ &= 1 + 4\sum_{n=1}^{\infty}\frac{\chi_{1,7}(n)nq^n}{1-q^n}, \end{aligned} \tag{1.6}$$

and are connected to one other by the identity [25, Chapter 21, Entry 5(i)]

$$(x+y+z)^2 = x^2 + y^2 + z^2. \tag{1.7}$$

This identity results from the Klein quartic equation, with symmetric form [16]

$$xy + xz + yz = 0. \tag{1.8}$$

Many formulations for modular forms of prime level in terms of theta quotients appear in the work of Klein and Ramanujan. Because symmetry in these constructions may appear incidental, no unified study of symmetric modular parameterizations or extensions to other settings have been undertaken. In the present work, bases for graded rings of modular forms of prime level p with $5 \leq p \leq 19$ are introduced that parameterize modular forms in a symmetric way. The coefficient symmetry is a hallmark of a special collection of theta quotients generating the graded ring of modular forms on

$$\Gamma_1(p) = \left\{ \begin{pmatrix} a & b \\ c & d \end{pmatrix} \in \mathrm{PSL}(2,\mathbb{Z}) \,\middle|\, c \equiv 0,\ a \equiv d \equiv 1 \pmod{p} \right\}. \tag{1.9}$$

The generators are cyclically permuted up to a change of sign under action by

$$\Gamma_0(p) = \left\{ \begin{pmatrix} a & b \\ c & d \end{pmatrix} \in \mathrm{PSL}(2,\mathbb{Z}) \,\middle|\, c \equiv 0 \pmod{p} \right\}. \tag{1.10}$$

Balanced polynomial representations for modular forms on $\Gamma_0(p)$ result from a cyclic permutative action on generators for the graded rings of modular forms for $\Gamma_1(p)$ induced by modular transformation formulas. In addition to the list of permuted generators of level five and seven already presented, a corresponding set of $\Gamma_0(11)$-permuted generators for the graded ring of forms on $\Gamma_1(11)$ will be given by

$$\frac{(q^4, q^7, q^{11}, q^{11}; q^{11})_\infty}{(q, q^{10}, q^2, q^9; q^{11})_\infty}, \tag{1.11a}$$

$$q\frac{(q^2, q^9, q^{11}, q^{11}; q^{11})_\infty}{(q^5, q^6, q, q^{10}; q^{11})_\infty}, \tag{1.11b}$$

$$q^2\frac{(q, q^{10}, q^{11}, q^{11}; q^{11})_\infty}{(q^3, q^8, q^5, q^6; q^{11})_\infty}, \tag{1.11c}$$

$$q\frac{(q^5,q^6,q^{11},q^{11};q^{11})_\infty}{(q^3,q^8,q^4,q^7;q^{11})_\infty}, \tag{1.12a}$$

$$q\frac{(q^3,q^8,q^{11},q^{11};q^{11})_\infty}{(q^2,q^9,q^4,q^7;q^{11})_\infty}. \tag{1.12b}$$

A similarly permuted set of generators for the graded ring of forms for $\Gamma_1(13)$ is

$$\frac{(q^6,q^7,q^{13},q^{13};q^{13})_\infty}{(q,q^{12},q^3,q^{10};q^{13})_\infty}, \tag{1.13a}$$

$$q^2\frac{(q^3,q^{10},q^{13},q^{13};q^{13})_\infty}{(q^5,q^8,q^6,q^7;q^{13})_\infty}, \tag{1.13b}$$

$$q\frac{(q^5,q^8,q^{13},q^{13};q^{13})_\infty}{(q^3,q^{10},q^4,q^9;q^{13})_\infty}, \tag{1.13c}$$

$$q\frac{(q^4,q^9,q^{13},q^{13};q^{13})_\infty}{(q^2,q^{11},q^5,q^8;q^{13})_\infty}, \tag{1.14a}$$

$$q\frac{(q,q^2,q^{11},q^{13};q^{13})_\infty}{(q,q^{12},q^4,q^9;q^{13})_\infty}, \tag{1.14b}$$

$$q^2\frac{(q,q^{12},q^{13},q^{13};q^{13})_\infty}{(q^2,q^{11},q^6,q^7;q^{13})_\infty}. \tag{1.14c}$$

A set of $\Gamma_0(17)$-permuted generators for the graded ring of forms on $\Gamma_1(17)$ is

$$\frac{(q^8,q^9,q^{17},q^{17};q^{17})_\infty}{(q^2,q^{15},q^3,q^{14};q^{17})_\infty}, \tag{1.15a}$$

$$q\frac{(q^3,q^{14},q^{17},q^{17};q^{17})_\infty}{(q,q^{16},q^5,q^{12};q^{17})_\infty}, \tag{1.15b}$$

$$q^3\frac{(q,q^{16},q^{17},q^{17};q^{17})_\infty}{(q^4,q^{13},q^6,q^{11};q^{17})_\infty}, \tag{1.15c}$$

$$q\frac{(q^6,q^{11},q^{17},q^{17};q^{17})_\infty}{(q^2,q^{15},q^7,q^{10};q^{17})_\infty}, \tag{1.16a}$$

$$q^3\frac{(q^2,q^{15},q^{17},q^{17};q^{17})_\infty}{(q^5,q^{12},q^8,q^9;q^{17})_\infty}, \tag{1.16b}$$

$$q\frac{(q^5,q^{12},q^{17},q^{17};q^{17})_\infty}{(q^3,q^{14},q^4,q^{13};q^{17})_\infty}, \tag{1.16c}$$

$$q\frac{(q^4,q^{13},q^{17},q^{17};q^{17})_\infty}{(q,q^{16},q^7,q^{10};q^{17})_\infty}, \tag{1.17a}$$

$$q^2\frac{(q^7,q^{10},q^{17},q^{17};q^{17})_\infty}{(q^6,q^{11},q^8,q^9;q^{17})_\infty}. \tag{1.17b}$$

Finally, a set of permuted generators for the graded ring of forms for $\Gamma_1(19)$ is

$$\frac{(q^8,q^{11},q^9,q^{10},q^{19},q^{19};q^{19})_\infty}{(q^3,q^{16},q^4,q^{15},q^5,q^{14};q^{19})_\infty}, \quad q^2\frac{(q^4,q^{15},q^5,q^{14},q^{19},q^{19};q^{19})_\infty}{(q^2,q^{17},q^7,q^{12},q^8,q^{11};q^{19})_\infty}, \tag{1.18}$$

$$q\frac{(q^2,q^{17},q^7,q^{12},q^7,q^{12};q^{19})_\infty}{(q,q^{18},q^4,q^{15},q^6,q^{13};q^{19})_\infty}, \quad q^2\frac{(q,q^{18},q^6,q^{13},q^{19},q^{19};q^{19})_\infty}{(q^2,q^{17},q^3,q^{16},q^9,q^{10};q^{19})_\infty}, \tag{1.19}$$

$$q\frac{(q^3,q^{16},q^9,q^{10},q^{19},q^{19};q^{19})_\infty}{(q,q^{18},q^5,q^{14},q^8,q^{11};q^{19})_\infty}, \quad q^2\frac{(q^5,q^{14},q^8,q^{11},q^{19},q^{19};q^{19})_\infty}{(q^4,q^{15},q^7,q^{12},q^9,q^{10};q^{19})_\infty}, \tag{1.20}$$

$$q\frac{(q^4,q^{15},q^7,q^{12},q^{19},q^{19};q^{19})_\infty}{(q^2,q^{17},q^5,q^{14},q^6,q^{13};q^{19})_\infty}, \quad q\frac{(q^2,q^{17},q^6,q^{13},q^{19},q^{19};q^{19})_\infty}{(q,q^{18},q^3,q^{16},q^7,q^{12};q^{19})_\infty}, \tag{1.21}$$

$$q^5\frac{(q,q^{18},q^3,q^{16},q^{19},q^{19};q^{19})_\infty}{(q^6,q^{13},q^8,q^{11},q^9,q^{10};q^{19})_\infty}. \tag{1.22}$$

The products above generate the graded algebra of modular forms on $\Gamma_1(p)$ for $1 \le p \le 19$. Any modular form on a subgroup containing $\Gamma_0(p)$ may be expressed as a polynomial in the parameters so that each coefficient agrees in absolute value with that of any other obtained through a faithful cyclic permutation of the parameters.

In Section 2, a canonical basis of weight one forms on $\Gamma_1(p)$ is constructed in terms of certain linear combinations of the weight $k = 1$ twisted Eisenstein series

$$E_{\chi,k}(\tau) = 1 + \frac{2}{L(1-k,\chi)}\sum_{n=1}^{\infty}\chi(n)\frac{n^{k-1}q^n}{1-q^n}, \tag{1.23}$$

where $L(1-k,\chi)$ is the analytic continuation of the associated Dirichlet L-series. The Eisenstein series of weight one generate the graded algebra of modular forms for $\Gamma_1(p)$. Moreover, the generating sums are cyclically permuted up to change of sign under action by $\Gamma_0(p)$. In Section 3, we show that the Eisenstein sums equal the theta quotients from (1.4)–(1.22). In Section 4, generating functions are constructed for enumerating p-core partitions from products of elements in each generating set. A new convolution representation for the number of p-core partitions is given in Corollary 4.2 for $5 \leq p \leq 19$. Coupled systems of differential equations are formulated for the generators of each graded ring of modular forms. The systems are invariant under action by $\Gamma_0(p)$. This generalizes the differential systems of level five and seven derived in [15, 16] and explains the common coefficients characterizing each coupled system.

Theorem 1.1. *Let* $E_2(q) = 1 - 24\sum_{n=1}^{\infty}\frac{nq^n}{1-q^n}$ *and* $\mathscr{P} = E_2(q^5)$. *Then*

$$q\frac{d}{dq}A = \frac{1}{60}A(-5A^{10} - 66A^5B^5 + 7B^{10} + 5\mathscr{P}), \tag{1.24}$$

$$q\frac{d}{dq}B = \frac{1}{60}B(-5B^{10} + 66A^5B^5 + 7A^{10} + 5\mathscr{P}), \tag{1.25}$$

$$q\frac{d}{dq}\mathscr{P} = \frac{5}{12}(\mathscr{P}^2 - B^{20} + 12B^{15}A^5 - 14B^{10}A^{10} - 12B^5A^{15} - A^{20}). \tag{1.26}$$

Theorem 1.2. *Let* $\mathcal{P}(q) = E_2(q^7)$. *Then*

$$q\frac{d}{dq}x = \frac{x}{12}(5y^2 + 5z^2 - 7x^2 + 20yz + 52xy + 7\mathcal{P}), \tag{1.27}$$

$$q\frac{d}{dq}y = \frac{y}{12}(5z^2 + 5x^2 - 7y^2 + 20xz + 52yz + 7\mathcal{P}), \tag{1.28}$$

$$q\frac{d}{dq}z = \frac{z}{12}(5x^2 + 5y^2 - 7z^2 + 20xy + 52xz + 7\mathcal{P}), \tag{1.29}$$

$$q\frac{d}{dq}\mathcal{P}(q) = \frac{7}{12}(\mathcal{P}^2 - x^4 - 4x^3y - 12xy^3 - y^4 - 12x^3z - 4y^3z - 4xz^3 - 12yz^3 - z^4).$$

Theorem 4.5 extends the differential equations from Theorems 1.1 and 1.2 to higher prime levels. These systems are analogous to those for modular parameters of lower levels [12–14, 24] and to Ramanujan's differential system for Eisenstein series [26]

$$q\frac{dE_2}{dq} = \frac{E_2^2 - E_4}{12}, \quad q\frac{dE_4}{dq} = \frac{E_2E_4 - E_6}{3}, \quad q\frac{dE_6}{dq} = \frac{E_2E_6 - E_4^2}{2}, \tag{1.30}$$

where the normalized Eisenstein series $E_k = E_k(q)$ for $\mathrm{PSL}(2,\mathbb{Z})$ are defined by

$$E_{2k}(q) = 1 + \frac{2}{\zeta(1-2k)} \sum_{n=1}^{\infty} \frac{n^{2k-1}q^n}{1-q^n}, \tag{1.31}$$

and where ζ is the analytic continuation of the Riemann ζ-function. In fact, the last equations of Theorem 1.1 and 1.2 are equivalent to the first equation of (1.30).

2. Symmetric Representations via Eisenstein Sums

Let $\mathcal{M}_k(\Gamma)$ denote the vector space of weight k modular forms for $\Gamma \leq \mathrm{PSL}(2,\mathbb{Z})$. If $E_k(\Gamma)$ and $\mathcal{S}_k(\Gamma)$ denote the Eisenstein space and space of cusp forms, respectively, then

$$\mathcal{M}_k(\Gamma) = E_k(\Gamma) \oplus \mathcal{S}_k(\Gamma).$$

Since the space of cusp forms of weight one for $\Gamma_1(p)$ is trivial when $1 \leq p \leq 19$, the set of $(p-1)/2$ weight one Eisenstein series of odd primitive character modulo p from (1.23) forms a basis for $\mathcal{M}_1(\Gamma_1(p))$ (cf. [4, 29; 8, Theorem 4.8.1]). To show that the products from Section 1 generate the graded algebra of modular forms for $\Gamma_1(p)$ with the claimed symmetry, we formulate bases of weight one forms as linear combinations of Eisenstein series so that the spanning elements are cyclically permuted under action by $\Gamma_0(p)$. The Eisenstein series of weight one generate the graded ring of modular forms for $\Gamma_1(p)$, and the spanning elements coincide with the products from (1.4)–(1.22).

Theorem 2.1. *For primes $5 \leq p \leq 19$, define $\mathcal{E}_0 := \mathcal{E}_{0,p}$ by*

$$\mathcal{E}_0(\tau) := \mathcal{E}_{0,p}(\tau) := \frac{2}{p-1} \sum_{\chi(-1)=-1} E_{\chi,k}(\tau). \tag{2.1}$$

Let a denote the smallest positive primitive root modulo p, and define

$$\langle a\rangle \mathcal{E}_0(\tau) = \frac{2}{p-1} \sum_{\chi(-1)=-1} \chi(a) E_{\chi,1}(\tau). \tag{2.2}$$

Then

$$\mathcal{M}_1(\Gamma_1(p)) = \bigoplus_{k=0}^{(p-3)/2} \mathbb{C}\langle a^k\rangle(\mathcal{E}_0). \tag{2.3}$$

Proof. Since the Eisenstein series of weight one for $\Gamma_0(p)$ form a basis for $\mathcal{M}_1(\Gamma_1(p))$, it suffices to show that the matrix connecting the two sets of modular forms is invertible. The orthogonality of the Dirichlet characters modulo p may be used to derive

$$\sum_{\chi(-1)=-1} \chi(m)\overline{\chi}(n) = \begin{cases} \pm(p-1)/2, & m \equiv \pm n \pmod{p}, \\ 0, & m \not\equiv \pm n \pmod{p}. \end{cases} \tag{2.4}$$

Therefore, if the Dirichlet characters $\{\chi_s\}_{s=0}^{(p-3)/2}$ are odd, the rows of

$$(B)_{k,s} = \chi_s(a^k), \quad 0 \le k, s \le (p-3)/2, \tag{2.5}$$

are orthogonal under the standard Hermitian inner product. Hence, B satisfies

$$B(E_{1,\chi_1}(\tau), \dots, E_{1,\chi_{(p-1)/2}}(\tau))^T = (\langle a^0\rangle(\mathcal{E}_0), \dots, \langle a^{(p-3)/2}\rangle(\mathcal{E}_0))^T. \tag{2.6}$$

and is an invertible linear transformation corresponding to a change of basis. □

We will show in Theorem 2.4 that the basis from Theorem 2.1 is permuted under action by $\Gamma_0(p)$. Before doing this, we extend the bases of weight one forms for $\Gamma_1(p)$ from Theorem 2.1 to generators for the graded algebra of positive integer weight modular forms for $\Gamma_1(p)$. The next result from [3, 28] shows that for prime levels $N \ge 5$, it suffices to generate modular forms up to weight three. The situation is better for principal congruence subgroups of level N, where weight one suffices [17].

Lemma 2.2. *Denote by $\mathcal{M}_k(\Gamma)$ the $\mathbb{C}$-vector space of weight k modular forms for the congruence subgroup Γ, and let $\mathcal{M}(\Gamma) = \bigoplus_{k=1}^{\infty} \mathcal{M}_k(\Gamma)$ be the corresponding graded ring.*

(1) *For $N \ge 5$, any algebra containing $\mathcal{M}_k(\Gamma_1(N))$, for $k \le 3$, contains $\mathcal{M}(\Gamma_1(N))$.*
(2) *For $N \ge 3$, any algebra containing $\mathcal{M}_1(\Gamma(N))$ contains $\mathcal{M}(\Gamma(N))$.*

By Lemma 2.2, it suffices to demonstrate the existence of $\dim \mathcal{M}_k(\Gamma_1(p))$ linearly independent monomials of degree $k = 2, 3$ in the generators. The dimensions of $\mathcal{M}_k(\Gamma_1(p))$ for $k = 2, 3$ are given below:

	$p=5$	$p=7$	$p=11$	$p=13$	$p=17$	$p=19$
$\dim \mathcal{M}_2(\Gamma_1(p))$	3	5	10	13	20	24
$\dim \mathcal{M}_3(\Gamma_1(p))$	4	7	15	20	32	39

Theorem 2.3. *For primes* $5 \le p \le 19$, *the set* $\{\langle a^k\rangle(\mathcal{E}_p)\}_{k=0}^{(p-3)/2}$ *generates* $\mathcal{M}(\Gamma_1(p))$.

Proof. Since,

$$\mathcal{E}_{0,5}(\tau) = 1 + O(q), \quad \langle 2\rangle(\mathcal{E}_{0,5}) = q + O(q^2), \tag{2.7}$$

any set of distinct monomials of degree k forms a linearly independent set of modular forms of weight k for $\Gamma_1(5)$. Hence, the three monomials in (2.7) of degree two and the four monomials of degree three form a basis for $\mathcal{M}_k(\Gamma_1(5))$, $k = 2, 3$ respectively. For the higher levels $p = 7, 11, 13, 17, 19$, we introduce the complete homogeneous symmetric polynomial in n variables of degree k,

$$h_k(\vec{x}) = h_k(x_1, x_2, \ldots, x_n) = \sum_{1\le i_1\le i_2\le\cdots\le i_k\le n} x_{i_1}x_{i_2}\ldots x_{i_k}. \tag{2.8}$$

For each fixed p, k, and $1 \le i, j \le \binom{k+\frac{p-1}{2}-1}{k}$, let $\mathcal{H}(i,j)$ denote the coefficient of q^j in the q-expansion of the ith monomial of

$$h_k(\mathcal{E}_0, \mathcal{E}_1, \ldots, \mathcal{E}_{\frac{p-3}{2}}) \tag{2.9}$$

under lexicographic ordering of the monomials of $h_k(\vec{x})$. Define the matrix $H_{p,k} = \{\mathcal{H}(i,j)\}$. For each prime $5 \le p \le 19$, a computer algebra system may be used to show

$$\text{Rank}(H_{p,k}) = \dim \mathcal{M}_k(\Gamma_1(p)), \quad k = 1, 2, 3. \tag{2.10}$$

This establishes that a subset of the terms in the polynomial (2.9) span $\mathcal{M}_k(\Gamma_1(p)$ for $k = 2, 3$. The proof of Theorem 2.3 may be completed by applying Lemma 2.2. □

For large values of p, the verification of (2.10) is nontrivial. In particular, the case $p = 19$, $k = 3$ requires knowledge of the first 165 terms in the q-expansions of the 165 monomials of degree 3 in the parameters from (1.18)–(1.22). This calculation and the rank of the corresponding matrix may be verified with the aid of a computer algebra system. Explicit bases for $\mathcal{M}_k(\Gamma_1(p))$ may be constructed by row reducing $H_{k,p}$ and selecting the $\dim \mathcal{M}_k(\Gamma_1(p))$ monomials corresponding to linear independent rows.

The next theorem shows that the Eisenstein sum generators above for $\mathcal{M}(\Gamma_1(p))$ are permuted up to change of sign under action by $\Gamma_0(p)$.

Theorem 2.4. *For* $5 \le p \le 19$, *define* $a \in \mathbb{N}$ *to be the smallest primitive root* mod p *and*

$$\mathcal{E}_j(\tau) := \mathcal{E}_{j,p}(\tau) = \pm\langle a\rangle^j \mathcal{E}_0(\tau), \tag{2.11}$$

where the sign in equation (2.11) is chosen so that the first coefficient in the q-expansion is 1. If arithmetic is performed modulo $(p-1)/2$ *on the subscripts, and* $\vec{\epsilon}$ *is given below,*

$$\langle a\rangle \mathcal{E}_j = \epsilon_j \mathcal{E}_{j+1}. \tag{2.12}$$

p	$\vec{\epsilon}$	p	$\vec{\epsilon}$
5	$(1,-1)$	7	$(1,1,-1)$
11	$(1,-1,1,-1,-1)$	13	$(1,1,-1,-1,-1,1)$
17	$(1,-1,1,-1,1,-1,-1,-1)$	19	$(1,1,-1,1,-1,1,-1,-1,-1)$

Proof. Let $c \in (\mathbb{Z}/p\mathbb{Z})^*$. The claimed formulas follow directly from the definition of $\langle c\rangle(\mathcal{E}_{0,p})$. It is readily verified that for each $c \in (\mathbb{Z}/p\mathbb{Z})^*$, the first nonzero coefficient of $\langle c\rangle(\mathcal{E}_{0,p})$ is either 1 or -1. Denote by C_p the set of $c \in (\mathbb{Z}/p\mathbb{Z})^*$ such that the first nonzero coefficient of $\langle c\rangle(\mathcal{E}_{0,p})$ is 1. The elements of C_p are listed below:

p	C_p	p	C_p
5	1, 2	7	1, 2, 3
11	1, 2, 3, 5, 7	13	1, 2, 3, 4, 5, 7
17	1, 2, 3, 5, 7, 8, 11, 13	19	1, 2, 3, 4, 5, 7, 9, 11, 13

By referring to the table we see that $\epsilon_1 = 1$, and for $j \ge 1$,

$$\mathcal{E}_j = (-1)^{\lambda_p(a^j)} \cdot \epsilon_1 \cdots \epsilon_{j-1}\mathcal{E}_{j+1}, \quad \lambda_p(n) = \begin{cases} 0, & \overline{n} \in C_p, \\ 1, & \overline{n} \notin C_p, \end{cases} \tag{2.13}$$

where $\overline{n}$ denotes the congruence class modulo p. The claimed cyclic permutative action of $\langle\cdot\rangle$ on the Eisenstein sums follows from (2.13). $\square$

Since $\Gamma_1(p) \subset \Gamma_0(p)$, any modular form on a subgroup containing $\Gamma_0(p)$ may be expressed in terms of linearly independent monomials of fixed degree generated by $\{\mathcal{E}_1, \ldots, \mathcal{E}_{(p-1)/2}\}$. Theorem 2.4 demonstrates that such a polynomial has the property that the coefficient of each monomial agrees in absolute value with the coefficient of any other monomial obtained by the given cyclic permutations of the parameters.

3. Theta Quotient Generators

We now show that the Eisenstein sums may be systematically represented as quotients of theta functions and the Dedekind eta function, $\eta(\tau) = q^{1/24}(q;q)_\infty$, a weight 1/2 modular form for $\mathrm{SL}(2,\mathbb{Z})$ with multiplier given explicitly by [22, p. 51]. We employ fundamental properties of the Jacobi theta function

$$\theta_1(z|q) = -iq^{1/8} \sum_{n=-\infty}^{\infty} (-1)^n q^{n(n+1)/2} e^{(2n+1)iz}, \tag{3.1}$$

an odd function of z with a simple zero at the origin such that [30, p. 489]

$$\frac{\theta_1'}{\theta_1}(z|q) = \cot z + 4 \sum_{n=1}^{\infty} \frac{q^n}{1-q^n} \sin 2nz \tag{3.2}$$

$$= i - 2i \sum_{n=1}^{\infty} \frac{q^n e^{2iz}}{1-q^n e^{2iz}} + 2i \sum_{n=0}^{\infty} \frac{q^n e^{-2iz}}{1-q^n e^{-2iz}}. \tag{3.3}$$

Our subsequent calculations require the following easily verified functional equations

$$\theta_1(z+n\pi) = (-1)^n \theta_1(z|q), \quad \theta_1(z+n\pi\tau|q) = (-1)^n q^{-n^2/2} e^{-2inz} \theta_1(z|q) \tag{3.4}$$

and the Jacobi Triple Product identity [30]

$$\theta_1(z|q) = -iq^{1/8} e^{iz} (q;q)_\infty (qe^{2iz};q)_\infty (e^{-2iz};q)_\infty. \tag{3.5}$$

In particular, we will make use of the special case

$$\theta_1(r\pi\tau|q^p) = iq^{-r} q^{p/8} q^{r/2} (q^r, q^{p-r}, q^p; q^p)_\infty. \tag{3.6}$$

By differentiating (3.5) at the origin, we obtain a product formulation for the derivative

$$\theta_1'(q) := \lim_{z\to 0} \frac{\theta_1(z|q)}{z} = 2q^{1/8}(q;q)_\infty^3. \tag{3.7}$$

To show that $\Gamma_0(p)$ acts as indicated on the generating theta quotients, it will be convenient to apply transformation formulas for special values of the Jacobi theta functions (3.1) expressed in terms of theta constants of characteristic $[\epsilon, \epsilon'] \in \mathbb{R}^2$ [10]

$$\theta\begin{bmatrix}\epsilon\\ \epsilon'\end{bmatrix}(z,\tau) = \sum_{n\in\mathbb{Z}} \exp 2\pi i \left\{\frac{1}{2}\left(n+\frac{\epsilon}{2}\right)^2\tau + \left(n+\frac{\epsilon}{2}\right)\left(z+\frac{\epsilon'}{2}\right)\right\}. \quad (3.8)$$

Theorem 3.1 ([10, pp. 215–219]). *For odd positive integers $k \geq 3$, let*

$$\mathcal{V}_k(\tau) = \left[\theta\begin{bmatrix}(k-2)/k\\ 1\end{bmatrix}(k\tau), \theta\begin{bmatrix}(k-4)/k\\ 1\end{bmatrix}(k\tau), \ldots, \theta\begin{bmatrix}1/k\\ 1\end{bmatrix}(k\tau)\right]^T. \quad (3.9)$$

Then $\mathcal{V}_k$ is a vector-valued form of weight $1/2$ *on* $\mathrm{PSL}(2,\mathbb{Z})$ *inducing a representation*

$$\begin{aligned}\pi_k : \mathrm{PSL}(2,\mathbb{Z}) \to \mathrm{PGL}((k-1)/2, \mathbb{C}) \quad &\textit{via } \mathcal{V}_k(\gamma\tau)\\ &= (\gamma_{21}\tau + \gamma_{22})^{1/2}\pi_k(\gamma)\mathcal{V}_k(\tau)\end{aligned}$$

determined by the images of generators for $\mathrm{SL}(2,\mathbb{Z})$, $S = (0,-1;1,0)$, $T = (1,1;0,1)$,

$$\begin{aligned}\mathcal{V}_N(T\tau) &= \mathcal{V}_N(\tau+1) = \pi_N(T)\mathcal{V}_N(\tau),\\ \mathcal{V}_N(S\tau) &= \mathcal{V}_N(-1/\tau) = \tau^{1/2}\pi_N(S)\mathcal{V}_N(\tau),\end{aligned}$$

where the matrices $\pi_N(S)$ and $\pi_N(T)$ have (ℓ, j)th entry, for $1 \leq \ell, j \leq (N-1)/2$,

$$\{\pi_N(T)\}_{(\ell,j)} = \begin{cases}\exp\left(\frac{(N-2\ell)^2\pi i}{4N}\right), & \ell = j,\\ 0 & \textit{else},\end{cases} \quad (3.10)$$

$$\{\pi_N(S)\}_{(\ell,j)} = \frac{(1+e^{\frac{(2j-N)(N-2\ell)}{k}\pi i})\exp\left(\frac{(j(-2N+4\ell+2)+N^2-2(N+1)\ell)}{2N}\pi i\right)}{\sqrt{iN}}. \quad (3.11)$$

Equivalent transformation formulas figure prominently in Klein's representation of the automorphism group of the icosahedron [9, 19, 20]; in the septic extension to the Klein quartic [21]; as well as in Klein's level 11 analysis of the symmetries of the Klein cubic [18]. Our formulation of the Klein quartic via the quadratic relation (1.8) may be derived from a classical theta function identity [30, p. 518]

$$\left(\frac{\theta_1'}{\theta_1}\right)'(x|q) - \left(\frac{\theta_1'}{\theta_1}\right)'(y|q) = \frac{\theta_1'(0|q)^2\theta_1(x-y|q)\theta_1(x+y|q)}{\theta_1^2(x|q)\theta_1^2(y|q)}. \tag{3.12}$$

Theorem 3.2 provides the fundamental connection between the Eisenstein series sums above and the products (1.4)–(1.22) from Section 1. In particular, for each prime, the sum of weight one Eisenstein series is a theta quotient.

Theorem 3.2. *Define $E_{\chi,k}(\tau)$ as in (1.23). For each prime $5 \le p \le 19$, let*

$$\mathcal{E}_{0,p}(\tau) = \frac{2}{p-1} \sum_{\chi(-1)=-1} E_{\chi,1}(\tau), \tag{3.13}$$

where the sum is over the odd primitive Dirichlet characters modulo p. Then

$$\mathcal{E}_{0,5}(\tau) = \frac{(q;q)_\infty^2}{(q,q^4;q^5)_\infty^5}, \quad \mathcal{E}_{0,7}(\tau) = \frac{(q^3,q^4,q^7,q^7;q^7)_\infty}{(q,q^6;q^7)_\infty^2}, \tag{3.14}$$

$$\mathcal{E}_{0,11}(\tau) = \frac{(q^4,q^7,q^{11},q^{11};q^{11})_\infty}{(q,q^{10},q^2,q^9;q^{11})_\infty},$$

$$\mathcal{E}_{0,13}(\tau) = \frac{(q^6,q^7,q^{13},q^{13};q^{13})_\infty}{(q,q^{12},q^3,q^{10};q^{13})_\infty}, \tag{3.15}$$

$$\mathcal{E}_{0,17}(\tau) = \frac{(q^8,q^9,q^{17},q^{17};q^{17})_\infty}{(q^2,q^{15},q^3,q^{14};q^{17})_\infty},$$

$$\mathcal{E}_{0,19}(\tau) = \frac{(q^8,q^{11},q^9,q^{10},q^{19},q^{19};q^{19})_\infty}{(q^3,q^{16},q^4,q^{15},q^5,q^{14};q^{19})_\infty}. \tag{3.16}$$

Proof. To prove each identity, we use the fact that the sum of the residues of an elliptic function on its period parallelogram is zero [30]. The challenge lies in writing down the relevant elliptic functions. We begin by proving the leftmost equation of (3.14). Let

$$f_5(z) = \frac{e^{-2iz}\theta_1^3(z-\pi\tau|q^5)}{\theta_1^2(z|q^5)\theta_1(z+2\pi\tau|q^5)}. \tag{3.17}$$

Apply (3.4) to verify that $f_5(z)$ is an elliptic function with periods π and $5\pi\tau$. From corresponding properties of the Jacobi theta function, observe that $f_5(z)$ has a simple pole at $z=-2\pi\tau$ and a double pole at $z=0$. The residue of $f_5(z)$ at $z=-2\pi\tau$ is

$$\lim_{z\to-2\pi\tau}\frac{(z+2\pi\tau)}{\theta_1(z+2\pi\tau|q^5)}\lim_{z\to-2\pi\tau}\frac{e^{-2iz}\theta_1^3(z-\pi\tau|q^5)}{\theta_1^2(z|q^5)}$$

$$=\frac{q^2}{\theta_1'(q^5)}\cdot\frac{\theta_1^3(-3\pi\tau|q^5)}{\theta_1^2(-2\pi\tau|q^5)}. \tag{3.18}$$

The residue of $f_5(z)$ at $z=0$ is

$$\lim_{z\to0}(z^2f_5(z))'=\lim_{z\to0}(z^2f(z))\left(\frac{2}{z}+\frac{f_5'(z)}{f_5(z)}\right) \tag{3.19}$$

$$=\left(\lim_{z\to0}\frac{z^2}{\theta_1^2(z|q^5)}\right)\left(\lim_{z\to0}\frac{e^{-2iz}\theta_1^3(z-\pi\tau|q^5)}{(z+2\pi\tau|q^5)}\right)$$

$$\times\left(\lim_{z\to0}\frac{2}{z}+\frac{f_5'(z)}{f_5(z)}\right) \tag{3.20}$$

$$=\frac{-1}{\theta_1'(q^5)^2}\left(\frac{\theta_1^3(\pi\tau|q^5)}{\theta_1(2\pi\tau|q^5)}\right)\left(\lim_{z\to0}\frac{2}{z}+\frac{f_5'(z)}{f_5(z)}\right). \tag{3.21}$$

Since the sum of the residues of $f_5(z)$ is zero, we obtain from (3.6)

$$2i\frac{(q^2,q^3,q^5;q^5)_\infty^2}{(q,q^4,q^5;q^5)_\infty^3}(q^5;q^5)_\infty^3=\lim_{z\to0}\frac{2}{z}+\frac{f_5'(z)}{f_5(z)}. \tag{3.22}$$

By applying identities (3.2)–(3.3), and the Laurent expansion for $\cot z$, we derive

$$\lim_{z\to0}\frac{2}{z}+\frac{f_5'(z)}{f_5(z)}=\lim_{z\to0}\left(\frac{2}{z}-2\frac{\theta_1'}{\theta_1}(z|q^5)\right)-2i$$

$$-3\frac{\theta_1'}{\theta_1}(\pi\tau|q^5)-\frac{\theta_1'}{\theta_1}(2\pi\tau|q^5) \tag{3.23}$$

$$=-2i-3\frac{\theta_1'}{\theta_1}(\pi\tau|q^5)-\frac{\theta_1'}{\theta_1}(2\pi\tau|q^5)$$

$$=2i+\sum_{n=1}^{\infty}\frac{c_nq^n}{1-q^n}, \tag{3.24}$$

where, from (3.3), $\{c_n\}_{n=1}^{\infty}$ is a periodic sequence modulo five such that

$$c_1 = 6i, \quad c_2 = 2i, \quad c_3 = -2i, \quad c_4 = -6i, \quad c_5 = 0. \tag{3.25}$$

If we denote the two odd primitive Dirichlet characters modulo five by

$$\langle \chi_{2,5}(n) \rangle_{n=0}^{4} = \langle 0, 1, i, -i, -1 \rangle, \quad \langle \chi_{4,5}(n) \rangle_{n=0}^{4} = \langle 0, 1, -i, i, -1 \rangle, \tag{3.26}$$

then, since, for χ nonprincipal modulo p, we may write [8, pp. 136–137],

$$L(\chi, 0) = \sum_{n=0}^{p-1} \chi(n) \left(\frac{1}{2} - \frac{n}{p} \right), \tag{3.27}$$

it follows from (3.26) and (3.27) that

$$c_n = \frac{2i\chi_{2,5}(n)}{L(\chi_{2,5}, 0)} + \frac{2i\chi_{4,5}(n)}{L(\chi_{4,5}, 0)}. \tag{3.28}$$

Therefore, from (3.13) and identities (3.24), (3.28), and (1.23),

$$\lim_{z \to 0} \frac{2}{z} + \frac{f_5'(z)}{f_5(z)} = 2i\mathcal{E}_5(q). \tag{3.29}$$

This completes the proof of the leftmost equation of (3.14). A formulation of rightmost equation of (3.14) from (3.30) is given in [23, equation (3.23)]. For prime levels $7 \leq p \leq 19$, the claimed product expansions for the Eisenstein sums $\mathcal{E}_{0,p}(\tau)$ may obtained by applying the residue theorem with the elliptic functions $f_p(z)$ of period $\pi, p\pi\tau$, defined by

$$f_7(z) = e^{2iz} \frac{\theta_1^2(z + \pi\tau | q^7)\theta_1(z + 2\pi\tau | q^7)}{\theta_1^2(z | q^7)\theta_1(z - 3\pi\tau | q^7)}, \tag{3.30}$$

$$f_{11}(z) = e^{-2iz} \frac{\theta_1(z - 2\pi\tau | q^{11})\theta_1(z - 3\pi\tau | q^{11})\theta_1(z - 5\pi\tau | q^{11})}{\theta_1^2(z | q^{11})\theta_1(z + \pi\tau | q^{11})}, \tag{3.31}$$

$$f_{13}(z) = e^{-2iz}\frac{\theta_1(z-3\pi\tau|q^{13})\theta_1(z-4\pi\tau|q^{13})\theta_1(z-5\pi\tau|q^{13})}{\theta_1^2(z|q^{13})\theta_1(z+\pi\tau|q^{13})}, \tag{3.32}$$

$$f_{17}(z) = e^{-2iz}\frac{\theta_1(z-3\pi\tau|q^{17})\theta_1(z-5\pi\tau|q^{17})\theta_1(z-7\pi\tau|q^{17})}{\theta_1^2(z|q^{17})\theta_1(z+2\pi\tau|q^{17})}, \tag{3.33}$$

$$f_{19}(z) = e^{-2iz}\frac{\theta_1(z-4\pi\tau|q^{19})\theta_1(z-5\pi\tau|q^{19})\theta_1(z-7\pi\tau|q^{19})}{\theta_1^2(z|q^{19})\theta_1(z+3\pi\tau|q^{19})}, \tag{3.34}$$

each constructed by writing $\mathcal{E}_{0,p}(\tau)$ in terms of $(\theta_1'/\theta_1)(k\pi\tau|q^p)$, $1 \le k \le (p-1)/2$. □

To demonstrate that the normalized Eisenstein sums from Theorem 2.4 equal the products (1.4)–(1.22) from Section 1, we compute the $\Gamma_0(p)$ orbit of the quotients from Theorem 3.2 under action by the diamond operator. From the transformation formulas in Theorem 3.1, it follows that each basis element of level p for $\mathcal{M}_1(\Gamma_1(p))$ from Theorem 2.1 is representable as a quotient of modified theta constants. The product representations follow from the triple product representation [10, p. 141]

$$\theta\begin{bmatrix} m/n \\ 1 \end{bmatrix}(n\tau) = \exp\left(\frac{\pi i m}{2n}\right) q^{m^2/(8n)} \times (q^{(n-m)/2};q^n)_\infty (q^{(n+m)/2};q^n)_\infty (q^n;q^n)_\infty. \tag{3.35}$$

The theta quotients are constructed by writing the product formulations of Theorem 2.1 in terms of theta constants and determining their orbit under $\langle a \rangle$, for $a \in (\mathbb{Z}/p\mathbb{Z})^*$.

Theorem 3.3. *Define for odd k,*

$$\varphi_{k,\ell}(\tau) = \theta\begin{bmatrix} \frac{2\ell-1}{k} \\ 1 \end{bmatrix}(0, k\tau), \quad 1 \le \ell \le \frac{k-1}{2}, \tag{3.36}$$

and, for $[b_1, \ldots, b_{(p-1)/2}] \in \mathbb{Z}^{(p-1)/2}$, *denote*

$$\mathfrak{T}_p[b_1, \ldots, b_{(p-1)/2}](\tau) = \eta^3(p\tau) \prod_{k=1}^{(p-1)/2} \exp\left(-\frac{\pi i b_k(2k-1)}{2p}\right) \varphi_{p,k}^{b_k}(\tau). \tag{3.37}$$

The bases for $\mathcal{M}_1(\Gamma_1(p))$ from Theorem 2.1 have the theta quotient representations:

Level, p	*Basis for $\mathcal{M}_1(\Gamma_1(p))$*
5	$\mathcal{E}_0 = \mathfrak{T}_5[2,-3]$, $\mathcal{E}_1 = \langle 2\rangle(\mathcal{E}_0) = \mathfrak{T}_5[-3,2]$
7	$\mathcal{E}_0 = \mathfrak{T}_7[1,0,-2]$, $\mathcal{E}_1 = \langle 3\rangle(\mathcal{E}_0) = \mathfrak{T}_7[0,-2,1]$, $\mathcal{E}_2 = \langle 2\rangle(\mathcal{E}_7) = \mathfrak{T}_7[-2,1,0]$
11	$\mathcal{E}_0 = \mathfrak{T}_{11}[0,1,0,-1,-1]$, $\mathcal{E}_1 = \langle 2\rangle(\mathcal{E}_0) = \mathfrak{T}_{11}[-1,0,0,1,-1]$ $\mathcal{E}_2 = \langle 7\rangle(\mathcal{E}_0) = \mathfrak{T}_{11}[-1,0,-1,0,1]$, $\mathcal{E}_3 = \langle 3\rangle(\mathcal{E}_0) = \mathfrak{T}_{11}[1,-1,-1,0,0]$ $\mathcal{E}_4 = \langle 5\rangle(\mathcal{E}_0) = \mathfrak{T}_{11}[0,-1,1,-1,0]$
13	$\mathcal{E}_0 = \mathfrak{T}_{13}[1,0,0,-1,0,-1]$, $\mathcal{E}_1 = \langle 2\rangle(\mathcal{E}_0) = \mathfrak{T}_{13}[-1,-1,0,1,0,0]$, $\mathcal{E}_2 = \langle 4\rangle(\mathcal{E}_0) = \mathfrak{T}_{11}[0,1,-1,-1,0,0]$, $\mathcal{E}_3 = \langle 5\rangle(\mathcal{E}_0) = \mathfrak{T}_0[0,-1,1,0,-1,0]$, $\mathcal{E}_4 = \langle 3\rangle(\mathcal{E}_0) = \mathfrak{T}_{13}[0,0,-1,0,1,-1]$ $\mathcal{E}_5 = \langle 7\rangle(\mathcal{E}_0) = \mathfrak{T}_{13}[-1,0,0,0,-1,1]$
17	$\mathcal{E}_0 = \mathfrak{T}_{17}[1,0,0,0,0,-1,-1,0]$, $\mathcal{E}_1 = \langle 3\rangle(\mathcal{E}_0) = \mathfrak{T}_{17}[0,0,0,-1,0,1,0,-1]$, $\mathcal{E}_2 = \langle 8\rangle(\mathcal{E}_0) = \mathfrak{T}_{17}[0,0,-1,0,-1,0,0,1]$ $\mathcal{E}_3 = \langle 7\rangle(\mathcal{E}_0) = \mathfrak{T}_{17}[0,-1,1,0,0,0,-1,0]$, $\mathcal{E}_4 = \langle 13\rangle(\mathcal{E}_0) = \mathfrak{T}_{17}[-1,0,0,-1,0,0,1,0]$, $\mathcal{E}_5 = \langle 5\rangle(\mathcal{E}_0) = \mathfrak{T}_{17}[0,0,0,1,-1,-1,0,0]$, $\mathcal{E}_6 = \langle 2\rangle(\mathcal{E}_0) = \mathfrak{T}_{17}[0,-1,0,0,1,0,0,-1]$, $\mathcal{E}_7 = \langle 11\rangle(\mathcal{E}_0) = \mathfrak{T}_{17}[-1,1,-1,0,0,0,0,0]$
19	$\mathcal{E}_0 = \mathfrak{T}_{19}[1,1,0,0,-1,-1,-1,0,0]$, $\mathcal{E}_1 = \langle 2\rangle(\mathcal{E}_0) = \mathfrak{T}_{19}[0,-1,-1,0,1,1,0,-1,0]$, $\mathcal{E}_2 = \langle 4\rangle(\mathcal{E}_0) = \mathfrak{T}_{19}[0,0,1,-1,0,-1,0,1,-1]$, $\mathcal{E}_3 = \langle 11\rangle(\mathcal{E}_0) = \mathfrak{T}_{19}[-1,0,0,1,0,0,-1,-1,1]$ $\mathcal{E}_4 = \langle 3\rangle(\mathcal{E}_0) = \mathfrak{T}_{19}[1,-1,0,0,-1,0,1,0,-1]$, $\mathcal{E}_5 = \langle 13\rangle(\mathcal{E}_0) = \mathfrak{T}_{19}[-1,1,-1,0,1,-1,0,0,0]$ $\mathcal{E}_6 = \langle 7\rangle(\mathcal{E}_0) = \mathfrak{T}_{19}[0,0,1,-1,-1,1,0,-1,0]$ $\mathcal{E}_7 = \langle 5\rangle(\mathcal{E}_0) = \mathfrak{T}_{19}[0,0,-1,1,0,0,-1,1,-1]$, $\mathcal{E}_8 = \langle 9\rangle(\mathcal{E}_0) = \mathfrak{T}_{19}[-1,-1,0,-1,0,0,1,0,1]$

Proof. The theta quotient representations for $\mathcal{E}_{0,p}(\tau) = \langle 1\rangle(\mathcal{E}_0)(\tau)$ may be deduced from the product representations proved in Theorem 2.1. Transformation formula for these theta quotients under $\Gamma_0(p)$, in turn, may be deduced from corresponding modular transformation formulas for $\eta(\tau)$ from [22, p. 51] and those for vectors of modified theta constants $\mathcal{V}_N(\tau)$ under generators for the full modular group from Theorem 3.1. For each prime p, we may deduce the product representations for each normalized Eisenstein series sum from modular transformation formulas for these building blocks. We illustrate the general procedure with $p=5$. From Theorem 3.2 and (3.35),

$$\langle 1\rangle(\mathcal{E}_{0,5}) = \frac{(q;q)_\infty^2}{(q,q^4;q^5)_\infty^5} = \eta^3(5\tau)\frac{e^{-2\pi i/10}\varphi_{5,1}^2(\tau)}{e^{-9\pi i/10}\varphi_{5,2}^3(\tau)} = \mathfrak{T}_5[2,-3](\tau). \quad (3.38)$$

A set of generators for $\Gamma_0(5)$ is given by

$$T=\begin{pmatrix}1 & 1\\ 0 & 1\end{pmatrix}, \quad \alpha=\begin{pmatrix}2 & -1\\ 5 & -2\end{pmatrix}, \quad \beta=\begin{pmatrix}3 & -2\\ 5 & -3\end{pmatrix}. \quad (3.39)$$

The generators for $\Gamma_0(5)$ may be given in terms of those for the full modular group

$$\alpha = TST^2ST^3S, \quad \beta = TST^3ST^2S. \quad (3.40)$$

Transformation matrices for the vectors of modified theta constants may be computed from their images in $\mathrm{PGL}((p-1)/2,\mathbb{C})$ via the representation π_p given in (3.10)–(3.11)

$$\pi_5(T)=\begin{pmatrix}e^{\frac{9\pi i}{20}} & 0\\ 0 & e^{\frac{\pi i}{20}}\end{pmatrix}, \quad \pi_5(\alpha)=\begin{pmatrix}0 & e^{\frac{\pi i}{20}}\\ -e^{\frac{9\pi i}{20}} & 0\end{pmatrix},$$

$$\pi_5(\beta)=\begin{pmatrix}0 & e^{\frac{\pi i}{4}}\\ -e^{\frac{\pi i}{4}} & 0\end{pmatrix}. \quad (3.41)$$

Hence, by (3.38), and the modular transformation formula for $\eta(\tau)$, we deduce that up to a constant multiple, C,

$$\begin{aligned}\langle 2\rangle(\mathcal{E}_{0,5}) &= (5\tau-3)^{-1}\mathfrak{T}_5[2,-3](\beta\tau)\\ &= C\eta^3(5\tau)\frac{\varphi_{5,2}^2(\tau)}{\varphi_{5,1}^3(\tau)} = C\frac{e^{6\pi i/10}}{e^{3\pi i/10}}q + O(q^2).\end{aligned} \quad (3.42)$$

On the other hand, if $\chi_{2,5}$ and $\chi_{4,5}$ denote the odd characters modulo five,

$$\langle 2\rangle(\mathcal{E}_{0,5}) = \chi_{2,5}(2)E_{\chi_{2,5},1}(\tau) + \chi_{4,5}(2)E_{\chi_{4,5},1}(\tau) = q + O(q^2). \quad (3.43)$$

Therefore, $C=e^{-3\pi i/10}$, and so

$$\begin{aligned}\langle 2\rangle(\mathcal{E}_{0,5}) &= e^{-3\pi i/10}\eta^3(5\tau)\frac{\varphi_{5,2}^2(\tau)}{\varphi_{5,1}^3(\tau)} = \mathfrak{T}_5[-3,2](\tau)\\ &= q\frac{(q;q)_\infty^2}{(q^2,q^3;q^5)_\infty^5}.\end{aligned} \quad (3.44)$$

For higher levels, $7 \leq p \leq 19$, we similarly use the fact that the image of $\Gamma_0(p)$ under the representation π_p defined by Theorem 3.1 is a matrix with a single nonzero entry in each row and column. We obtain the theta quotient representations of the bases for $\mathcal{M}_1(\Gamma_1(p))$ from those for $\mathcal{E}_{0,p}(\tau)$. In each case, the theta quotients are permuted according to the image of π_p and the transformation formula for Eisenstein series on $\Gamma_0(p)$ is applied

$$(\gamma_{21}\tau + \gamma_{22})^{-k} E_{k,\chi}(\gamma\tau) = \chi(\gamma_{22}) E_{k,\chi}(\tau), \quad \gamma = \begin{pmatrix} \gamma_{11} & \gamma_{12} \\ \gamma_{21} & \gamma_{22} \end{pmatrix} \in \Gamma_0(p).$$

We then compare the first nonzero entry in the resulting q-expansions. By repeating this process with an independent set of generators for $\Gamma_0(p)$, we ultimately obtain the linearly independent sets of theta quotient representations claimed in Theorem 3.3. □

4. Symmetric Representations for Modular Forms of Level p

This section is devoted to applications of the symmetric representations for modular forms in terms of the permuted generators for $\mathcal{M}(\Gamma_1(p))$. Theorem 4.1 presents a uniform parameterization for an important class of combinatorial generating functions, that of t-cores. For a given partition λ, each square in the Young diagram representation for λ defines a *hook* consisting of that square, all the squares to the right of that square, and all the squares below that square. The *hook number* of a given square is the number of squares in that hook. A partition λ is said to be a t-core if it has no hook numbers that are multiples of t. The generating function for the number of t-cores is [11]

$$\sum_{n=0}^{\infty} c_t(n) q^n = \frac{(q^{tn}; q^{tn})_\infty^t}{(q;q)_\infty}. \tag{4.1}$$

Series from the first three cases of Theorem 4.1 play a fundamental role in the derivation of Ramanujan's famous congruences for the partition function modulo $5, 7, 11$ [2, 27].

Theorem 4.1. *Let be defined by $\mathcal{E}_k$ by Theorem 2.4, and let $\delta_p = \frac{p^2-1}{24}$. Then, for each prime p with $5 \leq p \leq 19$,*

$$q^{\delta_p} \frac{(q^p; q^p)_\infty^p}{(q;q)_\infty} = \prod_{k=0}^{\frac{p-3}{2}} \mathcal{E}_{k,p}(\tau). \tag{4.2}$$

Proof. The claim (4.2) for each level p may be deduced by utilizing the product representations, given explicitly by (1.4)–(1.22), for the generators from Theorem 3.3. □

From Theorem 4.1, we deduce new divisor sum representations for p-cores, p prime, $5 \leq p \leq 19$, in terms of L-function values for odd Dirichlet characters at the origin.

Corollary 4.2. *Let δ_p be defined by Theorem 4.1. For primes $5 \leq p \leq 19$, and $n \geq \delta_p$,*

$$\pm c_p(n-\delta_p) = \left(\frac{2}{p-1}\right)^{\frac{p-3}{2}} \sum_{r_1+\cdots+r_{(p-3)/2}=n} \left(\prod_{k=1}^{\frac{p-3}{2}} \sum_{d_k \mid r_k} \ell_p(a^{k+1}\cdot d_k)\right)$$
$$+ \left(\frac{2}{p-1}\right)^{\frac{p-1}{2}} \sum_{r_1+\cdots+r_{(p-1)/2}=n} \left(\prod_{k=1}^{\frac{p-1}{2}} \sum_{d_k \mid r_k} \ell_p(a^{k}\cdot d_k)\right), \tag{4.3}$$

where the sign in (4.3) *is* $+1$ *except when* $p = 17, 19$. *Here* $r_i \geq 1$, *and*

$$\ell_p(d) = 2 \sum_{\substack{\chi(-1)=-1 \\ primitive \bmod p}} \frac{\chi(d)}{L(0,\chi)}, \qquad d \in \mathbb{Z}. \tag{4.4}$$

Proof. For a the minimal positive primitive root modulo p and λ as defined in (2.13),

$$(-1)^{\lambda(a^k)}\mathcal{E}_{k,p}(\tau) = \delta_{0,k} + \frac{2}{p-1}\sum_{n=1}^{\infty}\left(\sum_{d\mid n}\sum_{\chi(-1)=-1}\frac{2\chi(a^k d)}{L(0,\chi)}\right)q^n,$$
$$\delta_{ij} = \begin{cases} 1, & i=j, \\ 0, & i \neq j. \end{cases} \tag{4.5}$$

Equation (4.3) follows by considering the Cauchy product of coefficients of the series from (4.2). In particular, the first line of (4.3) comes from the case $k = 0$ in (4.5). □

For the primes $p = 5, 7, 13$, the Eisenstein series of weight two and trivial character modulo p have common representations in terms of the squares of the generators.

Theorem 4.3. *If $\chi_{1,p}$ is the principal character modulo p, and $\mathcal{E}_{0,p}$ is given by* (2.11),

$$1+6\sum_{n=1}^{\infty}\frac{\chi_{1,5}(n)nq^n}{1-q^n}=\mathcal{E}_{0,5}^2+\mathcal{E}_{1,5}^2,$$

$$1+4\sum_{n=1}^{\infty}\frac{\chi_{1,7}(n)nq^n}{1-q^n}=\mathcal{E}_{0,7}^2+\mathcal{E}_{1,7}^2+\mathcal{E}_{2,7}^2, \tag{4.6}$$

$$1+2\sum_{n=1}^{\infty}\frac{\chi_{1,13}(n)nq^n}{1-q^n}=\mathcal{E}_{0,13}^2+\mathcal{E}_{1,13}^2+\mathcal{E}_{2,13}^2+\mathcal{E}_{3,13}^2+\mathcal{E}_{4,13}^2+\mathcal{E}_{5,13}^2. \tag{4.7}$$

We next derive coupled systems of differential equations for the theta quotients of level p and explain the common coefficients within the differential systems from Theorems 1.1 and 1.2. To interpret and extend the phenomena from Theorems 1.1 and 1.2 to higher levels, we apply the following standard result with $h=pk/12$.

Lemma 4.4 (**[1, Lemma 2.1]**). *Suppose that p is a prime and that f is a meromorphic modular form of weight k on $\Gamma_0(p)$. If h is any constant, then the function*

$$F_f:=q\frac{d}{dq}f+\left\{\frac{\frac{k}{12}-h}{p-1}\cdot pE_2(p\tau))+\frac{h-\frac{pk}{12}}{p-1}\cdot E_2(\tau)\right\}\cdot f\in\mathcal{M}_{k+2}(\Gamma_0(p)). \tag{4.8}$$

The proof of Lemma 4.4 from [1] shows that the lemma remains true for the subgroup $\Gamma_1(p)$ of $\Gamma_0(p)$. Moreover, if $f,g\in\mathcal{M}_k(\Gamma_1(p))$, and $\gamma\in\Gamma_0(p)$ with $\langle\gamma\rangle(f)=g$, then $\langle\gamma\rangle(F_f)=F_g$. Thus, for $0\le i\le(p-3)/2$, $(q\frac{d}{dq}\mathcal{E}_i(\tau))/\mathcal{E}_i(\tau)-\frac{pE_2(p\tau)}{12}\in\mathcal{M}_2(\Gamma_1(p))$. Bases of weight two forms are encoded by monomials from Theorem 4.5. The theorem subsumes differential equations of level 5 and 7 from Theorems 1.1–1.2 and new coupled systems for levels $11\le p\le 19$. Each equation is permuted under action by $\Gamma_0(p)$.

Theorem 4.5. *Let $\mathcal{E}_{k,p}$ be given as in* (2.11) *and define*

$$\mathcal{F}_5(x_0,x_1)=-5x_0^2+66x_0x_1+7x_1^2, \tag{4.9}$$

$$\mathcal{F}_7(x_0,x_1,x_2)=-7x_0^2+5x_2^2+5x_1^2-20x_2x_1+52x_0x_2, \tag{4.10}$$

$$\begin{aligned}\mathcal{F}_{11}(x_0,x_1,\ldots,x_4)=&-11x_0^2+x_1^2+13x_3^2+x_4^2+x_2^2+34x_0x_3\\&+42x_1x_4-40x_1x_2+38x_3x_2-10x_4x_2,\end{aligned} \tag{4.11}$$

$$\begin{aligned}\mathcal{F}_{13}(x_0,x_1,\ldots,x_5) =& -13x_0^2+11x_1^2-x_4^2-x_2^2-x_3^2+11x_5^2+16x_0x_1\\ &+38x_0x_3+2x_1x_4-20x_1x_2+40x_1x_3\\ &-8x_1x_5+14x_2x_5, \qquad (4.12)\end{aligned}$$

$$\begin{aligned}\mathcal{F}_{17}(x_0,x_1,\ldots,x_7) =& -17x_0^2+19x_6^2+7x_1^2-5x_5^2-5x_3^2-5x_2^2-5x_7^2+19x_4^2\\ &-12x_6x_1+54x_6x_3+12x_3x_2+30x_6x_7-42x_3x_7\\ &-60x_6x_4+12x_1x_4-54x_3x_4-12x_7x_4, \qquad (4.13)\end{aligned}$$

$$\begin{aligned}\mathcal{F}_{19}(x_0,x_1,\ldots,x_8) =& -19x_0^2+41x_1^2+5x_4^2-7x_2^2-7x_7^2+5x_6^2\\ &+5x_8^2-7x_3^2-7x_5^2+16x_1x_7-36x_4x_7\\ &+8x_2x_7+32x_7x_6+12x_2x_8-16x_7x_8\\ &+28x_7x_3-12x_6x_3-4x_1x_5+68x_4x_5\\ &+16x_7x_5-28x_3x_5. \qquad (4.14)\end{aligned}$$

Then the generators satisfy a system of $(p-1)/2$ differential equations subsumed by

$$\frac{12}{\mathcal{E}_j}q\frac{d}{dq}\mathcal{E}_j = \mathcal{F}_p(\langle a^j\rangle\mathcal{E}_0, \langle a^j\rangle\mathcal{E}_1, \ldots, \langle a^j\rangle\mathcal{E}_{(p-3)/2}) + pE_2(p\tau), \qquad (4.15)$$

where $0 \le j \le (p-3)/2$ and a is the minimal positive primitive root modulo p.

A differential system for a full set of generators for $\mathcal{M}(\Gamma_1(p))$ may be formulated from (4.15) and the cyclic permutation of Theorem 2.4. For example, when $p = 11$, we may apply Theorem 2.4 to deduce a differential system in terms of the polynomial $\mathcal{F}_{11}$ defined by (4.11)

$$\frac{12}{\mathcal{E}_0}q\frac{d}{dq}\mathcal{E}_0 = \mathcal{F}_{11}(\mathcal{E}_0, \mathcal{E}_1, \mathcal{E}_2, \mathcal{E}_3, \mathcal{E}_4) + 11E_2(q^{11}), \qquad (4.16)$$

$$\frac{12}{\mathcal{E}_1}q\frac{d}{dq}\mathcal{E}_1 = \mathcal{F}_{11}(\mathcal{E}_1, -\mathcal{E}_2, \mathcal{E}_3, -\mathcal{E}_4, -\mathcal{E}_0) + 11E_2(q^{11}), \qquad (4.17)$$

$$\frac{12}{\mathcal{E}_2}q\frac{d}{dq}\mathcal{E}_2 = \mathcal{F}_{11}(-\mathcal{E}_2, -\mathcal{E}_3, -\mathcal{E}_4, \mathcal{E}_0, -\mathcal{E}_1) + 11E_2(q^{11}), \qquad (4.18)$$

$$\frac{12}{\mathcal{E}_3}q\frac{d}{dq}\mathcal{E}_3 = \mathcal{F}_{11}(-\mathcal{E}_3, \mathcal{E}_4, \mathcal{E}_0, \mathcal{E}_1, \mathcal{E}_2) + 11E_2(q^{11}), \qquad (4.19)$$

$$\frac{12}{\mathcal{E}_4}q\frac{d}{dq}\mathcal{E}_4 = \mathcal{F}_{11}(\mathcal{E}_4, -\mathcal{E}_0, \mathcal{E}_1, -\mathcal{E}_2, \mathcal{E}_3) + 11E_2(q^{11}). \qquad (4.20)$$

In each case, the differential equation for $\mathcal{E}_0$ may be derived from the q-expansion for $\mathcal{E}_0'/\mathcal{E}_0$ and from the terms of $\mathcal{F}_p$ comprising a basis for $\mathcal{M}_2(\Gamma_1(p))$. The remaining equations are induced by the action of $(\mathbb{Z}/p\mathbb{Z})^*$. A differential equation for $E_2(q^p)$ may be formulated from (1.30) by expressing $E_4(q^p)$ in terms of generators for $\mathcal{M}(\Gamma_1(p))$.

References

[1] S. Ahlgren, The theta-operator and the divisors of modular forms on genus zero subgroups, *Math. Res. Lett.* **10**(5–6) (2003) 787–798.

[2] B. C. Berndt and K. Ono, Ramanujan's unpublished manuscript on the partition and tau functions with proofs and commentary, *Sém. Lothar. Combin.* **42** (1999) Art. B42c, 63 pp. (electronic). The Andrews Festschrift (Maratea, 1998).

[3] L. A. Borisov and P. E. Gunnells, Toric modular forms of higher weight, *J. Reine Angew. Math.* **560** (2003) 43–64.

[4] K. Buzzard, Computing weight one modular forms over $\mathbb{C}$ and $\overline{\mathbb{F}}_p$, in *Computations with Modular Forms*, Contributions in Mathematical and Compututational Sciences, Vol. 6 (Springer, Cham, 2014), pp. 129–146.

[5] R. Charles, T. Huber and A. Mendoza, Parameterizations for quintic Eisenstein series, *J. Number Theory* **133**(1) (2013) 195–214.

[6] S. Cooper and P. C. Toh, Quintic and septic Eisenstein series, *Ramanujan J.* **19**(2) (2009) 163–181.

[7] P. Deligne and M. Rapoport, Les schémas de modules de courbes elliptiques, in *Modular Functions of One Variable, II*, Lecture Notes in Mathematics, Vol. 349 (Springer, Berlin, 1973), pp. 143–316.

[8] F. Diamond and J. Shurman, *A First Course in Modular Forms*, Graduate Texts in Mathematics, Vol. 228 (Springer, New York, 2005).

[9] W. Duke, Continued fractions and modular functions, *Bull. Amer. Math. Soc. (N.S.)* **42**(2) (2005) 137–162 (electronic).

[10] H. M. Farkas and I. Kra, *Theta Constants, Riemann Surfaces and the Modular Group*, Graduate Studies in Mathematics, Vol. 37 (American Mathematical Society, Providence, RI, 2001). An introduction with applications to uniformization theorems, partition identities and combinatorial number theory.

[11] F. Garvan, D. Kim and D. Stanton, Cranks and t-cores, *Invent. Math.* **101**(1) (1990) 1–17.

[12] H. Hahn, Eisenstein series associated with $\Gamma_0(2)$, *Ramanujan J.* 15 (2008) 235–257.

[13] T. Huber, Coupled systems of differential equations for modular forms of level n, in *Ramanujan Rediscovered: Proceedings of a Conference on Elliptic Functions, Partitions, and q-Series in memory of K. Venkatachaliengar*, Bangalore, 1–5 June, 2009 (The Ramanujan Mathematical Society, Banglore, 2009), pp. 139–146.

[14] T. Huber, Differential equations for cubic theta functions, *Int. J. Number Theory* **7**(7) (2011) 1945–1957.

[15] T. Huber, A theory of theta functions to the quintic base, *J. Number Theory* **134** (2014) 49–92.

[16] T. Huber and D. Lara, Differential equations for septic theta functions, *Ramanujan J.* **38** (2015) 211. https://doi.org/10.1007/s11139-014-9588-1.

[17] K. Khuri-Makdisi, Moduli interpretation of Eisenstein series, *Int. J. Number Theory* **8**(3) (2012) 715–748.

[18] F. Klein, Über die transformation elfter ordnung der elliptischen functionen, in *Gesammelte Mathematische Abhandlungen III* (J. Springer, Berlin, 1923), pp. 140–165.

[19] F. Klein, *Lectures on the Icosahedron and the Solution of Equations of the Fifth Degree*, rev. edn. (Dover Publications, Inc., New York, 1956). Translated into English by George Gavin Morrice.

[20] F. Klein, *Vorlesungen über das Ikosaeder und die Auflösung der Gleichungen vom fünften Grade* (Birkhäuser Verlag, Basel, 1993). Reprint of the 1884 original, edited, with an introduction and commentary by Peter Slodowy.

[21] F. Klein, On the order-seven transformation of elliptic functions, in *The Eightfold Way*, Mathematical Sciences Research Institude Publications, Vol. 35 (Cambridge University Press, Cambridge, 1999), pp. 287–331. Translated from the German and with an introduction by Silvio Levy.

[22] M. Knopp, *Modular Functions in Analytic Number Theory* (Markham Publishing Co., Chicago, IL, 1970).

[23] Z.-G. Liu, On certain identities of Ramanujan, *J. Number Theory* **83**(1) (2000) 59–75.

[24] R. Maier, Nonlinear differential equations satisfied by certain classical modular forms, *Manuscripta Math.* (2010) 1–42.

[25] S. Ramanujan, *Notebooks*, Vols. 1 and 2 (Tata Institute of Fundamental Research, Bombay, 1957).

[26] S. Ramanujan, On certain arithmetical functions [*Trans. Cambridge Philos. Soc.* **22**(9) (1916) 159–184], in *Collected Papers of Srinivasa Ramanujan* (AMS Chelsea Publication, Providence, RI, 2000), pp. 136–162.

[27] S. Ramanujan, Some properties of $p(n)$, the number of partitions of n [*Proc. Cambridge Philos. Soc.* **19** (1919) 207–210], in *Collected Papers of Srinivasa Ramanujan* (AMS Chelsea Publication, Providence, RI, 2000), pp. 210–213.

[28] N. Rustom, Generators of graded rings of modular forms, *J. Number Theory* **138** (2014) 97–118.

[29] W. Stein, *Modular Forms: A Computational Approach*, Graduate Studies in Mathematics, Vol. 79 (American Mathematical Society, Providence, RI, 2007). With an appendix by Paul E. Gunnells.

[30] E. T. Whittaker and G. N. Watson, *A Course of Modern Analysis. An Introduction to the General Theory of Infinite Processes and of Analytic Functions: With an Account of the Principal Transcendental Functions*, 4th edn. (Reprinted) (Cambridge University Press, New York, 1962).

[31] D. Zagier, Introduction to modular forms, in *From Number Theory to Physics* (Springer, Berlin, 1992), pp. 238–291.

Chapter 18

Some Smallest Parts Functions from Variations of Bailey's Lemma

Chris Jennings-Shaffer

Department of Mathematics, Oregon State University
Corvallis, OR 97331, USA
jennichr@math.oregonstate.edu

We construct new smallest parts partition functions and smallest parts crank functions by considering variations of Bailey's lemma and conjugate Bailey pairs. The functions we introduce satisfy simple linear congruences modulo 3 and 5. We introduce and give identities for two four variable q-hypergeometric functions; these functions specialize to some of our new spt-crank-type functions as well as many known spt-crank-type functions.

Keywords: Number theory; partitions; Bailey's lemma; Bailey pairs; conjugate Bailey pairs; congruences; ranks; cranks.

Mathematics Subject Classification 2010: 11P81, 11P83

1. Introduction

In the recent study of ranks and cranks for smallest parts partition it has become apparent that Bailey pairs and Bailey's lemma are inherent to this study. One can review the articles [9, 17, 21, 26] to see this is the case. We will demonstrate another method in which this occurs. We recall a partition of an integer n is a nonincreasing sequence of positive integers that sum to n. We let $p(n)$ denote the number of partitions of n; as an example $p(5) = 7$ since the partitions of 5 are 5, $4+1$, $3+2$, $3+1+1$, $2+2+1$, $2+1+1+1$, and $1+1+1+1+1$. There are many functions related to $p(n)$, one of them is spt (n), the smallest parts partition function. The smallest parts partition function was introduced by Andrews in [6] as a weighted count on the partitions of n, by counting each partition by the number of times the smallest part appears. From the partitions of 5 we see that spt $(5) = 14$.

Both $p(n)$ and spt (n) satisfy certain linear congruences, as do many functions related to partitions. In particular the partition function satisfies $p(5n+4) \equiv 0 \pmod 5$, $p(7n+5) \equiv 0 \pmod 7$, and $p(11n+6) \equiv 0 \pmod{11}$ and the smallest parts function satisfies $\operatorname{spt}(5n+4) \equiv 0 \pmod 5$, $\operatorname{spt}(7n+5) \equiv 0 \pmod 7$, and $\operatorname{spt}(13n+6) \equiv 0 \pmod{13}$. The congruences for $p(n)$ were first observed by Ramanujan and now have a plethora of proofs, a few of which can be found in [10, 12, 18, 27]. The congruences for spt (n) were first established by Andrews when he introduced the function in [6]. Focusing on spt (n), we note that Andrews original congruences were met with a storm of articles establishing various facts about the smallest parts function and new congruences. To name just a few of these congruences, in [25] Ono established for $\ell \geq 5$ prime and $\left(\frac{-\delta}{\ell}\right) = 1$ that

$$\operatorname{spt}\left(\frac{\ell^2(\ell n+\delta)+1}{24}\right) \equiv 0 \pmod{\ell},$$

in [19] Garvan established for $a, b, c \geq 1$ and with δ_a, λ_b, γ_c respectively denoting the least nonnegative reciprocals of 24 modulo 5^a, 7^b, and 13^c that

$$\begin{aligned}
\operatorname{spt}(5^a n+\delta_a) &\equiv 0 \pmod{5^{\lfloor \frac{a+1}{2} \rfloor}},\\
\operatorname{spt}(7^b n+\lambda_b) &\equiv 0 \pmod{7^{\lfloor \frac{b+1}{2} \rfloor}},\\
\operatorname{spt}(13^c n+\gamma_c) &\equiv 0 \pmod{13^{\lfloor \frac{c+1}{2} \rfloor}},
\end{aligned}$$

and in [1] Ahlgren, Bringmann, and Lovejoy established for $\ell \geq 5$ prime, $m \geq 1$, and $\left(\frac{-n}{\ell}\right) = 1$ that

$$\operatorname{spt}\left(\frac{\ell^2 n+1}{24}\right) \equiv 0 \pmod{\ell^m}.$$

The smallest parts function is of interest not just for its congruences and elegant combinatorial description, but also for its modular properties. While the generating function for $p(n)$ is a modular form of weight $-1/2$, the generating function for spt (n) is instead one of two pieces of the so-called holomorphic part of a weight $3/2$ harmonic Maass form [13, 25]. Maass forms are of great recent interest in number theory. One application of

Maass forms is that they give new explanations of Ramanujan's mock theta conjectures [2, 16]. Other smallest parts functions were also found to be related to Maass forms in [14].

Here we are interested in studying a wide array of functions whose generating functions have a similar form to that of $\operatorname{spt}(n)$. We further restrict our attention to those that satisfy simple linear congruences, like $\operatorname{spt}(5n+4) \equiv 0 \pmod 5$, that can be explained by a so-called spt-crank. The first spt-crank was given by Andrews, Garvan, and Liang in [9], however the idea of using partition statistics to explain partition congruences originated with Dyson's rank conjectures [10, 15]. This topic experienced a resurgence beginning with Garvan's vector crank [18] and the Andrews–Garvan crank [8]. The idea of a rank or crank is to define a statistic that yields a refinement of a congruence. This is best illustrated with an example. The Dyson rank of a partition is defined as the largest part minus the number of parts. The partition function satisfies the congruence $p(5n+4) \equiv 0 \pmod 5$. If one groups the partitions of $5n+4$ according to the value of their rank reduced modulo 5, then one has five sets of equal size. This can be seen with the partitions of 4 in the following table:

partition	rank	rank (mod 5)
4	3	3
$3+1$	1	1
$2+2$	0	0
$2+1+1$	-1	4
$1+1+1+1$	-3	2

For a given partition like function with a congruence, one can attempt to find a statistic to explain the congruence. However, the statistic may not have such an elegant and simple definition as the Dyson rank of a partition.

The goal of this chapter is to demonstrate a method by which we can find and introduce new partitions functions while simultaneously obtaining a crank function for them. From this method we select those functions that satisfy simple linear congruences and by which the crank gives a proof of the congruences. In Section 2 we introduce these new functions, state various identities and congruences, and prove a few preliminary results. In Section 3 we describe with an example our method for finding these new functions. In Sections 4 and 5 we prove the results stated in Section 2. Finally in Section 6 we give a few concluding remarks.

2. Preliminaries and Statement of Results

Throughout this chapter we use the standard product notation,

$$\begin{aligned}(z;q)_n &= \prod_{j=0}^{n-1}(1-zq^j), \quad (z;q)_\infty = \prod_{j=0}^{\infty}(1-zq^j),\\ (z_1,\ldots,z_k;q)_n &= (z_1;q)_n\ldots(z_k;q)_n\,,\\ (z_1,\ldots,z_k;q)_\infty &= (z_1;q)_\infty\ldots(z_k;q)_\infty\,.\end{aligned}$$

To begin we define we define two generic functions,

$$F(\rho_1,\rho_2,z;q) = \frac{(q;q)_\infty}{(z,z^{-1},\rho_1,\rho_2;q)_\infty}\sum_{n=1}^{\infty}\frac{\left(z,z^{-1},\rho_1,\rho_2;q\right)_n\left(\frac{q}{\rho_1\rho_2}\right)^n}{(q;q)_{2n}},$$

$$G(\rho_1,\rho_2,z;q) = \frac{(q;q)_\infty}{(z,z^{-1},\rho_1,\rho_2;q)_\infty}\sum_{n=1}^{\infty}\frac{\left(z,z^{-1},\rho_1,\rho_2;q\right)_n\left(\frac{q^2}{\rho_1\rho_2}\right)^n}{(q;q)_{2n}}.$$

We would also like to let $\rho_2\to\infty$ in $F(\rho_1,\rho_2,z;q)$ and $G(\rho_1,\rho_2,z;q)$, however this requires a slight alteration. In particular we let

$$\begin{aligned}F(\rho,z;q) &= \lim_{\rho_2\to\infty}(\rho_2;q)_\infty F(\rho,\rho_2,z,q)\\ &= \frac{(q;q)_\infty}{(\rho,z,z^{-1};q)_\infty}\sum_{n=1}^{\infty}\frac{\left(z,z^{-1},\rho;q\right)_n(-1)^n q^{\frac{n(n+1)}{2}}\rho^{-n}}{(q;q)_{2n}},\\ G(\rho,z;q) &= \lim_{\rho_2\to\infty}(\rho_2;q)_\infty G(\rho,\rho_2,z,q)\\ &= \frac{(q;q)_\infty}{(\rho,z,z^{-1};q)_\infty}\sum_{n=1}^{\infty}\frac{\left(z,z^{-1},\rho;q\right)_n(-1)^n q^{\frac{n(n+3)}{2}}\rho^{-n}}{(q;q)_{2n}}.\end{aligned}$$

The special cases of these functions we are interested in are

$$\begin{aligned}S_{G1}(z,q) &= G(q,q^2,z;q^2)\\ &= \frac{(q^2;q^2)_\infty}{(z,z^{-1},q,q^2;q^2)_\infty}\sum_{n=1}^{\infty}\frac{\left(z,z^{-1},q,q^2;q^2\right)_n q^n}{(q^2;q^2)_{2n}},\end{aligned}$$

$$\begin{aligned}
S_{G2}(z,q) &= G(iq^{1/2},-iq^{1/2},z;q)\\
&= \frac{(q;q)_\infty}{(z,z^{-1};q)_\infty(-q;q^2)_\infty}\sum_{n=1}^{\infty}\frac{(z,z^{-1};q)_n(-q;q^2)_n q^n}{(q;q)_{2n}},\\
S_{F1}(z,q) &= F(-q,z;q)\\
&= \frac{(q;q)_\infty}{(z,z^{-1},-q;q)_\infty}\sum_{n=1}^{\infty}\frac{(z,z^{-1},-q;q)_n q^{\frac{n(n-1)}{2}}}{(q;q)_{2n}},\\
S_{G3}(z,q) &= G(q,z;q)\\
&= \frac{(q;q)_\infty}{(z,z^{-1},q;q)_\infty}\sum_{n=1}^{\infty}\frac{(z,z^{-1},q;q)_n(-1)^n q^{\frac{n(n+1)}{2}}}{(q;q)_{2n}}.
\end{aligned}$$

Additionally, we define three functions of a similar form that we will find are related to the conjugate Bailey pair identities (1.7), (1.9), and (1.12) of [23],

$$\begin{aligned}
S_{L7}(z;q) &= \frac{(-q;q)_\infty}{(z,z^{-1};q^2)_\infty}\sum_{n=1}^{\infty}\frac{(z,z^{-1};q^2)_n q^{2n}}{(-q;q)_{2n}},\\
S_{L9}(z;q) &= \frac{(q;q^2)_\infty}{(z,z^{-1};q)_\infty}\sum_{n=1}^{\infty}\frac{(z,z^{-1};q)_n q^n}{(q;q^2)_n},\\
S_{L12}(z;q) &= \frac{(-q;q^2)_\infty^2}{(z,z^{-1};q^2)_\infty}\sum_{n=1}^{\infty}\frac{(z,z^{-1};q^2)_n q^{2n}}{(-q;q^2)_n(-q;q^2)_{n+1}}.
\end{aligned}$$

These seven functions are our spt-crank functions. By setting $z=1$ and simplifying the products, we obtain our smallest parts functions.

$$\begin{aligned}
S_{G1}(q) &= \sum_{n=1}^{\infty}\mathrm{spt}_{G1}(n)\,q^n = \sum_{n=1}^{\infty}\frac{q^n(q^{4n+2};q^2)_\infty}{(q^{2n},q^{2n},q^{2n+1},q^{2n+2};q^2)_\infty}\\
&= \sum_{n=1}^{\infty}\frac{q^n}{(1-q^{2n})^2(q^{2n+1};q)_\infty}\cdot\frac{1}{(q^{2n+2};q^2)_\infty}\cdot\frac{1}{(q^{2n+2};q^2)_n},\\
S_{G2}(q) &= \sum_{n=1}^{\infty}\mathrm{spt}_{G2}(n)\,q^n = \sum_{n=1}^{\infty}\frac{q^n(q^{2n+1};q)_\infty}{(q^n;q)_\infty^2(-q^{2n+1};q^2)_\infty}\\
&= \sum_{n=1}^{\infty}\frac{q^n}{(1-q^n)^2(q^{n+1};q)_n(q^{2n+2};q^2)_\infty}\cdot\frac{1}{(q^{n+1};q)_n(q^{4n+2};q^4)_\infty},
\end{aligned}$$

$$S_{F1}(q) = \sum_{n=1}^{\infty} \mathrm{spt}_{F1}(n)\, q^n = \sum_{n=1}^{\infty} \frac{q^{\frac{n(n-1)}{2}} \left(q^{2n+1};q\right)_\infty}{(q^n, q^n, -q^{n+1};q)_\infty}$$

$$= \frac{1}{(1-q)^2 \left(q^2;q^2\right)_\infty} + \frac{q}{(1-q^2)^2(1-q^3)\left(q^4;q^2\right)_\infty}$$

$$+ \sum_{n=3}^{\infty} \frac{q^n}{(1-q^n)^2 \left(q^{n+1};q\right)_n \left(q^{2n+2};q^2\right)_\infty} \cdot q^{\frac{n(n-3)}{2}},$$

$$S_{G3}(q) = \sum_{n=1}^{\infty} \mathrm{spt}_{G3}(n)\, q^n = \sum_{n=1}^{\infty} \frac{(-1)^n q^{\frac{n(n+1)}{2}} \left(q^{2n+1};q\right)_\infty}{(q^n, q^n, q^{n+1};q)_\infty}$$

$$= \sum_{n=1}^{\infty} \frac{(-1)^n q^n}{(1-q^n)^2 \left(q^{n+1};q\right)_\infty} \cdot \frac{q^{\frac{n(n-1)}{2}}}{\left(q^{n+1};q\right)_\infty} \cdot \frac{1}{\left(q^{n+1};q\right)_n},$$

$$S_{L7}(q) = \sum_{n=1}^{\infty} \mathrm{spt}_{L7}(n)\, q^n = \sum_{n=1}^{\infty} \frac{q^{2n}\left(-q^{2n+1};q\right)_\infty}{\left(q^{2n};q^2\right)_\infty^2}$$

$$= \sum_{n=1}^{\infty} \frac{q^{2n}}{(1-q^{2n})^2 \left(q^{2n+2};q^2\right)_\infty} \cdot \frac{\left(-q^{2n+1};q\right)_\infty}{\left(q^{2n+2};q^2\right)_\infty},$$

$$S_{L9}(q) = \sum_{n=1}^{\infty} \mathrm{spt}_{L9}(n)\, q^n = \sum_{n=1}^{\infty} \frac{q^n\left(q^{2n+1};q^2\right)_\infty}{\left(q^n;q\right)_\infty^2}$$

$$= \sum_{n=1}^{\infty} \frac{q^n}{(1-q^n)^2 \left(q^{n+1};q\right)_\infty} \cdot \frac{1}{\left(q^{n+1};q\right)_n \left(q^{2n+2};q^2\right)_\infty},$$

$$S_{L12}(q) = \sum_{n=1}^{\infty} \mathrm{spt}_{L12}(n)\, q^n = \sum_{n=1}^{\infty} \frac{q^{2n}\left(-q^{2n+1}, -q^{2n+3};q^2\right)_\infty}{\left(q^{2n};q^2\right)_\infty^2}$$

$$= \sum_{n=1}^{\infty} \frac{q^{2n}\left(-q^{2n+1};q^2\right)_\infty}{(1-q^{2n})^2 \left(q^{2n+2};q^2\right)_\infty} \cdot \frac{\left(-q^{2n+3};q^2\right)_\infty}{\left(q^{2n+2};q^2\right)_\infty}.$$

We now give the combinatorial interpretations of these functions. We recall that an overpartition is a partition where the first occurrence of a part may be overlined. For example the overpartitions of 3 are 3, $\overline{3}$, $2+1$, $2+\overline{1}$, $\overline{2}+1$, $\overline{2}+\overline{1}$, $1+1+1$, and $\overline{1}+1+1$. We say a vector $(\pi_1,\ldots,\pi_k)$, where each π_i is a partition or overpartition, is a vector partition of n if altogether the parts of the π_i sum to n. For a partition, or overpartition, π we let $s(\pi)$ denote the smallest part of a π, $\mathrm{spt}(\pi)$ the number of times $s(\pi)$ appears, and $\ell(\pi)$ the largest part of π. We use the convention that the empty partition has smallest part ∞ and largest part 0. Rather than interpret these functions in the order of their definitions, we begin with the simplest.

We see $\mathrm{spt}_{L7}(n)$ is the number of occurrences of the smallest part in the partition pairs (π_1, π_2) of n, where π_1 is a partition into even parts, π_2 is an overpartition with all nonoverlined parts even, and $s(\pi_1) < s(\pi_2)$. We see $\mathrm{spt}_{L9}(n)$ is the number of occurrences of the smallest part in the partition pairs (π_1, π_2) of n, where $s(\pi_1) < s(\pi_2)$ and all parts of π_2 larger than $2s(\pi_2)$ must be even. We see $\mathrm{spt}_{L12}(n)$ is the number of occurrences of the smallest part in the partition pairs (π_1, π_2) of n, where the odd parts of π_1 do not repeat, the odd parts of π_2 do not repeat, $s(\pi_1)$ is even, $s(\pi_1) < s(\pi_2)$, and the odd parts of π_2 are at least $s(\pi_1)+3$. We see $\mathrm{spt}_{G2}(n)$ is the number of occurrences of the smallest part in the partition pairs (π_1, π_2) of n, where $s(\pi_1) < s(\pi_2)$, the parts of π_1 larger than $2s(\pi_1)$ must be even, the parts of π_2 larger than $2s(\pi_1)$ must be 2 modulo 4, and π_2 has no parts in the interval $(2s(\pi_1), 4s(\pi_1)+2)$.

To interpret $\mathrm{spt}_{G1}(n)$, we first note that

$$\frac{q^n}{(1-q^{2n})^2} = q^n + 2q^{3n} + 3q^{5n} + 4q^{7n} + \cdots .$$

We see $\mathrm{spt}_{G1}(n)$ is a weighted count on certain partition triples (π_1, π_2, π_3) of n. Here the restrictions are $\mathrm{spt}(\pi_1)$ is odd, π_1 has no parts in the interval $(s(\pi_1), 2s(\pi_1)+1)$, π_2 and π_3 are partitions with only even parts, $2s(\pi_1) < s(\pi_2)$, $2s(\pi_1) < s(\pi_3)$, and $\ell(\pi_3) \le 4s(\pi_1)$. These partitions tripled are weighted by $\frac{\mathrm{spt}(\pi_1)+1}{2}$, rather than by just $\mathrm{spt}(\pi_1)$.

We see $\mathrm{spt}_{G3}(n)$ is a weighted count on certain partition triples (π_1, π_2, π_3) of n. Here the restrictions are π_2 is a partition where the parts $1, 2, \ldots, s(\pi_1)-1$ appear exactly once and $s(\pi_1)$ does not appear as a part of π_2 (with the understanding that this only means $s(\pi_2) \ge 2$ when $s(\pi_1) = 1$), $s(\pi_1) < s(\pi_3)$, and $\ell(\pi_3) \le 2s(\pi_1)$. These partition triples are weighted by $(-1)^{s(\pi_1)}\mathrm{spt}(\pi_1)$.

We view $\mathrm{spt}_{F1}(n)$ as a sum of three functions. The first is a weighted count on the partitions π of n, where 1 may appear as a part but all other parts are even. For this first function these partitions are weighted by one more than the number of times the part 1 appears. The second function is a weighted count on the partitions π of n, where 1 and 3 may appear as parts but all other parts are even and larger than 2, and 1 must appear an odd number of times. For this second function these partitions are weighted by $\frac{\mathrm{spt}(\pi)+1}{2}$, and we note $\mathrm{spt}(\pi)$ is the number of ones. The third function is a weighted count on the number of partition pairs (π_1, π_2) of n, where $s(\pi_1) \geq 3$, the parts of π_1 larger than $2s(\pi_1)$ must be even, and π_2 consists of exactly one copy of $2, 3, \ldots, s(\pi_1) - 2$ (with the understanding that π_2 is the empty partition when $s(\pi_1) = 3$). For this third function these partition pairs are weighted by $\mathrm{spt}(\pi_1)$.

Theorem 2.1. *For $n \geq 0$, we have the following congruences:*

$$\begin{aligned}
\mathrm{spt}_{G1}(3n) &\equiv 0 \pmod{3},\\
\mathrm{spt}_{G2}(3n) &\equiv 0 \pmod{3},\\
\mathrm{spt}_{G2}(3n+2) &\equiv 0 \pmod{3},\\
\mathrm{spt}_{L7}(3n+1) &\equiv 0 \pmod{3},\\
\mathrm{spt}_{L9}(3n+2) &\equiv 0 \pmod{3},\\
\mathrm{spt}_{L12}(3n) &\equiv 0 \pmod{3},\\
\mathrm{spt}_{F1}(10n+9) &\equiv 0 \pmod{5},\\
\mathrm{spt}_{G3}(5n+3) &\equiv 0 \pmod{5}.
\end{aligned}$$

To prove the congruences of Theorem 2.1 we will prove certain identities for the spt-crank two variable series. To explain this, for $X = L_i, F_1, G_i$ we write

$$S_X(z,q) = \sum_{n=1}^{\infty} \sum_{m=-\infty}^{\infty} M_X(m,n) z^m q^n.$$

Whatever each $M_X(m,n)$ is counting is the spt-crank associated to the function $\mathrm{spt}_X(n)$. We define the additional functions

$$M_X(k,t,n) = \sum_{m \equiv k \pmod{t}} M_X(m,n).$$

For now we consider just $S_{L7}(z,q)$, as the explanations for the other six functions are identical. Since $S_{L7}(q) = S_{L7}(1,q)$, we have that

$$\mathrm{spt}_{L7}(n) = \sum_{k=0}^{t-1} M_{L7}(k,t,n).$$

Next with ζ_t a tth root of unity, we have

$$S_{L7}(\zeta_t, q) = \sum_{n=1}^{\infty} \left(\sum_{k=0}^{t-1} M_{L7}(k,t,n) \zeta_t^k \right) q^n.$$

When t is prime and ζ_t is primitive, the minimal polynomial for ζ_t is $1+x+x^2+\cdots+x^{t-1}$. So if the coefficient of q^N in $S_{L7}(\zeta_t,q)$ is zero, then

$$\begin{aligned}M_{L7}(0,t,N) &= M_{L7}(1,t,N) = M_{L7}(2,t,N)\\ &= \cdots = M_{L7}(t-1,t,N) = \frac{1}{t}\operatorname{spt}_{L7}(N).\end{aligned}$$

Since the $M_{L7}(k,t,n)$ are integers, we clearly have $\operatorname{spt}_{L7}(N)\equiv 0 \pmod t$.

That is to say, one way to prove $\operatorname{spt}_{L7}(3n+1)\equiv 0 \pmod 3$ is to instead prove the stronger result that $M_{L7}(0,3,3n+1) = M_{L7}(1,3,3n+1) = M_{L7}(2,3,3n+1)$. We show these values of the spt-crank are equal by showing the coefficient of q^{3n+1} in $S_{L7}(\zeta_3,q)$ is zero. In Section 4 we prove that the coefficients of q^{3n+1} in $S_{L7}(\zeta_3,q)$, q^{3n+2} in $S_{L9}(\zeta_3,q)$, q^{3n} in $S_{L12}(\zeta_3,q)$, q^{3n} in $S_{G1}(\zeta_3,q)$, q^{3n} in $S_{G2}(\zeta_3,q)$, q^{3n+2} in $S_{G2}(\zeta_3,q)$, q^{10n+9} in $S_{F1}(\zeta_5,q)$, and q^{5n+3} in $S_{G3}(\zeta_5,q)$ are all zero. This establishes the following theorem and Theorem 2.1 as a corollary.

Theorem 2.2. *For $n\ge 0$, the spt-cranks satisfy the following equalities:*

$$\begin{aligned}M_{L7}(0,3,3n+1) &= M_{L7}(1,3,3n+1) = M_{L7}(2,3,3n+1),\\ M_{L9}(0,3,3n+2) &= M_{L9}(1,3,3n+2) = M_{L9}(2,3,3n+2),\\ M_{L12}(0,3,3n) &= M_{L12}(1,3,3n) = M_{L12}(2,3,3n),\\ M_{G1}(0,3,3n) &= M_{G1}(1,3,3n) = M_{G1}(2,3,3n),\\ M_{G2}(0,3,3n) &= M_{G2}(1,3,3n) = M_{G2}(2,3,3n),\\ M_{G2}(0,3,3n+2) &= M_{G2}(1,3,3n+2) = M_{G2}(2,3,3n+2),\\ M_{F1}(0,5,10n+9) &= M_{F1}(1,5,10n+9) = M_{F1}(2,5,10n+9)\\ &= M_{F1}(3,5,10n+9) = M_{F1}(4,5,10n+9),\\ M_{G3}(0,5,5n+3) &= M_{G3}(1,5,5n+3) = M_{G3}(2,5,5n+3)\\ &= M_{G3}(3,5,5n+3) = M_{G3}(4,5,5n+3).\end{aligned}$$

The main tools to prove Theorem 2.2 are the following identities. We note these are identities for all values of z, not just for z being a specific root of unity.

Theorem 2.3. *The spt-crank generating functions can be expressed as the following series:*

$$\begin{aligned}S_{L7}(z;q) = \frac{1}{(1+z)\left(q^2,z,z^{-1};q^2\right)_\infty}&\sum_{j=1}^{\infty}\sum_{n=0}^{\infty}(1-z^j)(1-z^{j-1})\\ &\times z^{1-j}(-1)^{j+1}(1+q^{2j-1})q^{j(j-1)+\frac{n(n-1)}{2}+2jn},\end{aligned} \tag{2.1}$$

$$S_{L9}(z;q) = \frac{1}{(1+z)(q,z,z^{-1};q)_\infty} \times \sum_{j=1}^{\infty}\sum_{n=0}^{\infty}(1-z^j)(1-z^{j-1})z^{1-j}(-1)^{j+n+1} \times (1-q^{2j-1})q^{\frac{j(j-1)}{2}+n^2+2jn}, \tag{2.2}$$

$$S_{L12}(z;q) = \frac{1}{(1+z)\,(q^2,z,z^{-1};q^2)_\infty} \times \sum_{j=1}^{\infty}\sum_{n=0}^{\infty}(1-z^j)(1-z^{j-1})z^{1-j}(-1)^{j+1} \times (1+q^{2j-1})q^{j(j-1)+n^2+2jn}, \tag{2.3}$$

$$F(\rho_1,\rho_2,z;q) = \sum_{j=1}^{\infty} \frac{(1-z^j)(1-z^{j-1})z^{1-j}(-1)^{j+1}q^{j(j-1)/2} \times (\rho_1,\rho_2;q)_{j-1}\left(\frac{q^{j+1}}{\rho_1},\frac{q^{j+1}}{\rho_2};q\right)_\infty}{(1+z)\rho_1^{j-1}\rho_2^{j-1}(z,z^{-1},\rho_1,\rho_2,\frac{q}{\rho_1\rho_2};q)_\infty} \times \left(1-\frac{q^j}{\rho_1}-\frac{q^j}{\rho_2}+\frac{q^{3j-1}}{\rho_1}+\frac{q^{3j-1}}{\rho_2}-q^{4j-2}\right), \tag{2.4}$$

$$G(\rho_1,\rho_2,z;q) = \sum_{j=1}^{\infty} \frac{(1-z^j)(1-z^{j-1})z^{1-j}(-1)^{j+1}(1-q^{2j-1}) \times q^{\frac{j(j+1)}{2}-1}(\rho_1,\rho_2;q)_{j-1}\left(\frac{q^{j+1}}{\rho_1},\frac{q^{j+1}}{\rho_2};q\right)_\infty}{(1+z)\rho_1^{j-1}\rho_2^{j-1}(z,z^{-1},\rho_1,\rho_2,\frac{q^2}{\rho_1\rho_2};q)_\infty}, \tag{2.5}$$

$$F(\rho,z;q) = \sum_{j=1}^{\infty} \frac{(1-z^j)(1-z^{j-1})z^{1-j}q^{(j-1)^2}(\rho;q)_{j-1}\left(\frac{q^{j+1}}{\rho};q\right)_\infty}{(1+z)\rho^{j-1}(z,z^{-1},\rho;q)_\infty} \times \left(1-\frac{q^j}{\rho}+\frac{q^{3j-1}}{\rho}-q^{4j-2}\right), \tag{2.6}$$

$$G(\rho,z;q) = \sum_{j=1}^{\infty} \frac{(1-z^j)(1-z^{j-1})z^{1-j}(1-q^{2j-1})q^{j(j-1)} \times (\rho;q)_{j-1}\left(\frac{q^{j+1}}{\rho};q\right)_\infty}{(1+z)\rho^{j-1}(z,z^{-1},\rho;q)_\infty}, \tag{2.7}$$

$$S_{G1}(z,q) = \frac{1}{(1+z)\,(z,z^{-1},q;q^2)_\infty} \times \sum_{j=1}^{\infty}(1-z^j)(1-z^{j-1})z^{1-j}(-1)^{j+1} \times (1+q^{2j-1})q^{(j-1)^2}, \tag{2.8}$$

$$S_{G2}(z,q) = \frac{1}{(1+z)\,(z,z^{-1},q;q)_\infty} \times \sum_{j=-\infty}^{\infty} \frac{(1-z^j)(1-z^{j-1})z^{1-j}(-1)^{j+1}q^{\frac{j(j-1)}{2}}}{(1+q^{2j-1})}, \tag{2.9}$$

$$S_{F1}(z,q) = \frac{1}{(1+z)\,(z,z^{-1};q)_\infty} \times \sum_{j=-\infty}^{\infty} (1-z^j)(1-z^{j-1})z^{1-j}(-1)^{j+1}q^{j^2-3j+2}(1+q^{j-1}) \tag{2.10}$$

$$= \frac{(z^{-1}q^2,zq^2,q^2;q^2)_\infty}{(zq,z^{-1}q;q)_\infty} + \frac{(z^{-1}q,zq,q^2;q^2)_\infty}{(z,z^{-1};q)_\infty} - \frac{(q,q,q^2;q^2)_\infty}{(z,z^{-1};q)_\infty}, \tag{2.11}$$

$$S_{G3}(z,q) = \frac{1}{(1+z)\,(z,z^{-1};q)_\infty} \times \sum_{j=1}^{\infty}(1-z^j)(1-z^{j-1})z^{1-j}(1-q^{2j-1})q^{(j-1)^2}. \tag{2.12}$$

We note (2.6) and (2.7) follow by taking limits in (2.4) and (2.5) as in the definitions of $F(\rho,z;q)$ and $G(\rho,z;q)$. It is worth pointing out that $S_{F1}(z,q)$ reducing to products tells us that $S_{F1}(z,q)$ will be a modular form when z is a root of unity. However, the other spt-cranks with single series representations do not appear to reduce to products. As such these functions will likely instead be false theta functions when z is a root of unity. The double series identities also have another interesting form.

Theorem 2.4. *The following spt-crank functions can be written as Hecke–Rogers double sums, in particular,*

$$S_{L7}(z;q) = \frac{1}{(1+z)\,(q^2,z,z^{-1};q^2)_\infty} \times \sum_{j=0}^{\infty}\sum_{n=-j}^{j} (1-z^{j-|n|+1})(1-z^{j-|n|})z^{|n|-j} \times (-1)^{j+n}q^{j(j+1)-\frac{n(n-1)}{2}}, \tag{2.13}$$

$$S_{L9}(z;q) = \frac{1}{(1+z)\,(q,z,z^{-1};q)_\infty}$$

$$\times \sum_{j=0}^{\infty} \sum_{n=-\lfloor j/2 \rfloor}^{\lfloor j/2 \rfloor} (1 - z^{j-2|n|+1})(1 - z^{j-2|n|}) z^{2|n|-j}$$
$$\times (-1)^{j+n} q^{\frac{j(j+1)}{2} - n(n-1)}, \tag{2.14}$$

$$S_{L12}(z;q) = \frac{1}{(1+z)\,(q^2, z, z^{-1}; q^2)_\infty}$$
$$\times \sum_{j=0}^{\infty} \sum_{n=-j}^{j} (1 - z^{j-|n|+1})(1 - z^{j-|n|}) z^{|n|-j} (-1)^{j+n} q^{j(j+1)+n}. \tag{2.15}$$

Hecke–Rogers double sums of this form recently arose in [20] for the Dyson rank of partitions, the Dyson rank of overpartitions, and the M_2-rank of partitions without repeated odd parts. The identities for those ranks lead to Hecke–Rogers series for certain related spt functions as well. One simple point to notice about the double series in Theorem 2.3 is that summation indices are independent, but the power of q is a quadratic in j and n with a cross term jn, whereas the double series in Theorem 2.4 have summation indices that are dependent but the power of q is a quadratic without a cross term.

Now that we have stated our results, we describe the q-series techniques we need to prove these identities. We will use Lemma 4.1 of [20], which is

$$\frac{(1+z)\left(z, z^{-1}; q\right)_n}{(q;q)_{2n}} = \sum_{j=-n}^{n+1} \frac{(-1)^{j+1}(1 - q^{2j-1}) z^j q^{\frac{j(j-3)}{2}+1}}{(q;q)_{n+j}\,(q;q)_{n-j+1}}. \tag{2.16}$$

We recall a pair of sequences (α, β) is a Bailey pair relative to (a, q) if

$$\beta_n = \sum_{k=0}^{n} \frac{\alpha_k}{(q;q)_{n-k}\,(aq;q)_{n+k}}.$$

Some authors suppress the dependence of q in the definition of a Bailey pair, however we will find this additional notation necessary. We recall a limiting case of Bailey's lemma states if (α, β) is a Bailey pair relative to (a, q) then

$$\sum_{n=0}^{\infty} (\rho_1, \rho_2; q)_n \left(\frac{aq}{\rho_1 \rho_2}\right)^n \beta_n = \frac{(aq/\rho_1, aq/\rho_2; q)_\infty}{(aq, \frac{aq}{\rho_1\rho_2}; q)_\infty}$$
$$\times \sum_{n=0}^{\infty} \frac{(\rho_1, \rho_2; q)_n \left(\frac{aq}{\rho_1\rho_2}\right)^n \alpha_n}{(aq/\rho_1, aq/\rho_2; q)_n}. \tag{2.17}$$

Bailey pairs and Bailey's lemma have a rich and varied history. They were introduced by Bailey in [11] in reproving the Rogers–Ramanujan identities and giving more identities of that type. Slater, a student of Bailey, established what is now a standard list of Bailey pairs in [29, 30] and used them to prove a massive list of Roger–Ramanujan type identities. Rather than say too much of their history, recent developments, and applications, we refer the reader to [4, Chapter 3] and the survey articles [5, 24, 31].

Lemma 2.5. *If (α, β) is a Bailey pair relative to (a, q), then*

$$\sum_{n=0}^{\infty} \left(aq;q^2\right)_n q^n \beta_n = \frac{1}{\left(aq^2;q^2\right)_\infty (q;q)_\infty} \sum_{r,n\geq 0} (-a)^n q^{n^2+2rn+r+n} \alpha_r, \tag{2.18}$$

$$\begin{aligned} &\sum_{n=0}^{\infty} (\rho_1\sqrt{a}, \rho_2\sqrt{a};q)_n \left(\frac{q}{\rho_1\rho_2}\right)^n \beta_n(a,q) \\ &\quad = \frac{(\sqrt{a}q/\rho_1, \sqrt{a}q/\rho_2;q)_\infty}{(aq, \frac{q}{\rho_1\rho_2};q)_\infty} \\ &\qquad \times \sum_{n=0}^{\infty} \frac{(\rho_1\sqrt{a}, \rho_2\sqrt{a};q)_n (\frac{q}{\rho_1\rho_2})^n \alpha_n(a,q)}{(\sqrt{a}q/\rho_1, \sqrt{a}q/\rho_2;q)_n}, \end{aligned} \tag{2.19}$$

$$\begin{aligned} &\sum_{n=0}^{\infty} (\rho_1\sqrt{a/q}, \rho_2\sqrt{a/q};q)n \left(\frac{q^2}{\rho_1\rho_2}\right)^n \beta_n(a,q) \\ &\quad = \frac{(\sqrt{a}q^{\frac{3}{2}}/\rho_1, \sqrt{a}q^{\frac{3}{2}}/\rho_2;q)_\infty}{(aq, \frac{q^2}{\rho_1\rho_2};q)_\infty} \\ &\qquad \times \sum_{n=0}^{\infty} \frac{(\rho_1\sqrt{a/q}, \rho_2\sqrt{a/q};q)_n \left(\frac{q^2}{\rho_1\rho_2}\right)^n \alpha_n(a,q)}{(\sqrt{a}q^{\frac{3}{2}}/\rho_1, \sqrt{a}q^{\frac{3}{2}}/\rho_2;q)_n}. \end{aligned} \tag{2.20}$$

If (α, β) is a Bailey pair relative to (a^2q^2, q^2), then

$$\sum_{n=0}^{\infty} (aq;q)_n q^{2n} \beta_n = \frac{(aq;q)_\infty}{\left(a^2q^4;q^2\right)_\infty} \sum_{r,n\geq 0} \frac{q^{\frac{n(n+1)}{2}+2nr+2r+n} a^n}{1-aq^{2r+1}} \alpha_r. \tag{2.21}$$

If (α, β) is a Bailey pair relative to (a^2, q), then

$$\sum_{n=0}^{\infty} \frac{\left(a^2;q\right)_{2n} q^n}{(a, aq;q)_n} \beta_n = \frac{1}{(q, aq, aq;q)_\infty} \sum_{r,n\geq 0} \frac{(-a)^n q^{\frac{n(n+1)}{2}+nr+r}(1+a)}{1+aq^r} \alpha_r. \tag{2.22}$$

Proof. Equations (2.18), (2.21), and (2.22) are exactly (1.9), (1.7), and (1.12) of [23]. We find (2.19) follows from (2.17) by letting $\rho_1 \mapsto \rho_1\sqrt{a}$ and $\rho_2 \mapsto \rho_2\sqrt{a}$ and (2.20) follows from (2.17) by letting $\rho_1 \mapsto \rho_1\sqrt{a/q}$ and $\rho_2 \mapsto \rho_2\sqrt{a/q}$. □

We only need the following two Bailey pairs relative to (a, q),

$$\beta_n^*(a,q) = \frac{1}{(aq,q;q)_n}, \quad \alpha_n^*(a,q) = \begin{cases} 1, & n = 0, \\ 0, & n \geq 1, \end{cases} \tag{2.23}$$

$$\beta_n^{**}(a,q) = \frac{1}{(aq^2,q;q)_n}, \quad \alpha_n^{**}(a,q) = \begin{cases} 1, & n = 0, \\ -aq, & n = 1, \\ 0, & n \geq 2. \end{cases} \tag{2.24}$$

That these are Bailey pairs relative to (a, q) follows immediately from the definition of a Bailey pair. We next explain how our spt-cranks were found.

3. The General Idea

Bailey's lemma comes from Bailey's transform, which states that if

$$\beta_n = \sum_{k=0}^{n} \alpha_k u_{n-k} v_{n+k}, \quad \gamma_n = \sum_{k=n}^{\infty} \delta_k u_{k-n} v_{k+n}, \tag{3.1}$$

then

$$\sum_{n=0}^{\infty} \beta_n \delta_n = \sum_{n=0}^{\infty} \alpha_n \gamma_n.$$

If in (3.1) we use $u_n = 1/(q;q)_n$ and $v_n = 1/(aq;q)_n$, then we see the condition on α and β is exactly that of being a Bailey pair. We refer to (γ, δ) as a conjugate Bailey pair. While the idea of a conjugate Bailey pair is clearly built into the theory of Bailey pairs, they were almost completely ignored until Schilling and Warnaar brought attention to them in [28]. A simple statement about conjugate Bailey pairs is that for each conjugate Bailey pair, we have a new version of Bailey's lemma.

Our method relies on applying these "new" versions of Bailey's lemma. This was born out of the proofs of [20, 21]. In [21] we defined an spt-crank-type function to be a series of the form

$$\frac{P(q)}{(z, z^{-1}; q)_\infty} \sum_{n=1}^{\infty} (z, z^{-1}; q)_n \, q^n \beta_n,$$

where $P(q)$ is some infinite product and β comes from a Bailey pair relative to $(1, q)$. In that article and others, we chose a Bailey pair of Slater [29, 30], applied Bailey's lemma to the spt-crank-type function, and obtained a generalized Lambert series and an infinite product to work with. Instead, here we take an spt-crank function to be of the form

$$S_A(z,q) = \frac{P(q)}{(z,z^{-1};q)_\infty} \sum_{n=1}^{\infty} \left(z,z^{-1};q\right)_n q^n A_n(q),$$

and do not worry about A occurring as part of a Bailey pair, we only require that it be a function of q and n. We instead ask how should we choose A so that we can transform the spt-crank function with a conjugate Bailey pair identity using one of the Bailey pairs (α^*, β^*) or $(\alpha^{**}, \beta^{**})$. This method leads to the series identities in Theorem 2.3.

Suppose one takes a fixed conjugate Bailey pair and the resulting Bailey's lemma type identity. To be explicit we take the identity (1.9) from [23],

$$\sum_{n=0}^{\infty} \left(aq;q^2\right)_n q^n \beta_n = \frac{1}{\left(aq^2;q^2\right)_\infty (q;q)_\infty} \sum_{r,n\geq 0} (-a)^n q^{n^2+2rn+r+n} \alpha_r, \quad (3.2)$$

where (α, β) is a Bailey pair relative to (a, q). First we apply (2.16) to $S_A(z, q)$ to find that

$$(1+z)\left(z,z^{-1};q\right)_\infty S_A(z,q) = P(q) \sum_{n=1}^{\infty} A_n(q) q^n (q;q)_{2n} \times \sum_{j=-n}^{n+1} \frac{(-1)^{j+1}(1-q^{2j-1}) z^j q^{\frac{j(j-3)}{2}+1}}{(q;q)_{n+j}(q;q)_{n-j+1}}.$$

We find that the coefficients of z^j and z^{1-j} in $(1+z)\left(z,z^{-1};q\right)_\infty S_A(z,q)$ are equal, due to $S_A(z,q)$ being symmetric in z and z^{-1}. For $j \geq 2$ we find the coefficient of z^j in $(1+z)\left(z,z^{-1};q\right)_\infty S_A(z,q)$ is given by

$$P(q)(-1)^{j+1}(1-q^{2j-1}) q^{\frac{j(j-3)}{2}+1} \sum_{n=j-1}^{\infty} \frac{A_n(q) q^n (q;q)_{2n}}{(q;q)_{n+j}(q;q)_{n-j+1}}$$

$$= P(q)(-1)^{j+1}(1-q^{2j-1}) q^{\frac{j(j-3)}{2}+1} \sum_{n=0}^{\infty} \frac{A_{n+j-1}(q) q^{n+j-1} (q;q)_{2n+2j-2}}{(q;q)_{n+2j-1}(q;q)_n}$$

$$= P(q)(-1)^{j+1} q^{\frac{j(j-1)}{2}} \sum_{n=0}^{\infty} \frac{A_{n+j-1}(q) q^n \left(q^{2j-1};q\right)_{2n}}{\left(q^{2j},q;q\right)_n}$$

$$= P(q)(-1)^{j+1}q^{\frac{j(j-1)}{2}}\sum_{n=0}^{\infty} A_{n+j-1}(q)q^n \left(q^{2j-1};q\right)_{2n} \beta_n^*(q^{2j-1};q)$$

$$= P(q)(-1)^{j+1}q^{\frac{j(j-1)}{2}}\sum_{n=0}^{\infty} A_{n+j-1}(q)q^n \left(q^{2j-1};q\right)_{2n} \beta_n^{**}(q^{2j-2};q).$$

The coefficient of z is derived in the same way, but because n will start at 1, rather than 0, we find the coefficient of z is given by

$$P(q)\sum_{n=0}^{\infty} A_n(q)q^n\,(q;q)_{2n}\,\beta_n^*(q;q) - A_0(q)P(q)$$

$$= P(q)\sum_{n=0}^{\infty} A_n(q)q^n\,(q;q)_{2n}\,\beta_n^{**}(1;q) - A_0(q)P(q).$$

If we are to next apply (3.2) with (α^*, β^*), then we must choose A so that

$$A_{n+j-1}(q)q^n \left(q^{2j-1};q\right)_{2n} = C_j(q)\left(q^{2j};q^2\right)_n q^n,$$

where C is a function dependent on j, but not dependent on n. Since

$$\frac{\left(q^{2j};q^2\right)_n}{(q^{2j-1};q)_{2n}} = \frac{\left(q^2;q^2\right)_{n+j-1}}{(q;q)_{2(n+j-1)}}\frac{(q;q)_{2j-2}}{\left(q^2;q^2\right)_{j-1}},$$

we see a reasonable choice for A is

$$A_n(q) = \frac{\left(q^2;q^2\right)_n}{(q;q)_{2n}} = \frac{1}{\left(q;q^2\right)_n}.$$

If instead we want apply (3.2) with $(\alpha^{**}, \beta^{**})$, then we must choose A so that

$$A_{n+j-1}(q)q^n \left(q^{2j-1};q\right)_{2n} = C_j(q)\left(q^{2j-1};q^2\right)_n q^n.$$

Here we find a reasonable choice for A is

$$A_n(q) = \frac{\left(q;q^2\right)_n}{(q;q)_{2n}} = \frac{1}{\left(q^2;q^2\right)_n}.$$

Which of these two choices should we use? Both are valid and lead to different functions. Using (α^*, β^*) we would likely consider the function

$$\frac{(q;q^2)_\infty}{(z,z^{-1};q)_\infty}\sum_{n=1}^{\infty}\frac{(z,z^{-1};q)_n q^n}{(q;q^2)_n}$$

and using (α^{**},β^{**}) we would use

$$\frac{(q^2;q^2)_\infty}{(z,z^{-1};q)_\infty}\sum_{n=1}^{\infty}\frac{(z,z^{-1};q)_n q^n}{(q^2;q^2)_n}.$$

However, after elementary rearrangements we recognize the latter as the overpartition spt-crank function of [17] and we have nothing new. With the former, which we earlier called $S_{L9}(z,q)$, we reduce the products, set $z=1$, and see if we observe any congruences on a computer. Given the empirical evidence of the congruence $\mathrm{spt}_{L9}(3n+2)$, we then check if the coefficients of q^{3n+2} of $S_{L9}(\zeta_3,q)$ appear to be zero. These coefficients do appear to be zero and so we have a new candidate spt function and spt-crank function to work with. We now apply (3.2) with (α^*,β^*), sum the powers of z, and after a bit of work we arrive at the identity for $S_{L9}(z,q)$ in Theorem 2.3.

We repeat this process with the other conjugate Bailey pair identities of [23], as well as with the classical form of Bailey's lemma (2.17). After determining choices of $A_n(q)$ for (α^*,β^*) and (α^{**},β^{**}), we see what functions they reduce to when $z=1$, and check for congruences. Doing so gives many functions previously studied and many functions without apparent congruences. The functions without apparent congruences will still satisfy identities like those in Theorem 2.3, however there are far too many to go through unless we have a vested interest.

4. Proof of Theorems 2.3 and 2.4

Proof of (2.1). We use the rearrangements described in the previous section with (2.21) applied to $\beta^*(q^{2j-2},q^2)$. We find the coefficient of z^j, for $j\ge 2$, of $(1+z)\left(z,z^{-1};q^2\right)_\infty S_{L7}(z;q)$ is given by

$$(-q;q)_\infty\sum_{n=j-1}^{\infty}\frac{(-1)^{j+1}(1-q^{4j-2})q^{2n+j(j-3)+2}\left(q^2;q^2\right)_{2n}}{(-q;q)_{2n}\left(q^2;q^2\right)_{n+j}\left(q^2;q^2\right)_{n-j+1}}$$

$$=\frac{(-1)^{j+1}(1+q^{2j-1})q^{j(j-1)}}{(q^2;q^2)_\infty}\sum_{n=0}^{\infty}q^{\frac{n(n-1)}{2}+2jn}.$$

When $j = 1$ we must subtract $(-q;q)_\infty$ from this. By summing the powers of z we then find that

$$\begin{aligned}&(1+z)\left(z,z^{-1};q^2\right)_\infty S_{L7}(z;q)\\&= -(1+z)(-q;q)_\infty + \frac{1}{(q^2q^2)_\infty}\\&\times\sum_{j=1}^{\infty}(z^j+z^{1-j})(-1)^{j+1}(1+q^{2j-1})q^{j(j-1)}\sum_{n=0}^{\infty}q^{\frac{n(n-1)}{2}+2jn}.\end{aligned}$$

However, we note the left-hand side is zero when $z = 1$, and so

$$(-q;q)_\infty = \frac{1}{(q^2q^2)_\infty}\sum_{j=1}^{\infty}(-1)^{j+1}(1+q^{2j-1})q^{j(j-1)}\sum_{n=0}^{\infty}q^{\frac{n(n-1)}{2}+2jn}.$$

Noting $z^j + z^{1-j} - 1 - z = (1-z^j)(1-z^{j-1})z^{1-j}$, we then have

$$\begin{aligned}&(1+z)\left(z,z^{-1};q^2\right)_\infty S_{L7}(z;q)\\&= \frac{1}{(q^2;q^2)_\infty}\sum_{j=1}^{\infty}(1-z^j)(1-z^{j-1})z^{1-j}(-1)^{j+1}(1+q^{2j-1})q^{j(j-1)}\\&\times\sum_{n=0}^{\infty}q^{\frac{n(n-1)}{2}+2jn},\end{aligned}$$

which immediately implies (2.1). □

Proof of (2.2). We use the rearrangements described in the previous section with (2.18) applied to $\beta^*(q^{2j-1},q)$. We find the coefficient of z^j, for $j \geq 2$, of $(1+z)\left(z,z^{-1};q\right)_\infty S_{L9}(z;q)$ is given by

$$\begin{aligned}&\left(q;q^2\right)_\infty\sum_{n=j-1}^{\infty}\frac{(-1)^{j+1}(1-q^{2j-1})q^{n+j(j-3)/2+1}(q;q)_{2n}}{(q;q^2)_n(q;q)_{n+j}(q;q)_{n-j+1}}\\&= \frac{(-1)^{j+1}(1-q^{2j-1})q^{j(j-1)/2}}{(q;q)_\infty}\sum_{n=0}^{\infty}(-1)^n q^{n^2+2jn}.\end{aligned}$$

When $j = 1$ we must subtract $\left(q;q^2\right)_\infty$ from this. By summing the powers of z we then find that

$$\begin{aligned}&(1+z)\left(z,z^{-1};q\right)_\infty S_{L9}(z;q)\\&= -(1+z)\left(q;q^2\right)_\infty + \frac{1}{(q;q)_\infty}\sum_{j=1}^{\infty}(z^j+z^{1-j})(-1)^{j+1}\\&\times(1-q^{2j-1})q^{j(j-1)/2}\sum_{n=0}^{\infty}(-1)^n q^{n^2+2jn}.\end{aligned}$$

However, we note the left-hand side is zero when $z=1$, and so

$$\left(q;q^2\right)_\infty = \frac{1}{(qq)_\infty}\sum_{j=1}^{\infty}(-1)^{j+1}(1-q^{2j-1})q^{j(j-1)/2}\sum_{n=0}^{\infty}(-1)^n q^{n^2+2jn}.$$

We then have

$$\begin{aligned}&(1+z)\left(z,z^{-1};q\right)_\infty S_{L9}(z;q)\\ &\quad=\frac{1}{(q;q)_\infty}\sum_{j=1}^{\infty}(1-z^j)(1-z^{j-1})z^{1-j}(-1)^{j+1}\\ &\qquad\times(1-q^{2j-1})q^{j(j-1)/2}\sum_{n=0}^{\infty}(-1)^n q^{n^2+2jn},\end{aligned}$$

which immediately implies (2.2). □

Proof of (2.3). We use the rearrangements described in the previous section with (2.22) applied to $\beta^*(q^{4j-2},q^2)$. We find the coefficient of z^j, for $j\geq 2$, of $(1+z)\left(z,z^{-1};q\right)_\infty S_{L12}(z;q)$ is given by

$$\begin{aligned}&\left(q;q^2\right)_\infty^2\sum_{n=j-1}^{\infty}\frac{(-1)^{j+1}(1-q^{4j-2})q^{2n+j(j-3)+2}\left(q^2;q^2\right)_{2n}}{(q;q^2)_n\,(q;q^2)_{n+1}\,(q^2;q^2)_{n+j}\,(q^2;q^2)_{n-j+1}}\\ &\quad=\frac{(-1)^{j+1}(1-q^{2j-1})q^{j(j-1)}}{(q^2;q^2)_\infty}\sum_{n=0}^{\infty}(-1)^n q^{n^2+2jn}.\end{aligned}$$

When $j=1$ we must subtract $\left(q,q^3;q^2\right)_\infty$ from this. By summing the powers of z we then find that

$$\begin{aligned}&(1+z)\left(z,z^{-1};q\right)_\infty S_{L12}(z;q)\\ &\quad=-(1+z)\left(q,q^3;q^2\right)_\infty+\frac{1}{(q^2;q^2)_\infty}\sum_{j=1}^{\infty}(z^j+z^{1-j})(-1)^{j+1}\\ &\quad\times(1-q^{2j-1})q^{j(j-1)}\sum_{n=0}^{\infty}(-1)^n q^{n^2+2jn}.\end{aligned}$$

However, we note the left-hand side is zero when $z=1$, and so

$$\left(q,q^3;q^2\right)_\infty=\frac{1}{(q^2q^2)_\infty}\sum_{j=1}^{\infty}(-1)^{j+1}(1-q^{2j-1})q^{j(j-1)}\sum_{n=0}^{\infty}(-1)^n q^{n^2+2jn}.$$

We then have

$$
\begin{aligned}
&(1+z)\left(z,z^{-1};q^2\right)_\infty S_{L12}(z;q)\\
&=\frac{1}{(q^2;q^2)_\infty}\sum_{j=1}^{\infty}(1-z^j)(1-z^{j-1})z^{1-j}(-1)^{j+1}(1-q^{2j-1})q^{j(j-1)}\\
&\quad\times\sum_{n=0}^{\infty}(-1)^n q^{n^2+2jn},
\end{aligned}
$$

which immediately implies (2.3). □

Proof of (2.4). We use the rearrangements described in the previous section with (2.19) applied to $\beta^{**}(q^{2j-2},q)$. We find the coefficient of z^j, for $j\geq 2$, of $(1+z)\left(z,z^{-1};q\right)_\infty F(\rho_1,\rho_2,z;q)$ is given by

$$
\begin{aligned}
&\frac{(q;q)_\infty}{(\rho_1,\rho_2;q)_\infty}\sum_{n=j-1}^{\infty}\frac{(\rho_1,\rho_2;q)_n\left(\frac{q}{\rho_1\rho_2}\right)^n(-1)^{j+1}(1-q^{2j-1})q^{\frac{j(j-3)}{2}+1}}{(q;q)_{n+j}\,(q;q)_{n-j+1}}\\
&=\frac{(-1)^{j+1}q^{j(j-1)/2}\,(\rho_1,\rho_2;q)_{j-1}\left(q^{j+1}/\rho_1,q^{j+1}/\rho_2;q\right)_\infty}{\rho_1^{j-1}\rho_2^{j-1}(\rho_1,\rho_2,\frac{q}{\rho_1\rho_2};q)_\infty}\\
&\quad\times(1-q^j/\rho_1-q^j/\rho_2+q^{3j-1}/\rho_1+q^{3j-1}/\rho_2-q^{4j-2}).
\end{aligned}
$$

When $j=1$ we must subtract $(q;q)_\infty/(\rho_1,\rho_2;q)_\infty$ from this. By summing the powers of z we then find that

$$
\begin{aligned}
&(1+z)\left(z,z^{-1};q\right)_\infty F(\rho_1,\rho_2,z;q)\\
&=-(1+z)\frac{(q;q)_\infty}{(\rho_1,\rho_2;q)_\infty}\\
&\quad+\sum_{j=1}^{\infty}(z^j+z^{1-j})\frac{(-1)^{j+1}q^{j(j-1)/2}\,(\rho_1,\rho_2;q)_{j-1}\left(\frac{q^{j+1}}{\rho_1},\frac{q^{j+1}}{\rho_2};q\right)_\infty}{\rho_1^{j-1}\rho_2^{j-1}(\rho_1,\rho_2,\frac{q}{\rho_1\rho_2}q)_\infty}\\
&\quad\times\left(1-\frac{q^j}{\rho_1}-\frac{q^j}{\rho_2}+\frac{q^{3j-1}}{\rho_1}+\frac{q^{3j-1}}{\rho_2}-q^{4j-2}\right).
\end{aligned}
$$

However, we note the left-hand side is zero when $z=1$, and so

$$
\begin{aligned}
\frac{(q;q)_\infty}{(\rho_1,\rho_2;q)_\infty}&=\sum_{j=1}^{\infty}\frac{(-1)^{j+1}q^{j(j-1)/2}\,(\rho_1,\rho_2;q)_{j-1}\left(\frac{q^{j+1}}{\rho_1},\frac{q^{j+1}}{\rho_2};q\right)_\infty}{\rho_1^{j-1}\rho_2^{j-1}(\rho_1,\rho_2,\frac{q}{\rho_1\rho_2};q)_\infty}\\
&\quad\times\left(1-\frac{q^j}{\rho_1}-\frac{q^j}{\rho_2}+\frac{q^{3j-1}}{\rho_1}+\frac{q^{3j-1}}{\rho_2}-q^{4j-2}\right).
\end{aligned}
$$

We then have

$$\begin{aligned}
&(1+z)\left(z,z^{-1};q\right)_\infty F(\rho_1,\rho_2,z;q)\\
&=\sum_{j=1}^{\infty}\frac{\begin{array}{c}(1-z^j)(1-z^{j-1})z^{1-j}(-1)^{j+1}q^{j(j-1)/2}\left(\rho_1,\rho_2;q\right)_{j-1}\\ \times(\frac{q^{j+1}}{\rho_1},\frac{q^{j+1}}{\rho_2};q)_\infty\end{array}}{\rho_1^{j-1}\rho_2^{j-1}(\rho_1,\rho_2,\frac{q}{\rho_1\rho_2};q)_\infty}\\
&\quad\times\left(1-\frac{q^j}{\rho_1}-\frac{q^j}{\rho_2}+\frac{q^{3j-1}}{\rho_1}+\frac{q^{3j-1}}{\rho_2}-q^{4j-2}\right),
\end{aligned}$$

which immediately implies (2.4). □

Proof of (2.5). We use the rearrangements described in the previous section with (2.20) applied to $\beta^*(q^{2j-1},q)$. We find the coefficient of z^j, for $j\geq 2$, of $(1+z)\left(z,z^{-1};q\right)_\infty G(\rho_1,\rho_2,z;q)$ is given by

$$\begin{aligned}
&\frac{(q;q)_\infty}{(\rho_1,\rho_2;q)_\infty}\sum_{n=j-1}^{\infty}\frac{(\rho_1,\rho_2;q)_n\,(\frac{q^2}{\rho_1\rho_2})^n(-1)^{j+1}(1-q^{2j-1})q^{\frac{j(j-3)}{2}+1}}{(q;q)_{n+j}\,(q;q)_{n-j+1}}\\
&=\frac{(-1)^{j+1}(1-q^{2j-1})q^{j(j+1)/2-1}\left(\rho_1,\rho_2;q\right)_{j-1}(q^{j+1}/\rho_1,q^{j+1}/\rho_2;q)_\infty}{\rho_1^{j-1}\rho_2^{j-1}(\rho_1,\rho_2,\frac{q^2}{\rho_1\rho_2};q)_\infty}.
\end{aligned}$$

When $j=1$ we must subtract $(q;q)_\infty/(\rho_1,\rho_2;q)_\infty$ from this. By summing the powers of z we then find that

$$\begin{aligned}
&(1+z)\left(z,z^{-1};q\right)_\infty G(\rho_1,\rho_2,z;q)\\
&=-(1+z)\frac{(q;q)_\infty}{(\rho_1,\rho_2;q)_\infty}+\sum_{j=1}^{\infty}(z^j+z^{1-j})\\
&\times\frac{(-1)^{j+1}(1-q^{2j-1})q^{j(j+1)/2-1}\left(\rho_1,\rho_2;q\right)_{j-1}(q^{j+1}/\rho_1,q^{j+1}/\rho_2 q)_\infty}{\rho_1^{j-1}\rho_2^{j-1}(\rho_1,\rho_2,\frac{q^2}{\rho_1\rho_2}q)_\infty}.
\end{aligned}$$

However, we note the left-hand side is zero when $z=1$, and so

$$\frac{(q;q)_\infty}{(\rho_1,\rho_2;q)_\infty}=\sum_{j=1}^{\infty}\frac{\begin{array}{c}(-1)^{j+1}(1-q^{2j-1})q^{j(j+1)/2-1}\left(\rho_1,\rho_2;q\right)_{j-1}\\ \times(q^{j+1}/\rho_1,q^{j+1}/\rho_2;q)_\infty\end{array}}{\rho_1^{j-1}\rho_2^{j-1}(\rho_1,\rho_2,\frac{q^2}{\rho_1\rho_2};q)_\infty}.$$

We note this is the same product as in the previous proof, but a rather different series. Regardless, we then have

$$
\begin{aligned}
&(1+z)\left(z,z^{-1};q\right)_\infty G(\rho_1,\rho_2,z;q)\\
&=\sum_{j=1}^{\infty}\frac{\begin{array}{c}(1-z^j)(1-z^{j-1})z^{1-j}(-1)^{j+1}(1-q^{2j-1})q^{j(j+1)/2-1}\\ \times(\rho_1,\rho_2;q)_{j-1}(q^{j+1}/\rho_1,q^{j+1}/\rho_2;q)_\infty\end{array}}{\rho_1^{j-1}\rho_2^{j-1}(\rho_1,\rho_2,\frac{q^2}{\rho_1\rho_2};q)_\infty},
\end{aligned}
$$

which immediately implies (2.5). □

Proof of (2.8). With $q\mapsto q^2$, $\rho_1=q$, and $\rho_2=q^2$ in (2.5), we have that

$$
\begin{aligned}
S_{G1}(z,q)&=\frac{1}{(1+z)\left(z,z^{-1};q^2\right)_\infty}\\
&\quad\times\sum_{j=1}^{\infty}\frac{\begin{array}{c}(1-z^j)(1-z^{j-1})z^{1-j}(-1)^{j+1}(1-q^{4j-2})q^{j(j+1)-2}\\ \times\left(q,q^2;q\right)_{j-1}\left(q^{2j+1},q^{2j};q^2\right)_\infty\end{array}}{q^{3j-3}\left(q,q^2,q;q^2\right)_\infty}\\
&=\frac{1}{(1+z)\left(z,z^{-1},q;q^2\right)_\infty}\\
&\quad\times\sum_{j=1}^{\infty}(1-z^j)(1-z^{j-1})z^{1-j}(-1)^{j+1}(1+q^{2j-1})q^{(j-1)^2},
\end{aligned}
$$

which is (2.8). □

Proof of (2.9). With $\rho_1=iq^{1/2}$, and $\rho_2=-iq^{1/2}$ in (2.5), we have that

$$
\begin{aligned}
S_{G2}(z,q)&=\frac{1}{(1+z)\left(z,z^{-1};q\right)_\infty}\sum_{j=1}^{\infty}(1-z^j)(1-z^{j-1})z^{1-j}(-1)^{j+1}\\
&\quad\times(1-q^{2j-1})q^{\frac{j(j+1)}{2}-1}(iq^{1/2},-iq^{1/2}q)_{j-1}\\
&\quad\times\frac{\left(-iq^{j+1/2},iq^{j+1/2};q\right)_\infty}{q^{j-1}\left(iq^{1/2},-iq^{1/2},q;q\right)_\infty}\\
&=\frac{1}{(1+z)\left(z,z^{-1};q\right)_\infty}
\end{aligned}
$$

$$\times \sum_{j=1}^{\infty} \frac{\begin{array}{c}(1-z^j)(1-z^{j-1})z^{1-j}(-1)^{j+1}(1-q^{2j-1})q^{\frac{j(j-1)}{2}} \\ \times \left(-q;q^2\right)_{j-1}\left(-q^{2j+1};q^2\right)_\infty\end{array}}{\left(-q;q^2\right)_\infty (q;q)_\infty}$$

$$= \frac{1}{(1+z)\left(z,z^{-1},q;q\right)_\infty}$$

$$\times \sum_{j=1}^{\infty} \frac{(1-z^j)(1-z^{j-1})z^{1-j}(-1)^{j+1}(1-q^{2j-1})q^{\frac{j(j-1)}{2}}}{(1+q^{2j-1})}$$

$$= \frac{1}{(1+z)\left(z,z^{-1},q;q\right)_\infty} \sum_{j=-\infty}^{\infty} \frac{(1-z^j)(1-z^{j-1})z^{1-j}(-1)^{j+1}q^{\frac{j(j-1)}{2}}}{(1+q^{2j-1})},$$

which is (2.9). □

Proof of (2.10). With $\rho = -q$ in (2.6), we have that

$$S_{F1}(z,q) = \frac{1}{(1+z)\left(z,z^{-1};q\right)_\infty}$$

$$\times \sum_{j=1}^{\infty} \frac{\begin{array}{c}(1-z^j)(1-z^{j-1})z^{1-j}(-1)^{j+1} \\ \times q^{(j-1)^2}\left(-q;q\right)_{j-1}\left(-q^j;q\right)_\infty\end{array}}{q^{j-1}\left(-q;q\right)_\infty}$$

$$\times (1+q^{j-1}-q^{3j-2}-q^{4j-2})$$

$$= \frac{1}{(1+z)\left(z,z^{-1};q\right)_\infty} \sum_{j=1}^{\infty} (1-z^j)(1-z^{j-1})z^{1-j}(-1)^{j+1}$$

$$\times (1+q^{j-1}-q^{3j-2}-q^{4j-2})q^{j^2-3j+2}$$

$$= \frac{1}{(1+z)\left(z,z^{-1};q\right)_\infty}$$

$$\times \sum_{j=-\infty}^{\infty} (1-z^j)(1-z^{j-1})z^{1-j}(-1)^{j+1}(1+q^{j-1})q^{j^2-3j+2},$$

which is (2.10). Next we will use the Jacobi triple product identity [3, Theorem 2.8],

$$\sum_{j=-\infty}^{\infty} (-1)^j t^j q^{j^2} = \left(tq, t^{-1}q, q^2; q^2\right)_\infty .$$

We have that

$$
\begin{aligned}
&(1+z)\left(z,z^{-1};q\right)_\infty S_{F1}(z,q)\\
&= -q^2\sum_{j=-\infty}^{\infty}(-1)^j z^j q^{j^2-3j} - q\sum_{j=-\infty}^{\infty}(-1)^j z^j q^{j^2-2j} - zq^2\\
&\quad\times\sum_{j=-\infty}^{\infty}(-1)^j z^{-j} q^{j^2-3j} - zq\sum_{j=-\infty}^{\infty}(-1)^j z^{-j} q^{j^2-2j}\\
&\quad+(1+z)q^2\sum_{j=-\infty}^{\infty}(-1)^j q^{j^2-3j} + (1+z)\sum_{j=-\infty}^{\infty}(-1)^j q^{j^2-2j}\\
&= -q^2\left(zq^{-2},z^{-1}q^4,q^2;q^2\right)_\infty - q\left(zq^{-1},z^{-1}q^3,q^2;q^2\right)_\infty\\
&\quad- zq^2\left(z^{-1}q^{-2},zq^4,q^2;q^2\right)_\infty\\
&\quad- zq\left(z^{-1}q^{-1},zq^3,q^2;q^2\right)_\infty + (1+z)q^2\left(q^{-2},q^4,q^2;q^2\right)_\infty\\
&\quad+(1+z)q\left(q^{-1},q^3,q^2;q^2\right)_\infty\\
&= (1-z^2)\left(z^{-1},zq^2,q^2;q^2\right)_\infty + (1+z)q\left(z^{-1}q,zq,q^2;q^2\right)_\infty\\
&\quad-(1+z)\left(q,q,q^2;q^2\right)_\infty,
\end{aligned}
$$

where the last equality follows from multiple applications of $\left(t,qt^{-1};q\right)_\infty = -t\left(tq,t^{-1};q\right)_\infty$. We see that (2.11) follows from the above after diving by $(1+z)\left(z,z^{-1};q\right)_\infty$ and elementary simplifications. □

Proof of (2.12). With $\rho = q$ in (2.9), we have that

$$
\begin{aligned}
S_{G3}(z,q) &= \frac{1}{(1+z)\left(z,z^{-1};q\right)_\infty}\\
&\quad\times\sum_{j=1}^{\infty}\frac{(1-z^j)(1-z^{j-1})z^{1-j}(1-q^{2j-1})q^{j(j-1)}(q;q)_{j-1}\left(q^j;q\right)_\infty}{q^{j-1}\left(q;q\right)_\infty}\\
&= \frac{1}{(1+z)\left(z,z^{-1};q\right)_\infty}\\
&\quad\times\sum_{j=1}^{\infty}(1-z^j)(1-z^{j-1})z^{1-j}(1-q^{2j-1})q^{(j-1)^2},
\end{aligned}
$$

which is (2.12). □

Proof of Theorem 2.4. The proofs of (2.13), (2.14), and (2.15) are all a rearrangement of the series in Theorem 2.3. We describe the rearrangements here and then proceed with the calculations. First we reverse the order of summation and expand the double series into a sum of two double series. Second we replace n by $-1-n$ in the second double series. Third we rewrite the summands in both double series in a common form and obtain a double series that is bilateral in n. Fourth we replace j by $j-|n|+1$, $j-2|n|+1$, and $j-|n|+1$ for (2.13), (2.14), and (2.15) respectively. Lastly we exchange the order of summation to obtain the identity. Since these calculations are so similar, we only write out the details for $S_{L7}(z,q)$.

By (2.1) we have that

$$
\begin{aligned}
&(1+z)\left(q^2,z,z^{-1};q^2\right)_\infty S_{L7}(z,q)\\
&=\sum_{n=0}^{\infty}\sum_{j=1}^{\infty}(1-z^j)(1-z^{j-1})z^{1-j}(-1)^{j+1}q^{j(j-1)+\frac{n(n-1)}{2}+2jn}\\
&\quad+\sum_{n=0}^{\infty}\sum_{j=1}^{\infty}(1-z^j)(1-z^{j-1})z^{1-j}(-1)^{j+1}q^{j(j+1)-1+\frac{n(n-1)}{2}+2jn}\\
&=\sum_{n=0}^{\infty}\sum_{j=1}^{\infty}(1-z^j)(1-z^{j-1})z^{1-j}(-1)^{j+1}q^{j(j-1)+\frac{n(n-1)}{2}+2jn}\\
&\quad+\sum_{n=-\infty}^{-1}\sum_{j=1}^{\infty}(1-z^j)(1-z^{j-1})z^{1-j}(-1)^{j+1}q^{j(j-1)+\frac{n(n+3)}{2}-2jn}\\
&=\sum_{n=-\infty}^{\infty}\sum_{j=1}^{\infty}(1-z^j)(1-z^{j-1})z^{1-j}(-1)^{j+1}q^{j(j-1)+\frac{n(n+1)}{2}-|n|+2j|n|}\\
&=\sum_{n=-\infty}^{\infty}\sum_{j=|n|}^{\infty}(1-z^{j-|n|+1})(1-z^{j-|n|})z^{|n|-j}(-1)^{j+n}q^{j(j+1)-\frac{n(n-1)}{2}}\\
&=\sum_{j=0}^{\infty}\sum_{n=-j}^{j}(1-z^{j-|n|+1})(1-z^{j-|n|})z^{|n|-j}(-1)^{j+n}q^{j(j+1)-\frac{n(n-1)}{2}},
\end{aligned}
$$

which implies (2.13). □

5. Proof of Theorem 2.2

To prove Theorem 2.2, we need to show the coefficients of the following terms are zero: q^{3m+1} in $S_{L7}(\zeta_3,q)$, q^{3m+2} in $S_{L9}(\zeta_3,q)$, q^{3m} in $S_{L12}(\zeta_3,q)$,

q^{3m} in $S_{G1}(\zeta_3, q)$, q^{3m} in $S_{G2}(\zeta_3, q)$, q^{3m+2} in $S_{G2}(\zeta_3, q)$, q^{10m+9} in $S_{F1}(\zeta_5, q)$, and q^{5m+3} in $S_{G3}(\zeta_5, q)$.

We first note that $\left(q, \zeta_3 q, \zeta_3^{-1} q; q\right)_\infty = \left(q^3; q^3\right)_\infty$. By (2.1) we have that

$$S_{L7}(z; q) = \frac{-3\zeta_3}{(q^6; q^6)_\infty} \sum_{j=1}^{\infty} \sum_{n=0}^{\infty} (1 - \zeta_3^j)(1 - \zeta_3^{j-1})\zeta_3^{1-j}(-1)^{j+1} \times (1 + q^{2j-1}) q^{j(j-1) + \frac{n(n-1)}{2} + 2jn}.$$

We note that the terms in the series are zero except when $j \equiv 2 \pmod 3$. However when $j \equiv 2 \pmod 3$, one finds that $(1 + q^{2j-1})q^{j(j-1)+\frac{n(n-1)}{2}+2jn}$ contributes only terms of the form q^{3m} and q^{3m+2}. Thus $S_{L7}(\zeta_3, q)$ has no nonzero terms of the form q^{3m+1}. Using (2.2), (2.3), and (2.9) we find the same reasoning shows the coefficients of q^{3m+2} in $S_{L9}(\zeta_3, q)$, q^{3m} in $S_{L12}(\zeta_3, q)$, q^{3m} in $S_{G2}(\zeta_3, q)$, and q^{3m+2} in $S_{G2}(\zeta_3, q)$ are all zero.

Next by (2.9) we have that

$$S_{G1}(z, q) = \frac{-3\zeta_3 \left(q^2; q^2\right)_\infty}{\left(q^6; q^6\right)_\infty \left(q; q^2\right)_\infty} \times \sum_{j=-\infty}^{\infty} (1 - z^j)(1 - z^{j-1}) z^{1-j} (-1)^{j+1} (1 + q^{2j-1}) q^{(j-1)^2}. \tag{5.1}$$

By Gauss [3, Corollary 2.10] we have

$$\frac{\left(q^2; q^2\right)_\infty}{\left(q; q^2\right)_\infty} = \sum_{n=0}^{\infty} q^{\frac{n(n+1)}{2}},$$

and so $\frac{(q^2;q^2)_\infty}{(q;q^2)_\infty}$ has only terms of the form q^{3m} and q^{3m+1}. In (5.1), the terms in the series are zero except when $j \equiv 2 \pmod 3$. However when $j \equiv 2 \pmod 3$, one finds that $(1 + q^{2j-1})q^{(j-1)^2}$ contributes only terms of the form q^{3m+1}. Thus $S_{G1}(\zeta_3, q)$ has no nonzero terms of the form q^{3m}.

For $S_{F1}(\zeta_5, q)$ and $S_{G3}(\zeta_5, q)$, we first note that Lemma 3.9 of [18] is

$$\frac{1}{\left(\zeta_5 q, \zeta_5^{-1} q; q\right)_\infty} = \frac{1}{\left(q^5, q^{20}; q^{25}\right)_\infty} + \frac{(\zeta_5 + \zeta_5^{-1})q}{\left(q^{10}, q^{15}; q^{25}\right)_\infty}. \tag{5.2}$$

Also with the Jacobi triple product identity one can easily deduce that

$$\left(\zeta_5 q^2, \zeta_5^{-1} q^2, q^2; q^2\right)_\infty = \left(q^{20}, q^{30}, q^{50}; q^{50}\right)_\infty + (\zeta_5^2 + \zeta_5^3) q^2 \left(q^{10}, q^{40}, q^{50}; q^{50}\right)_\infty,$$

$$\begin{aligned}\left(\zeta_5 q, \zeta_5^{-1} q, q^2; q^2\right)_\infty &= \left(q^{25}, q^{25}, q^{50}; q^{50}\right)_\infty \\ &\quad - (\zeta_5 + \zeta_5^4) q \left(q^{15}, q^{35}, q^{50}; q^{50}\right)_\infty \\ &\quad + (\zeta_5^2 + \zeta_5^3) q^4 \left(q^5, q^{45}, q^{50}; q^{50}\right)_\infty, \\ \left(q, q, q^2; q^2\right)_\infty &= \left(q^{25}, q^{25}, q^{50}; q^{50}\right)_\infty - 2q \left(q^{15}, q^{35}, q^{50}; q^{50}\right)_\infty \\ &\quad + 2q^4 \left(q^5, q^{45}, q^{50}; q^{50}\right)_\infty.\end{aligned}$$

By (2.11) and the above, we have that

$$\begin{aligned}S_{F1}(\zeta_5, q) &= \frac{\left(\zeta_5^{-1} q^2, \zeta_5 q^2, q^2; q^2\right)_\infty}{\left(\zeta_5 q, \zeta_5^{-1} q; q\right)_\infty} + \frac{\left(\zeta_5^{-1} q, \zeta_5 q, q^2; q^2\right)_\infty}{\left(\zeta_5, \zeta_5^{-1}; q\right)_\infty} - \frac{\left(q, q, q^2; q^2\right)_\infty}{\left(\zeta_5, \zeta_5^{-1}; q\right)_\infty} \\ &= \frac{\left(q^{20}, q^{30}, q^{50}; q^{50}\right)_\infty}{\left(q^5, q^{20}; q^{25}\right)_\infty} + (\zeta_5^2 + \zeta_5^3) q^5 \frac{\left(q^5, q^{45}, q^{50}; q^{50}\right)_\infty}{\left(q^{10}, q^{15}; q^{25}\right)_\infty} \\ &\quad + (\zeta_5 + \zeta_5^4) q \frac{\left(q^{20}, q^{30}, q^{50}; q^{50}\right)_\infty}{\left(q^{10}, q^{15}; q^{25}\right)_\infty} \\ &\quad + q \frac{\left(q^{15}, q^{35}, q^{50}; q^{50}\right)_\infty}{\left(q^5, q^{20}; q^{25}\right)_\infty} + (\zeta_5^2 + \zeta_5^3) q^2 \frac{\left(q^{10}, q^{40}, q^{50}; q^{50}\right)_\infty}{\left(q^5, q^{20}; q^{25}\right)_\infty} \\ &\quad + (\zeta_5 + \zeta_5^4) q^2 \frac{\left(q^{15}, q^{35}, q^{50}; q^{50}\right)_\infty}{\left(q^{10}, q^{15}; q^{25}\right)_\infty} \\ &\quad - q^3 \frac{\left(q^{10}, q^{40}, q^{50}; q^{50}\right)_\infty}{\left(q^{10}, q^{15}; q^{25}\right)_\infty} + (\zeta_5^3 + \zeta_5^2 - 1) q^4 \frac{\left(q^5, q^{45}, q^{50}; q^{50}\right)_\infty}{\left(q^5, q^{20}; q^{25}\right)_\infty}.\end{aligned}$$

However, we see that

$$\frac{\left(q^5, q^{45}, q^{50}; q^{50}\right)_\infty}{\left(q^5, q^{20}; q^{25}\right)_\infty} = \frac{\left(q^5, q^{45}, q^{50}; q^{50}\right)_\infty}{\left(q^5, q^{20}, q^{30}, q^{45}; q^{50}\right)_\infty} = \frac{\left(q^{50}; q^{50}\right)_\infty}{\left(q^{20}, q^{30}; q^{50}\right)_\infty},$$

and so while $S_{F1}(\zeta_5, q)$ does have terms of the form q^{5m+4}, it has no terms of the form q^{10m+9}.

By (2.12) we have that

$$\begin{aligned}S_{G3}(z, q) &= \frac{1}{(1+\zeta_5)(1-\zeta_5)(1-\zeta_5^{-1})\left(\zeta q, \zeta^{-1} q; q\right)_\infty} \\ &\quad \times \sum_{j=1}^{\infty} (1-z^j)(1-z^{j-1}) z^{1-j} (1-q^{2j-1}) q^{(j-1)^2}.\end{aligned}$$

However by (5.2), $\frac{1}{(\zeta q,\zeta^{-1}q;q)_\infty}$ contributes only terms of the form q^{5n} and q^{5n+1}. In the series, the terms are zero except when $j \equiv 2,3,4 \pmod 5$. One can verify when $j \equiv 2,3,4 \pmod 5$ that $(1-q^{2j-1})q^{(j-1)^2}$ contributes only terms of the form q^{5n+1} and q^{5n+4}. Thus $S_{G3}(\zeta_5,q)$ has no nonzero terms of the form q^{5m+3}.

6. Remarks

Here we have demonstrated that new spt-crank functions arise from conjugate Bailey pair identities and variations of Bailey's lemma, when applied carefully to the two generic Bailey pairs (α^*,β^*) and (α^{**},β^{**}). Previously we saw that spt-crank functions arise from applying the classical form of Bailey's lemma to a series

$$\frac{P(q)}{(z,z^{-1};q)_\infty}\sum_{n=1}^{\infty}\left(z,z^{-1};q\right)_n q^n\beta_n,$$

where (α,β) is a Bailey pair relative to $(1,q)$. There are spt-crank functions that appear in both forms. This was first noticed in [21]. In these cases we started with spt-crank functions defined by the above series and then applied the techniques of this chapter. It does appear one can work in the opposite direction as well. In particular it would appear that $S_{F1}(z,q)$ also comes from an spt-crank function in the form

$$\frac{(q;q)_\infty}{(z,z^{-1},-q;q)_\infty}\sum_{n=1}^{\infty}\left(z,z^{-1};q\right)_n q^n\beta_n,$$

with the Bailey pair relative to $(1,q)$ given by

$$\beta_n=\frac{q^{\frac{n(n-3)}{2}}(-q;q)_n}{(q;q)_{2n}},\qquad \alpha_n=\begin{cases}1 & \text{if } n=0,\\ (-1)^{\frac{n-1}{2}}q^{\frac{n^2-4n-1}{4}}(1-q^{2n}) & \text{if } n \text{ is odd},\\ (-1)^{\frac{n}{2}}q^{\frac{n^2-2n}{4}}(1+q^n) & \text{if } n \text{ is even}.\end{cases}$$

With this we could approach $S_{F1}(z,q)$ by applying Bailey's lemma with $\rho_1=z$, $\rho_2=z^{-1}$ and obtaining a difference of a generalized Lambert series and an infinite product we could dissect at roots of unity. However we would first need to verify that the above is indeed a Bailey pair, which we should be able to easily prove along the same lines as the Bailey pairs from group C of [29].

The functions F and G are fairly general, so we would expect other specializations are of interest as well. Actually many specializations are functions that were previously studied. To list just a few, $G(-z^{1/2}, -z^{-1/2}, z^{1/2}; -q)$ is the M2spt crank function $S2(z,q)$ from [17], $G(q,q,z;q^2)$ is the $\overline{\text{spt2}}$ crank function $S(z,q)$ from [22], $G(-q^{1/2}, q^{1/2}, z; q)$ is $S_{E2}(z,q)$ and $G(-q,z;q)$ is $S_{C5}(z,q)$ from [21]. Additionally, the function $\text{spt}_{G2}(n)$ was independently introduced as $\overline{\text{spt}}_{\omega}(n)$ in [7].

References

[1] S. Ahlgren, K. Bringmann and J. Lovejoy, ℓ-Adic properties of smallest parts functions, *Adv. Math.* **228**(1) (2011) 629–645.

[2] N. Andersen, Vector-valued modular forms and the mock theta conjectures, *Res. Number Theory* **2** (2016) Art. 32, 14 pp.

[3] G. E. Andrews, *The Theory of Partitions*, Encyclopedia of Mathematics and its Applications, Vol. 2 (Addison-Wesley, Reading, MA, 1976).

[4] G. E. Andrews, *q-Series: Their Development and Application in Analysis, Number Theory, Combinatorics, Physics, and Computer Algebra*, CBMS Regional Conference Series in Mathematics, Vol. 66, Published for the Conference Board of the Mathematical Sciences, Washington, DC (American Mathematical Society, Providence, RI, 1986).

[5] G. E. Andrews, Bailey's transform, lemma, chains and tree, in *Special Functions 2000: Current Perspective and Future Directions*, NATO Science Series II. Mathematics, Physics and Chemistry, Vol. 30 (Kluwer Academic Publishers, Dordrecht, 2001), pp. 1–22.

[6] G. E. Andrews, The number of smallest parts in the partitions of n, *J. Reine Angew. Math.* **624** (2008) 133–142.

[7] G. E. Andrews, A. Dixit, D. Schultz and A. J. Yee, Overpartitions related to the mock theta function $\omega(q)$, preprint (2016).

[8] G. E. Andrews and F. G. Garvan, Dyson's crank of a partition, *Bull. Amer. Math. Soc. (N.S.)* **18**(2) (1988) 167–171.

[9] G. E. Andrews, F. G. Garvan and J. Liang, Combinatorial interpretations of congruences for the spt-function, *Ramanujan J.* **29**(1–3) (2012) 321–338.

[10] A. O. L. Atkin and P. Swinnerton-Dyer, Some properties of partitions, *Proc. London Math. Soc.* (3) **4** (1954) 84–106.

[11] W. N. Bailey, Identities of the Rogers–Ramanujan type, *Proc. London Math. Soc.* (2) **50** (1948) 1–10.

[12] B. C. Berndt, Ramanujan's congruences for the partition function modulo 5, 7, and 11, *Int. J. Number Theory* **3**(3) (2007) 349–354.

[13] K. Bringmann, On the explicit construction of higher deformations of partition statistics, *Duke Math. J.* **144**(2) (2008) 195–233.

[14] K. Bringmann, J. Lovejoy and R. Osburn, Automorphic properties of generating functions for generalized rank moments and Durfee symbols, *Int. Math. Res. Notices* **2010**(2) (2010) 238–260.

[15] F. J. Dyson, Some guesses in the theory of partitions, *Eureka* **8** (1944) 10–15.
[16] A. Folsom, A short proof of the mock theta conjectures using Maass forms, *Proc. Amer. Math. Soc.* **136**(12) (2008) 4143–4149.
[17] F. Garvan and C. Jennings-Shaffer, The spt-crank for overpartitions, *Acta Arith.* **166**(2) (2014) 141–188.
[18] F. G. Garvan, New combinatorial interpretations of Ramanujan's partition congruences mod 5, 7 and 11, *Trans. Amer. Math. Soc.* **305**(1) (1988) 47–77.
[19] F. G. Garvan, Congruences for Andrews' spt-function modulo powers of 5, 7 and 13, *Trans. Amer. Math. Soc.* **364**(9) (2012) 4847–4873.
[20] F. G. Garvan, Universal mock theta functions and two-variable Hecke–Rogers identities, *Ramanujan J.* **36**(1–2) (2015) 267–296.
[21] F. G. Garvan and C. Jennings-Shaffer, Exotic Bailey–Slater spt-functions II: Hecke–Rogers-type double sums and Bailey pairs from groups A, C, E, *Adv. Math.* **299** (2016) 605–639.
[22] C. Jennings-Shaffer, Another SPT crank for the number of smallest parts in overpartitions with even smallest part, *J. Number Theory* **148** (2015) 196–203.
[23] J. Lovejoy, Ramanujan-type partial theta identities and conjugate Bailey pairs, *Ramanujan J.* **29**(1–3) (2012) 51–67.
[24] J. McLaughlin, A. V. Sills and P. Zimmer, Rogers–Ramanujan–Slater type identities, *Electron. J. Combin.* **15** (2008).
[25] K. Ono, Congruences for the Andrews spt function, *Proc. Natl. Acad. Sci. USA* **108**(2) (2011) 473–476.
[26] A. E. Patkowski, A strange partition theorem related to the second Atkin–Garvan moment, *Int. J. Number Theory* **11**(7) (2015) 2191–2197.
[27] S. Ramanujan, Congruence properties of partitions, *Math. Z.* **9**(1–2) (1921) 147–153.
[28] A. Schilling and S. O. Warnaar, Conjugate Bailey pairs: from configuration sums and fractional-level string functions to Bailey's lemma, in *Recent Developments in Infinite-Dimensional Lie algebras and Conformal Field Theory, Contemporary Mathematics*, Vol. 297 (American Mathematical Society, Providence, RI, 2002), pp. 227–255.
[29] L. J. Slater, A new proof of Rogers's transformations of infinite series, *Proc. London Math. Soc.* (2) **53** (1951) 460–475.
[30] L. J. Slater, Further identities of the Rogers–Ramanujan type, *Proc. London Math. Soc.* (2) **54** (1952) 147–167.
[31] S. O. Warnaar, 50 years of Bailey's lemma, in *Algebraic Combinatorics and Applications* (Springer, Berlin, 2001), pp. 333–347.

Chapter 19

Dual Addition Formulas Associated with Dual Product Formulas

Tom H. Koornwinder

Korteweg-de Vries Institute, University of Amsterdam
P.O. Box 94248, 1090 GE Amsterdam, The Netherlands
t.h.koornwinder@uva.nl

Dedicated to Mourad E. H. Ismail on the occasion of his 70th birthday

We observe that the linearization coefficients for Gegenbauer polynomials are the orthogonality weights for Racah polynomials with special parameters. Then it turns out that the linearization sum with such a Racah polynomial as extra factor inserted, can also be evaluated. The corresponding Fourier–Racah expansion is an addition-type formula which is dual to the well-known addition formula for Gegenbauer polynomials. The limit to the case of Hermite polynomials of this dual addition formula is also considered. Similar results as for Gegenbauer polynomials, although only formal, are given by taking the Ruijsenaars–Hallnäs dual product formula for Gegenbauer functions as a starting point and by working with Wilson polynomials.

Keywords: Addition formula for Gegenbauer polynomials; dual addition formula; Racah polynomials.

Mathematics Subject Classification 2010: 33C45, 33-03, 01A55

1. Introduction

A prototype for an addition formula for a family of special orthogonal polynomials is the *addition formula for Legendre polynomials* [17, (18.18.9)]

$$\begin{aligned} &P_n(\cos\theta_1\cos\theta_2+\sin\theta_1\sin\theta_2\cos\phi)\\ &\quad= P_n(\cos\theta_1)P_n(\cos\theta_2)+2\sum_{k=1}^{n}\frac{(n-k)!\,(n+k)!}{2^{2k}(n!)^2}\\ &\quad\quad\times(\sin\theta_1)^k P_{n-k}^{(k,k)}(\cos\theta_1)\,(\sin\theta_2)^k P_{n-k}^{(k,k)}(\cos\theta_2)\cos(k\phi). \qquad (1.1)\end{aligned}$$

The right-hand side is the Fourier-cosine expansion of the left-hand side as a function of ϕ. Integration with respect to ϕ over $[0,\pi]$ gives the constant term in this expansion, [17, (18.17.6)]

$$P_n(\cos\theta_1)P_n(\cos\theta_2)=\frac{1}{\pi}\int_0^{\pi}P_n(\cos\theta_1\cos\theta_2+\sin\theta_1\sin\theta_2\cos\phi)\,d\phi, \qquad (1.2)$$

which is known as the *product formula for Legendre polynomials.*

A formula dual to (1.2) is the *linearization formula for Legendre polynomials*, see [17, (18.18.22)] for $\lambda=\frac{1}{2}$ together with [17, (18.7.9)] and see [17, (5.2.4)] for the shifted factorial $(a)_n$. It reads

$$\begin{aligned} P_l(x)P_m(x)=&\sum_{j=0}^{\min(l,m)}\frac{(\frac{1}{2})_j(\frac{1}{2})_{l-j}(\frac{1}{2})_{m-j}(l+m-j)!}{j!\,(l-j)!\,(m-j)!\,(\frac{3}{2})_{l+m-j}}\\ &\times(2(l+m-2j)+1)\,P_{l+m-2j}(x). \qquad (1.3)\end{aligned}$$

On several occasions, during his lectures at conferences, Richard Askey raised the problem to find an addition-type formula associated with (1.3) in a similar way as the addition formula (1.1) is associated with the product formula (1.2). It is the purpose of the present paper to give such a formula, more generally associated with the linearization formula for Gegenbauer polynomials, and also a (formal) addition-type formula associated with the dual product formula for Gegenbauer functions which was recently given by Hallnäs and Ruijsenaars [9, (4.17)].

In order to get a better feeling for what a dual addition formula should look like, we first rewrite (1.1) and (1.2) by substituting $z=\sin\theta_1\sin\theta_2\cos\phi$, $x=\cos\theta_1$, $y=\cos\theta_2$, and putting $\cos(k\phi)=T_k(\cos\phi)$ (T_k is a Chebyshev polynomial [17, (18.5.1)]). Assume that $x,y\in[-1,1]$

and $z \in \big[-\sqrt{1-x^2}\,\sqrt{1-y^2}, \sqrt{1-x^2}\,\sqrt{1-y^2}\,\big]$. We obtain

$$\begin{aligned} P_n(z+xy) = P_n(x)P_n(y) + 2\sum_{k=1}^{n} &\frac{(n-k)!\,(n+k)!}{2^{2k}(n!)^2} \\ &\times (1-x^2)^{\frac{1}{2}k} P_{n-k}^{(k,k)}(x)\,(1-y^2)^{\frac{1}{2}k} P_{n-k}^{(k,k)}(y) \\ &\times T_k\left(\frac{z}{\sqrt{1-x^2}\,\sqrt{1-y^2}}\right) \end{aligned} \tag{1.4}$$

and

$$P_n(x)P_n(y) = \frac{1}{\pi}\int_{-\sqrt{1-x^2}\,\sqrt{1-y^2}}^{\sqrt{1-x^2}\,\sqrt{1-y^2}} \frac{P_n(z+xy)}{\sqrt{(1-x^2)(1-y^2)-z^2}}\,dz. \tag{1.5}$$

The rewritten addition formula (1.4) expands the left-hand side as a function of z in terms of Chebyshev polynomials of dilated argument, where the dilation factor depends on x, y, and the product formula (1.5) recovers the constant term in this expansion by integration with respect to the weight function over the orthogonality interval for these dilated Chebyshev polynomials. The linearization formula (1.3) very much looks as a dual formula with respect to (1.5). If we can recognize the coefficients in the sum in (1.3) as weights for some finite system of orthogonal polynomials, then the dual addition formula associated with (1.3) should be the corresponding orthogonal expansion of $P_{l+m-2j}(x)$ as a function of j.

It is not so easy to recognize the coefficients in (1.3) as weights for known orthogonal polynomials, but a strong hint was provided by the Hallnäs–Ruijsenaars dual product formula [9, (4.17)] for Gegenbauer functions, which can be rewritten as (6.2). There the weight function in the integral is clearly the weight function for Wilson polynomials [11, Section 9.1] with suitable parameters. This suggests that in (1.3) we should have weights of Racah polynomials [11, Section 9.2], which are the discrete analogues of the Wilson polynomials. Indeed, this turns out to work, see (4.2) (more generally for Gegenbauer polynomials), and a nice expansion (4.6) in terms of these Racah polynomials can be derived, which is the dual addition formula for Gegenbauer polynomials predicted by Askey. In a remark at the end of Section 4, we point to a paper by Koelink *et al.* [12] from 2013 which already has the dual addition formula in disguised form.

For better comparison of the dual results in Section 4, we state the addition formula for Gegenbauer polynomials and related formulas in Section 3. There we also mention the quite unknown paper [1] by Allé from

1865 which already gives the addition formula for Gegenbauer polynomials, much earlier than Gegenbauer's paper [7] from 1874.

In Section 5 we obtain a limit of the dual addition formula for Gegenbauer polynomials corresponding to the limit from Gegenbauer to Hermite polynomials. The resulting formulas for Hermite polynomials are well known. Remarkable is that the orthogonality for special Racah polynomials tends in the limit to a (well-known) biorthogonality for shifted factorials.

In Section 2 we introduce the needed special functions. The paper concludes in Section 7 with a list of possible follow-up work on dual addition formulas.

2. Preliminaries

2.1. Jacobi, Gegenbauer and Hermite polynomials

We will work with renormalized *Jacobi polynomials* [11, Section 9.8]

$$R_n^{(\alpha,\beta)}(x) := \frac{P_n^{(\alpha,\beta)}(x)}{P_n^{(\alpha,\beta)}(1)} = {}_2F_1\left(\begin{matrix}-n, n+\alpha+\beta+1\\ \alpha+1\end{matrix}; \tfrac{1}{2}(1-x)\right), \tag{2.1}$$

where $P_n^{(\alpha,\beta)}(1) = (\alpha+1)_n/n!$ and $R_n^{(\alpha,\beta)}(1) = 1$ (see [17, (16.2.1)] for the definition of a ${}_pF_q$ hypergeometric series). These are orthogonal polynomials on the interval $[-1,1]$ with respect to the weight function $(1-x)^\alpha(1+x)^\beta$ $(\alpha,\beta > -1)$. In particular, for $\alpha = \beta$, the polynomials are called *Gegenbauer polynomials*, then the weight function is even and we have

$$R_n^{(\alpha,\alpha)}(-x) = (-1)^n R_n^{(\alpha,\alpha)}(x).$$

The precise orthogonality relation is

$$\begin{aligned}&\int_{-1}^{1} R_m^{(\alpha,\alpha)}(x)\, R_n^{(\alpha,\alpha)}(x)\, (1-x^2)^\alpha\, dx = h_n^{(\alpha,\alpha)}\, \delta_{m,n},\\ &h_n^{(\alpha,\alpha)} = \frac{2^{2\alpha+1}\Gamma(\alpha+1)^2}{\Gamma(2\alpha+2)} \frac{n+2\alpha+1}{2n+2\alpha+1} \frac{n!}{(2\alpha+2)_n}.\end{aligned} \tag{2.2}$$

From (2.1) we see that

$$R_n^{(\alpha,\beta)}(x) = \frac{(n+\alpha+\beta+1)_n}{2^n(\alpha+1)_n}\, x^n + \text{terms of lower degree.} \tag{2.3}$$

The connection of $R_n^{(\alpha,\alpha)}$ with the usual $C_n^{(\lambda)}$ notation for Gegenbauer polynomials is

$$R_n^{(\alpha,\alpha)}(x) = \frac{n!}{(\alpha+1)_n} P_n^{(\alpha,\alpha)}(x) = \frac{n!}{(2\alpha+1)_n} C_n^{(\alpha+\frac{1}{2})}(x).$$

From [17, (18.5.10)] we have the power series

$$R_n^{(\alpha,\alpha)}(x) = \frac{n!}{(2\alpha+1)_n} \sum_{k=0}^{[\frac{1}{2}n]} \frac{(-1)^k (\alpha+\frac{1}{2})_{n-k}}{k!\,(n-2k)!}\,(2x)^{n-2k}. \tag{2.4}$$

We will need the difference formula

$$R_n^{(\alpha,\alpha)}(x) - R_{n-2}^{(\alpha,\alpha)}(x) = \frac{n+\alpha-\frac{1}{2}}{\alpha+1}\,(x^2-1)\,R_{n-2}^{(\alpha+1,\alpha+1)}(x) \quad (n \geq 2). \tag{2.5}$$

Proof of (2.5). More generally, let $w(x) = w(-x)$ be an even weight function on $[-1,1]$, let $p_n(x) = k_n x^n + \cdots$ be orthogonal polynomials on $[-1,1]$ with respect to the weight function $w(x)$, and let $q_n(x) = k_n' x^n + \cdots$ be orthogonal polynomials on $[-1,1]$ with respect to the weight function $w(x)(1-x^2)$. Assume that p_n and q_n are normalized by $p_n(1) = 1 = q_n(1)$. Let $n \geq 2$. Then $p_n(x) - p_{n-2}(x)$ vanishes for $x = \pm 1$. Hence $(p_n(x) - p_{n-2}(x))/(1-x^2)$ is a polynomial of degree $n-2$. It is seen immediately that x^k $(k < n-2)$ is orthogonal to this polynomial with respect to the weight function $w(x)(1-x^2)$ on $[-1,1]$. We conclude that

$$p_n(x) - p_{n-2}(x) = \frac{k_n}{k_{n-2}'}(x^2-1)q_{n-2}(x) \quad (n \geq 2).$$

Now specialize to $w(x) = (1-x^2)^\alpha$ and use (2.3). □

Hermite polynomials [11, Section 9.15] are orthogonal polynomials H_n on $(-\infty,\infty)$ with respect to the weight function e^{-x^2} and normalized such that $H_n(x) = 2^n x^n$ + terms of lower degree. From [17, (18.5.13)] we have the power series

$$H_n(x) = n! \sum_{k=0}^{n} \frac{(-1)^k}{k!\,(n-2k)!}\,(2x)^{n-2k}. \tag{2.6}$$

It follows from (2.4) and (2.6) that

$$\begin{aligned} &\lim_{\alpha\to\infty} \alpha^{\frac{1}{2}n} R_n^{(\alpha,\alpha)}(\alpha^{-\frac{1}{2}}x) = 2^{-n} H_n(x), \\ &\lim_{\alpha\to\infty} \alpha^{\mu n} R_n^{(\alpha,\alpha)}(\alpha^{-\mu}x) = x^n \quad \left(\mu < \frac{1}{2}\right), \end{aligned} \tag{2.7}$$

in particular,

$$\lim_{\alpha\to\infty} R_n^{(\alpha,\alpha)}(x) = x^n. \tag{2.8}$$

2.2. Racah polynomials

We will consider *Racah polynomials* [11, Section 9.2]

$$R_n\big(x(x+\gamma+\delta+1);\alpha,\beta,\gamma,\delta\big) := {}_4F_3\left(\begin{matrix}-n,n+\alpha+\beta+1,-x,x+\gamma+\delta+1\\ \alpha+1,\beta+\delta+1,\gamma+1\end{matrix};1\right) \tag{2.9}$$

for $\gamma=-N-1$, where $N\in\{1,2,\ldots\}$, and for $n\in\{0,1,\ldots,N\}$. These are orthogonal polynomials on the finite quadratic set $\{x(x+\gamma+\delta+1)\mid x\in\{0,1,\ldots,N\}\}$:

$$\sum_{x=0}^{N}(R_mR_n)(x(x+\gamma+\delta+1);\alpha,\beta,\gamma,\delta)\,w_{\alpha,\beta,\gamma,\delta}(x) = h_{n;\alpha,\beta,\gamma,\delta}\delta_{m,n}\quad (m,n\in\{0,1,\ldots,N\}) \tag{2.10}$$

with

$$w_{\alpha,\beta,\gamma,\delta}(x)=\frac{(\alpha+1)_x(\beta+\delta+1)_x(\gamma+1)_x(\gamma+\delta+1)_x}{(-\alpha+\gamma+\delta+1)_x(-\beta+\gamma+1)_x(\delta+1)_x\,x!}\,\frac{\gamma+\delta+1+2x}{\gamma+\delta+1}, \tag{2.11}$$

$$\frac{h_{n;\,\alpha,\beta,\gamma,\delta}}{h_{0;\,\alpha,\beta,\gamma,\delta}}=\frac{\alpha+\beta+1}{\alpha+\beta+2n+1}\,\frac{(\beta+1)_n(\alpha+\beta-\gamma+1)_n(\alpha-\delta+1)_n\,n!}{(\alpha+1)_n(\alpha+\beta+1)_n(\beta+\delta+1)_n(\gamma+1)_n}, \tag{2.12}$$

$$h_{0;\,\alpha,\beta,\gamma,\delta}=\sum_{x=0}^{N}w_{\alpha,\beta,\gamma,\delta}(x)=\frac{(\alpha+\beta+2)_N(-\delta)_N}{(\alpha-\delta+1)_N(\beta+1)_N}. \tag{2.13}$$

Clearly $R_n(0;\alpha,\beta,\gamma,\delta)=1$ while, by (2.9) and the Saalschütz formula [17, (16.4.3)], we can evaluate the Racah polynomial for $x=N$:

$$R_n(N\delta;\alpha,\beta,\gamma,\delta)=\frac{(\beta+1)_n(\alpha-\delta+1)_n}{(\alpha+1)_n(\beta+\delta+1)_n}. \tag{2.14}$$

The backward shift operator equation [11, (9.2.8)] can be rewritten as

$$\begin{aligned}&w_{\alpha,\beta,\gamma,\delta}(x)R_n(x(x+\gamma+\delta+1);\alpha,\beta,\gamma,\delta)\\&=\frac{\gamma+\delta+2}{\gamma+\delta+2+2x}\,w_{\alpha+1,\beta+1,\gamma+1,\delta}(x)\\&\quad\times R_{n-1}\big(x(x+\gamma+\delta+2);\alpha+1,\beta+1,\gamma+1,\delta\big)\end{aligned}$$

$$- \frac{\gamma+\delta+2}{\gamma+\delta+2x}\, w_{\alpha+1,\beta+1,\gamma+1,\delta}(x-1)$$
$$\times R_{n-1}\big((x-1)(x+\gamma+\delta+1);\alpha+1,\beta+1,\gamma+1,\delta\big). \quad (2.15)$$

This holds for $x = 0, \ldots, N$. For $x = 0$, (2.15) remains true if we put the second term on the right equal to 0, while for $x = N$ the first term on the right can be assumed to vanish. In this last case, the identity (2.15) can be checked by using (2.14) and (2.11).

Hence, for a function f on $\{0, 1, \ldots, N\}$ we have

$$\sum_{x=0}^{N} w_{\alpha,\beta,\gamma,\delta}(x) R_n\big(x(x+\gamma+\delta+1);\alpha,\beta,\gamma,\delta\big)\, f(x)$$
$$= \sum_{x=0}^{N-1} \frac{\gamma+\delta+2}{\gamma+\delta+2+2x} w_{\alpha+1,\beta+1,\gamma+1,\delta}(x)$$
$$\times R_{n-1}(x(x+\gamma+\delta+2);\alpha+1,\beta+1,\gamma+1,\delta)(f(x)-f(x+1)). \quad (2.16)$$

2.3. Jacobi and Gegenbauer functions

In [9, (4.3)] Hallnäs and Ruijsenaars define a *conical function*

$$F(g;r,2k) := \left(\frac{\pi}{4}\right)^{\frac{1}{2}} \frac{\Gamma(g+ik)\Gamma(g-ik)}{\Gamma(g)(2\sinh r)^{g-\frac{1}{2}}}\, P^{\frac{1}{2}-g}_{ik-\frac{1}{2}}(\cosh r) \quad (r>0,\ \mathrm{Re}\, g>0). \quad (2.17)$$

Here the P-function is the *associated Legendre function of the first kind* which is expressed by [17, (14.3.6) and (15.1.2)] as Gauss hypergeometric function:

$$P^{\mu}_{\nu}(x) = \frac{1}{\Gamma(1-\mu)} \left(\frac{x+1}{x-1}\right)^{\frac{1}{2}\mu} {}_2F_1\left(\begin{matrix}\nu+1,-\nu\\ 1-\mu\end{matrix}; \tfrac{1}{2}-\tfrac{1}{2}x\right) \quad (x>1). \quad (2.18)$$

Substitute (2.18) in (2.17) and also use Euler's transformation formula [17, (15.8.1)] and Legendre's duplication formula [17, (5.5.5)]. Then we obtain

$$F(g;r,2k) = \frac{\Gamma(g+ik)\Gamma(g-ik)}{2\Gamma(2g)}\, {}_2F_1\left(\begin{matrix}g+ik,g-ik\\ g+\frac{1}{2}\end{matrix}; -\sinh^2 \tfrac{1}{2}r\right). \quad (2.19)$$

By [17, (15.9.11)] this can be written in terms of *Jacobi functions* [14, 15]

$$\phi_\lambda^{(\alpha,\beta)}(t) := {}_2F_1\left(\begin{matrix}\frac{1}{2}(\alpha+\beta+1+i\lambda), \frac{1}{2}(\alpha+\beta+1-i\lambda)\\ \alpha+1\end{matrix}; -\sinh^2 t\right) \quad (t \in \mathbb{R}) \tag{2.20}$$

(called *Gegenbauer functions* if $\alpha = \beta$) as

$$F(g;r,2k) = \frac{\Gamma(g+ik)\Gamma(g-ik)}{2\Gamma(2g)}\,\phi_{2k}^{(g-\frac{1}{2},g-\frac{1}{2})}(\tfrac{1}{2}r).$$

By [14, (2.8)]

$$\phi_{2\lambda}^{(\alpha,\alpha)}(t) = \phi_\lambda^{(\alpha,-\frac{1}{2})}(2t), \tag{2.21}$$

this becomes

$$F(g;r,2k) = \frac{\Gamma(g+ik)\Gamma(g-ik)}{2\Gamma(2g)}\,\phi_k^{(g-\frac{1}{2},-\frac{1}{2})}(r) \tag{2.22}$$

or equivalently,

$$F(\alpha+\tfrac{1}{2};t,2\lambda) = \frac{\Gamma(\alpha+\frac{1}{2}+i\lambda)\Gamma(\alpha+\frac{1}{2}-i\lambda)}{2\Gamma(2\alpha+1)}\,\phi_\lambda^{(\alpha,-\frac{1}{2})}(t). \tag{2.23}$$

Note that, by (2.20), $\phi_\lambda^{(\alpha,\beta)}(0) = 1$. From [15, (6.1)] we have

$$|\phi_\lambda^{(\alpha,\beta)}(t)| \le 1 \quad (\alpha \ge \beta \ge -\tfrac{1}{2},\ t \in \mathbb{R},\ |\operatorname{Im}\lambda| \le \alpha+\beta+1). \tag{2.24}$$

The contiguous relation

$${}_2F_1\left(\begin{matrix}a,b\\ c\end{matrix};z\right) - {}_2F_1\left(\begin{matrix}a-1,b+1\\ c\end{matrix};z\right) = \frac{(b-a+1)z}{c}\,{}_2F_1\left(\begin{matrix}a,b+1\\ c+1\end{matrix};z\right),$$

which follows immediately by substitution of the power series for the three ${}_2F_1$ functions, can be rewritten by (2.20) in terms of Jacobi functions:

$$\frac{\phi_{\lambda-i}^{(\alpha,\beta)}(t) - \phi_{\lambda+i}^{(\alpha,\beta)}(t)}{i\lambda} = \frac{\sinh^2 t}{\alpha+1}\,\phi_\lambda^{(\alpha+1,\beta)}(t). \tag{2.25}$$

2.4. Wilson polynomials

Wilson polynomials [11, Section 9.1] are defined by

$$\frac{W_n(x^2;a,b,c,d)}{(a+b)_n(a+c)_n(a+d)_n} := {}_4F_3\left(\begin{matrix}-n,n+a+b+c+d-1,a+ix,a-ix\\ a+b,a+c,a+d\end{matrix};1\right). \tag{2.26}$$

We need these polynomials with parameters $\pm i\lambda \pm i\mu + \frac{1}{2}\alpha + \frac{1}{4}$ ($\alpha > -\frac{1}{2}$, $\lambda, \mu \in \mathbb{R}$). Then the orthogonality relation becomes [11, (9.1.2)]

$$\begin{aligned}\frac{1}{4\pi}\int_{-\infty}^{\infty}(W_mW_n)(\nu^2;\pm i\lambda \pm i\mu + \tfrac{1}{2}\alpha + \tfrac{1}{4})\left|\frac{\Gamma\big(i\nu \pm i\lambda \pm i\mu + \frac{1}{2}\alpha + \frac{1}{4}\big)}{\Gamma(2i\nu)}\right|^2 d\nu \\ = \frac{\Gamma(\alpha+\frac{1}{2})^2\,|\Gamma(n+\alpha+\frac{1}{2}+2i\lambda)|^2\,|\Gamma(n+\alpha+\frac{1}{2}+2i\mu)|^2}{\Gamma(2n+2\alpha+1)} \\ \times (n+2\alpha)_n\, n!\,\delta_{m,n}.\end{aligned} \tag{2.27}$$

Here and later $\pm i\lambda \pm i\mu + \frac{1}{2}\alpha + \frac{1}{4}$ in the parameter list means four elements in the list with the four possibilities given by the two $\pm$ signs. Similarly $\Gamma\big(i\nu \pm i\lambda \pm i\mu + \frac{1}{2}\alpha + \frac{1}{4}\big)$ stands for a product of four Gamma functions. We also wrote $(4\pi)^{-1}\int_{-\infty}^{\infty}$ instead of the usual $(2\pi)^{-1}\int_0^{\infty}$, which is allowed because the integrand is an even function of ν.

The backward shift operator equation [11, (9.1.9)] can be rewritten as

$$\begin{aligned}\frac{\Gamma(a\pm ix)\cdots\Gamma(d\pm ix)}{\Gamma(\pm 2ix)}W_n(x^2;a,b,c,d) \\ = \frac{\Gamma(a+\frac{1}{2}\pm i(x+\frac{1}{2}i))\cdots\Gamma(d+\frac{1}{2}\pm i(x+\frac{1}{2}i))}{2i(x+\frac{1}{2}i)\,\Gamma(\pm 2i(x+\frac{1}{2}i))} \\ \times W_n((x+\tfrac{1}{2}i)^2;a+\tfrac{1}{2},b+\tfrac{1}{2},c+\tfrac{1}{2},d+\tfrac{1}{2}) \\ - \frac{\Gamma(a+\frac{1}{2}\pm i(x-\frac{1}{2}i))\cdots\Gamma(d+\frac{1}{2}\pm i(x-\frac{1}{2}i))}{2i(x-\frac{1}{2}i)\,\Gamma(\pm 2i(x-\frac{1}{2}i))} \\ \times W_n((x-\tfrac{1}{2}i)^2;a+\tfrac{1}{2},b+\tfrac{1}{2},c+\tfrac{1}{2},d+\tfrac{1}{2}).\end{aligned} \tag{2.28}$$

3. The Addition Formula for Gegenbauer Polynomials

In this section we briefly review the addition formula for Gegenbauer polynomials and formulas associated with it. This extends the discussion of the Legendre case in Section 1. As a reference, see for instance [2, Section 9.8]. We use the notation (2.1).

Product formula ($\alpha > -\frac{1}{2}$)

$$R_n^{(\alpha,\alpha)}(x)R_n^{(\alpha,\alpha)}(y) = \frac{\Gamma(\alpha+1)}{\Gamma(\alpha+\frac{1}{2})\Gamma(\frac{1}{2})} \times\int_{-1}^{1} R_n^{(\alpha,\alpha)}\left(xy+(1-x^2)^{\frac{1}{2}}(1-y^2)^{\frac{1}{2}}t\right)(1-t^2)^{\alpha-\frac{1}{2}}\,dt. \tag{3.1}$$

Addition formula

$$R_n^{(\alpha,\alpha)}(xy+(1-x^2)^{\frac{1}{2}}(1-y^2)^{\frac{1}{2}}t) = \sum_{k=0}^{n}\binom{n}{k}\frac{\alpha+k}{\alpha+\frac{1}{2}k}\,\frac{(n+2\alpha+1)_k(2\alpha+1)_k}{2^{2k}(\alpha+1)_k^2} \times(1-x^2)^{\frac{1}{2}k}R_{n-k}^{(\alpha+k,\alpha+k)}(x)\,(1-y^2)^{\frac{1}{2}k}R_{n-k}^{(\alpha+k,\alpha+k)}(y)\,R_k^{(\alpha-\frac{1}{2},\alpha-\frac{1}{2})}(t). \tag{3.2}$$

Addition formula for $t = 1$

$$R_n^{(\alpha,\alpha)}(xy+(1-x^2)^{\frac{1}{2}}(1-y^2)^{\frac{1}{2}}) = \sum_{k=0}^{n}\binom{n}{k}\frac{\alpha+k}{\alpha+\frac{1}{2}k}\,\frac{(n+2\alpha+1)_k(2\alpha+1)_k}{2^{2k}(\alpha+1)_k^2} \times(1-x^2)^{\frac{1}{2}k}R_{n-k}^{(\alpha+k,\alpha+k)}(x)\,(1-y^2)^{\frac{1}{2}k}R_{n-k}^{(\alpha+k,\alpha+k)}(y). \tag{3.3}$$

For $x=\cos\theta_1$, $y=\cos\theta_2$ the left-hand side takes the form $R_n^{(\alpha,\alpha)}\big(\cos(\theta_1-\theta_2)\big)$.

Addition formula for $t = 1$, $x = y$

$$1=\sum_{k=0}^{n}\binom{n}{k}\frac{\alpha+k}{\alpha+\frac{1}{2}k}\,\frac{(n+2\alpha+1)_k(2\alpha+1)_k}{2^{2k}(\alpha+1)_k^2}\,(1-x^2)^k\big(R_{n-k}^{(\alpha+k,\alpha+k)}(x)\big)^2. \tag{3.4}$$

This shows in particular that $|R_n^{(\alpha,\alpha)}(x)| \le 1$ if $x \in [-1,1]$ and $\alpha > -\frac{1}{2}$, [17, (18.14.1)]. This is also well known by several other methods, including as a corollary of (3.1).

Limit to Hermite polynomials

In the addition formula (3.2) replace x by $\alpha^{-\frac{1}{2}}x$, t by $\alpha^{-\frac{1}{2}}t$, multiply both sides of (3.2) by $\alpha^{\frac{1}{2}n}$ and let $\alpha \to \infty$. By (2.7) and (2.8) we obtain

$$H_n\big(xy + (1-y^2)^{\frac{1}{2}}t\big) = \sum_{k=0}^{n} \binom{n}{k} H_{n-k}(x)\, H_k(t)\, (1-y^2)^{\frac{1}{2}k}\, y^{n-k}. \tag{3.5}$$

This is the case $n = 2$ of [5, 10.13(40)]. The corresponding limit of the product formula (3.1) is

$$H_n(x)\, y^n = \frac{1}{\sqrt{\pi}} \int_{-\infty}^{\infty} H_n\big(xy + (1-y^2)^{\frac{1}{2}}t\big)\, e^{-t^2}\, dt. \tag{3.6}$$

History of the addition formula for Gegenbauer polynomials

The addition formula (3.2) for Gegenbauer polynomials is usually ascribed to Gegenbauer [7] in (1874). However, it is already stated and proved by Allé [1] in 1865. The subsequent proofs by Gegenbauer [7, 8] in 1874 and 1893, and by Heine [10, p. 455] in (1878) do not mention Allé's result.

4. The Dual Addition Formula for Gegenbauer Polynomials

The linearization formula for Gegenbauer polynomials, see [3, (5.7)], can be written as

$$\begin{aligned} R_l^{(\alpha,\alpha)}(x) R_m^{(\alpha,\alpha)}(x) = \frac{l!\, m!}{(2\alpha+1)_l (2\alpha+1)_m} &\sum_{j=0}^{\min(l,m)} \frac{l+m+\alpha+\frac{1}{2}-2j}{\alpha+\frac{1}{2}} \\ &\times \frac{(\alpha+\frac{1}{2})_j (\alpha+\frac{1}{2})_{l-j} (\alpha+\frac{1}{2})_{m-j} (2\alpha+1)_{l+m-j}}{j!\, (l-j)!\, (m-j)!\, (\alpha+\frac{3}{2})_{l+m-j}} \\ &\times R_{l+m-2j}^{(\alpha,\alpha)}(x). \end{aligned} \tag{4.1}$$

As mentioned in [4, (4.18)], Rogers already gave the analogous linearization formula for q-ultraspherical polynomials in 1895 and observed (4.1) as a special case. Then (4.1) was independently given by Dougall in 1919 without proof. See [3, p. 40] for a discussion of further treatments of (4.1). See also [2, Theorem 6.8.2] and the proof and discussion following the theorem.

From now on assume $\alpha > -\frac{1}{2}$. Then the linearization coefficients in (4.1) are nonnegative (as they are in the degenerate case $\alpha = -\frac{1}{2}$). We also assume, without loss of generality, that $l \geq m$.

It is rather hidden in (4.1) that the linearization coefficients are special cases of orthogonality weights (2.11) for Racah polynomials. But indeed, a further rewriting of (4.1) and substitution of (2.11) and (2.13) gives

$$R_l^{(\alpha,\alpha)}(x) R_m^{(\alpha,\alpha)}(x) = \sum_{j=0}^{m} \frac{w_{\alpha-\frac{1}{2},\alpha-\frac{1}{2},-m-1,-l-\alpha-\frac{1}{2}}(j)}{h_{0;\,\alpha-\frac{1}{2},\alpha-\frac{1}{2},-m-1,-l-\alpha-\frac{1}{2}}} \times R_{l+m-2j}^{(\alpha,\alpha)}(x) \quad (l \geq m). \tag{4.2}$$

This identity can be considered as giving the constant term of an expansion of $R_{l+m-2j}^{(\alpha,\alpha)}(x)$ as a function of j in terms of Racah polynomials

$$R_n\big(j(j-l-m-\alpha-\tfrac{1}{2});\, \alpha-\tfrac{1}{2}, \alpha-\tfrac{1}{2}, -m-1, -l-\alpha-\tfrac{1}{2}\big) = {}_4F_3\left(\begin{matrix} -n, n+2\alpha, -j, j-l-m-\alpha-\frac{1}{2} \\ \alpha+\frac{1}{2}, -l, -m \end{matrix}; 1\right). \tag{4.3}$$

The general terms of this expansion will be obtained by evaluating the sum

$$S_{n,l,m}^{(\alpha)}(x) := \sum_{j=0}^{m} w_{\alpha-\frac{1}{2},\alpha-\frac{1}{2},-m-1,-l-\alpha-\frac{1}{2}}(j)\, R_{l+m-2j}^{(\alpha,\alpha)}(x) \times R_n\big(j(j-l-m-\alpha-\tfrac{1}{2});\; \alpha-\tfrac{1}{2}, \alpha-\tfrac{1}{2}, -m-1, -l-\alpha-\tfrac{1}{2}\big), \tag{4.4}$$

where we still assume $l \geq m$ and where $n \in \{0, \ldots, m\}$.

Theorem 4.1. *The sum (4.4) can be evaluated as*

$$S_{n,l,m}^{(\alpha)}(x) = \frac{(2\alpha+1)_{l+n}(2\alpha+1)_{m+n}(\alpha+\frac{1}{2})_{l+m}}{2^{2n}(\alpha+\frac{1}{2})_l(\alpha+\frac{1}{2})_m(2\alpha+1)_{l+m}(\alpha+1)_n^2} \times (x^2-1)^n\, R_{l-n}^{(\alpha+n,\alpha+n)}(x)\, R_{m-n}^{(\alpha+n,\alpha+n)}(x). \tag{4.5}$$

Proof. By (4.4), (2.16) and (2.5) we obtain the recurrence

$$S_{n,l,m}^{(\alpha)} = \frac{-l-m-\alpha+\frac{1}{2}}{\alpha+1}(1-x^2)\, S_{n-1,l-1,m-1}^{(\alpha+1)}.$$

Iteration gives

$$S_{n,l,m}^{(\alpha)} = \frac{(-l-m-\alpha+\frac{1}{2})_n}{(\alpha+1)_n} (1-x^2)^n S_{0,l-n,m-n}^{(\alpha+n)}.$$

Now use (4.4), (4.1) and (2.13). □

As an immediate corollary, by the orthogonality relation (2.10) for Racah polynomials and by substitution of (2.12) and (2.13), we obtain the following theorem.

Theorem 4.2 (Dual addition formula). *For $l \geq m$ and for $j \in \{0, \ldots, m\}$, there is the expansion*

$$\begin{aligned} R_{l+m-2j}^{(\alpha,\alpha)}(x) = \sum_{n=0}^{m} &\frac{\alpha+n}{\alpha+\frac{1}{2}n} \frac{(-l)_n(-m)_n(2\alpha+1)_n}{2^{2n}(\alpha+1)_n^2\, n!} \\ &\times (x^2-1)^n R_{l-n}^{(\alpha+n,\alpha+n)}(x)\, R_{m-n}^{(\alpha+n,\alpha+n)}(x) \\ &\times R_n\big(j(j-l-m-\alpha-\tfrac{1}{2}); \\ &\quad \alpha-\tfrac{1}{2}, \alpha-\tfrac{1}{2}, -m-1, -l-\alpha-\tfrac{1}{2}\big). \end{aligned} \tag{4.6}$$

In particular, for $j=0$,

$$\begin{aligned} R_{l+m}^{(\alpha,\alpha)}(x) = \sum_{n=0}^{m} &\frac{\alpha+n}{\alpha+\frac{1}{2}n} \frac{(-l)_n(-m)_n(2\alpha+1)_n}{2^{2n}(\alpha+1)_n^2\, n!} \\ &\times (x^2-1)^n R_{l-n}^{(\alpha+n,\alpha+n)}(x)\, R_{m-n}^{(\alpha+n,\alpha+n)}(x), \end{aligned} \tag{4.7}$$

and for $j=m$ we obtain by (2.14) that

$$\begin{aligned} R_{l-m}^{(\alpha,\alpha)}(x) = \sum_{n=0}^{m} &\binom{m}{n} \frac{\alpha+n}{\alpha+\frac{1}{2}n} \frac{(l+2\alpha+1)_n(2\alpha+1)_n}{2^{2n}(\alpha+1)_n^2} \\ &\times (1-x^2)^n R_{l-n}^{(\alpha+n,\alpha+n)}(x)\, R_{m-n}^{(\alpha+n,\alpha+n)}(x), \end{aligned} \tag{4.8}$$

which is dual to (3.3) and which has a further specialization to

$$\begin{aligned} 1 = \sum_{n=0}^{m} &\binom{m}{n} \frac{\alpha+n}{\alpha+\frac{1}{2}n} \frac{(m+2\alpha+1)_n(2\alpha+1)_n}{2^{2n}(\alpha+1)_n^2} \\ &\times (1-x^2)^n \big(R_{m-n}^{(\alpha+n,\alpha+n)}(x)\big)^2. \end{aligned} \tag{4.9}$$

Note that (4.9) coincides with formula (3.4), which is a specialization of the addition formula (3.2) for Gegenbauer polynomials.

Remark 4.3. It follows from (4.4) and (2.3) that

$$\int_{-1}^{1} S_{n,l,m}^{(\alpha)}(x)\, R_{l+m-2j}^{(\alpha,\alpha)}(x)\,(1-x^2)^\alpha\,dx$$
$$= w_{\alpha-\frac{1}{2},\alpha-\frac{1}{2},-m-1,-l-\alpha-\frac{1}{2}}(j)\, h_{l+m-2j}^{(\alpha,\alpha)}$$
$$\times R_n\big(j(j-l-m-\alpha-\tfrac{1}{2});\,\alpha-\tfrac{1}{2},\alpha-\tfrac{1}{2},-m-1,-l-\alpha-\tfrac{1}{2}\big). \tag{4.10}$$

By (4.5) and (4.3) we can rewrite this as

$$\int_{-1}^{1} R_{l-n}^{(\alpha+n,\alpha+n)}(x)\, R_{m-n}^{(\alpha+n,\alpha+n)}(x)\, R_{l+m-2j}^{(\alpha,\alpha)}(x)\,(1-x^2)^{\alpha+n}\,dx$$
$$= \text{const.}\ {}_4F_3\left(\begin{matrix}-n,n+2\alpha,-j,j-l-m-\alpha-\frac{1}{2}\\ \alpha+\frac{1}{2},-l,-m\end{matrix};1\right)$$
$$= \text{const.}\ {}_4F_3\left(\begin{matrix}-m+n,-m-n-2\alpha,j-m,l-j+\alpha+\frac{1}{2}\\ -m,-m-\alpha+\frac{1}{2},l-m+1\end{matrix};1\right),$$

where the second equality follows by twofold application of Whipple's identity [2, Theorem 3.3.3] and where the constants can be given as explicit, elementary, but somewhat tedious expressions. It turns out that the second ${}_4F_3$ evaluation of the integral above precisely matches the formula given by Koelink *et al.* [12, (2.6]. Just put there (without loss of generality) $k=0$ and replace α,β,n,m,t by $\alpha+n-\frac{1}{2},\alpha+\frac{1}{2},l-n,m-n,j-n$, respectively. So, in a sense, the dual addition formula for Gegenbauer polynomials was already derived there in disguised form.

5. A Limit to Hermite Polynomials

We will do a rescaling in the dual addition formula (4.6) such that we can take the limit for $\alpha\to\infty$. For this purpose observe that the Racah polynomial (4.3) (where $l \geq m \geq \max(j,n)$) has limits

$$\begin{aligned}
&\lim_{\alpha\to\infty} \alpha^{-j} R_n\big(j(j-l-m-\alpha-\tfrac{1}{2});\\
&\qquad \alpha-\tfrac{1}{2},\alpha-\tfrac{1}{2},-m-1,-l-\alpha-\tfrac{1}{2}\big) = \frac{2^j(-n)_j}{(-l)_j(-m)_j},\\
&\lim_{\alpha\to\infty} \alpha^{-n} R_n\big(j(j-l-m-\alpha-\tfrac{1}{2});\\
&\qquad \alpha-\tfrac{1}{2},\alpha-\tfrac{1}{2},-m-1,-l-\alpha-\tfrac{1}{2}\big) = \frac{2^n(-j)_n}{(-l)_n(-m)_n}.
\end{aligned} \tag{5.1}$$

Otherwise said, $R_n\big(j(j-l-m-\alpha-\frac{1}{2});\ \alpha-\frac{1}{2},\alpha-\frac{1}{2},-m-1,-l-\alpha-\frac{1}{2}\big) = O(\alpha^{\min(n,j)})$ as $\alpha\to\infty$ with the order constant given in (5.1).

Now, in (4.6), replace x by $\alpha^{-\frac{1}{2}}x$, multiply both sides by $\alpha^{\frac{1}{2}(l+m-2j)}$ and let $\alpha\to\infty$. By (2.7) and (5.1) we obtain

$$\begin{aligned}
&2^j(-l)_j(-m)_j H_{l+m-2j}(x)\\
&\quad = \sum_{n=j}^{m} \frac{(-n)_j}{n!}\,(-2)^n(-l)_n(-m)_n\, H_{l-n}(x)H_{m-n}(x) \quad (l\geq m),
\end{aligned} \tag{5.2}$$

which may be called the dual addition formula for Hermite polynomials. Formula (5.2) for arbitrary j is equivalent to its case $j=0$,

$$H_{l+m}(x) = \sum_{n=0}^{m} \frac{(-2)^n(-l)_n(-m)_n}{n!}\, H_{l-n}(x)H_{m-n}(x) \quad (l\geq m), \tag{5.3}$$

and this is precisely [5, 10.13(36)].

Next we want to consider the limit as $\alpha\to\infty$ of (4.5) with $S^{(\alpha)}_{n,l,m}$ given by (4.4). Recall that (4.5) together with (4.4) is the dual of (4.6) in the sense of Fourier–Racah inversion. Observe from (2.11)–(2.13) that

$$\lim_{\alpha\to\infty} \alpha^j w_{\alpha-\frac{1}{2},\alpha-\frac{1}{2},-m-1,-l-\alpha-\frac{1}{2}}(j) = \frac{(-l)_j(-m)_j}{2^j j!}, \tag{5.4}$$

$$\lim_{\alpha\to\infty} \alpha^{-n} h_{n;\alpha-\frac{1}{2},\alpha-\frac{1}{2},-m-1,-l-\alpha-\frac{1}{2}} = \frac{2^n n!}{(-l)_n(-m)_n}. \tag{5.5}$$

In (4.5), replace x by $\alpha^{-\frac{1}{2}}x$, multiply both sides by $\alpha^{\frac{1}{2}(l+m-2n)}$ and let $\alpha \to \infty$. By (2.7), (5.1) and (5.4) we obtain

$$\sum_{j=n}^{m} \frac{(-j)_n}{j!} 2^j (-l)_j (-m)_j \, H_{l+m-2j}(x) = (-2)^n (-l)_n (-m)_n \, H_{l-n}(x) H_{m-n}(x) \quad (l \geq m). \tag{5.6}$$

Again, as with (5.2), formula (5.6) for arbitrary n is equivalent to its case $n = 0$,

$$\sum_{j=0}^{m} \frac{2^j (-l)_j (-m)_j}{j!} H_{l+m-2j}(x) = H_l(x) H_m(x) \quad (l \geq m), \tag{5.7}$$

and this is precisely the linearization formula [3, p. 42] for Hermite polynomials.

Just as with (4.6) and (4.5), the identities (5.2) and (5.6) can be obtained from each other by a Fourier type inversion. This no longer involves an orthogonal system as the Racah polynomials but a biorthogonal system implied by the *biorthogonality relation* (see [20, Section 2.1])

$$\sum_{j=0}^{\infty} \frac{(-n)_j}{n!} \frac{(-j)_k}{k!} = \delta_{n,k} \,. \tag{5.8}$$

Note that the above sum in fact runs from $j = k$ to n. The biorthogonality (5.8) is also a limit case of the Racah orthogonality relation (2.10). Indeed, replace $\alpha, \beta, \gamma, \delta$ by $\alpha - \frac{1}{2}, \alpha - \frac{1}{2}, -m - 1, -l - \alpha - \frac{1}{2}$, multiply both sides of (2.10) by α^{-n}, let $\alpha \to \infty$, and use (5.1), (5.4) and (5.5). It is quite remarkable that a biorthogonal (and essentially nonorthogonal) system can be obtained as a limit case of an orthogonal system. Of course, before the limit it taken, the orthogonal system already has to be prepared as a biorthogonal system by rescaling.

6. The Dual Addition Formula for Gegenbauer Functions

The dual product formula for the functions (2.17) is given in [9, (4.17)] as

$$F(g; r, 2p) F(g; r, 2q) = \frac{1}{8\pi} \int_0^{\infty} F(g; r, 2k) \times \frac{\prod_{\delta_1, \delta_2, \delta_3 = +,-} \Gamma(\frac{1}{2}(g + i\delta_1 p + i\delta_2 q + i\delta_3 k))}{\Gamma(g)^2 \prod_{\delta = +,-} \Gamma(i\delta k) \Gamma(g + i\delta k)} \, dk, \tag{6.1}$$

where $g \in (0,\infty)$ and $r,p,q \in \mathbb{R}$. The formula is obtained there as a limit case of a similar formula for a q-analogue (or relativistic analogue) of the Gegenbauer function. By (2.22) we can rewrite (6.1) as

$$\frac{\Gamma(\alpha+\frac{1}{2})^2\,|\Gamma(\alpha+\frac{1}{2}+2i\lambda)|^2\,|\Gamma(\alpha+\frac{1}{2}+2i\mu)|^2}{\Gamma(2\alpha+1)}\,\phi_{2\lambda}^{(\alpha,-\frac{1}{2})}(t)\,\phi_{2\mu}^{(\alpha,-\frac{1}{2})}(t)$$

$$=\frac{1}{4\pi}\int_{-\infty}^{\infty}\phi_{2\nu}^{(\alpha,-\frac{1}{2})}(t)\left|\frac{\Gamma\big(i\nu\pm i\lambda\pm i\mu+\frac{1}{2}\alpha+\frac{1}{4}\big)}{\Gamma(2i\nu)}\right|^2 d\nu, \tag{6.2}$$

where $\alpha > -\frac{1}{2}$ and $t,\lambda,\mu \in \mathbb{R}$. Note that the integral in (6.2) converges absolutely by (2.24) and by estimates for the Gamma quotient using [17, (5.5.5) and (5.11.12)] and, from [17, (5.5.3)],

$$|\Gamma(\tfrac{1}{2}+i\nu)|^2=\frac{\pi}{\cosh(\pi\nu)}\,.$$

The cases $\alpha = 0$ and $\frac{1}{2}$ of (6.2) were earlier given by Mizony [16].

We recognize the weight function in the integrand of (6.2) as the weight function in the orthogonality relation (2.27) for Wilson polynomials with parameters $\pm i\lambda \pm i\mu + \frac{1}{2}\alpha + \frac{1}{4}$. The case $t = 0$ of (6.2) coincides with the case $m = n = 0$ of (2.27).

Similarly as the sum (4.4) is suggested by formula (4.2), we are led by formula (6.2) to try to evaluate the integral

$$I_n^\alpha(\lambda,\mu) := \frac{1}{4\pi}\int_{-\infty}^{\infty}\phi_{2\nu}^{(\alpha,-\frac{1}{2})}(t)\,W_n(\nu^2;\pm i\lambda\pm i\mu+\tfrac{1}{2}\alpha+\tfrac{1}{4})$$
$$\times\left|\frac{\Gamma\big(i\nu\pm i\lambda\pm i\mu+\frac{1}{2}\alpha+\frac{1}{4}\big)}{\Gamma(2i\nu)}\right|^2 d\nu. \tag{6.3}$$

Here and further in this section we will only work formally. We will not bother about convergence, moving of integration contours and justification of Fourier–Wilson inversion. Certainly this should be repaired later.

By (2.28) and by shifting integration contours we get

$$I_n^\alpha(\lambda,\mu) = \frac{1}{4\pi}\int_{-\infty}^{\infty}\frac{\phi_{2\nu-i}^{(\alpha,-\frac{1}{2})}(t)-\phi_{2\nu+i}^{(\alpha,-\frac{1}{2})}(t)}{2i\nu}$$
$$\times\,W_{n-1}(\nu^2;\pm i\lambda\pm i\mu+\tfrac{1}{2}(\alpha+1)+\tfrac{1}{4})$$

$$\times \left| \frac{\Gamma\big(i\nu \pm i\lambda \pm i\mu + \frac{1}{2}(\alpha+1) + \frac{1}{4}\big)}{\Gamma(2i\nu)} \right|^2 d\nu = \frac{\sinh^2 t}{\alpha+1}\, I_{n-1}^{\alpha+1}(\lambda,\mu),$$

where the last equality follows by (2.25). Iteration gives

$$I_n^\alpha(\lambda,\mu) = \frac{(\sinh t)^{2n}}{(\alpha+1)_n}\, I_0^{\alpha+n}(\lambda,\mu).$$

Hence, by (6.3) and (6.2),

$$\begin{aligned}
&\frac{1}{4\pi}\int_{-\infty}^{\infty} \phi_{2\nu}^{(\alpha,-\frac{1}{2})}(t)\, W_n(\nu^2; \pm i\lambda \pm i\mu + \tfrac{1}{2}\alpha + \tfrac{1}{4}) \\
&\qquad \times \left| \frac{\Gamma\big(i\nu \pm i\lambda \pm i\mu + \frac{1}{2}\alpha + \frac{1}{4}\big)}{\Gamma(2i\nu)} \right|^2 d\nu \\
&\quad = \frac{\Gamma(\alpha+\frac{1}{2})^2\, |\Gamma(n+\alpha+\frac{1}{2}+2i\lambda)|^2\, |\Gamma(n+\alpha+\frac{1}{2}+2i\mu)|^2}{\Gamma(2n+2\alpha+1)} \\
&\qquad \times \frac{(\sinh t)^{2n}}{(\alpha+1)_n}\, \phi_{2\lambda}^{(\alpha+n,-\frac{1}{2})}(t)\, \phi_{2\mu}^{(\alpha+n,-\frac{1}{2})}(t). \qquad (6.4)
\end{aligned}$$

By (2.27) there corresponds, at least formally, to (6.4) the orthogonal expansion

$$\begin{aligned}
\phi_{2\nu}^{(\alpha,-\frac{1}{2})}(t) &= \sum_{k=0}^{\infty} \frac{(\sinh t)^{2k}}{(\alpha+1)_k (k+2\alpha)_k\, k!}\phi_{2\lambda}^{(\alpha+k,-\frac{1}{2})}(t)\, \phi_{2\mu}^{(\alpha+k,-\frac{1}{2})}(t) \\
&\quad \times W_k(\nu^2; \pm i\lambda \pm i\mu + \tfrac{1}{2}\alpha + \tfrac{1}{4}).
\end{aligned}$$

Equivalently, by (2.21), we can write what we call the *dual addition formula for Gegenbauer functions*:

$$\begin{aligned}
\phi_{4\nu}^{(\alpha,\alpha)}(t) &= \sum_{k=0}^{\infty} \frac{(\sinh 2t)^{2k}}{(\alpha+1)_k (k+2\alpha)_k\, k!}\phi_{4\lambda}^{(\alpha+k,\alpha+k)}(t)\, \phi_{4\mu}^{(\alpha+k,\alpha+k)}(t) \\
&\quad \times W_k(\nu^2; \pm i\lambda \pm i\mu + \tfrac{1}{2}\alpha + \tfrac{1}{4}). \qquad (6.5)
\end{aligned}$$

7. Further Perspective

The results of this paper suggest much further work. The author will only discuss here the polynomial case. A very obvious thing to do is to imitate the approach of Section 4 for q-ultraspherical polynomials, starting with their linearization formula [2, (10.11.10)], which goes back to Rogers (1895). Quite probably, the q-Racah polynomials will pop up there. The results should be compared with the Rahman–Verma product and addition formula [19] for q-ultraspherical polynomials. It would be very interesting to find a dual addition formula for the addition formula for Jacobi polynomials [13], starting with the linearization formula [18].

It would be quite challenging to search for an addition formula on a higher level (for (q-)Racah polynomials) which has both the addition formula and the dual addition formula for Gegenbauer polynomials as a limit case. The recent formula for the linearization coefficients of Askey–Wilson polynomials by Foupouagnigni *et al.* [6, Theorem 21], which involves four summations, does not give much hope for a quick answer to this problem.

Very important will also be to give a group theoretic interpretation for the dual addition formula for Gegenbauer polynomials, for instance when $\alpha = 0$. Possibly this can be done in the context of tensor algebras associated with the group SU(2).

Acknowledgments

The author wants to thank Erik Koelink (Nijmegen) for pointing out that (4.5) would possibly be equivalent to [12, (2.6)] and for suggesting to consider the Hermite limit of the dual addition formula. Josef Hofbauer (Vienna) was very helpful in providing material about the history of the addition formula for Gegenbauer polynomials.

References

[1] M. Allé, Über die Eigenschaften derjenigen Gattung von Functionen, welche in der Entwicklung von $(1 - 2qx + q^2)^{-m/2}$ nach aufsteigenden Potenzen von q auftreten, und über Entwicklung des Ausdruckes $\{1 - 2q[\cos\theta\cos\theta' + \sin\theta\sin\theta'\cos(\psi - \psi')] + q^2\}^{-m/2}$, *Sitz. Math. Natur. Kl. Akad. Wiss. Wien* (*Abt. II*) **51** (1865) 429–458.

[2] G. E. Andrews, R. Askey and R. Roy, *Special Functions* (Cambridge University Press, 1999).

[3] R. Askey, *Orthogonal Polynomials and Special Functions*, Regional Conference Series in Applied Mathematics, Vol. 21 (SIAM, 1975).

[4] R. Askey and M. E. H. Ismail, A generalization of ultraspherical polynomials, in *Studies in Pure Mathematics*, ed. P. Erdös (Birkhäuser, 1983), pp. 55–78.

[5] A. Erdélyi, *Higher Transcendental Functions*, Vol. 2 (McGraw-Hill, 1953).

[6] M. Foupouagnigni, W. Koepf and D. D. Tcheutia, Connection and linearization coefficients of the Askey–Wilson polynomials, *J. Symbolic Comput.* **53** (2013) 96–118.

[7] L. Gegenbauer, Über einige bestimmte Integrale, *Sitz. Math. Natur. Kl. Akad. Wiss. Wien* (*Abt. II*) **70** (1874) 433–443.

[8] L. Gegenbauer, Das Additionstheorem der Functionen $C_n^\nu(x)$, *Sitz. Math. Natur. Kl. Akad. Wiss. Wien* (*Abt. IIa*) **102** (1893) 942–950.

[9] M. Hallnäs and S. Ruijsenaars, Product formulas for the relativistic and nonrelativistic conical functions, Preprint (2015); `arXiv:1508.07191v1 [math.CA]`, to appear in *Representation theory, Special Functions and Painlevé Equation*, ASPM (Advanced Studies in Pure Mathematics).

[10] E. Heine, *Handbuch der Kugelfunctionen. Theorie und Anwendungen. Erster Band, zweite umgearbeitete und vermehrte Auflage* (G. Reimer, Berlin, 1878).

[11] R. Koekoek, P. A. Lesky and R. F. Swarttouw, *Hypergeometric Orthogonal Polynomials and their q-Analogues* (Springer, 2010).

[12] E. Koelink, M. van Pruijssen and P. Román, Matrix-valued orthogonal polynomials associated to $(\mathrm{SU}(2) \times \mathrm{SU}(2), \mathrm{diag})$, II, *Publ. Res. Inst. Math. Sci.* **49** (2013) 271–312.

[13] T. H. Koornwinder, The addition formula for Jacobi polynomials I. Summary of results, *Indag. Math.* **34** (1972) 188–191.

[14] T. Koornwinder, A new proof of a Paley–Wiener type theorem for the Jacobi transform, *Ark. Mat.* **13** (1975) 145–159.

[15] T. H. Koornwinder, Jacobi functions and analysis on noncompact semisimple Lie groups, in *Special Functions: Group Theoretical Aspects and Applications* (Reidel, Dordrecht, 1984), pp. 1–85.

[16] M. Mizony, Algèbres et noyaux de convolution sur le dual sphérique d'un groupe de Lie semi-simple, non compact et de rang 1, *Publ. Dép. Math.* (*Lyon*) **13** (1976) 1–14.

[17] F. W. J. Olver *et al.*, *NIST Handbook of Mathematical Functions* (Cambridge University Press, 2010); `http://dlmf.nist.gov`.

[18] M. Rahman, A non-negative representation of the linearization coefficients of the product of Jacobi polynomials, *Canad. J. Math.* **33** (1981) 915–928.

[19] M. Rahman and A. Verma, Product and addition formulas for the continuous q-ultraspherical polynomials, *SIAM J. Math. Anal.* **17** (1986) 1461–1474.

[20] J. Riordan, *Combinatorial Identities* (Wiley, 1968).

Chapter 20

Holonomic Tools for Basic Hypergeometric Functions

Christoph Koutschan[*,‡] **and Peter Paule**[†,§]

[*]*Johann Radon Institute for Computational and Applied Mathematics (RICAM), Austrian Academy of Sciences*
Altenberger Straße 69, A-4040 Linz, Austria
[†]*Research Institute for Symbolic Computation (RISC)*
Johannes Kepler University
Altenberger Straße 69, A-4040 Linz, Austria
[‡]*christoph.koutschan@ricam.oeaw.ac.at*
[§]*peter.paule@risc.jku.at*

Dedicated to Professor Mourad Ismail at the occasion of his 70th birthday

With the exception of q-hypergeometric summation, the use of computer algebra packages implementing Zeilberger's "holonomic systems approach" in a broader mathematical sense is less common in the field of q-series and basic hypergeometric functions. A major objective of this chapter is to popularize the usage of such tools also in these domains. Concrete case studies showing software in action introduce to the basic techniques. An application highlight is a new computer-assisted proof of the celebrated Ismail–Zhang formula, an important q-analog of a classical expansion formula of plane waves in terms of Gegenbauer polynomials.

Keywords: Basic hypergeometric function; holonomic function; q-holonomic recurrence; symbolic summation; creative telescoping.

Mathematics Subject Classification 2010: 33F10, 05A19, 05A30, 13P10, 39A13, 68W30, 33D15, 33D45

1. Introduction

Quoting Knuth [11, p. 62] the *Concrete Tetrahedron* [10] is "sort of the sequel to *Concrete Mathematics* [3]." Indeed, presenting algorithmic ideas in connection with the symbolic treatment of combinatorial sums, recurrences, and generating functions, it can be viewed as an algorithmic supplement to [3] directed at an audience using computer algebra. Most of the methods under consideration fit into the "holonomic systems approach to special functions identities" notably pioneered by Zeilberger [18].

The authors of this chapter feel that in contrast to applications in the domain of classical hypergeometric functions, the use of such methods and tools is less common in the field of q-series and basic hypergeometric functions. A major objective of this chapter is to popularize the holonomic systems approach also in these domains. In order to illustrate some of the basic (pun intended!) techniques, concrete case studies of software in action are given. To this end, computer algebra packages written in Mathematica are used. These packages are freely downloadable (upon password request) by following the instructions at http://www.risc.uni-linz.ac.at/research/combinat/software.

The chapter is structured as follows: In Section 2 the objects of a computational case study are q-versions of modified Lommel polynomials introduced by Ismail in [4]. To derive properties of this polynomial family, we apply q-holonomic computer algebra tools for guessing, generalized telescoping, and the execution of closure properties.

Section 3 introduces to the algebraic language and concepts needed for an algorithmic treatment of functions defined by mixed (q-)difference-differential equations. Following Zeilberger's holonomic systems approach, special functions are described by (generators of) annihilating ideals in operator algebras. Special function operations like addition, multiplication, integration, or summation are lifted to operations on (the generators of) these ideals. The HolonomicFunctions package implements this algebraic/algorithmic framework. Gegenbauer polynomials are used to show some basic features of the software.

Employing the algorithmic machinery described, Section 4 presents a new computer-assisted proof of the celebrated Ismail–Zhang formula from [7]. This important identity is a q-analog of a classical expansion formula of plane waves in terms of Gegenbauer polynomials and involving Bessel functions. Ingredients of the q-analog are the basic exponential function $\mathcal{E}_q(x; i\omega)$, also introduced by Ismail and Zhang in [7], as well as Jackson's second q-analog of the Bessel functions $J_\nu(x)$ and q-Gegenbauer polynomials.

We want to mention explicitly that the task of computing an annihilating operator for the series side of the Ismail–Zhang formula is leading to the frontiers of what is computationally feasible today. To compute the operator **annSumRHS** in In[43], we had to use recent algorithmic developments [13] as well as human inspection and trial and error to determine suitable denominators in a decisive preprocessing step.

2. Basic Bessel Functions and q-Lommel Polynomials

We begin with basic Bessel functions considered by Ismail in [4]:

$$J_\nu^{(1)}(x;q) = \frac{(q^{\nu+1};q)_\infty}{(q;q)_\infty} \sum_{n=0}^{\infty} \frac{(-1)^n (x/2)^{\nu+2n}}{(q;q)_n (q^{\nu+1};q)_n}, \quad 0 < q < 1,$$

where

$$(a;q)_0 = 1, \quad (a;q)_n = \prod_{j=0}^{n-1} (1 - aq^j), \quad (a;q)_\infty = \prod_{j=0}^{\infty} (1 - aq^j).$$

After opening a Mathematica session we load Riese's package [15] which implements a q-version of Zeilberger's "fast" Algorithm [17]:

In[1]:= << **RISC\`qZeil\`**

Package q-Zeilberger version 4.50 written by Axel Riese Copyright 1992–2009, Research Institute for Symbolic Computation (RISC), Johannes Kepler University, Linz, Austria

The package provides the q-rising factorials via the **qPochhammer** command, i.e., **qPochhammer**$[a, q, k] := (a;q)_k$ and **qPochhammer**$[a, q] := (a;q)_\infty$. For better readability we set

In[2]:= $(a\_)_{k\_}$:= **qPochhammer**$[a, q, k]$

A recurrence for $J_\nu^{(1)}(x;q)$ (=: SUM[ν]) is computed as follows:

In[3]:= **qZeil**$\left[\frac{(q^{\nu+1})_\infty (-1)^n (x/2)^{\nu+2n}}{(q)_\infty (q)_n (q^{\nu+1})_n}, \{n, 0, \infty\}, \nu, 2\right]$

qZeil::natbounds : Assuming appropriate convergence.

Out[3]= $\mathrm{SUM}[\nu] = q^{1-\nu}(-\mathrm{SUM}[\nu-2]) - \frac{2(1-q^{1-\nu})\,\mathrm{SUM}[\nu-1]}{x}$

This corresponds exactly to (1.18), $k=1$, in [4]. Setting $r_1^{(\nu)}(x) := 2(1-q^\nu)x$ we rewrite the previous output Out[3] as

$$q^\nu \,\mathrm{SUM}[\nu+1] = r_1^{(\nu)}\left(\frac{1}{x}\right)\mathrm{SUM}[\nu] - \mathrm{SUM}[\nu-1]. \tag{2.1}$$

By iterating this recurrence, one produces a sequence $r_1^{(\nu)}(x), r_2^{(\nu)}(x), r_3^{(\nu)}(x), \ldots$ of polynomials such that

$$q^\nu q^{\nu+1}\,\mathrm{SUM}[\nu+2] = r_2^{(\nu)}\left(\frac{1}{x}\right)\mathrm{SUM}[\nu] - r_1^{(\nu+1)}\left(\frac{1}{x}\right)\mathrm{SUM}[\nu-1], \tag{2.2}$$

$$q^\nu q^{\nu+1} q^{\nu+2}\,\mathrm{SUM}[\nu+3] = r_3^{(\nu)}\left(\frac{1}{x}\right)\mathrm{SUM}[\nu] - r_2^{(\nu+1)}\left(\frac{1}{x}\right)\mathrm{SUM}[\nu-1], \tag{2.3}$$

and so on. In other words, setting $r_0^{(\nu)}(x) := 1$, the polynomial sequence $\left(r_n^{(\nu)}(x)\right)_{n\geqslant 0}$ determined this way satisfies the relation

$$q^{n\nu+n(n-1)/2} J_{\nu+n}^{(1)}(x;q) = r_n^{(\nu)}\left(\frac{1}{x}\right) J_\nu^{(1)}(x;q) - r_{n-1}^{(\nu+1)}\left(\frac{1}{x}\right) J_{\nu-1}^{(1)}(x;q), \quad n \geqslant 1. \tag{2.4}$$

This is recurrence (1.19) for $k=1$ in [4]. As noted in [4], the polynomials $r_n^{(\nu)}(x)$ are q-versions of the modified Lommel polynomials. The goal of the present case study is to illustrate how computer algebra tools can be used to find out more about the polynomials $r_n^{(\nu)}(x)$.

2.1. Guessing a q-Holonomic Recurrence

First, by iterating recurrence (2.1) as in (2.2) and (2.3), we compute the seven initial polynomials $1, r_1^{(\nu)}(x), r_2^{(\nu)}(x), \ldots, r_6^{(\nu)}(x)$ and store them in a list (not shown in full detail here for space reasons):

In[4]:= **rpolys** $=$ $\{1, 2(1-q^\nu)x, (4-4q^\nu-4q^{1+\nu}+4q^{1+2\nu})x^2 - q^\nu, -2(-1+q^{1+\nu})x(-q^\nu - q^{1+\nu} + 4x^2 - 4q^\nu x^2 - 4q^{2+\nu}x^2 + 4q^{2+2\nu}x^2), \ldots\}$;

As described in [9], the package

In[5]:= **<< RISC´qGeneratingFunctions´**

qGeneratingFunctions Package version 1.9.1
written by Christoph Koutschan
Copyright 2006–2015, Research Institute for Symbolic Computation (RISC),
Johannes Kepler University, Linz, Austria

can be used to guess a recursive pattern for the sequence $\left(r_n^{(\nu)}(x)\right)_{n\geqslant 0}$. To do so, we execute

In[6]:= **srec = QREGuess[rpolys, $s[n]$]**

QREGuess::data : Not enough data. The result might be wrong.

Out[6]= $\{-2qx\, s[n-1]\left(q-q^{n+\nu}\right)+s[n-2]q^{n+\nu}+q^2 s[n]=0,$
$s[0]=1, s[1]=2x\left(1-q^{\nu}\right)\}$

Ignoring the warning, and observing that when using as input more than seven polynomials the guessed recurrence remains stable, the output (i.e., the automatic guess) can be interpreted as a conjecture (it corresponds to (1.20) for $k=1$ in [4]).

Conjecture 2.1. *Let $s_n^{(\nu)}(x), n\geqslant 0$, be the sequence uniquely defined by the recurrence in* Out[6]. *Then $r_n^{(\nu)}(x)=s_n^{(\nu)}(x)$ for all $n\geqslant 0$.*

Definition 2.2. A sequence $(a_n)_{n\geqslant 0}$ that satisfies a linear recurrence with coefficients being polynomials in q^n with coefficients in a field $\mathbb{K}(q)$ is called q-holonomic.

In (computational) applications the coefficient field $\mathbb{K}(q)$ is a rational function field; usually $\mathbb{K}$ is a transcendental extension $\mathbb{K}=\mathbb{Q}(a,b,c,\ldots)$ of $\mathbb{Q}$ containing parameters a, b, c, and so on. In our example, $\mathbb{K}=\mathbb{Q}(q^{\nu})$.

As pointed out in [4], Conjecture 2.1 can be proved by straightforward induction. In order to introduce various aspects of computer algebra, we present an algorithmic proof exploiting *holonomic closure properties.*

2.2. Proof of Conjecture 2.1 and q-Holonomic Closure Properties

Iterating recurrence (2.1) as in (2.2) and (2.3) uniquely determines the polynomials $r_n^{(\nu)}$. Hence to prove Conjecture 2.1 it suffices to prove

$$q^{n\nu+n(n-1)/2}J_{\nu+n}^{(1)}(x;q)=s_n^{(\nu)}\left(\frac{1}{x}\right)J_{\nu}^{(1)}(x;q)$$
$$-s_{n-1}^{(\nu+1)}\left(\frac{1}{x}\right)J_{\nu-1}^{(1)}(x;q),\quad n\geqslant 1,\quad (2.5)$$

where $s_n^{(\nu)}(x)$ is the sequence uniquely defined by the recurrence Out[6] together with the initial values $s_0^{(\nu)}:=1=r_0^{(\nu)}$ and $s_1^{(\nu)}(x):=2(1-q^{\nu})x=r_1^{(\nu)}(x)$.

First we call the **qZeil** package to obtain a recurrence with respect to n for the left side of (2.5):

$$\text{In[7]:= } \mathbf{recLHS = qZeil}\left[q^{n\nu+n((n-1)/2)}\frac{(q^{\nu+n+1})_\infty(-1)^j(x/2)^{\nu+n+2j}}{(q)_j(q^{\nu+n+1})_j},\right.$$
$$\left.\{j,0,\infty\},n,2,\{\nu\}\right]\ /.\ n\to n+2$$

qZeil::natbounds : Assuming appropriate convergence.

$$\text{Out[7]= } \mathrm{SUM}[n+2]=\frac{2\mathrm{SUM}[n+1]\left(1-q^{n+\nu+1}\right)}{x}-\mathrm{SUM}[n]q^{n+\nu}$$

In what follows it will be convenient to work directly with operators. To this end we load

In[8]:= << **RISC´HolonomicFunctions´**

HolonomicFunctions Package version 1.7.1 (09-Oct-2013) written by Christoph Koutschan Copyright 2007–2013, Research Institute for Symbolic Computation (RISC), Johannes Kepler University, Linz, Austria

The following procedure writes the recurrence **recLHS**, which is Out[7], into an operator **opLHS** which annihilates $q^{n\nu+n(n-1)/2}J^{(1)}_{n+\nu}(x;q)$, the left-hand side of (2.5):

In[9]:= **opLHS = ToOrePolynomial[recLHS, SUM[*n*], OreAlgebra [QS[N, *q*^*n*]]]**

$$\text{Out[9]= } S_{N,q}^2+\left(\frac{2Nq^{\nu+1}}{x}-\frac{2}{x}\right)S_{N,q}+Nq^\nu$$

The notation becomes clear by comparison to **recLHS**: N stands for q^n, and $S_{N,q}$ denotes the shift operator with respect to n. For instance, $S_{N,q}F(n)=F(n+1)$, $S_{N,q}N=qNS_{N,q}$, and $S^2_{N,q}F(n)=F(n+2)$.

Calling the same procedure from the **HolonomicFunctions** package, we obtain operator forms of the recurrences of the two parts on the right-hand side of (2.5):

In[10]:= **srec1 = srec[[1]] /. {*x* → 1/*x*, *n* → *n* + 2}**

$$\text{Out[10]= } -\frac{2qs[n+1]\left(q-q^{n+\nu+2}\right)}{x}+s[n]q^{n+\nu+2}+q^2s[n+2]=0$$

This recurrence for the $s_n^{(\nu)}(1/x)$ then is rewritten as an annihilating operator of $s_n^{(\nu)}\left(\frac{1}{x}\right)J_\nu^{(1)}(x;q)$:

In[11]:= **op1RHS = {ToOrePolynomial[srec1, $s[n]$, OreAlgebra [QS[N, q^n]]]}**

Out[11]= $\left\{q^2 S_{N,q}^2 + \left(\frac{2Nq^{\nu+3}}{x} - \frac{2q^2}{x}\right) S_{N,q} + Nq^{\nu+2}\right\}$

Analogously we obtain an annihilating operator of $-s_{n-1}^{(\nu+1)}\left(\frac{1}{x}\right) J_\nu^{(1)}(x;q)$:

In[12]:= **srec2 = srec[[1]] /. {$x \to 1/x, n \to n+2, \nu \to \nu+1$}**

Out[12]= $-\frac{2qs[n+1]\left(q - q^{n+\nu+3}\right)}{x} + s[n]q^{n+\nu+3} + q^2 s[n+2] = 0$

In[13]:= **op2RHS = {ToOrePolynomial[srec2, $s[n]$, OreAlgebra [QS[N, q^n]]]}**

Out[13]= $\left\{q^2 S_{N,q}^2 + \left(\frac{2Nq^{\nu+4}}{x} - \frac{2q^2}{x}\right) S_{N,q} + Nq^{\nu+3}\right\}$

By constructively utilizing the holonomic closure properties, we compute an operator annihilating $s_n^{(\nu)}\left(\frac{1}{x}\right) J_\nu^{(1)}(x;q) - s_{n-1}^{(\nu+1)}\left(\frac{1}{x}\right) J_\nu^{(1)}(x;q)$:

In[14]:= **opRHS = DFinitePlus[op1RHS, op2RHS][[1]] // Factor**

Out[14]= $x^2 S_{N,q}^4 + 2(q+1)x\left(Nq^{\nu+3} - 1\right) S_{N,q}^3 + q\big(4N^2 q^{2\nu+5} + Nx^2 q^{\nu+1} + Nx^2 q^{\nu+2} - 4Nq^{\nu+2} - 4Nq^{\nu+3} + 4\big) S_{N,q}^2 + 2N(q+1)xq^{\nu+2}\left(Nq^{\nu+2} - 1\right) S_{N,q} + N^2 x^2 q^{2\nu+3}$

As often when applying holonomic closure properties this operator is not equal to **opLHS**, but a left multiple of it:

$$\mathbf{opRHS} = \left(x^2 S_{N,q}^2 + 2q(Nq^{\nu+3} - 1)xS_{N,q} + Nq^{\nu+3}x^2\right) \mathbf{opLHS}$$

With the **HolonomicFunctions** package this factorization can be found as follows:

In[15]:= **LMultiple = OreReduce[opRHS, {opLHS}, Extended $\to$ True]**

Out[15]= $\left\{0, 1, \left\{x^2 S_{N,q}^2 + 2qx\left(Nq^{\nu+3} - 1\right) S_{N,q} + Nx^2 q^{\nu+3}\right\}\right\}$

In[16]:= **OreTimes[LMultiple[[3, 1]], opLHS]**

Out[16]= $x^2 S_{N,q}^4 + 2(q+1)x\left(Nq^{\nu+3} - 1\right) S_{N,q}^3 + q\big(4N^2 q^{2\nu+5} + Nx^2 q^{\nu+1} + Nx^2 q^{\nu+2} - 4Nq^{\nu+2} - 4Nq^{\nu+3} + 4\big) S_{N,q}^2 + 2N(q+1)xq^{\nu+2}\left(Nq^{\nu+2} - 1\right) S_{N,q} + N^2 x^2 q^{2\nu+3}$

Summarizing, we have shown that both sides of (2.5) satisfy a recurrence of order 4 with respect to n. Hence the proof of Conjecture 2.1 is completed by verifying (2.5) for $n = 1, 2, 3, 4$ which amounts to checking $r_n^{(\nu)}(x) = s_n^{(\nu)}(x)$ for $n = 0, 1, 2, 3$.

2.3. q-Holonomic Functions and Generalized Telescoping

In order to gain more insight into the structure of the polynomials $s[n] := s_n^{(\nu)}(x)$ we look at the generating function

$$F(t) := \sum_{n=0}^{\infty} s[n]t^n = \sum_{j=0}^{\infty} s_n^{(\nu)}(x)t^n.$$

Taking as input the recurrence **srec**, which including the initial values is a unique presentation of $\left(s_n^{(\nu)}(x)\right)_{n\geqslant 0}$, we first derive a q-difference equation for $F(t)$ by calling a procedure from the **qGeneratingFunctions** package:

In[17]:= **qDiffEq = QRE2SE[srec, s[n], F[t]]**

Out[17]= $\{tq^{\nu}(t+2x)F[qt] + F[t](1-2tx) - 1 = 0, \langle 1\rangle[F[t]] = 1\}$

Here $\langle 1\rangle[F[t]] = 1$ stands for $F[0] = 1$. In view of the q-shift operator $S_{t,q}^k F(t) = F(q^k t)$, q-difference equations like this are also called q-shift equations.

Definition 2.3. A function $F(t)$ that satisfies a linear q-difference equation with coefficients being polynomials in t with coefficients in a field $\mathbb{K}(q)$ is called q-holonomic.

As noted in [4] **qDiffEq**, which is Out[17], can be iterated to obtain an explicit series representation for $F(t)$:

$$\begin{aligned}(1-2tx)F(t) &= 1 - 2xtq^{\nu}\left(1 + \frac{1}{2}\frac{t}{x}\right)F(qt) \\ &= 1 - 2xtq^{\nu}\frac{1+\frac{1}{2}\frac{t}{x}}{1-2txq} + q(2xtq^{\nu})^2 \\ &\quad \times \frac{1+\frac{1}{2}\frac{t}{x}}{1-2txq}\frac{1+\frac{1}{2}\frac{t}{x}q}{1-2txq^2}F(q^2t) + \cdots .\end{aligned}$$

In the limit this iteration process results in [4, (3.3)]:

$$F(t) = \sum_{j=0}^{\infty} \frac{(-2xtq^{\nu})^j \left(-\frac{1}{2}\frac{t}{x}\right)_j}{(2xt)_{j+1}} q^{j(j-1)/2}. \tag{2.6}$$

The series presentation (2.6) can be used to derive a q-hypergeometric sum representation for the $s_n^{(\nu)}(x)$ using two versions of the q-binomial theorem, the finite version by Gauss and the infinite one by Heine; see [4, (3.3) and (3.6)].

To illustrate further functionalities of the **HolonomicFunctions** package, we derive a q-difference equation for the right-hand side of (2.6):

In[18]:= $\mathbf{CreativeTelescoping}\left[\frac{(-2xtV)^j(-\frac{1}{2}t/x)_j}{(2xt)_{j+1}}q^{j(j-1)/2},\right.$ $\left.\mathbf{QS}[J,q^j]-1,\mathbf{QS}[t,q^T]\right]$ /. $V\to q^\nu$ // **Factor**

Out[18]= $\{\{-tq^\nu(t+2x)S_{t,q}+(2tx-1)\},\{2tx-1\}\}$

Setting

$$g_j(t):=\frac{(-2xtq^\nu)^j\left(-\frac{1}{2}\frac{t}{x}\right)_j}{(2xt)_{j+1}}q^{j(j-1)/2}$$

and $\Delta_j F(j):=F(j+1)-F(j)$, the output constitutes the solution of a generalized telescoping problem and means that

$$q^\nu t(t+2x)g_j(qt)-(2tx-1)g_j(t)=\Delta_j(2tx-1)g_j(t).$$

Summing this from $j=0$ to $j=\infty$, and setting the right-hand side of (2.6) to $G(t)$, gives

$$q^\nu t(t+2x)G(qt)-(2tx-1)G(t)=(2tx-1)\left(g_\infty(t)-\frac{1}{1-2xt}\right).$$

Noting that $g_\infty(t)=0$ (as a formal power series in q, or analytically taking $|q|<1$) we obtain **qDiffEq**; i.e., $G(t)$ satisfies the same q-difference equation as for $F(t)$.

As in the case $q=1$ [10, Theorem 7.1] there is a simple but important connection between q-holonomic generating functions and their coefficient sequences.

Theorem 2.4.

$$F(t)=\sum_{n=0}^{\infty}a_nt^n \text{ is } q\text{-holonomic} \iff \left(a_n\right)_{n\geqslant 0} \text{ is } q\text{-holonomic.}$$

One direction of the theorem has been exploited above when deriving **qDiffEq** from **srec**. The inverse direction, this means, to compute from the q-difference equation for $F(t)$ a q-recurrence for the coefficient polynomials $s(n)=s_n^{(\nu)}(x)$, is done as follows:

In[19]:= **QSE2RE[qDiffEq, F[t], s[n]]**

Out[19]= $\{q^2s[n]=2qx\,s[n-1]\left(q-q^{n+\nu}\right)-s[n-2]q^{n+\nu},s[0]=1,$ $s[1]=-2x\left(q^\nu-1\right)\}$

The output is **srec**, the q-recurrence Out[6] for the modified q-Lommel polynomials.

3. Interlude: Annihilating Ideals of Operators

In order to state, in an algebraic language, the concepts that are introduced in this section, and for writing mixed (q-)difference-differential equations in a concise way, the following operator notation is employed: let D_x denote the partial derivative operator with respect to x (x is then called a *continuous variable*), S_n the forward shift operator with respect to n (n is then called a *discrete variable*), and $S_{t,q}$ the q-shift operator with respect to t. More precisely, these operator symbols act on a function f by

$$D_x f = \frac{\partial f}{\partial x}, \quad S_n f = f\big|_{n\to n+1}, \quad \text{and} \quad S_{t,q} f = f\big|_{t\to qt}.$$

The operator notation allows us to translate linear homogeneous (q-)difference-differential equations into polynomials in the operator symbols D_x, S_n, $S_{t,q}$, etc., with coefficients in some field $\mathbb{F}$. Typically, $\mathbb{F}$ is a rational function field in the variables x, n, t, q, etc. For example, the equation

$$\frac{\partial}{\partial x} f(k, n+1, x, y) + n\frac{\partial}{\partial y} f(k, n, x, y) + x f(k+1, n, x, y) - f(k, n, x, y) = 0$$

turns into $Pf(k, n, x, y) = 0$, where P is the operator $D_x S_n + nD_y + xS_k - 1$. An example in the q-case is the annihilating operator **opLHS**, given in Out[9], for $q^{n\nu+n(n-1)/2} J^{(1)}_{n+\nu}(x;q)$,

$$J := S^2_{N,q} + \left(-\frac{2}{x} + \frac{2Nq^{\nu+1}}{x}\right) S_{N,q} + Nq^{\nu}, \tag{3.1}$$

which is a polynomial in the q-shift operator $S_{N,q}$ whose coefficients are elements of the rational function field $\mathbb{Q}(q, q^{\nu}, N, x)$ where $N = q^n$. Note that in general the ring $\mathbb{F}[D_x, S_n, S_{t,q}, \ldots]$ is not commutative: coefficients from $\mathbb{F}$ do not commute with the "variables" D_x, S_n, $S_{t,q}$, etc. For instance, for some $a(x, n, t) \in \mathbb{F}$ one has

$$\begin{aligned} D_x \cdot a(x, n, t) &= a(x, n, t) \cdot D_x + \frac{\partial}{\partial x} a(x, n, t), \\ S_n \cdot a(x, n, t) &= a(x, n+1, t) \cdot S_n, \\ S_{t,q} \cdot a(x, n, t) &= a(x, n, qt) \cdot S_{t,q}. \end{aligned}$$

Such noncommutative rings of operators are called *Ore algebras*; more precise definitions and properties of such algebras can be found in [12].

Example 3.1. We demonstrate how arithmetic operations in an Ore algebra can be used to compute the polynomials $r_i^{(\nu)}$ for $i = 1, 2, 3, \ldots$. For this purpose let us convert the recurrence Out[3] into an operator:

In[20]:= $\mathbf{op = Factor[ToOrePolynomial[SUM[\nu] == -q^{1-\nu}}$
$\mathbf{SUM[\nu-2] - 2(1 - q^{1-\nu}\, SUM[\nu-1])/x,}$
$\mathbf{SUM[\nu], OreAlgebra[QS[V, q^{\nu}]]]]}$

Out[20]= $S_{V,q}^2 + \frac{2(qV-1)}{qVx} S_{V,q} + \frac{1}{qV}$

Then iterating this recurrence according to (2.2) and (2.3) corresponds to reducing the operator $V(qV)\cdots(q^{i-1}V)S_{V,q}^i$, which encodes the left-hand side, with the previously defined **op**; the result is an operator that corresponds to the right-hand side. Its leading coefficient is precisely the desired $r_i^{(\nu+1)}(1/x)$ (note that the symbol ** stands for noncommutative multiplication).

In[21]:= $\mathbf{Table[LeadingCoefficient[OreReduce[(q^{i(i-1)/2}V^{i-1})}$
$\mathbf{{**}\, QS[V, q^{\nu}]^i,\ \{op\ /.\ x \to 1/x\}]]\ /.\ V \to V/q, \{i, 1, 4\}]}$

Out[21]= $\{1, -2(V-1)x, 4qV^2x^2 - 4qVx^2 - 4Vx^2 - V + 4x^2,$
$-2x(qV-1)(4q^2V^2x^2 - 4q^2Vx^2 - qV - 4Vx^2 - V + 4x^2)\}$

We define the *annihilator* (with respect to some Ore algebra $\mathbb{O}$) of a function f by:

$$\mathrm{Ann}_{\mathbb{O}}(f) := \{P \in \mathbb{O} \mid Pf = 0\}.$$

It can easily be seen that $\mathrm{Ann}_{\mathbb{O}}(f)$ is a left ideal in $\mathbb{O}$. Every left ideal $I \subseteq \mathrm{Ann}_{\mathbb{O}}(f)$ is called an *annihilating ideal* for f. For example, the operator J given in (3.1) is an element of $\mathrm{Ann}_{\mathbb{O}}\big(q^{n\nu+n(n-1)/2}J_{n+\nu}^{(1)}(x;q)\big)$ with $\mathbb{O} = \mathbb{F}[S_{N,q}] = \mathbb{Q}(q, q^{\nu}, N, x)[S_{N,q}]$. Actually, it is the unique (up to multiplication by elements from $\mathbb{F}$) generator of that principal left ideal.

Definition 3.2. Let $\mathbb{O} = \mathbb{F}[\ldots]$ be an Ore algebra. A function f is called *∂-finite* with respect to $\mathbb{O}$ if $\mathbb{O}/\mathrm{Ann}_{\mathbb{O}}(f)$ is a finite-dimensional $\mathbb{F}$-vector space. The dimension of this vector space is called the (*holonomic*) *rank* of f with respect to $\mathbb{O}$.

Example 3.3. The following **HolonomicFunctions** procedure delivers the annihilator of the Gegenbauer (also called ultraspherical) polynomials $C_m^{(\nu)}(x)$:

In[22]:= $\mathbf{annG = Annihilator[GegenbauerC[m, \nu, x],}$
$\mathbf{\{S[m], S[\nu], Der[x]\}]}$

Out[22]= $\{2\nu S_{\nu} - xD_x + (-m - 2\nu),\ (m+1)S_m + (1 - x^2)D_x + (-mx - 2\nu x),$
$(x^2 - 1)D_x^2 + (2\nu x + x)D_x + (-m^2 - 2m\nu)\}$

It means that, these three elements generate $I := \operatorname{Ann}_{\mathbb{O}}\big(C_m^{(\nu)}(x)\big)$ as a left ideal in the operator algebra $\mathbb{O} = \mathbb{F}[S_m, S_\nu, D_x]$ with $\mathbb{F} = \mathbb{Q}(m, \nu, x)$. Their leading monomials are S_ν, S_m, and D_x^2, which shows that the function $C_m^{(\nu)}(x)$ is ∂-finite with respect to $\mathbb{O}$ and in particular:

$$\operatorname{rank}_{\mathbb{O}}\big(C_m^{(\nu)}(x)\big) = \dim_{\mathbb{F}}\big(\mathbb{O}/I\big) = 2.$$

In the holonomic systems approach, the data structure for representing functions is an annihilating ideal (given by a finite set of generators) plus initial values. When working with (left) ideals, we make use of *(left) Gröbner bases* [1, 8] which are an important tool for executing certain operations (e.g., the ideal membership test) in an algorithmic way.

For functions annihilated by univariate operators from the Ore algebras $\mathbb{F}[S_n]$ or $\mathbb{F}[D_x]$ or $\mathbb{F}[S_{N,q}]$, the notions of ∂-finite and (q-)holonomic coincide. Despite being closely related to being ∂-finite, for functions annihilated by multivariate Ore operators the definition of holonomic is much more technical. In general, the holonomic property reflects certain elimination properties of annihilating operators which are required for summation and integration of special functions.

Without proof we state the following theorem about *closure properties* of ∂-finite functions; its proof can be found in [12, Chapter 2.3]. We remark that all of them are algorithmically executable, and the algorithms work with the above-mentioned data structure.

Theorem 3.4. *Let $\mathbb{O}$ be an Ore algebra and let f and g be ∂-finite with respect to $\mathbb{O}$ of rank r and s, respectively. Then we have the following:*

(i) *$f + g$ is ∂-finite of rank $\leqslant r + s$.*

(ii) *$f \cdot g$ is ∂-finite of rank $\leqslant rs$.*

(iii) *f^2 is ∂-finite of rank $\leqslant r(r+1)/2$.*

(iv) *Pf is ∂-finite of rank $\leqslant r$ for any $P \in \mathbb{O}$.*

(v) *$f|_{x\to A(x,y,\ldots)}$ is ∂-finite of rank $\leqslant rd$ if $x, y, \ldots$ are continuous variables and if the algebraic function A satisfies a polynomial equation of degree d.*

(vi) *$f|_{n\to A(n,k,\ldots)}$ is ∂-finite of rank $\leqslant r$ if A is an integer-linear expression in the discrete variables $n, k, \ldots$.*

The bounds on the ranks are generically sharp. For example, the operator **opRHS** annihilating the right-hand side of (2.5) has been computed by exploiting ∂-finite closure properties in the spirit of Theorem 3.4. We continue with Theorem 3.5 which establishes the closure of holonomic functions with respect to sums and integrals; for its proof, we once again refer to [12, 18].

Theorem 3.5. *Let the function f be holonomic with respect to D_x (respectively, S_n). Then also $\int_a^b f\,\mathrm{d}x$ (respectively, $\sum_{n=a}^b f$) is holonomic.*

Example 3.6. We continue the discussion from Example 3.3 by again considering the Gegenbauer polynomials

$$C_m^{(\nu)}(x) := \sum_{k=0}^{m} F[x, m, \nu, k], \tag{3.2}$$

where

$$\text{In[23]:= } \mathbf{F[x\_, m\_, \nu\_, k\_] := \frac{P[2\nu, m]}{m!} \frac{P[-m, k]\, P[m + 2\nu, k]}{P[\nu + 1/2, k]\, k!)} \left(\frac{1 - x}{2}\right)^k}$$

with

$$\text{In[24]:= } \mathbf{P[x\_, k\_] := Pochhammer[x, k]}$$

This time we want to derive the annihilator of $C_m^{(\nu)}(x)$ from its definition (3.2). For this purpose, we compute annihilating operators of the hypergeometric term $F[x, m, \nu, k]$ in telescoping form:

$$\text{In[25]:= } \mathbf{CreativeTelescoping[F[x, m, \nu, k], S[k] - 1, \{Der[x], S[m]\}]\ //\ Factor}$$

$$\text{Out[25]= } \Big\{\{-(x-1)(x+1)D_x + (m+1)S_m + x(-(m+2\nu)),$$
$$(m+2)S_m^2 - 2x(m+\nu+1)S_m + (m+2\nu)\},$$
$$\left\{-\frac{k(2k+2\nu-1)}{k-m-1}, \frac{2k(2k+2\nu-1)(m+\nu+1)}{(k-m-2)(k-m-1)}\right\}\Big\}$$

The output has to be interpreted as follows:

$$\begin{aligned}&\big(-(-1+x)(1+x)D_x + (m+1)S_m - (m+2\nu)x\big)F[x, m, \nu, k]\\ &\quad = (S_k - 1)\frac{k(2k+2\nu-1)}{k-m-1} F[x, m, \nu, k]\end{aligned} \tag{3.3}$$

and

$$\begin{aligned}&\big((m+2)S_m^2 - 2(m+\nu+1)xS_m + (m+2\nu)\big)F[x, m, \nu, k]\\ &\quad = -(S_k - 1)\frac{2k(m+\nu+1)(2k+2\nu-1)}{(k-m-2)(k-m-1)} F[x, m, \nu, k].\end{aligned} \tag{3.4}$$

Note that the relations (3.3) and (3.4) can be easily verified (even without using a computer). To compute them, the package **HolonomicFunctions** employs noncommutative Gröbner bases; the monomial order is deduced from the order in which the operators are given. Indeed, by changing in the input the order of **Der**$[x]$ and **S**$[m]$, one obtains a different result:

In[26]:= **CreativeTelescoping[F[x, m, ν, k], S[k] − 1, {S[m], Der[x]}] // Factor**

Out[26]= $\Big\{\{(m+1)S_m - (x-1)(x+1)D_x + x(-(m+2\nu)),$

$- (x-1)(x+1)D_x^2 - (2\nu+1)xD_x + m(m+2\nu)\},$

$\Big\{-\dfrac{k(2k+2\nu-1)}{k-m-1}, -\dfrac{k(2k+2\nu-1)}{x-1}\Big\}\Big\}$

This repeats (3.3), but computes another purely differential relation

$$\big(-(-1+x)(1+x)D_x^2 - (2\nu+1)xD_x + m(m+2\nu)\big)F[x,m,\nu,k]$$
$$= (S_k - 1)\frac{k(2k+2\nu-1)}{x-1}F[x,m,\nu,k]. \tag{3.5}$$

Summing (3.3), (3.4), and (3.5) with respect to k from 0 to ∞ gives the well-known shift/differential relations for the Gegenbauer polynomials.

These computations were done in the operator algebra $\mathbb{O} = \mathbb{F}[S_m, S_k, D_x]$ with $\mathbb{F} = \mathbb{Q}(\nu, m, k, x)$. Let us include in addition the shift operator S_ν:

In[27]:= **CreativeTelescoping[F[x, m, ν, k], S[k] − 1, {S[m], S[ν], Der[x]}]**

Out[27]= $\Big\{\Big\{2\nu S_\nu - xD_x + (-m-2\nu), (m+1)S_m + (1-x^2)D_x + (-mx-2\nu x),$

$(1-x^2)D_x^2 + (-2\nu x - x)D_x + (m^2+2m\nu)\},$

$\Big\{-\dfrac{k}{x-1}, \dfrac{-2k^2-2k\nu+k}{k-m-1}, \dfrac{-2k^2-2k\nu+k}{x-1}\Big\}\Big\}$

We see that summing the resulting telescoping relations with respect to k from 0 to ∞, gives the generators of the annihilating ideal **annG** computed in Out[22].

Finally we note that for the q-case we need to consider the Gegenbauer polynomials in the (equivalent) form:

$$C_m^{(\nu)}\big(\cos(\theta)\big) := \sum_{k=0}^{m} G[x,m,\nu,k],$$

where

In[28]:= $\boldsymbol{G[x\_, m\_, \nu\_, k\_] := \dfrac{P[\nu,k]\,P[\nu,m-k]}{k!(m-k)!}A^{m-2k}}$

with $x = \cos(\theta)$, $A = e^{i\theta}$ and $P[\nu, k]$ being defined as the Pochhammer symbol $(\nu)_k$, as in In[6]. In the q-context we will be interested to compute annihilating operators containing shifts in m and ν, and, as above, in telescoping form with respect to $S_k - 1$. More precisely, in the next section we will compute a q-version of

In[29]:= **CreativeTelescoping[$G[x, m, \nu, k]$, S[k] − 1, {S[m], S[ν]}] // Factor**

$$\text{Out[29]=}\ \Big\{\{A(A^2+1)(m+1)S_m - (A-1)^2(A+1)^2\nu S_\nu - 2A^2(m+2\nu),$$

$$-(A-1)^2(A+1)^2\nu(\nu+1)S_\nu^2 + \nu\big(A^4m + A^4\nu + A^4 - 2A^2m - 6A^2\nu$$

$$-4A^2 + m + \nu + 1\big)S_\nu + A^2(m+2\nu)(m+2\nu+1)\},$$

$$\Big\{\frac{A^2k(k-m-\nu)\big(A^2k - A^2m + A^2\nu - A^2 - k + m + \nu + 1\big)}{\nu(k-m-1)},$$

$$(A^2k(-k+m+\nu)\big(A^2k^2 - A^2km + A^2k\nu - A^2m\nu - k^2 + km + k\nu$$

$$+2k - m - \nu - 1)\big)\Big/(\nu(\nu+1))\Big\}\Big\}$$

4. The Ismail–Zhang Formula

An important classical expansion formula is the expansion of the plane wave in terms of ultraspherical polynomials $C_m^{(\nu)}(x)$, also called Gegenbauer polynomials:

$$e^{irx} = \left(\frac{2}{r}\right)^{\nu} \Gamma(\nu) \sum_{m=0}^{\infty} i^m (\nu+m)\, J_{\nu+m}(r)\, C_m^{(\nu)}(x).$$

Ismail and Zhang [7, (3.32)] had found the following q-analog of this formula:

$$\mathcal{E}_q(x; i\omega) = \frac{(q;q)_\infty\, \omega^{-\nu}}{(q^\nu;q)_\infty\, (-q\omega^2;q^2)_\infty}$$
$$\times \sum_{m=0}^{\infty} i^m (1-q^{\nu+m})\, q^{m^2/4} J_{\nu+m}^{(2)}(2\omega;q)\, C_m(x;q^\nu|q), \quad (4.1)$$

where $J_{\nu+m}^{(2)}(2\omega;q)$ is Jackson's q-Bessel function defined by

$$J_\nu^{(2)}(z;q) = \frac{(q^{\nu+1};q)_\infty}{(q;q)_\infty} \sum_{n=0}^{\infty} q^{(\nu+n)n} \frac{(-1)^n (z/2)^{\nu+2n}}{(q;q)_n (q^{\nu+1};q)_n}.$$

In the Ismail–Zhang formula (4.1), Jackson's second q-analog of the Bessel function $J_\nu(z)$ appears; the remaining ingredients, the basic exponential function $\mathcal{E}_q(x; i\omega)$ and the q-Gegenbauer polynomials $C_m(x; q^\nu|q)$, are explained subsequently. There are several proofs of the Ismail–Zhang formula; see the books [5, 16] for references and for the embedding of the formula in a broader context. In this section we present a new, computer-assisted proof of (4.1).

4.1. The Basic Exponential Function

The basic exponential function $\mathcal{E}_q(x; i\omega)$, as well as its more general version $\mathcal{E}_q(x, y; i\omega)$, was introduced by Ismail and Zhang [7] and satisfies numerous important and also beautiful properties. For illustrative reasons we choose to introduce $\mathcal{E}_q(x; i\omega)$ via the basic cosine and sine functions: For $x = \cos(\theta)$ and $|\omega| < 1$ we define

$$\mathcal{E}_q(x; i\omega) := C_q(x; \omega) + i\, S_q(x; \omega),$$

where the basic cosine function $C_q(x; \omega)$ is defined as

$$C_q(x; \omega) := \frac{(-\omega^2; q^2)_\infty}{(-q\omega^2; q^2)_\infty} \sum_{j=0}^{\infty} \frac{(-qe^{2i\theta}; q^2)_j\, (-qe^{-2i\theta}; q^2)_j}{(q; q^2)_j\, (q^2; q^2)_j} (-\omega^2)^j,$$

and the basic sine function $S_q(x; \omega)$ as

$$\begin{aligned} S_q(x; \omega) : &= \frac{(-\omega^2; q^2)_\infty}{(-q\omega^2; q^2)_\infty} \frac{2q^{1/4}\omega}{1-q} \cos(\theta) \\ &\times \sum_{j=0}^{\infty} \frac{(-qe^{2i\theta}; q^2)_j\, (-qe^{-2i\theta}; q^2)_j}{(q^3; q^2)_j\, (q^2; q^2)_j} (-\omega^2)^j. \end{aligned}$$

It is not difficult to check that

$$\lim_{q \to 1-} C_q\big(x; \omega(1-q)/2\big) = \cos(\omega x),$$

$$\lim_{q \to 1-} S_q\big(x; \omega(1-q)/2\big) = \sin(\omega x).$$

In the following we shall use the abbreviation $A = e^{i\theta}$, as before, and the following short-hand notation for the **qPochhammer** command:

In[30]:= **qP = qPochhammer;**

Consequently, the input **qP**$[-w^2, q^2]$ stands for $(-\omega^2; q^2)_\infty$, and **qP**$[-qA^2, q^2, j]$ for $(-qe^{2i\theta}; q^2)_j$. The continuous q-ultraspherical (q-Gegenbauer) polynomials $C_m(x; q^\nu|q)$, $x = \cos(\theta)$, are defined as

$$C_m(\cos\theta; \beta|q) := \sum_{k=0}^{m} \frac{(\beta; q)_k\, (\beta; q)_{m-k}}{(q; q)_k\, (q; q)_{m-k}} e^{i(m-2k)\theta}.$$

To prove (4.1) we compute annihilating operators representing q-difference equations for the left- and right-hand sides, respectively. First we derive a q-shift equation for the q-cosine. This is done analogously to the treatment of the right-hand side of (2.6):

In[31]:= **CreativeTelescoping**

$$\left[\frac{\mathbf{qP}[-\omega^2, q^2]}{\mathbf{qP}[-q\omega^2, q^2]} \frac{\mathbf{qP}[-qA^2, q^2, j]\, \mathbf{qP}[-q/A^2, q^2, j]}{\mathbf{qP}[q, q^2, j]\, \mathbf{qP}[q^2, q^2, j]}(-\omega^2)^j, \mathbf{QS}[J, q^j] - 1, \mathbf{QS}[\omega, q^w]\right] \ // \ \mathbf{Factor}$$

$$\text{Out[31]=} \left\{\{A^2(q^2\omega^2+1)S^2_{\omega,q} + (A^4q^2\omega^2 - A^2q - A^2 + q^2\omega^2)S_{\omega,q} + A^2q(q\omega^2+1)\}, \left\{\frac{A^2(J-1)(J+1)(J^2-q)(q\omega^2+1)}{\omega^2+1}\right\}\right\}$$

Denoting by $c_j(\omega)$ the summand in the q-cosine series and in view of $S^j_{\omega,q}f(\omega) = f(q^j\omega)$, this output means that

$$\begin{aligned} A^2(q^2\omega^2+1)c_j(q^2\omega) &+ (A^4q^2\omega^2 - A^2q - A^2 + q^2\omega^2)c_j(q\omega) \\ &+ A^2q(q\omega^2+1)c_j(\omega) \\ &= -\Delta_j \frac{A^2(J-1)(J+1)(J^2-q)(q\omega^2+1)}{\omega^2+1}c_j(\omega), \end{aligned}$$

where $J = q^j$. Summing the right-hand side from $j = 0$ to $j = \infty$ gives

$$-\frac{qA^2(1+q\omega^2)}{1+\omega^2}c_\infty(\omega) + \frac{A^2(-1+q^0)(1+q^0)(q^0-q)(1+q\omega^2)}{1+\omega^2}c_0(\omega) = 0.$$

Hence

In[32]:= **annCos = %[[1]]**

$$\text{Out[32]=} \{A^2(q^2\omega^2+1)S^2_{\omega,q} + (A^4q^2\omega^2 - A^2q - A^2 + q^2\omega^2)S_{\omega,q} + A^2q(q\omega^2+1)\}$$

annihilates $C_q(x;\omega)$. An annihilator, respectively q-difference equation, for the q-sine is derived analogously:

In[33]:= **Factor** [**CreativeTelescoping**

$$\left[\mathbb{i}\frac{\mathbf{qP}[-\omega^2, q^2]\, 2q^{1/4}\omega}{\mathbf{qP}[-q\omega^2, q^2](1-q)}\mathbf{Cos}[\theta] \frac{\mathbf{qP}[-q^2A^2, q^2, j]\, \mathbf{qP}[-q^2/A^2, q^2, j]}{\mathbf{qP}[q^3, q^2, j]\, \mathbf{qP}[q^2, q^2, j]}(-\omega^2)^j, \mathbf{QS}[J, q^j] - 1, \mathbf{QS}[\omega, q^w]\right]\Bigg]$$

$$\text{Out[33]=} \left\{\{A^2(q^2\omega^2+1)S^2_{\omega,q} + (A^4q^2\omega^2 - A^2q - A^2 + q^2\omega^2)S_{\omega,q} + A^2q(q\omega^2+1)\}, \left\{\frac{A^2(J-1)(J+1)q(J^2q-1)(q\omega^2+1)}{\omega^2+1}\right\}\right\}$$

In[34]:= **annSin = %[[1]]**

Out[34]= $\{A^2(q^2\omega^2+1)S_{\omega,q}^2+(A^4q^2\omega^2-A^2q-A^2+q^2\omega^2)S_{\omega,q}+A^2q(q\omega^2+1)\}$

Finally we exploit the q-holonomic closure properties; more precisely, in view of Theorem 3.4(i) we "add" the q-difference equations for the q-cosine function and i times the q-sine function to obtain a q-difference equation for $\mathcal{E}_q(x;i\omega)$. The latter is the generator of the annihilating ideal of $\mathcal{E}_q(x;i\omega)$:

In[35]:= **annLHS = DFinitePlus[annCos, annSin]**

Out[35]= $\{(A^2q^2\omega^2+A^2)S_{\omega,q}^2+(A^4q^2\omega^2-A^2q-A^2+q^2\omega^2)S_{\omega,q}+(A^2q^2\omega^2+A^2q)\}$

The result is not surprising: since $C_q(x;\omega)$ and $S_q(x;\omega)$ satisfy the same q-difference equation (compare **annSin** with **annCos**), also their linear combination satisfies the same equation. Conversely, the operator above annihilates any linear combination

$$c_1C_q(x;\omega)+c_2S_q(x;\omega)$$

where c_1 and c_2 are constants, i.e., independent of ω. The order of **annLHS** is 2, hence one derives explicit expressions for the c_i by picking the coefficients of ω^0 and ω^1 in $\mathcal{E}_q(x;i\omega)$, respectively:

$$c_1=1,\quad c_2=\frac{2q^{1/4}}{1-q}\cos(\theta). \tag{4.2}$$

Summarizing, $\mathcal{E}_q(x;i\omega)$ is uniquely determined by the q-difference operator **annLHS** and the initial values (4.2).

4.2. An Annihilator for the Ismail–Zhang Series

To compute an annihilating operator for the right-hand side of (4.1), we algorithmically exploit ∂-finite, respectively, q-holonomic, closure properties as described in Section 3. Let us first compute generators of the ideal of operators annihilating $C_m(\cos\theta;q^\nu|q)$. Recall $x=\cos(\theta)$ and $A=e^{i\theta}$. In addition, we will use the abbreviations

$$K=q^k,\quad M=q^m,\quad N=q^n,\quad V=q^\nu,\ \text{and}\quad \omega=q^w.$$

In[36]:= **annqGegenbauer =**
$$\mathbf{CreativeTelescoping}\left[\frac{\mathbf{qP}[q^\nu,q,k]\,\mathbf{qP}[q^\nu,q,m-k]}{\mathbf{qP}[q,q,k]\,\mathbf{qP}[q,q,m-k]}A^{m-2k},\ \mathbf{QS}[\mathbf{K},q^k]-1,\{\mathbf{QS}[M,q^m],\mathbf{QS}[V,q^\nu],\mathbf{QS}[\omega,q^w]\}\right][[1]]\ //$$
Factor

Out[36]= $\{S_{\omega,q}-1,-A(A^2+1)V(Mq-1)S_{M,q}+(V-1)(A^2-V)(A^2V-1)S_{V,q}$
$+A^2(V+1)(MV^2-1),(V-1)(qV-1)(A^2-qV)(A^2qV-1)S_{V,q}^2$

$$-(V-1)\big(A^4Mq^2V^2-A^4qV-A^2Mq^3V^3-A^2Mq^2V^3+A^2q+A^2 + Mq^2V^2-qV\big)S_{V,q}-A^2q\big(MV^2-1\big)\big(MqV^2-1\big)\}$$

The algorithmic method to compute **annqGegenbauer** follows the creative telescoping strategy described in Section 3. In contrast to the $q=1$ case, here we also include the shift with respect to ω which in the output gives rise to an additional, trivial generator $S_{\omega,q}-1$. This is done in order to be able to execute all required closure property computations in one common operator algebra. For further details see [12, 14]; there is also an on-line description of the **CreativeTelescoping** procedure in the **HolonomicFunctions** package:

In[37]:= **?CreativeTelescoping**

> CreativeTelescoping[f, delta, {op1, ..., opk}] or CreativeTelescoping[ann, delta, {op1, ..., opk}] computes creative telescoping relations for the given function f (resp. the given ∂-finite ideal ann annihilating some function f). In particular it returns {{q1, ..., qm}, {r1, ..., rm}}, two lists of OrePolynomials such that qj + delta*rj is in the annihilator of f for all $1 \leqslant j \leqslant m$. The polynomials qj form a Groebner basis in the rational Ore algebra with generators op1, ..., opk whereas the rj's live in the Ore algebra with generators Join[OreOperators[delta], {op1, ..., opk}] (resp. the Ore algebra of ann). For summation (w.r.t. n) set delta to S[n]−1 or Delta[n], and in the q-case to QS[qn,q^n]−1; for integration (w.r.t. x) set delta to Der[x].

In an analogous fashion we compute generators of the annihilating ideal of the q-Bessel function $J_\nu^{(2)}(2\omega;q)$:

In[38]:= **annqBesselJ =**

$$\mathbf{CreativeTelescoping}\left[q^{(\nu+n)n}\frac{\mathbf{qP}[q^{\nu+1},q]\,(-1)^n\omega^{\nu+2n}}{\mathbf{qP}[q,q]\,\mathbf{qP}[q,q,n]\,\mathbf{qP}[q^{\nu+1},q,n]},\ \mathbf{QS}[\mathrm{N},q^n]-1,\{\mathbf{QS}[M,q^m],\mathbf{QS}[V,q^\nu],\mathbf{QS}[\omega,q^w]\}\right][[1]]$$

Out[38]= $\{(-V\omega-\omega)S_{V,q}+(q\omega^4+q\omega^2+\omega^2+1)S_{\omega,q}+(\omega^2-V),S_{M,q}-1,$
$(q^5V\omega^4+q^3V\omega^2+q^2V\omega^2+V)S_{\omega,q}^2+(q^2V\omega^2+qV\omega^2-V^2-1)S_{\omega,q}+V\}$

The annihilating ideal of $h_1(\omega,m,n):=i^m(1-q^{\nu+m})$ is obtained as follows:

In[39]:= **annh1 = Annihilator**$[\mathbb{i}^m(1-q^{\nu+m}),\{\mathbf{QS}[M,q^m],\mathbf{QS}[V,q^\nu],\mathbf{QS}[\omega,q^w]\}]$

Out[39]= $\{S_{\omega,q}-1,(MV-1)S_{V,q}+(1-MqV),(MV-1)S_{M,q}+(i-iMqV)\}$

Note that in fact it is trivial to compute the generators of this ideal, just consider the quotients

$$\frac{h_1(q\omega,m,n)}{h_1(\omega,m,n)},\quad \frac{h_1(\omega,m+1,n)}{h_1(\omega,m,n)},\quad \text{and}\quad \frac{h_1(\omega,m,n+1)}{h_1(\omega,m,n)},$$

which, after simplification, yield rational functions, whose numerators and denominators appear as coefficients in the first-order operators of Out[39]. The reason for this is that $h_1(\omega,m,n)$ is actually a q-hypergeometric term. When trying to compute the annihilating ideal of $h_2(m) := q^{m^2/4}$ by the **Annihilator** command, the package is trapped by the factor $\frac{1}{4}$ in the exponent and delivers the fourth-order operator $S_{M,q}^4 - q^4M^2$. Although this is correct, in the sense that it is a left multiple of the minimal-order annihilating operator, it is not the operator we wish to work with. Instead, we write down the annihilator of $h_2(m)$ by hand, and convert its generators to Ore polynomials that live in the same Ore algebra as **annh1**:

In[40]:= $\mathbf{annh2 = ToOrePolynomial[\{QS[V, q^\nu] - 1, QS[\omega, q^w] - 1, QS[M, q^m]^2 - qM\}, OreAlgebra[QS[M, q^m], QS[V, q^\nu], QS[\omega, q^w]]]}$

Out[40]= $\{S_{V,q} - 1, S_{\omega,q} - 1, S_{M,q}^2 - Mq\}$

The annihilating ideal of $h_1(\omega,m,n)h_2(m) = i^m(1-q^{\nu+m})q^{m^2/4}$ is obtained by applying the closure property "multiplication", see Theorem 3.4(ii):

In[41]:= **annh1h2 = DFiniteTimes[annh1, annh2]**

Out[41]= $\{S_{\omega,q}-1, (MV-1)S_{V,q}+(1-MqV), (MV-1)S_{M,q}^2+(M^2q^3V-Mq)\}$

We continue by applying the same closure property again, in order to obtain an annihilating ideal of $i^m(1-q^{\nu+m})q^{m^2/4}J_{\nu+m}^{(2)}(2\omega;q)C_m(\cos\theta;q^\nu|q)$; here we use the previously computed annihilators **annqBesselJ** and **annqGegenbauer** of the q-Bessel function and the q-Gegenbauer polynomials, respectively, together with discrete substitution as described in Theorem 3.4(vi):

In[42]:= $\mathbf{annSmnd = DFiniteTimes[annh1h2, DFiniteSubstitute[annqBesselJ, \{\nu \to \nu + m\}], annqGegenbauer];}$

The output list consists of three annihilating operators, and it would require about two pages to display them. The output **annSmnd** is of the form:

$$\{(1+q^2\omega^2+q^3\omega^2+q^5\omega^4)MVS_{\omega,q}^2$$
$$+(qMV\omega^2+q^2MV\omega^2-M^2V^2-1)S_{\omega,q}$$

$$
\begin{aligned}
&+MV, (-A^2MV\omega^2 + \cdots - q^5A^2M^5V^9\omega^2)S_{V,q}^2 \\
&+(A^2MV\omega + \cdots + q^6A^2M^5V^8\omega^5)S_{V,q}S_{\omega,q} \\
&+(-qA^2M^2V^2\omega - \cdots + q^5A^2M^5V^8\omega^3)S_{V,q} \\
&+(-A^2 + \cdots + q^7A^2M^6V^8\omega^4)S_{\omega,q} + (A^2MV - \cdots + q^6A^2M^6V^8\omega^2), \\
&\quad(-A^2V^2\omega^2 + \cdots + q^4A^2M^5V^5\omega^2)S_{M,q}^2 \\
&+(qA^2MV\omega - \cdots - q^4A^2M^4V^7\omega^5)S_{V,q}S_{\omega,q} \\
&+(-q^2A^2M^2V^2\omega + \cdots - q^3A^2M^4V^7\omega^3)S_{V,q} \\
&+(-A^2 + \cdots - q^6A^2M^5V^7\omega^4)S_{\omega,q} \\
&+(A^2MV - \cdots - q^5A^2M^5V^7\omega^2)\}
\end{aligned}
\tag{4.3}
$$

The next step in the process of constructing an annihilating ideal of the right-hand side of (4.1) consists in "doing the sum"

$$\sum_{m=0}^{\infty} i^m (1 - q^{\nu+m})\, q^{m^2/4} J_{\nu+m}^{(2)}(2\omega; q)\, C_m(\cos\theta; q^\nu|q).$$

However, applying the **CreativeTelescoping** command, as we did before, does not deliver any result within a reasonable amount of time. By inspecting the leading monomials of **annSmnd**—they are $S_{\omega,q}^2$, $S_{V,q}^2$, and $S_{M,q}^2$—we find that the holonomic rank of **annSmnd** is 8, which is relatively large and which explains the failure of the first attempt. Luckily there exists an algorithm [13] that is more efficient in such situations, but whose drawback is that sometimes it is not able to deduce the correct denominator of the output. The current summation problem is such an example, and therefore we give the correct denominator with an additional option (it can be found by looking at the leading coefficients of **annSmnd** plus some trial and error). The computation then takes about 20 seconds and for better readability we suppress parts of the output.

In[43]:= **annSumRHS = FindCreativeTelescoping[annSmnd,**
QS[M, q^m] − 1,
Denominator → $(M^2V^2 - 1)(q^2M^2V^2 - 1)(q^4M^2V^2 - 1)$]

Out[43]= $\Big\{\{(V-1)S_{V,q} + \omega, (-q^5A^2\omega^4 - q^3A^2\omega^2 - q^2A^2\omega^2 - A^2)S_{\omega,q}^2 +$
$(-q^2A^4V\omega^2 + qA^2V + A^2V - q^2V\omega^2)S_{\omega,q} - qA^2V^2\}, \{\cdots\}\Big\}$

Finally we obtain the annihilating ideal of the right-hand side of (4.1):

In[44]:= **annRHS = DFiniteTimes[annSumRHS[[1]], Annihilator[**
qP[q, q]/(ω^νqP[q^ν, q] qP[$(-q)\omega^2, q^2$]),
{QS[V, q^ν], QS[ω, q^w]}]]

Out[44]= $\{S_{V,q}-1,(q^2A^2\omega^2+A^2)S_{\omega,q}^2+(q^2A^4\omega^2-qA^2-A^2+q^2\omega^2)S_{\omega,q}$
$+(q^2A^2\omega^2+qA^2)\}$

Comparison with the left-hand side of (4.1):

In[45]:= **annLHS**

Out[45]= $\{(q^2A^2\omega^2+A^2)S_{\omega,q}^2+(q^2A^4\omega^2-qA^2-A^2+q^2\omega^2)S_{\omega,q}+(q^2A^2\omega^2+qA^2)\}$

To complete the proof we have to incorporate initial conditions. To this end we convert the q-shift equation to an equivalent version which is in the format of a q-differential equation. This is supported by the command **QSE2DE** from the **qGeneratingFunctions** package; in order to invoke it, we need to convert the operator **annLHS** in Out[45] to a standard q-shift equation:

In[46]:= **qSeq = ApplyOreOperator[annLHS, $f[\omega]$] /. q^(a_. m + b_.) → $q^b\omega^a$**

Out[46]= $\{(q^2A^2\omega^2+qA^2)f[\omega]+(q^2A^4\omega^2-qA^2-A^2+q^2\omega^2)f[q\omega]$
$+(q^2A^2\omega^2+A^2)f[q^2\omega]\}$

In[47]:= **QSE2DE[qSeq, $f[\omega]$]**

Out[47]= $\{q(A^2+1)^2f[\omega]+q(q-1)(A^4+qA^2+A^2+1)\omega f'[\omega]$
$+(q-1)^2(q^2\omega^2+1)A^2f''[\omega]=0\}$

In this equivalent form, $f'(\omega):=D_qf(\omega)$ refers to the q-derivative defined on (formal) power series as

$$D_q\sum_{n=0}^{\infty}a_n\omega^n:=\sum_{n=1}^{\infty}a_n\frac{q^n-1}{q-1}\omega^{n-1}.$$

Now our proof can be completed as follows: let $l(\omega)$ and $r(\omega)$ denote the left and right sides of (4.1), respectively. Above we have shown that both $l(\omega)$ and $r(\omega)$ satisfy the q-differential equation Out[47]. So what is left to show is that

$$l(0)=r(0)=1\quad\text{and}\quad l'(0)=r'(0)=\frac{2iq^{1/4}\cos(\theta)}{1-q}.$$

But this, in view of the definition of D_q, amounts to comparing the coefficients of ω^0 and ω^1, respectively, in the Taylor expansions of $l(\omega)$ and $r(\omega)$. Owing to the definitions of the functions involved and noticing that $\omega^{-\nu}$ is cancelling out, this task is an easy verification.

5. Conclusion

As expressed in Section 1, a major objective of this chapter is to popularize the holonomic systems approach in the field of q-series and basic hypergeometric functions. In the case studies we presented, RISC software written in Mathematica was used. With respect to q-summation one can find various packages written in Maple or in other computer algebra systems; with regard to the more general q-holonomic setting (operator algebras, non-commutative Gröbner basis methods, etc.) we point explicitly to the Maple package **Mgfun** by Chyzak [2].

We want to conclude with a few remarks on the fact that computing an annihilating operator for the series side of the Ismail–Zhang formula (4.1) is leading to the frontiers of what is computationally feasible today. As already pointed out, ongoing research is trying to push frontiers further by the design of new constructive methods like [13]. Formulas like (4.1) or the very well-poised basic hypergeometric series ${}_{10}W_9$ in [6], which has been successfully treated by the **HolonomicFunctions** package, provide excellent challenges and inspirations for such algorithmic developments.

Acknowledgments

The first author was supported by the Austrian Science Fund (FWF): DK W1214 and SFB F50-11. The second author was supported by the Austrian Science Fund (FWF): SFB F50-06.

References

[1] B. Buchberger, Ein Algorithmus zum Auffinden der Basiselemente des Restklassenrings nach einem nulldimensionalen Polynomideal, Ph.D. thesis, University of Innsbruck, Austria (1965).

[2] F. Chyzak, Fonctions holonomes en calcul formel, Ph.D. thesis, École polytechnique (1998).

[3] R. L. Graham, D. E. Knuth and O. Patashnik, *Concrete Mathematics*, 2nd edn. (Addison-Wesley, Reading, MA, 1994).

[4] M. E. H. Ismail, The zeros of basic Bessel functions, the functions $J_{\nu+ax}(x)$, and associated orthogonal polynomials, *J. Math. Anal. Appl.* **86**(1) (1982) 1–19.

[5] M. E. H. Ismail, *Classical and Quantum Orthogonal Polynomials in One Variable*, Encyclopedia of Mathematics and its Applications, Vol. 98 (Cambridge University Press, Cambridge, 2005).

[6] M. E. H. Ismail, E. M. Rains and D. Stanton, Orthogonality of very well-poised series, Unpublished manuscript (2015); http://www.math.umn.edu/~stant001/PAPERS/10W9Feb2015.pdf.

[7] M. E. H. Ismail and R. Zhang, Diagonalization of certain integral operators, *Adv. Math.* **109**(1) (1994) 1–33.

[8] A. Kandri-Rody and V. Weispfenning, Non-commutative Gröbner bases in algebras of solvable type, *J. Symbolic Comput.* **9**(1) (1990) 1–26.

[9] M. Kauers and C. Koutschan, A Mathematica package for q-holonomic sequences and power series, *Ramanujan J.* **19**(2) (2009) 137–150.

[10] M. Kauers and P. Paule, *The Concrete Tetrahedron*, Text & Monographs in Symbolic Computation, (Springer Wien, 2011).

[11] D. E. Knuth and E. G. Daylight, *Algorithmic Barriers Falling: P=NP?* (Lonely Scholar, 2014).

[12] C. Koutschan, Advanced applications of the holonomic systems approach, Ph.D. thesis, Research Institute for Symbolic Computation (RISC), Johannes Kepler University, Linz, Austria (2009).

[13] C. Koutschan, A fast approach to creative telescoping, *Math. Comput. Sci.* **4**(2–3) (2010) 259–266.

[14] C. Koutschan, HolonomicFunctions (user's guide), Technical Report 10-01, RISC Report Series, Johannes Kepler University, Linz, Austria (2010).

[15] P. Paule and A. Riese, A Mathematica q-analogue of Zeilberger's algorithm based on an algebraically motivated approach to q-hypergeometric telescoping, in *Special Functions, q-Series and Related Topics*, eds. M. E. H. Ismail, D. R. Masson and M. Rahman, Fields Institute Communications, Vol. 14 (American Mathematical Society, 1997), pp. 179–210.

[16] S. K. Suslov, *An Introduction to Basic Fourier Series*, Developments in Mathematics, Vol. 9 (Kluwer Academic Publishers, Dordrecht, 2003).

[17] D. Zeilberger, A fast algorithm for proving terminating hypergeometric identities, *Discrete Math.* **80**(2) (1990) 207–211.

[18] D. Zeilberger, A holonomic systems approach to special functions identities. *J. Comput. Appl. Math.* **32**(3) (1990) 321–368.

Chapter 21

A Direct Evaluation of an Integral of Ismail and Valent

Alexey Kuznetsov

Department of Mathematics and Statistics, York University
4700 Keele Street, Toronto, ON, Canada M3J 1P3
kuznetsov@mathstat.yorku.ca

We give a direct evaluation of a curious integral identity, which follows from the work of Ismail and Valent on the Nevanlinna parameterization of solutions to a certain indeterminate moment problem.

Keywords: Cauchy Residue Theorem; Jacobi elliptic functions; theta functions; indeterminate moment problem; Nevanlinna parameterization.

Mathematics Subject Classification 2010: 33E05, 30E05

1. Introduction and the Main Results

At the *International Conference on Orthogonal Polynomials and q-Series*, which was held in Orlando in May 2015 in celebration of the 70th birthday of Mourad Ismail, Dennis Stanton gave a plenary talk titled "A small slice of Mourad's work". One of the topics in that talk was about "the mystery integral of Mourad Ismail": a curious integral that has first appeared in the paper by Ismail and Valent [3] (see also [1] for a special case). To present this integral, we fix $k \in (0,1)$ and denote by

$$K(k) = \frac{\pi}{2} \times {}_2F_1(\tfrac{1}{2}, \tfrac{1}{2}; 1; k^2)$$

the complete elliptic integral of the first kind. We also denote $K = K(k)$, $k' = \sqrt{1-k^2}$ and $K' = K(k')$. The "mystery integral", which Dennis Stanton referred to in his talk, is the following one:

$$\frac{1}{2}\int_{\mathbb{R}} \frac{\mathrm{d}x}{\cos(\sqrt{x}K) + \cosh(\sqrt{x}K')} = 1. \tag{1.1}$$

This is essentially formula (1.16) in [3] after correcting the typo — an extra factor of $1/2$ multiplying $\sqrt{x}$.

The identity (1.1) is indeed rather unusual and mysterious. First of all, there is a free parameter k that affects in a non-trivial way the integrand in the left-hand side, but the right-hand side stays constant. The appearance of the complete elliptic integral in combination with trigonometric and hyperbolic functions is also uncommon. The integrand looks very simple (after all, it only has two elementary trigonometric functions and well-known complete elliptic integrals), but this simplicity is deceptive. In fact, this is the most striking feature of this integral: it is not at all clear how to prove such a result directly. The identity (1.1) follows as a byproduct of explicit computations related to Nevanlinna parameterization in a certain indeterminate moment problem, see [3]. The goal of this paper is to evaluate the integral in (1.1) directly without using the theory of the indeterminate moment problem.

Our main result is the following theorem, which gives a more general statement than (1.1). As we will see later, this theorem allows one to compute explicitly all moments of the probability measure appearing in (1.1). In what follows, we will be working with Jacobi elliptic functions; we refer the reader to [5, Chap. 22] for their definition and various properties.

Theorem 1.1. *Assume that $k \in (0,1)$ and denote $k' = \sqrt{1-k^2}$, $K = K(k)$ and $K' = K(k')$. Then for $u \in \mathbb{C}$ satisfying $|\mathrm{Re}(u)| < K$ and $|\mathrm{Im}(u)| < K'$, we have*

$$\frac{1}{2}\int_{\mathbb{R}} \frac{\sin(\sqrt{x}u)}{\sqrt{x}} \times \frac{\mathrm{d}x}{\cos(\sqrt{x}K) + \cosh(\sqrt{x}K')} = \frac{\mathrm{sn}(u,k)}{\mathrm{cd}(u,k)}. \tag{1.2}$$

Proof. Our plan is to establish (1.2) for $u = v(K + \mathrm{i}K')/2$ with $v \in (-1,1)$ and then apply an analytic continuation argument to extend this result to other values of u. Thus, we fix $v \in (-1,1)$ and we denote

$$I := \frac{1}{2}\int_{\mathbb{R}} \frac{\sin(\sqrt{x}v(K+\mathrm{i}K')/2)}{\sqrt{x}} \times \frac{\mathrm{d}x}{\cos(\sqrt{x}K) + \cosh(\sqrt{x}K')}. \tag{1.3}$$

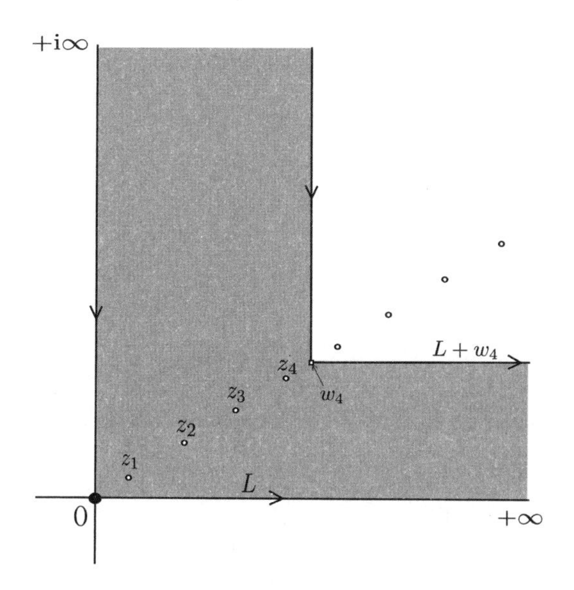

Fig. 1. The contour of integration L that is used in formula (1.5), the poles at points $\{z_n\}_{n\geq 1}$ and the shifted contour of integration $L + w_N$ when $N = 4$.

Our first step is to change the variable of integration $x = (2z/K)^2$ in (1.3). This implies $z = K\sqrt{x}/2$, and the original contour of integration $\mathbb{R} \ni x$ is mapped into the contour $L \ni z$, where L consists of two half-lines $(+\mathrm{i}\infty, 0] \cup [0, +\infty)$, see Fig. 1. This contour is traversed in the direction $+\mathrm{i}\infty \to 0 \to +\infty$. After this change of variables, we obtain

$$I = \frac{2}{K} \int_L \frac{\sin(zv(1+\tau))\mathrm{d}z}{\cos(2z) + \cos(2z\tau)}, \tag{1.4}$$

where we have denoted

$$\tau := \mathrm{i}\frac{K'}{K}.$$

Using trigonometric sum-to-product identity, we rewrite (1.4) in the form

$$I = \frac{1}{K} \int_L \frac{\sin(zv(1+\tau))\mathrm{d}z}{\cos(z(1+\tau))\cos(z(1-\tau))}. \tag{1.5}$$

Our second step is to compute the integral in (1.5) via Cauchy's Residue Theorem. Note that the integrand in (1.5) is a meromorphic function that has only simple poles. Only the poles lying in the first quadrant are of importance to us, and these are given by

$$z_n := \pi(n - 1/2)\frac{1}{1-\tau}, \quad n \in \mathbb{N}.$$

We also introduce the following notation:

$$w_n := \pi n \frac{1}{1-\tau} \quad \text{and} \quad t := \frac{1+\tau}{1-\tau}.$$

It is clear that the points z_n all lie on a ray in the first quadrant, and w_n are the midpoints between z_n and z_{n+1}, see Fig. 1.

We choose $N \in \mathbb{N}$ and we shift the contour of integration $L \mapsto L + w_N$, apply the Cauchy Residue Theorem and take into account the residues at $z = z_n$, $1 \le n \le N$:

$$I = 2\pi \mathrm{i} \times \frac{1}{K} \sum_{n=1}^{N} r_n(v) + I_N, \tag{1.6}$$

where we have denoted

$$I_N := \frac{1}{K} \int_{L+w_N} \frac{\sin(vz(1+\tau))\mathrm{d}z}{\cos(z(1+\tau))\cos(z(1-\tau))}, \tag{1.7}$$

and

$$\begin{aligned} r_n(v) &:= \operatorname{Res}\left(\frac{\sin(vz(1+\tau))}{\cos(z(1+\tau))\cos(z(1-\tau))} \,\middle|\, z = z_n \right) \\ &= \frac{(-1)^n}{1-\tau} \frac{\sin(\pi(n-1/2)tv)}{\cos(\pi(n-1/2)t)}. \end{aligned} \tag{1.8}$$

Note that shifting the contour of integration is justified, since the integrand decays exponentially fast as $z \to \infty$ in the area between the two contours L and $L + w_N$ (this is the gray area in Fig. 1).

Next, we plan to show that $I_N \to 0$ as $N \to +\infty$. By changing the variable of integration $z = w_N + w$, we rewrite (1.7) in the form

$$I_N = \frac{(-1)^N}{K} \int_L \frac{\sin(vw(1+\tau) + \pi N t v)\mathrm{d}w}{\cos(w(1+\tau) + \pi N t)\cos(w(1-\tau))}. \tag{1.9}$$

Note that $\arg(1+\tau) \in (0, \pi/2)$ and $\arg(1-\tau) \in (-\pi/2, 0)$, therefore

$$\operatorname{Im}(t) > 0 \quad \text{and} \quad \operatorname{Im}(w(1+\tau)) > 0, \quad \text{for all } w \in L.$$

The above conditions imply $\mathrm{Im}(w(1+\tau)+\pi Nt)>10$ for all $w\in L$ and all N large enough. Using the following trivial estimates

$$|\sin(a)|<\exp(\mathrm{Im}(a)), \qquad \text{for } a\in\mathbb{C} \text{ such that } \mathrm{Im}(a)>0,$$
$$|\cos(a)|>\exp(\mathrm{Im}(a))/10, \qquad \text{for } a\in\mathbb{C} \text{ such that } \mathrm{Im}(a)>10,$$

we conclude that for all $z\in L$ and all N large enough, we have

$$\begin{aligned}&\left|\frac{\sin(vz(1+\tau)+\pi Ntv)}{\cos(z(1+\tau)+\pi Nt)}\right|\\ &\quad<10\exp((v-1)\,\mathrm{Im}(z(1+\tau))+\pi N\,\mathrm{Im}(t)(v-1))\\ &\quad<10\exp(\pi N\,\mathrm{Im}(t)(v-1)).\end{aligned} \tag{1.10}$$

Combining (1.9) and (1.10), we obtain

$$|I_N|\le\frac{10}{K}\exp(\pi N\,\mathrm{Im}(t)(v-1))\times\int_L\frac{|\mathrm{d}z|}{|\cos(z(1-\tau))|},$$

and the right-hand side converges to zero as $N\to+\infty$ (recall that $v-1<0$ and $\mathrm{Im}(t)>0$). The above result and formula (1.6) imply the following identity:

$$I=\frac{2\pi\mathrm{i}}{K(1-\tau)}\sum_{n=1}^{\infty}(-1)^n\frac{\sin(\pi(n-1/2)tv)}{\cos(\pi(n-1/2)t)}. \tag{1.11}$$

Our third step is to express the infinite sum in (1.11) in terms of Jacobi elliptic functions. Formula (22.11.5) in [2] tells us that

$$\pi\sum_{n=1}^{\infty}(-1)^n\frac{\sin(\pi(n-1/2)z)}{\cosh(\pi(n-1/2)K'/K)}=-Kkk'\frac{\mathrm{sn}(Kz,k)}{\mathrm{dn}(Kz,k)}.$$

Using this result and formulas (22.2.4) and (22.2.6) in [2], which express Jacobi elliptic functions in terms of theta functions and formula (22.2.2) in [2], which express the constants k, k' and K in terms of theta functions, we arrive at the following expression:

$$I=-\frac{\pi\mathrm{i}}{K(1-\tau)}\theta_2(0,t)\theta_4(0,t)\frac{\theta_1(\pi tv/2,t)}{\theta_3(\pi\pi tv/2,t)}. \tag{1.12}$$

Here, $\theta_i(z,t)$ (with $z,t\in\mathbb{C}$ and $\mathrm{Im}(t)>0$) are the four theta functions, as defined in [2, formulas (20.2.1)–(20.2.4)].

Our plan is to apply transformations of theta functions $\theta_i(\cdot,t)$ with respect to the parameter t so that we obtain an expression involving $\theta_i(\cdot,\tau)$.

This will be done in four steps, and the sequence of transformations is summarized as follows:

$$t = \frac{1+\tau}{1-\tau} \longmapsto t_1 := t+1 = \frac{2}{1-\tau} \longmapsto t_2 := t_1/2 = \frac{1}{1-\tau}$$
$$\longmapsto t_3 := -1/t_2 = \tau - 1 \longmapsto t_4 := t_3 + 1 = \tau.$$

Transformation 1, $t \longmapsto t_1$: We apply formulas (20.7.26)–(20.7.29) in [2] to the expression in (1.12) and obtain

$$I = -\frac{\pi}{K(1-\tau)}\theta_2(0,t_1)\theta_3(0,t_1)\frac{\theta_1(\pi t v/2, t_1)}{\theta_4(\pi t v/2, t_1)}. \tag{1.13}$$

Transformation 2, $t_1 \longmapsto t_2$: We apply formulas (20.7.11)–(20.7.12) in [2] to the expression in (1.13) and obtain

$$I = -\frac{\pi}{2K(1-\tau)}\theta_2(0,t_2)^2\frac{\theta_1(\pi t v/4, t_2)\theta_2(\pi t v/4, t_2)}{\theta_3(\pi t v/4, t_2)\theta_4(\pi t v/4, t_2)}. \tag{1.14}$$

Transformation 3, $t_2 \longmapsto t_3$: We apply formulas (20.7.30)–(20.7.33) in [2] to the expression in (1.14) and obtain

$$I = -\frac{\pi}{2K}\theta_4(0,t_3)^2\frac{\theta_1(\pi t v t_3/4, t_3)\theta_4(\pi t v t_3/4, t_3)}{\theta_3(\pi t v t_3/4, t_3)\theta_2(\pi t v t_3/4, t_3)}. \tag{1.15}$$

Transformation 4, $t_3 \longmapsto \tau$: Finally, we apply formulas (20.7.26)–(20.7.29) in [2] to the expression in (1.15) and obtain

$$I = -\frac{\pi}{2K}\theta_3(0,\tau)^2\frac{\theta_1(\pi t v t_3/4, \tau)\theta_3(\pi t v t_3/4, \tau)}{\theta_4(\pi t v t_3/4, \tau)\theta_2(\pi t v t_3/4, \tau)}. \tag{1.16}$$

Now, we check that $\pi t v t_3/4 = -\pi v(1+\tau)/4$, we apply formulas (22.2.4) and (22.2.8) in [2] and rewrite the expression in (1.16) in terms of Jacobi elliptic functions, which gives us

$$I = \frac{\mathrm{sn}(vK(1+\tau)/2, k)}{\mathrm{cd}(vK(1+\tau)/2, k)}.$$

Recalling our definition of I in (1.3), we see that we have established formula (1.2) for $u = v(K + \mathrm{i}K')/2$ with $v \in (-1, 1)$.

As the final step, we need to show that (1.2) holds true in the bigger region $D := \{u \in \mathbb{C} : |\mathrm{Re}(u)| < K, |\mathrm{Im}(u)| < K'\}$. This is easy to achieve by analytic continuation. Indeed, the integral in the left-hand side of (1.2) converges absolutely and uniformly for all u on compact subsets of D, thus this integral defines an analytic function on D. The right-hand side of (1.2) is also analytic in D (one can check this by locating the poles of $\mathrm{sn}(u,k)$ and the roots of $\mathrm{cd}(u,k)$; see [2, Tables 22.4.1 and 22.4.2]). Thus, by analytic continuation, the identity (1.2) is valid not only for $u = v(K+\mathrm{i}K')/2$ with $v \in (-1,1)$, but also for all $u \in D$. $\square$

By expanding both sides in (1.2) in Taylor series in u, we obtain the following result. The mystery integral identity (1.1) follows by setting $n=0$ in formula (1.17) below.

Corollary 1.2. *With the notation of Theorem* 1.1, *we have*

$$\frac{1}{2}\int_{\mathbb{R}} \frac{x^n \mathrm{d}x}{\cos(\sqrt{x}K)+\cosh(\sqrt{x}K')} = (-1)^n \times \left.\frac{\mathrm{d}^{2n+1}}{\mathrm{d}u^{2n+1}} \frac{\mathrm{sn}(u,k)}{\mathrm{cd}(u,k)}\right|_{u=0}, \quad (1.17)$$

for $n \geq 0$.

2. A More General Version of the Mystery Integral

Ismail and Valent [3] compute explicitly the functions $D(x)$ and $B(x)$ appearing in the Nevanlinna parameterization of the indeterminate moment problem, which has the same moments as in (1.17). These functions are as follows:

$$D(x) = -\frac{4}{\pi}\sin(\sqrt{x}K/2)\sinh(\sqrt{x}K'/2),$$

$$B(x) = \frac{2}{\pi}\ln(k/k')\sin(\sqrt{x}K/2)\sinh(\sqrt{x}K'/2) + \cos(\sqrt{x}K/2)\cosh(\sqrt{x}K'/2)$$

(see [3, formulas (4.16) and (4.17)]). For $t \in \mathbb{R}$ and $\gamma > 0$, we define

$$w(x;t,\gamma) := \frac{\gamma/\pi}{(D(x)-tB(x))^2+\gamma^2 B(x)^2}, \quad x \in \mathbb{R}.$$

As was proved in [3] using the Nevanlinna parameterization and the theory of indeterminate moment problems, the measures $w(x;t,\gamma)\mathrm{d}x$ have the same moments for all $t \in \mathbb{R}$ and $\gamma > 0$. One can check (after some tedious algebraic computations) that for all $x \in \mathbb{R}$,

$$w(x;t^*,\gamma^*) = \frac{1/2}{\cos(\sqrt{x}K)+\cosh(\sqrt{x}K')},$$

provided that

$$\gamma^* = \frac{4}{\pi(1+C^2)}, \quad t^* = -C\gamma^*, \quad \text{and} \quad C = \frac{2}{\pi}\ln(k/k').$$

Combining this fact with Corollary 1.2, we obtain the following result, which generalizes Theorem 1.1.

Theorem 2.1. *Assume that $t \in \mathbb{R}$, $\gamma > 0$, $k \in (0,1)$ and denote $k' = \sqrt{1-k^2}$, $K = K(k)$ and $K' = K(k')$. Then for $u \in \mathbb{C}$ satisfying $|\text{Re}(u)| < K$ and $|\text{Im}(u)| < K'$, we have*

$$\int_{\mathbb{R}} \frac{\sin(\sqrt{x}u)}{\sqrt{x}} \times w(x;t,\gamma)\text{d}x = \frac{\text{sn}(u,k)}{\text{cd}(u,k)}. \tag{2.1}$$

It turns out that the integral in (2.1) also can be evaluated directly, without using the Nevanlinna parameterization or the theory of the indeterminate moment problem. To do this, one only needs to show directly that the measures $w(x;t,\gamma)\text{d}x$ have the same moments. We plan to present this approach in a more general setting in the forthcoming paper [4].

Acknowledgments

This research was supported by the Natural Sciences and Engineering Research Council of Canada. We would like to thank Mourad Ismail for the helpful discussions.

References

[1] Y. Chen and H. M. E. Ismail, Some indeterminate moment problems and Freud-like weights, *Construct. Approx.* **14**(3) (1998) 439–458.

[2] NIST Digital Library of Mathematical Functions, http://dlmf.nist.gov/, Release 1.0.11 of 2016-06-08. Online companion to F. W. J. Olver, D. W. Lozier, R. F. Boisvert and C. W. Clark (eds.), *NIST Handbook of Mathematical Functions* (Cambridge University Press, New York, 2010).

[3] M. E. H. Ismail and G. Valent, On a family of orthogonal polynomials related to elliptic functions, *Illinois J. Math.* **42**(2) (1998) 294–312.

[4] A. Kuznetsov, Constructing measures with identical moments, preprint (2016).

[5] F. W. J. Olver, D. W. Lozier, R. F. Boisvert and C. W. Clark (eds.), *NIST Handbook of Mathematical Functions* (Cambridge University Press, New York, 2010).

Chapter 22

Algebraic Generating Functions for Gegenbauer Polynomials

Robert S. Maier

Departments of Mathematics and Physics, University of Arizona
Tucson, AZ 85721, USA
rsm@math.arizona.edu

It is shown that several of Brafman's generating functions for the Gegenbauer polynomials are algebraic functions of their arguments if the Gegenbauer parameter differs from an integer by one-fourth or one-sixth. Two examples are given, which come from recently derived expressions for associated Legendre functions with tetrahedral or octahedral monodromy. It is also shown that if the Gegenbauer parameter is restricted as stated, the Poisson kernel for the Gegenbauer polynomials can be expressed in terms of complete elliptic integrals. An example is given.

Keywords: Generating function; Gegenbauer polynomial; Legendre polynomial; Legendre function; Hypergeometric transformation; Poisson kernel.

Mathematics Subject Classification 2010: 33C45, 33C05, 33C20

1. Introduction

The Gegenbauer polynomials $C_n^\lambda(x)$, $n = 0, 1, 2, \ldots$, for any $\lambda \in \mathbb{C}$, are defined by

$$\sum_{n=0}^{\infty} C_n^\lambda(x) t^n = (1 - 2xt + t^2)^{-\lambda}. \tag{1.1}$$

That is, they have $R^{-2\lambda} := (1-2xt+t^2)^{-\lambda}$ as their ordinary generating function. (In the sequel, R will signify $(1-2xt+t^2)^{1/2}$, with $R=1$ when $t=0$.) When $\lambda=1/2$, the polynomials $C_n^\lambda(x)$ become the Legendre polynomials $P_n(x)$. They are specializations themselves: $C_n^\lambda(x)$ is proportional to the ultraspherical polynomial $P_n^{(\lambda-1/2,\lambda-1/2)}$, which is the $\alpha=\beta=\lambda-1/2$ case of the Jacobi polynomial $P_n^{(\alpha,\beta)}(x)$. If $\lambda \neq -1/2, -3/2, \ldots$, one can write

$$C_n^\lambda(x) = \frac{(2\lambda)_n}{n!}\, {}_2F_1\left(\begin{matrix} -n,\ n+2\lambda \\ \lambda+1/2 \end{matrix} \,\middle|\, \frac{1-x}{2}\right), \tag{1.2}$$

where $(a)_n := a(a+1)\cdots(a+n-1)$ is the Pochhammer symbol and ${}_2F_1$ is the Gauss hypergeometric function.

We were intrigued by a remark of Mourad Ismail, [1, Section 4.3], to the effect that not many generating functions for Jacobi polynomials are known, which are algebraic functions of their arguments (denoted t, x here). In this chapter, we show that for the Gegenbauer polynomials with the parameter λ differing by one-fourth or one-sixth from an integer, there are several distinct nonordinary generating functions that are algebraic and can be expressed in closed form. (A similar result in the more familiar case when λ is an integer or a half-odd-integer was previously known.) The simplest example is

$$\sum_{n=0}^{\infty} \frac{(-1/12)_n}{(1/2)_n}\, C_n^{1/4}(x) t^n = 2^{-1/4} R^{1/12} \left[\cosh(\xi/3) + \sqrt{\frac{\sinh\xi}{3\sinh(\xi/3)}}\,\right]^{1/4},$$
$$e^\xi := R^{-1}[1-(x-\sqrt{x^2-1})t], \tag{1.3}$$

Because $\sinh\xi$, $\sinh(\xi/3)$, and $\cosh(\xi/3)$ are algebraic functions of e^ξ, the right-hand side is an algebraic function of t, x. Why it is most easily written trigonometrically will become clear.

The algebraic generating functions derived below are specializations of two of Brafman's nonordinary (and in general, non-algebraic) generating functions [2], which appear in Theorems 2.1 and 2.2, and their respective extensions [3], which appear in Theorems 3.1 and 3.2. For additional light on his two generating functions, see [4, Chapter 17; 5, Chapter III, Section 4; 6]. The two extensions come with the aid of an identity appearing as Equation (4) in [3], which is generalized and placed in context in [7; 8, Section 4.1]. The generating functions of Brafman are usually expressed in terms of ${}_2F_1$, but can also be written in terms of the associated Legendre functions P_ν^μ, or their Ferrers counterparts P_ν^μ.

There are cases in which the Legendre function P_ν^μ, of degree ν and order μ, can be written in closed form, such as when $\mu = 0$ and $\nu = n \in \mathbb{Z}$, in which case P_ν^μ and P_ν^μ equal P_n. There are several other cases. It has been known since the early work of Schwarz [9] on the algebraicity of ${}_2F_1$ that if the ordered pair $(\nu, \mu) \in \mathbb{C}^2$ differs from $(\pm 1/6, \pm 1/4)$, $(\pm 1/4, \pm 1/3)$, or $(\pm 1/6, \pm 1/3)$ by an element of $\mathbb{Z}^2$, the functions P_ν^μ and P_ν^μ will be algebraic. (For an exposition focusing on ${}_2F_1$, see [10, Chapter VII; 11, Section 2.7.2; 12].) But simple, nonparametric representations of these algebraic functions were not known.

We recently obtained explicit trigonometric formulas [13] for the functions $P_{-1/6}^{-1/4}, \mathrm{P}_{-1/6}^{-1/4}$, and the second-kind Legendre function $Q_{-1/4}^{-1/3}$. In the appendix, the resulting simple formulas for $P_{-1/6}^{\pm 1/4}, \mathrm{P}_{-1/6}^{\pm 1/4}$ and $P_{-1/4}^{\pm 1/3}$, $\mathrm{P}_{-1/4}^{\pm 1/3}$ are given. In Sections 2 and 3, Brafman's four results are specialized with the aid of these representations and yield novel algebraic generating functions for the set $\{C_n^\lambda(x)\}_{n=0}^\infty$ when $\lambda \in \mathbb{Z} \pm 1/4$ and $\lambda \in \mathbb{Z} \pm 1/6$. That specializing Brafman's two generating functions yields interesting identities has been pointed out [6], but the present focus on algebraicity is new.

For the Gegenbauer polynomials $C_n^\lambda(x)$, the cases $\lambda \in \mathbb{Z} \pm 1/4$ and $\lambda \in \mathbb{Z} \pm 1/6$ have long been recognized as special. For instance, $C_n^{-1/4}(x)$ and $C_n^{-1/6}(x)$ have been expressed in terms of elliptic functions [14]. Also, polynomials of the form $C_n^{1/6}(x)$ and $C_n^{7/6}(x)$ have recently been used in applied mathematics in the modeling of wave scattering in the exterior of what is locally a right-angled wedge [15]. In Section 4, we point out that the Poisson kernel for each of the sets $\{C_n^{1/4}(x)\}_{n=0}^\infty$ and $\{C_n^{1/6}(x)\}_{n=0}^\infty$ is also special: it can be expressed in terms of the first and second complete elliptic integral functions, $K = K(m)$ and $E = E(m)$. (A similar elliptic formula in the case $\lambda = 1/2$, i.e., for a kernel arising from the Legendre set $\{P_n(x)\}_{n=0}^\infty$, was obtained by Watson [16].) Some final remarks appear in Section 5.

2. The First Gegenbauer Generating Function

The following theorem presents Brafman's first ${}_2F_1$-based generating function [2].

Theorem 2.1. *For any $\lambda \in \mathbb{C} \setminus \{0, -1/2, -1, \dots\}$ and $\gamma \in \mathbb{C}$, one has*

$$\sum_{n=0}^{\infty} \frac{(\gamma)_n}{(2\lambda)_n} C_n^\lambda(x) t^n = R^{-\gamma} {}_2F_1\left(\begin{matrix} \gamma, 2\lambda - \gamma \\ \lambda + 1/2 \end{matrix} \,\middle|\, \frac{R - 1 + xt}{2R} \right) \tag{2.1a}$$

$$= (1-xt)^{-\gamma}\, {}_2F_1\left(\begin{matrix}\gamma/2,\ \gamma/2+1/2\\ \lambda+1/2\end{matrix}\,\middle|\, 1-\left(\frac{R}{1-xt}\right)^2\right), \tag{2.1b}$$

as equalities between power series in t.

Note. Version (2.1b) is Brafman's, rewritten; (2.1a) follows by a quadratic transformation of ${}_2F_1$. When $\gamma = -N$, $N = 0, 1, 2, \ldots$, the series terminates, and the ${}_2F_1$ in (2.1a) is proportional to $C_N^{\lambda}((1-xt)/R)$ by (1.2).

The ${}_2F_1$ in (2.1a) can be rewritten as an associated Legendre (or Ferrers) function by (A.1); as is true of the ${}_2F_1$ in (2.1b) if $\gamma = 2\lambda - 1/2$. It must be remembered that $P_\nu^\mu(z), \mathrm{P}_\nu^\mu(z)$ are defined if, respectively, $z \notin (-\infty, 1]$ and $z \notin (-\infty, -1] \cup [1, \infty)$. The rewriting yields the following theorem.

Theorem 2.2. *For any $\mu \notin \{1/2, 1, 3/2, \ldots\}$, one has*

$$\sum_{n=0}^{\infty} \frac{(-\nu-\mu)_n}{(1-2\mu)_n} C_n^{1/2-\mu}(x) t^n$$
$$= 2^{-\mu}\,\Gamma(1-\mu)\, R^{\nu+\mu}[(z^2-1)^{\mu/2} P_\nu^\mu(z)]\Big|_{z=\frac{1-xt}{R}}, \tag{2.2a}$$

for any $\nu \in \mathbb{C}$. Moreover,

$$\sum_{n=0}^{\infty} \frac{(1/2-2\mu)_n}{(1-2\mu)_n} C_n^{1/2-\mu}(x) t^n$$
$$= 2^{-\mu}\,\Gamma(1-\mu)\,(1-xt)^{2\mu-1/2}[(z^2-1)^{\mu/2} P_{-1/4}^\mu(z)]\Big|_{z=2\left(\frac{R}{1-xt}\right)^2-1}. \tag{2.2b}$$

These hold as equalities between power series in t. For real z, they hold as stated when $z > 1$, and hold when $z \in (-1, 1)$ if P_ν^μ and z^2-1 are replaced by P_ν^μ and $1-z^2$.

Note. By examination, if x is real and t is real and of sufficiently small magnitude, then $z > 1$ (implying that the Legendre case is applicable) in (2.2a) when $|x| > 1$ and in (2.2b) when $|x| < 1$. Conversely, $z \in (-1, 1)$ (implying that the Ferrers case is applicable) in (2.2a) when $|x| < 1$ and in (2.2b) when $|x| > 1$. For the $|x| < 1$ case of (2.2a), cf. [17, Theorem 3].

Specializing parameters ν, μ in this theorem yields a number of interesting identities. As is summarized in the appendix, the associated Legendre function P_ν^μ and Ferrers function P_ν^μ can be written in terms of elementary functions in several cases. They are numbered here as the reducible case (i), the quasi-algebraic cases (ii[a]), (ii[b]), and the algebraic cases (iii), (iv[a]), (iv[b]).

In the reducible case (i), $\nu = -\mu + N$, $N = 0, 1, 2, \ldots$. The functions $P_\nu^\mu = P_{-\mu+N}^\mu$ and $\mathrm{P}_\nu^\mu = \mathrm{P}_{-\mu+N}^\mu$ can be expressed in terms of the polynomial $C_N^{\mu-1/2}$ (see (A.3)). This leads to the following theorem.

Theorem 2.3. *For any $\lambda \in \mathbb{C} \setminus \{0, -1/2, -1, \ldots\}$ and $N = 0, 1, 2, \ldots$, one has*

$$\sum_{n=0}^{N} \frac{(-N)_n}{(2\lambda)_n} C_n^\lambda(x) t^n = \frac{N!}{(2\lambda)_N} R^N C_N^\lambda\left(\frac{1 - xt}{R}\right), \tag{2.3}$$

$$\sum_{n=0}^{\infty} \frac{(2\lambda + N)_n}{(2\lambda)_n} C_n^\lambda(x) t^n = \frac{N!}{(2\lambda)_N} R^{-2\lambda - N} C_N^\lambda\left(\frac{1 - xt}{R}\right), \tag{2.4}$$

as equalities between power series in t.

Proof. To obtain (2.3), substitute (A.3) into (2.2a), and for (2.4) do the same, first using $P_\nu^\mu = P_{-\nu-1}^\mu$ (or $\mathrm{P}_\nu^\mu = \mathrm{P}_{-\nu-1}^\mu$). □

The finite identity (2.3) has been derived in another way by Miller (see [18, equation (4.11)]), and (2.4) is also known. The $N = 0$ case of (2.4) is of course the defining generating function (1.1) for the Gegenbauer polynomials. There are also identities resembling (2.3) and (2.4) that come from (2.2b).

In the first quasi-dihedral case (ii[a]), the degree ν is an integer, and P_ν^μ is therefore expressible in closed form; for instance, $P_{-1}^\mu(\coth \xi) = P_0^\mu(\coth \xi)$ equals $\Gamma(1-\mu)^{-1} e^{\mu\xi}$, or equivalently,

$$P_{-1}^\mu(z) = P_0^\mu(z) = \Gamma(1-\mu)^{-1} \left[(z+1)/(z-1)\right]^{\mu/2}. \tag{2.5}$$

Setting $\nu = -1, 0$ in (2.2a) yields the pair

$$\sum_{n=0}^{\infty} \frac{(\lambda + 1/2)_n}{(2\lambda)_n} C_n^\lambda(x) t^n = R^{-1} \left(\frac{1 + R - xt}{2}\right)^{1/2 - \lambda}, \tag{2.6a}$$

$$\sum_{n=0}^{\infty} \frac{(\lambda - 1/2)_n}{(2\lambda)_n} C_n^\lambda(x) t^n = \left(\frac{1 + R - xt}{2}\right)^{1/2 - \lambda}. \tag{2.6b}$$

The identity (2.6a) is a well-known alternative generating function for the Gegenbauer polynomials. But its companion (2.6b) is less well known, though a generalization to Jacobi polynomials was found by Carlitz [19];

as is the fact that a closed form can be computed whenever the coefficient $(\lambda+1/2)_n/(2\lambda)_n$ is replaced by $(\lambda+k+1/2)_n/(2\lambda)_n$, with $k\in\mathbb{Z}$.

In the second quasi-dihedral case (ii[b]), the order μ is a half-odd-integer, which in (2.2a) and (2.2b) means that the Gegenbauer parameter $\lambda = 1/2-\mu$ must be an integer. This is the fairly straightforward trigonometric (e.g., Chebyshev) case, and the resulting identities are not given here.

The focus here is on the octahedral case (iii), when $(\nu,\mu)\in\mathbb{Z}^2 + (\pm1/6,\pm1/4)$, and the tetrahedral cases (iv[a]) and (iv[b]), when $(\nu,\mu)\in\mathbb{Z}^2+(\pm1/4,\pm1/3)$ and $(\pm1/6,\pm1/3)$. The functions $P_\nu^\mu, \mathrm{P}_\nu^\mu$ are then algebraic (see the appendix).

Theorem 2.4. *The Gegenbauer generating function*

$$\sum_{n=0}^{\infty}\frac{(\gamma)_n}{(2\lambda)_n}C_n^\lambda(x)t^n$$

is algebraic (1) *if* $\lambda\in\mathbb{Z}\pm1/4$ *with* $\gamma-\lambda\in\mathbb{Z}\pm1/3$; *or* (2) *if* $\lambda\in\mathbb{Z}\pm1/6$ *with* $\gamma-\lambda\in\mathbb{Z}\pm\{1/3,1/4\}$.

Proof. Claim 1 comes by restricting (2.2a) to the octahedral case (iii), and claim 2 by restricting (2.2a) to the tetrahedral cases (iv[a]),(iv[b]). □

For the octahedral case (iii) and the tetrahedral case (iv[a]), the fundamental algebraic formulas are (A.7) and (A.9), where $P_{-1/6}^{\pm1/4}(\cosh\xi)$, $\mathrm{P}_{-1/6}^{\pm1/4}(\cos\theta)$ and $P_{-1/4}^{\pm1/3}(\coth\xi)$, $\mathrm{P}_{-1/4}^{\pm1/3}(\tanh\xi)$ are given in terms of trigonometric functions: hyperbolic ones of ξ and circular ones of θ. In effect, they are given in terms of e^ξ or $e^{i\theta}$. But in Theorem 2.2, the Legendre/Ferrers argument z equals $1-xt$ (in (2.2a)) or $2\left[R/(1-xt)\right]^2-1$ (in (2.2b)). The following table adapts (A.7) and (A.9) to the needs of Theorem 2.2.

$z=(1-xt)/R$		
$z=\cosh\xi$	$\Longrightarrow$	$e^\xi = R^{-1}\left[1-(x-\sqrt{x^2-1})t\right]$
$z=\cos\theta$	$\Longrightarrow$	$e^{\mathrm{i}\theta} = R^{-1}\left[1-(x-\mathrm{i}\sqrt{1-x^2})t\right]$
$z=\coth\xi$	$\Longrightarrow$	$e^\xi = t\sqrt{x^2-1}/(1-R-xt)$
$z=\tanh\xi$	$\Longrightarrow$	$e^\xi = t\sqrt{1-x^2}/(-1+R+xt)$
$z=2\left[R/(1-xt)\right]^2-1$		
$z=\cosh\xi$	$\Longrightarrow$	$e^{\xi/2} = (1-xt)/\left[R-t\sqrt{1-x^2}\right]$
$z=\cos\theta$	$\Longrightarrow$	$e^{\mathrm{i}\theta/2} = (1-xt)/\left[R-\mathrm{i}t\sqrt{x^2-1}\right]$
$z=\coth\xi$	$\Longrightarrow$	$e^\xi = R/\left(t\sqrt{1-x^2}\right)$
$z=\tanh\xi$	$\Longrightarrow$	$e^\xi = R/\left(t\sqrt{x^2-1}\right)$

By combining the $(\nu,\mu)=(-1/6,1/4)$ case of (2.2a) with (A.7a), aided by the first line of this table, one readily derives an explicit, octahedrally algebraic generating function for the set $\{C_n^{1/4}(x)\}_{n=0}^{\infty}$. It appeared in Section 1 as equation (1.3). One also derives a tetrahedrally algebraic generating function for $\{C_n^{1/6}(x)\}_{n=0}^{\infty}$,

$$\begin{aligned}
&\sum_{n=0}^{\infty}\frac{(-1/12)_n}{(1/3)_n}\,C_n^{1/6}(x)t^n\\
&\quad=2^{-7/12}\,3^{-3/8}R^{1/12}(\sinh\xi)^{-1/3}\big[\sqrt{\sqrt{3}+1}\,f_+ + \sqrt{\sqrt{3}-1}\,f_-\big],\\
&\qquad\text{when } e^{\xi}:=t\sqrt{x^2-1}/(1-R-xt)\\
&\quad=2^{-7/12}\,3^{-3/8}R^{1/12}(\cosh\xi)^{-1/3}\big[\sqrt{\sqrt{3}+1}\,g_+ + \sqrt{\sqrt{3}-1}\,g_-\big],\\
&\qquad\text{when } e^{\xi}:=t\sqrt{1-x^2}/(-1+R+xt),
\end{aligned}\tag{2.7}$$

by combining the $(\nu,\mu)=(-1/4,1/3)$ case of (2.2a) with (A.9a) and (A.9b), aided by the third and fourth lines of the table. Here, the functions $f_\pm = f_\pm(\coth\xi)$ and $g_\pm = g_\pm(\tanh(\xi))$ are algebraic in e^{ξ} and are defined in (A.8a) and (A.8b). The two right-hand sides of (2.7) are equivalent, but when the argument x is real, they are most useful when, respectively, $|x|>1$ and $|x|<1$. Additional explicitly algebraic generating functions that arise from (2.2a) or (2.2b) can be worked out.

The preceding results, which stemmed from Theorem 2.1, can be generalized because Brafman's first generating function can be hypergeometrically extended. The extension uses the identities appearing in [3, equations (4) and (11)]. The latter, which is a specialization of the former to Gegenbauer polynomials, can be restated as follows.

Lemma 2.5. *For any $\lambda\in\mathbb{C}$, and parameters $c_1,\ldots,c_p$, $d_1,\ldots,d_q$ and $u\in\mathbb{C}$ for which the below ${}_{p+1}F_q$ coefficients are defined, one has*

$$\begin{aligned}
&\sum_{n=0}^{\infty}{}_{p+1}F_q\left(\begin{matrix}-n,\,c_1,\ldots,\,c_p\\ d_1,\ldots,d_q\end{matrix}\,\middle|\,u\right)C_n^{\lambda}(x)t^n\\
&\quad=R^{-2\lambda}\sum_{n=0}^{\infty}\frac{(c_1)_n\cdots(c_p)_n}{(d_1)_n\cdots(d_q)_n}\,C_n^{\lambda}\left(\frac{x-t}{R}\right)\left(\frac{-tu}{R}\right)^n,
\end{aligned}$$

as an equality between power series in t.

Note. In [3], the argument of the C_n^λ on the right is written as w, which is defined in equation (1) of that work to equal $2(x-t)/R$. The "2" is easily seen to be erroneous and has been removed. This identity is proved by series rearrangement, once each C_n^λ has been expressed hypergeometrically: not as in (1.2), but in a form that incorporates a quadratic transformation.

By applying the $p = q = 1$ case of the lemma to the statement of Theorem 2.1, one readily obtains the following corollaries of (2.1a) and (2.1b). Here and below, U signifies $[1-2(1-u)xt+(1-u)^2t^2]^{1/2}$, with $U = 1$ when $t = 0$. Hence, U interpolates between R at $u = 0$ and unity at $u = 1$. Similarly, $R^2 + u(x-t)t$ interpolates between R^2 and $1 - xt$. It should be noted that to make (2.8b) resemble (2.1b) as closely as possible, Euler's transformation of ${}_2F_1$ has been applied to its right-hand side.

Theorem 2.6. *For any* $\lambda \in \mathbb{C} \setminus \{0, -1/2, -1, \ldots\}$ *and* $\gamma \in \mathbb{C}$, *and arbitrary* u, *one has*

$$\sum_{n=0}^{\infty} {}_2F_1\left(\begin{matrix}-n,\ 2\lambda-\gamma\\ 2\lambda\end{matrix}\,\middle|\, u\right) C_n^\lambda(x)t^n$$

$$= U^{\gamma-2\lambda}R^{-\gamma}\,{}_2F_1\left(\begin{matrix}\gamma,\ 2\lambda-\gamma\\ \lambda+1/2\end{matrix}\,\middle|\, \frac{UR-[R^2+u(x-t)t]}{2UR}\right) \tag{2.8a}$$

$$= U^{2\gamma-2\lambda}[R^2+u(x-t)t]^{-\gamma}$$

$$\times\,{}_2F_1\left(\begin{matrix}\gamma/2,\ \gamma/2+1/2\\ \lambda+1/2\end{matrix}\,\middle|\, 1-\left[\frac{UR}{R^2+u(x-t)t}\right]^2\right), \tag{2.8b}$$

where the equalities are between power series in t.

Note. When $\gamma = -N$, $N = 0, 1, 2, \ldots$, the series terminates, and the ${}_2F_1$ in (2.8a) is proportional to $C_N^\lambda\left([R^2+u(x-t)t]/UR\right)$.

Theorem 2.6 is not merely a corollary of Theorem 2.1, but an extension. It reduces to the previous theorem when $u = 1$, because the left-hand ${}_2F_1$ then equals $(\gamma)_n/(2\lambda)_n$ by the Chu–Vandermonde formula. Rewriting each right-hand ${}_2F_1$ in Theorem 2.6 as an associated Legendre function yields the following, which is an extension of Theorem 2.2. (In (2.8b), the rewriting is possible only if $\gamma = 2\lambda - 1/2$.)

Theorem 2.7. *For any $\mu \notin \{1/2, 1, 3/2, \dots\}$, and arbitrary u, one has*

$$\sum_{n=0}^{\infty} {}_2F_1\left(\begin{matrix} -n,\, \nu-\mu+1 \\ 1-2\mu \end{matrix} \,\middle|\, u \right) C_n^{1/2-\mu}(x) t^n$$
$$= 2^{-\mu}\,\Gamma(1-\mu)\, U^{-\nu+\mu-1} R^{\nu+\mu} [(z^2-1)^{\mu/2}\, P_\nu^\mu(z)] \Big|_{z=\frac{R^2+u(x-t)t}{UR}}, \tag{2.9a}$$

for any $\nu \in \mathbb{C}$. Moreover,

$$\sum_{n=0}^{\infty} {}_2F_1\left(\begin{matrix} -n,\, 1/2 \\ 1-2\mu \end{matrix} \,\middle|\, u \right) C_n^{1/2-\mu}(x) t^n$$
$$= 2^{-\mu}\,\Gamma(1-\mu)$$
$$\times U^{-2\mu} \left[R^2 + u(x-t)t\right]^{2\mu-1/2}$$
$$\times [(z^2-1)^{\mu/2}\, P_{-1/4}^\mu(z)] \Big|_{z=2[\frac{UR}{R^2+u(x-t)t}]^2-1}. \tag{2.9b}$$

These hold as equalities between power series in t. For real z, they hold as stated when $z>1$ and hold when $z \in (-1,1)$ if P_ν^μ and z^2-1 are replaced by P_ν^μ and $1-z^2$.

By specializing Theorem 2.7, one can immediately extend the preceding results on closed-form generating functions and algebraicity to include the free parameter u. The following specializations of (2.9a), based on (A.3), are extensions of the identities (2.3) and (2.4) of Theorem 2.3; for the latter, cf. [6, equation (4.14)].

Theorem 2.8. *For any $\lambda \in \mathbb{C} \setminus \{0, -1/2, -1, \dots\}$ and $N = 0, 1, 2, \dots$, and arbitrary u, one has*

$$\sum_{n=0}^{\infty} {}_2F_1\left(\begin{matrix} -n,\, 2\lambda+N \\ 2\lambda \end{matrix} \,\middle|\, u \right) C_n^{\lambda}(x) t^n$$
$$= \frac{N!}{(2\lambda)_N} U^{-2\lambda-N} R^N\, C_N^\lambda\left(\frac{R^2+u(x-t)t}{UR} \right),$$

$$\sum_{n=0}^{\infty} {}_2F_1\left(\begin{matrix} -n,\, -N \\ 2\lambda \end{matrix} \,\middle|\, u \right) C_n^{\lambda}(x) t^n$$
$$= \frac{N!}{(2\lambda)_N} U^{N} R^{-2\lambda-N}\, C_N^\lambda\left(\frac{R^2+u(x-t)t}{UR} \right),$$

as equalities between power series in t.

The extension of Theorem 2.4, which was a consequence of (2.2a), arises similarly as a consequence of its extension (2.9a).

Theorem 2.9. *The Gegenbauer generating function*

$$\sum_{n=0}^{\infty} {}_2F_1\left(\begin{matrix}-n,\, 2\lambda-\gamma\\ 2\lambda\end{matrix}\,\middle|\, u\right) C_n^{\lambda}(x)t^n$$

is algebraic, for arbitrary u, (1) *if* $\lambda \in \mathbb{Z} \pm 1/4$ *with* $\gamma - \lambda \in \mathbb{Z} \pm 1/3$; *or* (2) *if* $\lambda \in \mathbb{Z} \pm 1/6$ *with* $\gamma - \lambda \in \mathbb{Z} \pm \{1/3, 1/4\}$.

Explicit expressions for these u-dependent generating functions, which are algebraic in u as well as in t, x, can be computed with some effort from Theorem 2.7 if one exploits the fundamental formulas (A.7) and (A.9).

3. The Second Gegenbauer Generating Function

The following theorem presents Brafman's second ${}_2F_1$-based generating function [2].

Theorem 3.1. *For any* $\lambda \in \mathbb{C} \setminus \{0, -1/2, -1, \ldots\}$ *and* $\gamma \in \mathbb{C}$, *one has*

$$\begin{aligned}\sum_{n=0}^{\infty} &\frac{(\gamma)_n(2\lambda-\gamma)_n}{(2\lambda)_n(\lambda+1/2)_n} C_n^{\lambda}(x)t^n \\ &= {}_2F_1\left(\begin{matrix}\gamma,\, 2\lambda-\gamma\\ \lambda+1/2\end{matrix}\,\middle|\, \frac{1-R-t}{2}\right) {}_2F_1\left(\begin{matrix}\gamma,\, 2\lambda-\gamma\\ \lambda+1/2\end{matrix}\,\middle|\, \frac{1-R+t}{2}\right)\end{aligned} \tag{3.1a}$$

$$\begin{aligned}&= (1-2xt)^{-\gamma} \\ &\quad\times {}_2F_1\left(\begin{matrix}\gamma/2,\, \gamma/2+1/2\\ \lambda+1/2\end{matrix}\,\middle|\, 1-\frac{1}{(R+t)^2}\right) \\ &\quad\times {}_2F_1\left(\begin{matrix}\gamma/2,\, \gamma/2+1/2\\ \lambda+1/2\end{matrix}\,\middle|\, 1-\frac{1}{(R-t)^2}\right),\end{aligned} \tag{3.1b}$$

as equalities between power series in t.

Note. Version (3.1a) is Brafman's; (3.1b) follows by quadratically transforming each ${}_2F_1$. When $\gamma = -N$, $N = 0, 1, 2, \ldots$, the series terminates, and the right-hand side of (3.1a) is proportional to $C_N^{\lambda}(R+t)C_N^{\lambda}(R-t)$. Irrespective of γ, the Legendre (i.e., $\lambda = 1/2$) case is of special interest. The C_n^{λ} then reduces to the Legendre polynomial P_n, and each ${}_2F_1$ in (3.1a) is proportional to the Legendre function $P_{-\gamma}$.

By applying the $p = q = 2$ case of Lemma 2.5 to the statement of Theorem 3.1, one readily obtains the following corollaries of (3.1a) and (3.1b); for the former, cf. [3].

Theorem 3.2. *For any $\lambda \in \mathbb{C} \setminus \{0, -1/2, -1, \dots\}$ and $\gamma \in \mathbb{C}$, and arbitrary u, one has*

$$\sum_{n=0}^{\infty} {}_3F_2\left(\begin{matrix} -n,\, \gamma,\, 2\lambda-\gamma \\ 2\lambda,\, \lambda+1/2 \end{matrix} \,\middle|\, u\right) C_n^{\lambda}(x) t^n$$

$$= R^{-2\lambda}\, {}_2F_1\left(\begin{matrix} \gamma,\, 2\lambda-\gamma \\ \lambda+1/2 \end{matrix} \,\middle|\, \frac{R-U+ut}{2R}\right) {}_2F_1\left(\begin{matrix} \gamma,\, 2\lambda-\gamma \\ \lambda+1/2 \end{matrix} \,\middle|\, \frac{R-U-ut}{2R}\right) \tag{3.2a}$$

$$= (U^2 - u^2t^2)^{-\gamma} R^{2\gamma-2\lambda}$$

$$\times\, {}_2F_1\left(\begin{matrix} \gamma/2,\, \gamma/2+1/2 \\ \lambda+1/2 \end{matrix} \,\middle|\, 1 - \frac{R^2}{(U-ut)^2}\right)$$

$$\times\, {}_2F_1\left(\begin{matrix} \gamma/2,\, \gamma/2+1/2 \\ \lambda+1/2 \end{matrix} \,\middle|\, 1 - \frac{R^2}{(U+ut)^2}\right), \tag{3.2b}$$

where the equalities are between power series in t.

Note. The Legendre case of (3.2b), i.e., that of $\lambda = 1/2$ with γ unrestricted, was derived before Brafman by Rice [20], who attributed its $u=1$ subcase to Bateman [21] (cf. Rice's equations (2.11) and (2.14), and Bateman's (4.3)).

Theorem 3.2 does not reduce easily to Theorem 3.1 (for instance, by setting $u = 1$, as in Theorem 2.6). It is better described as a corollary than as an extension.

The ${}_2F_1$ functions in Theorems 3.1 and 3.2 are familiar: they have the same parameters as in Theorems 2.1 and 2.6, to which Theorems 3.1 and 3.2 are analogous. Rewriting Theorem 3.1 in terms of associated Legendre or Ferrers functions yields the following, which is analogous to Theorem 2.2. (For the first identity (3.3a), cf. [17, Theorem 2].)

Theorem 3.3. *For any $\mu \notin \{1/2, 1, 3/2, \dots\}$, one has*

$$\sum_{n=0}^{\infty} \frac{(-\nu-\mu)_n (1+\nu-\mu)_n}{(1-2\mu)_n (1-\mu)_n} C_n^{1/2-\mu}(x) t^n$$

$$= 2^{-2\mu}\, \Gamma(1-\mu)^2\, \mathcal{F}_\nu^\mu(R+t)\, \mathcal{F}_\nu^\mu(R-t), \tag{3.3a}$$

for any $\nu \in \mathbb{C}$. *Moreover,*

$$\sum_{n=0}^{\infty} \frac{(1/2-2\mu)_n(1/2)_n}{(1-2\mu)_n(1-\mu)_n} C_n^{1/2-\mu}(x)t^n$$
$$= 2^{-2\mu}\,\Gamma(1-\mu)^2\,(1-2xt)^{2\mu-1/2}$$
$$\times\, \mathfrak{F}^{\mu}_{-1/4}\left(\frac{2}{(R-t)^2}-1\right) \mathfrak{F}^{\mu}_{-1/4}\left(\frac{2}{(R+t)^2}-1\right). \tag{3.3b}$$

These hold as equalities between power series in t. *The function* $\mathfrak{F}^{\mu}_{\nu}(z)$ *is defined as* $(z^2-1)^{\mu/2}\, P^{\mu}_{\nu}(z)$ *or* $(1-z^2)^{\mu/2}\, \mathrm{P}^{\mu}_{\nu}(z)$, *which are equivalent. (For real* z, *the Legendre definition should be used if* $z>1$ *and the Ferrers definition if* $z \in (-1,1)$.)

The rewriting of Theorem 3.2 in terms of associated Legendre functions proceeds similarly. By specializing parameters in the two rewritten theorems, one can derive a number of interesting identities. For example, in the reducible case (i), when $\nu = -\mu + N$, $N = 0, 1, 2, \ldots$, one can derive analogues of Theorems 2.3 and 2.8.

The focus here is on the octahedral and tetrahedral algebraic cases. From the algebraic formulas in the appendix, substituted into the rewritten theorems, one deduces the following from their first halves; again, as in the last section.

Theorem 3.4. *The Gegenbauer generating functions*

$$\sum_{n=0}^{\infty} \frac{(\gamma)_n(2\lambda-\gamma)_n}{(2\lambda)_n(\lambda+1/2)_n} C_n^{\lambda}(x)t^n, \quad \sum_{n=0}^{\infty} {}_3F_2\left(\begin{matrix} -n,\ \gamma,\ 2\lambda-\gamma \\ 2\lambda,\ \lambda+1/2 \end{matrix} \,\middle|\, u\right) C_n^{\lambda}(x)t^n,$$

are algebraic (1) *if* $\lambda \in \mathbb{Z} \pm 1/4$ *with* $\gamma-\lambda \in \mathbb{Z}\pm 1/3$; *or* (2) *if* $\lambda \in \mathbb{Z}\pm 1/6$ *with* $\gamma-\lambda \in \mathbb{Z}\pm\{1/3, 1/4\}$. *In the latter generating function,* u *is arbitrary: the algebraicity is in* x, t *and* u.

4. The Poisson Kernel

The Poisson kernel for a set of orthogonal polynomials $\{p_n(x)\}_{n=0}^{\infty}$ plays a major role in approximation theory. It is a bilinear generating function of the form

$$K_t(x,y) = \sum_{n=0}^{\infty} h_n\, p_n(x) p_n(y)\, t^n, \tag{4.1}$$

where the normalization coefficients h_n would be absent if the polynomials were orthonormal and not merely orthogonal. The Poisson kernel for the Gegenbauer polynomials can be expressed in terms of ${}_2F_1$, as can a slightly simpler companion function [22–24]. Let $x = \cos\theta$ and $y = \cos\phi$, which are appropriate when $x, y \in (-1, 1)$, and define

$$\tilde{z} = \frac{-4\,t\sin\theta\sin\phi}{1 - 2t\cos(\theta-\phi) + t^2}, \tag{4.2a}$$

$$z = \frac{4\,t^2\sin^2\theta\sin^2\phi}{(1 - 2t\cos\theta\cos\phi + t^2)^2}, \tag{4.2b}$$

which are related by $z = [\tilde{z}/(2-\tilde{z})]^2$. The Poisson kernel for $\{C_n^\lambda(x)\}_{n=0}^\infty$ is

$$\begin{aligned}
&\sum_{n=0}^{\infty} \frac{\lambda+n}{\lambda}\,\frac{n!}{(2\lambda)_n}\,C_n^\lambda(x)C_n^\lambda(y)\,t^n \\
&\quad = \frac{1-t^2}{[1-2t\cos(\theta-\phi)+t^2]^{\lambda+1}}\,{}_2F_1\left(\begin{matrix}\lambda,\ \lambda+1\\ 2\lambda\end{matrix}\,\middle|\,\tilde{z}\right) \\
&\quad = \frac{1-t^2}{(1-2t\cos\theta\cos\phi+t^2)^{\lambda+1}}\,{}_2F_1\left(\begin{matrix}(\lambda+1)/2,\ (\lambda+2)/2\\ \lambda+1/2\end{matrix}\,\middle|\,z\right)
\end{aligned} \tag{4.3}$$

and its companion is

$$\begin{aligned}
&\sum_{n=0}^{\infty} \frac{n!}{(2\lambda)_n}\,C_n^\lambda(x)C_n^\lambda(y)\,t^n \\
&\quad = \frac{1}{[1-2t\cos(\theta-\phi)+t^2]^{\lambda}}\,{}_2F_1\left(\begin{matrix}\lambda,\ \lambda\\ 2\lambda\end{matrix}\,\middle|\,\tilde{z}\right) \\
&\quad = \frac{1}{(1-2t\cos\theta\cos\phi+t^2)^{\lambda}}\,{}_2F_1\left(\begin{matrix}\lambda/2,\ (\lambda+1)/2\\ \lambda+1/2\end{matrix}\,\middle|\,z\right).
\end{aligned} \tag{4.4}$$

In each of (4.3) and (4.4), the two right-hand sides are related by a quadratic hypergeometric transformation. Equation (4.3) can be obtained from equation (4.4) by applying the operator $\lambda^{-1}t^{-\lambda+1}\frac{\mathrm{d}}{\mathrm{d}t}\circ t^\lambda$ to both sides.

When the parameter λ is an integer (e.g., in the Chebyshev case), the Poisson kernel and its companion are elementary functions. When $\lambda = 1/2$, so that $\{C_n^\lambda(x)\}_{n=0}^\infty$ are the Legendre polynomials, Watson [16] expressed the companion in terms of the first complete elliptic integral function, $K = K(m)$. In principle, this can be done when λ is any half-odd-integer.

It does not seem to have been remarked that when λ differs by an integer by one-fourth or one-sixth, expressions in terms of complete elliptic integrals can also be obtained. This is implied by the pattern of hypergeometric parameters in (4.4), as will be explained. The focus is on the cases $\lambda = 1/4$ and $\lambda = 1/6$; the general $\lambda \in \mathbb{Z} \pm 1/4$ and $\lambda \in \mathbb{Z} \pm 1/6$ cases can be handled by applying the contiguity relations of ${}_2F_1$.

The Gauss hypergeometric ODE satisfied by the function ${}_2F_1(a, b; c; z)$ has singular points at $z = 0, 1, \infty$, with respective characteristic exponents $0, 1-c$; $0, c-a-b$; and a, b. The respective exponent *differences* are $1-c$, $c-a-b$, and $b-a$, and differences are significant only up to sign. It is well known that if a Gauss ODE has an unordered set of (up-to-sign) exponent differences $\{\frac{1}{2}, \delta, \delta\}$, it admits of a quadratic transformation to one with differences $\{\delta, \delta, 2\delta\}$, and the solutions of the two ODEs will correspond. (For instance, in each of (4.3) and (4.4) the first right-hand side comes from the second in this way.) One may write $\{\frac{1}{2}, \delta, \delta\} \sim \{\delta, \delta, 2\delta\}$. There are hypergeometric transformations of higher order than the quadratic (see [10, Section 25], *inter alia*). In particular, there are sextic ones that arise as compositions of quadratic and cubic ones, the action of which is summarized by $\{\frac{1}{2}, \frac{1}{3}, \delta\} \sim \{2\delta, 2\delta, 2\delta\}$.

The exponent differences for the ${}_2F_1$s in the two right-hand sides of (4.4) are respectively $1 - 2\lambda, 0, 0$ and $1/2 - \lambda, 0, 1/2$. If $\lambda = 1/6$, the second of these triples (when unordered) is $\{\frac{1}{2}, \frac{1}{3}, 0\} \sim \{0, 0, 0\}$. If $\lambda = 1/4$, the first is $\{\frac{1}{2}, 0, 0\} \sim \{0, 0, 0\}$. In other words, a sextic and a quadratic transformation will respectively convert the ${}_2F_1$s in the two right-hand sides of (4.4) to solutions of a Gauss ODE with exponent differences $0, 0, 0$. This is the Gauss ODE with parameters $a = b = 1/2$, $c = 1$, one of the solutions of which is ${}_2F_1(1/2, 1/2; 1; \tilde{z})$. (Here, $\tilde{z}$ signifies the new independent variable, which is determined by the hypergeometric transformation.) But ${}_2F_1(1/2, 1/2; 1; \tilde{z})$ equals $(2/\pi)K(\tilde{z})$, and the full solution space of the ODE is spanned by $K(\tilde{z})$ and $K'(\tilde{z}) := K(1 - \tilde{z})$.

The expressions resulting from the just-described reduction procedure are somewhat inelegant, but better ones can be obtained heuristically. In fact, one can begin with the Poisson kernel itself, rather than its companion. The case $\lambda = 1/4$ is illustrative. When $\lambda = 1/4$, the ${}_2F_1$ in the first right-hand side of (4.3) is of the form ${}_2F_1(1/4, 5/4; 1/2; w)$, where $w = \tilde{z}$. An explicit formula for this function can be found in the database of Roach [25], which is currently available at `www.planetquantum.com`. The formula is

$$\frac{\Gamma(1/4)^2}{2\sqrt{\pi}}\,{}_2F_1\left(\begin{matrix}1/4,\,5/4\\1/2\end{matrix}\,\middle|\,w\right)$$
$$=\frac{2\sqrt{w}}{1-w}\,E(\tilde{w}_+)-\frac{2\sqrt{w}}{1-w}\,E(\tilde{w}_-)+\frac{1}{1+\sqrt{w}}\,K(\tilde{w}_+)+\frac{1}{1-\sqrt{w}}\,K(\tilde{w}_-), \tag{4.5}$$

where $\tilde{w}_\pm = (1 \pm \sqrt{w})/2$. The presence of the second complete elliptic integral, $E = E(m)$, for which the exponent differences are $1, 1, 0$ and not $0, 0, 0$, can be attributed to an application of the contiguous relations of ${}_2F_1$.

5. Final Remarks

It has been shown that if the Gegenbauer parameter λ differs by one-fourth or one-sixth from an integer, there are several generating functions for the Gegenbauer polynomials $\{C_n^\lambda\}_{n=0}^\infty$ that are algebraic. These are special cases of Brafman's generating functions, including the extended ones with an additional free parameter (denoted u here). Gegenbauer polynomials with λ restricted as stated were shown to be special in another way: the Poisson kernel computed from them can be expressed with the aid of hypergeometric transformations in terms of complete elliptic integrals.

The results on algebraicity are consequences of Schwarz's classification of the algebraic cases of the Gauss function ${}_2F_1$, and the explicit examples of algebraic generating functions came from recently developed closed-form expressions for certain algebraic ${}_2F_1$s with tetrahedral and octahedral monodromy; i.e., tetrahedral and octahedral associated Legendre functions [13]. This is because the ${}_2F_1$s in Brafman's generating functions admit of quadratic transformations, so that in essence, they are Legendre functions.

Generalizations can be considered. One matter worthy of investigation is the applicability of icosahedral ${}_2F_1$s, which although algebraic cannot be expressed in terms of radicals. Some parametric formulas for them are known [26], and may yield manageable parameterizations of the consequent algebraic generating functions for Gegenbauer polynomials.

The generalization from Gegenbauer to Jacobi polynomials is also worth pursuing. It follows readily from Schwarz's classification that certain special cases of Brafman's second generating function, generalized to non-Gegenbauer Jacobi polynomials but still expressed in terms of ${}_2F_1$, are algebraic functions of their arguments. However, the Jacobi generalization of his first generating function is known to involve the Appell function F_4, i.e., a bivariate hypergeometric function. A full classification of the algebraic cases of the first generating function, generalized to Jacobi polynomials, will require results on the algebraicity of Appell functions; and the same is true of the Poisson kernel.

Finally, it should be mentioned that Brafman's extension procedure, leading to identities parameterized by u, is not the only one that can be applied to Gegenbauer generating functions. By exploiting the connection formula for Gegenbauer polynomials, Cohl and collaborators [27, 28] have obtained novel extensions of the defining relation (1.1), as equation (11) of [27], and of Brafman's second identity (3.1a), as equation (26) of [28]. For suitably chosen parameter values, such extensions will be algebraic.

Appendix. Associated Legendre Functions in Closed Form

The associated Legendre function $P_\nu^\mu(z)$ of degree $\nu \in \mathbb{C}$ and order $\mu \in \mathbb{C}$ is defined in terms of the Gauss function ${}_2F_1$ by

$$P_\nu^\mu(z) = \frac{2^\mu}{\Gamma(1-\mu)}(z^2-1)^{-\mu/2}\,{}_2F_1\left(\begin{matrix}-\nu-\mu,\ 1+\nu-\mu\\ 1-\mu\end{matrix}\,\middle|\,\frac{1-z}{2}\right). \tag{A.1}$$

The Ferrers function P_ν^μ is defined similarly, with $1-z^2$ replacing z^2-1. By convention, $P_\nu^\mu(z)$ and $\mathrm{P}_\nu^\mu(z)$ are defined and analytic on the complex z-plane, with the respective omissions of the cut $(-\infty,1]$ and the cut-pair $(-\infty,-1]\cup[1,\infty)$. When $\mu = 1,2,\ldots$, (A.1) must be taken in a limiting sense. In the singular case when $\nu = 0,1,2,\ldots$ and $\mu-\nu$ is a positive integer, P_ν^μ and P_ν^μ are identically zero.

On their respective domains, $P_\nu^{\pm\mu}$ and $\mathrm{P}_\nu^{\pm\mu}$ span the two-dimensional solution space of the associated Legendre ODE, except when $(\nu,\mu)\in\mathbb{Z}^2$. This space can also be viewed as the span of P_ν^μ and Q_ν^μ, respectively P_ν^μ and Q_ν^μ, where Q_ν^μ and Q_ν^μ are the associated Legendre and Ferrers functions of the second kind. (Again, singular cases are excepted.) The function $P_\nu^\mu(z)$ is singled out as the element $f(z)$ of the solution space with

$$f(z) \sim \frac{2^{\mu/2}}{\Gamma(1-\mu)}(z-1)^{-\mu/2}, \qquad z \to 1 \tag{A.2}$$

as asymptotic behavior.

For all $\nu, \mu \in \mathbb{C}$, it follows from (A.1) that $P^{\mu}_{-\nu-1} = P^{\mu}_{\nu}$ and $\mathrm{P}^{\mu}_{-\nu-1} = \mathrm{P}^{\mu}_{\nu}$. Also, the ordered pair (ν, μ) can be displaced by any element of $\mathbb{Z}^2$, for either P^{μ}_{ν} or P^{μ}_{ν}, by applying an appropriate differential operator. (See [13, Section 6]). Such "ladder operators", which increment and decrement ν and/or μ, come from the contiguity relations of ${}_2F_1$.

There are several cases when the functions $P_{\nu}, \mathrm{P}^{\mu}_{\nu}$ are elementary; or to put it more broadly, when all solutions of the associated Legendre ODE can be reduced to quadratures [12]. These include the case when the ODE, or the equivalent ODE satisfied by the ${}_2F_1$ in (A.1), is "reducible" (see [11, Section 2.2]); and certain algebraic cases, when the ODE has a finite projective monodromy group (see [11, Section 2.7.2; 10, Chapter VII]). In the algebraic cases, this group as a subgroup of the Möbius group may be dihedral, tetrahedral, octahedral, or icosahedral, but the last of these possibilities does not lead to radical expressions. The remaining cases are numbered (i)–(iv) here.

The reducible case is: (i) when $\nu = -\mu + N$, $N = 0, 1, 2, \ldots$, which is also called the degenerate or Gegenbauer case. It follows from (1.2) and (A.1) that when $\mu \neq \frac{1}{2}, 1, \frac{3}{2}, \ldots$,

$$P^{\mu}_{-\mu+N}(z) = \frac{2^{\mu}}{\Gamma(1-\mu)} \frac{N!}{(1-2\mu)_N} (z^2-1)^{-\mu/2} \, C_N^{1/2-\mu}(z), \tag{A.3}$$

with the same holding if $P^{\mu}_{-\mu+N}$ and $z^2 - 1$ are replaced by $\mathrm{P}^{\mu}_{-\mu+N}$ and $1 - z^2$.

Of the algebraic cases, the simplest is (ii[a]): the first dihedral case, more accurately called 'cyclic', when the degree ν is an integer. The basic formulas are

$$P^{\mu}_{0}(\coth \xi) = \Gamma(1-\mu)^{-1} \, e^{\mu\xi}, \tag{A.4a}$$

$$\mathrm{P}^{\mu}_{0}(\tanh \xi) = \Gamma(1-\mu)^{-1} \, e^{\mu\xi}. \tag{A.4b}$$

Actually, $P^{\mu}_0(z), \mathrm{P}^{\mu}_0(z)$ are algebraic in z only if μ is rational; for general μ, the term "quasi-cyclic" can be used. For any nonzero $\nu \in \mathbb{Z}$, $P^{\mu}_{\nu}, \mathrm{P}^{\mu}_{\nu}$ are computed from $P^{\mu}_0, \mathrm{P}^{\mu}_0$ by applying ladder operators that shift the degree.

There is also (ii[b]): the second dihedral case, when the order μ is a half-odd-integer. The basic formulas are

$$P_\nu^{1/2}(\cosh\xi) = \sqrt{\frac{2}{\pi}}\,\frac{\cosh\left[(\nu+1/2)\xi\right]}{\sqrt{\sinh\xi}}, \tag{A.5a}$$

$$\mathrm{P}_\nu^{1/2}(\cos\theta) = \sqrt{\frac{2}{\pi}}\,\frac{\cos\left[(\nu+1/2)\xi\right]}{\sqrt{\sin\theta}}, \tag{A.5b}$$

which define algebraic functions $P_\nu^{1/2}, \mathrm{P}_\nu^{1/2}$ only if ν is rational; for general ν, this case is only quasi-dihedral. For any half-odd-integer μ other than 1/2, $P_\nu^\mu, \mathrm{P}_\nu^\mu$ are computed from $P_\nu^{1/2}, \mathrm{P}_\nu^{1/2}$ by applying ladder operators that shift the order.

The cases recently examined [13] include (iii): the octahedral case, when $(\nu,\mu) \in \mathbb{Z}^2 + (\pm 1/6, \pm 1/4)$. For this, define algebraic functions $h_\pm, k_\pm$ trigonometrically by

$$h_\pm(\cosh\xi) = \left\{ (\sinh\xi)^{-1}\left[\pm\cosh(\xi/3) + \sqrt{\frac{\sinh\xi}{3\sinh(\xi/3)}}\right]\right\}^{1/4}, \tag{A.6a}$$

$$k_\pm(\cos\theta) = \left\{ (\sin\theta)^{-1}\left[\cos(\theta/3) \pm \sqrt{\frac{\sin\theta}{3\sin(\theta/3)}}\right]\right\}^{1/4}. \tag{A.6b}$$

Then the basic formulas are

$$P_{-1/6}^{\pm 1/4}(\cosh\xi) = 3^{(3/8)(1\mp 1)}\,\Gamma(1\mp 1/4)^{-1}\,h_\pm(\cosh\xi), \tag{A.7a}$$

$$\mathrm{P}_{-1/6}^{\pm 1/4}(\cos\theta) = 3^{(3/8)(1\mp 1)}\,\Gamma(1\mp 1/4)^{-1}\,k_\pm(\cos\theta). \tag{A.7b}$$

(The plus formulas were derived in [13] and the minus formulas follow from them, the normalization factors coming from the condition (A.2).) Ladder operators can be applied to these basic formulas, as needed.

The other case recently examined [13] is (iv[a]): the first tetrahedral case, when $(\nu,\mu) \in \mathbb{Z}^2 + (\pm 1/4, \pm 1/3)$. For this, define algebraic functions $f_\pm, g_\pm$ trigonometrically by

$$f_\pm(\coth\xi) = \left\{ (\sinh\xi)\left[\pm\cosh(\xi/3) + \sqrt{\frac{\sinh\xi}{3\sinh(\xi/3)}}\right]\right\}^{1/4}, \tag{A.8a}$$

$$g_\pm(\tanh\xi) = \left\{ (\cosh\xi)\left[\pm\sinh(\xi/3) + \sqrt{\frac{\cosh\xi}{3\cosh(\xi/3)}}\right]\right\}^{1/4}. \tag{A.8b}$$

Then the basic formulas are

$$P_{-1/4}^{\pm 1/3}(\coth\xi) = 2^{1/2\mp 3/4}\,3^{-3/8}\,\Gamma(1\mp 1/3)^{-1} \times \left[\sqrt{\sqrt{3}\pm 1}\,f_+ \pm \sqrt{\sqrt{3}\mp 1}\,f_-\right](\coth\xi), \tag{A.9a}$$

$$\mathrm{P}_{-1/4}^{\pm 1/3}(\tanh\xi) = 2^{1/2\mp 3/4}\,3^{-3/8}\,\Gamma(1\mp 1/3)^{-1} \times \left[\pm\sqrt{\sqrt{3}\pm 1}\,g_+ + \sqrt{\sqrt{3}\mp 1}\,g_-\right](\tanh\xi). \tag{A.9b}$$

(It was shown in [13] that $Q_{-1/4}^{-1/3}$ is a multiple of f_-, and f_+ is an independent solution of the same associated Legendre ODE; so $P_{-1/4}^{\pm 1/3}$ must be linear combinations of f_+, f_-, and the coefficients shown in (A.9a) can be deduced with some effort from the condition (A.2).) Ladder operators can be applied to these basic formulas, as needed.

There remains (iv[b]): the second tetrahedral case, when $(\nu, \mu) \in \mathbb{Z}^2 + (\pm 1/6, \pm 1/3)$. This case is related to the first tetrahedral one by a quadratic hypergeometric transformation, but the resulting formulas are complicated and are not given here.

References

[1] M. E. H. Ismail, *Classical and Quantum Orthogonal Polynomials in One Variable*, Encyclopedia of Mathematics and its Applications, Vol. 98 (Cambridge University Press, Cambridge, 2005). With two chapters by W. Van Assche.

[2] F. Brafman, Generating functions of Jacobi and related polynomials, *Proc. Amer. Math. Soc.* **2**(6) (1951) 942–949.

[3] F. Brafman, An ultraspherical generating function, *Pacific J. Math.* **7**(3) (1957) 1319–1323.

[4] E. D. Rainville, *Special Functions* (Macmillan, New York, 1960).

[5] E. B. McBride, *Obtaining Generating Functions*, Springer Tracts in Natural Philosophy, Vol. 21 (Springer, New York, 1971).

[6] B. Viswanathan, Generating functions for ultraspherical functions, *Canad. J. Math.* **20** (1968) 120–134.

[7] H. M. Srivastava, An extension of the Hille–Hardy formula, *Math. Comp.* **23** (1969) 305–311.

[8] H. M. Srivastava and H. L. Manocha, *A Treatise on Generating Functions* (Halsted Press, New York, 1984).

[9] H. A. Schwarz, Ueber diejenigen Fälle, in welchen die Gaussische hypergeometrische Reihe eine algebraische Function ihres vierten Elementes darstellt, *J. Reine Angew. Math.* **75** (1873) 292–335.

[10] E. G. C. Poole, *Introduction to the Theory of Linear Differential Equations* (Oxford University Press, Oxford, 1936).

[11] A. Erdélyi, W. Magnus, F. Oberhettinger and F. G. Tricomi (eds.), *Higher Transcendental Functions* (McGraw–Hill, New York, 1953–55).

[12] T. Kimura, On Riemann's equations which are solvable by quadratures, *Funkcial. Ekvac.* **12** (1969/70) 269–281.

[13] R. S. Maier, Legendre functions of fractional degree: Transformations and evaluations, *Proc. Roy. Soc. London Ser. A* **472** (2016) 20160097, 29 pp.

[14] L. Koschmieder, Über besondere Jacobische Polynome, *Math. Z.* **8**(1–2) (1920) 123–137.

[15] C. M. Linton, Accurate solution to scattering by a semi-circular groove, *Wave Motion* **46**(3) (2009) 200–209.

[16] G. N. Watson, Notes on generating functions of polynomials: (3) Polynomials of Legendre and Gegenbauer, *J. London Math. Soc.* **S1-8**(3) (1933) 289–292.

[17] H. S. Cohl and C. MacKenzie, Generalizations and specializations of generating functions for Jacobi, Gegenbauer, Chebyshev and Legendre polynomials with definite integrals, *J. Class. Anal.* **3**(1) (2013) 17–33.

[18] W. Miller, Jr., Special functions and the complex Euclidean group in 3-space. III, *J. Math. Phys.* **9**(9) (1968) 1434–1444.

[19] L. Carlitz, Some generating functions for the Jacobi polynomials, *Boll. Un. Mat. Ital.* (3) **16** (1961) 150–155.

[20] S. O. Rice, Some properties of ${}_3F_2(-n, n+1, \zeta;\ 1, p;\ v)$, *Duke Math. J.* **6**(1) (1940) 108–119.

[21] H. Bateman, Spheroidal and bipolar coördinates, *Duke Math. J.* **4**(1) (1938) 39–50.

[22] L. Weisner, Group-theoretic origin of certain generating functions, *Pacific J. Math.* **5** (1955) 1033–1039.

[23] G. Gasper and M. Rahman, Positivity of the Poisson kernel for the continuous q-ultraspherical polynomials, *SIAM J. Math. Anal.* **14**(2) (1983) 409–420.

[24] T. H. Koornwinder, Additions to the formula lists in "Hypergeometric orthogonal polynomials and their q-analogues" by Koekoek, Lesky and Swarttouw, preprint (2015); arXiv:1401.0815 [math.CA].

[25] K. B. Roach, Hypergeometric function representations, in *ISSAC '96: Proceedings of the 1996 International Symposium on Symbolic and Algebraic Computation*, ed. Y. N. Lakshman (Association for Computing Machinery (ACM), New York, 1996), pp. 301–308.

[26] R. Vidūnas, Darboux evaluations of algebraic Gauss hypergeometric functions, *Kyushu J. Math.* **67** (2013) 249–280.

[27] H. S. Cohl, On a generalization of the generating function for Gegenbauer polynomials, *Integral Transforms Spec. Funct.* **24**(10) (2013) 807–816.

[28] H. S. Cohl, C. MacKenzie and H. Volkmer, Generalizations of generating functions for hypergeometric orthogonal polynomials with definite integrals, *J. Math. Anal. Appl.* **407**(2) (2013) 211–225.

Chapter 23

q-Analogues of Two Product Formulas of Hypergeometric Functions by Bailey

Michael J. Schlosser

Fakultät für Mathematik, Universität Wien
Oskar-Morgenstern-Platz 1, A-1090 Vienna, Austria
michael.schlosser@univie.ac.at
http://www.mat.univie.ac.at/~schlosse

Dedicated to Mourad E. H. Ismail

We use Andrews' q-analogues of Watson's and Whipple's ${}_3F_2$ summation theorems to deduce two formulas for products of specific basic hypergeometric functions. These constitute q-analogues of corresponding product formulas for ordinary hypergeometric functions given by Bailey. The first formula was obtained earlier by Jain and Srivastava by a different method.

Keywords: Basic hypergeometric series; product formulas.

Mathematics Subject Classification 2010: 33D15.

1. Introduction

We refer to Slater's text [9] for an introduction to hypergeometric series and to Gasper and Rahman's text [5] for an introduction to basic hypergeometric series, whose notations we follow. Throughout, we assume $|q| < 1$ and $|z| < 1$.

In [1], Andrews proved the following two theorems.

Theorem 1.1.

$$
{}_4\phi_3\left[\begin{matrix} a, b, c^{\frac{1}{2}}, -c^{\frac{1}{2}} \\ (abq)^{\frac{1}{2}}, -(abq)^{\frac{1}{2}}, c \end{matrix} ; q, q\right] = a^{\frac{n}{2}} \frac{(aq, bq, cq/a, cq/b; q^2)_\infty}{(q, abq, cq, cq/ab; q^2)_\infty}, \tag{1.1}
$$

where $b = q^{-n}$ and n is a nonnegative integer.

Theorem 1.2.

$$
{}_4\phi_3\left[\begin{matrix} a, q/a, c^{\frac{1}{2}}, -c^{\frac{1}{2}} \\ -q, e, cq/e \end{matrix} ; q, q\right] = q^{\binom{n+1}{2}} \frac{(ea, eq/a, caq/e, cq^2/ae; q^2)_\infty}{(e, cq/e; q)_\infty}, \tag{1.2}
$$

where $a = q^{-n}$ and n is a nonnegative integer.

By a standard polynomial argument, (1.2) also holds when a is a complex variable, but $c = q^{-2n}$ with n being a nonnegative integer. (This is the case we will make use of.)

Theorems 1.1 and 1.2 are q-analogues of Watson's and of Whipple's ${}_3F_2$ summation theorems, listed as equations (III.23) and (III.24) in [9, p. 245], respectively.

2. Two Product Formulas for Basic Hypergeometric Functions

We now have the following two product formulas which are derived using Theorems 1.1 and 1.2. The first one in Theorem 2.1 was already given earlier by Jain and Srivastava [7, equation (4.9)] (as Slobodan Damjanović has kindly pointed out to the author after seeing an earlier version of this note), who established the result by specializing a general reduction formula for double basic hypergeometric series. The second formula in Theorem 2.2 appears to be new.

Theorem 2.1.

$$
{}_2\phi_1\left[\begin{matrix} a, -a \\ a^2 \end{matrix} ; q, z\right] {}_2\phi_1\left[\begin{matrix} b, -b \\ b^2 \end{matrix} ; q, -z\right] = {}_4\phi_3\left[\begin{matrix} ab, -ab, abq, -abq \\ a^2q, b^2q, a^2b^2 \end{matrix} ; q^2, z^2\right]. \tag{2.1}
$$

Theorem 2.2.

$$
{}_2\phi_1\left[\begin{matrix} a, q/a \\ -q \end{matrix} ; q, z\right] {}_2\phi_1\left[\begin{matrix} b, q/b \\ -q \end{matrix} ; q, -z\right]
$$

$$
= \sum_{j=0}^{\infty} \frac{(q^{2-j}/ab, aq^{1-j}/b; q^2)_j}{(q^2; q^2)_j} q^{\binom{j}{2}} (bz)^j \tag{2.2a}
$$

$$= {}_4\phi_3\left[\begin{matrix} ab, q^2/ab, aq/b, bq/a \\ -q^2, q, -q \end{matrix}; q^2, z^2\right]$$

$$-\frac{(a-b)(1-q/ab)}{1-q^2}\, z$$

$$\times\, {}_4\phi_3\left[\begin{matrix} abq, q^3/ab, aq^2/b, bq^2/a \\ -q^2, q^3, -q^3 \end{matrix}; q^2, z^2\right]. \tag{2.2b}$$

Sketch of proofs. To prove Theorem 2.1, compare coefficients of z^n. The resulting identity is equivalent to Theorem 1.1. The proof of Theorem 2.2 is similar. Comparison of coefficients of z^n gives an identity which is equivalent to Theorem 3.2 (where in the latter theorem, the restriction $a = q^{-n}$ is replaced by $c = q^{-2n}$ as mentioned). The second identity in equation (2.2) follows from splitting the sum over j into two parts depending on the parity of j. (This is motivated by the particular numerator factors in the j-th summand.) The technical details — elementary manipulation of q-shifted factorials — are routine and thus omitted. □

Theorem 2.1 is a q-analogue of Bailey's formula in [2, equation (2.11), p. 246]:

$${}_1F_1\left[\begin{matrix} a \\ 2a \end{matrix}; z\right] {}_1F_1\left[\begin{matrix} b \\ 2b \end{matrix}; -z\right] = {}_2F_3\left[\begin{matrix} \frac{1}{2}(a+b), \frac{1}{2}(a+b+1) \\ a+\frac{1}{2}, b+\frac{1}{2}, a+b \end{matrix}; \frac{1}{4}z\right]. \tag{2.3}$$

To obtain (2.3) from Theorem 2.1, replace (a, b, z) by $(q^a, q^b, (1-q)z/2)$, and let $q \to 1$.

Similarly, Theorem 2.2 is a q-analogue of Bailey's formula in [2, equation (2.08), p. 245]:

$${}_2F_0\left[\begin{matrix} a, 1-a \\ - \end{matrix}; z\right] {}_2F_0\left[\begin{matrix} b, 1-b \\ - \end{matrix}; -z\right]$$

$$= {}_4F_1\left[\begin{matrix} \frac{1}{2}(1+a-b), \frac{1}{2}(1-a+b), \frac{1}{2}(a+b), \frac{1}{2}(2-a-b) \\ \frac{1}{2} \end{matrix}; 4z^2\right]$$

$$-(a-b)(a+b-1)\, z$$

$$\times\, {}_4F_1\left[\begin{matrix} \frac{1}{2}(2+a-b), \frac{1}{2}(2-a+b), \frac{1}{2}(1+a+b), \frac{1}{2}(3-a-b) \\ \frac{3}{2} \end{matrix}; 4z^2\right]. \tag{2.4}$$

To obtain (2.4) from Theorem 2.2, replace (a, b, z) by $(q^a, q^b, 2z/(1-q))$ and let $q \to 1$.

3. Related Results in the Literature

A different product formula for basic hypergeometric functions was established by Srivastava [10, equation (21)] (see also [11, equation (3.13)]):

$$ {}_2\phi_1\left[\begin{matrix} a, b \\ -ab \end{matrix}; q, z\right] {}_2\phi_1\left[\begin{matrix} a, b \\ -ab \end{matrix}; q, -z\right] = {}_4\phi_3\left[\begin{matrix} a^2, b^2, ab, abq \\ a^2b^2, -ab, -abq \end{matrix}; q^2, z^2\right]. \tag{3.1} $$

This formula is a q-extension of Bailey's formula in [2, equation (2.08), p. 245] (or, equivalently, of an identity recorded by Ramanujan [8, Chapter 13, Entry 24]).

Finally, we mention that in 1941, Jackson [6] had derived the identity

$$ {}_2\phi_1\left[\begin{matrix} a^2, b^2 \\ a^2b^2q \end{matrix}; q^2, z\right] {}_2\phi_1\left[\begin{matrix} a^2, b^2 \\ a^2b^2q \end{matrix}; q^2, qz\right] = {}_4\phi_3\left[\begin{matrix} a^2, b^2, ab, -ab \\ a^2b^2, abq^{\frac{1}{2}}, -abq^{\frac{1}{2}} \end{matrix}; q, z\right], \tag{3.2} $$

which is a q-analogue of Clausen's formula of 1828,

$$ \left({}_2F_1\left[\begin{matrix} a, b \\ a+b+\frac{1}{2} \end{matrix}; z\right]\right)^2 = {}_3F_2\left[\begin{matrix} 2a, 2b, a+b \\ 2a+2b, a+b+\frac{1}{2} \end{matrix}; z\right]. \tag{3.3} $$

Another q-analogue of Clausen's formula was delivered by Gasper in [4]. While it has the advantage that it expresses a square of a basic hypergeometric series as a basic hypergeometric series, it only holds, provided the series terminate:

$$ \left({}_4\phi_3\left[\begin{matrix} a, b, aby, ab/y \\ abq^{\frac{1}{2}}, -abq^{\frac{1}{2}}, -ab \end{matrix}; q, q\right]\right)^2 = {}_5\phi_4\left[\begin{matrix} a^2, b^2, ab, aby, ab/y \\ a^2b^2, abq^{\frac{1}{2}}, -abq^{\frac{1}{2}}, -ab \end{matrix}; q, q\right]. \tag{3.4} $$

See [5, Sec. 8.8] for a nonterminating extension of (3.4) and related identities.

Acknowledgments

I would like to thank George Gasper for his interest and for informing me of the papers [10, 11] by Srivastava. I am especially indebted to Slobodan Damjanović for pointing out that Theorem 2.1 was already given by Jain and Srivastava [7, equation (4.9)]. This work was partly supported by FWF Austrian Science Fund grant F50-08.

References

[1] G. E. Andrews, On q-analogues of the Watson and Whipple summations, *SIAM J. Math. Anal.* **7**(3) (1976) 332–336.

[2] W. N. Bailey, Products of generalized hypergeometric series, *Proc. London Math. Soc.* **S2–28**(1) (1928) 242–254.

[3] T. Clausen, Ueber die Fälle, wenn die Reihe von der Form ... ein Quadrat von der Form ... hat, *J. Reine Angew. Math.* **3** (1828) 89–91.

[4] G. Gasper, q-Extensions of Clausen's formula and of the inequalities used by de Branges in his proof of the Bieberbach, Robertson and Milin conjectures, *SIAM J. Math. Anal.* **20** (1989) 1019–1034.

[5] G. Gasper and M. Rahman, *Basic Hypergeometric Series*, 2nd edn., Encyclopedia of Mathematics and Its Applications, Vol. 96 (Cambridge University Press, Cambridge, 2004).

[6] F. H. Jackson, Certain q-identities, *Quart. J. Math.* (*Oxford*) **12** (1941) 167–172.

[7] V. K. Jain and H. M. Srivastava, q-Series identities and reducibility of basic double hypergeometric functions, *Canad. J. Math.* **38** (1986) 215–231.

[8] S. Ramanujan, *Notebooks of Srinivasa Ramanujan*, Vols. I and II (Tata Institute of Fundamental Research, Bombay, 1957).

[9] L. J. Slater, *Generalized Hypergeometric Functions* (Cambridge University Press, Cambridge, 1966).

[10] H. M. Srivastava, Some formulas of Srinivasa Ramanujan involving products of hypergeometric functions, *Indian J. Math.* **29** (1987) 91–100.

[11] H. M. Srivastava, Srinivasa Ramanujan and generalized basic hypergeometric functions, *Serdica Bulgariacae Math. Publ.* **19** (1993) 191–197.

This page intentionally left blank

Chapter 24

Summation Formulas for Noncommutative Hypergeometric Series

Michael J. Schlosser

Fakultät für Mathematik, Universität Wien
Oskar-Morgenstern-Platz 1, A-1090 Wien, Austria
michael.schlosser@univie.ac.at

Dedicated to Mourad E. H. Ismail

We establish several summation formulas for hypergeometric and basic hypergeometric series involving noncommutative parameters and argument. These results were inspired by a paper of J. A. Tirao [*Proc. Natl. Acad. Sci. USA* **100**(14) (2003) 8138–8141].

Keywords: Noncommutative hypergeometric series; noncommutative basic hypergeometric series.

Mathematics Subject Classification 2010: 33C20, 33C99, 33D15, 33D99

1. Introduction

Hypergeometric series with noncommutative parameters and argument, in the special case involving square matrices, have been studied by a number of researchers including (in alphabetical order) Aldenhoven, Durán, Duval, Grünbaum, de la Iglesia, Iliev, Koelink, Ovsienko, Pacharoni, de los Ríos, Tirao, and others. See [1, 5, 8, 10–15, 22] for some selected papers.

The subject of hypergeometric series involving matrices is closely related to, and partly overlapping, the theory of orthogonal matrix polynomials. (For a number of examples, see Sinap and Van Assche's paper [20].) The study of the latter was initiated by Krein [17] and subsequently has experienced a steady development. Whereas a good amount of theory of orthogonal matrix polynomials has already been worked out, see, e.g., [6, 7, 22], it seems as appropriate to study noncommutative hypergeometric series (involving not only matrices but also more generally arbitrary noncommutative parameters of some unit ring, or, in the case of infinite series, of some Banach algebra) from an entirely elementary point of view. This includes the search for identities for noncommutative hypergeometric and noncommutative basic hypergeometric series, extending their classical commutative versions which can be found, for instance, in the standard textbooks of Bailey [3], Slater [21], Gasper and Rahman [9], and of Andrews, Askey, and Roy [2].

This paper contains some results of our search which we hope will be the starting point of a systematic study towards a theory of identities for noncommutative hypergeometric series and their basic analogues (q-analogues). The special types of noncommutative hypergeometric series we are considering were inspired by a recent paper of Tirao [22]. To be precise, we consider noncommutative hypergeometric series of types I and II. Tirao's matrix extension of the Gauss hypergeometric function belongs to type I, according to our terminology in Section 2. We would like to stress that by "noncommutative" we do not mean "q-commutative" or "quasi-commutative", i.e., involving a relation like $yx = qxy$. For some results on the latter, see the papers by Koornwinder [16] and Volkov [23] and the references therein. Unless we specify explicit commutation relations (which will sometimes happen) our parameters, elements of an abstract noncommutative unit ring, are understood *not* to commute with each other.

Our paper is organized as follows. In Section 2 we first recall the classical definitions for (q-)shifted factorials and (basic) hypergeometric series, and then define their noncommutative versions. In Section 3 we prove by induction a couple of lemmas containing simple addition formulas for the noncommutative (Q-)shifted factorials. Sections 4–6 are the heart pieces of our paper. Here we derive noncommutative extensions of several important terminating summations for hypergeometric and basic hypergeometric series, in particular of the (q-)Chu–Vandermonde summations, the (q-) Pfaff–Saalschütz summation, and of Dougall's ${}_7F_6$ summation. Concerning the latter summation, we were unfortunately not able to establish a noncommutative Q-analogue. In other words, the problem of finding a noncommutative Q-Dougall summation (or Jackson summation) is still open. The summations in Sections 4–6 are proved by entirely elementary means, namely by induction. The situation is quite different in Section 7 where, now working in some abstract Banach algebra, we give some nonterminating identities, in particular, two Q-Gauss summations (derived using a formal argument) as conjectures, and further two nonterminating Q-binomial theorems which we prove using functional equations. Finally, in Section 8 we indicate two ways for obtaining even more identities for noncommutative (basic) hypergeometric series, one of them is "telescoping", the other is "reversing all products".

The results of this paper provide (to the best of our knowledge) a first collection of identities for noncommutative hypergeometric and noncommutative basic hypergeometric series. We were honestly surprised when finding the identities in this paper which extend some of the most important summation formulas in the theory of (basic) hypergeometric series to noncommuting parameters. A continuation of our program may include the derivation of yet other summations but also noncommutative analogues of some of the classical transformation formulas (as listed in [3, Appendix III; 9, Appendix III]). Another issue left open is the study of the "type II" Gauss hypergeometric function and the "type I" and "type II" Q-Gauss hypergeometric functions from the view-point of the second-order (Q-)differential equations they (presumably) satisfy when considered in an appropriate analytic setting, in the spirit of Tirao's [22] illuminating investigation of the (what we call) ordinary (i.e. "non-Q") "type I" case. While this paper offers an elementary approach (the terminating summations are obtained by two applications of induction in each instance) and focuses on explicit summation formulas only, we feel that in order to gain more insight the subject of noncommutative (Q-)hypergeometric series should also be investigated from a broader perspective, connecting it to other theories (in combinatorics, representation theory, physics, etc.) where possible and appropriate. We have strong confidence that our summation formulas will be useful in the theory of (Q-)orthogonal matrix polynomials (e.g., to show that certain "principal" specializations of these polynomials factor) and may even provide motivation for defining new selected families of (Q-)orthogonal matrix polynomials. Furthermore, it seems not too farfetched to expect that our identities will have applications in some noncommutative models of mathematical physics (see also [8]).

2. Preliminaries

2.1. Classical (commutative) hypergeometric and basic hypergeometric series

The standard references for hypergeometric series and basic hypergeometric series are [21] and [9], respectively.

Define the *shifted factorial* for all integers k by the following quotient of gamma functions,

$$(a)_k := \frac{\Gamma(a+k)}{\Gamma(a)}.$$

Further, the (ordinary) *hypergeometric* ${}_{r+1}F_r$ *series* is defined as

$${}_{r+1}F_r\left[\begin{matrix}a_1, a_2, \ldots, a_{r+1} \\ b_1, b_2, \ldots, b_r\end{matrix}; z\right] := \sum_{k=0}^{\infty} \frac{(a_1)_k \cdots (a_{r+1})_k}{(b_1)_k \cdots (b_r)_k} \frac{z^k}{k!}. \tag{2.1}$$

Let q be a complex number such that $0 < |q| < 1$. Define the *q-shifted factorial* for all integers k (including infinity) by

$$(a;q)_k := \prod_{j=1}^{k}(1 - aq^j).$$

We write

$${}_{r+1}\phi_r\left[\begin{matrix}a_1, a_2, \ldots, a_{r+1} \\ b_1, b_2, \ldots, b_r\end{matrix}; q, z\right] := \sum_{k=0}^{\infty} \frac{(a_1;q)_k \ldots (a_{r+1};q)_k}{(b_1;q)_k \ldots (b_r;q)_k} \frac{z^k}{(q;q)_k} \tag{2.2}$$

to denote the *basic hypergeometric* ${}_{r+1}\phi_r$ *series.* In (2.1) and (2.2), $a_1, \ldots, a_{r+1}$ are called the *upper parameters,* $b_1, \ldots, b_r$ the *lower parameters,* z the *argument,* and (in (2.2)) q the *base* of the series. The ${}_{r+1}\phi_r$ series in (2.2) reduces to the ${}_{r+1}F_r$ series in (2.1) after first replacing all parameters a_i by q^{a_i} and b_i by q^{b_i} and then letting $q \to 1$. This possibility of taking limits to obtain ordinary hypergeometric series from basic hypergeometric series is not shared by the noncommutative versions of (2.1) and (2.2) which we will define in Section 2.2.

The hypergeometric ${}_{r+1}F_r$ series terminates if one of the upper parameters, say a_{r+1}, is of the form $-n$, for a nonnegative integer n. On the other hand, the basic hypergeometric ${}_{r+1}\phi_r$ series terminates if one of the upper parameters, say a_{r+1}, is of the form q^{-n}, for a nonnegative integer n. See [21, p. 45; 9, p. 25] for the criteria of when the hypergeometric, respectively basic hypergeometric, series converge if they do not terminate.

The classical theories of hypergeometric basic hypergeometric series contain several important summation and transformation formulas involving ${}_{r+1}F_r$ and ${}_{r+1}\phi_r$ series. Many of these summation theorems require that the parameters satisfy the condition of being either balanced and/or very-well-poised. An ${}_{r+1}F_r$ hypergeometric series is called *balanced* (or 1-*balanced*) if $b_1 + \cdots + b_r = a_1 + \cdots + a_{r+1} + 1$ and $z = 1$. More generally, it is called k-*balanced* if $b_1 + \cdots + b_r = a_1 + \cdots + a_{r+1} + k$ and $z = 1$. Similarly, an ${}_{r+1}\phi_r$ basic hypergeometric series is called *balanced* if $b_1 \cdots b_r = a_1 \cdots a_{r+1} q$ and $z = q$. An ${}_{r+1}F_r$ series is called *well-poised* if $a_1 + 1 = a_2 + b_1 = \cdots = a_{r+1} + b_r$ and is *very-well-poised* if in addition $a_2 = \frac{a_1}{2} + 1$. Note that this choice of a_2 entails that the factor

$$\frac{\frac{a_1}{2} + k}{\frac{a_1}{2}}$$

appears in a very-well-poised series. Similarly, an ${}_{r+1}\phi_r$ basic hypergeometric series is called *well-poised* if $a_1 q = a_2 b_1 = \cdots = a_{r+1} b_r$ and is *very-well-poised* if in addition $a_2 = -a_3 = q\sqrt{a_1}$. Here this choice of a_2 and a_3 entails that the factor

$$\frac{1 - a_1 q^{2k}}{1 - a_1}$$

appears in a very-well-poised basic series. In both cases (ordinary and basic), the parameter a_1 is referred to as the *special parameter* of the very-well-poised series.

2.2. Noncommutativity

Let R be a unit ring (i.e., a ring with a multiplicative identity). Throughout this chapter, the elements of R shall be denoted by capital letters $A, B, C, \ldots$. In general these elements do not commute with each other; however, we may sometimes specify certain commutation relations explicitly. We denote the identity by I and the zero element by O. Whenever a multiplicative inverse element exists for any $A \in R$, we denote it by A^{-1}. (Since R is a unit ring, we have $AA^{-1} = A^{-1}A = I$.) On the other hand, as we shall implicitly assume that all the expressions which appear are well defined, whenever we write A^{-1} we assume its existence. For instance, in (2.5) and (2.6) we assume that $C_i + jI$ is invertible for all $1 \leq i \leq r$, $0 \leq j < k$.

An important special case is when R is the ring of $n \times n$ square matrices (our notation is certainly suggestive with respect to this interpretation), or, more generally, one may view R as a space of some abstract operators.

Let $\mathbb{Z}$ be the set of integers. For $l, m \in \mathbb{Z} \cup \{\pm\infty\}$ we define the noncommutative product as follows:

$$\prod_{j=l}^{m} A_j = \begin{cases} 1, & m = l-1, \\ A_l A_{l+1} \dots A_m, & m \geq l, \\ A_{l-1}^{-1} A_{l-2}^{-1} \dots A_{m+1}^{-1}, & m < l-1. \end{cases} \tag{2.3}$$

Note that

$$\prod_{j=l}^{m} A_j = \prod_{j=m+1}^{l-1} A_{m+l-j}^{-1}, \tag{2.4}$$

for all $l, m \in \mathbb{Z} \cup \{\pm\infty\}$. We will make use of (2.4) at the end of Section 4 when reversing the order of summation of a series and pulling out factors.

Let $k \in \mathbb{Z}$. We define the generalized *noncommutative shifted factorial of type I* by

$$\left\lceil \begin{matrix} A_1, A_2, \dots, A_r \\ C_1, C_2, \dots, C_r \end{matrix} ; Z \right\rfloor_k := \prod_{j=1}^{k} \left[\left(\prod_{i=1}^{r} (C_i + (k-j)I)^{-1} (A_i + (k-j)I) \right) Z \right], \tag{2.5}$$

and the *noncommutative shifted factorial of type II* by

$$\left\lfloor \begin{matrix} A_1, A_2, \dots, A_r \\ C_1, C_2, \dots, C_r \end{matrix} ; Z \right\rceil_k := \prod_{j=1}^{k} \left[\left(\prod_{i=1}^{r} (C_i + (j-1)I)^{-1} (A_i + (j-1)I) \right) Z \right]. \tag{2.6}$$

Note the unusual usage of brackets ("floors" and "ceilings" are intermixed) on the left-hand sides of (2.5) and (2.6) which is intended to suggest that the products involve noncommuting factors in a prescribed order. In both cases, the product, read from left to right, starts with a denominator factor. The brackets in the form "$\lceil - \rfloor$" are intended to denote that the factors are *falling*, while in "$\lfloor - \rceil$" that they are *rising*.

If $Z = I$, we write

$$\left\lceil \begin{matrix} A_1, A_2, \dots, A_r \\ C_1, C_2, \dots, C_r \end{matrix} ; I \right\rfloor_k = \left\lceil \begin{matrix} A_1, A_2, \dots, A_r \\ C_1, C_2, \dots, C_r \end{matrix} \right\rfloor_k, \tag{2.7}$$

and

$$\left\lfloor \begin{matrix} A_1, A_2, \ldots, A_r \\ C_1, C_2, \ldots, C_r \end{matrix} ; I \right\rceil_k = \left\lfloor \begin{matrix} A_1, A_2, \ldots, A_r \\ C_1, C_2, \ldots, C_r \end{matrix} \right\rceil_k, \tag{2.8}$$

for simplicity in notation.

We define the *noncommutative hypergeometric series of type I* by

$${}_{r+1}F_r \left\lceil \begin{matrix} A_1, A_2, \ldots, A_{r+1} \\ C_1, C_2, \ldots, C_r \end{matrix} ; Z \right\rfloor := \sum_{k \geq 0} \left\lceil \begin{matrix} A_1, A_2, \ldots, A_{r+1} \\ C_1, C_2, \ldots, C_r, I \end{matrix} ; Z \right\rfloor_k, \tag{2.9}$$

and the *noncommutative hypergeometric series of type II* by

$${}_{r+1}F_r \left\lfloor \begin{matrix} A_1, A_2, \ldots, A_{r+1} \\ C_1, C_2, \ldots, C_r \end{matrix} ; Z \right\rceil := \sum_{k \geq 0} \left\lfloor \begin{matrix} A_1, A_2, \ldots, A_{r+1} \\ C_1, C_2, \ldots, C_r, I \end{matrix} ; Z \right\rceil_k. \tag{2.10}$$

In each case, the series terminates if one of the upper parameters A_i is of the form $-nI$. The situation is more delicate if the series is nonterminating. In this case we shall assume that R is a Banach algebra with norm $\| \cdot \|$. Then the series converges in R if $\|Z\| < 1$. If $\|Z\| = 1$ the series may converge in R for some particular choice of upper and lower parameters. Exact conditions depend on the Banach algebra R.

Throughout this paper, Q will be a parameter which commutes with any of the other parameters appearing in the series. (For instance, a central element such as $Q = qI$, a scalar multiple of the unit element in R, for $qI \in R$, trivially satisfies this requirement.)

Let $k \in \mathbb{Z}$. The generalized *noncommutative Q-shifted factorial of type I* is defined by

$$\left\lceil \begin{matrix} A_1, A_2, \ldots, A_r \\ C_1, C_2, \ldots, C_r \end{matrix} ; Q, Z \right\rfloor_k := \prod_{j=1}^{k} \left[\left(\prod_{i=1}^{r} (I - C_i Q^{k-j})^{-1} (I - A_i Q^{k-j}) \right) Z \right]. \tag{2.11}$$

Similarly, the generalized *noncommutative Q-shifted factorial of type II* is defined by

$$\left\lfloor \begin{matrix} A_1, A_2, \ldots, A_r \\ C_1, C_2, \ldots, C_r \end{matrix} ; Q, Z \right\rceil_k := \prod_{j=1}^{k} \left[\left(\prod_{i=1}^{r} (I - C_i Q^{j-1})^{-1} (I - A_i Q^{j-1}) \right) Z \right]. \tag{2.12}$$

Formally, one may let $k \to \infty$ (or, if one desires, even $k \to -\infty$).

We define the *noncommutative basic hypergeometric series of type I* by

$$
{}_{r+1}\phi_r \left\lceil \begin{matrix} A_1, A_2, \ldots, A_{r+1} \\ C_1, C_2, \ldots, C_r \end{matrix}; Q, Z \right\rfloor := \sum_{k\geq 0} \left\lceil \begin{matrix} A_1, A_2, \ldots, A_{r+1} \\ C_1, C_2, \ldots, C_r, Q \end{matrix}; Q, Z \right\rfloor_k, \tag{2.13}
$$

and the *noncommutative basic hypergeometric series of type II* by

$$
{}_{r+1}\phi_r \left\lfloor \begin{matrix} A_1, A_2, \ldots, A_{r+1} \\ C_1, C_2, \ldots, C_r \end{matrix}; Q, Z \right\rceil := \sum_{k\geq 0} \left\lfloor \begin{matrix} A_1, A_2, \ldots, A_{r+1} \\ C_1, C_2, \ldots, C_r, Q \end{matrix}; Q, Z \right\rceil_k. \tag{2.14}
$$

We also refer to the respective series as (*noncommutative*) *Q-hypergeometric series.* In each case, the series terminates if one of the upper parameters A_i is of the form Q^{-n}. If the series does not terminate, then (implicitly assuming that R is a Banach algebra with norm $\|\cdot\|$) it converges if $\|Z\| < 1$.

Note that the factors in the generalized noncommutative (Q-)shifted factorials are strongly interlaced, e.g.,

$$
\left\lfloor \begin{matrix} A, B \\ C, D \end{matrix}; Z \right\rceil_2 = C^{-1}AD^{-1}BZ(C+I)^{-1}(A+I)(D+I)^{-1}(B+I)Z.
$$

It is this interlacing which is mainly responsible that the noncommutative hypergeometric series considered in this paper can be summed in closed form. This is maybe best understood by regarding the general procedure for proving (all) the terminating identities in this chapter, namely induction: A particular factor of the summand is usually rewritten such that the original sum is split into two sums. After shifting the index of summation in one of the sums some factors can be pulled out and a similar sum remains to which the inductive hypothesis applies. (See Sections 4–6 for several demonstrations of this procedure.) The (Q-)shifted factorials of types I and II have been defined exactly in a way that induction can be successfully applied for proving the respective summations.

Remark 2.1. Tirao's [22] matrix extension of the Gauss hypergeometric function corresponds to the special case of our noncommutative ${}_2F_1$ series of type I when the parameters are $n \times n$ matrices over the complex numbers $\mathbb{C}$ and, in addition, the argument Z is a diagonal matrix zI with $z \in \mathbb{C}$. Restated in terms of the notation introduced in this section, Tirao essentially shows (among other results) that [22, Theorem 2]: *if*

$A, B, C, F_0 \in R$ and $C + jI$ is invertible for all nonnegative integers j, then

$$F(z) = {}_2F_1\left[\begin{matrix}A, B\\ C\end{matrix}; zI\right] F_0 \tag{2.15}$$

is analytic on $|z| < 1$ with values in R, and $F(z)$ is a solution of the hypergeometric equation

$$z(1-z)F'' + [C - z(I + A + B)]F' - ABF = O \tag{2.16}$$

such that $F(0) = F_0$, and conversely any solution of F analytic at $z = 0$ is of this form. He further shows (see [22, Corollary 3]) that the matrix-valued Jacobi polynomials introduced by Grünbaum [10] can be expressed in terms of hypergeometric functions of the above type (2.15), thereby giving an explicit example within the theory of matrix-valued orthogonal polynomials initiated by Krein [17].

A study of the ${}_2F_1$ function of type II in (2.10) and the ${}_2\phi_1$ functions of types I and II, in (2.13) and (2.14), respectively, in terms of the second-order (Q-)differential equations they satisfy was carried out by us (see [4]). (Similar but different results were obtained by Román and Simondi in [18].)

In this chapter, we are concerned with the (elementary) derivation of explicit summation formulas.

3. Elementary Identities for Noncommutative (Q-)shifted Factorials

Here we provide a couple of lemmas which will be utilized for proving the summations in Sections 4–6. These lemmas concern addition formulas for noncommutative (Q-)shifted factorials of type I and type II. Throughout we assume n to be a nonnegative integer. All formulas are proved in the same manner, by induction on n.

Lemma 3.1. *Let A and C be noncommutative parameters of some unit ring, and let n be a nonnegative integer. Then we have the following addition formula for shifted factorials of type I:*

$$\left[\begin{matrix}C-A\\ C\end{matrix}\right]_n - \left[\begin{matrix}C-A\\ C+I\end{matrix}\right]_n C^{-1}A = \left[\begin{matrix}C-A\\ C\end{matrix}\right]_{n+1}. \tag{3.1}$$

Further, we have the following addition formula for shifted factorials of type II:

$$\left\lfloor \begin{matrix} C-A+I \\ C \end{matrix} \right\rceil_n (C-A+nI)^{-1} - C^{-1}A \left\lfloor \begin{matrix} C-A+I \\ C+I \end{matrix} \right\rceil_n (C-A+nI)^{-1}$$

$$= \left\lfloor \begin{matrix} C-A+I \\ C \end{matrix} \right\rceil_{n+1} (C-A+(n+1)I)^{-1}. \tag{3.2}$$

Proof. We start with (3.1). For $n = 0$, (3.1) is easily verified. Assume the identity is true for all nonnegative integers less than a fixed positive integer n. Then we rewrite the left-hand side of (3.1) as

$$\left\lceil \begin{matrix} C-A \\ C \end{matrix} \right\rfloor_n - \left\lceil \begin{matrix} C-A \\ C+I \end{matrix} \right\rfloor_n C^{-1}A$$

$$= \left\lceil \begin{matrix} C-A+I \\ C+I \end{matrix} \right\rfloor_{n-1} C^{-1}(C-A) - \left\lceil \begin{matrix} C-A \\ C+I \end{matrix} \right\rfloor_n C^{-1}A.$$

We have simply pulled out the last two factors from the first term. Now we can apply the inductive hypothesis (with C replaced by $C+I$) to the first term to transform the last expression into

$$\left(\left\lceil \begin{matrix} C-A+I \\ C+I \end{matrix} \right\rfloor_n + \left\lceil \begin{matrix} C-A+I \\ C+2I \end{matrix} \right\rfloor_{n-1} (C+I)^{-1}A \right)$$

$$\times C^{-1}(C-A) - \left\lceil \begin{matrix} C-A \\ C+I \end{matrix} \right\rfloor_n C^{-1}A$$

$$= \left\lceil \begin{matrix} C-A \\ C \end{matrix} \right\rfloor_{n+1} + \left\lceil \begin{matrix} C-A+I \\ C+2I \end{matrix} \right\rfloor_{n-1} (C+I)^{-1}A$$

$$\times C^{-1}(C-A) - \left\lceil \begin{matrix} C-A \\ C+I \end{matrix} \right\rfloor_n C^{-1}A.$$

What remains to be shown is that in this sum of three terms the last two cancel each other. This is equivalent to

$$AC^{-1}(C-A) = (C-A)C^{-1}A, \tag{3.3}$$

which is immediately verified since both sides equal $A - AC^{-1}A$. Hence, we have established (3.1).

Next, we prove (3.2). The $n = 0$ case is trivial. Next, assume that the formula is true for all nonnegative integers less than a fixed positive integer n. Then we rewrite the left-hand side of (3.2) as

$$\begin{bmatrix} C-A+I \\ C \end{bmatrix}_n (C-A+nI)^{-1} - C^{-1}A \begin{bmatrix} C-A+I \\ C+I \end{bmatrix}_n (C-A+nI)^{-1}$$

$$= C^{-1}(C-A+I) \begin{bmatrix} C-A+2I \\ C+I \end{bmatrix}_{n-1} (C-A+nI)^{-1}$$

$$- C^{-1}A \begin{bmatrix} C-A+I \\ C+I \end{bmatrix}_n (C-A+nI)^{-1}.$$

We have simply pulled out the first two factors from the first term. Now we can apply the inductive hypothesis (with C replaced by $C+I$) to the first term to transform the last expression into

$$C^{-1}(C-A+I) \left(\begin{bmatrix} C-A+2I \\ C+I \end{bmatrix}_n (C-A+(n+1)I)^{-1} \right.$$

$$\left. + (C+I)^{-1}A \begin{bmatrix} C-A+2I \\ C+2I \end{bmatrix}_{n-1} (C-A+nI)^{-1} \right)$$

$$- C^{-1}A \begin{bmatrix} C-A+I \\ C+I \end{bmatrix}_n (C-A+nI)^{-1}.$$

What remains to be shown is that in the resulting sum of three terms the last two cancel each other, which is equivalent to

$$(C-A+I)(C+I)^{-1}A = A(C+I)^{-1}(C-A+I).$$

But this is simply the $C \mapsto C+I$ case of (3.3). Thus, we have established (3.2). $\square$

Lemma 3.2. *Let A and C be noncommutative parameters of some unit ring, and suppose that Q commutes both with A and C. Further, let n be a nonnegative integer. Then we have the following addition formulas for*

Q-shifted factorials of type I:

$$\left\lceil \begin{matrix} CA^{-1} \\ C \end{matrix} ; Q, A \right\rfloor_n - \left\lceil \begin{matrix} CA^{-1} \\ CQ \end{matrix} ; Q, A \right\rfloor_n (I - C)^{-1}(I - A)$$

$$= \left\lceil \begin{matrix} CA^{-1} \\ C \end{matrix} ; Q, A \right\rfloor_{n+1}, \tag{3.4}$$

and

$$\left\lceil \begin{matrix} A^{-1}C \\ C \end{matrix} ; Q, I \right\rfloor_n - \left\lceil \begin{matrix} A^{-1}C \\ CQ \end{matrix} ; Q, I \right\rfloor_n (I - C)^{-1}(I - A)A^{-1}CQ^n$$

$$= \left\lceil \begin{matrix} A^{-1}C \\ C \end{matrix} ; Q, I \right\rfloor_{n+1}. \tag{3.5}$$

Further, we have the following addition formulas for Q-shifted factorials of type II:

$$\left\lfloor \begin{matrix} CA^{-1}Q \\ C \end{matrix} ; Q, A \right\rceil_n (A - CQ^n)^{-1}$$

$$- (I - C)^{-1}(I - A) \left\lfloor \begin{matrix} CA^{-1}Q \\ CQ \end{matrix} ; Q, A \right\rceil_n (A - CQ^n)^{-1}$$

$$= \left\lfloor \begin{matrix} CA^{-1}Q \\ C \end{matrix} ; Q, A \right\rceil_{n+1} (A - CQ^{n+1})^{-1}, \tag{3.6}$$

and

$$\left\lfloor \begin{matrix} A^{-1}CQ \\ C \end{matrix} ; Q, I \right\rceil_n (I - A^{-1}CQ^n)^{-1}$$

$$- (I - C)^{-1}(I - A)A^{-1}CQ^n \left\lfloor \begin{matrix} A^{-1}CQ \\ CQ \end{matrix} ; Q, I \right\rceil_n (I - A^{-1}CQ^n)^{-1}$$

$$= \left\lfloor \begin{matrix} A^{-1}CQ \\ C \end{matrix} ; Q, I \right\rceil_{n+1} (I - A^{-1}CQ^{n+1})^{-1}. \tag{3.7}$$

Proof. We restrict ourselves to proving (3.6), the proofs of (3.4), (3.5) and (3.7) being similar. For $n = 0$, (3.6) is easily verified. Assume the identity is true for all nonnegative integers less than a fixed positive integer n. Then we rewrite the left-hand side of (3.6) as

$$\begin{aligned}
&\left[\begin{matrix} CA^{-1}Q \\ C \end{matrix};Q,A\right]_n (A-CQ^n)^{-1} \\
&\quad -(I-C)^{-1}(I-A)\left[\begin{matrix} CA^{-1}Q \\ CQ \end{matrix};Q,A\right]_n (A-CQ^n)^{-1} \\
&= (I-C)^{-1}(A-CQ)\left[\begin{matrix} CA^{-1}Q^2 \\ CQ \end{matrix};Q,A\right]_{n-1} (A-CQ^n)^{-1} \\
&\quad -(I-C)^{-1}(I-A)\left[\begin{matrix} CA^{-1}Q \\ CQ \end{matrix};Q,A\right]_n (A-CQ^n)^{-1}.
\end{aligned}$$

We have simply pulled out the first two factors from the first term. Now we can apply the inductive hypothesis (with C replaced by CQ) to the first term to transform the last expression into

$$\begin{aligned}
&(I-C)^{-1}(A-CQ)\left(\left[\begin{matrix} CA^{-1}Q^2 \\ CQ \end{matrix};Q,A\right]_n (A-CQ^{n+1})^{-1}\right. \\
&\quad \left. +(I-CQ)^{-1}(I-A)\left[\begin{matrix} CA^{-1}Q^2 \\ CQ^2 \end{matrix};Q,A\right]_{n-1} (A-CQ^n)^{-1}\right) \\
&\quad -(I-C)^{-1}(I-A)\left[\begin{matrix} CA^{-1}Q \\ CQ \end{matrix};Q,A\right]_n (A-CQ^n)^{-1}.
\end{aligned}$$

What remains to be shown is that in the resulting sum of three terms the last two cancel each other, which is equivalent to

$$(A-CQ)(I-CQ)^{-1}(I-A) = (I-A)(I-CQ)^{-1}(A-CQ). \tag{3.8}$$

However, splitting the factor $(A-CQ)$ on each side of (3.8) into two terms as $(I-CQ)-(I-A)$, both sides can be reduced to the same expression, namely

$$(I-A)-(I-A)(I-CQ)^{-1}(I-A),$$

which immediately establishes (3.6). □

Lemma 3.3. *Let A, B and C be noncommutative parameters of some unit ring, and suppose that the sum $A+B-C$ commutes each with A, B and C.*

Further, let n be a nonnegative integer. Then we have the following addition formula for shifted factorials of type I:

$$\begin{aligned}&\begin{bmatrix} C-B, C-A \\ C, C-A-B \end{bmatrix}_n - \begin{bmatrix} C-B, C-A \\ C+I, C-A-B-I \end{bmatrix}_n \\ &\quad \times C^{-1}A(A+B-C+(1-n)I)^{-1}B(A+B-C-nI)^{-1} \\ &\quad \times (A+B-C+I) = \begin{bmatrix} C-B, C-A \\ C, C-A-B \end{bmatrix}_{n+1}. \end{aligned} \tag{3.9}$$

Further, we have the following addition formula for shifted factorials of type II:

$$\begin{aligned}&\begin{bmatrix} C-B+I, C-A+I \\ C, C-A-B \end{bmatrix}_n (C-A+nI)^{-1}(C-B+nI)^{-1} \\ &\quad - C^{-1}A(A+B-C+(1-n)I)^{-1} \\ &\quad \times B(A+B-C-nI)^{-1}(A+B-C+I) \\ &\quad \times \begin{bmatrix} C-B+I, C-A+I \\ C+I, C-A-B-I \end{bmatrix}_n (C-A+nI)^{-1}(C-B+nI)^{-1} \\ &= \begin{bmatrix} C-B+I, C-A+I \\ C, C-A-B \end{bmatrix}_{n+1} (C-A+(n+1)I)^{-1} \\ &\quad \times (C-B+(n+1)I)^{-1}. \end{aligned} \tag{3.10}$$

Proof. We prove (3.10) by induction on n and leave the proof of (3.9) (which is similar) to the reader. For $n = 0$, (3.10) is easily verified. Assume the identity is true for all nonnegative integers less than a fixed positive integer n. Then we rewrite the left-hand side of (3.10) as

$$\begin{aligned}&\begin{bmatrix} C-B+I, C-A+I \\ C, C-A-B \end{bmatrix}_n (C-A+nI)^{-1}(C-B+nI)^{-1} \\ &\quad - C^{-1}A(A+B-C+(1-n)I)^{-1} \\ &\quad \times B(A+B-C-nI)^{-1}(A+B-C+I) \\ &\quad \times \begin{bmatrix} C-B+I, C-A+I \\ C+I, C-A-B-I \end{bmatrix}_n (C-A+nI)^{-1}(C-B+nI)^{-1} \end{aligned}$$

$$
\begin{aligned}
&= C^{-1}(C-B+I)(C-A-B)^{-1}(C-A+I) \\
&\quad \times \begin{bmatrix} C-B+2I, C-A+2I \\ C+I, C-A-B+I \end{bmatrix}_{n-1} (C-A+nI)^{-1}(C-B+nI)^{-1} \\
&\quad - C^{-1}A(A+B-C+(1-n)I)^{-1} \\
&\quad \times B(A+B-C-nI)^{-1}(A+B-C+I) \\
&\quad \times \begin{bmatrix} C-B+I, C-A+I \\ C+I, C-A-B-I \end{bmatrix}_{n} (C-A+nI)^{-1}(C-B+nI)^{-1}.
\end{aligned}
$$

We have simply pulled out the first four factors from the first term. Now we can apply the inductive hypothesis (with C replaced by $C+I$) to the first term to transform the last expression into

$$
\begin{aligned}
&C^{-1}(C-B+I)(C-A-B)^{-1}(C-A+I) \\
&\quad \times \left(\begin{bmatrix} C-B+2I, C-A+2I \\ C+I, C-A-B+I \end{bmatrix}_{n} (C-A+(n+1)I)^{-1} \right. \\
&\quad \times (C-B+(n+1)I)^{-1} + (C+I)^{-1}A(A+B-C+(1-n)I)^{-1} \\
&\quad \times B(A+B-C-nI)^{-1}(A+B-C) \\
&\quad \times \left. \begin{bmatrix} C-B+2I, C-A+2I \\ C+2I, C-A-B \end{bmatrix}_{n-1} (C-A+nI)^{-1}(C-B+nI)^{-1} \right) \\
&\quad - C^{-1}A(A+B-C+(1-n)I)^{-1} \\
&\quad \times B(A+B-C-nI)^{-1}(A+B-C+I) \\
&\quad \times \begin{bmatrix} C-B+I, C-A+I \\ C+I, C-A-B-I \end{bmatrix}_{n} (C-A+nI)^{-1}(C-B+nI)^{-1}.
\end{aligned}
$$

What remains to be shown is that in the resulting sum of three terms the last two cancel each other, which is equivalent to

$$
\begin{aligned}
&(C-B+I)(C-A+I)(C+I)^{-1}AB \\
&\quad = AB(C+I)^{-1}(C-B+I)(C-A+I). \qquad (3.11)
\end{aligned}
$$

However, splitting the factor $(C-B+I)(C-A+I)$ on each side of (3.11) into two terms as $AB + (C-A-B+I)(C+I)$ (which can be done since $A(C-B) = (C-B)A$), both sides can be reduced to the same expression, namely

$$AB(C+I)^{-1}AB + (C-A-B+I)AB,$$

which immediately establishes (3.10). □

Lemma 3.4. *Let A, B and C be noncommutative parameters of some unit ring, and suppose that Q commutes each with A, B and C. Further, assume that the product $BC^{-1}A$ commutes each with A, B and C. Moreover, let n be a nonnegative integer. Then we have the following addition formula for Q-shifted factorials of type I:*

$$\begin{aligned}&\left\lfloor \begin{matrix} CB^{-1}, A^{-1}C \\ C, A^{-1}CB^{-1} \end{matrix}; Q, I \right\rfloor_n - \left\lfloor \begin{matrix} CB^{-1}, A^{-1}C \\ CQ, A^{-1}CB^{-1}Q^{-1} \end{matrix}; Q, I \right\rfloor_n \\ &\quad \times (I-C)^{-1}(I-A)(I-BC^{-1}AQ^{1-n})^{-1}(I-B) \\ &\quad \times (I-BC^{-1}AQ^{-n})^{-1}(I-BC^{-1}AQ) = \left\lfloor \begin{matrix} CB^{-1}, A^{-1}C \\ C, A^{-1}CB^{-1} \end{matrix}; Q, I \right\rfloor_{n+1}. \end{aligned} \tag{3.12}$$

Further, we have the following addition formula for Q-shifted factorials of type II:

$$\begin{aligned}&\left\lceil \begin{matrix} CB^{-1}Q, A^{-1}CQ \\ C, A^{-1}CB^{-1} \end{matrix}; Q, I \right\rceil_n (I-A^{-1}CQ^n)^{-1}(I-CB^{-1}Q^n)^{-1} \\ &\quad - (I-C)^{-1}(I-A)(I-BC^{-1}AQ^{1-n})^{-1}(I-B) \\ &\quad \times (I-BC^{-1}AQ^{-n})^{-1}(I-BC^{-1}AQ) \\ &\quad \times \left\lceil \begin{matrix} CB^{-1}Q, A^{-1}CQ \\ CQ, A^{-1}CB^{-1}Q^{-1} \end{matrix}; Q, I \right\rceil_n (I-A^{-1}CQ^n)^{-1}(I-CB^{-1}Q^n)^{-1} \\ &= \left\lceil \begin{matrix} CB^{-1}Q, A^{-1}CQ \\ C, A^{-1}CB^{-1} \end{matrix}; Q, I \right\rceil_{n+1} (I-A^{-1}CQ^{n+1})^{-1}(I-CB^{-1}Q^{n+1})^{-1}. \end{aligned} \tag{3.13}$$

Proof. We prove (3.13) by induction on n and leave the proof of (3.12) (which is similar) to the reader. For $n=0$, (3.13) is easily verified. Assume the identity is true for all nonnegative integers less than a fixed positive integer n. Then we rewrite the left-hand side of (3.13) as

$$\begin{aligned}
&\left[\begin{matrix}CB^{-1}Q, A^{-1}CQ\\ C, A^{-1}CB^{-1}\end{matrix};Q,I\right]_n (I-A^{-1}CQ^n)^{-1}(I-CB^{-1}Q^n)^{-1}\\
&\quad -(I-C)^{-1}(I-A)(I-BC^{-1}AQ^{1-n})^{-1}(I-B)\\
&\quad \times (I-BC^{-1}AQ^{-n})^{-1}(I-BC^{-1}AQ)\\
&\quad \times \left[\begin{matrix}CB^{-1}Q, A^{-1}CQ\\ CQ, A^{-1}CB^{-1}Q^{-1}\end{matrix};Q,I\right]_n (I-A^{-1}CQ^n)^{-1}(I-CB^{-1}Q^n)^{-1}\\
&= (I-C)^{-1}(I-CB^{-1}Q)(I-A^{-1}CB^{-1})^{-1}(I-A^{-1}CQ)\\
&\quad \times \left[\begin{matrix}CB^{-1}Q^2, A^{-1}CQ^2\\ CQ, A^{-1}CB^{-1}Q\end{matrix};Q,I\right]_{n-1} (I-A^{-1}CQ^n)^{-1}(I-CB^{-1}Q^n)^{-1}\\
&\quad -(I-C)^{-1}(I-A)(I-BC^{-1}AQ^{1-n})^{-1}(I-B)\\
&\quad \times (I-BC^{-1}AQ^{-n})^{-1}(I-BC^{-1}AQ)\\
&\quad \times \left[\begin{matrix}CB^{-1}Q, A^{-1}CQ\\ CQ, A^{-1}CB^{-1}Q^{-1}\end{matrix};Q,I\right]_n (I-A^{-1}CQ^n)^{-1}(I-CB^{-1}Q^n)^{-1}.
\end{aligned}$$

We have simply pulled out the first four factors from the first term. Now we can apply the inductive hypothesis (with C replaced by CQ) to the first term to transform the last expression into

$$\begin{aligned}
&(I-C)^{-1}(I-CB^{-1}Q)(I-A^{-1}CB^{-1})^{-1}(I-A^{-1}CQ)\\
&\quad \times \Bigg(\left[\begin{matrix}CB^{-1}Q^2, A^{-1}CQ^2\\ CQ, A^{-1}CB^{-1}Q^{-1}\end{matrix};Q,I\right]_n\\
&\quad \times (I-A^{-1}CQ^{n+1})^{-1}(I-CB^{-1}Q^{n+1})^{-1} + (I-CQ)^{-1}(I-A)\\
&\quad \times (I-BC^{-1}AQ^{1-n})^{-1}(I-B)(I-BC^{-1}AQ^{-n})^{-1}(I-BC^{-1}A)\\
&\quad \times \left[\begin{matrix}CB^{-1}Q^2, A^{-1}CQ^2\\ CQ^2, A^{-1}CB^{-1}\end{matrix};Q,I\right]_{n-1}\\
&\quad \times (I-A^{-1}CQ^n)^{-1}(I-CB^{-1}Q^n)^{-1}\Bigg) - (I-C)^{-1}(I-A)\\
&\quad \times (I-BC^{-1}AQ^{1-n})^{-1}(I-B)(I-BC^{-1}AQ^{-n})^{-1}(I-BC^{-1}AQ)\\
&\quad \times \left[\begin{matrix}CB^{-1}Q, A^{-1}CQ\\ CQ, A^{-1}CB^{-1}Q^{-1}\end{matrix};Q,I\right]_n (I-A^{-1}CQ^n)^{-1}(I-CB^{-1}Q^n)^{-1}.
\end{aligned}$$

What remains to be shown is that in the resulting sum of three terms the last two cancel each other, which is equivalent to

$$\begin{aligned}(I - CB^{-1}Q)&(I - A^{-1}CQ)(I - CQ)^{-1}(I - A)(I - B)\\ &= (I - A)(I - B)(I - CQ)^{-1}(I - CB^{-1}Q)(I - A^{-1}CQ). \qquad (3.14)\end{aligned}$$

However, splitting the factor $(I - CB^{-1}Q)(I - A^{-1}CQ)$ on each side of (3.14) into two terms as $(I-A)(I-B)A^{-1}CB^{-1}Q+(I-A^{-1}CB^{-1}Q)(I-CQ)$ (which can be done since $CB^{-1}A^{-1} = A^{-1}CB^{-1}$, etc.), both sides can be reduced to the same expression, namely

$$\begin{aligned}(I - A)(I - B)&(I - CQ)^{-1}(I - A)(I - B)A^{-1}CB^{-1}Q\\ &+ (I - A^{-1}CB^{-1}Q)(I - A)(I - B),\end{aligned}$$

which immediately establishes (3.13). □

Lemma 3.5. *Let A, B, C and D be noncommutative parameters of some unit ring. Assume that the commutation relations* (6.6) *hold. Further, let n be a nonnegative integer. Then we have the following addition formula for shifted factorials of type I.*

$$\begin{aligned}&\begin{bmatrix} A-C-D+I, A-B-D+I, A+I, A-B-C+I \\ A-C+I, A-B+I, A-D+I, A-B-C-D+I \end{bmatrix}_n\\ &\quad - \begin{bmatrix} A-C-D+I, A-B-D+I, A+3I, A-B-C+I \\ A-C+2I, A-B+2I, A-D+2I, A-B-C-D \end{bmatrix}_n\\ &\quad \times (B+C+D-A-(n+1)I)^{-1}(B+C+D-A)\\ &\quad \times (A+(n+1)I)^{-1}(A+I)\\ &\quad \times (A+(n+2)I)^{-1}(A+2I)(B+C+D-A-nI)^{-1}\\ &\quad \times (2A-B-C-D+(2+2n)I)(A-C+I)^{-1}\\ &\quad \times B(A-B+I)^{-1}C(A-D+I)^{-1}D\\ &= \begin{bmatrix} A-C-D+I, A-B-D+I, A+I, A-B-C+I \\ A-C+I, A-B+I, A-D+I, A-B-C-D+I \end{bmatrix}_{n+1}. \end{aligned} \qquad (3.15)$$

Remark 3.6. An equivalent, almost identical formula holds for shifted factorials of type II, see also Remark 6.3.

Proof of Lemma 3.5. We prove (3.15) by induction on n. For $n = 0$, (3.15) is easily verified. Assume the identity is true for all nonnegative integers less than a fixed positive integer n. Then we rewrite the left-hand side of (3.15) as

$$
\begin{aligned}
&\left[\begin{matrix} A-C-D+I, A-B-D+I, A+I, A-B-C+I \\ A-C+I, A-B+I, A-D+I, A-B-C-D+I \end{matrix}\right|_n \\
&\quad - \left[\begin{matrix} A-C-D+I, A-B-D+I, A+3I, A-B-C+I \\ A-C+2I, A-B+2I, A-D+2I, A-B-C-D \end{matrix}\right|_n \\
&\quad \times (B+C+D-A-(n+1)I)^{-1}(B+C+D-A) \\
&\quad \times (A+(n+1)I)^{-1}(A+I)(A+(n+2)I)^{-1}(A+2I) \\
&\quad \times (B+C+D-A-nI)^{-1}(2A-B-C-D+(2+2n)I) \\
&\quad \times A-C+I)^{-1}B(A-B+I)^{-1}C(A-D+I)^{-1}D \\
&\quad = \left[\begin{matrix} A-C-D+2I, A-B-D+2I, A+2I, A-B-C+2I \\ A-C+2I, A-B+2I, A-D+2I, A-B-C-D+2I \end{matrix}\right|_{n-1} \\
&\quad \times (A-C+I)^{-1}(A-C-D+I)(A-B+I)^{-1}(A-B-D+I) \\
&\quad \times (A-D+I)^{-1}(A+I)(A-B-C-D+I)^{-1}(A-B-C+I) \\
&\quad - \left[\begin{matrix} A-C-D+I, A-B-D+I, A+3I, A-B-C+I \\ A-C+2I, A-B+2I, A-D+2I, A-B-C-D \end{matrix}\right|_n \\
&\quad \times (B+C+D-A-(n+1)I)^{-1}(B+C+D-A) \\
&\quad \times (A+(n+1)I)^{-1}(A+I)(A+(n+2)I)^{-1}(A+2I) \\
&\quad \times (B+C+D-A-nI)^{-1}(2A-B-C-D+(2+2n)I) \\
&\quad \times (A-C+I)^{-1}B(A-B+I)^{-1}C(A-D+I)^{-1}D.
\end{aligned}
$$

We have simply pulled out the first eight factors from the first term. Now we can apply the inductive hypothesis (with A replaced by $A + I$) to the first term to transform the last expression into

$$
\begin{aligned}
&\Bigg(\left[\begin{matrix} A-C-D+2I, A-B-D+2I, A+2I, A-B-C+2I \\ A-C+2I, A-B+2I, A-D+2I, A-B-C-D+2I \end{matrix}\right|_n \\
&\quad + \left[\begin{matrix} A-C-D+2I, A-B-D+2I, A+4I, A-B-C+2I \\ A-C+3I, A-B+3I, A-D+3I, A-B-C-D+I \end{matrix}\right|_{n-1}
\end{aligned}
$$

$$
\begin{aligned}
&\times (B+C+D-A-(n+1)I)^{-1}(B+C+D-A-I)\\
&\times (A+(n+1)I)^{-1}(A+2I)(A+(n+2)I)^{-1}\\
&\times (A+3I)(B+C+D-A-nI)^{-1}(2A-B-C-D+(2+2n)I)\\
&\times (A-C+2I)^{-1}B(A-B+2I)^{-1}C(A-D+2I)^{-1}D\Big)\\
&\times (A-C+I)^{-1}(A-C-D+I)(A-B+I)^{-1}(A-B-D+I)\\
&\times (A-D+I)^{-1}(A+I)(A-B-C-D+I)^{-1}(A-B-C+I)\\
&-\left[\begin{matrix}A-C-D+I, A-B-D+I, A+3I, A-B-C+I\\ A-C+2I, A-B+2I, A-D+2I, A-B-C-D\end{matrix}\right]_n\\
&\times (B+C+D-A-(n+1)I)^{-1}(B+C+D-A)(A+(n+1)I)^{-1}\\
&\times (A+I)(A+(n+2)I)^{-1}(A+2I)(B+C+D-A-nI)^{-1}\\
&\times (2A-B-C-D+(2+2n)I)(A-C+I)^{-1}\\
&\times B(A-B+I)^{-1}C(A-D+I)^{-1}D.
\end{aligned}
$$

What remains to be shown is that in the resulting sum of three terms the last two cancel each other, which is equivalent to

$$
\begin{aligned}
BCD&(A-D+2I)^{-1}(A-C+I)^{-1}(A-B+I)^{-1}\\
&\times (A-C-D+I)(A-B-D+I)(A-B-C+I)\\
&= (A-C-D+I)(A-B-D+I)(A-B-C+I)\\
&\times (A-D+2I)^{-1}(A-C+I)^{-1}(A-B+I)^{-1}BCD.
\end{aligned}
$$

This follows from the fact that the three products BCD, $(A-B+I)(A-C+I)(A-D+2I)$ and $(A-C-D+I)(A-B-D+I)(A-B-C+I)$ mutually commute, which can be readily verified using (6.6) and (6.8). □

4. Chu–Vandermonde Summations

The classical Chu–Vandermonde summation formula (cf. [21, Appendix (III.4)]) sums a terminating ${}_2F_1$ series with unit argument:

$$
{}_2F_1\left[\begin{matrix}a, -n\\ c\end{matrix}; 1\right] = \frac{(c-a)_n}{(c)_n}. \tag{4.1}
$$

We provide two noncommutative extensions of (4.1).

Theorem 4.1. *Let A, C be noncommutative parameters of some unit ring, and let n be a nonnegative integer. Then we have the following summation for a noncommutative hypergeometric series of type I:*

$$ {}_2F_1\left\lceil \begin{matrix} A, -nI \\ C \end{matrix} ; I \right\rfloor = \left\lceil \begin{matrix} C-A \\ C \end{matrix} \right\rfloor_n . \tag{4.2}$$

Further, we have the following summation for a noncommutative hypergeometric series of type II:

$$ {}_2F_1\left\lfloor \begin{matrix} A, -nI \\ C \end{matrix} ; I \right\rceil = \left\lfloor \begin{matrix} C-A+I \\ C \end{matrix} \right\rceil_n (C-A+nI)^{-1}(C-A). \tag{4.3}$$

Note that the right-hand side of (4.3) may also be written as

$$\left\lfloor \begin{matrix} C-A+I \\ C \end{matrix} \right\rceil_{n-1} (C+(n-1)I)^{-1}(C-A);$$

however, in the simple case $n=0$ it is a little bit easier to see that the sum reduces correctly to I when one uses the expression on the right-hand side of (4.3).

Proof of Theorem 4.1. Both identities (4.2) and (4.3) are readily proved by induction on n. We start with the proof of the first summation (4.2).

For $n=0$ the formula is trivial. We assume the summation is true up to a fixed n. To prove (4.2) for $n+1$, use the elementary identity

$$\left\lceil \begin{matrix} -(n+1)I \\ I \end{matrix} \right\rfloor_k = \left\lceil \begin{matrix} -nI \\ I \end{matrix} \right\rfloor_k [I-((-n-1+k)I)^{-1}kI]$$

to obtain

$$\begin{aligned} {}_2F_1\left\lceil \begin{matrix} A, -(n+1)I \\ C \end{matrix} ; I \right\rfloor &= \sum_{k=0}^{n} \left\lceil \begin{matrix} A, -nI \\ C, I \end{matrix} \right\rfloor_k [I-((-n-1+k)I)^{-1}kI] \\ &= {}_2F_1\left\lceil \begin{matrix} A, -nI \\ C \end{matrix} ; I \right\rfloor - {}_2F_1\left\lceil \begin{matrix} A+I, -nI \\ C+I \end{matrix} ; I \right\rfloor C^{-1}A \\ &= \left\lceil \begin{matrix} C-A \\ C \end{matrix} \right\rfloor_n - \left\lfloor \begin{matrix} C-A \\ C+I \end{matrix} \right\rceil_n C^{-1}A = \left\lfloor \begin{matrix} C-A \\ C \end{matrix} \right\rceil_{n+1}, \end{aligned}$$

the penultimate equation due to the inductive hypothesis, the last equation due to Lemma 3.1, equation (3.1).

We turn now to the proof of the second identity (4.3). For $n = 0$ the formula is trivial. We assume the summation is true up to a fixed n. To prove (4.3) for $n+1$, use the elementary identity

$$\begin{bmatrix} -(n+1)I \\ I \end{bmatrix}_k = \begin{bmatrix} -nI \\ I \end{bmatrix}_k [I - ((-n-1+k)I)^{-1}kI]$$

to obtain

$$\begin{aligned}
{}_2F_1\begin{bmatrix} A, -(n+1)I \\ C \end{bmatrix}; I\end{bmatrix} &= \sum_{k=0}^{n} \begin{bmatrix} A, -nI \\ C, I \end{bmatrix}_k [I - ((-n-1+k)I)^{-1}kI] \\
&= {}_2F_1\begin{bmatrix} A, -nI \\ C \end{bmatrix}; I\end{bmatrix} - C^{-1}A\, {}_2F_1\begin{bmatrix} A+I, -nI \\ C+I \end{bmatrix}; I\end{bmatrix} \\
&= \begin{bmatrix} C-A+I \\ C \end{bmatrix}_n (C-A+nI)^{-1}(C-A) \\
&\quad - C^{-1}A \begin{bmatrix} C-A+I \\ C+I \end{bmatrix}_n (C-A+nI)^{-1}(C-A) \\
&= \begin{bmatrix} C-A+I \\ C \end{bmatrix}_{n+1} (C-A+(n+1)I)^{-1}(C-A),
\end{aligned}$$

the penultimate equation due to the inductive hypothesis, the last equation due to Lemma 3.1, equation (3.2). □

The following summation is a q-analogue of (4.1) (cf. [9, Appendix (II.6)]):

$$ {}_2\phi_1\begin{bmatrix} a, q^{-n} \\ c \end{bmatrix}; q, q\end{bmatrix} = \frac{(c/a;q)_n}{(c;q)_n} a^n. \tag{4.4}$$

We provide the following noncommutative extensions of (4.4).

Theorem 4.2. *Let A and C be noncommutative parameters of some unit ring, and suppose that Q commutes both with A and C. Further, let n be a nonnegative integer. Then we have the following summation for a noncommutative hypergeometric series of type I:*

$$ {}_2\phi_1\begin{bmatrix} A, Q^{-n} \\ C \end{bmatrix}; Q, Q\end{bmatrix} = \begin{bmatrix} CA^{-1} \\ C \end{bmatrix}; Q, A\end{bmatrix}_n. \tag{4.5}$$

Further, we have the following summation for a noncommutative hypergeometric series of type II:

$$ {}_2\phi_1\left[\begin{matrix}A, Q^{-n}\\ C\end{matrix}; Q, Q\right] = \left[\begin{matrix}CA^{-1}Q\\ C\end{matrix}; Q, A\right]_n (A - CQ^n)^{-1}(A - C). \qquad (4.6)$$

Note that the right-hand side of (4.6) may also be written as

$$\left[\begin{matrix}CA^{-1}Q\\ C\end{matrix}; Q, A\right]_{n-1} (I - CQ^{n-1})^{-1}(A - C);$$

however, as in (4.3) it is in the simple case $n = 0$ easier to see that the sum reduces correctly to I when one uses the expression on the right-hand side of (4.6).

Proof of Theorem 4.2. We prove (4.5) by induction on n (and leave (4.6) to the reader). For $n = 0$ the formula is trivial. Assume the formula is true up to a fixed n. To prove it for $n + 1$, use the elementary identity

$$\left[\begin{matrix}Q^{-(n+1)}\\ Q\end{matrix}; Q, I\right]_k = \left[\begin{matrix}Q^{-n}\\ Q\end{matrix}; Q, I\right]_k [I - Q^{-n-1}(I - Q^{-n-1+k})^{-1}(I - Q^k)]$$

to obtain

$$\begin{aligned}
&{}_2\phi_1\left[\begin{matrix}A, Q^{-(n+1)}\\ C\end{matrix}; Q, Q\right]\\
&= \sum_{k=0}^{n}\left[\begin{matrix}A, Q^{-n}\\ C, Q\end{matrix}; Q, Q\right]_k [I - Q^{-n-1}(I - Q^{-n-1+k})^{-1}(I - Q^k)]\\
&= {}_2\phi_1\left[\begin{matrix}A, Q^{-n}\\ C\end{matrix}; Q, Q\right] - {}_2\phi_1\left[\begin{matrix}AQ, Q^{-n}\\ CQ\end{matrix}; Q, Q\right] Q^{-n}(I - C)^{-1}(I - A)\\
&= \left[\begin{matrix}CA^{-1}\\ C\end{matrix}; Q, A\right]_n - \left[\begin{matrix}CA^{-1}\\ CQ\end{matrix}; Q, A\right]_n (I - C)^{-1}(I - A),
\end{aligned}$$

the last equation due to the inductive hypothesis. Now apply Lemma 3.1, equation (3.4), which completes the proof of (4.5). □

Here is another q-analogue of (4.1) (cf. [9, Appendix (II.7)]):

$$ {}_2\phi_1\left[\begin{matrix} a, q^{-n} \\ c \end{matrix}; q, \frac{cq^n}{a}\right] = \frac{(c/a;q)_n}{(c;q)_n}. \tag{4.7} $$

We provide the following noncommutative extensions of (4.7).

Theorem 4.3. *Let A and C be noncommutative parameters of some unit ring, and suppose that Q commutes both with A and C. Further, let n be a nonnegative integer. Then we have the following summation for a noncommutative hypergeometric series of type I:*

$$ {}_2\phi_1\left\lceil\begin{matrix} A, Q^{-n} \\ C \end{matrix}; Q, A^{-1}CQ^n\right\rfloor = \left\lceil\begin{matrix} A^{-1}C \\ C \end{matrix}; Q, I\right\rfloor_n. \tag{4.8} $$

Further, we have the following summation for a noncommutative hypergeometric series of type II:

$$ {}_2\phi_1\left\lfloor\begin{matrix} A, Q^{-n} \\ C \end{matrix}; Q, A^{-1}CQ^n\right\rceil $$

$$ = \left\lfloor\begin{matrix} A^{-1}CQ \\ C \end{matrix}; Q, I\right\rceil_n (I - A^{-1}CQ^n)^{-1}(I - A^{-1}C). \tag{4.9} $$

Proof. We prove (4.8) by induction on n (and leave (4.9) to the reader). For $n = 0$ the formula is trivial. Assume the formula is true up to a fixed n. To prove it for $n+1$, use the elementary identity

$$ \left\lceil\begin{matrix} Q^{-(n+1)} \\ Q \end{matrix}; Q, Q^{n+1}\right\rfloor_k = \left\lceil\begin{matrix} Q^{-n} \\ Q \end{matrix}; Q, Q^n\right\rfloor_k [I - (I - Q^{-n-1+k})^{-1}(I - Q^k)] $$

to obtain

$$ {}_2\phi_1\left\lceil\begin{matrix} A, Q^{-(n+1)} \\ C \end{matrix}; Q, A^{-1}CQ^{n+1}\right\rfloor $$

$$ = \sum_{k=0}^{n} \left\lceil\begin{matrix} A, Q^{-n} \\ C, Q \end{matrix}; Q, A^{-1}CQ^n\right\rfloor_k [I - (I - Q^{-n-1+k})^{-1}(I - Q^k)] $$

$$ = {}_2\phi_1\left\lceil\begin{matrix} A, Q^{-n} \\ C \end{matrix}; Q, A^{-1}CQ^n\right\rfloor - {}_2\phi_1\left\lceil\begin{matrix} AQ, Q^{-n} \\ CQ \end{matrix}; Q, A^{-1}CQ^n\right\rfloor $$

$$ \times (I - C)^{-1}(I - A)A^{-1}CQ^n $$

$$= \left[\begin{matrix} A^{-1}C \\ C \end{matrix} ; Q, I\right]_n - \left[\begin{matrix} A^{-1}C \\ CQ \end{matrix} ; Q, I\right]_n (I-C)^{-1}(I-A)A^{-1}CQ^n,$$

the last equation due to the inductive hypothesis. Now apply Lemma 3.1, equation (3.5), and the proof of (4.8) is complete. □

There are two ways to obtain Theorem 4.3 from Theorem 4.2 directly, namely by *inverting the base* $Q \to Q^{-1}$, or by *reversing the sum*. We give some details for both of these possibilities, but restrict ourselves to the derivation of (4.8) from (4.5). The details of deriving (4.9) from (4.6) by inverting the base or by reversing the sum are similar and left to the reader.

(1) *Inverting the base.* It is easy to verify that

$$\left[\begin{matrix} A \\ C \end{matrix} ; Q^{-1}, Z\right]_k = C^{-1} \left[\begin{matrix} A^{-1} \\ C^{-1} \end{matrix} ; Q, AZC^{-1}\right]_k C. \tag{4.10}$$

Now since

$${}_2\phi_1 \left[\begin{matrix} A, Q^n \\ C \end{matrix} ; Q^{-1}, Q^{-1}\right] = \left[\begin{matrix} CA^{-1} \\ C \end{matrix} ; Q^{-1}, A\right]_n$$

by (4.5), we deduce from (4.10) that

$$C^{-1}\, {}_2\phi_1 \left[\begin{matrix} A^{-1}, Q^{-n} \\ C^{-1} \end{matrix} ; Q, AC^{-1}Q^n\right] C = C^{-1} \left[\begin{matrix} AC^{-1} \\ C^{-1} \end{matrix} ; Q, I\right]_n C.$$

Now simply replace A by A^{-1} and C by C^{-1} and then multiply both sides by C^{-1} from the left and by C from the right to get (4.8).

(2) *Reversing the sum.* Using

$$\begin{aligned}\prod_{j=1}^{n-k} B_{n-k-j} &= \prod_{j=1}^{-k} B_{n-k-j} \prod_{j=1-k}^{n-k} B_{n-k-j} \\ &= \prod_{j=1-k}^{0} B_{n-1+j}^{-1} \prod_{j=1}^{n} B_{n-j} \\ &= \prod_{j=1}^{k} B_{n-1-k+j}^{-1} \prod_{j=1}^{n} B_{n-j},\end{aligned}$$

which is derived using (2.4), we readily deduce

$$\left[\begin{matrix} A \\ C \end{matrix} ; Q, I\right]_{n-k} = A^{-1} \left[\begin{matrix} C^{-1}Q^{1-n} \\ A^{-1}Q^{1-n} \end{matrix} ; Q, CA^{-1}\right]_k A \left[\begin{matrix} A \\ C \end{matrix} ; Q, I\right]_n .$$

Similarly,

$$\begin{bmatrix} Q^{-n} \\ Q \end{bmatrix}; Q, I \Bigg]_{n-k} = \begin{bmatrix} Q^{-n} \\ Q \end{bmatrix}; Q, I \Bigg]_{k} \begin{bmatrix} Q^{-n} \\ Q \end{bmatrix}; Q, I \Bigg]_{n} Q^{(n+1)k}$$

$$= \begin{bmatrix} Q^{-n} \\ Q \end{bmatrix}; Q, I \Bigg]_{k} (-1)^n Q^{(n+1)k-\binom{n+1}{2}}.$$

Hence we have from (4.5), by reversing the sum on the left hand side,

$$A^{-1}{}_2\phi_1 \left[\begin{matrix} C^{-1}Q^{1-n}, Q^{-n} \\ A^{-1}Q^{1-n} \end{matrix}; Q, CA^{-1}Q^n \right] A(-1)^n Q^{-\binom{n}{2}} \begin{bmatrix} A \\ C \end{bmatrix}; Q, I \Bigg]_n$$

$$= \begin{bmatrix} CA^{-1} \\ C \end{bmatrix}; Q, A \Bigg]_n .$$

Performing the simultaneous substitutions $A \mapsto C^{-1}Q^{1-n}$, $C \mapsto A^{-1}Q^{1-n}$, and putting some factors to the other side gives

$${}_2\phi_1 \left[\begin{matrix} A, Q^{-n} \\ C \end{matrix}; Q, A^{-1}CQ^n \right]$$

$$= C^{-1} \left[\begin{matrix} A^{-1}C \\ A^{-1}Q^{1-n} \end{matrix}; Q, C^{-1}Q^{1-n} \right]_n \left[\begin{matrix} C^{-1}Q^{1-n} \\ A^{-1}Q^{1-n} \end{matrix}; Q, I \right]_n^{-1} C(-1)^n Q^{\binom{n}{2}}. \tag{4.11}$$

We want to show that the right-hand side of (4.11) reduces to the right-hand side of (4.8). First, we compute

$$\left[\begin{matrix} C^{-1}Q^{1-n} \\ A^{-1}Q^{1-n} \end{matrix}; Q, I \right]_n^{-1} = C \left[\begin{matrix} A \\ C \end{matrix}; Q, A^{-1}C \right]_n C^{-1}. \tag{4.12}$$

Further,

$$\left[\begin{matrix} A^{-1}C \\ A^{-1}Q^{1-n} \end{matrix}; Q, C^{-1}Q^{1-n} \right]_n$$

$$= (-1)^n Q^{\binom{n}{2}} \prod_{j=1}^{n} [A(I - AQ^{j-1})^{-1}(I - A^{-1}CQ^{n-j})C^{-1}]. \tag{4.13}$$

Thus, applying (4.12) and (4.13) to the right-hand side of (4.11) and equating the result to the right-hand side of (4.8), we have established (4.8) once we have shown the following lemma.

Lemma 4.4.

$$\prod_{j=1}^{n} [C^{-1}A(I - AQ^{j-1})^{-1}(I - A^{-1}CQ^{n-j})]$$
$$\times \left[\begin{matrix} A \\ C \end{matrix}; Q, A^{-1}C\right]_n = \left[\begin{matrix} A^{-1}C \\ C \end{matrix}; Q, I\right]_n . \qquad (4.14)$$

Proof. We proceed by induction on n. For $n = 0$ the identity is trivial. Observe that for $n = 1$ the statement amounts to

$$C^{-1}A(I - A)^{-1}(I - A^{-1}C)(I - C)^{-1}(I - A)A^{-1}C$$
$$= (I - C)^{-1}(I - A^{-1}C). \qquad (4.15)$$

This is easily verified by splitting the factor $(I - A^{-1}C)(I - C)^{-1}$ on the left-hand side of (4.15) into two terms as $I - (I - A)A^{-1}C(I - C)^{-1}$, and simplifying the expression to $I-(I-C)^{-1}(I-A)A^{-1}C$ which clearly equals $(I - C)^{-1}(I - A^{-1}C)$.

Assume now that (4.14) is true for all nonnegative integers up to a fixed n. Then

$$\prod_{j=1}^{n} [C^{-1}A(I - AQ^{j-1})^{-1}(I - A^{-1}CQ^{n-j})] \left[\begin{matrix} A \\ C \end{matrix}; Q, A^{-1}C\right]_n$$
$$= C^{-1}A(I - A)^{-1}(I - A^{-1}CQ^{n-1})$$
$$\times \prod_{j=1}^{n-1} [C^{-1}A(I - AQ^{j})^{-1}(I - A^{-1}CQ^{n-1-j})]$$
$$\times \left[\begin{matrix} AQ \\ CQ \end{matrix}; Q, A^{-1}C\right]_{n-1} (I - C)^{-1}(I - A)A^{-1}C$$
$$= C^{-1}A(I - A)^{-1}(I - A^{-1}CQ^{n-1}) \left[\begin{matrix} A^{-1}C \\ CQ \end{matrix}; Q, I\right]_{n-1}$$
$$\times (I - C)^{-1}(I - A)A^{-1}C, \qquad (4.16)$$

the last equation due to the $A \mapsto AQ$, $C \mapsto CQ$ case of the inductive hypothesis. The proof is complete after application of Lemma 4.5. □

Lemma 4.5.

$$C^{-1}A(I-A)^{-1}(I-A^{-1}CQ^{n-1})\left[\begin{matrix}A^{-1}C\\ CQ\end{matrix};Q,I\right]_{n-1}$$
$$\times(I-C)^{-1}(I-A)A^{-1}C=\left[\begin{matrix}A^{-1}C\\ C\end{matrix};Q,I\right]_{n}. \tag{4.17}$$

Proof. We proceed by induction on n. For $n=1$, (4.17) is simply (4.15). Assume now that (4.17) is true for all positive integers up to a fixed n. Then

$$C^{-1}A(I-A)^{-1}(I-A^{-1}CQ^{n})\left[\begin{matrix}A^{-1}C\\ CQ\end{matrix};Q,I\right]_{n}(I-C)^{-1}(I-A)A^{-1}C$$
$$=C^{-1}Q^{-1}A(I-A)^{-1}(I-A^{-1}CQ^{n})\left[\begin{matrix}A^{-1}CQ\\ CQ^{2}\end{matrix};Q,I\right]_{n-1}$$
$$\times(I-CQ)^{-1}(I-A^{-1}C)(I-C)^{-1}(I-A)A^{-1}CQ$$
$$=\left[\begin{matrix}A^{-1}CQ\\ CQ\end{matrix};Q,I\right]_{n-1}C^{-1}Q^{-1}A(I-A)^{-1}$$
$$\times(I-A^{-1}C)(I-C)^{-1}(I-A)A^{-1}CQ, \tag{4.18}$$

the last equation due to the $C\mapsto CQ$ case of the inductive hypothesis. Finally, an application of (4.15) simplifies the last expression to

$$\left[\begin{matrix}A^{-1}CQ\\ CQ\end{matrix};Q,I\right]_{n-1}(I-C)^{-1}(I-A^{-1}C),$$

which is equal to the generalized Q-shifted factorial on the right-hand side of (4.17). □

5. Pfaff–Saalschütz Summations

The classical Pfaff–Saalschütz summation formula (cf. [21, Appendix (III.2)]) sums a terminating balanced ${}_3F_2$ series:

$${}_3F_2\left[\begin{matrix}a,b,-n\\ c,a+b-c+1-n\end{matrix};1\right]=\frac{(c-a)_n(c-b)_n}{(c)_n(c-a-b)_n}. \tag{5.1}$$

In order to derive a noncommutative extension of (5.1), one should at least be able to extend its $n = 1$ special case, which is

$$1 - \frac{ab}{c(a+b-c)} = \frac{(c-a)(c-b)}{c(c-a-b)}.$$

This is no problem indeed, as we have

$$I - C^{-1}A(A+B-C)^{-1}B = C^{-1}(C-B)(C-A-B)^{-1}(C-A), \tag{5.2}$$

for noncommutative parameters A, B, C, as one immediately verifies. In fact,

$$\begin{aligned} I - C^{-1}A(A+B-C)^{-1}B &= C^{-1}[C - A(A+B-C)^{-1}B] \\ &= C^{-1}[C - ((A+B-C) - (B-C))(A+B-C)^{-1}B] \\ &= C^{-1}[C - B - (C-B)(A+B-C)^{-1}B] \\ &= C^{-1}(C-B)[I - (A+B-C)^{-1}B] \\ &= C^{-1}(C-B)(C-A-B)^{-1}(C-A). \end{aligned}$$

Despite of the $n = 1$ case of the Pfaff–Saalschütz summation to perfectly extend to the noncommutative setting, we were nevertheless not able to extend (5.2) to a full noncommutative Pfaff–Saalschütz summation with arbitrary noncommutative parameters A, B and C. For the case of general n, we need a restriction on the sum $D = A + B - C$, namely that it must commute with the other elements A, B and C (e.g., $D = dI$).

Here are two noncommutative extensions of (5.1).

Theorem 5.1. *Let A, B and C be noncommutative parameters of some unit ring, and assume that the sum $A + B - C$ commutes each with A, B and C. Further, let n be a nonnegative integer. Then we have the following summation for a noncommutative hypergeometric series of type I:*

$$ {}_3F_2\left\lceil \begin{matrix} A, B, -nI \\ C, A+B-C+(1-n)I \end{matrix} ; I \right\rfloor = \left\lceil \begin{matrix} C-B, C-A \\ C, C-A-B \end{matrix} \right\rfloor_n . \tag{5.3}$$

Further, we have the following summation for a noncommutative hypergeometric series of type II:

$$ {}_3F_2\left\lfloor \begin{matrix} A, B, -nI \\ C, A+B-C+(1-n)I \end{matrix} ; I \right\rceil = \left\lfloor \begin{matrix} C-B+I, C-A+I \\ C, C-A-B \end{matrix} \right\rceil_n \times (C-A+nI)^{-1}(C-B+nI)^{-1}(C-B)(C-A). \tag{5.4}$$

Note that the right-hand side of (5.4) may also be written as

$$\begin{bmatrix} C-B+I, C-A+I \\ C, C-A-B \end{bmatrix}_{n-1} (C+(n-1)I)^{-1}$$
$$\times (C-A-B+(n-1)I)^{-1}(C-B)(C-A).$$

Proof of Theorem 5.1. We prove (5.4) by induction on n, leaving the proof of (5.3) (which is similar) to the reader. For $n=0$ the formula is trivial. Assume the formula is true up to a fixed n. To prove it for $n+1$, use the elementary identity

$$\begin{bmatrix} -(n+1)I \\ A+B-C-nI \end{bmatrix}_k = \begin{bmatrix} -nI \\ A+B-C+(1-n)I \end{bmatrix}_k [I-(A+B-C-nI)^{-1}$$
$$\times (A+B-C+I)((-n-1+k)I)^{-1}kI]$$

to obtain

$$\begin{aligned}
&{}_3F_2\begin{bmatrix} A, B, -(n+1)I \\ C, A+B-C-nI \end{bmatrix}; I \\
&= \sum_{k=0}^{n} \begin{bmatrix} A, B, -nI \\ C, A+B-C+(1-n)I, I \end{bmatrix}_k [I-(A+B-C-nI)^{-1} \\
&\times (A+B-C+I)((-n-1+k)I)^{-1}kI] \\
&= {}_3F_2\begin{bmatrix} A, B, -nI \\ C, A+B-C+(1-n)I \end{bmatrix}; I - C^{-1}A(A+B-C+(1-n)I)^{-1} \\
&\times B(A+B-C-nI)^{-1}(A+B-C+I) \\
&\times {}_3F_2\begin{bmatrix} A+I, B+I, -nI \\ C+I, A+B-C+(2-n)I \end{bmatrix}; I = \begin{bmatrix} C-B+I, C-A+I \\ C, C-A-B \end{bmatrix}_n \\
&\times (C-A+nI)^{-1}(C-B+nI)^{-1}(C-B)(C-A) - C^{-1} \\
&\times A(A+B-C+(1-n)I)^{-1}B(A+B-C-nI)^{-1}(A+B-C+I) \\
&\times \begin{bmatrix} C-B+I, C-A+I \\ C+I, C-A-B-I \end{bmatrix}_n (C-A+nI)^{-1} \\
&\times (C-B+nI)^{-1}(C-B)(C-A),
\end{aligned}$$

the last equation due to the inductive hypothesis. We are done after application of Lemma 3.3, equation (3.10). □

The q-analogue of (5.1) is the following summation formula (cf. [9, Appendix (II.12)]):

$$ {}_3\phi_2\left[\begin{matrix} a, b, q^{-n} \\ c, abq^{1-n}/c \end{matrix}; q, q\right] = \frac{(c/a;q)_n (c/b;q)_n}{(c;q)_n (c/ab;q)_n}. \tag{5.5} $$

In order to derive a noncommutative extension of (5.5), one should at least be able to extend its $n = 1$ special case, which is

$$ 1 - \frac{(1-a)(1-b)}{(1-c)(1-ab/c)} = \frac{(1-c/a)(1-c/b)}{(1-c)(1-c/ab)}. $$

This is no problem indeed, as we have

$$ \begin{aligned} & I - (I-C)^{-1}(I-A)(I-BC^{-1}A)^{-1}(I-B) \\ & = (I-C)^{-1}(I-CB^{-1})(I-A^{-1}CB^{-1})^{-1}(I-A^{-1}C), \end{aligned} \tag{5.6} $$

for noncommutative parameters A, B, C, as one immediately verifies. In fact,

$$ \begin{aligned} & I - (I-C)^{-1}(I-A)(I-BC^{-1}A)^{-1}(I-B) \\ & = (I-C)^{-1}[I - C - (I-A)(I-BC^{-1}A)^{-1}(I-B)] \\ & = (I-C)^{-1}[I - C - ((I-BC^{-1}A) \\ & \quad -(I-BC^{-1})A)(I-BC^{-1}A)^{-1}(I-B)] \\ & = (I-C)^{-1}[I - C - (I-B) \\ & \quad -(I-CB^{-1})BC^{-1}A(I-BC^{-1}A)^{-1}(I-B)] \\ & = (I-C)^{-1}(I-CB^{-1})[B + (I-A^{-1}CB^{-1})^{-1}(I-B)] \\ & = (I-C)^{-1}(I-CB^{-1})(I-A^{-1}CB^{-1})^{-1}[B - A^{-1}C + (I-B)] \\ & = (I-C)^{-1}(I-CB^{-1})(I-A^{-1}CB^{-1})^{-1}(I-A^{-1}C). \end{aligned} $$

Alternatively, the identity (5.6) can be directly derived from (5.2) by the substitution

$$ (A, B, C) \mapsto (BC^{-1}(I-A), I-B, BC^{-1}(I-C)). $$

Despite of the $n = 1$ case of the q-Pfaff–Saalschütz summation to perfectly extend to the noncommutative setting, we were nevertheless not able to extend (5.6) to a full noncommutative q-Pfaff–Saalschütz summation with arbitrary noncommutative parameters A, B and C. For the case of general n, we need a restriction on the product $D = BC^{-1}A$, namely that it must commute with the other elements A, B and C (e.g., $D = dI$).

Here are two noncommutative extensions of (5.5).

Theorem 5.2. *Let A, B and C be noncommutative parameters of some unit ring, and suppose that Q commutes each with A, B and C. Moreover, assume that the product $BC^{-1}A$ commutes each with A, B and C. Further, let n be a nonnegative integer. Then we have the following summation for a noncommutative basic hypergeometric series of type I:*

$$ {}_3\phi_2\left[\begin{matrix} A, B, Q^{-n} \\ C, BC^{-1}AQ^{1-n} \end{matrix}; Q, Q\right| = \left[\begin{matrix} CB^{-1}, A^{-1}C \\ C, A^{-1}CB^{-1} \end{matrix}; Q, I\right|_n . \tag{5.7}$$

Further, we have the following summation for a noncommutative basic hypergeometric series of type II:

$$ \begin{aligned} &{}_3\phi_2\left|\begin{matrix} A, B, Q^{-n} \\ C, BC^{-1}AQ^{1-n} \end{matrix}; Q, Q\right] \\ &= \left|\begin{matrix} CB^{-1}Q, A^{-1}CQ \\ C, A^{-1}CB^{-1} \end{matrix}; Q, I\right]_n (I - A^{-1}CQ^n)^{-1} \\ &\quad \times (I - CB^{-1}Q^n)^{-1}(I - CB^{-1})(I - A^{-1}C). \end{aligned} \tag{5.8}$$

Note that the right-hand side of (5.8) may also be written as

$$ \begin{aligned} &\left|\begin{matrix} CB^{-1}Q, A^{-1}CQ \\ C, A^{-1}CB^{-1} \end{matrix}; Q, I\right]_{n-1} (I - CQ^{n-1})^{-1} \\ &\quad \times (I - A^{-1}CB^{-1}Q^{n-1})^{-1}(I - CB^{-1})(I - A^{-1}C). \end{aligned}$$

Proof of Theorem 5.2. We prove (5.8) by induction on n. We leave the proof of (5.7) (which is similar) to the reader. For $n = 0$ the formula is trivial. Assume the formula is true up to a fixed n. To prove it for $n + 1$, use the elementary identity

$$ \begin{aligned} &\left|\begin{matrix} Q^{-(n+1)} \\ BC^{-1}AQ^{-n} \end{matrix}; Q, I\right]_k \\ &= \left|\begin{matrix} Q^{-n} \\ BC^{-1}AQ^{1-n} \end{matrix}; Q, I\right]_k [I - (I - BC^{-1}AQ^{-n})^{-1} \\ &\quad \times (I - BC^{-1}AQ)(I - Q^{-n-1+k})^{-1}(I - Q^k)] \end{aligned}$$

to obtain

$$
\begin{aligned}
&{}_3\phi_2\left[\begin{matrix}A,B,Q^{-(n+1)}\\ C,BC^{-1}AQ^{-n}\end{matrix};Q,Q\right]\\
&=\sum_{k=0}^{n}\left[\begin{matrix}A,B,Q^{-n}\\ C,BC^{-1}AQ^{1-n},Q\end{matrix};Q,Q\right]_k [I-(I-BC^{-1}AQ^{-n})^{-1}\\
&\quad\times(I-BC^{-1}AQ)(I-Q^{-n-1+k})^{-1}(I-Q^k)]\\
&={}_3\phi_2\left[\begin{matrix}A,B,Q^{-n}\\ C,BC^{-1}AQ^{1-n}\end{matrix};Q,Q\right]-(I-C)^{-1}(I-A)\\
&\quad\times(I-BC^{-1}AQ^{1-n})^{-1}(I-B)(I-BC^{-1}AQ^{-n})^{-1}(I-BC^{-1}AQ)\\
&\quad\times{}_3\phi_2\left[\begin{matrix}AQ,BQ,Q^{-n}\\ CQ,BC^{-1}AQ^{2-n}\end{matrix};Q,Q\right]=\left[\begin{matrix}CB^{-1}Q,A^{-1}CQ\\ C,A^{-1}CB^{-1}\end{matrix};Q,I\right]_n\\
&\quad\times(I-A^{-1}CQ^n)^{-1}(I-CB^{-1}Q^n)^{-1}(I-CB^{-1})(I-A^{-1}C)\\
&\quad-(I-C)^{-1}(I-A)(I-BC^{-1}AQ^{1-n})^{-1}(I-B)\\
&\quad\times(I-BC^{-1}AQ^{-n})^{-1}(I-BC^{-1}AQ)\left[\begin{matrix}CB^{-1}Q,A^{-1}CQ\\ CQ,A^{-1}CB^{-1}Q^{-1}\end{matrix};Q,I\right]_n\\
&\quad\times(I-A^{-1}CQ^n)^{-1}(I-CB^{-1}Q^n)^{-1}(I-CB^{-1})(I-A^{-1}C),
\end{aligned}
$$

the last equation due to the inductive hypothesis. We are done after application of Lemma 3.4, equation 3.13. □

6. Very-Well-Poised ${}_7F_6$ Summations

Dougall's summation formula (cf. [21, Appendix (III.14)]) sums a terminating very-well-poised 2-balanced ${}_7F_6$ series:

$$
\begin{aligned}
&{}_7F_6\left[\begin{matrix}\frac{a}{2}+1,a,b,c,d,2a-b-c-d+1+n,-n\\ \frac{a}{2},a-b+1,a-c+1,a-d+1,b+c+d-a-n,a+n+1\end{matrix};1\right]\\
&=\frac{(a+1)_n(a-b-c+1)_n(a-b-d+1)_n(a-c-d+1)_n}{(a-b+1)_n(a-c+1)_n(a-d+1)_n(a-b-c-d+1)_n}. \qquad (6.1)
\end{aligned}
$$

In order to derive a noncommutative extension of (6.1), one should at least be able to extend its $n=1$ special case, which (with a replaced by $a-1$) is

$$1-\frac{bcd(2a-b-c-d)}{(a-b)(a-c)(a-d)(b+c+d-a)}$$
$$=\frac{a(a-b-c)(a-b-d)(a-c-d)}{(a-b)(a-c)(a-d)(a-b-c-d)}. \tag{6.2}$$

A noncommutative extension of (6.2) is indeed available when one assumes that A, B, C, and D are parameters such that A, B and D mutually commute while C does not commute with any of the other parameters. We then have

$$I-(A-C)^{-1}B(A-B)^{-1}C(B+C+D-A)^{-1}D(A-D)^{-1}$$
$$\times(2A-B-C-D)=(A-C)^{-1}(A-B-D)(A-B)^{-1}$$
$$\times(A-C-D)(A-B-C-D)^{-1}A(A-D)^{-1}(A-B-C). \tag{6.3}$$

We already know from (5.2) that

$$I-G^{-1}E(E+F-G)^{-1}F=G^{-1}(G-F)(G-E-F)^{-1}(G-E) \tag{6.4}$$

holds, for noncommuting parameters E, F and G. Now let

$$E=BC,$$
$$F=D(2A-B-C-D),$$
$$G=(A-B)(A-C),$$

while assuming that A, B, C, and D are parameters such that A, B and D mutually commute while C does not commute with any of the other parameters. Then one readily verifies that

$$G-E=A(A-B-C),$$
$$G-F=(A-B-D)(A-C-D),$$
$$G-E-F=(A-D)(A-B-C-D).$$

After these substitutions into (6.4) and rearranging the order of some factors (which do not involve C), we immediately establish (6.3).

However, rather than (6.3), we prefer to consider its variant which is obtained by replacing D by $2A-B-C-D$, for convenience. The result is the following.

Under the assumption that A, B, C, and D are (noncommuting) parameters such that A, B and $C+D$ mutually commute, there holds

$$I-(A-C)^{-1}B(A-B)^{-1}C(A-D)^{-1}(2A-B-C-D)$$
$$\times(B+C+D-A)^{-1}D=(A-C)^{-1}(A-C-D)(A-B)^{-1}$$
$$\times(A-B-D)(A-D)^{-1}A(A-B-C-D)^{-1}(A-B-C). \quad (6.5)$$

As a matter of fact, we were not able to extend (6.3) (or the equivalent (6.5)) to a noncommutative extension of Dougall's summation (6.1), valid for general n, without introducing more commutation relations. What we require in the general case (in place of the weaker requirement that A, B and $C+D$ shall mutually commute) are the following commutation relations:

$$A \quad \text{commutes with} \quad B,\ C,\ D, \quad (6.6a)$$
$$B+C+D \quad \text{commutes with} \quad B,\ C,\ D, \quad (6.6b)$$

and the following "rotation relations":

$$BCD=CDB=DBC. \quad (6.6c)$$

It is easy to check that (6.6b) implies

$$BC-CB=CD-DC=DB-BD, \quad (6.7)$$

which, using Lie brackets, writes elegantly as

$$[B,C]=[C,D]=[D,B].$$

Of course, these Lie products need not to be O (the zero element of the unit ring R). Suppose that E commutes each with B, C and D. Since it follows from (6.6c) that

$$BCD \text{ commutes with } BC+BD+CD,$$

we immediately deduce from

$$(B+E)(C+E)(D+E)=BCD+(BC+BD+CD)E+(B+C+D)E^2+E^3,$$

that (6.6b) and (6.6c) imply that

$$BCD \text{ commutes with } (B+E)(C+E)(D+E). \quad (6.8)$$

Remark 6.1. One may wonder whether the above commutation relations (6.6) allow any room for noncommutativity, where B, C, D do not already mutually commute. The following nontrivial example of A, B, C, D realizing (6.6) has been kindly communicated to us (essentially in the given form) by Hjalmar Rosengren.

Consider R to be the ring of (3×3)-matrices over the complex numbers $\mathbb{C}$ with unit element I and zero element O. Let $\omega = e^{2\pi i/3}$, a cubic root of unity. Suppose $a, b, c, d \in \mathbb{C}$ with $b \neq 0$, $c \neq 0$. Let $A = aI$,

$$B = \begin{pmatrix} d & b & 0 \\ b^{-1} & d & c \\ 0 & -\omega^2 c^{-1} & d \end{pmatrix}, \quad C = \begin{pmatrix} d & \omega b & 0 \\ \omega^2 b^{-1} & d & \omega c \\ 0 & -\omega c^{-1} & d \end{pmatrix},$$

$$D = \begin{pmatrix} d & \omega^2 b & 0 \\ \omega b^{-1} & d & \omega^2 c \\ 0 & -c^{-1} & d \end{pmatrix}.$$

Then the matrices A, B, C and D satisfy (6.6): Of course, (6.6a) is trivially satisfied. Observe that B, C, D do not mutually commute since

$$X = BC - CB = CD - DC = DB - BD = (\omega^2 - \omega)\begin{pmatrix} 1 & 0 & 0 \\ 0 & \omega & 0 \\ 0 & 0 & \omega^2 \end{pmatrix} \neq O.$$

Further observe that with the above choice of B, C and D one has $C = X^{-1}BX$, $D = XBX^{-1}$, hence $XC = BX$, $DX = XB$, and $CX = XD$. It is straightforward to compute $B + C + D = 3dI$, thus (6.6b) is satisfied. Finally, since $BCD = (CB + X)D = CBD + XD = C(DB - X) + XD = CDB - CX + XD = CDB$, etc., one immediately deduces (6.6c).

We are ready to present our noncommutative extension of (6.1).

Theorem 6.2. *Let A, B, C and D be noncommutative parameters of some unit ring, and assume that the relations (6.6) hold. Further, let n be a nonnegative integer. Then we have the following summation for a noncommutative hypergeometric series of type I (and II).*

$$\begin{aligned} &{}_7F_6\left[\begin{matrix} \tfrac{1}{2}A + I, A, B, C, D, 2A - B - C - D + (n+1)I, \\ \tfrac{1}{2}A, A + (n+1)I, A - C + I, A - B + I, A - D + I, \end{matrix}\right. \\ &\qquad\qquad \left.\begin{matrix} -nI \\ B + C + D - A - nI \end{matrix}; I\right] \\ &= \left[\begin{matrix} A - C - D + I, A - B - D + I, A + I, A - B - C + I \\ A - C + I, A - B + I, A - D + I, A - B - C - D + I \end{matrix}\right]_n. \end{aligned} \tag{6.9}$$

Remark 6.3. The above summation holds for both types, I and II, of noncommutative series and shifted factorials. More precisely, we could switch the type I brackets to type II brackets on either side (or on both sides) of (6.9) and the formula would be still valid. This is a consequence of the conditions (6.6) (from which we extracted, in particular, (6.8)). Nevertheless, for brevity of display we write (6.9) using type I brackets only.

Proof of Theorem 6.2. We prove (6.9) by induction on n. For $n = 0$ the formula is trivial. Assume the formula is true up to a fixed n. To prove it for $n+1$, use the elementary identity

$$\left[\begin{matrix} 2A-B-C-D+(n+2)I, -(n+1)I \\ A+(n+2)I, B+C+D-A-(n+1)I \end{matrix}\right]_k$$

$$= \left[\begin{matrix} 2A-B-C-D+(n+1)I, -nI \\ A+(n+1)I, B+C+D-A-nI \end{matrix}\right]_k$$

$$\times [I-(2A-B-C-D+(n+1)I)^{-1}(2A-B-C-D+(2n+2)I)$$

$$\times (B+C+D-A-(n+1)I)^{-1}(B+C+D-A)$$

$$\times (A+(n+1+k)I)^{-1}(A+kI)((-n-1+k)I)^{-1}kI]$$

to obtain

$${}_7F_6\left[\begin{matrix} \frac{1}{2}A+I, A, B, C, D, 2A-B-C-D+(n+2)I, \\ \frac{1}{2}A, A+(n+2)I, A-C+I, A-B+I, A-D+I, \end{matrix}\right.$$

$$\left.\begin{matrix} -(n+1)I \\ B+C+D-A-(n+1)I \end{matrix}; I\right]$$

$$= \sum_{k=0}^{n} \left[\begin{matrix} \frac{1}{2}A+I, A, B, C, D, 2A-B-C-D+(n+1)I, \\ \frac{1}{2}A, A+(n+1)I, A-C+I, A-B+I, A-D+I, I, \end{matrix}\right.$$

$$\left.\begin{matrix} -nI \\ B+C+D-A-nI \end{matrix}\right]_k$$

$$\times [I-(2A-B-C-D+(n+1)I)^{-1}(2A-B-C-D+(2n+2)I)$$

$$\times (B+C+D-A-(n+1)I)^{-1}(B+C+D-A)$$

$$\times (A+(n+1+k)I)^{-1}(A+kI)((-n-1+k)I)^{-1}kI]$$

$$= {}_7F_6\left[\begin{matrix}\frac{1}{2}A+I, A, B, C, D, 2A-B-C-D+(n+1)I,\\ \frac{1}{2}A, A+(n+1)I, A-C+I, A-B+I, A-D+I,\end{matrix}\right.$$

$$\left.\begin{matrix}-nI\\ B+C+D-A-nI\end{matrix};I\right]$$

$$-{}_7F_6\left[\begin{matrix}\frac{1}{2}A+2I, A+2I, B+I, C+I, D+I,\\ 2A-B-C-D+(n+2)I,\\ \frac{1}{2}A+I, A+(n+3)I, A-C+2I, A-B+2I,\\ A-D+2I,\end{matrix}\right.$$

$$\left.\begin{matrix}-nI\\ B+C+D-A-(n-1)I\end{matrix};I\right]$$

$$\times (B+C+D-A-(n+1)I)^{-1}(B+C+D-A)(A+(n+1)I)^{-1}$$
$$\times (A+I)(A+(n+2)I)^{-1}(A+2I)(B+C+D-A-nI)^{-1}$$
$$\times (2A-B-C-D+(2+2n)I)$$
$$\times (A-C+I)^{-1}B(A-B+I)^{-1}C(A-D+I)^{-1}D$$

$$= \left[\begin{matrix}A-C-D+I, A-B-D+I, A+I, A-B-C+I\\ A-C+I, A-B+I, A-D+I, A-B-C-D+I\end{matrix}\right]_n$$

$$-\left[\begin{matrix}A-C-D+I, A-B-D+I, A+3I, A-B-C+I\\ A-C+2I, A-B+2I, A-D+2I, A-B-C-D\end{matrix}\right]_n$$

$$\times (B+C+D-A-(n+1)I)^{-1}(B+C+D-A)(A+(n+1)I)^{-1}$$
$$\times (A+I)(A+(n+2)I)^{-1}(A+2I)(B+C+D-A-nI)^{-1}$$
$$\times (2A-B-C-D+(2+2n)I)$$
$$\times (A-C+I)^{-1}B(A-B+I)^{-1}C(A-D+I)^{-1}D,$$

the last equation due to the inductive hypothesis. We are done after application of Lemma 3.5. □

A natural question is as follows: if one can derive a noncommutative terminating ${}_5F_4$ summation (extending [21, Appendix (III.13)]) as a special case from Theorem 6.2. The answer is negative. If, say, $D = A - C + I$, then since D and C would commute, (6.7) would imply that all variables commute so one just obtains the usual (commutative) terminating ${}_5F_4$ summation.

We now turn to the question of deriving a Q-analogue of Theorem 6.2. The q-analogue of (6.1) is the following summation formula (cf. [9, Appendix (II.22)]):

$$
\begin{aligned}
&{}_8\phi_7\left[\begin{matrix} qa^{\frac{1}{2}}, qa^{-\frac{1}{2}}, a, b, c, d, a^2q^{n+1}/bcd, q^{-n} \\ a^{\frac{1}{2}}, a^{-\frac{1}{2}}, aq/b, aq/c, aq/d, bcdq^{-n}/a, aq^{n+1} \end{matrix}; q, q\right] \\
&\quad = \frac{(aq;q)_n(aq/bc;q)_n(aq/bd;q)_n(aq/cd;q)_n}{(aq/b;q)_n(aq/c;q)_n(aq/d;q)_n(aq/bcd;q)_n}. \qquad (6.10)
\end{aligned}
$$

In order to derive a noncommutative extension of (6.10), one should at least be able to extend its $n = 1$ special case, which (with a replaced by a/q) is

$$
\begin{aligned}
&1 - \frac{(1-b)(1-c)(1-d)(1-a^2/bcd)}{(1-a/b)(1-a/c)(1-a/d)(1-bcd/a)} \\
&\quad = \frac{(1-a)(1-a/bc)(1-a/bd)(1-a/cd)}{(1-a/b)(1-a/c)(1-a/d)(1-a/bcd)}. \qquad (6.11)
\end{aligned}
$$

A noncommutative extension of (6.11) is indeed available when one assumes that A, B, C, and D are parameters such that A, B and D mutually commute while C does not commute with any of the other parameters. We then have

$$
\begin{aligned}
&I - (I - C^{-1}A)^{-1}(I - B)(I - AB^{-1})^{-1}(I - B^{-1}CB) \\
&\qquad \times (I - DA^{-1}CB)^{-1}(I - D)(I - AD^{-1})^{-1}(I - AB^{-1}D^{-1}C^{-1}A) \\
&\quad = (I - C^{-1}A)^{-1}(I - AB^{-1}D^{-1})(I - AB^{-1})^{-1}(I - C^{-1}AD^{-1}) \\
&\qquad \times (I - B^{-1}C^{-1}AD^{-1})^{-1}(I - A)(I - AD^{-1})^{-1}(I - B^{-1}C^{-1}A). \qquad (6.12)
\end{aligned}
$$

This is a consequence of (5.6) and the identity

$$
\begin{aligned}
&I - D^{-1}(I - BA^{-1})^{-1}(I - B)(I - DA^{-1}CB)^{-1}D(I - A^{-1}CBDA^{-1}) \\
&\quad = D^{-1}(I - BA^{-1})^{-1}(I - C^{-1}AD^{-1}) \\
&\qquad \times (I - B^{-1}C^{-1}AD^{-1})^{-1}(I - A^{-1})D, \qquad (6.13)
\end{aligned}
$$

where A, B and D mutually commute while C does not commute with any of the other parameters.

Since our attempts to extend (6.12) to a noncommutative extension of (6.10) for general n failed, we omit writing out the technical details of the proofs of (6.13) and (6.12). When trying to establish a Q-analogue of

Theorem 6.2, we had no problem with applying induction, when assuming that A and DCB commute with all other parameters. However, for the last step in the corresponding Q-analogue of Lemma 3.5 to work out, we would need that the three products $(I - B)(I - B^{-1}CB)(I - D)$, $(I - AB^{-1}Q)(I - AC^{-1}Q)(I - AD^{-1}Q^2)$ and $(I - AC^{-1}D^{-1}Q)(I - AD^{-1}B^{-1}Q)(I - AB^{-1}C^{-1}Q)$ mutually commute. This means that additional conditions on the parameters are needed. We were not able to specify any reasonable additional conditions on A, B, C and D satisfying all of the above but where A, B, C and D do not already mutually commute.

The question remains whether there exists any more suitable noncommutative extension of (6.11) than (6.12), for the purpose of deriving a full Q-analogue of Theorem 6.2.

7. Summations for Nonterminating Series

The noncommutative (basic) hypergeometric series in Sections 4–6 terminate due to the occurrence of the upper parameter $-nI$ (or Q^{-n}) where n is a nonnegative integer. Assuming that (the unit ring) R is a Banach algebra with norm $\| \cdot \|$, we may obtain summations for nonterminating noncommutative (basic) hypergeometric series by (possibly substituting some variables and) formally considering the term-wise limit as $n \to \infty$. However, the validity of this procedure which involves the interchange of limit and summation would need to be justified case by case. The main problem appears with convergence, especially for noncommutative "ordinary" hypergeometric series, for which precise conditions would need to be given (as the argument in the summations of interest is often I). Furthermore, in the limit a noncommutative generalized gamma function (which would reduce to a product of quotients of gamma functions in the case of commuting parameters) would be needed which we did not (yet) define. This problem does not appear in the basic Q-case where infinite Q-shifted factorials make perfectly sense assuming $\|Q\| < 1$. Wherever we consider infinite Q-shifted factorials we shall implicitly assume such Q.

The following two identities are obtained from Theorem 5.2 by taking the limit $n \to \infty$ on each side, termwise on the left-hand sides. These are conjectured noncommutative Q-Gauss summations, extending [9, Appendix (II.8)].

Conjecture 7.1. *Let A, B and C be noncommutative parameters of some Banach algebra, and suppose that Q commutes each with A, B and C.*

Further, assume that the product $BC^{-1}A$ commutes each with A, B and C. Moreover, assume that $\|A^{-1}CB^{-1}\| < 1$. Then we have the following summation for a noncommutative basic hypergeometric series of type I:

$$ {}_2\phi_1\left\lceil\begin{matrix}A,B\\C\end{matrix};Q,A^{-1}CB^{-1}\right\rfloor = \left\lceil\begin{matrix}CB^{-1},A^{-1}C\\C,A^{-1}CB^{-1}\end{matrix};Q,I\right\rfloor_\infty . \tag{7.1}$$

Further, we have the following summation for a noncommutative basic hypergeometric series of type II:

$$ {}_2\phi_1\left\lfloor\begin{matrix}A,B\\C\end{matrix};Q,A^{-1}CB^{-1}\right\rceil = \left\lfloor\begin{matrix}CB^{-1}Q,A^{-1}CQ\\C,A^{-1}CB^{-1}\end{matrix};Q,I\right\rceil_\infty \times (I-CB^{-1})(I-A^{-1}C). \tag{7.2}$$

Next we give two noncommutative extensions of the nonterminating q-binomial theorem (cf. [9, II.3]).

Theorem 7.2. *Let A and C be noncommutative parameters of some Banach algebra, and suppose that Q commutes both with A and C. Further, assume that $\|Z\| < 1$. Then we have the following summation for a noncommutative hypergeometric series of type I:*

$$ {}_1\phi_0\left\lceil\begin{matrix}A\\-\end{matrix};Q,Z\right\rfloor = \left\lfloor\begin{matrix}AZ\\Z\end{matrix};Q,I\right\rceil_\infty . \tag{7.3}$$

Further, we have the following summation for a noncommutative hypergeometric series of type II:

$$ {}_1\phi_0\left\lfloor\begin{matrix}A\\-\end{matrix};Q,Z\right\rceil = \left\lceil\begin{matrix}AZQ^{-1}\\Z\end{matrix};Q,I\right\rfloor_\infty (I-AZQ^{-1})^{-1}. \tag{7.4}$$

Note that with the notation defined in Section 8.2, (7.4) can be written more compactly as

$$ {}_1\phi_0\left\lfloor\begin{matrix}A\\-\end{matrix};Q,Z\right\rceil = {}^{\sim}\left\lfloor\begin{matrix}AZ\\Z\end{matrix};Q,I\right\rceil_\infty . \tag{7.5}$$

Identities (7.3) and (7.4) can be obtained from Theorem 4.2 by performing the substitution $C \mapsto Q^{1-n}Z^{-1}$, taking the limit $n \to \infty$ on each side of the respective identities, and multiplying each side by Z^{-1} from the left and by Z from the right. Since we did not justify taking termwise limits, we provide an independent proof.

Proof of Theorem 7.2. We prove (7.3), leaving the proof of (7.4) to the reader.

Let $f(A, Z)$ denote the series on the left-hand side of (7.3). We make use of the two simple identities

$$Z = AZQ^k + (I - AQ^k)Z, \tag{7.6a}$$

$$I = Q^k + (I - Q^k), \tag{7.6b}$$

to obtain two functional equations for f. First, (7.6a) gives

$$Zf(A, Z) = AZf(A, ZQ) + f(AQ, Z)(I - A)Z, \tag{7.7}$$

while (7.6b) gives

$$f(A, Z) = f(A, ZQ) + f(AQ, Z)(I - A)Z. \tag{7.8}$$

Combining (7.7) and (7.8), one immediately has

$$f(A, Z) = f(A, ZQ) + Zf(A, Z) - AZf(A, ZQ),$$

or equivalently

$$(I - Z)f(A, Z) = (I - AZ)f(A, ZQ),$$

thus

$$f(A, Z) = (I - Z)^{-1}(I - AZ)f(A, ZQ). \tag{7.9}$$

Iteration of (7.9) gives the result since $f(A, O) = I$ by definition of f. □

For generalizations of Theorem 7.2 to noncommutative extensions of Ramanujan's ${}_1\psi_1$ summation, see [19].

8. More Identities

8.1. Telescoping

By iterating any of the simple identities (5.2), (5.6), (6.3) or (6.12), indefinite summations involving general parameters can be derived. For example, using (5.6) one has

$$\begin{aligned}
&\sum_{k=0}^{n} \prod_{j=0}^{k-1} [(I - C_j)^{-1}(I - A_j)(I - B_j C_j^{-1} A_j)^{-1}(I - B_j)] \\
&\qquad \times (I - C_k)^{-1}(I - C_k B_k^{-1})(I - A_k^{-1} C_k B_k^{-1})^{-1}(I - A_k^{-1} C_k) \\
&\quad = \sum_{k=0}^{n} \left[\prod_{j=0}^{k-1} [(I - C_j)^{-1}(I - A_j)(I - B_j C_j^{-1} A_j)^{-1}(I - B_j)] \right.
\end{aligned}$$

$$- \prod_{j=0}^{k} [(I - C_j)^{-1}(I - A_j)(I - B_j C_j^{-1} A_j)^{-1}(I - B_j)] \Bigg]$$

$$= I - \prod_{j=0}^{n} [(I - C_j)^{-1}(I - A_j)(I - B_j C_j^{-1} A_j)^{-1}(I - B_j)], \qquad (8.1)$$

for arbitrary noncommutative parameters A_j, B_j, C_j, where $0 \leq j \leq n$.

One can now obtain a basic (or even multi-basic) identity in a natural way by setting $A_j = AQ^j$, etc. and similarly for B_j and C_j. We leave the details to the reader.

8.2. Reversing all products

For all identities given in this paper, one may obtain new ones by simply reversing all the products (of elements of the unit ring R) simultaneously on each side of the respective identities. This is clearly an involution.[a]

For instance, with the new definitions

$$\widetilde{\begin{bmatrix} A_1, A_2, \dots, A_r \\ C_1, C_2, \dots, C_r \end{bmatrix}; Z}_k := \prod_{j=1}^{k} \left(Z \prod_{i=1}^{r} (A_i + (j-1)I)(C_i + (j-1)I)^{-1} \right), \qquad (8.2)$$

and

$$\widetilde{{}_{r+1}F_r} \left[\begin{matrix} A_1, A_2, \dots, A_{r+1} \\ C_1, C_2, \dots, C_r \end{matrix}; Z \right] := \sum_{k \geq 0} \widetilde{\begin{bmatrix} A_1, A_2, \dots, A_{r+1} \\ C_1, C_2, \dots, C_r, I \end{bmatrix}; Z}_k \qquad (8.3)$$

for reversed (or "transposed") versions of generalized noncommutative shifted factorials and noncommutative hypergeometric series of type I

---

[a] For a more solid construction, assume R to be a Banach $*$-algebra, which is a Banach algebra R over a field K equipped with an involutive antiautomorphism $* : R \longrightarrow R$ satisfying the following properties for every $X, Y \in R$:

$$\begin{aligned} &(X^*)^* = X, \quad \text{i.e., the map } * \text{ is an involution,} \\ &(X+Y)^* = X^* + Y^*, \\ &(XY)^* = Y^* X^*, \\ &\|X^*\| = \|X\|, \end{aligned}$$

such that the restriction $* : K \longrightarrow K$ is an involutive automorphism (since K is commutative).

In particular, the ring of complex $n \times n$ square matrices is such a Banach $*$-algebra, with the map $*$ being the adjoint operator, X^* being the conjugate transpose of X.

"Reversing all products" can thus be formalized by starring the respective expressions.

(compare with (2.5) and (2.9); similar definitions can be made for type II and in the basic cases), and writing

$$\widetilde{\left[\begin{matrix} A_1, A_2, \ldots, A_r \\ C_1, C_2, \ldots, C_r \end{matrix}; I\right]}_k = \widetilde{\left[\begin{matrix} A_1, A_2, \ldots, A_r \\ C_1, C_2, \ldots, C_r \end{matrix}\right]}_k \tag{8.4}$$

for brevity, we have the following noncommutative Chu–Vandermonde summation.

Theorem 8.1. *Let A, C be noncommutative parameters of some unit ring. Then*

$${}_2\widetilde{F}_1\left[\begin{matrix} A, -nI \\ C \end{matrix}; I\right] = \widetilde{\left[\begin{matrix} C-A \\ C \end{matrix}\right]}_n . \tag{8.5}$$

This is a simple consequence of Theorem 4.1, obtained by reversing all products. Similarly, all the other identities involving noncommutative parameters appearing in this chapter have reversed versions. We do not write them out explicitly.

Acknowledgments

The author was supported by Austrian Science Fund FWF, grants P17563-N13 and F50-08, and partly by EC's IHRP Programme, grant HPRN-CT-2001-00272 "Algebraic Combinatorics in Europe". The author would like to thank George Gasper and Hjalmar Rosengren for their comments. We are especially indebted to the latter for providing an explicit matrix solution to a particular set of algebraic equations (see Remark 6.1). The main material in this paper stems from an earlier preprint which was put online (`arXiv:math/0411136`) more than a dozen of years ago but was never regularly published. The present version is a slightly updated version of that preprint and has been dedicated to Mourad Ismail for his continuous support and interest.

References

[1] N. Aldenhoven, E. Koelink and A. M. de los Ríos, Matrix-valued little q-Jacobi polynomials, *J. Approx. Theory* **193** (2015) 164–183.

[2] G. E. Andrews, R. Askey and R. Roy, *Special Functions*, Encyclopedia of Mathematics and Its Applications, Vol. 71 (Cambridge University Press, Cambridge, 1999).

[3] W. N. Bailey, *Generalized Hypergeometric Series* (Cambridge University Press, Cambridge, 1935).

[4] A. Conflitti and M. J. Schlosser, Noncommutative hypergeometric and basic hypergeometric equations, *J. Nonlinear Math. Phys.* **17**(4) (2010) 429–443.
[5] A. J. Durán and F. A. Grünbaum, Orthogonal matrix polynomials satisfying second-order differential equations, *Int. Math. Res. Notices* **10** (2004) 461–484.
[6] A. J. Durán and P. López-Rodríguez, Orthogonal matrix polynomials, in *Loredo Lectures on Orthogonal Polynomials and Special Functions*, Advances in the Theory of Special Functions and Applications (Nova Science Publication, New York, 2004), pp. 13–44.
[7] A. J. Durán and W. Van Assche, Orthogonal matrix polynomials and higher order recurrence relations, *Lin. Algebra Appl.* **219** (1995) 261–280.
[8] C. Duval and V. Ovsienko, Projectively equivariant quantization and symbol calculus: noncommutative hypergeometric functions, *Lett. Math. Phys.* **57**(1) (2001) 61–67.
[9] G. Gasper and M. Rahman, *Basic Hypergeometric Series*, 2nd edn., Encyclopedia of Mathematics and Its Applications, Vol. 96 (Cambridge University Press, Cambridge, 2004).
[10] F. A. Grünbaum, Matrix valued Jacobi polynomials, *Bull. Sci. Math.* **127**(3) (2003) 207–214.
[11] F. A. Grünbaum and M. D. de la Iglesia, Matrix-valued orthogonal polynomials related to $\mathrm{SU}(n+1)$, their algebra of differential operators, and the corresponding curves, *Experiment. Math.* **16**(2) (2007) 189–207.
[12] F. A. Grünbaum and P. Iliev, A noncommutative version of the bispectral problem, *J. Comput. Appl. Math.* **161**(1) (2003) 99–118.
[13] F. A. Grünbaum, I. Pacharoni and J. A. Tirao, A matrix valued solution to Bochner's problem, *J. Phys. A* **34**(48) (2001) 10647–10656.
[14] F. A. Grünbaum, I. Pacharoni and J. A. Tirao, Matrix valued spherical functions associated to the complex projective plane, *J. Funct. Anal.* **188**(2) (2002) 350–441.
[15] F. A. Grünbaum, I. Pacharoni and J. A. Tirao, Matrix valued spherical functions associated to the three dimensional hyperbolic space, *Internat. J. Math.* **13**(7) (2002) 727–784.
[16] T. H. Koornwinder, Special functions and q-commuting variables, in *Special Functions, q-Series, and Related Topics*, eds. M. E. H. Ismail, D. R. Masson and M. Rahman, Fields Institute Communications, Vol. 14 (American Mathematical Society, 1997), pp. 131–166.
[17] M. G. Krein, Fundamental aspects of the representation theory of Hermitian operators with deficiency index (m, m), *Ukrain. Mat. Ž.* **1** (1949) 3–66; *Amer. Math. Soc. Transl.* **97**(2) (1970) 75–143.
[18] P. Román and S. Simondi, The generalized matrix valued hypergeometric equation, *Internat. J. Math.* **21**(2) (2010) 145–155.
[19] M. J. Schlosser, Noncommutative extensions of Ramanujan's ${}_1\psi_1$ summation, *Electron. Trans. Numer. Anal.* **24** (2006) 94–102.
[20] A. Sinap and W. Van Assche, Orthogonal matrix polynomials and applications, *J. Comput. Appl. Math.* **66**(1–2) (1996) 27–52.

[21] L. J. Slater, *Generalized Hypergeometric Functions* (Cambridge University Press, Cambridge, 1966).

[22] J. A. Tirao, The matrix-valued hypergeometric equation, *Proc. Natl. Acad. Sci. USA* **100**(14) (2003) 8138–8141.

[23] A. Y. Volkov, Noncommutative hypergeometry, *Comm. Math. Phys.* **258**(2) (2005) 257–273.

Chapter 25

Asymptotics of Generalized Hypergeometric Functions

Y. Lin* **and R. Wong**†

Department of Mathematics, South China University of Technology
Guangzhou, China
†*Department of Mathematics, City University of Hong Kong*
Kowloon, Hong Kong

The asymptotic behavior of the generalized hypergeometric function ${}_pF_q(z)$ depends critically on the values of the two nonnegative integers p and q. There are four cases to be considered; namely, (i) $p = q + 1$, (ii) $p = q$, (iii) $p = q - 1$ and (iv) $p \leq q - 2$. The purpose of this chapter is to provide a self-contained and easy-to-follow presentation of rigorous derivations of the asymptotic expansions of this function in all four cases.

Keywords: Asymptotic expansion; generalized hypergeometric function.

Mathematics Subject Classification 2010: 41A60, 33C20

1. Introduction

With the increasing amount of activities in random matrix theory, the once dormant Meijer G-function again appears frequently in recent research articles; see, e.g., [3, 4, 7, 8]. Motivated by this observation, Beals and Szmigielski wrote an expository paper on this function, which appeared in the popular *Notices of Amer. Math. Soc.* [1]. In their paper, they expressed that this is a remarkable family of functions of one variable. Following [1], the present authors also wrote an expository paper with a focus on the asymptotics of these functions for large values of the variable. But, in one of the cases considered in [9], we only show that the Meijer G-function can be expressed as a linear combination of the generalized hypergeometric function ${}_pF_q$, and indicate that the asymptotic expansions of the latter functions can be found in the NIST Handbook [10]. It is only when we started to look for the proofs of the results listed in [10, pp. 411 and 412], we realized that the derivations of these asymptotic expansions are very complicated. The proofs of these results in the existing literature are either too long or, in some sense, not quite complete. For example, the article by Braaksma [5] is over one hundred pages, the paper of Wright [14] gives only the results but no proof, and the arguments in the book of Paris and Wood [12] seem to have places where more details are needed (see, e.g., the paragraph running from p. 49 into p. 50). For this reason, we intend to write an article which is self-contained, completely rigorous and reasonably easy to read; this chapter is the outcome of this effort.

Throughout this chapter, we will adopt the standard notations used in the NIST Handbook [10]. Let $a_1, a_2, \ldots, a_p$ and $b_1, b_2, \ldots, b_q$ be real or complex parameters, and p and q be nonnegative integers. Assume also that none of parameters b_k is a nonpositive integer when $1 \leq k \leq q$. The generalized hypergeometry function is defined by

$$ {}_pF_q(z) = {}_pF_q\left(\begin{matrix} a_1, \ldots, a_p \\ b_1, \ldots, b_q \end{matrix}; z\right) = \sum_{n=0}^{\infty} \frac{(a_1)_n \cdots (a_p)_n}{(b_1)_n \cdots (b_q)_n} \frac{z^n}{n!}, \tag{1.1} $$

where the Pochhammer symbol $(a)_n$ is used to represent the shifted factorial

$$ (a)_n = a(a+1)\cdots(a+n-1) = \frac{\Gamma(a+n)}{\Gamma(a)}. \tag{1.2} $$

Note that when one or more of parameters a_k is a nonpositive integer the series (1.1) terminates, and the generalized hypergeometric function is a polynomial in z. The ratio test shows that the radius of convergence of the series (1.1) is zero if $p > q+1$, while the radius of convergence is 1 if $p = q+1$ and infinite if $p \leq q$. Therefore, we assume that $p \leq q+1$, and none of the a_k is a nonpositive integer throughout the following discussion. For further properties and applications of the generalized hypergeometric function, we refer to [2].

When $z \neq 0$, the integral representation of ${}_pF_q(z)$ is given by

$$\begin{aligned} H(z) &:= \left(\prod_{k=1}^{p}\Gamma(a_k)\Big/\prod_{k=1}^{q}\Gamma(b_k)\right){}_pF_q(z) \\ &= \frac{1}{2\pi i}\int_C\left(\prod_{k=1}^{p}\Gamma(a_k+s)\Big/\prod_{k=1}^{q}\Gamma(b_k+s)\right)\Gamma(-s)(-z)^s ds \\ &= \frac{1}{2\pi i}\int_C h(s)(-z)^s ds, \end{aligned} \tag{1.3}$$

where the contour of integration separates the poles of the factors $\Gamma(a_k+s)$, $1 \le k \le p$, from those of $\Gamma(-s)$. There are two possible choices for C:

(i) C goes from $-i\infty$ to $i\infty$. The integral converges if $q < p+1$ and $|\arg(-z)| < (p+1-q)\pi/2$.
(ii) C is a loop that starts at infinity on a line parallel to the positive real axis, encircles all the poles of $\Gamma(-s)$ once in the negative sense, and returns to infinity on another line parallel to the positive real axis. The integral converges for all $z \neq 0$ if $p < q+1$, and for $0 < |z| < 1$ if $p = q+1$.

Convergence of the integral in (1.3) can be established by using Stirling's formula and the reflection formula for the gamma function. The details can be found in [11].

2. Vertical Line of Integration

Assume $a_j - a_k \neq 0, \pm 1, \pm 2, \ldots$, for $j, k = 1, \ldots, p, j \neq k$. Furthermore, without loss of generality, we suppose that $\operatorname{Re} a_p \le \cdots \le \operatorname{Re} a_2 \le \operatorname{Re} a_1$. Let w be a positive number such that $-w < -\operatorname{Re} a_1 - \delta$, where $\delta > 0$ is a small number. Furthermore, let K_j be the largest nonnegative integer strictly less than $w - \operatorname{Re} a_j$ for $j = 1, \ldots, n$. Then, the contour C in (1.3) can be moved to the left so that we have by Cauchy's theorem

$$H(z) = \sum_{j=1}^{p}\sum_{n=0}^{K_j} \operatorname{Res}\{h(s)(-z)^s : s = -a_j - n\} + R_w(z), \tag{2.1}$$

where the remainder $R_w(z)$ is given by

$$R_w(z) = \frac{1}{2\pi i}\int_{-w-i\infty}^{-w+i\infty} h(s)(-z)^s ds. \tag{2.2}$$

Since $\operatorname{Res}_{s=-n}\Gamma(s) = (-1)^n/n!$, we have

$$\begin{aligned}&\operatorname{Res}\{h(s)(-z)^s : s = -a_j - n\}\\&= \frac{\Gamma(a_j+n)\prod_{k=1,k\neq j}^{p}\Gamma\left(a_k - a_j - n\right)}{\prod_{k=1}^{q}\Gamma(b_k - a_j - n)}\frac{(-1)^n}{n!}(-z)^{-a_j-n}.\end{aligned} \tag{2.3}$$

To estimate the remainder $R_w(z)$ in (2.2), we let $s = -w + it$. With $\theta = \arg(-z)$, we have

$$|R_w(z)| \le \frac{1}{2\pi}|z|^{-w}\int_{-\infty}^{\infty}|h(-w+it)|e^{-\theta t}dt. \tag{2.4}$$

Using Stirling's formula, it can be shown that for large t, there is a positive constant C such that

$$|h(-w+it)| \le C|t|^{\beta_1}e^{-\beta_2|t|}, \tag{2.5}$$

where $\beta_2 = (p+1-q)\pi/2$ and

$$\beta_1 = \operatorname{Re}\left(\sum_{k=1}^{p}a_k - \sum_{k=1}^{q}b_k - 1\right) + (q+1-p)\left(w+\frac{1}{2}\right).$$

The integral in (2.4) is independent of $|z|$, and converges when $|\arg(-z)| < (p+1-q)\pi/2$. Hence,

$$R_w(z) = \mathcal{O}(|z|^{-w}). \tag{2.6}$$

Coupling this with (2.1) and (2.3) gives

$$\begin{aligned}H(z) =&\sum_{j=1}^{p}\sum_{n=0}^{K_j}\frac{\Gamma(a_j+n)\prod_{k=1,k\neq j}^{p}\Gamma\left(a_k - a_j - n\right)}{\prod_{k=1}^{q}\Gamma(b_k - a_j - n)}\\&\times\frac{(-1)^n}{n!}(-z)^{-a_j-n} + \mathcal{O}(z^{-w})\\&= Q_w(z) + \mathcal{O}(z^{-w})\end{aligned} \tag{2.7}$$

for z in the sector $|\arg(-z)| < (p+1-q)\pi/2$, where

$$\begin{aligned}Q_w(z) := \sum &\text{ residues of } h(s)(-z)^s \text{ at the points}\\&s = -a_j - n, \text{ with } \operatorname{Re} s > -w.\end{aligned} \tag{2.8}$$

3. Inverse Factorial Expansions

Before proceeding to the discussion of the case when the contour C in (1.3) is a loop (i.e., case (ii) mentioned in Section 1), we need some results concerning inverse factorial expansions. The following statement is given in [6, 13] with no proof.

Lemma 3.1. *Let $\mu > 0$, and a and b denote arbitrary complex numbers. Then, we have*

$$\frac{1}{\mu(s+a)} = \sum_{n=0}^{\infty} \frac{(b-\mu a)_n}{(\mu s+b)_{n+1}}, \tag{3.1}$$

which uniformly converges for $|\mu(s+a)+1| > 1$, or simply $\mathrm{Re}(s+a) > 0$.

Proof.

First, we assume that

$$\frac{1}{\mu(s+a)} = \sum_{k=0}^{\infty} \frac{a_k}{(\mu s+b)_{k+1}} \tag{3.2}$$

is convergent for some a_k. By letting $s \to \infty$, it is easily seen that $a_0 = 1$.

To apply induction on k, we assume that $a_k = (b-\mu a)_k$, $k = 0, 1, \ldots, n$. We shall prove $a_{n+1} = (b-\mu a)_{n+1}$. Note that

$$\begin{aligned}\frac{1}{\mu(s+a)} &= \frac{1}{\mu s+b} + \cdots + \frac{(b-\mu a)_n}{(\mu s+b)_{n+1}} + \frac{a_{n+1}}{(\mu s+b)_{n+2}} \\ &\quad + \sum_{k=n+2}^{\infty} \frac{a_k}{(\mu s+b)_{k+1}}.\end{aligned} \tag{3.3}$$

Multiplying by $(\mu s+b)_{n+2}$ on both sides of (3.3), we obtain

$$\begin{aligned}\frac{(\mu s+b)_{n+2}}{\mu(s+a)} &= (\mu s+b+1)_{n+1} + (b-\mu a)(\mu s+b+2)_n \\ &\quad + \cdots + (b-\mu a)_n(\mu s+b+n+1) + a_{n+1} \\ &\quad + \sum_{j=1}^{\infty} \frac{a_{n+1+j}}{(\mu s+b+n+2)_j},\end{aligned} \tag{3.4}$$

from which it follows that

$$\begin{aligned}&\frac{1}{\mu(s+a)}\Big[(\mu s+b)_{n+2} - \mu(s+a)(\mu s+b+1)_{n+1} \\ &\quad -\mu(s+a)(b-\mu a)(\mu s+b+2)_n \\ &\quad - \cdots - \mu(s+a)(b-\mu a)_n(\mu s+b+n+1)\Big] \\ &= a_{n+1} + \sum_{j=1}^{\infty} \frac{a_{n+1+j}}{(\mu s+b+n+2)_j}.\end{aligned} \tag{3.5}$$

Note that

$$(b-\mu a)_j \cdot (\mu s+b+j)_{k+1} - (b-\mu a)_j \cdot \mu(s+a)(\mu s+b+j+1)_k$$
$$= (b-\mu a)_{j+1} \cdot (\mu s+b+j+1)_k. \tag{3.6}$$

Coupling this with (3.5) gives

$$\frac{(b-\mu a)_{n+1}(\mu s+b+n+1)}{\mu(s+a)} = a_{n+1} + \sum_{j=1}^{\infty} \frac{a_{n+1+j}}{(\mu s+b+n+2)_j}. \tag{3.7}$$

Taking $s \to +\infty$ yields

$$a_{n+1} = (b-\mu a)_{n+1}. \tag{3.8}$$

By Gauss's test,

$$\frac{1}{\mu(s+a)} = \sum_{n=0}^{\infty} \frac{(b-\mu a)_n}{(\mu s+b)_{n+1}} \tag{3.9}$$

converges uniformly for $|\mu s+\mu a+1| > 1$. □

Define

$$h_0(s) := \prod_{k=1}^{p} \Gamma(a_k+s) \Bigg/ \left(\Gamma(s+1) \prod_{k=1}^{q} \Gamma(b_k+s) \right). \tag{3.10}$$

Note that this function is related to the function $h(s)$ in (1.3). We next use Riney's method [13] to establish the following expansion; see also [11, pp. 39 and 41].

Lemma 3.2. *Let $\kappa := q+1-p > 0$ and define*

$$\alpha = \sum_{k=1}^{q} b_k - \sum_{k=1}^{p} a_k + \frac{1}{2}(p-q+2). \tag{3.11}$$

Then, there exist A_0, $c_1, c_2, \ldots$, independent of s, such that

$$h_0(s)\kappa^{-\kappa s} = A_0 \left(\sum_{j=0}^{N-1} \frac{c_j}{\Gamma(\kappa s+\alpha+j)} + \frac{r_N(s)}{\Gamma(\kappa s+\alpha+N)} \right), \tag{3.12}$$

where the remainder function $r_N(s)$ is analytic in s except at the point $s = -a_j - n$, $n = 0, 1, 2, \ldots$ $(1 \le j \le p)$, and

$$r_N(s) = \mathcal{O}(1) \tag{3.13}$$

as $s \to \infty$ uniformly for $|\arg s| < \pi - \varepsilon$. Here

$$A_0 = (2\pi)^{\frac{1}{2}(p-q)} \kappa^{\alpha - \frac{1}{2}}, \tag{3.14}$$

and c_j satisfy the recursion formula

$$c_j = -\frac{1}{j\kappa^\kappa}\sum_{k=0}^{j-1} c_k e(j,k) \tag{3.15}$$

with $c_0 = 1$,

$$e(j,k) = \sum_{\ell=1}^{q+1} \frac{\prod_{r=1}^{p}(a_r - b_\ell)}{\prod_{r=1,r\neq\ell}^{q+1}(b_r - b_\ell)} \frac{\Gamma(\alpha - \kappa b_\ell + \kappa + j)}{\Gamma(\alpha - \kappa b_\ell + k)}, \tag{3.16}$$

and $b_{q+1} = 1$.

Proof. By the asymptotic expansion of the gamma function, we have

$$h_0(s)\Gamma(\kappa s + \alpha) = A_0\kappa^{\kappa s}\left(1 + \sum_{j=1}^{N-1} A_j s^{-j} + \mathcal{O}(s^{-N})\right) \tag{3.17}$$

as $s \to \infty$ for $|\arg s| \leq \pi - \varepsilon$. Note that as $s \to \infty$

$$\sum_{j=1}^{N-1} A_j s^{-j} = \frac{c_1}{\kappa s + \alpha} + \frac{c_2}{(\kappa s + \alpha)_2} + \cdots + \frac{c_{N-1}}{(\kappa s + \alpha)_{N-1}} + \mathcal{O}(s^{-N}), \tag{3.18}$$

where c_j will be determined later. From (3.17) and (3.18), it follows that

$$\begin{aligned} h_0(s)\kappa^{-\kappa s} &= \frac{A_0}{\Gamma(\kappa s + \alpha)}\left(1 + \sum_{j=1}^{N-1}\frac{c_j}{(\kappa s + \alpha)_j} + \mathcal{O}(s^{-N})\right) \\ &= A_0\left(\sum_{j=0}^{N-1}\frac{c_j}{\Gamma(\kappa s + \alpha + j)} + \frac{\mathcal{O}(1)}{\Gamma(\kappa s + \alpha + N)}\right) \end{aligned} \tag{3.19}$$

with $c_0 = 1$.

Now, we wish to determine c_j for $j \geq 1$. First, we note that

$$h_0(s+1) = \frac{\prod_{k=1}^{p}(s + a_k)}{\prod_{k=1}^{q+1}(s + b_k)} h_0(s) = \sum_{\ell=1}^{q+1}\frac{D_\ell}{s + b_\ell} h_0(s), \tag{3.20}$$

where $b_{q+1} = 1$ and

$$D_\ell = \frac{\prod_{k=1}^{p}(a_k - b_\ell)}{\prod_{k=1,k\neq\ell}^{q+1}(b_k - b_\ell)}. \tag{3.21}$$

It follows from (3.19) that

$$h_0(s+1) = A_0\kappa^{\kappa s}\left(\sum_{\ell=1}^{q+1}\frac{D_\ell}{s+b_\ell}\right)\left\{\sum_{k=0}^{N-1}\frac{c_k}{\Gamma(\kappa s+\alpha+k)}+\frac{\mathcal{O}(1)}{\Gamma(\kappa s+\alpha+N)}\right\}$$
$$= A_0\kappa^{\kappa s}\left\{\sum_{\ell=1}^{q+1}\sum_{k=0}^{N-1}\frac{c_k D_\ell}{(s+b_\ell)\Gamma(\kappa s+\alpha+k)}+\frac{\mathcal{O}(1)}{\Gamma(\kappa s+\alpha+\kappa+N)}\right\}. \tag{3.22}$$

Here, we have used the fact that $\sum_{\ell=1}^{q+1} D_\ell/(s+b_\ell) \sim s^{-\kappa}$ on account of (3.20) and $s^\kappa\Gamma(\kappa s+\alpha+N) = \Gamma(\kappa s+\alpha+\kappa+N)\mathcal{O}(1)$ as $s \to \infty$.

By using Lemma 3.1 with $a = b_\ell$ and $b = \alpha + k$, we have

$$\sum_{\ell=1}^{q+1}\sum_{k=0}^{N-1}\frac{c_k D_\ell}{(s+b_\ell)\Gamma(\kappa s+\alpha+k)} = \sum_{k=0}^{N-1} c_k \sum_{\ell=1}^{q+1}\frac{D_\ell}{(s+b_\ell)\Gamma(\kappa s+\alpha+k)}$$
$$= \sum_{k=0}^{N-1} c_k \sum_{\ell=1}^{q+1}\left(D_\ell\cdot\kappa\sum_{n=0}^{\infty}\frac{(\alpha+k-\kappa b_\ell)_n}{\Gamma(\kappa s+\alpha+k+n+1)}\right)$$
$$= \kappa\sum_{k=0}^{N-1} c_k\left[\sum_{n=0}^{\infty}\frac{1}{\Gamma(\kappa s+\alpha+k+n+1)}\cdot\left(\sum_{\ell=1}^{q+1} D_\ell(\alpha+k-\kappa b_\ell)_n\right)\right]. \tag{3.23}$$

Clearly,

$$\sum_{\ell=1}^{q+1} D_\ell(\alpha+k-\kappa b_\ell)_n = \sum_{\ell=1}^{q+1} D_\ell \sum_{j=0}^{n} b_{n,j}(-b_\ell)^j$$
$$= \sum_{j=0}^{n} b_{n,j}\left(\sum_{\ell=1}^{q+1} D_\ell\cdot(-b_\ell)^j\right) \tag{3.24}$$

with $b_{n,n} = \kappa^n$ and $b_{n,n-1} = \kappa^{n-1}\{n(\alpha+k)+n(n-1)/2\}$. Since $\prod_{k=1}^{p}(s+a_k)/\prod_{k=1}^{q+1}(s+b_k) = \sum_{\ell=1}^{q+1} D_\ell/(s+b_\ell)$ from (3.20), comparing the coefficients of s^{-1} gives

$$\sum_{\ell=1}^{q+1} D_\ell\cdot(-b_\ell)^j = \begin{cases} 0, & j \le \kappa-2, \\ 1, & j = \kappa-1, \\ \frac{1}{2}(1-\kappa)-\alpha, & j = \kappa. \end{cases} \tag{3.25}$$

Here, we have made use of the fact that $b_{q+1} = 1$.

Let

$$e(j,k) := \sum_{\ell=1}^{q+1} D_\ell \frac{\Gamma(\alpha - \kappa b_\ell + j + \kappa)}{\Gamma(\alpha + k - \kappa b_\ell)}; \tag{3.26}$$

then,

$$\sum_{\ell=1}^{q+1} D_\ell (\alpha + k - \kappa b_\ell)_n = e(n + k - \kappa, k). \tag{3.27}$$

From (3.24) and (3.25), it follows that

$$\sum_{\ell=1}^{q+1} D_\ell (\alpha + k - \kappa b_\ell)_n = 0, \quad n \le \kappa - 2, \tag{3.28}$$

and

$$e(j,k) = \begin{cases} 0, & j \le k-2, \\ \kappa^{\kappa-1}, & j = k-1, \\ k \cdot \kappa^\kappa, & j = k. \end{cases} \tag{3.29}$$

From (3.23), (3.27) and (3.29), it follows that

$$\begin{aligned} \sum_{\ell=1}^{q+1}\sum_{k=0}^{N-1} \frac{c_k D_\ell}{(s+b_\ell)\Gamma(\kappa s + \alpha + k)} &= \kappa \sum_{k=0}^{N-1} c_k \sum_{n=\kappa-1}^{\infty} \frac{e(n+k-\kappa,k)}{\Gamma(\kappa s + \alpha + k + n + 1)} \\ &= \kappa \sum_{k=0}^{N-1} c_k \sum_{j=k}^{\infty} \frac{e(j-1,k)}{\Gamma(\kappa s + \alpha + \kappa + j)} \\ &= \kappa \sum_{j=0}^{N-1} \frac{1}{\Gamma(\kappa s + \alpha + \kappa + j)} \sum_{k=0}^{j} c_k e(j-1,k) \\ &\quad + \kappa \sum_{j=N}^{\infty} \frac{1}{\Gamma(\kappa s + \alpha + \kappa + j)} \\ &\quad \times \sum_{k=0}^{N-1} c_k e(j-1,k); \end{aligned} \tag{3.30}$$

thus,

$$h_0(s+1) = A_0 \kappa^{\kappa s+1} \left\{ \sum_{j=0}^{N-1} \frac{1}{\Gamma(\kappa s + \alpha + \kappa + j)} \sum_{k=0}^{j} c_k e(j-1,k) + \frac{\mathcal{O}(1)}{\Gamma(\kappa s + \alpha + \kappa + N)} \right\}. \tag{3.31}$$

On the other hand, we have from (3.19)

$$h_0(s+1) = A_0 \kappa^{\kappa s+\kappa} \left\{ \sum_{j=0}^{N-1} \frac{c_j}{\Gamma(\kappa s+\alpha+\kappa+j)} + \frac{\mathcal{O}(1)}{\Gamma(\kappa s+\alpha+\kappa+N)} \right\}. \tag{3.32}$$

Comparing (3.31) and (3.32) gives

$$\kappa^{\kappa-1} c_j = \sum_{k=0}^{j} c_k e(j-1,k) = \sum_{k=0}^{j-1} c_k e(j-1,k) + \kappa^{\kappa-1} c_j; \tag{3.33}$$

thus, $\sum_{k=0}^{j} c_k e(j,k) = 0$. By (3.29) with $j=k$, we have

$$c_j = -\frac{1}{j\kappa^{\kappa}} \sum_{k=0}^{j-1} c_k e(j,k), \tag{3.34}$$

which is exactly (3.15).

Coupling (3.21) and (3.26), we obtain (3.16). □

4. Loop Contour

Next, we wish to study the asymptotic behavior of

$$H(z) := \frac{1}{2\pi i} \int_C h(s)(-z)^s ds, \tag{4.1}$$

where $h(s)$ is defined in (1.3) and C is chosen as the contour shown in case (ii) in Section 1. Here we follow the method used by Braaksma [5].

Recall

$$\kappa := q+1-p, \tag{4.2}$$

and let w, l be positive numbers so that

$$\begin{aligned} -w \neq \operatorname{Re}(a_k - 1 - n), &\quad k = 1,\dots,p;\ n = 0,1,2,\dots, \\ l > \max\{|\operatorname{Im} a_k|, |\operatorname{Im} b_j| : &\quad k = 1,\dots,\kappa;\ j = 1,\dots q\}. \end{aligned} \tag{4.3}$$

Recall $h_0(s)$ given in (3.10), and define

$$h_1(s) := -\frac{\pi}{\sin \pi s}; \tag{4.4}$$

hence, $h(s) = h_0(s)h_1(s)$. Now we need some approximations for the function $h_1(s)$.

Lemma 4.1. *For* $\operatorname{Im} s > 0$, *we have*

$$h_1(s) = 2\pi i \sum_{k=0}^{\infty} e^{(2k+1)\pi i s}; \tag{4.5}$$

while for $\operatorname{Im} s < 0$, *we have*

$$h_1(s) = -2\pi i \sum_{k=0}^{\infty} e^{-(2k+1)\pi i s}. \tag{4.6}$$

Furthermore, for $\operatorname{Im} s \geq \varepsilon > 0$, *there exists a positive number* K *so that*

$$\left| \left(h_1(s) - 2\pi i \sum_{k=0}^{j-1} e^{(2k+1)\pi i s} \right) \cdot e^{-(2j+1)\pi i s} \right| \leq K \tag{4.7}$$

while for $\operatorname{Im} s \leq -\varepsilon, \varepsilon > 0$,

$$\left| \left(h_1(s) + 2\pi i \sum_{k=0}^{j-1} e^{-(2k+1)\pi i s} \right) \cdot e^{(2j+1)\pi i s} \right| \leq K. \tag{4.8}$$

Proof. It is easily seen that

$$\frac{-\pi}{\sin(\pi s)} = \frac{-2\pi i}{e^{i\pi s} - e^{-i\pi s}} = \frac{2\pi i}{1 - e^{2i\pi s}} e^{i\pi s} = 2\pi i e^{i\pi s} \sum_{k=0}^{\infty} e^{2ki\pi s}$$

for $\operatorname{Im} s > 0$. Moreover, we have

$$\begin{aligned} &\left| \left(h_1(s) - 2\pi i \sum_{k=0}^{j-1} e^{(2k+1)\pi i s} \right) \cdot e^{-(2j+1)\pi i s} \right| \\ &= \left| 2\pi i \sum_{k=j}^{\infty} e^{(2k+1)\pi i s} \cdot e^{-(2j+1)\pi i s} \right| \\ &= \left| 2\pi i \sum_{k=j}^{\infty} e^{2(k-j)\pi i s} \right| \leq K \end{aligned} \tag{4.9}$$

for $\operatorname{Im} s \geq \varepsilon$.

Similarly, we obtain (4.6) and (4.8). □

The integral in (1.3) can be rewritten as

$$H(-z) = \left(\prod_{k=1}^{p} \Gamma(a_k) \Big/ \prod_{k=1}^{q} \Gamma(b_k) \right) {}_pF_q(-z) = \frac{1}{2\pi i} \int_C h(s) z^s ds \tag{4.10}$$

with $z \neq 0$. As in case (i) discussed in Section 2, the loop contour C can be moved to the left so that we have by Cauchy's theorem

$$H(-z) = Q_w(-z) + \frac{1}{2\pi i}\int_L h_0(s)h_1(s)z^s ds - \frac{1}{2\pi i}\int_{L_1} h_0(s)h_1(s)z^s ds, \tag{4.11}$$

where $Q_w(z)$ is as given in (2.8), L is a contour which runs from $s = -w$ to $s = -w + il$ and then to $s = \infty + il$, and L_1 runs from $s = -w$ to $s = -w - il$ and then to $s = \infty - il$. Next, we need some estimates for the integrals.

Lemma 4.2. *Let L, L_1, w, l be defined as above, and let*

$$f(s) = \mathcal{O}(1)/\Gamma(\mu s + c) \tag{4.12}$$

as $s \to \infty$. Here, μ is a positive constant and c is a complex number. The integral

$$\int_L f(s)z^s ds \tag{4.13}$$

converges for all $z \neq 0$, and the contour L can be deformed into the vertical straight line Re $s = -w$ *from $-w$ to $-w + i\infty$ if*

$$\arg z > \tfrac{1}{2}\mu\pi. \tag{4.14}$$

Moreover, there exists a positive constant ε such that

$$\int_L f(s)z^s ds = \mathcal{O}(z^{-w}) \tag{4.15}$$

if $\arg z \geq \frac{1}{2}\mu\pi + \varepsilon$.

Similarly, the integral

$$\int_{L_1} f(s)z^s ds \tag{4.16}$$

converges for all $z \neq 0$, and the contour L_1 can be deformed into the vertical straight line Re $s = -w$ *from $-w$ to $-w - i\infty$ if*

$$\arg z < -\tfrac{1}{2}\mu\pi. \tag{4.17}$$

Moreover, there exists a positive constant ε such that

$$\int_{L_1} f(s)z^s ds = \mathcal{O}(z^{-w}) \tag{4.18}$$

for $\arg z \leq -\frac{1}{2}\mu\pi - \varepsilon$.

Proof. Let

$$s = w' + re^{\theta i}, \tag{4.19}$$

where $w' = -w + il$, $r = |s - w'|$ and $\theta = \arg(s - w')$. By using Stirling's formula, we have

$$\begin{aligned} |f(s)z^s| &= \mathcal{O}(1)|z|^{-w}(\mu r)^{\mathrm{Re}\, c'} \\ &\quad \times \exp\{-\mu\cos\theta r\ln(\mu r) + [\mu\cos\theta + \cos\theta\ln|z| \\ &\quad + (\mu\theta - \arg z)\sin\theta]r\} \end{aligned} \tag{4.20}$$

as $s \to \infty$, where $c' = 1/2 - c - \mu w'$. If $\arg z > \mu\pi/2$, it is readily seen that $f(s)z^s \to 0$ uniformly for $0 \le \theta \le \pi/2$. Hence, the contour L can be deformed into the vertical line Re $s = w$ from $s = -w$ to $s = w + i\infty$. By the same argument as in case (i) discussed in Section 2, we have

$$\int_L f(s)z^s ds = \int_{-w}^{-w+i\infty} f(s)z^s ds = \mathcal{O}(z^{-w}) \tag{4.21}$$

for $\arg z \ge \frac{1}{2}\mu\pi + \varepsilon$. Similarly, we can prove (4.16) and (4.18). □

Then, we define

$$\sigma = \lfloor \tfrac{1}{4}\kappa + 1 \rfloor. \tag{4.22}$$

It follows from (4.11) that

$$\begin{aligned} H(-z) &= Q_w(-z) + \sum_{k=0}^{\sigma-1}\int_L h_0(s)e^{(2k+1)\pi i s}z^s ds \\ &\quad + \frac{1}{2\pi i}\int_L h_0(s)\left[h_1(s) - 2\pi i\sum_{k=0}^{\sigma-1} e^{(2k+1)\pi i s}\right] z^s ds \\ &\quad - \frac{1}{2\pi i}\int_{L_1} h_0(s)\left[h_1(s) + 2\pi i\sum_{k=0}^{\sigma-1} e^{-(2k+1)\pi i s}\right] z^s ds \\ &\quad + \sum_{k=0}^{\sigma-1}\int_{L_1} h_0(s)e^{-(2k+1)\pi i s}z^s ds \end{aligned} \tag{4.23}$$

for $z \ne 0$. From (4.22), it is readily seen that

$$\begin{aligned} &\arg(ze^{(2\sigma+1)\pi i}) \ge 2\sigma\pi \ge \tfrac{1}{2}\kappa\pi + \varepsilon, \quad \text{and} \\ &\arg(ze^{-(2\sigma+1)\pi i}) \le -2\sigma\pi \le -\tfrac{1}{2}\kappa\pi - \varepsilon \end{aligned} \tag{4.24}$$

for $-\pi \le \arg z \le \pi$. Using Lemmas 3.2, 4.1 and 4.2, we have

$$\begin{aligned}
&\int_L h_0(s)\left[h_1(s) - 2\pi i \sum_{k=0}^{\sigma-1} e^{(2k+1)\pi i s}\right] z^s ds \\
&= \int_L h_0(s)\kappa^{-\kappa s}\left[h_1(s) - 2\pi i \sum_{k=0}^{\sigma-1} e^{(2k+1)\pi i s}\right] \\
&\quad \times e^{-(2\sigma+1)\pi i s}(\kappa^\kappa z e^{(2\sigma+1)\pi i})^s ds \\
&= \mathcal{O}(z^{-w}),
\end{aligned} \tag{4.25}$$

and

$$\begin{aligned}
&\int_{L_1} h_0(s)\left[h_1(s) + 2\pi i \sum_{k=0}^{\sigma-1} e^{-(2k+1)\pi i s}\right] z^s ds \\
&= \int_{L_1} h_0(s)\kappa^{-\kappa s}\left[h_1(s) + 2\pi i \sum_{k=0}^{\sigma-1} e^{-(2k+1)\pi i s}\right] \\
&\quad \times e^{(2\sigma+1)\pi i s}(\kappa^\kappa z e^{-(2\sigma+1)\pi i})^s ds \\
&= \mathcal{O}(z^{-w}),
\end{aligned} \tag{4.26}$$

for $-\pi \le \arg z \le \pi$.

Define for $z \neq 0$ and $\kappa > 0$

$$F(z) := \int_L h_0(s) z^s ds \quad \text{and} \quad \widetilde{F}(z) := \int_{L_1} h_0(s) z^s ds. \tag{4.27}$$

Then, it follows that

$$H(-z) = Q_w(-z) + \sum_{k=0}^{\sigma-1} F(ze^{(2k+1)\pi i}) + \sum_{k=0}^{\sigma-1} \widetilde{F}(ze^{-(2k+1)\pi i}) + \mathcal{O}(z^{-w}), \tag{4.28}$$

for $-\pi \le \arg z \le \pi$.

Now, we need to derive estimates for $F(z)$ and $\widetilde{F}(z)$. First, we consider $F(z)$ in the sector $0 \le \arg z \le \kappa\pi$. On account of the identity

$$\frac{1}{2i} \frac{e^{\pi i(\kappa s+\alpha)} - e^{-\pi i(\kappa s+\alpha)}}{\sin(\pi(\kappa s+\alpha))} = 1, \tag{4.29}$$

we have

$$
\begin{aligned}
F(z) &= \int_L h_0(s) \cdot \frac{e^{\pi i(\kappa s+\alpha)} - e^{-\pi i(\kappa s+\alpha)}}{2i\sin(\pi(\kappa s+\alpha))} z^s ds \\
&= \frac{i}{2}\int_L h_0(s)\frac{e^{-\pi i(\kappa s+\alpha)}}{\sin(\pi(\kappa s+\alpha))} z^s ds - \frac{i}{2}\int_L h_0(s)\frac{e^{\pi i(\kappa s+\alpha)}}{\sin(\pi(\kappa s+\alpha))} z^s ds \\
&= \frac{i}{2}\int_L h_0(s)\frac{e^{-\pi i(\kappa s+\alpha)}}{\sin(\pi(\kappa s+\alpha))} z^s ds - \frac{i}{2}\int_L h_0(s)\frac{e^{\pi i(\kappa s+\alpha)}}{\sin(\pi(\kappa s+\alpha))} z^s ds \\
&\quad -\frac{i}{2}\int_{L_1} h_0(s)\frac{e^{-\pi i(\kappa s+\alpha)}}{\sin(\pi(\kappa s+\alpha))} z^s ds + \frac{i}{2}\int_{L_1} h_0(s)\frac{e^{-\pi i(\kappa s+\alpha)}}{\sin(\pi(\kappa s+\alpha))} z^s ds.
\end{aligned} \tag{4.30}
$$

Since

$$
\left|\frac{e^{\mp\pi i\kappa s}}{\sin(\pi(\kappa s+\alpha))}\right| \text{ is bound for } \pm\operatorname{Im} s \geq l, \tag{4.31}
$$

it follows from Lemmas 3.2 and 4.2 that for $0 \leq \arg z \leq \kappa\pi$,

$$
\begin{aligned}
&\int_L h_0(s)\frac{e^{\pi i(\kappa s+\alpha)}}{\sin(\pi(\kappa s+\alpha))} z^s ds \\
&\quad = \int_L h_0(s)\kappa^{-\kappa s}\frac{e^{\pi i\alpha}e^{-\pi i\kappa s}}{\sin(\pi(\kappa s+\alpha))}(\kappa^\kappa z e^{2\pi i\kappa})^s ds = \mathcal{O}(z^{-w}),
\end{aligned} \tag{4.32}
$$

and

$$
\begin{aligned}
&\int_{L_1} h_0(s)\frac{e^{-\pi i(\kappa s+\alpha)}}{\sin(\pi(\kappa s+\alpha))} z^s ds \\
&\quad = \int_{L_1} h_0(s)\kappa^{-\kappa s}\frac{e^{-\pi i\alpha}e^{\pi i\kappa s}}{\sin(\pi(\kappa s+\alpha))}(\kappa^\kappa z e^{-2\pi i\kappa})^s ds = \mathcal{O}(z^{-w}).
\end{aligned} \tag{4.33}
$$

Coupling these with (4.30) gives

$$
\begin{aligned}
F(z) &= \frac{i}{2}\left(\int_L - \int_{L_1}\right) h_0(s)\frac{e^{-\pi i(\kappa s+\alpha)}}{\sin(\pi(\kappa s+\alpha))} z^s ds + \mathcal{O}(z^{-w}) \\
&= \frac{i}{2}e^{-\pi i\alpha}\int_C \frac{h_0(s)}{\sin(\pi(\kappa s+\alpha))}(z e^{-\kappa\pi i})^s ds + \mathcal{O}(z^{-w}),
\end{aligned} \tag{4.34}
$$

for $0 \le \arg z \le \kappa\pi$. Similarly, we have

$$\widetilde{F}(z) = \frac{i}{2} e^{\pi i \alpha} \int_C \frac{h_0(s)}{\sin(\pi(\kappa s + \alpha))} (z e^{\kappa \pi i})^s ds + \mathcal{O}(z^{-w}), \tag{4.35}$$

for $-\kappa\pi \le \arg z \le 0$.

Lemma 4.3. *Let N be a positive integer, and let $F(z)$ and $\widetilde{F}(z)$ be given as in (4.27). Define*

$$E_N(z) := (2\pi)^{(p-q)/2} \kappa^{\alpha - 3/2} e^{\kappa z^{1/\kappa}} \times \left(\sum_{j=0}^{N-1} c_j (\kappa z^{1/\kappa})^{1-\alpha-j} + \mathcal{O}((\kappa z^{1/\kappa})^{1-\alpha-N}) \right). \tag{4.36}$$

As $z \to \infty$, we have

$$F(z) = E_N(z) + \mathcal{O}(z^{(1-\alpha-N)/\kappa}) + \mathcal{O}(z^{-w}) \tag{4.37}$$

for $0 \le \arg z \le \kappa\pi$. Similarly, as $z \to \infty$, we have

$$\widetilde{F}(z) = E_N(z) + \mathcal{O}(z^{(1-\alpha-N)/\kappa}) + \mathcal{O}(z^{-w}) \tag{4.38}$$

for $-\kappa\pi \le \arg z \le 0$.

Proof. By (4.34) and Lemma 3.2, we have

$$\begin{aligned} F(z) &= \frac{i}{2} e^{-\pi i \alpha} A_0 \left(\sum_{j=0}^{N-1} \int_C \frac{c_j}{\Gamma(\kappa s + \alpha + j) \sin(\pi(\kappa s + \alpha))} (\kappa^\kappa z e^{-\kappa \pi i})^s ds \right. \\ &\quad \left. + \int_C \frac{r_N(s)}{\Gamma(\kappa s + \alpha + N) \sin(\pi(\kappa s + a))} (\kappa^\kappa z e^{-\kappa \pi i})^s ds \right) + \mathcal{O}(z^{-w}) \\ &= \frac{i}{2\pi} e^{-\pi i \alpha} A_0 \sum_{j=0}^{N-1} (-1)^j c_j \int_C \Gamma(1 - \kappa s - \alpha - j) (\kappa^\kappa z e^{-\kappa \pi i})^s ds \\ &\quad + c' R_w(z) + \mathcal{O}(z^{-w}), \end{aligned} \tag{4.39}$$

where $c' = \frac{(-1)^N i}{2\pi} e^{-\pi i \alpha} A_0$ and the remainder is given by

$$R_w(z) = \int_C r_N(s) \Gamma(1 - \kappa s - \alpha - N)(\kappa^\kappa z e^{-\kappa \pi i})^s ds. \tag{4.40}$$

Since

$$\frac{1}{\kappa} z^{\beta/\kappa} e^{-z^{1/\kappa}} = \frac{1}{2\pi i} \int_{C_\beta} \Gamma(\beta - \kappa s) z^s ds, \tag{4.41}$$

which is valid for all $z \neq 0$, where C_β is a curve encircling the point $s = \beta$ (in the negative sense) with endpoints at infinity in $\operatorname{Re} s > 0$, from (4.39) we have

$$\begin{aligned} F(z) &= \frac{i}{2\pi} e^{-\pi i \alpha} A_0 \sum_{j=0}^{N-1} (-1)^j c_j \int_C \Gamma(1 - \kappa s - \alpha - j)(\kappa^\kappa z e^{-\kappa \pi i})^s ds \\ &\quad + c' R_w(z) + \mathcal{O}(z^{-w}) \\ &= - e^{-\pi i \alpha} \frac{A_0}{\kappa} \sum_{j=0}^{N-1} (-1)^j c_j \cdot (\kappa^\kappa z e^{-\kappa \pi i})^{(1-\alpha-j)/\kappa} \exp(-(\kappa^\kappa z e^{-\kappa \pi i})^{1/\kappa}) \\ &\quad + c' R_w(z) + \mathcal{O}(z^{-w}) \\ &= (2\pi)^{(p-q)/2} \kappa^{\alpha - 3/2} e^{\kappa z^{1/\kappa}} \sum_{j=0}^{N-1} c_j (\kappa z^{1/\kappa})^{1-\alpha-j} + c' R_w(z) + \mathcal{O}(z^{-w}), \end{aligned} \tag{4.42}$$

for $0 \leq \arg z \leq \kappa\pi$.

Next, we have to estimate the function $R_w(z)$ in (4.40). For $\arg z = \frac{1}{2}\kappa\pi + \varepsilon$, by using Lemma 4.2 we have

$$R_w(z) = \int_{-w-i\infty}^{-w+i\infty} r_N(s) \frac{\Gamma(3 - \kappa s - \alpha - N)}{(1 - \kappa s - \alpha - N)_2} (\kappa^\kappa z e^{-\kappa \pi i})^s ds. \tag{4.43}$$

Also, note that

$$\begin{aligned} &\Gamma(3 - \kappa s - \alpha - N)(\kappa^\kappa z e^{-\kappa \pi i})^s \\ &= (\kappa^\kappa z e^{-\kappa \pi i})^{(3-\alpha-N)/\kappa} \int_0^\infty t^{2-\kappa s - \alpha - N} e^{\kappa z^{1/\kappa} t} dt; \end{aligned} \tag{4.44}$$

hence,

$$\begin{aligned} R_w(z) = (\kappa^\kappa z e^{-\kappa \pi i})^{(3-\alpha-N)/\kappa} \int_{-w-i\infty}^{-w+i\infty} \frac{r_N(s)}{(1 - \kappa s - \alpha - N)_2} \\ \times \left(\int_0^\infty t^{2-\kappa s-\alpha-N} e^{\kappa z^{1/\kappa} t} dt \right) ds. \end{aligned} \tag{4.45}$$

By interchanging the order of integration, we have

$$R_w(z) = (-\kappa z^{1/\kappa})^{3-\alpha-N} \int_0^\infty r(t) e^{\kappa z^{1/\kappa} t} dt, \tag{4.46}$$

where

$$r(t) = \int_{-w-i\infty}^{-w+i\infty} \frac{r_N(s) t^{2-\kappa s-\alpha-N}}{(1-\kappa s-\alpha-N)_2} ds, \tag{4.47}$$

which converges for only $t > 0$. From (3.13), it is easily seen that

$$r_N(s)/(1-\kappa s-\alpha-N)_2 = \mathcal{O}(s^{-2}) \tag{4.48}$$

as $s \to \infty$, and

$$\begin{aligned} |r(t)| &\leq \int_{-w-i\infty}^{-w+i\infty} |r_N(s) t^{2-\kappa s-\alpha-N}/(1-s-\alpha-N)_2| \, |ds| \\ &\leq K t^{2+\kappa w-\operatorname{Re}\alpha-N} \end{aligned} \tag{4.49}$$

for $t > 0$. Note that here we can choose w and N such that $-w < (1 - \operatorname{Re}\alpha - N)/\kappa$.

If $t > 1$ then, for every positive number R, the part between $-w - iR$ and $-w + iR$ of the path of integration in (4.47) can be replaced by the half of the circle $|s + w| = R$ on the right-hand side of the line Re $s = -w$. Note that $r_N(s)$ has a finite number of poles with $\operatorname{Re} s > -w$; see Lemma 3.2. By Cauchy's theorem, we also have

$$\begin{aligned} r(t) &= \left(\int_{-w-i\infty}^{-w-iR} + \int_{C'} + \int_{-w+iR}^{-w+i\infty} \right) \frac{r_N(s) t^{2-\kappa s-\alpha-N}}{(1-\kappa s-\alpha-N)_2} ds \\ &\quad + \int_{A'} \frac{r_N(s) t^{2-\kappa s-\alpha-N}}{(1-\kappa s-\alpha-N)_2} ds, \end{aligned} \tag{4.50}$$

where C' is the right half of the circle $|s + w| = R$ joining $-w - iR$ and $-w + iR$, and A' is a finite contour which encloses all the poles of the integrand in the negative sense with $-w < \operatorname{Re} s < -w + R$. It follows from (4.48) that for $t > 1$

$$\begin{aligned} &\left| \int_{C'} \frac{r_N(s) t^{2-\kappa s-\alpha-N}}{(1-\kappa s-\alpha-N)_2} ds \right| \\ &\leq K \int_{-\pi/2}^{\pi/2} \left| \frac{r_N(-w+Re^{\theta i})}{(1+\kappa w-\kappa Re^{\theta i}-\alpha-N)_2} \right| t^{-\kappa R\cos\theta} R d\theta = \mathcal{O}(R^{-1}). \end{aligned} \tag{4.51}$$

By letting $R \to \infty$, (4.50) shows that for $t > 1$ the function $r(t)$ is equal to the sum of the residues of the poles of the integrand in (4.47) with $\operatorname{Re} s > -w$. The number of these poles is finite; hence, it follows that $r(t)$ can be analytically continued to all $t \neq 0$. Define

$$\widetilde{r}(t) := \int_A r_N(s) t^{2-\kappa s-\alpha-N} \frac{ds}{(1-\kappa s-\alpha-N)_2}, \tag{4.52}$$

where A is a finite contour in the half-plane $\operatorname{Re} s > -w$ which encloses all the poles of the integrand in (4.47) in the negative sense. It is easily seen that $\widetilde{r}(t)$ is analytic in t for $t \neq 0$ and $\widetilde{r}(t) = r(t)$ if $t > 1$. Moreover, we have

$$\widetilde{r}(t) = \mathcal{O}(t^{2+\kappa w-\operatorname{Re}\alpha-N}) \tag{4.53}$$

as $t \to \infty$ for $|\arg t| < \pi$.

Now, we rewrite (4.46) as

$$\begin{aligned} R_w(z) &= (-\kappa z^{1/\kappa})^{3-\alpha-N} \left(\int_0^1 r(t) e^{\kappa z^{1/\kappa} t} dt + \int_1^\infty \widetilde{r}(t) e^{\kappa z^{1/\kappa} t} dt \right) \\ &:= (-\kappa z^{1/\kappa})^{3-\alpha-N} \left(\tau_1(z) + \tau_2(z)\right), \end{aligned} \tag{4.54}$$

with $\arg z = \frac{1}{2}\kappa\pi + \varepsilon$. In view of (4.49), the function $\tau_1(z)$ can be analytically continued to all $z \neq 0$, and

$$\tau_1(z) = \mathcal{O}(e^{\kappa z^{1/\kappa}}) + \mathcal{O}(1) \tag{4.55}$$

as $z \to \infty$ for $0 \le \arg z \le \kappa\pi$. By the properties of $\widetilde{r}(t)$ and (4.53), we can rotate the horizontal path of integration of $\tau_2(z)$ to a ray $\arg t = \theta$ with $0 < \theta \le \pi - \delta, \delta > 0$. Indeed, for t in the sector $0 \le \arg t \le \theta$, we have

$$\frac{\pi}{2} + \frac{\varepsilon}{\kappa} \le \arg(z^{1/\kappa} t) = \frac{1}{\kappa} \arg z + \arg t = \frac{\pi}{2} + \frac{\varepsilon}{\kappa} + \arg t \le \frac{3\pi}{2} - \delta'$$

with $\delta' = \delta - \frac{\varepsilon}{\kappa} > 0$; that is, the function $e^{\kappa z^{1/\kappa} t}$ is exponentially small. Thus,

$$\tau_2(z) = \int_1^\infty \widetilde{r}(t) e^{\kappa z^{1/\kappa} t} dt = \int_1^{1+\infty e^{i\theta}} \widetilde{r}(t) e^{\kappa z^{1/\kappa} t} dt, \tag{4.56}$$

if $\arg z = \frac{1}{2}\kappa\pi + \varepsilon$. Choosing $\theta = \frac{\pi}{2} + \delta$, the function $e^{\kappa z^{1/\kappa} t}$ is exponentially small if $0 \le \arg z \le \kappa(\pi - 2\delta)$, and $\tau_2(z)$ can be analytically continued to this sector. Furthermore,

$$\begin{aligned} |\tau_2(z)| &= \left| \int_1^{1+\infty e^{i\theta}} \widetilde{r}(t) e^{\kappa z^{1/\kappa} t} dt \right| \\ &\le |e^{\kappa z^{1/\kappa}}| \int_0^\infty |\widetilde{r}(1+se^{i\theta})| |e^{\kappa z^{1/\kappa} s e^{i\theta}}| ds = \mathcal{O}(e^{\kappa z^{1/\kappa}}). \end{aligned} \tag{4.57}$$

The order estimate follows from the fact that the integral on the right is absolutely convergent. By choosing $\theta = \frac{\pi}{2} - \delta$, the same argument shows that $\tau_2(z)$ can be analytically continued to $2\kappa\delta \le \arg z \le \kappa\pi$, and the estimate in (4.57) also holds in this region. Therefore, we have extended $\tau_2(z)$ to the sector $0 \le \arg z \le \kappa\pi$, and established (4.57) in this region.

From (4.55) and (4.57), we have for $0 \le \arg z \le \kappa\pi$

$$\begin{aligned} R_w(z) &= (-\kappa z^{1/\kappa})^{3-\alpha-N}\,(\tau_1(z) + \tau_2(z)) \\ &= (-\kappa z^{1/\kappa})^{3-\alpha-N}(\mathcal{O}(e^{\kappa z^{1/\kappa}}) + \mathcal{O}(1)) \\ &= \mathcal{O}((\kappa z^{1/\kappa})^{3-\alpha-N})e^{\kappa z^{1/\kappa}} + \mathcal{O}((\kappa z^{1/\kappa})^{3-\alpha-N}). \end{aligned} \tag{4.58}$$

From (4.42) and (4.58), it follows that

$$\begin{aligned} F(z) &= (2\pi)^{(p-q)/2}\kappa^{\alpha-3/2}e^{\kappa z^{1/\kappa}}\left(\sum_{j=0}^{N-1} c_j(\kappa z^{1/\kappa})^{1-\alpha-j} + \mathcal{O}((\kappa z^{1/\kappa})^{3-\alpha-N})\right) \\ &\quad + \mathcal{O}((\kappa z^{1/\kappa})^{3-\alpha-N}) + \mathcal{O}(z^{-w}) \\ &= (2\pi)^{(p-q)/2}\kappa^{\alpha-3/2}e^{\kappa z^{1/\kappa}}\left(\sum_{j=0}^{N'+1} c_j(\kappa z^{1/\kappa})^{1-\alpha-j} + \mathcal{O}((\kappa z^{1/\kappa})^{1-\alpha-N'})\right) \\ &\quad + \mathcal{O}(z^{(1-\alpha-N')/\kappa}) + \mathcal{O}(z^{-w}) \\ &= (2\pi)^{(p-q)/2}\kappa^{\alpha-3/2}e^{\kappa z^{1/\kappa}}\left(\sum_{j=0}^{N'-1} c_j(\kappa z^{1/\kappa})^{1-\alpha-j} + \mathcal{O}((\kappa z^{1/\kappa})^{1-\alpha-N'})\right) \\ &\quad + \mathcal{O}(z^{(1-\alpha-N')/\kappa}) + \mathcal{O}(z^{-w}) \end{aligned} \tag{4.59}$$

as $z \to \infty$ for $0 \le \arg z \le \kappa\pi$. Here we have used the fact that

$$(\kappa z^{1/\kappa})^{1-\alpha-j} = \mathcal{O}((\kappa z^{1/\kappa})^{1-\alpha-N'}) \tag{4.60}$$

for $j = N'$ and $j = N' + 1$.

Similarly, we can prove (4.38). □

5. Summary of Results

Define

$$\nu := 1 - \alpha,$$

and recall that

$$\kappa = q + 1 - p, \quad \alpha = \sum_{k=1}^{q} b_k - \sum_{k=1}^{p} a_k + \frac{1}{2}(p - q + 2),$$

and

$$c_j = -\frac{1}{j\kappa^\kappa} \sum_{k=0}^{j-1} c_k e(j,k),$$

where $c_0 = 1$ and $e(j,k)$ is given by (3.16).

We are now ready to present the asymptotic expansions of the generalized hypergeometric function ${}_pF_q(z)$. We shall do this case-by-case, depending on the values of p and q.

5.1. Case $p = q + 1$

In this case, the radius of convergence of the series in (1.1) is 1, and the function ${}_pF_q(z)$ has a cut from 1 to $+\infty$ along the real line. The contour C in (1.3) is the vertical line from $-i\infty$ to $+i\infty$. This case has already been discussed in Section 2, and the result is

$$H(z) \sim \sum_{j=1}^{p} \sum_{n=0}^{\infty} \frac{\Gamma(a_j + n) \prod_{k=1, k\neq j}^{p} \Gamma(a_k - a_j - n)}{\prod_{k=1}^{q} \Gamma(b_k - a_j - n)} \frac{(-1)^n}{n!} (-z)^{-a_j - n} \tag{5.1}$$

as $z \to \infty$ in $|\arg(-z)| < \pi$.

5.2. Case $p = q$

In this case, we have $\kappa = 1$ and $\sigma = 1$; see (4.22). In particular, as $z \to \infty$, (2.7) can be rewritten as

$$H(z) \sim \sum_{j=1}^{p} \sum_{n=0}^{\infty} \frac{\Gamma(a_j + n) \prod_{k=1, k\neq j}^{p} \Gamma(a_k - a_j - n)}{\prod_{k=1}^{q} \Gamma(b_k - a_j - n)} \frac{(-1)^n}{n!} (-z)^{-a_j - n} \tag{5.2}$$

for z in the sector $|\arg(-z)| < \pi/2$.

From (4.28), we have for $-\pi \leq \arg(-z) \leq \pi$

$$H(z) = Q_w(z) + F(-ze^{\pi i}) + \widetilde{F}(-ze^{-\pi i}) + \mathcal{O}(z^{-w}). \tag{5.3}$$

Note that for $-\pi \le \arg(-z) \le 0$, we have $\arg(-ze^{-\pi i}) \le -\pi$, and by Lemma 4.2

$$\widetilde{F}(-ze^{-\pi i}) = \mathcal{O}(z^{-w}). \tag{5.4}$$

From (5.3) and Lemma 4.3, it follows that

$$H(z) = Q_w(z) + E_N(-ze^{\pi i}) + \mathcal{O}(z^{(1-\alpha-N)/\kappa}) + \mathcal{O}(z^{-w}), \tag{5.5}$$

for $-\pi \le \arg(-z) \le 0$.

Similarly, for $0 \le \arg(-z) \le \pi$, we have by Lemma 4.2

$$F(-ze^{\pi i}) = \mathcal{O}(z^{-w}), \tag{5.6}$$

and by Lemma 4.3

$$H(z) = Q_w(z) + E_N(-ze^{-\pi i}) + \mathcal{O}(z^{(1-\alpha-N)/\kappa}) + \mathcal{O}(z^{-w}). \tag{5.7}$$

Combining (5.5) and (5.7), we obtain as $z \to \infty$,

$$H(z) \sim \sum_{j=1}^{p}\sum_{n=0}^{\infty} \frac{\Gamma(a_j+n)\prod_{k=1,k\neq j}^{p}\Gamma(a_k-a_j-n)}{\prod_{k=1}^{q}\Gamma(b_k-a_j-n)} \frac{(-1)^n}{n!}(-z)^{-a_j-n}$$
$$+ e^z \sum_{j=0}^{\infty} c_j z^{\nu-j} \tag{5.8}$$

for $z \to \infty$ in $-\pi \le \arg(-z) \le \pi$.

5.3. Case $p = q - 1$

Here, we have $\kappa = q + 1 - p = 2$ and $\sigma = 1$. Then, (4.28) can be rewritten as

$$H(-z) = Q_w(-z) + F(ze^{\pi i}) + \widetilde{F}(ze^{-\pi i}) + \mathcal{O}(z^{-w}) \tag{5.9}$$

for $-\pi \le \arg z \le \pi$.

By Lemma 4.3, we have

$$H(-z) \sim \sum_{j=1}^{p}\sum_{n=0}^{\infty} \frac{\Gamma(a_j+n)\prod_{k=1,k\neq j}^{p}\Gamma(a_k-a_j-n)}{\prod_{k=1}^{q}\Gamma(b_k-a_j-n)} \frac{(-1)^n}{n!} z^{-a_j-n}$$
$$+(2\pi)^{-1/2}\kappa^{-\nu-1/2}e^{\kappa(ze^{\pi i})^{1/\kappa}} \sum_{j=0}^{\infty} c_j(\kappa(ze^{\pi i})^{1/\kappa})^{\nu-j}$$
$$+(2\pi)^{-1/2}\kappa^{-\nu-1/2}e^{\kappa(ze^{-\pi i})^{1/\kappa}} \sum_{j=0}^{\infty} c_j(\kappa(ze^{-\pi i})^{1/\kappa})^{\nu-j}, \tag{5.10}$$

as $z \to \infty$, for $-\pi \le \arg z \le \pi$.

5.4. Case $p \le q-2$

In this final case, we have $\kappa = q+1-p > 2$. Formula (4.28) gives

$$H(-z) = Q_w(-z) + \sum_{k=0}^{\sigma-1} F(ze^{(2k+1)\pi i}) + \sum_{k=0}^{\sigma-1} \widetilde{F}(ze^{-(2k+1)\pi i}) + \mathcal{O}(z^{-w}) \tag{5.11}$$

for $-\pi \le \arg z \le \pi$.

If $\kappa \ge 4$, we have from (4.22)

$$0 \le \arg(ze^{(2k+1)\pi i}) \le \arg(ze^{(2\sigma-1)\pi i}) \le \tfrac{1}{2}\kappa\pi + 2\pi \le \kappa\pi \tag{5.12}$$

and

$$0 \ge \arg(ze^{-(2k+1)\pi i}) \ge \arg(ze^{-(2\sigma-1)\pi i}) \ge -\tfrac{1}{2}\kappa\pi - 2\pi \ge -\kappa\pi \tag{5.13}$$

for $k = 0, \ldots, \sigma-1$. Hence,

$$0 \le \arg(ze^{(2k+1)\pi i}) \le \kappa\pi \quad \text{and} \quad -\kappa\pi \le \arg(ze^{-(2k+1)\pi i}) \le 0 \tag{5.14}$$

for $k = 0, \ldots, \sigma-1$. Coupling these with (5.11) and Lemma 4.3 gives

$$H(-z) = Q_w(-z) + \sum_{k=0}^{\sigma-1} E_N(ze^{(2k+1)\pi i}) + \sum_{k=0}^{\sigma-1} E_N(ze^{-(2k+1)\pi i}) + \mathcal{O}(z^{(1-\alpha-N)/\kappa}) + \mathcal{O}(z^{-w}). \tag{5.15}$$

Also, note that for $-\pi \le \arg z \le \pi$ and for $k = 1, \ldots, \sigma-1$, we have

$$0 \le \arg(ze^{\pi i}) < \arg(ze^{(2k+1)\pi i}) \le \kappa\pi \tag{5.16}$$

and $0 \le \arg(ze^{\pi i}) \le 2\pi \le \frac{1}{2}\kappa\pi$; thus, the terms in the series for $E_N(ze^{(2k+1)\pi i})$ contain the factor $e^{\kappa(ze^{(2k+1)\pi i})^{1/\kappa}}$, which satisfies

$$|\exp(\kappa(ze^{(2k+1)\pi i})^{1/\kappa})| < |\exp(\kappa(ze^{\pi i})^{1/\kappa})| \tag{5.17}$$

for $k = 1, \ldots, \sigma-1$, that is, $E_N(ze^{\pi i})$ is exponentially larger than $E_N(ze^{(2k+1)\pi i})$ as $z \to \infty$, for $k = 1, \ldots, \sigma-1$. Similarly, $E_N(ze^{-\pi i})$ is exponentially larger than $E_N(ze^{-(2k+1)\pi i})$ as $z \to \infty$, for $k = 1, \ldots, \sigma-1$. From (5.15), it follows that

$$H(-z) = Q_w(-z) + E_N(ze^{\pi i})[1 + \mathcal{O}(e^{-\delta|z|})] + E_N(ze^{-\pi i})[1 + \mathcal{O}(e^{-\delta|z|})] + \mathcal{O}(z^{(1-\alpha-N)/\kappa}) + \mathcal{O}(z^{-w}), \tag{5.18}$$

where δ is a positive number, and $E_N(ze^{\pi i}) + E_N(ze^{-\pi i})$ is exponentially large as $z \to \infty$.

If $\kappa = 3$, it follows from (4.22), (5.13) and Lemma 4.3 that $\sigma = 1$ and

$$H(-z) = Q_w(-z) + E_N(ze^{\pi i}) + E_N(ze^{-\pi i}) + \mathcal{O}(z^{(1-\alpha-N)/\kappa}) + \mathcal{O}(z^{-w}), \tag{5.19}$$

for $-\pi \le \arg z \le \pi$.

For $-\pi \le \arg z \le \pi/2 - \varepsilon$, we have

$$0 \le \arg(ze^{\pi i}) \le \tfrac{3}{2}\pi - \varepsilon; \tag{5.20}$$

thus, the term $\exp(\kappa(ze^{\pi i})^{1/\kappa})$ in the series for $E_N(ze^{\pi i})$ is exponentially large as $z \to \infty$. For $\pi/2 - \varepsilon \le \arg z \le \pi$, we have

$$-\tfrac{1}{2}\pi - \varepsilon \le \arg(ze^{-\pi i}) \le 0, \tag{5.21}$$

and the term $\exp(\kappa(ze^{-\pi i})^{1/\kappa})$ in the series for $E_N(ze^{-\pi i})$ is exponentially large as $z \to \infty$. Hence, for $-\pi \le \arg z \le \pi$, $E_N(ze^{\pi i}) + E_N(ze^{-\pi i})$ is exponentially large as $z \to \infty$.

Therefore, for $\kappa > 2$, we have from (5.18) and (5.19)

$$\begin{aligned} H(-z) \sim{} & (2\pi)^{(1-\kappa)/2}\kappa^{-\nu-1/2}e^{\kappa(ze^{\pi i})^{1/\kappa}}\sum_{j=0}^{\infty} c_j(\kappa(ze^{\pi i})^{1/\kappa})^{\nu-j} \\ & + (2\pi)^{(1-\kappa)/2}\kappa^{-\nu-1/2}e^{\kappa(ze^{-\pi i})^{1/\kappa}}\sum_{j=0}^{\infty} c_j(\kappa(ze^{-\pi i})^{1/\kappa})^{\nu-j}, \end{aligned} \tag{5.22}$$

as $z \to \infty$, for $-\pi \le \arg z \le \pi$.

Finally, it is worth noting that our results (5.1), (5.8), (5.10) and (5.22) are equivalent to formulas (16.11.6)–(16.11.9) in NIST handbook [10].

Acknowledgments

We would like to thank the referee for a careful reading of the chapter and making some constructive suggestions.

References

[1] R. Beals and J. Szmigielski, Meijer G-function: a gentle introduction, *Notices Amer. Math. Soc.* **60** (2013) 866–872.

[2] R. Beals and R. Wong, *Special Functions and Orthogonal Polynomials* (Cambridge University Press, Cambridge, 2010).

[3] M. Bertola and T. Bothner, Universality conjecture and results for a model of several coupled positive-definite matrices, *Comm. Math. Phys.* **337** (2015) 1077–1141.

[4] M. Bertola, M. Gekhtman and J. Szmigielski, Cauchy–Laguerre two-matrix model and the Meijer-G random point field, *Comm. Math. Phys.* **326** (2014) 111–144.

[5] B. L. J. Braaksma, Asymptotic expansions and analytic continuations for a class of Barnes-integrals, *Compositio Math.* **15** (1962–1964) 239–341.

[6] W. B. Ford, *The Asymptotic Developments of Functions Defined by Maclaurin Series,* University of Michigan Studies, Scientific Series, Vol. 11 (University of Michigam Press, 1936).

[7] P. J. Forrester, Eigenvalue statistics for product complex Wishart matrices, *J. Phys. A* **47** (2014) 345202, 22 pp.

[8] A. B. J. Kuijlaars and L. Zhang, Singular values of products of Ginibre random matrices, multiple orthogonal polynomials and hard edge scaling limits, *Comm. Math. Phys.* **332** (2014) 759–781.

[9] Y. Lin and R. Wong, Asymptotics of the Meijer G-functions, in *Modern Trends in Constructive Function Theory,* Contemporary Mathematics, Vol. 661 (American Mathematical Society, 2016), pp. 243–251.

[10] F. W. J. Olver, D. W. Lozier, R. F. Boisvert and C. W. Clark, *NIST Handbook of Mathematical Functions* (Cambridge University Press, Cambridge, 2010).

[11] R. B. Paris and D. Kaminski, *Asymptotics and Mellin–Barnes Integrals* (Cambridge University Press, Cambridge, 2001).

[12] R. B. Paris and A. D. Wood, *Asymptotics of High Order Differential Equations,* Pitman Research Notes in Mathematics, Vol. 129 (Longman, 1986).

[13] T. D. Riney, On the coefficients in asymptotic factorial expansions, *Proc. Amer. Math. Soc.* **7** (1956) 245–249.

[14] E. M. Wright, The asymptotic expansions of the generalized hypergeometric function, *Proc. London Math. Soc.* **46** (1940) 389–408.

This page intentionally left blank

Chapter 26

Mock Theta-Functions of the Third Order of Ramanujan in Terms of Appell–Lerch Series

Changgui Zhang
Laboratoire P. Painlevé (UMR – CNRS 8524), UFR Math.
Université de Lille 1, Cité scientifique
59655 Villeneuve d'Ascq cedex, France
zhang@math.univ-lille1.fr

In this chapter, six of seven mock theta-functions of the third order of Ramanujan are expressed in terms of Appell–Lerch series and generalized Mordell integrals.

Keywords: Mock theta-functions; stokes phenomenon; Appell–Lerch series; q-difference equations.

Mathematics Subject Classification 2010: 11F03, 33D15, 34M30, 39A13

1. Introduction

This work is devoted to the study of the third-order mock theta functions discovered by S. Ramanujan. Namely, we consider the following q-series that G. N. Watson have delivered at the LMS meeting of 14 November 1935 for his presidential address [32]:

$$f(q) = \sum_{n=0}^{\infty} \frac{q^{n^2}}{(1+q)^2(1+q^2)^2 \dots (1+q^n)^2},$$

$$\varphi(q) = \sum_{n=0}^{\infty} \frac{q^{n^2}}{(1+q^2)(1+q^4)\dots(1+q^{2n})},$$

$$\psi(q) = \sum_{n=1}^{\infty} \frac{q^{n^2}}{(1-q)(1-q^3)\dots(1-q^{2n-1})},$$

$$\chi(q) = \sum_{n=0}^{\infty} \frac{q^{n^2}}{(1-q+q^2)(1-q^2+q^4)\dots(1-q^n+q^{2n})},$$

$$\omega(q) = \sum_{n=0}^{\infty} \frac{q^{2n(n+1)}}{(1-q)^2(1-q^3)^2(1-q^5)^2\dots(1-q^{2n+1})^2},$$

$$\nu(q) = \sum_{n=0}^{\infty} \frac{q^{n(n+1)}}{(1+q)(1+q^3)(1+q^5)\dots(1+q^{2n+1})},$$

$$\rho(q) = \sum_{n=0}^{\infty} \frac{q^{2n(n+1)}}{(1+q+q^2)(1+q^3+q^6)(1+q^5+q^{10})\dots(1+q^{2n+1}+q^{4n+2})}.$$

In the paper *A Survey of Classical Mock Theta Functions* [17], Gordon and McIntosh gave four conditions for a holomorphic function in the unit disc to be (strong) mock theta. The results found in the present paper show that six of these seven functions satisfy the properties (i), (iii) and (iv) of this survey [17, pp. 98–99], and these properties are particularly considered by Andrews and Hickerson in [5]. The function to which our approaches cannot directly be applied is ψ, and we guess that this is only a technical problem.

Since 1930s, the mock theta-functions constitute a mathematical branch for what the research activities continue to motivate many people, in mathematics as in other fields of sciences; see, for example, [13, 33]. G. N. Watson has been a pioneer and leader in this field and both the transformation theory and the asymptotic analysis developed by him remain key elements needed inside the classic framework. In addition to all that, nobody can forget the influence for the development of this domain by several important works of Andrews [2, 3, 5]. For more details, we are really content to invite the interested reader to revisit [17] and the references therein [24–27, 42].

In this chapter, we would like to make use of the analytic theory of linear q-difference equations and, especially, the analysis around a singular point for obtaining information on the structure of solutions; see [30]. To see a general viewpoint of these topics related to our works [37–41], we start with some historical landmarks.

Hermite generalized elliptic functions and Appell–Lerch series — Hermite [18] first introduced three kinds of elliptic functions, the first one of what is the class of classic elliptic functions. These new kinds of elliptic functions can be identified as solutions of first-order Fuchsian or singular irregular q-difference equations. In [6, 7], Appell studied elementary decomposition of the elliptic functions of the third kind. Later, Lerch [20] studied several series with the view of writing a general elliptic function as linear combination of simple ones, where the factors may be theta functions. By this way, one finds Appell–Lerch series in the literature; see (3.3) for the definition.

Every nonramified Appell–Lerch series satisfies a nonhomogenous linear q-difference equation of the first order, and has simple poles along a given lattice (that is identified as q-spiral). Apart from a factor of theta function, this is exactly our sum-function along a spiral for the q-Euler series $\hat{E}(x;q)$, that is of the form ${}_2\varphi_0(q,0;q;x)$. See (1.1) for the definition of a general basic hypergeometric series. Thus, the summation of confluent basic hypergeometric series along a spiral yields a generalization of the nonramified Appell–Lerch series.

Mention also that the elementary decomposition problem that Appell considered for the third-order elliptic function is intimately related with the analytic classification of linear q-difference equations.

Stokes co-cycles, Mordell's integral and modular relations — In the analytic theory of differential equations, every irregular singularity is characterized by the existence of solutions having exponential growth near this point; see [8]. This is the same with linear q-difference equations, where the usual exponential growth is replaced with q-exponential growth; see [11, 30, 35]. The Jacobi theta function and the heat-kernel are two q-exponential functions, that are related with the well-known theta-modular formula [23]. This leads us to consider both the summation along q-spiral and the summation with heat-kernel, the last one being called Gq-summation.

For the simple but typical divergent power series $\hat{E}(x;q)$, its Gq-sum is also known as Mordell's integral [22]. This integral is efficient for proving the reciprocity law for generalized Gauss sums [10, Theorem 1.2.2, p. 13]. It should be interesting to notice that this same integral is used in the Riemann-Siegel integral formula for Riemann's zeta-function; see [14, p. 166].

Comparing two different sum-functions associated to a given power series is often called Stokes analysis [21, 29]. Our Stokes analysis consists first of considering the link between Appell–Lerch series and Mordell's integral. This shows that Appelle–Lerch series, combined together with Mordell's integral, is left invariant by the modular transformation modulo a Mordell's integral. In other words, this Stokes analysis give raise to modular-type relations.

Main step undertaken in the chapter — By writing the third-order mock theta-functions into the form of the degenerate q-hypergeometric series ${}_2\varphi_1(q,0;\alpha;q,\beta)$ as in (2.13), we are led, with the Ramanujan's ${}_1\psi_1$-summation, to a generalization of the Appell–Lerch series. This generalization is in fact a solution of linear q-difference equation with a singular point; see (3.4). What is important is the fact that, by an Euler infinite product, this functional equation can be transformed into an equation very similar to that for Appell–Lerch series. From the view of point of the analytic classification of linear q-difference equations, these functional equations have the same invariants. Thus, mock theta-functions can be expressed by means of Appell–Lerch series to what one needs only to add some correctional terms. One will see that the corrections can be formulated by Mordell-type integrals of the form

$$\int_0^\infty \frac{(\alpha\xi;q)_\infty}{1+\xi}\exp\left(\frac{\log^2(\frac{\xi}{x})}{2\log q}\right)\frac{d\xi}{\xi},\quad q\in(0,1).$$

Main results of the chapter — One will see that six of seven mock theta functions of the third order of Ramanujan can belong to the family of ${}_1\varphi_2(q;\frac{q}{a},\frac{q}{b};q,\frac{q}{ab})$'s, where a and b denote nonzero complex numbers that are not a positive power of q. See (2.5)–(2.9) and (2.11) in the below. For simplify, we will denote by $\mathcal{R}(a,b;q)$ this basic hypergeometric function. A key point will be to observe that $\mathcal{R}(a,b;q)$ can be expressed by means of the sum of the divergent q-hypergeometric series ${}_2\varphi_0(q,qb;q,x)$ at $x=\frac{a}{qb}$ along the q-spiral passing through a; see Theorem 2.4. This sum-function will be denoted as $L(x,\alpha,\mu;q)$; see (3.7).

The summation of any q-hypergeometric series, while divergent, is not unique. When $q\in(0,1)$, applying the Gq-summation defined in [35] gives a solution to the q-difference equation satisfied by ${}_2\varphi_0(q,qb;q,x)$. This solution will be denoted as $G(x,\alpha;q)$; see (3.11). Therefore, one has two different types of sum for the same power series. Comparing them allows one to interpret $\mathcal{R}(a,b;q)$ in terms of generalized Mordell integrals, like as in Theorem 4.1.

Theorem 3.6 plays a primary role in matter of the structure of mock theta-functions. This allows us to decompose, near $q = 1$, each mock theta-function as the linear combination of one Appell–Lerch series plus two Mordell integrals, one of which contains an Euler infinite product under the integration. In a future paper, it will be stated that, near a primitive root of order m, each mock theta-functions is related to $2m$ Mordell integrals.

Notational conventions — The symbol q will denote a complex number with $0 < |q| < 1$. We will often write $q = e^{2\pi i\tau}$ with $\tau \in \mathcal{H}$. As usual, one says that q tends radially to a root of unity $\zeta = e^{ir}$ with $r \in \mathbb{Q}$ if $qe^{-ir} \to 1$ in the open interval $(0,1)$. When $(0,1)$ can be replaced with some sector $\arg(1-x) \in (-\epsilon, \epsilon)$ of vertex at 1 with some $\epsilon > 0$, we will say that q tends almost radially to ζ in $\mathbb{D}$ and simply write $q \overset{a.r.}{\longrightarrow} \zeta$. At the same time, we will say that τ tends almost vertically to r in $\mathcal{H}$ and write $\tau \overset{a.v.}{\longrightarrow} r$.

We will denote by $\tilde{\mathbb{C}}^*$ the Riemann surface of logarithm, and by π the natural evaluation map on $\tilde{\mathbb{C}}^*$, this means that $\pi(x) = e^{\log x} \in \mathbb{C}^*$. For simplify, we will write x instead of $\pi(x)$ when no confusion is possible. As usual, let ${}_\ell\varphi_m$ denote the general basic hypergeometric series given as follows [16]:

$$
{}_\ell\varphi_m\left(\begin{matrix} a_1, \ldots, a_\ell \\ b_1, \ldots, b_m \end{matrix}\ ; q, x\right) = \sum_{n\geq 0} \frac{(a_1, \ldots, a_\ell; q)_n}{(q, b_1, \ldots, b_m; q)_n} \left((-1)^n q^{\frac{1}{2}n(n-1)}\right)^{m+1-\ell} x^n . \tag{1.1}
$$

Here, we will use the following notation: $(a;q)_0 = 1$,

$$
(a;q)_n = \prod_{\ell=0}^{n-1}(1 - aq^\ell), \quad (a_1, \ldots, a_m; q)_n = \prod_{k=1}^{m}(a_k;q)_n \tag{1.2}
$$

for $n \in \mathbb{Z}_{\geq 1} \cup \{\infty\}$ and a, $a_k \in \mathbb{C}$. It is well known that, generically, ${}_\ell\varphi_m$ has a finite (or infinite or zero, respectively) radius of convergence when $\ell = m+1$ (or $\ell \leq m$ or $\ell > m+1$, respectively).

2. Viewpoint of the Confluent Basic Hypergeometric Series

Let us start by introducing a *unified coding* in terms of generating series for these mock theta-functions. To each pair $(a,b) \in (\mathbb{C} \setminus q^{\mathbb{Z}}) \times (\mathbb{C} \setminus q^{\mathbb{Z}})$, we associate the power series $\mathcal{R}(a, b; q, x)$ as follows:

$$\mathcal{R}(a, b; q, x) = \sum_{n \geq 0} \frac{(a, b; q)_\infty}{(aq^{-n}, bq^{-n}; q)_\infty} \left(\frac{x}{q}\right)^n. \tag{2.1}$$

Since

$$(aq^{-n}; q)_\infty = (aq^{-n}; q)_n \, (a; q)_\infty \tag{2.2}$$

and

$$(aq^{-n}; q)_n = (-a)^n \, q^{\frac{1}{2}n(n-1)} \left(\frac{q}{a}; q\right)_n \tag{2.3}$$

for all nonzero complex a and nonnegative integer n, one can write $\mathcal{R}(a, b; q, x)$ into the following form:

$$\mathcal{R}(a, b; q, x) = \sum_{n=0}^{\infty} \frac{q^{n^2}}{(q/a, q/b\,; q)_n} \left(\frac{x}{ab}\right)^n. \tag{2.4}$$

Thus, the mock theta functions $f(q)$, ..., $\rho(q)$ can be rewritten into the following form:

$$f(q) = \mathcal{R}(-1, -1; q, 1), \quad \varphi(q) = \mathcal{R}(i, -i; q, 1), \tag{2.5}$$
$$\psi(q) = \mathcal{R}(\sqrt{q}, -\sqrt{q}; q, q) - 1, \quad \chi(q) = \mathcal{R}(-j, -j^2; q, 1), \tag{2.6}$$
$$\omega(q) = \mathcal{R}(q, q; q^2, 1) - 1, \quad \nu(q) = \mathcal{R}(\sqrt{q}\,i, -\sqrt{q}\,i; q, 1) - 1 \tag{2.7}$$

and

$$\rho(q) = \mathcal{R}(qj, qj^2; q^2, 1) - 1, \tag{2.8}$$

where $i = e^{\pi i/2}$ and $j = e^{2\pi i/3}$.

Apart from $\psi(q)$, every third-order mock theta function is associated to the special value of some power series $\mathcal{R}(a, b; q', x)$ taken at $x = 1$, where $q' = q$ or q^2. For the sake of brevity, we will write

$$\mathcal{R}(a, b; q) = \mathcal{R}(a, b; q, 1) \quad (a, \; b \in \mathbb{C}^* \setminus q^{\mathbb{Z}_{>0}}). \tag{2.9}$$

Remark 2.1. The seven cases included in (2.5)–(2.8) are the only cases of (a, b, q') satisfying the following conditions: $q' \in \{q, q^2\}$ and $(a, b) \in \left(\mathbb{C}^* \setminus q'^{\mathbb{Z}_{>0}}\right) \times \left(\mathbb{C}^* \setminus q'^{\mathbb{Z}_{>0}}\right)$ such that

$$ab \in \{1, q'\}, \quad a + b \in \mathbb{Z} \cup (q\mathbb{Z}), \quad |a| = |b|. \tag{2.10}$$

By (2.4), it follows that

$$\mathcal{R}(a,b;q,x) = {}_1\varphi_2\left(q; \frac{q}{a}, \frac{q}{b}; q, \frac{qx}{ab}\right),$$

which implies the following relation for $x=1$:

$$\mathcal{R}(a,b;q) = {}_1\varphi_2\left(q; \frac{q}{a}, \frac{q}{b}; q, \frac{q}{ab}\right). \tag{2.11}$$

Let $\mathcal{F}(a,b,t)$ be the function defined by N. J. Fine in [15, (1.1), p. 1] as follows:

$$\mathcal{F}(a,b,t) = {}_2\varphi_1(q, aq; bq; q, t).$$

By the relation given in [15, (12.3), p. 13], one finds that

$$(1-t)\, {}_2\varphi_1(q, 0; bq; q, t) = {}_1\varphi_2(q; bq, tq; q, qbt)\,. \tag{2.12}$$

In view of (2.11), replacing (b,t) with $(\frac{1}{b}, \frac{1}{a})$ in (2.12) yields that

$$\mathcal{R}(a,b;q) = \left(1-\frac{1}{a}\right) {}_2\varphi_1\left(q, 0; \frac{q}{b}; q, \frac{1}{a}\right). \tag{2.13}$$

Thus, $\mathcal{R}(a,b;q)$ is closely related to the basic hypergeometric series ${}_2\varphi_1(\alpha,\beta;\gamma;q,x)$. It should be emphasized that this last series in (2.13) is *degenerate* because of the fact that $\beta=0$.

The famous Ramanujan's ${}_1\psi_1$-summation formula can be expressed as follows [4, Theorem 10.5.1, p. 502; 16]:

$${}_1\psi_1(a;b;q,z) = \sum_{n\in\mathbb{Z}} \frac{(a;q)_n}{(b;q)_n}\, z^n = \frac{(az, q/az, q, b/a; q)_\infty}{(z, b/az, b, q/a; q)_\infty}, \tag{2.14}$$

where $|\frac{b}{a}| < |z| < 1$ and

$$(a;q)_{-n} = \frac{(a;q)_\infty}{(aq^{-n};q)_\infty} \tag{2.15}$$

for all positive integer n. The relations in (2.2) and (2.3) yield that

$$(a;q)_{-n} = (-a)^n\, q^{-\frac{1}{2}n(n-1)} \left(\frac{q}{a};q\right)_n.$$

By gathering together the terms of positive or null index of the above Laurent series in (2.14) and also that for the terms of negative index, one can rewrite this summation formula in the following manner:

$${}_2\varphi_1(a,q;b;q,z) + {}_2\varphi_1\left(\frac{q}{b}, q; \frac{q}{a}; q, \frac{b}{az}\right) = 1 + \frac{(az, q/az, q, b/a; q)_\infty}{(z, b/az, b, q/a; q)_\infty}. \tag{2.16}$$

The following result is proved in [37], and it allows one to understand how to have the "real or true" convergence of ${}_2\varphi_1$ towards sum-functions of ${}_2\varphi_0$ during the confluence process. There is a similar reasoning for the case of hypergeometric series ${}_2F_1$ and ${}_2F_0$; see [19, 28, 34].

Theorem 2.2. *Let α, $\beta \in \mathbb{C}$ and let ${}_2\varphi_0(\alpha,\beta;\mu,q,x)$ to be the sum-function of ${}_2\varphi_0(\alpha,\beta;q,x)$ along the q-spiral $[-\mu;q]$. If $\mu \in \mathbb{C}^* \setminus [1;q]$, then*

$$\lim_{\mathbb{N}\ni n\to\infty} {}_2\varphi_1\left(\alpha,\beta;\frac{q^{-n}}{\mu};q,\frac{q^{-n}}{\mu}x\right) = {}_2\varphi_0(\alpha,\beta;\mu,q,x), \tag{2.17}$$

the convergence being uniform on any compact of the domain $\mathbb{C}^* \setminus [\mu;q]$.

Proof. See [37, Theorem 3.1]. □

The summation process used in the above can be described as follows [31, 36]. For any power series $\hat{f}(x)$, one replaces each power x^n with $q^{\frac{1}{2}n(n-1)}\xi^n$ to define its so-called (formal) *q-Borel transform*, which will be usually denoted by $\hat{\mathcal{B}}_q\hat{f}$:

$$\hat{f} = \sum_{n\geq 0} a_n x^n \Longrightarrow \hat{\mathcal{B}}_q\hat{f} = \sum_{n\geq 0} a_n\, q^{\frac{1}{2}n(n-1)}\,\xi^n\,. \tag{2.18}$$

The sum-function along a q-spiral $[\mu] = [\mu;q] = \mu q^{\mathbb{Z}}$ that will be denoted as $\mathcal{S}^{[\mu]}\hat{f}$ is defined by the (analytic) *q-Laplace transform* of $\hat{\mathcal{B}}_q\hat{f}$ with θ-function:

$$\mathcal{S}^{[\mu]}\hat{f}(x) = \sum_{n\in\mathbb{Z}} \frac{\varphi(\mu q^n)}{\theta(\frac{\mu}{x}q^n)}, \quad \varphi = \hat{\mathcal{B}}_q\hat{f}, \tag{2.19}$$

where

$$\theta(x) = \theta(x;q) = \sum_{n\in\mathbb{Z}} q^{\frac{1}{2}n(n-1)}\,x^n\,. \tag{2.20}$$

Furthermore, by the triple product formula of Jacobi [4, Theorem 10.4.1, p. 497], it follows that

$$\theta(x;q) = \left(q,-x,-\frac{q}{x};q\right)_\infty, \tag{2.21}$$

which gives the zeros of θ in the complex plane: $\theta(x;q) = 0$ if and only if $x \in [-1;q]$. This leads to the observation that every sum-function defined by (2.19) will, in general, have simple poles $x \in [-\mu;q]$.

In Theorem 2.2 in the above, $\hat{f} = {}_2\varphi_0(\alpha,\beta;q,x)$, thus $\hat{\mathcal{B}}_q\hat{f} = {}_2\varphi_1(\alpha,\beta;0;q,-\xi)$. It follows that

$$ {}_2\varphi_0(\alpha,\beta;\mu,q,x) = \sum_{n\in\mathbb{Z}} \frac{\varphi(-\mu q^n)}{\theta(-\frac{\mu}{x}q^n)}, \quad \varphi(\xi) = {}_2\varphi_1(\alpha,\beta;0;q,-\xi). \tag{2.22} $$

Theorem 2.3. *Let* $\mu \in \mathbb{C}^* \setminus [1;q]$, $\alpha \in \mathbb{C} \setminus \left(q^{\mathbb{Z}_{>0}} \cup \{0\}\right)$, *and let* ${}_2\varphi_0(\alpha,0;\mu,q,x)$ *be as given by* (2.22) *with* $\beta = q$. *Then*

$$ {}_2\varphi_0(\alpha,q;\mu,q,x) = 1 - {}_2\varphi_1\left(0,q;\frac{q}{\alpha};q,\frac{1}{\alpha x}\right) + \Upsilon(\alpha;\mu,q,x), \tag{2.23} $$

where

$$ \Upsilon(\alpha;\mu,q,x) = \frac{(\alpha x/\mu, q\mu/\alpha x, 1/\mu\alpha, q\mu\alpha, q;q)_\infty}{(x/\mu, q\mu/x, 1/\alpha x, 1/\mu, q\mu, q/\alpha;q)_\infty}. \tag{2.24} $$

Proof. Put $a = \alpha$, $b = q^{-n}/\mu$, $z = q^{-n}x/\mu$ in (2.16), take $n \to \infty$ in $\mathbb{N}$, and apply the limit formula in (2.17) of Theorem 2.2. Then, one finds the following decomposition:

$$ {}_2\varphi_0(\alpha,q;\mu,q,x) = A + B, \quad A = 1 - {}_2\varphi_1\left(0,q;\frac{q}{\alpha};q,\frac{1}{\alpha x}\right) $$

and

$$ \begin{aligned} B &= \lim_{\mathbb{N}\ni n\to\infty} \frac{(\alpha x/\mu q^n, q, 1/\mu\alpha q^n;q)_\infty}{(x/\mu q^n, 1/\alpha x, 1/\mu q^n, q/\alpha;q)_\infty} \\ &= \frac{(\alpha x/\mu, q, 1/\mu\alpha;q)_\infty}{(x/\mu, 1/\alpha x, 1/\mu, q/\alpha;q)_\infty} \lim_{\mathbb{N}\ni n\to\infty} \frac{(q\mu/\alpha x, q\mu\alpha;q)_n}{(q\mu/x, q\mu;q)_n} \\ &= \Upsilon(\alpha;\mu,q,x). \end{aligned} $$

This implies the expression stated in (2.23). □

Theorem 2.3 can be viewed as a particular case of [37, Theorem 3.3], in which the sum-function ${}_2\varphi_0(\alpha,\beta;\mu,q,x)$ is expanded at $x = \infty$ for any pair of parameters (α,β). In (2.24), if $\mu = \alpha x \notin [1;q]$, then $\Upsilon(\alpha;\mu,q,x) = 0$; this, together with (2.23), implies that

$$ {}_2\varphi_0(\alpha,q;\alpha x,q,x) = 1 - {}_2\varphi_1\left(0,q;\frac{q}{\alpha};q,\frac{1}{\alpha x}\right). $$

Equivalently, it follows that

$$ {}_2\varphi_1(q,0;\gamma;q,x) = 1 - {}_2\varphi_0\left(q,\frac{q}{\gamma};\frac{1}{x},q,\frac{\gamma}{qx}\right). \tag{2.25} $$

This last relation can be seen as the expansion of the function ${}_2\varphi_1(q,0;\gamma;q,x)$ at $x=\infty$.

Theorem 2.4. *Let a, b and $\mathcal{R}(a,b;q)$ be as in (2.9). Then*

$$\mathcal{R}(a,b;q) = -\frac{1}{b}(1-a)(1-b)\,{}_2\varphi_0\left(q,qb;a,q,\frac{a}{qb}\right). \tag{2.26}$$

Proof. Putting $\gamma=\frac{q}{b}$ and $x=\frac{1}{a}$ into (2.25) yields that

$${}_2\varphi_1\left(q,0;\frac{q}{b};q,\frac{1}{a}\right) = 1 - {}_2\varphi_0\left(q,b;a,q,\frac{a}{b}\right).$$

Thus, the expression of $\mathcal{R}(a,b)$ in (2.13) becomes

$$\mathcal{R}(a,b;q) = \left(1-\frac{1}{a}\right)\left(1-{}_2\varphi_0\left(q,b;a,q,\frac{a}{b}\right)\right). \tag{2.27}$$

Letting $\alpha=q$, $\beta=b$, $\mu=a$ and $x=\frac{a}{b}$ in the definition of ${}_2\varphi_0(\alpha,\beta;\mu,q,x)$ in (2.22) gives that

$$1-{}_2\varphi_0\left(q,b;a,q,\frac{a}{b}\right) = \sum_{n\in\mathbb{Z}} \frac{\psi(-aq^n)}{\theta(-bq^n)}, \tag{2.28}$$

where

$$\psi(\xi) = 1-{}_2\varphi_1(q,b;0;q,-\xi) = (1-b)\xi\,{}_2\varphi_1(q,qb;0,q,-\xi).$$

This means that (2.28) can be read as follows:

$$\begin{aligned} 1-{}_2\varphi_0\left(q,b;a,q,\frac{a}{b}\right) &= (1-b)\sum_{n\in\mathbb{Z}} \frac{-aq^n}{\theta(-bq^n)}\,{}_2\varphi_1(q,qb;0,q,aq^n) \\ &= (1-b)\,\frac{a}{b}\sum_{n\in\mathbb{Z}} \frac{1}{\theta(-qbq^n)}\,{}_2\varphi_1(q,qb;0,q,aq^n) \\ &= (1-b)\,\frac{a}{b}\,{}_2\varphi_0\left(q,qb;a,q,\frac{a}{qb}\right). \end{aligned}$$

In this way, one deduces (2.26) from (2.27). □

3. Generalized Appell–Lerch Series and Mordell Integrals

In our recent work [39], the Appell–Lerch series $\mathrm{R}_1(x,\mu;q)$ are considered as being related with sum-functions of the following power series

$\sum_{n\geq 0}(-1)^n q^{\frac{1}{2}n(n-1)}x^n$, which is simply ${}_2\varphi_0(q,0;q,x)$. Namely, if one writes $L(x,\mu;q)={}_2\varphi_0(q,0;\mu,q,x)$, (2.22) implies that

$$L(x,\mu;q)=\frac{1}{\theta(-\frac{\mu}{x})}\sum_{n\in\mathbb{Z}}\frac{1}{1-\mu q^n}\left(-\frac{\mu}{x}\right)^n q^{\frac{1}{2}n(n-1)}. \tag{3.1}$$

Equivalently, it follows that

$$L(x,\mu;q)=\frac{1}{\theta(-\frac{\mu}{x})}\,\mathrm{R}_1\left(-\frac{\mu}{\sqrt{q}\,x},\mu;q\right), \tag{3.2}$$

where $\mathrm{R}_1(x,\mu;q)$ is the first-order Appell–Lerch series defined as follows:

$$\mathrm{R}_1(x,\mu;q)=\sum_{n\in\mathbb{Z}}\frac{q^{\frac{1}{2}n^2}}{1-\mu\,q^n}\,x^n\,. \tag{3.3}$$

More generally, let $\alpha\in\mathbb{C}$. By putting $\ell=2$ and $m=0$ into (1.1), the confluent basic hypergeometric series ${}_2\varphi_0(q,\alpha;q,x)$ can be expressed as follows:

$${}_2\varphi_0(q,\alpha;q,x)=\sum_{n\geq 0}(\alpha;q)_n\,q^{-\frac{1}{2}n(n-1)}\,(-x)^n\,.$$

From the relation

$$1-x\,{}_2\varphi_0\left(q,\alpha;q,\frac{x}{q}\right)=\sum_{n\geq 1}(\alpha;q)_n\,q^{-(n+1)n/2}\,(-x)^{n+1},$$

one finds easily that the power series ${}_2\varphi_0(q,\alpha;q,x)$ satisfies the following linear q-difference equation:

$$(1-\alpha x)\,y(x)+x\,y\left(\frac{x}{q}\right)=1\,. \tag{3.4}$$

Moreover, replacing y with any power series in (3.4) shows that ${}_2\varphi_0(q,\alpha;q,x)$ is the *unique* power series solution of this equation.

If one makes use of the function change $y=(q\alpha x;q)_\infty u$, then equation (3.4) takes the following form:

$$u(x)+x\,u\left(\frac{x}{q}\right)=\frac{1}{(\alpha x;q)_\infty}, \tag{3.5}$$

which has a *unique* power series solution. We shall apply to both sides of (3.5) the q-Borel transform defined by (2.18). Since

$$\hat{\mathcal{B}}_q\left(xu\left(\frac{x}{q}\right)\right)=\xi\hat{\mathcal{B}}_q(u),\quad \hat{\mathcal{B}}_q\left(\frac{1}{(\alpha x;q)_\infty}\right)=(-\alpha\xi;q)_\infty,$$

it follows that, if $\tilde{u}=\hat{\mathcal{B}}_q u$, then (3.5) is transformed as follows:

$$(1+\xi)\,\tilde{u}(\xi)=(-\alpha\xi;q)_\infty\,.$$

Thus, one gets that

$$\tilde{u}(\xi) = \frac{(-\alpha\xi; q)_\infty}{1+\xi}. \tag{3.6}$$

In order to keep in line with the notation $L(x, \mu; q)$ appearing in (3.1), we shall denote by $L(x, \alpha, \mu; q)$ the sum-function given by (2.19) with $\varphi = \tilde{u}(\xi)$ along the q-spiral $[-\mu; q]$, namely:

$$L(x, \alpha, \mu; q) = \sum_{n\in\mathbb{Z}} \frac{1}{\theta(-\frac{\mu}{x}q^n)} \frac{(\alpha\mu q^n; q)_\infty}{1-\mu q^n}. \tag{3.7}$$

By comparing this with (3.1), it is obvious to see that

$$L(x, 0, \mu; q) = L(x, \mu; q). \tag{3.8}$$

Let $\tilde{\mathbb{D}}^* = \pi^{-1}(\mathbb{D}^*) \subset \tilde{\mathbb{C}}^*$, where π denotes the natural evaluation map on $\tilde{\mathbb{C}}^*$. The following statement will be used several times in the whole paper.

Lemma 3.1. *Given $\nu \in \mathbb{Z}_{\geq 0}$, $r > 0$, $c \in \mathbb{C}^*$ and $q \in \tilde{\mathbb{D}}^*$, the functional equations*

$$x^\nu y\left(\frac{x}{q}\right) = c\, y(x), \quad y(xe^{2\pi i}) = y(x) \tag{3.9}$$

have no nontrivial common analytic solution for $x \in \tilde{\mathbb{C}}^*$ with $|x| < r$ provided that either $\nu > 0$ or $c \notin q^{\mathbb{Z}}$ for $\nu = 0$.

Proof. Let f be a common solution in $|x| < r$ for both equations in (3.9). The second equation implies that f can be represented by a convergent Laurent series $\sum_{n\in\mathbb{Z}} a_n x^n$. Using the first equation gives that $a_n q^{-n} = c\, a_{n+\nu}$ for $n \in \mathbb{Z}$, thus

$$a_{k\nu+\ell} = c^{-k} q^{-k(\ell+\frac{1}{2}(k-1)\nu)}\, a_\ell$$

for all $(k, \ell) \in \mathbb{Z}^2$. So, if $\nu > 0$, letting $k \to \pm\infty$ shows that this series converges in $0 < |x| < r$ only when $a_\ell = 0$ for all $\ell \in \mathbb{Z}$. If $\nu = 0$, the nonzero power series solution exists only when $c = q^{-n}$ for some $n \in \mathbb{Z}$, which case is excluded. In this way, one obtains Lemma 3.1. □

Theorem 3.2. *For any given $\mu \in \mathbb{C} \setminus [1; q]$, let ${}_2\varphi_0(q, \alpha; \mu, q, x)$ to be the function defined by (2.22) with $\beta = q$. One has the following properties.*

(1) *The sum-function ${}_2\varphi(q, \alpha; \mu, q, x)$ is the* unique *solution of* (3.4) *that admits only simple poles over the spiral $[\mu; q]$ and that is analytic in $\mathbb{C}^* \setminus [\mu; q]$.*

(2) *Moreover, if* $L(x,\alpha,\mu;q)$ *is given as in* (3.7), *then:*

$$ {}_2\varphi_0(q,\alpha;\mu,q,x)=(\alpha q x;q)_\infty\, L(x,\alpha,\mu;q)\,. \tag{3.10}$$

Proof. In order to simplify the notation, let $F(x)={}_2\varphi_0(q,\alpha;\mu,q,x)$.

(1) By [36, Theorem 1.5.3], it is shown that, for any generic q-spiral, the associated sum-function of a power series solution of a given linear q-difference equation, when this sum is well-defined, is also solution of the same functional equation. As is shown in the above, ${}_2\varphi_0(q,\alpha;q,x)$ satisfies (3.4), so it follows that its sum-function $F(x)$ is also solution of (3.4).

In view of the fact that any basic hypergeometric series ${}_2\varphi_1(\alpha,\beta;\gamma;q,x)$ is analytic in $\mathbb{C}\setminus q^{\mathbb{Z}_{\leq 0}}$, the connection formula (2.23) implies that $F(x)$, with simple poles in $[\mu;q]$, is analytic at least in $\mathbb{C}^*\setminus\big([\mu;q]\cup(\frac{1}{\alpha}q^{\mathbb{Z}_{\leq 0}})\big)$. Particularly, $F(x)$ is analytic at $x=\frac{q}{\alpha}$, thus writing (3.4) as

$$F(x)=\frac{1}{qx}\left(1-(1-q\alpha x)F(qx)\right)$$

implies that $F(x)$ is analytic at $x=\frac{1}{\alpha}$, and then it is analytic at $\frac{1}{q\alpha}$, $\frac{1}{q^2\alpha}$, etc. One finds finally that $F(x)$ is analytic in $\mathbb{C}^*\setminus[\mu;q]$.

In order to see the uniqueness, let $f(x)$ be any analytic solution of (3.4) in $\mathbb{C}^*\setminus[\mu;q]$ with only simple poles over the spiral $[\mu;q]$, and define $g(x)=F(x)-f(x)$. It follows that g is solution of the homogeneous equation of (3.4): $(1-\alpha x)g(x)+xg(\frac{x}{q})=0$. By dividing this last equation by $(\alpha x;q)_\infty$, one gets that

$$\frac{g(x)}{(\alpha q x;q)_\infty}=-x\frac{g(\frac{x}{q})}{(\alpha x;q)_\infty}\,.$$

Thus, if $h(x)=\theta(-\frac{\mu}{x})\,\frac{g(x)}{(\alpha q x;q)_\infty}$, then h is an analytic function for $x\in\mathbb{C}$ such that $|q\alpha x|<1$. Moreover, one can observe that h satisfies the following functional equation:

$$h(x)=-x\frac{\theta(-\frac{\mu}{x})}{\theta(-q\frac{\mu}{x})}\,h\left(\frac{x}{q}\right)=\mu\, h\left(\frac{x}{q}\right).$$

Thus, applying Lemma 3.1 implies that $h=0$ and then $g=0$, so $f=F$.

(2) In the above from (3.5) till (3.7), we proved that $(\alpha q x;q)_\infty$ $L(x,\alpha,\mu;q)$ is solution of (3.4). Moreover, $L(x,\alpha,\mu;q)$ is analytic for x near zero in $\mathbb{C}^*\setminus[\mu;q]$, (3.5) gives that $L(x,\alpha,\mu;q)$ is analytic in $\mathbb{C}^*\setminus\big([\mu;q]\cup(\frac{1}{\alpha}q^{\mathbb{Z}_{<0}})\big)$, which shows that $(\alpha q x;q)_\infty\, L(x,\alpha,\mu;q)$ has the same analyticity-domain as $F(x)$. Thus, the uniqueness stated in (1) implies immediately the equality in (3.10). This finishes the proof of Theorem 3.2. $\square$

In a similar way as what done for $G(x;q)$ in [39], we define, when $q \in (0,1)$,

$$G(x,\alpha;q) = \int_0^\infty \frac{(-\alpha\xi;q)_\infty}{1+\xi}\,\omega\left(\frac{\xi}{x};q\right)\frac{d\xi}{\xi}, \tag{3.11}$$

where the integration is taken along any half-line from 0 to infinity in the cut-plane $\mathbb{C}\setminus(-\infty,0]$ and where $\omega(t;q)$ is given for all t in the Riemann surface of logarithm by the following expression:

$$\omega(t;q) = \frac{1}{\sqrt{2\pi\ln 1/q}}\, e^{\frac{1}{2\ln q}((\log\frac{t}{\sqrt{q}})^2}, \quad t\in\tilde{\mathbb{C}}^*. \tag{3.12}$$

Analogously with (3.8), one can see that

$$G(x,0;q) = G(x;q) = \int_0^\infty \frac{1}{1+\xi}\,\omega\left(\frac{\xi}{x};q\right)\frac{d\xi}{\xi}, \tag{3.13}$$

where $G(x;q)$ is defined in [39, §2].

Remark 3.3. Both functions $L(x,\mu;q)$ and $G(x;q)$ satisfy the following q-difference equation:

$$x\,y\left(\frac{x}{q}\right) + y(x) = 1\,. \tag{3.14}$$

Indeed, letting $\alpha = 0$ into (3.4) yields directly the above functional equation (3.14).

One will say that $G(x,\alpha;q)$ is the *Gq-sum function* of the power series u determined by the q-difference equation in (3.5). By following a general reasoning about the Gq-summation method in [35], one can find that this function satisfies also (3.5).

Theorem 3.4. *The following relation holds for all* $q\in(0,1)$:

$$G(x,\alpha;q) = (\alpha;q)_\infty\, G(x;q) + \frac{\alpha}{(\alpha q x;q)_\infty}\,{}_3\varphi_2(q,\alpha,\alpha q x;0,0;q,q)\,. \tag{3.15}$$

Proof. By replacing $(-\alpha\xi;q)_\infty$ with $(1+\alpha\xi)(-\alpha q\xi;q)_\infty$ in the integral of (3.11), one can notice that

$$\begin{aligned} G(x,\alpha;q) &= (1-\alpha)\,G(x,\alpha q;q) + \alpha\int_0^\infty (-\alpha q\xi;q)_\infty\,\omega\left(\frac{\xi}{x};q\right)\frac{d\xi}{\xi} \\ &= (1-\alpha)\,G(x,\alpha q;q) + \frac{\alpha}{(\alpha q x;q)_\infty}\,. \end{aligned}$$

By iterating this last relation, it follows that

$$
\begin{aligned}
G(x,\alpha;q) &= (\alpha;q)_n\, G(x,\alpha q^n;q) + \sum_{k=0}^{n-1} \frac{\alpha q^k}{(\alpha q^{k+1}x;q)_\infty}\,(\alpha;q)_k \\
&= (\alpha;q)_n\, G(x,\alpha q^n;q) + \frac{\alpha}{(\alpha q x;q)_\infty} \sum_{k=0}^{n-1} (\alpha,\alpha q x;q)_k\, q^k
\end{aligned}
\tag{3.16}
$$

for any positive integer n.

As $n\to\infty$, one finds that

$$G(x,\alpha q^n;q)\to G(x,0;q)=G(x;q),$$

by (3.13). Moreover, by using the definition of ${}_\ell\varphi_m$ given in (1.1), one knows that

$$ {}_3\varphi_2(q,a,b;0,0;q,q) = \sum_{k\geq 0} (a,b;q)_k\, q^k\,. \tag{3.17}$$

Thus, letting $n\to\infty$ in (3.16) gives (3.15). □

As before, let $\tilde{\mathbb{D}}^* = \pi^{-1}(\mathbb{D}^*)\subset\tilde{\mathbb{C}}^*$. It is easy to see that $q\in\tilde{\mathbb{D}}^*$ if and only if $\Re(\log q)<0$. By using [39, Theorem 2.1], one finds that $G(x;q)$ is well defined and analytic for $(x,q)\in\tilde{\mathbb{C}}^*\times\tilde{\mathbb{D}}^*$.

Corollary 3.5. *The function $G(x,\alpha;q)$ can be extended into an analytic function for all $(x,\alpha,q)\in\tilde{\mathbb{C}}^*\times\mathbb{C}\times\tilde{\mathbb{D}}^*$ such that $\alpha x\notin q^{\mathbb{Z}_{<0}}$.*

Furthermore, given $(\alpha,q)\in\mathbb{C}\times\tilde{\mathbb{D}}^*$, the function $(q\alpha x;q)_\infty\, G(x,\alpha;q)$ represents an analytic solution of the q-difference equation (3.4) for $x\in\tilde{\mathbb{C}}^*$.

Proof. The analytic continuation of $G(x,\alpha;q)$ can be done with the help of the equality in (3.15), by noticing that $G(x;q)$ is analytic for $(x,q)\in\tilde{\mathbb{C}}^*\times\tilde{\mathbb{D}}^*$.

A direct computation shows that the function $(q\alpha x;q)_\infty\, G(x,\alpha;q)$ satisfies the same equation as ${}_2\varphi_0(q,\alpha;q,x)$. Thus, Corollary 3.5 follows immediately. □

From now on, the relation in (3.15) will be read in the following fashion for all $(x,\alpha,q)\in\tilde{\mathbb{C}}^*\times\mathbb{C}\times\tilde{\mathbb{D}}^*$:

$$(q\alpha x;q)_\infty\, G(x,\alpha;q) = (\alpha,\alpha q x;q)_\infty\, G(x;q) + \alpha\, {}_3\varphi_2(q,\alpha,\alpha q x;0,0;q,q)\,. \tag{3.18}$$

In fact, one has the following theorem.

Theorem 3.6. *Let $L(x,\alpha,\mu;q)$, $L(x,\mu;q)$, $G(x,\alpha;q)$ and $G(x;q)$ be as in (3.7), (3.8), (3.11) and (3.13), respectively. Then:*

$$L(x,\alpha,\mu;q) - G(x,\alpha;q) = (\alpha;q)_\infty \left(L(x,\mu;q) - G(x;q)\right). \tag{3.19}$$

Proof. Let

$$f(x) = L(x,\alpha,\mu;q) - (\alpha;q)_\infty L(x,\mu;q) \tag{3.20}$$

and

$$g(x) = G(x,\alpha;q) - (\alpha;q)_\infty G(x;q). \tag{3.21}$$

By (3.5), one finds that both $f(x)$ and $g(x)$ are solution of the following modified equation:

$$y(x) + xy(\frac{x}{q}) = \frac{1}{(\alpha x;q)_\infty} - (\alpha;q)_\infty. \tag{3.22}$$

In view of (3.6), applying to this last equation the q-Borel transform yields that

$$(1+\xi)\,\tilde{y} = (-\alpha\xi;q)_\infty - (\alpha;q)_\infty, \tag{3.23}$$

where $\tilde{y} = \hat{\mathcal{B}}_q y$. This implies that $\tilde{y}$ is analytic for all $\xi \in \mathbb{C}$, by observing that the right-hand side of this last equation is equal also to zero for $\xi = -1$. Thus, equation (3.22) admits an analytic solution in a neighborhood of zero in $\mathbb{C}$. Consequently, it follows that the summation along a spiral and the Gq-summation applied to this power series give the same function that equal to the classic sum; thus $f = g$. This finishes the proof. □

Remark 3.7. At each pole $x = \mu q^n$, where $n \in \mathbb{Z}$, it follows that

$$\text{Res}\left(L(x,\alpha,\mu;q) : x = \mu q^n\right) = (\alpha;q)_\infty \,\text{Res}\left(L(x,\mu;q) : x = \mu q^n\right). \tag{3.24}$$

Proof. This follows from the analyticity of f at zero; see (3.20). □

4. Mock Theta-Functions in Terms of Appell–Lerch Series and Mordell Integrals

We start with the following theorem.

Theorem 4.1. *Given $(a,b) \in (\mathbb{C}^* \setminus q^{\mathbb{Z}}) \times (\mathbb{C}^* \setminus q^{\mathbb{Z}_{>0}})$, let R denote the associated functions $\mathcal{R}(a,b;q)$ defined by (2.9). Then*

$$\begin{aligned}\mathcal{R}(q) &= -\frac{1}{b}(a,b;q)_\infty \left(L\left(\frac{a}{qb},a;q\right) - G\left(\frac{a}{qb};q\right)\right) \\ &\quad + \left(1-\frac{1}{b}\right)(a;q)_\infty\, G\left(\frac{a}{qb},qb;q\right).\end{aligned} \tag{4.1}$$

Proof. Letting $x = \frac{a}{qb}$, $\alpha = qb$ and $\mu = a$ into the equality (3.10) of Theorem 3.2 (2) yields that

$$ {}_2\varphi_0\left(q, qb; a, q, \frac{a}{qb}\right) = (qa;q)_\infty\, L\left(\frac{a}{qb}, qb, a; q\right). $$

Thus, considering Theorem 2.4 implies that

$$ \mathcal{R}(q) = -\frac{1}{b}(1-b)\,(a;q)_\infty\; L\left(\frac{a}{qb}, qb, a; q\right). \tag{4.2} $$

Besides, Theorem 3.6 says that

$$ L\left(\frac{a}{qb}, qb, a; q\right) = (qb;q)_\infty\,\left(L\left(\frac{a}{qb}, a; q\right) - G\left(\frac{a}{qb}; q\right)\right) + G\left(\frac{a}{qb}, qb; q\right). $$

Consequently, one gets immediately (4.1) from (4.2), and finishes the proof. □

In the right-hand side of (4.1), one needs to consider $\frac{a}{qb}$ as being an element of $\tilde{\mathbb{C}}^*$ for both terms $G(\frac{a}{qb};q)$ and $G(\frac{a}{qb}, qb; q)$. Indeed, by considering the integral of $G(x;q)$ in (3.11) with $\alpha = 0$ and applying the Residue theorem, one finds the following remark.

Remark 4.2. The following relation holds for $x \in \mathbb{C}^*$:

$$ G(xe^{-2\pi i}; q) - G(x;q) = 2\pi i\,\omega\left(\frac{e^{\pi i}}{x}; q\right). \tag{4.3} $$

By (3.1), the function $L(x,\mu;q)$ is well defined only when $\mu \notin q^{\mathbb{Z}}$. This requires that $a \notin q^{\mathbb{Z}}$ in the above identity in (4.1), due to the term $L(\frac{a}{qb}, a; q)$. The following statement may be used to complete this case.

Theorem 4.3. *If $\nu \in \mathbb{Z}_{\geq 0}$, $a = q^{-\nu}$ and $b \in \mathbb{C}^* \setminus q^{\mathbb{Z}_{>0}}$, then*

$$ \mathcal{R}(q^{-\nu}, b; q) = \frac{(q;q)_\nu}{(\frac{q}{b};q)_\infty}\, b^\nu - \sum_{n=1}^{\nu}(q^{-\nu}, b; q)_n\, q^n\,. \tag{4.4} $$

Proof. Indeed, let $\epsilon \sim 0$ and consider (4.1) with $a = q^{-\nu+\epsilon}$. One the one hand, since

$$ \begin{aligned} (q^{-\nu+\epsilon};q)_\infty &= (q^{-\nu+\epsilon};q)_\nu\,(1-q^\epsilon)\,(q^{1+\epsilon};q)_\infty \\ &= (-1)^\nu\, q^{\frac{1}{2}\nu(-\nu-1+2\epsilon)}\,(q^{1-\epsilon};q)_\nu\,(q^{1+\epsilon};q)_\infty\,(1-q^{-\epsilon}), \end{aligned} $$

using the definition of $L(x,\mu;q)$ in (3.1) yields that

$$\begin{aligned}
&\lim_{\epsilon\to 0}(q^{-\nu+\epsilon};q)_\infty\, L(\frac{q^{-\nu+\epsilon}}{qb},q^{-\nu+\epsilon};q)\\
&\quad = A_\nu \lim_{\epsilon\to 0}(1-q^{-\epsilon})\, L(\frac{q^{-\nu+\epsilon}}{qb},q^{-\nu+\epsilon};q)\\
&\quad = \frac{A_\nu}{\theta(-qb;q)}\,(-qb)^\nu\, q^{\frac{1}{2}\nu(\nu-1)},
\end{aligned}$$

where $A_\nu=(-1)^\nu q^{\frac{1}{2}\nu(-\nu-1)}\,(q;q)_\nu\,(q;q)_\infty$. This implies that

$$\begin{aligned}
&\lim_{a\to q^{-\nu}} -\frac{1}{b}(a,b;q)_\infty\left(L\left(\frac{a}{qb},a;q\right)-G\left(\frac{a}{qb};q\right)\right)\\
&\quad = \frac{(1-\frac{1}{b})\,(q;q)_\nu\,(q,qb;q)_\infty}{\theta(-qb;q)}\,b^\nu\\
&\quad = \frac{(q;q)_\nu}{(\frac{q}{b};q)_\infty}\,b^\nu,
\end{aligned} \tag{4.5}$$

where one used the triple formula (2.21) to get the second equality.

One the other hand, considering (3.18) yields that

$$\begin{aligned}
&(a;q)_\infty\, G\left(\frac{a}{qb},qb;q\right)\\
&\quad = (1-a)\left((qb,qa;q)_\infty\, G(x;q)+qb\,{}_3\varphi_2(q,qb,qa;0,0;q,q)\right).
\end{aligned}$$

Letting $a=q^{-\nu}$ in this equation implies that

$$(q^{-\nu};q)_\infty\, G\left(\frac{q^{-\nu}}{qb},qb;q\right)=(1-q^{-\nu})\,qb\,{}_3\varphi_2(q,qb,q^{1-\nu};0,0;q,q)\,.$$

As $(q^{-\nu};q)_n=0$ for $n>\nu$, using the definition of ${}_\ell\varphi_m$ in (1.1) with $(\ell,m)=(3,2)$ gives that

$$\left(1-\frac{1}{b}\right)qb\,(1-q^{-\nu})\,{}_3\varphi_2(q,qb,q^{1-\nu};0,0;q,q)=-\sum_{n=1}^{\nu}(q^{-\nu},b;q)_n\,q^n\,.$$

Thus, one obtains (4.4) by considering the limit for $a\to q^{-\nu}$ in both sides of (4.1), with the help of (4.5). □

Remark 4.4. When $\nu=0$ and $a=1$, Theorem 4.3 states the following identity:

$$\sum_{n\geq 0}\frac{q^{n^2}}{(q,\frac{q}{b};q)_n}\,(\frac{1}{b})^n=\frac{1}{(\frac{q}{b};q)_\infty}, \tag{4.6}$$

known since Cauchy; see [4, (10.9.2), p. 522].

In addition to the power series $\mathcal{R}(a,b;q,x)$ given in (2.1), let us consider the following Laurent series:

$$M(a,b;q,x)=\sum_{n=-\infty}^{\infty}\frac{(a,b;q)_\infty}{(aq^{-n},bq^{-n};q)_\infty}\left(\frac{x}{q}\right)^n. \tag{4.7}$$

By separating the terms with $n\geq 0$ and $n<0$ in this summation, it follows from (2.15) that

$$M(a,b;q,x)=\mathcal{R}(a,b;q,x)+r(a,b;q,x), \tag{4.8}$$

where $r(a,b;q,x)$ can be expressed as follows:

$$r(a,b;q,x)=\sum_{n\geq 1}(a,b;q)_n\left(\frac{q}{x}\right)^n. \tag{4.9}$$

When $x=1$, we will simply write

$$M(a,b;q)=M(a,b;q,1),\quad r(a,b;q)=r(a,b;q,1). \tag{4.10}$$

By using (1.1) for the definition of ${}_\ell\varphi_m$, (4.9) implies that

$$r(a,b;q)=\sum_{n=1}^{\infty}(a,b;q)_n\,q^n={}_3\varphi_2(q,a,b;0,0;q,q)-1\,. \tag{4.11}$$

Theorem 4.5. *Let $G(x;q)$ and $G(x,\alpha;q)$ be as in (3.13) and (3.11). Let $r(a,b;q)$ be as in (4.11). Then*

$$r(a,b;q)=\left(\frac{1}{b}-1\right)(a;q)_\infty\,G\left(\frac{a}{qb},qb;q\right)-\frac{1}{b}\,(a,b;q)_\infty\,G\left(\frac{a}{qb};q\right). \tag{4.12}$$

Consequently, the values of the series ${}_3\varphi_2(q,a,b;0,0;q)$ can always be expressed by means of (generalized) Mordell integrals.

Proof. Letting $x=\frac{a}{qb}$ and $\alpha=qb$ into (3.18) yields that

$$(qa;q)_\infty\,G\left(\frac{a}{qb},qb;q\right)=(qb,qa;q)_\infty\,G\left(\frac{a}{qb};q\right)+qb\,{}_3\varphi_2(q,qb,qa;0,0;q,q).$$

Besides, by (4.11), it follows that

$$r(a,b;q)=q(1-a)(1-b)\,{}_3\varphi_2(q,qa,qb;0,0;q,q).$$

Thus, one gets immediately (4.12). □

Corollary 4.6. *For all $(a,b)\in\mathbb{C}^{*2}\setminus(q^{\mathbb{Z}_{>0}})^2$,*

$$M(a,b;q)=-\frac{1}{b}\,(a,b;q)_\infty\,L\left(\frac{a}{qb},a;q\right). \tag{4.13}$$

Proof. This comes directly from Theorems 4.1–4.5. □

Important remark — Theorem 4.3 will be generalized, in a forthcoming paper, to the general case where $q = \rho\zeta$ with $\rho \in (0,1)$ and $\zeta^m = 1$ for some positive integer m.

References

[1] G. E. Andrews, Ramanujan's "lost" notebook. I. Partial θ-functions, *Adv. Math.* **41** (1981) 137–172.

[2] G. E. Andrews, Mordell integrals and Ramanujan's "lost" notebook, in *Analytic Number Theory*, ed. M. I. Knopp, Lecture Notes in Mathematics, Vol. 899 (Springer, 1981), pp. 10–48.

[3] G. E. Andrews, Mock theta functions, in *Theta functions*, Proceedings of Symposia in Pure Mathematics, Vol. 49, Part 2 (American Mathematical Society, Providence, RI, 1989), pp. 283–298.

[4] G. E. Andrews, R. Askey, R. Roy, *Special Functions*, Encyclopedia of Mathematics and its Applications, Vol. 71 (Cambridge University Press, Cambridge, 1999).

[5] G. E. Andrews and D. Hickewrson, Ramanujan "lost" notebook VII: The sixth order mock theta functions, *Adv. Math.* **73** (1989) 242–255.

[6] P. Appell, Sur les fonctions doublement périodiques de troisième espèce, *Ann. Sci. École Norm. Sup. Sér.* (3) **1** (1884) 135–164.

[7] P. Appell, Sur les fonctions doublement périodiques de troisième espèce, *Ann. Sci. École Norm. Sup. Sér.* (3) **3** (1884) 9–42.

[8] W. Balser, *Formal Power Series and Linear Systems of Meromorphic Ordinary Differential Equations*, Universitext (Springer, New York, 2000).

[9] B. C. Berndt, *Ramanujan's Notebooks. Part III* (Springer, New York, 1991).

[10] B. C. Berndt, R. J. Evans and K. S. Williams, *Gauss and Jacobi Sums*, (Wiley, New York, 1998).

[11] L. Di Vizio, J.-P. Ramis, J. Sauloy and C. Zhang, Équations aux q-différences, *Gaz. Math.* **96** (2003) 20–49.

[12] L. A. Dragonette, Some asymptotic formulae for the Mock theta series of Ramanujan, *Trans. Amer. Math. Soc.* **73** (1952) 474–500.

[13] F. J. Dyson, Missed opportunities, *Bull.* **78** (1972) 635–652.

[14] H. M. Edwards, *Riemann's Zeta-Function* (Dover Publication, 2001).

[15] N. J. Fine, *Basic Hypergeometric Series and Applications*, Mathematical Surveys and Monographs, Vol. 27 (American Mathematical Society, Providence, RI, 1988).

[16] G. Gasper and M. Rahman, *Basic Hypergeometric Series*, 2nd edn., Encyclopedia of Mathematics and its Applications, Vol. 96 (Cambridge University Press, Cambridge, 2004).

[17] B. Gordon and R. J. McIntosh, A survey of classical Mock theta function, in *Partitions, q-Series, and Modular Forms*, eds. K. Alladi and F. Garvan (Springer, 2012), pp. 95–144.

[18] C. Hermite, Remarques sur la décomposition en éléments simples des fonctions doublement périodiques, *Ann. Fac. Sci. Toulouse Sér.* (1) **2** (1888) pp. 1–12.

[19] C. Lambert and C. Rousseau, The Stokes phenomenon in the confluence of the hypergeometric equation using Riccati equation, *J. Differential Equations* **244** (2008) 2641–2664.

[20] M. Lerch, Bemerkungen zur Theorie der elliptischen Funktionen, *Jahrbuch Fortschr. Math.* **24** (1892) 442–445.

[21] B. Malgrange, Sommation des séries divergentes, *Exposition. Math.* **13**(2–3) (1995) 163–222.

[22] L. J. Mordell, The definite integral $\int_{-\infty}^{\infty} \frac{e^{ax^2+bx}}{e^{cx}+d} dx$ and the analytic theory of numbers, *Acta Math.* **61** (1933) 323–360.

[23] H. Rademacher, *Topics in Analytic Number Theory* (Springer, 1973).

[24] S. Ramanujan, Some definite integrals connected with Gauss's sums, *Messenger Math.* **44** (1915) 75–85; reprinted in *Collected Papers of Srinivasa Ramanujan* (Chelsea, 1962), pp. 59–67.

[25] S. Ramanujan, On certain infinite series, *Messenger Math.* **45** (1916) 11–15; reprinted in *Collected Papers of Srinivasa Ramanujan* (Chelsea, 1962), pp. 129–132.

[26] S. Ramanujan, On some definite integrals, *J. Indian Math. Soc.* **11** (1919) 81–87; reprinted in *Collected Papers of Srinivasa Ramanujan* (Chelsea, 1962), pp. 202–207.

[27] S. Ramanujan, *The Lost Notebook and Other Unpublished Papers: With an Introduction by George E. Andrews* (Springer, Berlin; Narosa Publishing House, New Delhi, 1988).

[28] J.-P. Ramis, Confluence et résurgence, *J. Fac. Sci. Univ. Tokyo Sect. IA Math.* **36** (1989) 703–716.

[29] J.-P. Ramis, Séries divergentes et théories asymptotiques, *Bull. Soc. Math. France* **121** (1993) 1–74.

[30] J. P. Ramis, J. Sauloy and C. Zhang, Local analytic classification of q-difference equations, to appear in *Astérique* **355** (2013) vi+151.

[31] J. P. Ramis and C. Zhang, Développements asymptotiques q-Gevrey et fonction thêta de Jacobi, *C. R. Acad. Sci. Paris, Ser. I* **335** (2002) 277–280.

[32] G. N. Watson, The final problem: An account of the mock theta functions. *J. London Math. Soc.* **11** (1936) 55–80.

[33] D. Zagier, Ramanujan's mock theta functions and their applications (after Zwegers and Ono-Bringmann), *Astérisque* **326** (2009) 143–164.

[34] C. Zhang, Confluence et phénomène de Stokes, *J. Math. Sci. Univ. Tokyo* **3** (1996) 91–107.

[35] C. Zhang, Développements asymptotiques q-Gevrey et séries Gq-sommables, *Ann. Inst. Fourier* **49** (1999) 227–261.

[36] C. Zhang, Une sommation discrète pour des équations aux q-différences linéaires et à coefficients analytiques: théorie générale et exemples, in

Differential Equations and the Stokes Phenomenon, eds. B. L. J. Braaksma *et al.* (World Scientific, 2002), pp. 309–329.

[37] C. Zhang, Remarks on some basic hypergeometric series, in *Theory and Applications of Special Functions*, Devolpments in Mathematics, Vol. 13 (Springer, New York, 2005), pp. 479-491.

[38] C. Zhang, On the modular behaviour of the infinite product $(1-x)(1-xq)(1-xq^2)(1-xq^3)\ldots$, *C. R. Acad. Sci. Paris, Ser. I* **349** (2011) 725–730. See also "A modular-type formula for the infinite product $(1-x)(1-xq)(1-xq^2)(1-xq^3)\ldots$", preprint version 2 (2011); arXiv:0905.1343.

[39] C. Zhang, On the mock-theta behavior of Appell–Lerch series, *C. R. Math. Acad. Sci. Paris* **353** (12)(2015), 1067–1073. See also "Appell–Lerch series viewed as mock theta functions", preprint (2015); https://hal.archives-ouvertes.fr/hal-01230050.

[40] C. Zhang, Only four Euler infinity products are theta-type functions, preprint (2015); https://hal.archives-ouvertes.fr/hal-01230057.

[41] S. Zhou, Shuang, Z. Luo and C. Zhang, On summability of formal solutions to a Cauchy problem and generalization of Mordell's theorem, *C. R. Math. Acad. Sci. Paris* **348** (2010) 753–758.

[42] S. P. Zwegers, Mock θ-functions and real analytic modular forms, in *q-Series with Applications to Combinatorics, Number Theory, and Physics*, Contemporary Mathematics, Vol. 291 (American Mathematical Society, Providence, RI, 2001), pp. 269–277.

Chapter 27

Certain Positive Semidefinite Matrices of Special Functions

Ruiming Zhang

College of Science, Northwest A&F University
Yangling, Shaanxi 712100, P. R. China
ruimingzhang@yahoo.com

Special functions are often defined as a Fourier or Laplace transform of a positive measure, and the positivity of the measure manifests as positive definiteness of certain matrices. The purpose of this expository note is to give a sample of such positive definite matrices to demonstrate this connection for some well-known special functions such as gamma, beta, hypergeometric, theta, elliptic, zeta and basic hypergeometric functions.

Keywords: Special functions; positive semidefinite matrices; special function inequalities.

Mathematics Subject Classification 2010: 14K25, 15A45, 26D15, 33B15, 33C20, 33D05, 33D15, 33E05

1. Introduction

Recall that for $n \in \mathbb{N}$ and $A = (a_{j,k})_{j,k=1}^{n}$, $a_{j,k} \in \mathbb{C}$, A is called positive semidefinite if and only if the quadratic form $\sum_{j,k=1}^{n} a_{j,k} z_j \overline{z_k} \geq 0$ for all $z_1, z_2, \ldots, z_n \in \mathbb{C}$, and it is positive definite if it is positive semidefinite and $\sum_{j,k=1}^{n} a_{j,k} z_j \overline{z_k} = 0$ implies that $z_1 = \cdots = z_n = 0$. Given two positive semidefinite $n \times n$ matrices,

$$A = (a_{j,k})_{j,k=1}^{n}, \quad B = (b_{j,k})_{j,k=1}^{n}, \quad a_{j,k}, b_{j,k} \in \mathbb{C},$$

it is well known that the Schur (Hadamard) product $A \circ B = (a_{j,k}b_{j,k})_{j,k=1}^{n}$ is also positive semidefinite, and it satisfies [4]

$$\det(A \circ B) \geq \det(A) \cdot \det(B).$$

Since all the minors of a positive semidefinite matrix are nonnegative, hence a positive semidefinite matrix of special function entries can yield many inequalities for special functions.

Given a positive measure $\mu(x)$ on the real line $\mathbb{R}$, we denote $\mathcal{H}$ as the Hilbert space of μ-square integrable functions,

$$\mathcal{H} = \left\{ f \,\middle|\, \int_{\mathbb{R}} |f(x)|^2 \, \mu(dx) < \infty \right\},$$

endowed with the usual inner product,

$$\langle f, g \rangle = \int_{\mathbb{R}} f(x)\overline{g(x)}\mu(dx), \quad f,\, g \in \mathcal{H}.$$

Let $\{f_n(x)\}_{n=0}^{N} \subset \mathcal{H}$, where N may be any nonnegative integer in $\mathbb{N}_0$ or equal to ∞, then the finite sections of the Gram matrices [1, 5]

$$G_n = (\langle f_j,\, f_k \rangle)_{j,k=0}^{n}, \quad n = 0, \ldots, N,$$

are positive semidefinite, and they are positive definite if $\{f_n(x)\}_{n=0}^{N} \subset \mathcal{H}$ are linearly independent.

In this article, we shall list some of the positive semidefinite matrices with special function entries. Our method to obtain positive semidefinite matrices is first to isolate an inner product structure associated with the special function, then choose a function set to compute the corresponding Gram matrices, and finally apply Schur product to the obtained more general positive semidefinite matrices. Even though the proofs are completely trivial, these positive semidefinite matrices sometimes may turn out to be very handy. In the following discussion, if any of the formulas below are not specifically referenced, it means that they can be found in [2, 6].

2. Main Results

Recall that the Jacobi θ_3-function is defined by

$$\theta_3(z, q) = \sum_{n=-\infty}^{\infty} q^{n^2} e^{2\pi i n v}, \quad z = e^{2\pi i v}, \ |q| < 1.$$

For $0 < q < 1$, define

$$\mu(x) = \sum_{n=-\infty}^{\infty} q^{n^2} \delta\left(x - n\right),$$

then $\mu(x)$ is a positive measure on $(-\infty,\infty)$. For $n\in\mathbb{N}$, $c_1,c_2,\ldots,c_n\in\mathbb{C}$ and $v_j\in\mathbb{C},\ j=1,\ldots,n$, we have

$$\begin{aligned}\int_{-\infty}^{\infty}\left|\sum_{j=1}^{n}c_je^{2\pi i v_j x}\right|^2 d\mu(x) &= \sum_{j,k=0}^{n}c_j\overline{c_k}\int_{-\infty}^{\infty}e^{2\pi i x(v_j-\overline{v_k})}d\mu(x)\\ &= \sum_{j,k=0}^{n}c_j\overline{c_k}\sum_{\ell=-\infty}^{\infty}q^{\ell^2}e^{2\pi i\ell(v_j-\overline{v_k})}\\ &= \sum_{j,k=0}^{n}c_j\overline{c_k}\theta_3(e^{2\pi i(v_j-\overline{v_k})},q)\geq 0.\end{aligned}$$

Thus, for $n\in\mathbb{N}$, $0<q<1$ and $v_j\in\mathbb{C},\ j=1,\ldots,n$, the matrix

$$(\theta_3(e^{2\pi i(v_j-\overline{v_k})},q))_{j,k=1}^{n} \tag{2.1}$$

is positive semi-definite. For

$$v_{j,\ell}\in\mathbb{C},\ 0<q_\ell<1,\ 1\leq j\leq n,\ 1\leq\ell\leq m,\quad m,\,n\in\mathbb{N},$$

by taking Schur product of the above matrix, we prove that

$$\left(\prod_{\ell=1}^{m}\theta_3(e^{2\pi i(v_{j,\ell}-\overline{v_{k,\ell}})},\ q_\ell)\right)_{j,k=1}^{n} \tag{2.2}$$

is also positive semidefinite.

The Jacobi elliptic function $\mathrm{dn}(2Kv)$ is defined by

$$\mathrm{dn}(2Kv)=\frac{\pi}{K}\sum_{n\in\mathbb{Z}}\frac{q^n}{1+q^{2n}}e^{2n\pi vi},$$

where

$$|q|<1,\quad K=\frac{\pi}{2}\theta_3^2(0,\,q),\quad qe^{2\pi|\Im(v)|}<1.$$

For $0<q<1$, let

$$\mu(x)=\frac{\pi}{K}\sum_{n\in\mathbb{Z}}\frac{q^n\delta(x-n)}{1+q^{2n}}$$

is a positive measure on $\mathbb{R}$. For $n\in\mathbb{N}$, $c_1,\,c_2,\ldots,c_n\in\mathbb{C}$ and

$$qe^{4\pi|\Im(v_j)|}<1,\quad v_j\in\mathbb{C},\ j=1,\ldots,n, \tag{2.3}$$

then the quadratic form

$$\int_{-\infty}^{\infty}\left|\sum_{j=1}^{n}c_je^{2\pi i v_j x}\right|^2 d\mu(x)=\sum_{j,k=1}^{n}c_j\overline{c_k}\mathrm{dn}\left(2K(v_j-\overline{v_k})\right)$$

is nonnegative. Hence, for

$$0 < q < 1,\ K = \frac{\pi}{2}\theta_3^2(0,\, q),\ qe^{4\pi|\Im(v_j)|} < 1,$$

the matrix

$$(\mathrm{dn}(2K(v_j - \overline{v_k})))_{j,k=1}^n \tag{2.4}$$

is positive semidefinite. By taking Schur product, we see that

$$\left(\prod_{\ell=1}^m \mathrm{dn}\left(2K_\ell\left(v_{j,\ell} - \overline{v_{k,\ell}}\right)\right)\right)_{j,k=1}^n, \quad m, n \in \mathbb{N} \tag{2.5}$$

is also positive semidefinite where

$$0 < q_j < 1,\ K_j = \frac{\pi}{2}\theta_3^2(0,\, q_j),\ q_je^{4\pi|\Im(v_{j,\ell})|} < 1,\ 1 \le j \le m.$$

The Riemann zeta function is defined as the analytic continuation of the Dirichlet series [2, 6]

$$\zeta(s) = \sum_{n=1}^{\infty} \frac{1}{n^s}, \quad \Re(s) > 1.$$

For $\Re(s) > 0$, it has the following integral representations:

$$\frac{1}{s-1} - \frac{\zeta(s)}{s} = \int_1^{\infty} u^{-s}\frac{\{u\}}{u}dx, \quad \Re(s) > 0,$$

where $0 \le \{x\} = x - \lfloor x \rfloor < 1$ is the fractional part of x.

Given $n \in \mathbb{N}$, for $c_j,\ s_j \in \mathbb{C},\ \Re(s_j) > 0,\ j = 1, \ldots, n$, we have

$$\int_1^{\infty} \left|\sum_{j=1}^n \frac{c_j}{u^{s_j}}\right|^2 \frac{\{u\}}{u}dx \ge 0.$$

Then for $n \in \mathbb{N}$, the matrix

$$\left(\frac{1}{s_j + \overline{s_k} - 1} - \frac{\zeta(s_j + \overline{s_k})}{s_j + \overline{s_k}}\right)_{j,k=1}^n \tag{2.6}$$

is positive semidefinite where $\Re(s_j) > 0,\ j = 1, \ldots, n$.

For $m,\ n \in \mathbb{N}$, by taking Schur product, we see the matrix

$$\left(\prod_{\ell=1}^m \left\{\frac{1}{s_{j,\ell} + \overline{s_{k,\ell}} - 1} - \frac{\zeta(s_{j,\ell} + \overline{s_{k,\ell}})}{s_{j,\ell} + \overline{s_{k,\ell}}}\right\}\right)_{j,k=1}^n \tag{2.7}$$

is also positive semidefinite where $\Re(s_{j,\ell}) > 0,\ j = 1, \ldots, n,\ \ell = 1, \ldots, m$.

Recall the Euler gamma function $\Gamma(z)$ is defined as the analytic continuation of integral,

$$\Gamma(z) = \int_0^\infty e^{-x} x^{z-1} dx, \quad \Re(z) > 0.$$

Then for $m, n \in \mathbb{N}$, by taking

$$f_k(x) = x^{z_k}, \ d\mu(x) = e^{-x} \frac{dx}{x} 1_{\{x>0\}},$$

then $\langle f_j, f_k \rangle = \Gamma(z_j + \overline{z_k})$, hence matrices

$$\left(\Gamma\left(z_j + \overline{z_k}\right)\right)_{j,k=1}^{n} \tag{2.8}$$

and

$$\left(\prod_{\ell=1}^{m} \Gamma\left(z_{j,\ell} + \overline{z_{k,\ell}}\right)\right)_{j,k=1}^{n} \tag{2.9}$$

are positive semidefinite for $\Re(z_j),\ \Re(z_{j,\ell}) > 0,\ 1 \le j \le n,\ 1 \le \ell \le m.$

For $\lambda > 0$ and $0 < \phi < \pi$, we have the following integral:

$$\frac{1}{2\pi} \int_{-\infty}^{\infty} e^{(2\phi-\pi)x} \left|\Gamma(\lambda + ix)\right|^2 dx = \frac{\Gamma(2\lambda)}{(2\sin\phi)^{2\lambda}},$$

which is the integral of the weight function for the Meixner–Pollaczek orthogonal polynomials. By taking function sequence $f_j(x) = e^{2\phi_j x}$, $j = 1, \ldots, n$ and $d\mu(x) = e^{-\pi x}|\Gamma(\lambda + ix)|^2 dx$, we see the matrix

$$\left(\frac{1}{\sin^{\lambda}(\phi_j + \phi_k)}\right)_{j,k=1}^{n} \tag{2.10}$$

is positive semidefinite where $\lambda > 0,\ \pi/2 > \phi_j > 0,\ j = 1, \ldots, n,\ n \in \mathbb{N}.$ By taking the Schur product, the matrix

$$\left(\frac{1}{\prod_{\ell=1}^{m} \sin^{\lambda_\ell}(\phi_{j,\ell} + \phi_{k,\ell})}\right)_{j,k=1}^{n} \tag{2.11}$$

is also positive semidefinite where

$$\frac{\pi}{2} > \phi_{j,\ell} > 0,\ \lambda_j > 0, \quad 1 \le j \le n, 1 \le \ell \le m, \quad m, n \in \mathbb{N}.$$

Similarly, from the Euler beta function,

$$B(p,q) = \int_0^1 x^p (1-x)^q \frac{dx}{x(1-x)}, \quad \Re(p), \Re(q) > 0,$$

we get that for $m,\, n \in \mathbb{N}$ the matrices, by considering

$$f_j(x) = x^{p_j}(1-x)^{q_j} \;\; d\mu(x) = \frac{dx}{x(1-x)} 1_{\{0<x<1\}},$$

we see the matrices

$$(B(p_j + \overline{p_k},\; q_j + \overline{q_k}))_{j,k=1}^{n}, \tag{2.12}$$

and

$$\left(\prod_{\ell=1}^{m} B\left(p_{j,\ell} + \overline{p_{k,\ell}},\; q_{j,\ell} + \overline{q_{k,\ell}}\right) \right)_{j,k=1}^{n} \tag{2.13}$$

are positive semidefinite where $\Re(p_j),\; \Re(p_{j,\ell}),\; \Re(q_j),\; \Re(q_{j,\ell}) > 0,\; 1 \le j \le n,\; 1 \le \ell \le m.$

The shifted factorial is defined by [1, 3, 5]

$$(a)_n = \frac{\Gamma(a+n)}{\Gamma(a)},\;\; (a_1, \ldots, a_m)_n = \prod_{k=1}^{m} (a_k)_n,$$

where $a,\, n,\, a_1, \ldots,\, a_m \in \mathbb{C}$. Then for $s+1 \ge r$ and $a_1, \ldots, a_r,\, b_1, \ldots, b_s \in \mathbb{C}$, the hypergeometric series is defined by

$${}_rF_s\left(\begin{matrix} a_1, \ldots,\, a_r \\ b_1, \ldots,\, b_s \end{matrix} \middle| \, z \right) = \sum_{n=0}^{\infty} \frac{(a_1, \ldots, a_r)_n}{(1,\, b_1, \ldots, b_s)_n} z^n,$$

where $z \in \mathbb{C}$ for $s+1 > r$ and $|z| < 1$ for $s+1 = r$. Given $m,\, n \in \mathbb{N}$ and nonnegative integers $r_\ell,\, s_\ell,\; 1 \le \ell \le m$, we assume that $a_{1,\ell}, \ldots, a_{r_\ell,\ell},\, b_{1,\ell}, \ldots, b_{s_\ell,\ell},\; 1 \le \ell \le m$ are so chosen such that

$$\frac{(a_{1,\ell}, \ldots, a_{r_\ell,\ell})_n}{(b_{1,\ell}, \ldots, b_{s_\ell,\ell})_n} \ge 0, \quad 1 \le \ell \le m,\, n \in \mathbb{N}_0.$$

Then the matrix

$$\left(\prod_{\ell=1}^{m} {}_{r_\ell}F_{s_\ell}\left(\begin{matrix} a_{1,\ell}, \ldots,\, a_{r_\ell,\ell} \\ b_{1,\ell}, \ldots,\, b_{s_\ell,\ell} \end{matrix} \middle| \, z_{j,\ell}\overline{z_{k,\ell}} \right) \right)_{j,k=1}^{n} \tag{2.14}$$

is positive semidefinite where for $1 \le \ell \le m$ we assume that $z_{j,\ell} \in \mathbb{C},\; 1 \le j \le n$ if $s_\ell + 1 > r_\ell$, and $|z_{j,\ell}| < 1,\; 1 \le j \le n$ if $s_\ell + 1 = r_\ell$.

From

$$(1 - 2^{1-s})\Gamma(s)\zeta(s) = \int_0^\infty u^s \frac{du}{u(e^u + 1)}, \quad \Re(s) > 0,$$

we see the matrix

$$\left(\prod_{\ell=1}^{m} \left(1 - 2^{1 - s_{j,\ell} - \overline{s_{k,\ell}}}\right) \Gamma(s_{j,\ell} + \overline{s_{k,\ell}}) \zeta(s_{j,\ell} + \overline{s_{k,\ell}}) \right)_{j,k=1}^{n} \tag{2.15}$$

is positive semidefinite where n, $m \in \mathbb{N}$ and $\Re(s_{j,\ell}) > 0,\ j = 1, \dots n, \ell = 1, \dots, m$.

From

$$(1-2^{1-s})\Gamma(s+1)\zeta(s) = \int_0^\infty u^s \frac{e^u du}{(e^u+1)^2}, \quad \Re(s) > 0,$$

we see that

$$\left(\prod_{\ell=1}^{m} \left(1-2^{1-s_{j,\ell}-\overline{s_{k,\ell}}}\right) \Gamma(s_{j,\ell}+\overline{s_{k,\ell}}+1)\zeta(s_{j,\ell}+\overline{s_{k,\ell}})\right)_{j,k=1}^{n} \tag{2.16}$$

is positive semidefinite where $\Re(s_{j,\ell}) > 0,\ j = 1, \dots n, \ell = 1, \dots, m,\ n, m \in \mathbb{N}$.

Because

$$\frac{\pi(s)_p\zeta(p+s)}{\sin \pi s} = \int_0^\infty ((-1)^{p-1}\psi^{(p)}(1+x))\frac{dx}{x^s}, \quad \Re(s) \in (0,1),\ p \in \mathbb{N},$$

where

$$\frac{(-1)^{p-1}\psi^{(p)}(1+x)}{p!} = \sum_{n=1}^{\infty} \frac{1}{(x+n)^{p+1}}$$

is positive on $(0, \infty)$. Hence, the matrix

$$\left(\prod_{\ell=1}^{m} \frac{(s_{j,\ell}+\overline{s_{k,\ell}})_{p_\ell}\zeta(p_\ell+s_{j,\ell}+\overline{s_{k,\ell}})}{\sin \pi(s_{j,\ell}+\overline{s_{k,\ell}})}\right)_{j,k=1}^{n} \tag{2.17}$$

is positive semidefinite where n, m, $p \in \mathbb{N}$ and $0 < \Re(s_{j,\ell}) < \frac{1}{2}$, $p_\ell \in \mathbb{N}$, $j = 1, \dots, n$, $\ell = 1, \dots, m$.

The Riemann Xi function

$$\Xi(z) = -\frac{(1+4z^2)}{8\pi^{(1+2iz)/4}}\Gamma\left(\frac{1+2iz}{4}\right)\zeta\left(\frac{1+2iz}{2}\right)$$

is an even entire function of genus 1. It satisfies

$$\Xi(z) = \int_{-\infty}^{\infty} e^{-itz}\phi(t)dt.$$

Thus, the matrix

$$\left(\prod_{\ell=1}^{m} \Xi\left(z_{j,\ell}-\overline{z_{k,\ell}}\right)\right)_{j,k=1}^{n} \tag{2.18}$$

is positive semidefinite where $n, m \in \mathbb{N}$ and $z_{j,\ell} \in \mathbb{C},\ j = 1, \ldots, n,\ \ell = 1, \ldots, m$.

The Hurwitz zeta function $\zeta(s,a)$ is defined as the analytic continuation of

$$\zeta(s,a) = \sum_{n=0}^{\infty} \frac{1}{(n+a)^s}, \quad \Re(s) > 1,\ -a \notin \mathbb{N}_0.$$

For $\Re(s) > 0,\ a > 0$ and $m, n \in \mathbb{N}$, from

$$\left(\frac{1}{a^s} + \frac{1}{(1+a)^s} + \frac{(1+a)^{1-s}}{s-1} - \zeta(s,a)\right) = s\int_1^{\infty} \frac{\{x\}dx}{(x+a)^{s+1}},$$

we see that

$$\left(\prod_{\ell=1}^{m}\left(\frac{a_\ell^{-s_{j,\ell}-\overline{s_{k,\ell}}} + (1+a_\ell)^{-s_{j,\ell}-\overline{s_{k,\ell}}} - \zeta(s_{j,\ell}+\overline{s_{k,\ell}}, a_\ell)}{s_{j,\ell}+\overline{s_{k,\ell}}} + \frac{(1+a_\ell)^{1-s_{j,\ell}-\overline{s_{k,\ell}}}(s_{j,\ell}+\overline{s_{k,\ell}})^{-1}}{(s_{j,\ell}+\overline{s_{k,\ell}}-1)}\right)\right)_{j,k=1}^{n} \tag{2.19}$$

is positive semidefinite for $\Re(s_{j,\ell}),\ a_\ell,\ 1 \le j \le n,\ 1 \le \ell \le m$.

Since for $\Re(s) > 0$ and $a > 0$, we have

$$\frac{\Gamma(s)}{4^s}\left(\zeta\left(s, \frac{a+1}{4}\right) - \zeta\left(s, \frac{a+3}{4}\right)\right) = \int_0^{\infty} x^s \frac{dx}{2xe^{ax}\cosh x}.$$

Then, the matrix

$$\left(\prod_{\ell=1}^{m}\Gamma(s_{j,\ell}+\overline{s_{k,\ell}})\left(\zeta\left(s_{j,\ell}+\overline{s_{k,\ell}}, \frac{a_\ell+1}{4}\right) - \zeta\left(s_{j,\ell}+\overline{s_{k,\ell}}, \frac{a_\ell+3}{4}\right)\right)\right)_{j,k=1}^{n} \tag{2.20}$$

is positive semidefinite for $\Re(s_{j,\ell}),\ a_\ell > 0,\ 1 \le j \le n,\ 1 \le \ell \le m$.

The Lerch's transcendent $\Phi(z,s,a)$ is defined as the analytic continuation of

$$\Phi(z,s,a) = \sum_{n=0}^{\infty} \frac{z^n}{(a+n)^s}, \quad -z \notin \mathbb{N}_0,\ |z| < 1;\ \Re(s) > 1,\ |z| = 1.$$

For $a > 0,\ z < 1$ and $\Re(s) > 0$, it has the following integral representation:

$$\Gamma(s)\Phi(z,\ s,\ a) = \int_0^{\infty} x^s \frac{dx}{xe^{ax}\left(1-ze^{-x}\right)}.$$

Then for $m, n \in \mathbb{N}$, the matrix

$$\left(\prod_{\ell=1}^{m} \Gamma(s_{j,\ell} + \overline{s_{k,\ell}})\Phi(z_\ell, s_{j,\ell} + \overline{s_{k,\ell}}, a_\ell)\right)_{j,k=1}^{n} \tag{2.21}$$

is positive semidefinite for $\Re(s_{j,\ell}),\ a_\ell > 0,\ z_\ell < 1,\ 1 \le j \le n, 1 \le \ell \le m$.

For $q \in (0,1)$ and $m \in \mathbb{N}$, let [1, 3, 5]

$$(z;\, q)_\infty = \prod_{n=0}^{\infty}(1 - zq^n),\ (z,\, q)_n = \frac{(z;\, q)_\infty}{(zq^n;\, q)_\infty}, \quad z, n \in \mathbb{C},$$

and

$$(z_1,\, z_2, \dots, z_m;\, q)_n = \prod_{j=1}^{m}(z_j;\, q)_n,\ z_j,\, n \in \mathbb{C}.$$

For $\alpha_1, \alpha_2, \alpha_3, \alpha_4 > 0$, a weaker form of the Askey–Wilson beta integral is

$$\int_0^\pi \frac{(1-q)^5(q;\, q)_\infty^6|(e^{2i\theta};\, q)_\infty|^2 d\theta}{|(q^{\alpha_1}e^{i\theta},\, q^{\alpha_2}e^{i\theta},\, q^{\alpha_3}e^{i\theta},\, q^{\alpha_4}e^{i\theta};\, q)_\infty|^2 2\pi}$$
$$= \frac{(1-q)^{2(\alpha_1+\alpha_2+\alpha_3+\alpha_4)}\prod_{1\le j<k\le 4}\Gamma_q(\alpha_j+\alpha_k)}{\Gamma_q(\alpha_1+\alpha_2+\alpha_3+\alpha_4)}.$$

By taking the function sequence,

$$u_k(\theta) = \frac{1}{|(q^{\alpha_{k,1}}e^{i\theta},\, q^{\alpha_{k,2}}e^{i\theta};\, q)_\infty|^2},$$

and the Schur product, we see the matrix

$$\left(\prod_{\ell=1}^{m} \frac{\Gamma_q(\alpha_{j,1,\ell} + \alpha_{k,1,\ell})\Gamma_q(\alpha_{j,2,\ell} + \alpha_{k,2,\ell})}{\Gamma_q(\alpha_{j,1,\ell} + \alpha_{j,2,\ell} + \alpha_{k,1,\ell} + \alpha_{k,2,\ell})}\right)_{j,k=1}^{n} \tag{2.22}$$

is positive semidefinite for

$$m,\, n \in \mathbb{N},\ \alpha_{k,1,\ell},\, \alpha_{k,2,\ell} > 0, \quad 1 \le k \le n, 1 \le \ell \le m.$$

For $r,\, s \in \mathbb{N}_0$, $0 < q < 1$, and $a_1, \dots, a_r,\ b_1, \dots, b_s \in \mathbb{C}$, let

$${}_rA_s^{(\alpha)}(a_1, \dots, a_r;\, b_1, \dots,\, b_s;\, q;\, z) = {}_r\phi_s\left(\begin{matrix} a_1, \dots,\, a_r \\ b_1, \dots,\, b_s \end{matrix}\,\middle|\, q,\, z\right)$$
$$= \sum_{n=0}^{\infty} \frac{(a_1, \dots, a_r;\, q)_n}{(b_1, \dots, b_s;\, q)_n} q^{\alpha n^2} z^n.$$

Then it is clear that for $0 < q < 1$ and $s+1 \geq r$, we have

$$
\begin{aligned}
&{}_rA_s^{((s+1-r)/2)}(a_1, \dots a_r;\ q,\ b_1, \dots,\ b_s\ ;\ q;\ (-1/\sqrt{q})^{s+1-r} z) \\
&= {}_r\phi_s\left(\begin{matrix} a_1, \dots, a_r \\ b_1, \dots, b_s \end{matrix}\,\middle|\, q,\ z\right),
\end{aligned}
$$

where the basic hypergeometric series ${}_r\phi_s$ is defined by [1, 3, 5]

$$
{}_r\phi_s\left(\begin{matrix} a_1, \dots, a_r \\ b_1, \dots, b_s \end{matrix}\,\middle|\, q,\ z\right) = \sum_{n=0}^{\infty} \frac{(a_1, \dots, a_r;\ q)_n}{(q, b_1, \dots, b_s;\ q)_n} (-q^{(n-1)/2})^{n(s+1-r)} z^n.
$$

For $n, m \in \mathbb{N}$ and $0 < q_\ell < 1$, $1 \leq \ell \leq m$, we assume that $z_{j,\ell} \in \mathbb{C}$, $1 \leq j \leq n$ when $\alpha_\ell > 0$, and when $\alpha_\ell = 0$, we restrict $z_{j,\ell}$, $1 \leq j \leq n$ inside an open disk with certain radius less than 1 to ensure the associated series converges. Furthermore, we also assume that

$$
\frac{(a_{1,\ell}, \dots, a_{r_\ell,\ell};\ q_\ell)_n}{(b_{1,\ell}, \dots, b_{s_\ell,\ell};\ q_\ell)_n} \geq 0, \quad n \in \mathbb{N}_0.
$$

Then the matrix

$$
\left(\prod_{\ell=1}^{m} {}_{r_\ell}A_{s_\ell}^{(\alpha_\ell)}\left(\begin{matrix} a_{1,\ell}, \dots, a_{r_\ell,\ell} \\ b_{1,\ell}, \dots, b_{s_\ell,\ell} \end{matrix}\,\middle|\, q_\ell,\ z_{j,\ell}\overline{z_{k,\ell}}\right)\right)_{j,k=1}^{n} \tag{2.23}
$$

are positive semidefinite.

For [3]

$$
q = e^{2\pi i\sigma},\ p = e^{2\pi i\tau}, \quad \Im(\sigma),\ \Im(\tau) > 0,
$$

let

$$
\theta(x;\ p) = (x,\ p/x;\ p)_\infty,\ \theta(x_1, \dots, x_m;\ p) = \prod_{k=1}^{m} \theta(x_k;\ p), \quad m \in \mathbb{N},
$$

$$
(a;\ q,\ p)_n = \begin{cases} \displaystyle\prod_{k=0}^{n-1} \theta(aq^k;\ p), & n \in \mathbb{N}, \\ 1, & n = 0, \\ 1\Big/\displaystyle\prod_{k=0}^{-n-1} \theta(aq^{n+k};\ p), & -n \in \mathbb{N}, \end{cases}
$$

and

$$
(a_1,\ a_2, \dots, a_m;\ q,\ p)_n = \prod_{k=1}^{m} (a_k;\ q,\ p)_n, \quad m \in \mathbb{N}.
$$

For $r,\ s \in \mathbb{N}_0$, the modular series ${}_rE_s$ and ${}_rG_s$ are defined by [3]

$$
\begin{aligned}
&{}_rE_s(a_1, \ldots, a_r;\ b_1, \ldots, b_s;\ q,\ p;\ A,\ z)\\
&\quad = {}_rE_s\left(\begin{matrix} a_1, \ldots,\ a_r \\ b_1, \ldots,\ b_s \end{matrix}\ \middle|\ q,\ p;\ A,\ z\right) = \sum_{n=0}^{\infty} \frac{(a_1,\ a_2, \ldots, a_r;\ q,\ p)_n}{(q,\ b_1, \ldots, a_s;\ q,\ p)_n} A_n z^n,
\end{aligned}
$$

and

$$
\begin{aligned}
&{}_rG_s\left(c_1, \ldots, c_r;\ d_1, \ldots, d_s;\ q,\ p;\ B,\ z\right)\\
&\quad = {}_rG_s\left(\begin{matrix} c_1, \ldots,\ c_r \\ d_1, \ldots,\ d_s \end{matrix}\ \middle|\ q,\ p;\ B,\ z\right) = \sum_{n=-\infty}^{\infty} \frac{(c_1, \ldots, c_r;\ q,\ p)_n}{(d_1, \ldots, d_s;\ q,\ p)_n} B_n z^n,
\end{aligned}
$$

where the sequences A_n and B_n are so-chosen to guarantee the above series converge in their appropriate domains.

Given $m \in \mathbb{N},\ 1 \le \ell \le m$, we let

$$
r_\ell,\ s_\ell \in \mathbb{N}_0,\ q_\ell = e^{-2\pi\sigma_\ell},\ p_\ell = e^{-2\pi\tau_\ell},\ \sigma_\ell,\ \tau_\ell > 0,
$$

and $a_{1,\ell}, \ldots, a_{r_\ell,\ell};\ b_{1,\ell}, \ldots, b_{s_\ell,\ell};\ c_{1,\ell}, \ldots, c_{r_\ell,\ell};\ d_{1,\ell}, \ldots, d_{s_\ell,\ell}$ and sequences $\{A_{n,\ell}\}_{n=0}^{\infty}$ and $\{B_{n,\ell}\}_{n=0}^{\infty}$ to ensure that

$$
\begin{aligned}
\frac{(a_{1,\ell}, \ldots, a_{r_\ell,\ell};\ q_\ell,\ p_\ell)_n}{(q_\ell,\ b_{1,\ell}, \ldots, a_{s_\ell,\ell};\ q_\ell,\ p_\ell)_n} A_{n,\ell} \ge 0, &\quad n \in \mathbb{N}_0,\\
\frac{(c_{1,\ell}, \ldots, c_{r_\ell,\ell};\ q_\ell,\ p_\ell)_n}{(d_{1,\ell}, \ldots, d_{s_\ell,\ell};\ q_\ell,\ p_\ell)_n} B_{n,\ell} \ge 0, &\quad n \in \mathbb{Z},
\end{aligned}
$$

and the series ${}_{r_\ell}E_{s_\ell}\ {}_{r_\ell}G_{s_\ell}$ are convergent on some symmetric open subsets $S_\ell \subset \mathbb{C}$ and $T_\ell \subset \mathbb{C}$ with respect to the complex conjugation, then the matrices

$$
\left(\prod_{\ell=1}^{m} {}_{r_\ell}E_{s_\ell}\left(\begin{matrix} a_{1,\ell}, \ldots,\ a_{r_\ell,\ell} \\ b_{1,\ell}, \ldots,\ b_{s_\ell,\ell} \end{matrix}\ \middle|\ q_\ell,\ p_\ell;\ A_\ell,\ z_{j,\ell}\overline{z_{k,\ell}}\right)\right)_{j,k=1}^{n}, \tag{2.24}
$$

and

$$
\left(\prod_{\ell=1}^{m} {}_{r_\ell}G_{s_\ell}\left(\begin{matrix} c_{1,\ell}, \ldots,\ c_{r_\ell,\ell} \\ d_{1,\ell}, \ldots,\ d_{s_\ell,\ell} \end{matrix}\ \middle|\ q_\ell,\ p_\ell;\ B_\ell,\ w_{j,\ell}\overline{w_{k,\ell}}\right)\right)_{j,k=1}^{n} \tag{2.25}
$$

are positive semidefinite where $z_{j,\ell} \in S_\ell,\ 1 \le j \le n,\ 1 \le \ell \le m$ and $w_{j,\ell} \in T_\ell,\ 1 \le j \le n,\ 1 \le \ell \le m$.

Acknowledgments

This work was partially supported by National Natural Science Foundation of China, grant No. 11371294 and Northwest A&F University of China.

References

[1] G. E. Andrews, R. A. Askey and R. Roy, *Special Functions* (Cambridge University Press, Cambridge, 1999).
[2] A. Erdelyi, *Higher Transcendental Functions*, Vols. I–III, ed. A. Erdelyi (McGraw-Hill, 1953). Reprinted by Krieger Publishing Co., Malabar, FL, 1981.
[3] G. Gasper and M. Rahman, *Basic Hypergeometric Series*, 2nd edn. (Cambridge University Press, Cambridge, 2004).
[4] R. A. Horn and C. R. Johnson, *Matrix Analysis* (Cambridge University Press, Cambridge, 1992).
[5] M. E. H. Ismail, *Classical and Quantum Orthogonal Polynomials in one Variable*, paperback edn. (Cambridge University Press, Cambridge, 2009).
[6] F. W. J. Olver, D. W. Lozier, R. F. Boisvert and C. W. Clark, *NIST Handbook of Mathematical Functions* (Cambridge University Press, 2010).

Index

A

addition formula, 382
addition formula for Gegenbauer polynomials, 381
addition formula for Legendre polynomials, 374
Ahmed, Shafique, 254, 259, 265
Al-Salam, Waleed, 3
Al-Salam–Carlitz II polynomials, 161
Al-Salam–Carlitz polynomials, 157–158, 161, 163–167
Allé, 383
Alladi, 239–240
Allaway, William, 3
ambiguity function, 122
Andrews, 240
Andrews–Gordon identities, 7, 10, 18
angular Fourier transform, 122
annihilating ideal, 403–404, 406, 410–413
annihilating operator, 398, 402, 404, 409, 412
annihilator, 403, 412
Apéry-like numbers, 173
Apéry-like sequences, 172
Appell function, 440
Appell's hypergeometric function, 185
Appell–Lerch series, 525
arithmetic–geometric mean (AGM), 171, 180
Askey, Richard A., 251, 254, 283
associated Legendre function, 379
asymptotic expansions, 497

B

Bailey pair, 343, 347, 354
Bailey's lemma, 343, 354
Banach algebra, 452–453, 457, 490
basic Bessel functions, 395
basic cosine function, 408
basic exponential function, 408
basic hypergeometric functions, 394, 415
basic hypergeometric series, 21, 446, 451, 454, 554
basic sine function, 408
Bateman, H., 435
Brafman, F., 425
binomial sums, 169
binomial theorem, 172
biorthogonality, 388
birth and death process polynomials, 252
bisection type method, 277
bispectral property, 142
block diagram for SAFT domain convolution, 129
bracket, 307, 310–317
bracket series, 310–317
Bressoud identities, 16

C

canonical, 285
canonical sign-pattern, 274, 286, 289
canonical sign-pattern matrix, 288
Carlitz, L., 429
characteristic exponents, 438
Charris, Jairo, 4
Chebyshev polynomials, 253, 374
Chebyshev-type weight, 150
chirp modulation, 128
Christoffel, Elwin Bruno, 100
Christoffel–Darboux formula, 144
Chu–Vandermonde formula, 432
classical, 157
Clausen's formula, 448
combinatorial, 394
combinatorial interpretation, 241
commutation relations, 452, 485
complete elliptic integral, 427, 437, 439
complete monotonicity, 263
computer algebra, 394, 415
conical function, 379
conjecture of Askey, 283
conjugate Bailey pair, 356
connection relations, 164, 167
contiguous relation, 438, 441
creative telescoping, 401, 406–407, 409, 411, 413
Cristoffel–Darboux formula, 149

D

Darboux, Jean Gaston, 100
Dedekind eta function, 329
definite integrals, 307–308, 310, 317
differential operator, 140, 145, 151
differentiation formula, 144
discrete time SAFT, 133
discrete top-down Markov problem, 267, 269, 278
discrete top-down problems, 270
divided differences, 259, 260–261
divisibility, 171, 173
Draux, Andre, 96
dual addition formula, 385
dual addition formula for Gegenbauer functions, 390
dual addition formula for Gegenbauer polynomials, 386
dual addition formula for Hermite polynomials, 387
dual product formula, 388

E

Elbert, Arpad, 251, 254–255, 259, 265
elementary Lagrange interpolation polynomials, 270, 276
elliptic integrals, 308
elliptic functions, 427
Euler beta function, 549
Euler gamma function, 310, 549
Euler's transformation, 432
extended Parseval identity, 228
extremal polynomial, 269–272, 276, 282

F

factorial, 156
Favard's result, 158, 163
Favard, Jean, 90
Fejer, Leopold, 2
Feldheim, Ervin, 2
Feynman diagrams, 307–309, 313, 315
Fourier *convolution theorem*, 126
Fractional Fourier Transform, 122
free parameters, 311–312, 316
free space propagation, 124
Freeden and Nashed, 228, 234
Freud polynomials, 252
Frobenius, Ferdinand Georg, 86–89, 94–95

G

Gantmacher, Felix, 86
Gatteschi, Luigi, 265
Gauss, 400
Gauss quadrature, 257
Gauss's theorem, 240
Gauss–Lobatto quadrature points, 285–287

Gauss–Weistrass summability, 226, 228
Gaussian circle problem, 213, 217
Gaussian quadrature, 163, 260, 276
Gautschi, Walter, 251, 265, 267, 289
Gegenbauer, 383
Gegenbauer functions, 380
Gegenbauer generating function, 427, 430
Gegenbauer polynomials, 376, 403, 406–407, 425
Gegenbauer, Leopold, 252
generalized hypergeometric function, 497
generalized Laguerre polynomials, 252
generating functions, 164, 167, 171, 174, 179
Gröbner basis, 404, 406, 415
guessing, 396
Gundelfinger, Simund, 86

H

Hager, William W., 269, 289
Hankel determinant polynomials, 85
Hankel, Hermann, 85–86, 89, 94–96
Hardy's Conjecture, 217–218
Hardy–Landau identities, 213
Hecke–Rogers double sums, 353–354
Hellmann–Feynman theorem, 252
Hermite functions, 120–121
Hermite polynomials, 120, 377, 383
higher monotonicity, 259
Hilbert, David, 93
holonomic, 175
holonomic closure properties, 397, 399, 404, 410, 412
holonomic functions, 173, 405
HolonomicFunctions package, 398–399, 401, 403, 405, 411, 415
holonomic systems approach, 394, 404, 415
hook number, 337
Hou, Hongyan, 289
Hurwitz zeta function, 552
hyperbolic transformation, 124
hypergeometric functions, 171
hypergeometric series, 445, 451–452, 454, 550
hypergeometric term, 405
hypergeometric transformation, 428, 436–439, 443

I

ILA conjecture, 252, 254–255, 259
incomplete monotonicity, 263
index of a representations, 308, 311, 313
indicator, 310
inequalities, 546
integral kernel, 145
integral operator, 140, 142, 145, 148, 150
interpolation polynomial, 271, 280, 282, 288
inverse factorial expansions, 501
Iohvidov, Iosif, 85, 87, 96–98
Ismail, Mourad E. H., 1, 251–254, 265–266, 289, 394, 407
Ismail–Letessier conjecture, 255, 259
Ismail–Zhang formula, 407, 415

J

Jackson's q-Bessel function, 407
Jacobi θ_3-function, 546
Jacobi elliptic function, 547
Jacobi functions, 380
Jacobi polynomials, 252, 256, 277, 279–281, 283–284, 286, 376, 426, 440
Jacobi theta functions, 329–330
Jacobi triple product identity, 10, 329, 365
Jacobi weight, 143
Jacobi weight function, 285
Jacobi, Cari Gustav, 86

K

Klein cubic, 330
Klein polynomials, 319
Klein quartic, 321, 330

Knuth, Donald, 394
Koelink, 386
Kronecker, Leopold, 85, 90–93

L

ladder operator, 441–443
Laforgia's conjecture, 253, 255, 259
Laforgia, Andrea, 251, 253, 265, 266
Lagrange interpolation formula, 271
Laguerre polynomails, 252
Landau, 139–140, 151
Lanzewizky, I. L., 2
lattice, 158
lattice points inside circles, 215, 218
Lebesgue constant, 273
Lebesgue function, 273
Legendre function, 425
Legendre polynomials, 171, 184, 253, 269, 273, 278, 287, 426, 434, 437
lens transformation, 124
Leopardi, Paul, 265
Lerch's transcendent, 552
Letessier, Jean, 251, 253–254, 266
Linear Canonical Transform, 122, 124
linear congruence, 344
linearization formula for Gegenbauer polynomials, 383
linearization formula for Legendre polynomials, 374
Lorch, Lee, 251, 266
Lucas congruences, 173, 175

M

Markov brothers' inequalities, 267
Markov's theorem, 252–253
Markov, Andrei A., 268
Markov, Vladimir A., 268
massless bubble diagram, 313
Mathematica, 394–395, 415
Matlab software for orthogonal polynomials, 255
matrix orthogonal polynomials, 145
matrix-valued orthogonal polynomials, 139
maximum property, 280
Mehler's formula, 121
Meijer G-function, 497
Meixner polynomials, 252
Meixner–Pollaczek orthogonal polynomials, 549
Mendeleev, Dmitri I., 268
method of brackets, 309–310, 315, 317
Minkowski's Theorem, 227
Mizony, 389
Möbius group, 441
mock theta-functions, 524
modified q-Lommel polynomials, 396
modular forms, 170–171, 174–176, 179
monodromy, 425, 441
monotonicity conjecture for zeros, 251
Mueller two-dimensional approach, 227–228
Muldoon, Martin E., 252–254, 259, 265–266
multinomial power, 315
multivariate Shannon-type sampling, 222

N

Nashed ad Walter, 234
Newton–Cotes formulas, 284
Newton–Cotes quadrature formula, 283
noncommutative hypergeometric series, 452–453, 456, 458, 471, 473–474, 479, 486, 490–491, 493
noncommutative parameters, 451, 459–461, 466, 468, 471–472, 479, 481–482, 484–486, 490–491, 493–494
normality of the largest bound, 275, 283, 286
number of divisors, 241

O

Offset Linear Canonical Transform, 127
Olver, Frank W. J., 266
optimal control, 269

Ore algebras, 402–404, 412
orthogonal collocation methods, 269
orthogonal polynomials, 2, 171
overpartitions, 17

P

∂-finite functions, 404
Paley–Wiener spaces, 231
partition weights, 241
partitions, 240, 248, 343
Pearson-type distributional equations, 163
periodic table of elements, 268
phase-free SAFT convolution, 132
phase-space transformations, 123
Pochhammer symbol, 407
Poisson kernel, 427, 436–440
Pollak, 139–140, 151
polynomials (or orthogonal polynomails)
  q-Hermite, 4
  Legendre, 1
  Pollaczek, 1
  Rogers, 4
  sieved, 4
positive semidefinite, 546–555
positivity of the Newton–Cotes formulas, 283
product formula, 382, 446
product formula for Legendre polynomials, 374
Property M, 277–278, 283, 285–286
Property M_k, 275, 288–289
Property M_n, 275

Q

q-Bessel function, 411–412
q-binomial coefficient, 9
q-binomial theorem, 9, 245, 400
q-cosine function, 409–410
q-derivative, 156
q-difference equation, 400–402, 409, 414, 525
q-Gegenbauer polynomials, 408, 412
qGeneratingFunctions package, 396, 400, 414
q-holonomic recurrence, 396
q-hypergeometric term, 412
q-Jackson integral, 157, 159
q-Lommel polynomials, 401
q-Pochhammer symbol, 156, 166
q-rising factorials, 395
q-series, 21
q-shift equation, 414
q-sine function, 409–410
q-Vandermonde sum, 12
qZeil package, 395, 398
quadratic transformation of hypergeometric functions, 253
quasi-definite, 157

R

Racah polynomials, 378, 384
Ramanujan's differential system, 325
Ramanujan's master theorem, 310
Ramanujan-type formulas for $1/\pi$, 171
random matrix theory, 497
Rao, Anil V., 289
recurrence relation, 158
reproducing kernel, 233
Richard Askey, 374
Riemann Xi function, 551
Riemann zeta function, 548
Rogers–Ramanujan identities, 7, 19, 21

S

SAFT convolution and product theorem, 129
SAFT convolution/filtering, 128
SAFT duality principle, 135
Schwarz, H. A., 427, 439, 440
Schwinger parameterization, 309
Shadrin, Aleksei, 268, 289
Shannon, 139–140
Shannon-type sampling theorem, 230

Siafarikas, Panayiotis, 251, 254–255, 259, 265
sign-pattern matrix, 273, 279, 281, 285
sign-pattern vector, 283
singular overpartitions, 19
Slepian, 139–140, 151
smallest parts, 343, 347
Sottas, G., 283
Special Affine Fourier Transformation, 123, 126
special functions, 546
Spigler, Renato, 253–254, 259, 265
spt-crank, 345, 347, 350, 357
Stanton, Dennis, 1
Stieltjes, Thomas Joannes, 89, 253
Sturm comparison theorem, 251
Sturmian methods, 253
sub-range Freud polynomials, 252
subnormality of the largest bound, 275
summation formula, 459–460, 478–479, 483
supercongruences, 173–174
supnormality of the largest bound, 275, 284
symmetric differential operator, 148
symmetric expansions, 21
symmetric second-order differential operator, 146
Szegö's formulations of the Sturm comparison theorem, 253
Szegö, Gabor, 2, 266

T

t-cores, 337
theta constants, 330
three-term recurrence relation, 163
time-and-band limiting, 140
time-band limiting, 139
time-frequency representations, 123
Toeplitz pattern, 279
top-down Markov problems, 268–269
triangular number, 242
twisted Eisenstein series, 323

U

ultraspherical polynomials, 251, 256–257, 259–260, 407, 426
unilateral basic hypergeometric series, 165

V

Viennot, Gerard, 4

W

Watson, G. N., 427, 437
weight function, 157
weighted partition identities, 239
Wigner distribution, 122
Wilson polynomials, 381, 389
Wilson, Jim, 1

Z

Zeilberger's algorithm, 395
Zeilberger, Doron, 394
zeros, 157
zeros of Freud polynomials, 252
zeros of Jacobi polynomials, 252
zeros of orthogonal polynomials, 257
Zhang, Ruiming, 266, 407

◎ 编辑手记

世界著名数学家 G. Choquet 曾指出：

> 数学中的基本结构可以和那种成批制造产品的机器相比，这种机器只能制造一种产品，但是短时间内可以制造出许多件，并且每一件都造得相当完美. 当今的年轻数学家应该首先在学习大型数学机器上多花些时间，也就是学习基本结构，进而从中受益.

本书就是这样一部与大型数学机器相关的英文数学专著，中文书名或可译为《正交多项式和 q – 级数的前沿》.

本书是一部文集，其主编有两位，一位是 M. 祖海尔 · 纳什德(M. Zuhair Nashed)，美国人，中佛罗里达大学教授，另一位是李欣(Xin Li)，美国人，1990 年加入中佛罗里达大学并任助理教授，2001 年起任教授，在数学、统计学、计算机视觉、教育等领域发表论文 70 余篇.

正如世界著名数学家 R. L. Wilder 所指出：

> 数学的普遍性看来是它在各个文化因素中最显著的特征. 但是有些数学还具有明显的民族特征. 长期以来，人们认为法国数学家偏爱函数论，英国数学家对应用感兴趣，德国数学家着重

于数学基础，意大利数学家感兴趣于几何，而美国的数学则以其抽象特征著称.然而，尽管有这些由于文化影响造成的差异，数学在今日还是可以看作具有绝大多数其他人类活动所没有的普遍性.

正如两位主编在前言中所介绍：

本书的灵感来自于“正交多项式与 q - 级数的国际会议”(2015年5月10日—12日在奥兰多的中佛罗里达大学举行)中的特邀演讲.本次会议是为纪念穆拉德·伊斯梅尔(Mourad Ismail)教授70岁生日而举办的.人们强烈地感觉到，准备一本重点介绍正交多项式、q - 级数和逼近理论、数论、组合学、应用和计算谐波分析中的相关主题的重要研究方向的书是很必要的.出席会议的世界科技出版公司的执行主编罗谢尔·科荣泽(Rochelle Kronzek)女士很赞同这个想法，并最终达成了出版协议.

本书共27章，其中探讨了正交多项式、q - 级数和涉及上述相关主题研究前沿的各种主题.我们努力写出一本能激励年轻研究人员，并供这些领域的其他几代研究人员进行愉快阅读的著作.

本书所涉内容相当广泛，版权编辑李丹女士为了方便读者阅读，特翻译了本书的目录如下：

1. Mourad Ismail
2. 二项 Andrews-Gordon-Bressoud 恒等式
3. 极良平衡基本超几何级数的对称展开
4. Hahn 差分算子的 Sturm-Liouville 理论
5. 实序列 Hankel 行列式问题的可解性
6. 特殊仿射 Fourier 变换的卷积与乘积定理
7. 矩阵正交多项式的时间和频带限制的进一步研究
8. 复参数 Al-Salam-Carlitz 多项式的正交性

9. 隐藏的 AGM,隐藏模块化

10. 正交多项式与 Painlevé 超越函数的渐近性

11. 从 Gauss 圆问题到多元 Shannon 抽样

12. 加权分拆恒等式和除数和

13. 关于超球多项式零点的 Ismail-Letessier-Askey 单调性猜想

14. 逼近论中的离散自顶向下 Markov 问题

15. 量子旋度的超对称性

16. 实验数学中的方括号方法

17. 平衡模参数化

18. Bailey 引理变分中的最小部分函数

19. 与对偶积公式相关的对偶加法公式

20. 基本超几何函数的完整工具

21. Ismail 和 Valent 积分的直接求值

22. Gegenbauer 多项式的代数生成函数

23. Bailey 超几何函数的两个乘积公式的 q - 类比

24. 非交换超几何级数的求和公式

25. 广义超几何函数的渐近性

26. 根据 Appell-Lerch 级数拟合 Ramanujan 的三阶 θ 函数

27. 特殊函数的若干正半定矩阵

本书内容相当有趣,比如第 11 章"从 Gauss 圆问题到多元 Shannon 抽样".

由离散的部分组成的最简单的图形 是平面正方形点格[①](图 1). 要得到这样的点格,我们需要在平面上画出面积为一单位的正方形的四个顶点,把正方形沿其一边的一个方向移出一边之长,画出所得的两个新顶点. 设想这样的步骤可以在一个方向上和它的相反方向上无止境地继续进行,这样我们在平面上就得到了由距离相等的两列点所组成的长条. 把长条在跟它垂直的方向上移出一边之长,画出得到的新顶点. 假定这个步骤又

① "点格"是指由点组成的"格子","格子点"是指点格中的任意点.

可以在两个方向上无止境地进行,这样就作出了全部的正方形点格.正方形点格还可以定义为在平面笛卡儿坐标系中整数坐标点的集合.

当然,由一个点格的四个顶点不仅可以组成正方形,也可以组成其他图形,例如平行四边形.容易知道,由平行四边形出发得到的点格与由正方形出发得到的相同,只要这个平行四边形不以格子点为顶点,而它的内部和边上不含任何其他格子点就行了(否则,用这种办法得出的点不能包括格子点的全体).试取任意的这样一个平行四边形,可以看出,它的面积等于原来正方形的面积(图1).关于这句话的严格证明,将在下文给出.

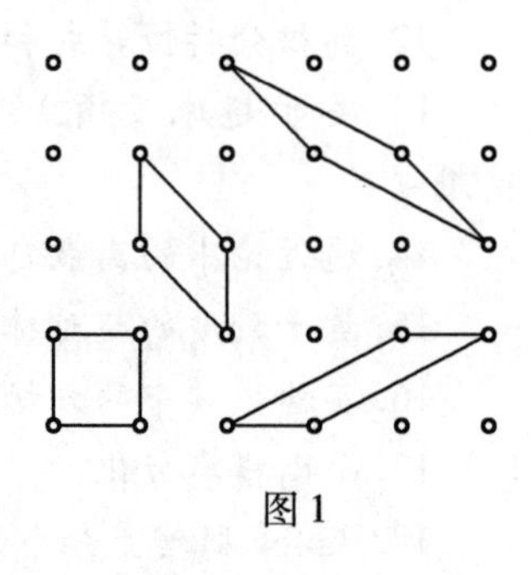
图1

尽管点格如此简单,但它却可以引起重要的数学研究,其中最早的是 Gauss 的研究,Gauss 试图在半径等于 r 的圆面中找出格子点的数目 $f(r)$,这里圆心是一格子点,而 r 是一整数. Gauss 凭实验找出许多对于不同 r 值的 $f(r)$ 值来,如

$$r=10,\ f(r)=317$$
$$r=20,\ f(r)=1\ 257$$
$$r=30,\ f(r)=2\ 821$$
$$r=100,\ f(r)=31\ 417$$
$$r=200,\ f(r)=125\ 629$$
$$r=300,\ f(r)=282\ 697$$

Gauss 研究函数 $f(r)$ 的目的是想借这个结果来计算 π 的近似值.每个基本正方形的面积,按假设都等于一个单位,因此 $f(r)$ 就等于被左下顶点在圆面之内或边上的所有正方形覆盖着的面积 F(图2).这样说来,$f(r)$ 与圆面面积 $r^2\pi$ 之差不超过与圆面相交的(包括计算进去的或未计算的)正方形面积的总和 $A(r)$,即

$$|f(r)-\pi r^2|\leqslant A(r)$$
$$\left|\frac{f(r)}{r^2}-\pi\right|\leqslant\frac{A(r)}{r^2}$$

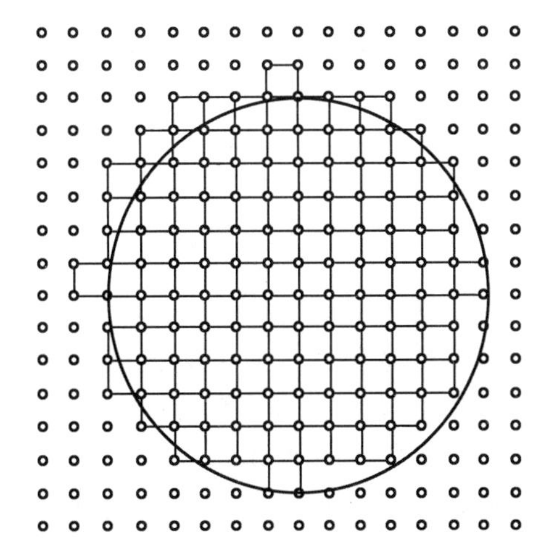

图 2

要求 $A(r)$ 的估值是不难的. 单位正方形的两点间的最大距离是$\sqrt{2}$,所以所有跟圆相交的正方形都落在一个圆环内,环的宽度为$2\sqrt{2}$,而夹在半径分别为$r+\sqrt{2}$和$r-\sqrt{2}$的两圆之间. 圆环的面积是

$$B(r) = [(r+\sqrt{2})^2 - (r-\sqrt{2})^2]\pi = 4\sqrt{2}\pi r$$

但 $A(r) < B(r)$,所以

$$\left|\frac{f(r)}{r^2} - \pi\right| < \frac{4\sqrt{2}\pi}{r}$$

由此再运用极限过程,就得到我们所要找的公式

$$\lim_{r\to\infty}\frac{f(r)}{r^2} = \pi \tag{1}$$

现在把 Gauss 求得的 $f(r)$ 的值代入上式,得出下列 π 的近似值($\pi = 3.141\ 59\cdots$)

$$r = 10, \frac{f(r)}{r^2} = 3.17$$

$$r = 20, \frac{f(r)}{r^2} = 3.142\ 5$$

$$r = 30, \frac{f(r)}{r^2} = 3.134$$

$$r = 100, \frac{f(r)}{r^2} = 3.141\,7$$

$$r = 200, \frac{f(r)}{r^2} = 3.140\,725$$

$$r = 300, \frac{f(r)}{r^2} = 3.141\,07$$

式(1)的另一个应用是证明凡是能产生正方形点格的任一平行四边形,它的面积都等于1. 为了证明,我们设想圆域内的每一个格子点都是一基本平行四边形的一顶点,并约定所有这些顶点在平行四边形相同的位置上. 让我们把平行四边形覆盖着的面积 F 跟圆面的面积比较一下. 这里也存在由半径为 $r+c$ 和 $r-c$ 的圆所作成的圆环面积 $B(r)$ 产生的微小误差,其中 c 是一基本平行四边形的两点间的最大距离而与 r 无关. 假定基本平行四边形的面积为 a,则面积 F 为 $a \cdot f(r)$,由此得出公式

$$|af(r) - r^2\pi| < B(r) = 4rc\pi$$

于是

$$\left|\frac{af(r)}{r^2} - \pi\right| < \frac{4c\pi}{r}$$

$$\lim_{r\to\infty}\frac{f(r)}{r^2} = \frac{\pi}{a}$$

我们已经证明过

$$\lim_{r\to\infty}\frac{f(r)}{r^2} = \pi$$

据此①,我们的断言 $a = 1$ 得证.

现在我们转来研究一般的"单位点格",即根据由单位正方形产生正方形点格的方法,由面积等于一单位的任意平行四边形产生的点格. 这里也是一样,不同的平行四边形可以产生相同的点格,但是这些平行四边形的面积必定等于一单位. 证法如同正方形点格的情形.

对任意这样一个单位点格来说,两个格子点间最短的距离 c

---

① 在这个证明中,也可以不必是圆,而是任何的一块平面,只要这块平面的边上被覆盖的那一部分对整个的这块平面来说,是任意地狭窄就可以.

是一个特征值．一方面，单位点格的 c 可以任意小，这只要考虑以 c 和 $\frac{1}{c}$ 为边长的矩形产生的点格就明白了．但是另一方面，c 显然不能庞大无边，否则点格就不是单位点格了．因此 c 必有一上界，今试决定这个上界．

从任一单位点格中任意选取距离最短（比如说，最短距离是 c）的两点（图3），通过这两点作一直线 g．按照点格的定义，在这条直线上应有无穷多的、间隔为 c 的其他点．在平行于 g 且与 g 的距离为 $\frac{1}{c}$ 的直线 h 上也应有无穷多的格子点，但在两条平行线 g 和 h 之间的区域内不应包含任何格子点．以上两项事实都是从所讨论的点格是单位点格推出来的．以 c 为半径、以 g 上的所有的格子点为圆心作圆，全部的圆将覆盖着一平面长条，这个长条的边界是一些圆弧．长条内部的任一点至少同一个格子点的距离小于 c，所以按照 c 的定义，这点不是格子点．因此 $\frac{1}{c}$ 必大于或等于长条边界线到 g 的最短距离．这个距离显然是以 c 为边长的等边三角形的高．于是就有

$$\frac{1}{c} \geqslant \frac{c}{2}\sqrt{3}$$

$$c \leqslant \sqrt{\frac{2}{\sqrt{3}}}$$

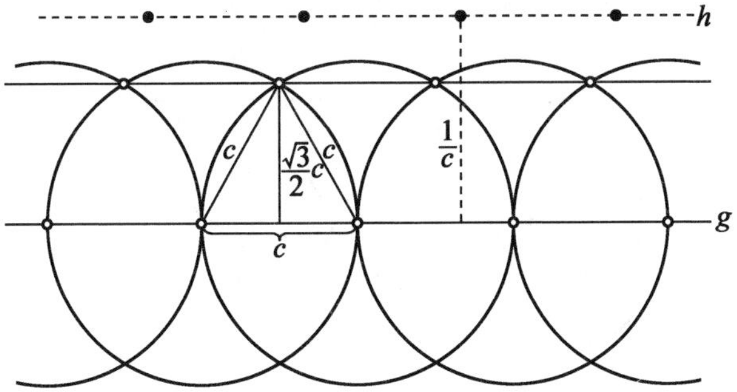

图3

$\sqrt{\frac{2}{\sqrt{3}}}$ 就是所求的 c 的上界,而且确实也有一个点格达到这个极值,因为,这样的点格可以用由两个等边三角形拼成的平行四边形产生出来.

单位点格经过膨胀或收缩后可以得到任意大小的点格. 设 a^2 是某个点格的基本平行四边形的面积, C 是两个格子点间的最短距离,那么应有

$$C \leqslant a \sqrt{\frac{2}{\sqrt{3}}}$$

等号当且仅当点格是由等边三角形组成的时候才成立. 所以,对一给定的最短距离,这样的点格含有最小的基本平行四边形. 但是,大的图形的面积近似地等于在那个区域里格子点数乘以基本平行四边形的面积,所以,在具有给定的最短距离的一切点格中,等边三角形点格(在给定的大区域里)含有最多的点.

若以每一格子点为圆心,以两个格子点间的最短距离的一半为半径作圆,则得到一组彼此相切但没有覆盖现象的圆. 这样作出来的圆组称为(正则的)圆形格子式堆积. 我们说一种圆形格子式堆积较另一种为紧密,如果前一种堆积能在相当大的已知区域内放进较多的圆. 据此得知,等边三角形点格产生最紧密的圆形格子式堆积(图 4).

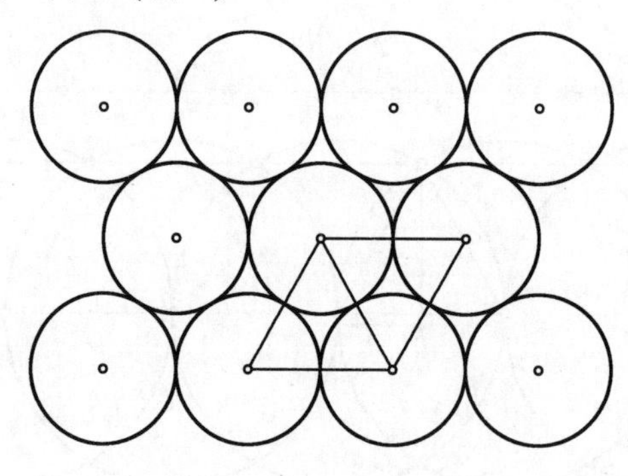

图 4

作为圆形格子式堆积的密度的度量,我们取包含在已知区域内各圆的总面积除以这个区域的面积. 对于充分大的区域来

说这个值显然近似等于一个圆的面积除以基本平行四边形的面积. 等边三角形点格给出密度的最大值

$$D = \frac{1}{2\sqrt{3}}\pi = 0.289\pi$$

点格在许多数论问题中都有用处.

1. 莱布尼兹级数

$\frac{\pi}{4} = 1 - \frac{1}{3} + \frac{1}{5} - \frac{1}{7} + \cdots$. 如同前面讲的,假定 $f(r)$ 代表在以 r 为半径、以一格子点为圆心的圆内,平面正方形单位点格的点的个数. 我们把圆心作为笛卡儿坐标的原点,并把格子点配以整数坐标. 这样 $f(r)$ 便是适合 $x^2 + y^2 \leqslant r^2$ 的所有整数偶 x, y 的个数. 因为 $x^2 + y^2$ 总表示整数 n,因此可以如此去求 $f(r)$:对每一整数 $n \leqslant r^2$,找出能以两个整数的平方和表示它的方法的个数,把所有这些分解法的个数相加. 在数论中有这样一个定理:一整数 n 表示为两个整数的平方和的方法的个数,等于 n 的具有 $4k+1$ 和 $4k+3$ 形式的因子的个数之差的四倍. 但在这种表示方法中,如 $n = a^2 + b^2$, $n = b^2 + a^2$, $n = (-a)^2 + b^2$ 等须认为是不同的分解式,因为它们对应不同的格子点. 这样说来,每种分解式可以导出 8 种分解式(但如 $a = \pm b$, $a = 0$, $b = 0$ 是例外). 作为本定理的例子,我们来看 $n = 65$ 这个数,这个数共有 4 个因子:1,5,13,65. 所有这些因子都可以写成 $4k+1$ 的形式,但是 $4k+3$ 形式的却一个也没有. 因此所求的差是 4,根据我们的定理,65 这个数可以写成 16 种两个数的平方之和(换句话说,以原点为中心、以 $\sqrt{65}$ 为半径的圆周通过 16 个格子点). 实际上,$65 = 1^2 + 8^2$, $65 = 4^2 + 7^2$,每个式子又可以写成 8 种形式.

根据这个定理,对于每个正整数 $n \leqslant r^2$,从形为 $4k+1$ 的因子的个数减去形为 $4k+3$ 的因子的个数,再将各差相加,则得出 $\frac{1}{4}(f(r) - 1)$. 不过,如果把加减的次序作适当的变更,可以简化许多. 这就是说,从所有的 $n \leqslant r^2$ 中形为 $4k+1$ 的因子数之和减去所有的形为 $4k+3$ 的因子数之和. 要决定第一个和,把形为

$4k+1$ 的各数按从小到大的次序写成 1,5,9,13,…,所有大于 r^2 的数一概不计. 每个这样的数累加几次不超过 r^2,在计算因子时,它就应该计算几次. 因此,1 应有$[r^2]$ 次,5 有$\left[\frac{r^2}{5}\right]$次,这里的$[a]$ 一般表示不超过 a 的最大整数. 所以,我们所求的 $4k+1$ 形式的因子总数是$[r^2]+\left[\frac{r^2}{5}\right]+\left[\frac{r^2}{9}\right]+\cdots$. 由符号$[a]$ 的定义,这个级数只要方括号中的分母一超过分子就中断了. 对于 $4k+3$ 形式的因子也可以同样地处理,从而得出这种因子总数为

$$\left[\frac{r^2}{3}\right]+\left[\frac{r^2}{7}\right]+\left[\frac{r^2}{11}\right]+\cdots$$

我们还要从第一个和中减去第二个和. 因为这两个和的项数都是有限的,所以级数的次序可以随意变更. 如用下法,则在过渡到极限 $r\to\infty$ 时较为方便. 我们把得出的结果写成下面的形式

$$\frac{1}{4}(f(r)-1)=[r^2]-\left[\frac{r^2}{3}\right]+\left[\frac{r^2}{5}\right]-\left[\frac{r^2}{7}\right]+\left[\frac{r^2}{9}\right]-\left[\frac{r^2}{11}\right]+\cdots$$

为了弄清楚这个级数什么时候中断,我们姑且假定 r 是奇数,这样级数一共有$\frac{r^2+1}{2}$项. 这个和中的符号加减相间,同时绝对值不增加. 因此,如果在$\left[\frac{r^2}{r}\right]=[r]=r$这项处就中断了,由此所产生的误差最多等于最后的一项 r,所以我们可以把这个误差写作 vr,这里的 v 是一个真分数. 如果我们要把留下的$\frac{1}{2}(r+1)$ 项的方括号去掉,结果每项的误差都小于 1,所以总的误差又可以写作 $v'r$,而 v' 是一个真分数. 于是,我们有

$$\frac{1}{4}(f(r)-1)=r^2-\frac{r^2}{3}+\frac{r^2}{5}-\frac{r^2}{7}+\cdots\pm r\pm vr\pm v'r$$

各项除以 r^2 后,得

$$\frac{1}{4}\left(\frac{f(r)}{r^2}-\frac{1}{r^2}\right)=1-\frac{1}{3}+\frac{1}{5}-\frac{1}{7}+\cdots\pm\frac{1}{r}\pm\frac{v+v'}{r}$$

让 r 无限增加(取所有的奇数值),则$\frac{f(r)}{r^2}$趋近于 π,这是容易证明的. 这样我们就导出了莱布尼兹级数

$$\frac{1}{4}\pi = 1 - \frac{1}{3} + \frac{1}{5} - \frac{1}{7} + \cdots$$

2. 二次形式的最小值

命

$$f(m,n) = am^2 + 2bmn + cn^2$$

是以实数a,b,c为系数且行列式$D = ac - b^2 = 1$的二次形式. 在这种情形下a不能等于零. 不妨假定$a > 0$. 众所周知,满足这些条件的$f(m,n)$是正定的,也就是说,对于所有的实数偶m,n,除$m = n = 0$之外,$f(m,n)$是正的. 以下我们要证明:不管如何选择系数a,b,c,只要它们适合条件$ac - b^2 = 1$且$a > 0$,总有两个不全为零的整数m,n,使$f(m,n) \leqslant \frac{2}{\sqrt{3}}$成立.

这句断言可以从单位点格中两点间的最短距离得知. 利用条件$D = 1$,再用通常的配方法,$f(m,n)$可写成

$$f(m,n) = \left(\sqrt{a}m + \frac{b}{\sqrt{a}}n\right)^2 + \left(\sqrt{\frac{1}{a}}n\right)^2$$

现在考虑平面笛卡儿坐标系中的坐标为

$$x = \sqrt{a}m + \frac{b}{\sqrt{a}}n$$

$$y = \sqrt{\frac{1}{a}}n$$

的点,这里的m和n取所有的整数值. 根据解析几何的初等定理知道这些点应该作成单位点格. 因为它们可以从正方形单位点格$x = m, y = n$经过行列式等于1的平面仿射变换

$$x = \sqrt{a}\xi + \frac{b}{\sqrt{a}}\eta$$

$$y = \sqrt{\frac{1}{a}}\eta$$

而得. 但是现在$f(m,n) = x^2 + y^2$,所以当m和n取所有的整数值时,$\sqrt{f(m,n)}$表示从原点到相应的格子点的距离. 点格中有一点P,可使这个距离不超过$\sqrt{\frac{2}{\sqrt{3}}}$. 由此对$P$的两个整数坐标$m$,

n 来说,我们就有

$$f(m,n) \leqslant \frac{2}{\sqrt{3}}$$

这就是要证明的.

这个结果可以用来解决通过有理数来逼近实数的问题. 设 a 是任一实数,我们考虑形式

$$f(m,n) = \left(\frac{an-m}{\varepsilon}\right)^2 + \varepsilon^2 n^2 = \frac{1}{\varepsilon^2}m^2 - 2\frac{a}{\varepsilon^2}mn + \left(\frac{a^2}{\varepsilon^2} + \varepsilon^2\right)n^2$$

这个形式的行列式是

$$D = \frac{1}{\varepsilon^2}\left(\frac{a^2}{\varepsilon^2} + \varepsilon^2\right) - \frac{a^2}{\varepsilon^4} = 1$$

这里的 ε 是任意正数. 我们可以找到两个数 m,n,使其满足不等式

$$\left(\frac{an-m}{\varepsilon}\right)^2 + \varepsilon^2 n^2 \leqslant \frac{2}{\sqrt{3}}$$

从这里显然可得两个不等式

$$\left|\frac{an-m}{\varepsilon}\right| \leqslant \sqrt{\frac{2}{\sqrt{3}}}, \ |\varepsilon n| \leqslant \sqrt{\frac{2}{\sqrt{3}}}$$

又得[①]

$$\left|a - \frac{m}{n}\right| \leqslant \frac{\varepsilon}{|n|}\sqrt{\frac{2}{\sqrt{3}}}, \ |n| \leqslant \frac{1}{\varepsilon}\sqrt{\frac{2}{\sqrt{3}}}$$

如果 a 不是有理数,那么第一个不等式的左边必不等于零. 所以假如给定的 ε 的值越来越小,则必得无穷多的这样的数偶 m,n,因为此时 $\left|a - \frac{m}{n}\right|$ 必无限减少. 用这种方法我们得到与无理数 a 任意逼近的有理数 $\frac{m}{n}$. 另一方面,借用第二个不等式可以消去 ε,从而得到

① 对于充分小的 ε,是可以用 n 除的,因为如果 n 等于零,那么不等式 $|an-m| \leqslant \varepsilon\sqrt{\frac{2}{\sqrt{3}}}$ 就不成立了.

$$\left|a-\frac{m}{n}\right| \leqslant \frac{2}{\sqrt{3}} \cdot \frac{1}{n^2}$$

这样一来,我们有了一个近似分数的序列,其近似的程度与分母的平方成正比例.这种近似值的分母不需要很大而近似程度就相当高.

3.闵可夫斯基定理

闵可夫斯基建立了一个关于点格的定理,这个定理虽然很简单,但是能够解决数论上许多用别的方法难以解决的问题.为了容易明白起见,这里我们不讲定理的一般形式,而只讨论一个特殊的情形,这种情形不但非常容易表述,而且从方法上说,已包含了主要的一切.这个定理是说:

如果以边长为 2 的正方形覆盖平面上任一已知单位点格,且使正方形的中心与一格子点重合,那么在这个正方形的内部或边上还有一个格子点.

要证明这一定理,设想在点格平面上划定了任意一个大的区域,譬如以大的 r 为半径、以一个格子点为圆心的圆的内部和圆周.对于在这个区域中的每一格子点都以这点为中心作一个以 s 为边长的正方形(图 5).现在要求不管选择 r 多么大,也没有两个正方形是覆盖着的,在这个要求之下,来估计一下边长 s.在所说的区域内有 $f(r)$ 个格子点,因为正方形不得互相覆盖,所以它们的总面积为 $s^2f(r)$.另一方面,这些正方形都落在较大的半径为 $r+2s$ 的同心圆内,因此,我们得到下面的不等式

$$s^2f(r) \leqslant \pi(r+2s)^2$$

或

$$s^2 \leqslant \frac{\pi r^2}{f(r)}\left(1+\frac{2s}{r}\right)^2$$

如果将 s 固定,让 r 无限增加,我们知道不等式的右方趋近于 1.所以,我们得出 s 的条件是

$$s \leqslant 1$$

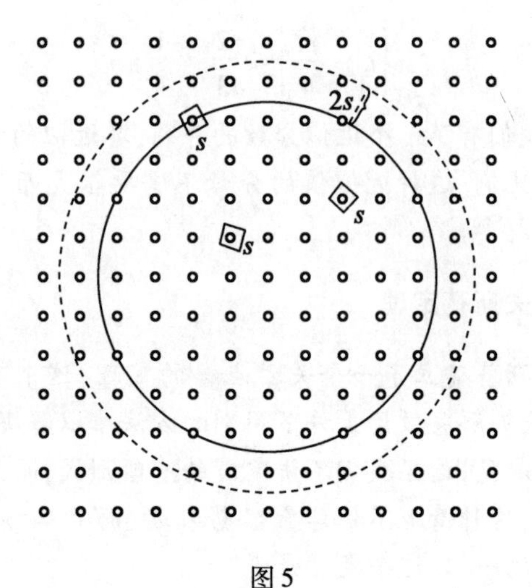

图5

因为两个正方形只可能有覆盖或不覆盖两种情形,由此可知,对于任意正数ε,不管它多么小,如果从边长为$1+\varepsilon$的正方形出发,必然得到覆盖的正方形. 直到现在,我们并没有假定正方形的相互位置,因此我们可以把正方形绕其中心做任何角度的转动. 让我们假定所有的正方形都平行地放着. 现取出以A和B为中心的两个互相覆盖的正方形a,b来看(按照题设a,b即是格子点),则线段AB的中点M必落在这两个正方形的内部(图6).

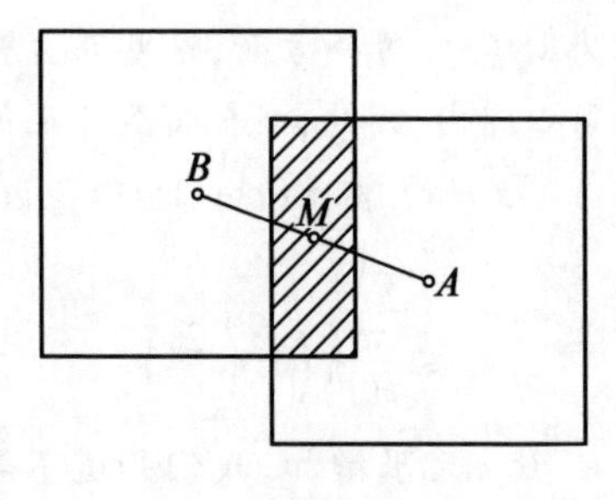

图6

为了简明起见,今后凡是遇到两个格子点连线的中点,例如M,我们一概用点格的"平分点"一词代表. 现在我们可以推出这样的结论:以一格子点为中心且以$1+\varepsilon$为边长的任一正方形a,

一定包含一平分点. 因为如以所有其他格子点为中心作一些正方形与 a 全等且同方向,则必有某些地方覆盖起来,又因为所有的正方形在这个图形上都有相同资格,所以 a 自己也必部分地被另一正方形 b 覆盖着,因此 a 必包含一平分点. 现在我们可以用反证法来完成定理的证明. 假如以一格子点 A 为中心、以 2 为边长的正方形的内部或边上再没有另外的格子点,那么我们可以把这个正方形在保持边的方向和中心位置不变的条件下稍微地扩大一下,使得扩大后的正方形 a' 的边长为 $2(1+\varepsilon)$,也不包含其他的格子点. 另一方面,我们把这个正方形也在保持边的方向和中心的条件下收缩到原边的一半,就得到以 A 为中心、以 $1+\varepsilon$ 为边长的正方形 a,刚才证明过这个正方形必包含一平分点 M. 这就是个矛盾,因为将 AM 延长一倍到 B,则 B 必是一格子点,而且从 a 和 a' 的相互位置来看,可知这个格子点必在 a' 之内(图 7).

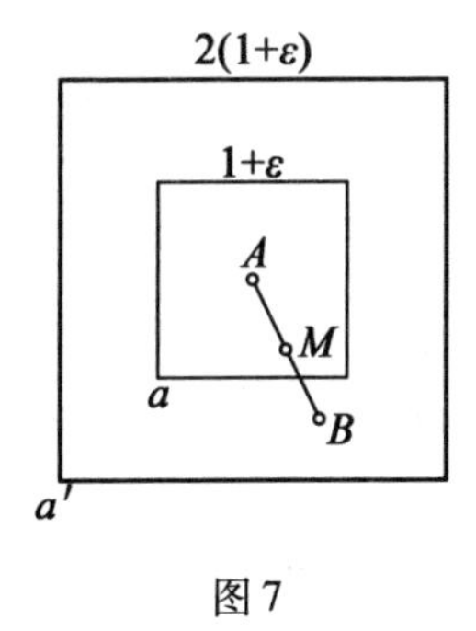

图 7

闵可夫斯基定理的一个有效应用是处理前面我们曾经讲过的用有理数逼近实数的问题. 我们的方法和之前的十分相似,可是得到了更好的结果. 借用已知的无理数 a,我们作点格、格子点的笛卡儿坐标为

$$x = \frac{an - m}{\varepsilon}, y = \varepsilon n$$

式中,m 和 n 取所有的整数值,ε 是任意的正数. 像以前一样,可以知道这个点格是单位点格.

图 8 表示点格中的一个基本平行四边形,假定 $0 < a < 1$. 我们作一正方形,以 2 为边长,中心在原点,且使其边平行于坐标

轴. 应用闵可夫斯基定理,这个正方形必包含另一格子点. 这个格子点由不全等于零的某两个整数 m 和 n 决定. 另一方面,在正方形的内部和边上的点的坐标都满足不等式 $|x|\leqslant 1$, $|y|\leqslant 1$. 因此, m, n 应满足下面的两个不等式

$$\left|\frac{an-m}{\varepsilon}\right|\leqslant 1,\ |\varepsilon n|\leqslant 1$$

或

$$\left|a-\frac{m}{n}\right|\leqslant\frac{\varepsilon}{|n|},\ |n|\leqslant\frac{1}{\varepsilon}$$

这就得出逼近 a 到任意准确程度的另一分数 $\frac{m}{n}$ 的序列. 消去 ε, 得

$$\left|a-\frac{m}{n}\right|\leqslant\frac{1}{n^2}$$

由此可见,闵可夫斯基定理证明了有逼近 a 的分数序列存在,它比在之前作出的分数序数更好,那里我们不过得到了近似式

$$\left|a-\frac{m}{n}\right|\leqslant\frac{2}{\sqrt{3}}\frac{1}{n^2}$$

这个结果比较弱,因为 $\frac{2}{\sqrt{3}}>1$.

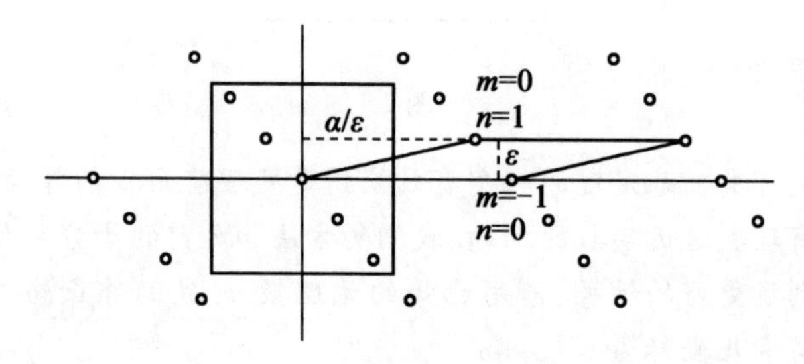

图 8

第 11 章从 Gauss 圆问题(1801) 出发,概述了导致 Shannon 抽样定理(1949) 当前推广的历史概况,证明了 Gauss 和 Shannon 的工作之间的桥梁是由几何格点理论中著名的 Hardy-Landau 恒等式的某些扩展构成的;特别感兴趣的是处理(比如) 与地球科学相关领域相对应的有限带宽函

数的问题;重点研究了所得多元基数级数的收敛性,以及多元 Shannon 抽样中欠抽样与超抽样的精确的显式说明;最后,在 Paley-Wiener 空间中展示了抽样展开(sampling expansion) 的路径,从而得到多元 sinc 型再生核.

1. 引言

在一条被广泛引用的格言中,"数学王子" 卡尔·弗里德里希·高斯(Carl Friedrich Gauss,1777—1855) 断言"数学是科学的皇后,而数论是数学的皇后."

此外,Gauss 在 *Eisenstein's Mathematische Abhandlungen*(1847)的导言中写道:

"高等算术(数论的古称) 提供给我们一座用之不竭的宝库,储满了有趣的真理,这些真理不是孤立的,而是最紧密地相互联系着. 伴随着这门科学的每一次成功发展,我们不断发现全新的,有时甚至是完全意想不到的起点. 算术理论的特殊魅力大多来源于我们由归纳法轻易获得的重要命题,这些命题拥有简洁的表达式,其证明却深埋于斯,在无数徒劳的努力之后才得以发掘;即便通过冗长的、人为的手段取得成功之后,更为清新自然的证明依然藏而不露."

所有这一切都在 Gauss 最著名的出版物中得到了很好的说明,即他的《算术研究》(*Disquisitiones Arithmethcae*),由 Gerhard Fleischer 于 1801 年出版的拉丁文原版. 这本书被描述为"数论的大宪章",这是很有道理的. Gauss 写这本书的时候只有 18 岁左右,这本书所展现的思想的深度和原创性尤其引人注目. 考虑到 Gauss 对现代数论的大范围影响,任何接近其影响的全面表述似乎都是站不住脚的. 不足为奇的是,有大量的文献关注 Gauss 的数论结果,他对现代科学的影响也是巨大的.

这篇文章的目的是解决"数学皇后" 的设置是否真的与现代抽样理论和设置有密切联系的问题. 事实上,我们的文章旨在表明,从圆问题开始,有一些重要的历史阶段构成了今天多元 Shannon 抽样定理的基础,引出了 Hardy-Landau 恒等式和关于圆内格点数目渐近性态的Hardy猜想,继续研究正则区域上的多元加权格点和,得出Shannon型抽样理论(最近由Freeden和Nashed提出).

2. 圆内格点

我们从数论的Gauss概念到现代抽样的桥梁开始,重述了关于圆绕原点O的半径为$N > \frac{\sqrt{2}}{2}$的圆$S_N^1 = \{x \in \mathbf{R}^2 : |x| = N\}$内格点数量的一些结果(图1).

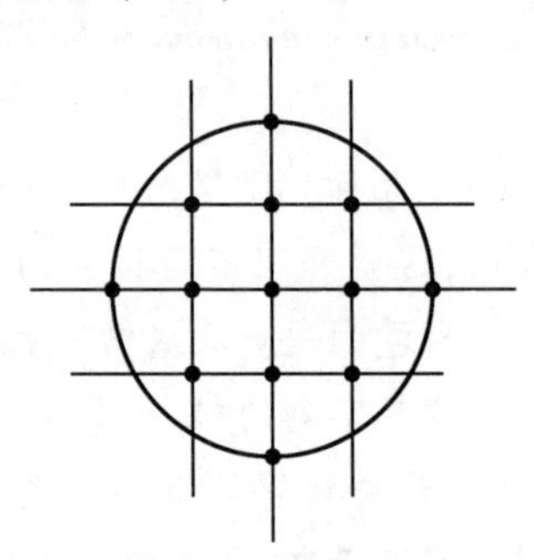

图1　圆内格点

确定圆盘$\overline{B_N^2} = \{x \in \mathbf{R}^2 : |x| \leqslant N\}$内部和边界上$\mathbf{Z}^2$的格点总数的问题,即

$$\#_{\mathbf{Z}^2}(\overline{B_N^2}) = \#\{(n_1, n_2)^{\mathrm{T}} \in \mathbf{Z}^2 : n_1^2 + n_2^2 \leqslant N^2\} \tag{2.1}$$

可追溯到Euler时期.用现代的术语,它可以等价地表示为一个和的形式

$$\#_{\mathbf{Z}^2}(\overline{B_N^2}) = \sum_{\substack{n_1^2 + n_2^2 \leqslant N^2 \\ (n_1, n_2)^T \in \mathbf{Z}^2}} 1 \tag{2.2}$$

Gauss发现了一种简单而有效的估计方法:将西北边缘作为格点与每个正方形相关联.

在$\overline{B_N^2}$内部具有格点的所有正方形的并集定义了具有面积

$$\| P_N^2 \| = \#_{\mathbf{Z}^2}(\overline{B_N^2}) \tag{2.3}$$

的多面体集P_N^2(图2).由于每个正方形的对角线长为$\sqrt{2}$,图2的

几何结构告诉我们

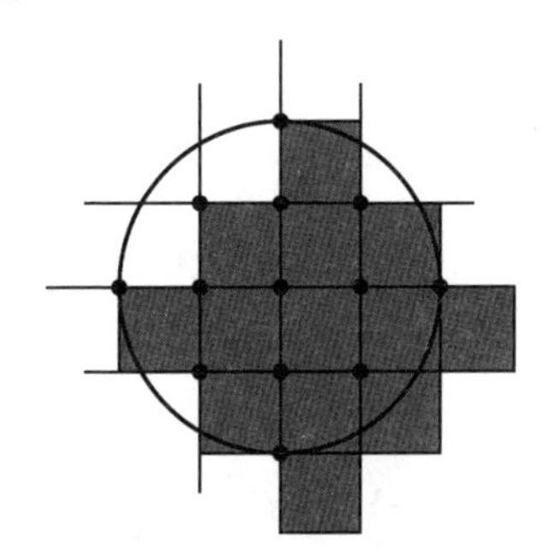

图 2　多面体集 P_N^2

$$\pi\left(N-\frac{\sqrt{2}}{2}\right)^2 \leqslant \#_{\mathbf{Z}^2}(\overline{B_N^2}) \leqslant \pi\left(N+\frac{\sqrt{2}}{2}\right)^2 \tag{2.4}$$

因此,$\#_{\mathbf{Z}^2}(\overline{B_N^2})-\pi N^2$ 除以 N 后对 $N\to\infty$ 有界,通常用 Landau 的 O 符号书写为

$$\#_{\mathbf{Z}^2}(\overline{B_N^2}) = \pi N^2 + O(N) \tag{2.5}$$

换句话说,$\overline{B_N^2}$ 中格点的数量等于该圆的面积加上边界的阶 (order) 的剩余. 特别地

$$\#_{\mathbf{Z}^2}(\overline{B_N^2}) \sim \pi N^2 \tag{2.6}$$

因此,确定无理数、超越数 π 的方法变得显而易见

$$\lim_{N\to\infty}\frac{\#_{\mathbf{Z}^2}(\overline{B_N^2})}{N^2} = \pi \tag{2.7}$$

Gauss 通过取 $N^2=100\ 000$ 来说明他的结果,在这种情况下,他计算了

$$\sum_{\substack{|g|^2\leqslant 100\ 000\\ g\in\mathbf{Z}^2}} 1 = 314\ 197 \tag{2.8}$$

此计算确定数 π 到小数点后最多 3 位数.

3. 圆问题与 Hardy 猜想

在 Landau 的 O 符号的命名法中,由 Gauss 得到的式(2.5),允许表示

$$\#_{\mathbf{Z}^2}(\overline{B_N^2}) = \pi N^2 + O(N) \tag{3.1}$$

所谓的圆问题,是关于确定边界

$$\alpha_2 = \inf\{\gamma: \#_{\mathbf{Z}^2}(\overline{B^2_{\sqrt{N}}}) = \pi N + O(N^\gamma)\} \tag{3.2}$$

的问题. 到目前为止,我们从(3.1)中知道 $\alpha_2 \leqslant \frac{1}{2}$. 对 Gauss 结果的改进是非常困难的,事实上,需要付出巨大的努力. 第一个显著的结果来自于 Sierpinski,他通过使用他的老师 Voronoi 的方法证明了

$$\#_{\mathbf{Z}^2}(\overline{B^2_{\sqrt{N}}}) = \pi N + O(N^{\frac{1}{3}}) \tag{3.3}$$

即 $\alpha_2 \leqslant \frac{1}{3}$. Sierpinski 的证明是初等的,它是几何和数论之间的联系. 今天,他的证明被大大缩短了.

通过使用指数和的高等方法(基于 Weyl,Chen 和其他人的工作),估计$\frac{1}{3}$可以在一定程度上得到加强. 它在 Kolesnik 的出版物中达到了顶峰,他用这些技巧取得了最显著的成果

$$\#_{\mathbf{Z}^2}(\overline{B^2_{\sqrt{N}}}) - \pi N = O(N^{\frac{139}{429}}) \tag{3.4}$$

Huxley 设计了一种全新的方法(不在这里讨论),他最有力的结论是估计

$$\#_{\mathbf{Z}^2}(\overline{B^2_{\sqrt{N}}}) - \pi N = O(N^{\frac{131}{416}}) \tag{3.5}$$

$\left(注意\frac{139}{429} = 0.324\,009\cdots, 而\frac{131}{416} = 0.314\,903\cdots\right)$.

表 1 在估计(3.7)中值 ε_2 的增量改进

0.250 000	Gauss
0.083 333…	Voronoi, Sierpinski
0.080 357…	Littlewood, A. Walfisz
0.079 268…	van der Corput
0.074 324…	Chen
0.074 009…	Kolesnik
0.064 903…	Huxley

Hardy 猜想断言

$$\#_{\mathbf{Z}^2}(\overline{B^2_{\sqrt{N}}}) - \pi N = O(N^{\frac{1}{4}+\varepsilon}) \tag{3.6}$$

对于每个 $\varepsilon > 0$ 成立,这个猜想似乎仍然是对未来工作的一个挑战. 然而,在 2007 年,S. Cappell 和 J. Shaneson 发表了一篇题为 *Some Problems in Number Theory* Ⅰ:*The Circle Problem* 的论文,证明了 $O(N^{\frac{1}{4}+\varepsilon})$ 的界,其中 $\varepsilon > 0$.

表 1 列出了圆问题上极限的量 ε_2 的不完全增量改进

$$\#_{\mathbf{Z}^2} \overline{B^2_{\sqrt{N}}} - \pi N = O(N^{\frac{1}{4}+\varepsilon_2}) \tag{3.7}$$

对于所有最近的改进,证明变得相当长,并在硬分析中使用了大型的机器.

总结我们关于圆内格点的结果,我们面临以下情况

$$\frac{1}{4} \leqslant \alpha_2 \leqslant \frac{1}{4} + \varepsilon_2 \tag{3.8}$$

且

$$\#_{\mathbf{Z}^2}(\overline{B^2_{\sqrt{N}}}) - \pi N \neq O(N^{\frac{1}{4}}) \tag{3.9}$$

$$\#_{\mathbf{Z}^2}(\overline{B^2_{\sqrt{N}}}) - \pi N = O(N^{\frac{1}{4}+\varepsilon_2}) \tag{3.10}$$

其中

$$0 < \varepsilon_2 \leqslant \frac{1}{4} \tag{3.11}$$

4. 圆问题的变体

如前所述,对于圆的 Gauss 格点问题,有许多观点可以表述它的变体. Landau 的文章 *Über die Gitterpunkte in einem Kreis* 的优点是指出了一些特别有趣的领域,例如下面列出的前两项:

(1) 一般的二维格

$$\Lambda = \{g = ng_1 + mg_2 : n, m \in \mathbf{Z}\} \tag{4.1}$$

其中 $g_1, g_2 \in \mathbf{R}^2$ 线性无关(图 3),可以用来代替单位(点)格 $\mathbf{Z}^2$.

(2) 余项可以表示为交替级数,称为 Hardy-Landau 级数,用 1 阶 Bessel 函数 J_1 表示

$$\sum_{\substack{|g| \leqslant N \\ g \in \Lambda}}{}' 1 = \frac{\pi N^2}{\| \mathcal{F}_\Lambda \|} + \lim_{R \to \infty} \frac{\pi N^2}{\| \mathcal{F}_\Lambda \|} \sum_{\substack{0 < |h| \leqslant R \\ h \in \Lambda^{-1}}} \frac{J_1(2\pi |h| N)}{\pi |h| N} \tag{4.2}$$

其中 Λ 为 $\mathbf{R}^2$ 中的任意格,且

$$\mathcal{F}_\Lambda = \left\{ x = x_1 g_1 + x_2 g_2 \in \mathbf{R}^2 : -\frac{1}{2} \leqslant x_i < \frac{1}{2}, i = 1,2 \right\} \tag{4.3}$$

是 $\Lambda \subset \mathbf{R}^2$ 的基本胞腔(fundamental cell),满足

$$\| \mathcal{F}_\Lambda \| = \sqrt{\det((g_i \cdot g_j)_{i,j=1,2})} \tag{4.4}$$

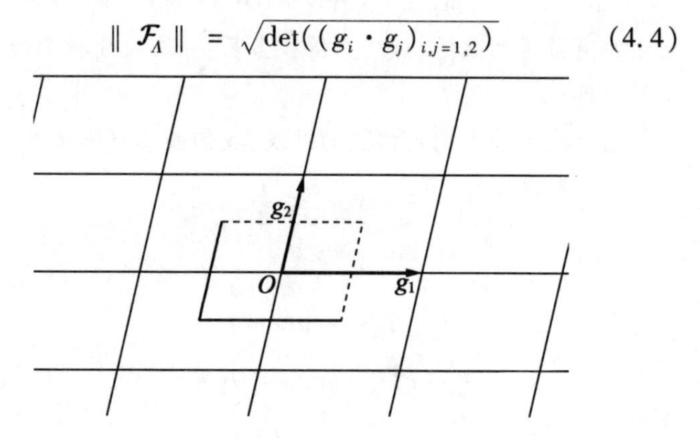

图3 由 $g_1, g_2 \in \mathbf{R}^2$ 生成的二维格 Λ

作为$\mathcal{F}_\Lambda$的面积 $\| \mathcal{F}_\Lambda \|$. 此外,我们遵循约定

$$\sum_{\substack{|g| \leqslant N \\ g \in \Lambda}}{}' 1 = \sum_{\substack{|g| < N \\ g \in \Lambda}} 1 + \frac{1}{2} \sum_{\substack{|g| = N \\ g \in \Lambda}} 1 \tag{4.5}$$

在格点理论中使用(注意,只有当存在格点 $g \in \Lambda$, $|g| = N$ 时,才会出现最后一个和).

(3) 格点可能会受到非常值权重(nonconstant weight) 的影响

$$\begin{aligned} &\sum_{\substack{|a+g| \leqslant N \\ g \in \Lambda}}{}' \mathrm{e}^{2\pi i y \cdot (a+g)} F(a+g) \\ &= \lim_{R \to \infty} \frac{1}{\| \mathcal{F}_\Lambda \|} \sum_{\substack{|h-y| \leqslant R \\ h \in \Lambda^{-1}}} \mathrm{e}^{2\pi i a \cdot h} \int_{\substack{|x| \leqslant N \\ x \in \mathbf{R}^2}} F(x) \mathrm{e}^{-2\pi i x \cdot (h-y)} \mathrm{d}y \end{aligned} \tag{4.6}$$

其中 dy 是体积元，$a,y\in\mathbf{R}^2$，F 在 $\overline{B_N^2}(N>0)$ 中二次连续可微，并且类似于式(4.5)，使用了以下约定

$$\sum_{\substack{|a+g|\leqslant N\\ g\in\Lambda}}{}'\cdots=\sum_{\substack{|a+g|<N\\ g\in\Lambda}}\cdots+\frac{1}{2}\sum_{\substack{|a+g|=N\\ g\in\Lambda}}\cdots \tag{4.7}$$

注意，对于 $F=1$，这个公式返回到

$$\mathrm{e}^{2\pi\mathrm{i}a\cdot y}\sum_{\substack{|g+a|\leqslant N\\ g\in\Lambda}}{}'\mathrm{e}^{2\pi\mathrm{i}g\cdot y}=\lim_{R\to\infty}\frac{\pi N^2}{\|\mathcal{F}_\Lambda\|}\sum_{\substack{|h-y|\leqslant R\\ h\in\Lambda^{-1}}}\mathrm{e}^{2\pi\mathrm{i}a\cdot h}\frac{J_1(2\pi|h-y|N)}{\pi|h-y|N} \tag{4.8}$$

对于 $a=y=0$，我们得到了经典的 Hardy-Landau 恒等式，即恒等式

$$\sum_{\substack{|g|\leqslant N\\ g\in\Lambda}}{}'1=\lim_{R\to\infty}\frac{\pi N^2}{\|\mathcal{F}_\Lambda\|}\sum_{\substack{|h|\leqslant R\\ h\in\Lambda^{-1}}}\frac{J_1(2\pi|h|N)}{\pi|h|N} \tag{4.9}$$

成立. 观察到 J_1 满足渐近关系 $J_1(r)=\dfrac{r}{2}+\cdots$，因此

$$\sum_{\substack{|g|\leqslant N\\ g\in\Lambda}}{}'1=\frac{\pi N^2}{\|\mathcal{F}_\Lambda\|}+\lim_{R\to\infty}\frac{\pi N^2}{\|\mathcal{F}_\Lambda\|}\sum_{\substack{0<|h|\leqslant R\\ h\in\Lambda^{-1}}}\frac{J_1(2\pi|h|N)}{\pi|h|N} \tag{4.10}$$

(4) 格点可以扩展到 $\mathbf{R}^2$ 中的特定区域，设 Λ 是 $\mathbf{R}^2$ 中的任意格. $\mathcal{G}\subset\mathbf{R}^2$ 是一个包含原点且具有边界曲线 ∂G 的凸区域，其正规域 ν 是连续可微的，并且其曲率是非零的. 假设 F 属于类 $C^{(2)}(\overline{\mathcal{G}})$，那么对于所有 $a,y\in\mathbf{R}^2$，级数

$$\sum_{g\in\Lambda}\mathrm{e}^{2\pi\mathrm{i}a\cdot g}\underbrace{\int_{\mathcal{G}}F(x)\mathrm{e}^{-2\pi\mathrm{i}x\cdot(g-y)}\mathrm{d}x}_{=F_{\hat{\mathcal{G}}}(g-y)} \tag{4.11}$$

在球面意义上收敛，并且我们有

$$\begin{aligned}&\frac{1}{\|\mathcal{F}_\Lambda\|}\sum_{\substack{a+h\in\overline{\mathcal{G}}\\ h\in\Lambda^{-1}}}{}'\mathrm{e}^{2\pi\mathrm{i}y\cdot(a+h)}F(a+h)\\&=\lim_{N\to\infty}\sum_{\substack{|g-y|\leqslant N\\ g\in\Lambda}}\mathrm{e}^{2\pi\mathrm{i}a\cdot g}\int_{\mathcal{G}}F(x)\mathrm{e}^{-2\pi\mathrm{i}x\cdot(g-y)}\mathrm{d}x\end{aligned} \tag{4.12}$$

(5) 对格$\Lambda \subset \mathbf{R}^q$,正则区域$\mathcal{G} \subset \mathbf{R}^q$,$q \geqslant 2$,以及$\overline{\mathcal{G}} = \mathcal{G} \cup \partial \mathcal{G}$上连续函数的推广可以用 Gauss 可和性来表达

$$\sum_{\substack{a+g \in \overline{\mathcal{G}} \\ g \in \Lambda}}{}' \mathrm{e}^{2\pi \mathrm{i} y \cdot (a+g)} F(a+g) = \lim_{\substack{\tau \to 0 \\ \tau > 0}} \frac{1}{\| \mathcal{F}_\Lambda \|} \sum_{h \in \Lambda^{-1}} \mathrm{e}^{-\tau \pi^2 h^2} \mathrm{e}^{2\pi \mathrm{i} h \cdot a} \cdot \int_{\mathcal{G}} F(x) \mathrm{e}^{-2\pi \mathrm{i} x \cdot (h-y)} \mathrm{d}x, a, y \in \mathbf{R}^q \tag{4.13}$$

其中,$\mathbf{R}^q$ 中的正则区域$\mathcal{G}$被理解为一个开连通集$\mathcal{G} \subset \mathbf{R}^q$,$q \geqslant 2$,其中:(i) 它的边界 $\partial \mathcal{G}$构成了维数为 $q-1$ 的可定向的、分段光滑的 Lipschitz 流形;(ii) 原点包含在$\mathcal{G}$中;(iii) $\mathcal{G}$将 $\mathbf{R}^q$ 唯一地划分为"内部空间" $\mathcal{G}$和"外部空间"$\mathbf{R}^q \setminus \overline{\mathcal{G}}$,$\overline{\mathcal{G}} = \mathcal{G} \cup \partial\mathcal{G}$.

显然,我们有

$$\sum_{\substack{a+g \in \overline{\mathcal{G}} \\ g \in \Lambda}}{}' \mathrm{e}^{2\pi \mathrm{i} y \cdot (a+g)} F(a+g) = \frac{1}{\| \mathcal{F}_\Lambda \|} \int_{\mathcal{G}} F(x) \mathrm{e}^{2\pi \mathrm{i} x \cdot y} \mathrm{d}x + \lim_{\substack{\tau \to 0 \\ \tau > 0}} \frac{1}{\| \mathcal{F}_\Lambda \|} \sum_{\substack{0 < |h| \leqslant R \\ h \in \Lambda^{-1}}} \mathrm{e}^{-\tau \pi^2 h^2} \mathrm{e}^{2\pi \mathrm{i} h \cdot a} \cdot \int_{\mathcal{G}} F(x) \mathrm{e}^{-2\pi \mathrm{i} x \cdot (h-y)} \mathrm{d}x, a, y \in \mathbf{R}^q \tag{4.14}$$

其中下列缩写形式一直被使用

$$\sum_{\substack{a+g \in \overline{\mathcal{G}} \\ g \in \Lambda}}{}' \cdots = \sum_{\substack{a+g \in \mathcal{G} \\ g \in \Lambda}} \cdots + \sum_{\substack{a+g \in \partial\mathcal{G} \\ g \in \Lambda}} \alpha_{\mathcal{G}}(a+g) \cdots \tag{4.15}$$

其中 $\alpha_{\mathcal{G}}(a+g)$ 表示 $\partial \mathcal{G}$在 $a+g$ 处所对(向) 的立体角(solid angle)(注意,作为与地球科学相关的正则区域,我们可以选择(实际的) 地球体或其部分的内部,地球科学相关表面的内部,如大地水准面、近似地形面等,也可以选择球体、椭球体、立方体、多面体等,根据上述定义被包括在内)(图 4 和图 5).

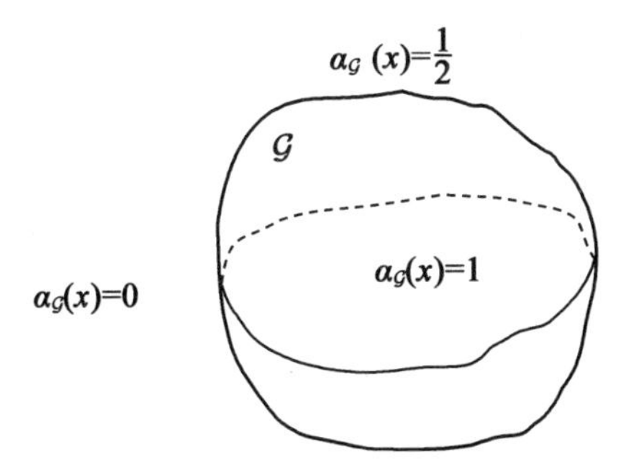

图 4　具有"光滑边界"的正则区域$\mathcal{G}$的表面 $\partial\mathcal{G}$在 $x\in\mathbf{R}^3$ 处对向的立体角

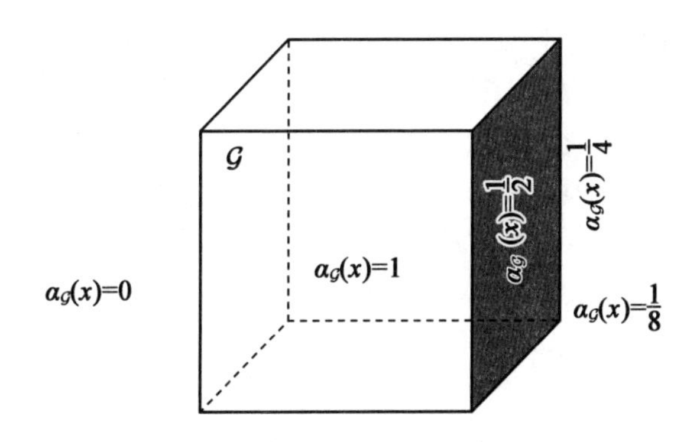

图 5　"非光滑"立方体$\mathcal{G}=(-1,1)^3$ 的表面 $\partial\mathcal{G}$在 $x\in\mathbf{R}^3$ 处对向的立体角

5. 多元 Shannon 型抽样

让我们继续观察,对于每个 $y\in\mathbf{R}^q$,级数

$$a\to\lim_{\substack{\tau\to0\\ \tau>0}}\lim_{N\to\infty}\frac{1}{\|\mathcal{F}_\Lambda\|}\sum_{\substack{|h-y|\leqslant N\\ h\in\Lambda^{-1}}}\mathrm{e}^{-\tau\pi^2h^2}\mathrm{e}^{2\pi\mathrm{i}a\cdot h}\underbrace{\int_{\mathcal{G}}F(x)\mathrm{e}^{-2\pi\mathrm{i}x\cdot(h-y)}\mathrm{d}x}_{F_{\hat{\mathcal{G}}}(h-y)}\tag{5.1}$$

以及有限和

$$a\to\sum_{\substack{a+g\in\overline{\mathcal{G}}\\ g\in\Lambda}}{}'\mathrm{e}^{2\pi\mathrm{i}y\cdot(a+g)}F(a+g),a\in\mathbf{R}^q\tag{5.2}$$

表明 Λ - 周期性,即作为变量 $a\in\mathbf{R}^q$ 的函数,它们关于格 $\Lambda\subset\mathbf{R}^q$ 是周期的. 因此,可以通过格点恒等式(4.13) 在正则区域$\mathcal{H}$

上的形式积分获得 Shannon 型抽样过程,该正则区域$\mathcal{H}$不一定等于正则区域$\mathcal{G}$

$$\begin{aligned}
&\int_{\mathcal{H}}\sum_{\substack{a+g\in\overline{\mathcal{G}}\\ g\in\Lambda}}{}' \mathrm{e}^{-2\pi\mathrm{i}y\cdot(a+g)}F(a+g)\,\mathrm{d}a\\
&=\sum_{\substack{(\mathcal{F}_\Lambda+\{g'\})\cap\mathcal{H}\neq\varnothing\\ g'\in\Lambda}}\underbrace{\int_{\mathcal{G}\cap\bigcup_{g\in\Lambda}(((\overline{\mathcal{H}}\cap(\mathcal{F}_\Lambda+\{g'\}))-\{g'\})+\{g\})}F(x)\mathrm{e}^{-2\pi\mathrm{i}y\cdot x}\mathrm{d}x}_{=F\hat{}_{\mathcal{G}\cap\bigcup_{g\in\Lambda}(((\overline{\mathcal{H}}\cap(\mathcal{F}_\Lambda+\{g'\}))-\{g'\})+\{g\})}(y)}\\
&=\lim_{\substack{\tau\to0\\ \tau>0}}\frac{1}{\|\mathcal{F}_\Lambda\|}\sum_{h\in\Lambda^{-1}}\mathrm{e}^{-\tau\pi^2h^2}\underbrace{\int_{\mathcal{G}}F(x)\mathrm{e}^{-2\pi\mathrm{i}h\cdot x}\mathrm{d}x}_{=F\hat{}_{\mathcal{G}}(h)}\underbrace{\int_{\mathcal{H}}\mathrm{e}^{2\pi\mathrm{i}a\cdot(h-y)}\mathrm{d}a}_{=K_{\mathcal{H}}(h-y)}
\end{aligned}\tag{5.3}$$

恒等式(5.3) 有许多有趣的性质. 例如,由于 Gauss 可和性,(5.3) 右侧基数型级数的收敛速度是指数增长的. 此外,根据所选正则区域$\mathcal{G},\mathcal{H}$的几何结构,可以通过恒等式左侧的傅里叶变换的有限和来明确分析超抽样和欠抽样的所有表现形式.

一些特殊的结构,即正则区域$\mathcal{H},\mathcal{G}$的特殊选择应该被更详细地研究,我们从$\overline{\mathcal{H}}\subset\mathcal{F}_\Lambda$和任意的$\mathcal{G}$开始. 在这种情况下,我们有

$$\begin{aligned}
&\sum_{\substack{(\mathcal{F}_\Lambda+\{g'\})\cap\mathcal{H}\neq\varnothing\\ g'\in\Lambda}}F\hat{}_{\mathcal{G}\cap\bigcup_{g\in\Lambda}(((\overline{\mathcal{H}}\cap(\mathcal{F}_\Lambda+\{g'\}))-\{g'\})+\{g\})}(y)\\
&=F\hat{}_{\mathcal{G}\cap\bigcup_{g\in\Lambda}(\mathcal{H}+\{g\})}(y)\\
&=\frac{1}{\|\mathcal{F}_\Lambda\|}\lim_{\substack{\tau\to0\\ \tau>0}}\sum_{h\in\Lambda^{-1}}\mathrm{e}^{-\tau\pi^2h^2}\underbrace{\int_{\mathcal{G}}F(x)\mathrm{e}^{-2\pi\mathrm{i}y\cdot x}\mathrm{d}x}_{=F\hat{}_{\mathcal{G}}(h)}\underbrace{\int_{\mathcal{H}}\mathrm{e}^{2\pi\mathrm{i}a\cdot(h-y)}\mathrm{d}a}_{=K_{\mathcal{H}}(h-y)}
\end{aligned}\tag{5.4}$$

因此,对于$\overline{\mathcal{H}}=\mathcal{F}_\Lambda$和任意的$\mathcal{G}$,我们得到

$$\begin{aligned}
&\sum_{\substack{(\mathcal{F}_\Lambda+\{g'\})\cap\mathcal{H}\neq\varnothing\\ g'\in\Lambda}}F\hat{}_{\mathcal{G}\cap\bigcup_{g\in\Lambda}(((\mathcal{H}\cap(\mathcal{F}_\Lambda+\{g'\}))-\{g'\})+\{g\})}(y)\\
&=F\hat{}_{\mathcal{G}}(y)\\
&=\frac{1}{\|\mathcal{F}_\Lambda\|}\lim_{\substack{\tau\to0\\ \tau>0}}\sum_{h\in\Lambda^{-1}}\mathrm{e}^{-\tau\pi^2h^2}\underbrace{\int_{\mathcal{G}}F(x)\mathrm{e}^{-2\pi\mathrm{i}h\cdot x}\mathrm{d}x}_{=F\hat{}_{\mathcal{G}}(h)}\underbrace{\int_{\mathcal{F}_\Lambda}\mathrm{e}^{2\pi\mathrm{i}a\cdot(h-y)}\mathrm{d}a}_{=K_{\mathcal{F}_\Lambda}(h-y)}
\end{aligned}\tag{5.5}$$

我们继续使用$\overline{\mathcal{G}}\subset\overline{\mathcal{F}_\Lambda}$和$\mathcal{H}$任意. 这就得到了恒等式

$$\sum_{\substack{(\mathcal{F}_\Lambda+\{g'\})\cap\mathcal{H}\neq\varnothing\\ g'\in\Lambda}} F_{\hat{\mathcal{G}}\cap\bigcup_{g\in\Lambda}(((\mathcal{H}\cap(\mathcal{F}_\Lambda+\{g'\}))-\{g'\})+\{g\})}(y)$$

$$=\sum_{\substack{(\mathcal{F}_\Lambda+\{g'\})\cap\mathcal{H}\neq\varnothing\\ g'\in\Lambda}} F_{\hat{\mathcal{G}}\cap((\mathcal{H}\cap(\mathcal{F}_\Lambda+\{g'\}))-\{g'\})}(y)$$

$$=\frac{1}{\|\mathcal{F}_\Lambda\|}\lim_{\substack{\tau\to 0\\ \tau>0}}\sum_{h\in\Lambda^{-1}}\mathrm{e}^{-\tau\pi^2h^2}\underbrace{\int_{\mathcal{G}}F(x)\mathrm{e}^{-2\pi\mathrm{i}h\cdot x}\mathrm{d}x}_{=F_{\hat{\mathcal{G}}}(h)}\underbrace{\int_{\mathcal{H}}\mathrm{e}^{2\pi\mathrm{i}a\cdot(h-y)}\mathrm{d}a}_{=K_{\mathcal{H}}(h-y)} \tag{5.6}$$

对于$\overline{\mathcal{G}},\overline{\mathcal{H}}\subset\overline{\mathcal{F}_\Lambda}$,我们有

$$\sum_{\substack{(\mathcal{F}_\Lambda+\{g'\})\cap\mathcal{H}\neq\varnothing\\ g'\in\Lambda}} F_{\hat{\mathcal{G}}\cap\bigcup_{g\in\Lambda}(((\mathcal{H}\cap(\mathcal{F}_\Lambda+\{g'\}))-\{g'\})+\{g\})}(y)$$

$$=F_{\hat{\mathcal{G}}\cap\mathcal{H}}(y)$$

$$=\frac{1}{\|\mathcal{F}_\Lambda\|}\lim_{\substack{\tau\to 0\\ \tau>0}}\sum_{h\in\Lambda^{-1}}\mathrm{e}^{-\tau\pi^2h^2}F_{\hat{\mathcal{G}}}(h)K_{\mathcal{H}}(h-y) \tag{5.7}$$

对于$\overline{\mathcal{G}}\subset\overline{\mathcal{H}}\subset\overline{\mathcal{F}_\Lambda}$,我们有

$$F_{\hat{\mathcal{G}}}(y)=\frac{1}{\|\mathcal{F}_\Lambda\|}\lim_{\substack{\tau\to 0\\ \tau>0}}\sum_{h\in\Lambda^{-1}}\mathrm{e}^{-\tau\pi^2h^2}F_{\hat{\mathcal{G}}}(h)K_{\mathcal{H}}(h-y) \tag{5.8}$$

对于$\overline{\mathcal{H}}\subset\overline{\mathcal{G}}\subset\overline{\mathcal{F}_\Lambda}$,我们有

$$F_{\hat{\mathcal{H}}}(y)=\frac{1}{\|\mathcal{F}_\Lambda\|}\lim_{\substack{\tau\to 0\\ \tau>0}}\sum_{h\in\Lambda^{-1}}\mathrm{e}^{-\tau\pi^2h^2}F_{\hat{\mathcal{G}}}(h)K_{\mathcal{H}}(h-y) \tag{5.9}$$

特别地,对于$\overline{\mathcal{G}}=\overline{\mathcal{H}}\subset\overline{\mathcal{F}_\Lambda}$,我们能够表达以下恒等式

$$F_{\hat{\mathcal{G}}}(y)=\frac{1}{\|\mathcal{F}_\Lambda\|}\lim_{\substack{\tau\to 0\\ \tau>0}}\sum_{h\in\Lambda^{-1}}\mathrm{e}^{-\tau\pi^2h^2}F_{\hat{\mathcal{G}}}(h)K_{\mathcal{G}}(h-y),y\in\mathbf{R}^q \tag{5.10}$$

显然,通过选择一个小的抽样密度,使得$\overline{\mathcal{F}_\Lambda}$覆盖原始信号$F_{\hat{\mathcal{G}}}$的紧支集$\overline{\mathcal{G}}$,用于重构的样本$F_{\hat{\mathcal{G}}}(h),h\in\Lambda^{-1}$的数量是高的,反之亦然.因此,在实际应用中,我们需要在抽样密度和样本总数之间找到一个折中点.这可以通过Λ的选择,使$\overline{\mathcal{F}_\Lambda}$紧紧覆盖$\overline{\mathcal{G}}$来实现.

对于$\overline{\mathcal{G}}\subset\overline{\mathcal{F}_\Lambda}\subset\overline{\mathcal{H}}$和由$g_1,\cdots,g_q\in\mathbf{R}^q$生成的格$\Lambda$,我们有明确的书写形式

$$F_{\hat{\mathcal{G}}}(y) = \lim_{\substack{\tau\to 0\\ \tau>0}} \sum_{h\in\Lambda^{-1}} e^{-\tau\pi^2 h^2} F_{\hat{\mathcal{G}}}(h) \cdot \frac{\sin(\pi(g_1\cdot(h-y)))}{\pi(g_1\cdot(h-y))}\cdot\cdots\cdot\frac{\sin(\pi g_q\cdot(h-y))}{\pi(g_q\cdot(h-y))} \tag{5.11}$$

换句话说,在指数收敛时,对于足够小的 $\tau > 0$,$F_{\hat{\mathcal{G}}}$ 可以用(5.11)右侧的级数表示,即

$$(F^{(\tau)})_{\hat{\mathcal{G}}}(y) \simeq \sum_{h\in\Lambda^{-1}} e^{-\tau\pi^2 h^2} F_{\hat{\mathcal{G}}}(h) \cdot \frac{\sin(\pi(g_1\cdot(h-y)))}{\pi(g_1\cdot(h-y))}\cdot\cdots\cdot\frac{\sin(\pi(g_q\cdot(h-y)))}{\pi(g_q\cdot(h-y))} \tag{5.12}$$

将格 Λ 替换为它的扩张格(dilated lattice)$\sigma\Lambda$,$\sigma > 1$,我们有

$$F_{\hat{\mathcal{G}}}(y) = \lim_{\substack{\tau\to 0\\ \tau>0}} \sigma^q \sum_{h\in\Lambda^{-1}} e^{-\tau\pi^2\left(\frac{h}{\sigma}\right)^2} F_{\hat{\mathcal{G}}}\left(\frac{h}{\sigma}\right) \cdot \frac{\sin\left(\pi\left(\sigma g_1\cdot\left(\frac{h}{\sigma}-y\right)\right)\right)}{\pi\left(\sigma g_1\cdot\left(\frac{h}{\sigma}-y\right)\right)}\cdot\cdots\cdot\frac{\sin\left(\pi\left(\sigma g_q\cdot\left(\frac{h}{\sigma}-y\right)\right)\right)}{\pi\left(\sigma g_q\cdot\left(\frac{h}{\sigma}-y\right)\right)} \tag{5.13}$$

Gauss-Weierstrass 可和性中的抽样的标准形式是通过取 $\overline{\mathcal{G}} = \overline{\mathcal{H}} = \overline{\mathcal{F}_\Lambda}$ 来提供的,即

$$F_{\hat{\mathcal{F}}_\Lambda}(y) = \frac{1}{\|\mathcal{F}_\Lambda\|} \lim_{\substack{\tau\to 0\\ \tau>0}} \sum_{h\in\Lambda^{-1}} e^{-\tau\pi^2 h^2} F_{\hat{\mathcal{F}}_\Lambda}(h) K_{\mathcal{F}_\Lambda}(h-y),\ y\in\mathbf{R}^q \tag{5.14}$$

备注5.1 给定任意格 $\Lambda\subset\mathbf{R}^q$ 和任意正则区域 $\mathcal{H}\subset\mathbf{R}^q$,我们能够找到一个常数 $\sigma\in\mathbf{R}$,使得 $\overline{\mathcal{H}}\subset\overline{\mathcal{F}_{\sigma\Lambda}}$ 尽可能紧. 在这种情况下,我们得到了

$$\begin{aligned}
&F_{\hat{\mathcal{G}}\cap\bigcup_{g\in\sigma\Lambda}(\mathcal{H}+\{g\})}(y)\\
&= \frac{1}{\|\mathcal{F}_{\sigma\Lambda}\|} \lim_{\substack{\tau\to 0\\ \tau>0}} \sum_{h\in(\sigma\Lambda)^{-1}} e^{-2\tau\pi^2 h^2} F_{\hat{\mathcal{G}}}(h) K_{\mathcal{H}}(h-y)\\
&= \frac{1}{\|\mathcal{F}_\Lambda\|} \lim_{\substack{\tau\to 0\\ \tau>0}} \frac{1}{\sigma^q} \sum_{h\in\Lambda^{-1}} e^{-\tau\pi^2\left(\frac{h}{\sigma}\right)^2} F_{\hat{\mathcal{G}}}\left(\frac{h}{\sigma}\right) K_{\mathcal{H}}\left(\frac{h}{\sigma}-y\right)
\end{aligned} \tag{5.15}$$

与关于格 Λ 的抽样相比,与 $\sigma\Lambda$ 相关的恒等式(5.15) 分别为值 $\sigma > 1$ 提供了上抽样或为值 $\sigma < 1$ 提供了下抽样. 特别重要的是 σ 的选择,使得 Λ 是 $\sigma\Lambda(\sigma > 1)$ 的子格,或者 $\sigma\Lambda(\sigma < 1)$ 是 Λ 的子格.

设$\mathcal{G}$,$\mathcal{H}$是正则区域,假设在$\overline{\mathcal{G}}$上 $F = 1$. 那么,对于所有 $y \in \mathbf{R}^q$,我们有

$$\int_{\mathcal{H}} \sum_{\substack{a+g \in \overline{\mathcal{G}} \\ g \in \Lambda}}{}' \mathrm{e}^{-2\pi \mathrm{i} y(a+g)} \mathrm{d}a$$
$$= \sum_{\substack{(\mathcal{F}_\Lambda + \{g'\}) \cap \mathcal{H} \neq \varnothing \\ g \in \Lambda}} \int_{\mathcal{G} \cap \bigcup_{g \in \Lambda}((\mathcal{H} \cap (\mathcal{F}_\Lambda + \{\gamma'\}) - \{g'\}) + \{g\})} \mathrm{e}^{-2\pi \mathrm{i} y \cdot x} \mathrm{d}x$$
$$= \frac{1}{\| \mathcal{F}_\Lambda \|} \lim_{\substack{\tau \to 0 \\ \tau > 0}} \sum_{h \in \Lambda^{-1}} \mathrm{e}^{-\tau \pi^2 h^2} \int_{\mathcal{G}} \mathrm{e}^{-2\pi \mathrm{i} h \cdot x} \mathrm{d}x \int_{\mathcal{H}} \mathrm{e}^{2\pi \mathrm{i} x \cdot (h-y)} \mathrm{d}x \tag{5.16}$$

使得,特别是对于$\mathcal{H} = \mathcal{G}$,有

$$\int_{\mathcal{G}} \sum_{\substack{a+g \in \overline{\mathcal{G}} \\ g \in \Lambda}}{}' \mathrm{e}^{-2\pi \mathrm{i} y(a+g)} \mathrm{d}a$$
$$= \sum_{\substack{(\mathcal{F}_\Lambda + \{g'\}) \cap \mathcal{G} \neq \varnothing \\ g \in \Lambda}} \int_{\mathcal{G} \cap \bigcup_{g \in \Lambda}((\mathcal{G} \cap (\mathcal{F}_\Lambda + \{\gamma'\}) - \{g'\}) + \{g\})} \mathrm{e}^{-2\pi \mathrm{i} y \cdot x} \mathrm{d}x$$
$$= \frac{1}{\| \mathcal{F}_\Lambda \|} \lim_{\substack{\tau \to 0 \\ \tau > 0}} \sum_{h \in \Lambda^{-1}} \mathrm{e}^{-\tau \pi^2 h^2} \int_{\mathcal{G}} \mathrm{e}^{-2\pi \mathrm{i} h \cdot x} \mathrm{d}x \int_{\mathcal{G}} \mathrm{e}^{2\pi \mathrm{i} x \cdot (h-y)} \mathrm{d}x \tag{5.17}$$

特别地,对于 $y = 0$,我们得到

$$\int_{\mathcal{H}} \sum_{\substack{a+g \in \overline{\mathcal{G}} \\ g \in \Lambda}}{}' 1 \mathrm{d}a$$
$$= \sum_{\substack{(\mathcal{F}_\Lambda + \{g'\}) \cap \mathcal{H} \neq \varnothing \\ g \in \Lambda}} \int_{\mathcal{G} \cap \bigcup_{g \in \Lambda}((\mathcal{H} \cap (\mathcal{F}_\Lambda + \{g'\}) - \{g'\}) + \{g\})} 1 \mathrm{d}x$$
$$= \frac{1}{\| \mathcal{F}_\Lambda \|} \lim_{\substack{\tau \to 0 \\ \tau > 0}} \sum_{h \in \Lambda^{-1}} \mathrm{e}^{-\tau \pi^2 h^2} \int_{\mathcal{G}} \mathrm{e}^{-2\pi \mathrm{i} h \cdot x} \mathrm{d}x \int_{\mathcal{H}} \mathrm{e}^{2\pi \mathrm{i} x \cdot h} \mathrm{d}x \tag{5.18}$$

并且,尤其是对$\mathcal{H} = \mathcal{G}$,有

$$\int_{\mathcal{G}} \sum_{\substack{a+g \in \overline{\mathcal{G}} \\ g \in \Lambda}}{}' 1 \mathrm{d}a$$

$$= \sum_{\substack{(\mathcal{F}_\Lambda+\{g'\})\cap\overline{\mathcal{G}}\neq\varnothing \\ g\in\Lambda}} \int_{\mathcal{G}\cap\bigcup_{g\in\Lambda}((\mathcal{G}\cap(\mathcal{F}_\Lambda+\{g'\})-\{g'\})+\{g\})} 1\mathrm{d}x$$

$$= \lim_{\substack{\tau\to 0\\ \tau>0}} \frac{1}{\|\mathcal{F}_\Lambda\|} \sum_{h\in\Lambda^{-1}} \mathrm{e}^{-\tau\pi^2 h^2} \left| \int_{\mathcal{G}} \mathrm{e}^{-2\pi i h\cdot x}\mathrm{d}x \right|^2 \tag{5.19}$$

恒等式(5.18)和(5.19)是在 Gauss-Weierstrass 框架内转换为涉及正则区域$\mathcal{G},\mathcal{H}\subset\mathbf{R}^q$ 的 Parseval 恒等式的规范准备.

备注5.2 公式(5.19)已经用于二维欧几里得空间$\mathbf{R}^2$. 事实上,Müller 注意到 Parseval 型恒等式

$$\int_{\mathcal{G}} \sum_{\substack{a+g\in\overline{\mathcal{G}}\\ g\in\mathbf{Z}^2}}{}' 1\mathrm{d}a = \sum_{h\in\mathbf{Z}^2} \left| \int_{\mathcal{G}} \mathrm{e}^{-2\pi i h\cdot x}\mathrm{d}x \right|^2 \tag{5.20}$$

对于所有对称(关于原点)和凸区域$\mathcal{G}\subset\mathbf{R}^2$ 成立. 式(5.20)的一个简单结果是不等式

$$\int_{\mathcal{G}} \sum_{\substack{a+g\in\overline{\mathcal{G}}\\ g\in\mathbf{Z}^2}}{}' 1\mathrm{d}a \geqslant \|\mathcal{G}\|^2 = \left(\int_{\mathcal{G}} \mathrm{d}x\right)^2 \tag{5.21}$$

这样,在假设$\|\mathcal{G}\|>4$下,同时考虑格$2\mathbf{Z}^2$,可以向 Minkowski 定理迈出重要的一步. 事实上,不等式

$$\int_{\mathcal{G}} \sum_{\substack{a+g\in\overline{\mathcal{G}}\\ g\in 2\mathbf{Z}^2}}{}' 1\mathrm{d}a \geqslant \frac{\|\mathcal{G}\|}{4}\|\mathcal{G}\| > \|\mathcal{G}\| \tag{5.22}$$

是保证一个对称(关于原点)和凸区域$\mathcal{G}\subset\mathbf{R}^2(\|\mathcal{G}\|>4)$包含与原点不同的$\mathbf{Z}^2$ 的格点的关键.

应当注意的是,Müller 的二维方法被 Freeden 和 Nashed 以各种方式推广. 例如,可以独立于欧几里得空间$\mathbf{R}^q(q\geqslant 1)$的维数来避免 Gauss-Weierstrass 可和性. (5.20)类型的 Parseval 恒等式可以扩展到任意格$\Lambda\subset\mathbf{R}^q$ 和正则区域$\mathcal{G}\subset\mathbf{R}^q$. 此外,可以包含任意连续的权函数来代替常数权函数.

定理5.3(Gauss-Weierstrass 可和性中的广义 Parseval 恒等式) 令Λ为$\mathbf{R}^q$ 中的格. 设$\mathcal{G},\mathcal{H}\subset\mathbf{R}^q$ 是任意正则区域. 假设F属于$C^{(0)}(\overline{\mathcal{G}})$,$G$属于$C^{(0)}(\overline{\mathcal{H}})$. 那么,我们有

$$\lim_{\substack{\tau\to 0\\ \tau>0}} \frac{1}{\|\mathcal{F}_\Lambda\|} \sum_{h\in\Lambda^{-1}} \mathrm{e}^{-\tau\pi^2 h^2} \int_{\mathcal{G}} F(x)\mathrm{e}^{-2\pi i x\cdot h}\mathrm{d}x\, \overline{\int_{\mathcal{H}} G(x)\mathrm{e}^{-2\pi i x\cdot h}\mathrm{d}x}$$

$$= \lim_{\substack{\tau\to 0\\ \tau>0}} \frac{1}{\|\mathcal{F}_\Lambda\|} \sum_{h\in\Lambda^{-1}} \mathrm{e}^{-\tau\pi^2 h^2} F_{\hat{\mathcal{G}}}(h)\, \overline{G_{\hat{\mathcal{H}}}(h)}$$

$$= \int_{\mathcal{H}} \sum_{\substack{a+g \in \overline{\mathcal{G}} \\ g \in \Lambda}}{}' F(a+g)\, \overline{G(a)}\, \mathrm{d}a \tag{5.23}$$

现在,如果$\overline{\mathcal{G}}$,$\overline{\mathcal{H}}$是$\overline{\mathcal{F}_\Lambda}$的子集,那么我们从广义 Parseval 恒等式中得到(参见定理 5.3)

$$\lim_{\substack{\tau \to 0 \\ \tau > 0}} \frac{1}{\| \mathcal{F}_\Lambda \|} \sum_{g \in \Lambda^{-1}} \mathrm{e}^{-\tau\pi^2 h^2} F_{\hat{\mathcal{G}}}(h)\, \overline{G_{\hat{\mathcal{H}}}(h)}$$

$$= \int_{\mathcal{G} \cap \mathcal{H}} F(a)\, \overline{G(a)}\, \mathrm{d}a \tag{5.24}$$

如果$\overline{\mathcal{G}} \subset \overline{\mathcal{H}} \subset \overline{\mathcal{F}_\Lambda}$,那么

$$\lim_{\substack{\tau \to 0 \\ \tau > 0}} \frac{1}{\| \mathcal{F}_\Lambda \|} \sum_{g \in \Lambda^{-1}} \mathrm{e}^{-\tau\pi^2 h^2} F_{\hat{\mathcal{G}}}(h)\, \overline{G_{\hat{\mathcal{H}}}(h)}$$

$$= \int_{\mathcal{G}} F(a)\, \overline{G(a)}\, \mathrm{d}a \tag{5.25}$$

因此,特殊构形$\overline{\mathcal{G}} = \overline{\mathcal{H}} = \overline{\mathcal{F}_\Lambda}$又回到了傅里叶理论中常规的 Parseval 恒等式(然而,在 Gauss-Weierstrass 可和性中).

根据定理5.3($\mathcal{H} = \mathcal{G}$),我们得到了 Gauss-Weierstrass 可和性中的以下推论.

推论 5.4(Gauss-Weierstrass 可和性中的 Parseval 型恒等式) 设Λ是$\mathbf{R}^q$中的格,$\mathcal{G} \subset \mathbf{R}^q$是正则区域. 假设$F,G$属于$C^{(0)}(\overline{\mathcal{G}})$,那么下面的 Parseval 恒等式的变体成立

$$\lim_{\substack{\tau \to 0 \\ \tau > 0}} \frac{1}{\| \mathcal{F}_\Lambda \|} \sum_{h \in \Lambda^{-1}} \mathrm{e}^{-\tau\pi^2 h^2} F_{\hat{\mathcal{G}}}(h)\, \overline{G_{\hat{\mathcal{G}}}(h)}$$

$$= \int_{\mathcal{G}} \sum_{\substack{a+g \in \overline{\mathcal{G}} \\ g \in \Lambda}}{}' F(a+g)\, \overline{G(a)}\, \mathrm{d}a \tag{5.26}$$

特别地,我们有

$$\lim_{\substack{\tau \to 0 \\ \tau > 0}} \frac{1}{\| \mathcal{F}_\Lambda \|} \sum_{h \in \Lambda^{-1}} \mathrm{e}^{-\tau\pi^2 h^2} | F_{\hat{\mathcal{G}}}(h) |^2$$

$$= \int_{\mathcal{G}} \sum_{\substack{a+g \in \overline{\mathcal{G}} \\ g \in \Lambda}}{}' F(a+g)\, \overline{F(a)}\, \mathrm{d}a \tag{5.27}$$

从数值角度看,(5.3) 右侧的基数级数的 Gauss 可和性非常重要;它使级数的快速计算成为可能. 尽管如此,Freeden 和 Nashed 表明(5.3) 在普通意义上是正确的.

推论 5.5(正则区域的 Parseval 型恒等式) 在定理 5.3 的假设下,我们有

$$\frac{1}{\|\mathcal{F}_\Lambda\|}\sum_{h\in\Lambda^{-1}}F_{\hat{\mathcal{G}}}(h)\,\overline{G_{\hat{\mathcal{H}}}(h)}=\int_{\mathcal{H}}\sum_{\substack{a+g\in\overline{\mathcal{G}}\\ g\in\Lambda}}{}'F(a+g)\,\overline{G(a)}\,\mathrm{d}a \tag{5.28}$$

特别地,如果$\overline{\mathcal{G}}=\overline{\mathcal{H}}$且 $F=G$,那么

$$\frac{1}{\|\mathcal{F}_\Lambda\|}\sum_{h\in\Lambda^{-1}}|F_{\hat{\mathcal{G}}}(h)|^2=\int_{\mathcal{G}}\sum_{\substack{a+g\in\overline{\mathcal{G}}\\ g\in\Lambda}}{}'F(a+g)\,\overline{F(a)}\,\mathrm{d}a \tag{5.29}$$

在选择$\overline{\mathcal{G}}=\overline{\mathcal{H}}\subset\overline{\mathcal{F}_\Lambda}$和 $F=G$ 时,我们发现

$$\frac{1}{\|\mathcal{F}_\Lambda\|}\sum_{h\in\Lambda^{-1}}|F_{\hat{\mathcal{G}}}(h)|^2=\int_{\mathcal{G}}|F(a)|^2\mathrm{d}a \tag{5.30}$$

值得注意的是,推论 5.5 也使我们能够表述普遍意义上的 Shannon 抽样定理,避免了 Gauss-Weierstrass 命名法(通过令 $G(x)=1(x\in\mathcal{H})$,用 $F(x)\mathrm{e}^{-2\pi iy\cdot x}$ 代替 $F(x)(x\in\mathcal{G})$).

定理 5.6(Shannon 型抽样定理) 设$\mathcal{G},\mathcal{H}$是 $\mathbf{R}^q$ 中的正则区域,假设 F 是类 $C^{(0)}(\overline{\mathcal{G}})$ 中的一元,那么

$$\int_{\mathcal{H}}\sum_{\substack{a+g\in\overline{\mathcal{G}}\\ g\in\Lambda}}{}'\mathrm{e}^{-2\pi iy\cdot(a+g)}F(a+g)\,\mathrm{d}a \tag{5.31}$$

$$=\sum_{\substack{(\mathcal{F}_\Lambda+\{g'\})\cap\mathcal{H}\neq\varnothing\\ g'\in\Lambda}}F_{\hat{\mathcal{G}}\cap\bigcup_{g\in\Lambda}(((\mathcal{H}\cap(\mathcal{F}_\Lambda+\{g'\}))-\{g'\})+\{g\})}(y)$$

$$=\frac{1}{\|\mathcal{F}_\Lambda\|}\sum_{h\in\Lambda^{-1}}F_{\hat{\mathcal{G}}}(h)K_{\mathcal{H}}(h-y) \tag{5.32}$$

对所有 $y\in\mathbf{R}^q$ 有效. 对于$\overline{\mathcal{G}}\subset\overline{\mathcal{H}}\subset\overline{\mathcal{F}_\Lambda}$,我们有

$$F_{\hat{\mathcal{G}}}(y)=\frac{1}{\|\mathcal{F}_\Lambda\|}\sum_{h\in\Lambda^{-1}}F_{\hat{\mathcal{G}}}(h)K_{\mathcal{H}}(h-y),\ y\in\mathbf{R}^q \tag{5.33}$$

最后,在假设$\overline{\mathcal{G}}=\overline{\mathcal{H}}\subset\overline{\mathcal{F}_\Lambda}$下,恒等式(5.33)蕴含

$$F_{\hat{\mathcal{G}}}(y)=\frac{1}{\|\mathcal{F}_\Lambda\|}\sum_{h\in\Lambda^{-1}}F_{\hat{\mathcal{G}}}(h)K_{\mathcal{G}}(h-y),\ y\in\mathbf{R}^q \tag{5.34}$$

备注 5.7 此外,对于$\overline{\mathcal{G}}\subset\overline{\mathcal{H}}=\overline{\mathcal{F}_\Lambda}$,我们得到

$$F_{\hat{\mathcal{G}}}(y)=\frac{1}{\|\mathcal{F}_\Lambda\|}\sum_{h\in\Lambda^{-1}}F_{\hat{\mathcal{G}}}(h)K_{\mathcal{F}_\Lambda}(h-y),y\in\mathbf{R}^q \tag{5.35}$$

明确地写出来,对于由 $g_1,\cdots,g_q\in\mathbf{R}^q$ 生成的格 Λ,我们得到

$$F_{\hat{\mathcal{G}}}(y)=\sum_{h\in\Lambda^{-1}}F_{\hat{\mathcal{G}}}(h)\frac{\sin(\pi(g_1\cdot(h-y)))}{\pi(g_1\cdot(h-y)))}\cdot\cdots\cdot\frac{\sin(\pi(g_q\cdot(h-y)))}{\pi(g_q\cdot(h-y))},y\in\mathbf{R}^q \tag{5.36}$$

$$F_{\hat{\mathcal{G}}}(y)=\frac{1}{\|\mathcal{F}_\Lambda\|}\sum_{h\in\Lambda^{-1}}F_{\hat{\mathcal{G}}}(h)K_{\mathcal{G}}(h-y) \tag{5.37}$$

事实上,恒等式(5.37)是Shannon抽样定理的多元变体,但现在用于多元正则区域$\mathcal{G}$. Shannon 抽样对信息论的主要影响是,它允许用离散的样本序列替换与$\mathcal{G}$相关的有限带宽信号 $F_{\hat{\mathcal{G}}}$,而不会丢失任何信息. 此外,它还指定了最低速率,即 Nyquist 速率,使其能够再现原始信号. 换句话说,Shannon 抽样为有限带宽函数的连续和离散版本提供了桥梁.

本文介绍的 Shannon 抽样定理的扩展在工程和物理学中有许多应用. 例如,在信号处理、数据传输、密码学、构造性近似及逆问题(比如天线问题)中.

6. Paley-Wiener 空间

我们可以更详细地研究抽样定理的许多扩展,最后将解释导致 Payley-Wiener 空间的一些方面.

我们将研究对象限制在正则区域$\mathcal{G}\subset\mathbf{R}^q(\overline{\mathcal{G}}\subset\overline{\mathcal{F}_\Lambda})$上,以便从逆格(inverse lattice)$F_{\hat{\mathcal{G}}}(h),h\in\Lambda^{-1}$,即

$$F_{\hat{\mathcal{G}}}(y)=\frac{1}{\|\mathcal{F}_\Lambda\|}\sum_{h\in\Lambda^{-1}}F_{\hat{\mathcal{G}}}(h)K_{\mathcal{G}}(h-y),y\in\mathbf{R}^q \tag{6.1}$$

$$K_{\mathcal{G}}(x-y)=\int_{\mathcal{G}}\mathrm{e}^{-2\pi ia\cdot(x-y)}\mathrm{d}a,x,y\in\mathbf{R}^q \tag{6.2}$$

的格点上的抽样信号中恢复连续信号

$$F_{\hat{\mathcal{G}}}(y)=\int_{\mathcal{G}}F(a)\mathrm{e}^{-2\pi ia\cdot y}\mathrm{d}a,y\in\mathbf{R}^q \tag{6.3}$$

我们能够在近似积分领域中推导出一些有趣的结果. 事实上,有

限带宽函数 $F_{\hat{\mathcal{G}}}$ 允许用格密度与逆格的点中的所有样本上的和的乘积来表示其在欧几里得空间 $\mathbf{R}^q$ 上的积分

$$\lim_{N\to\infty}\int_{\substack{|x|\leqslant N\\ x\in\mathbf{R}^q}} F_{\hat{\mathcal{G}}}(x)\,\mathrm{d}x = \frac{1}{\|\mathcal{F}_\Lambda\|}\sum_{h\in\Lambda^{-1}} F_{\hat{\mathcal{G}}}(h) \tag{6.4}$$

此外,Parseval 恒等式有效

$$\frac{1}{\|\mathcal{F}_\Lambda\|}\sum_{h\in\Lambda^{-1}} |F_{\hat{\mathcal{G}}}(h)|^2 = \int_{\mathcal{G}} |F(a)|^2\mathrm{d}a \tag{6.5}$$

从傅里叶理论可以得出

$$\int_{\mathcal{G}} |F(a)|^2\mathrm{d}a = \int_{\mathbf{R}^q} |F_{\hat{\mathcal{G}}}(y)|^2\mathrm{d}x \tag{6.6}$$

换句话说,如果 $F_{\hat{\mathcal{G}}}$,$\overline{\mathcal{G}}\subset\overline{\mathcal{F}_\Lambda}$,属于内积空间

$$C_{\hat{\mathcal{G}}}^{(0)} = \left\{y\mapsto\underbrace{\int_{\mathcal{G}} \mathrm{e}^{-2\pi\mathrm{i}a\cdot y}F(a)\,\mathrm{d}a}_{=F_{\hat{\mathcal{G}}}(y)} : F\in C^{(0)}(\overline{\mathcal{G}})\right\} \tag{6.7}$$

那么

$$\int_{\mathbf{R}^q} |F_{\hat{\mathcal{G}}}(y)|^2\mathrm{d}y = \frac{1}{\|\mathcal{F}_\Lambda\|}\sum_{h\in\Lambda^{-1}} |F_{\hat{\mathcal{G}}}(h)|^2 \tag{6.8}$$

将 Λ 替换为其逆格 Λ^{-1},我们发现

$$\int_{\mathbf{R}^q} |F_{\hat{\mathcal{G}}}(y)|^2\mathrm{d}y = \|\mathcal{F}_\Lambda\|\sum_{g\in\Lambda} |F_{\hat{\mathcal{G}}}(g)|^2 \tag{6.9}$$

仔细观察我们的方法,我们注意到 Shannon 抽样是在参考集 $C_{\hat{\mathcal{G}}}^{(0)}$ 上阐述的,它是相关 Paley-Wiener 空间

$$B_{\overline{\mathcal{G}}} = \left\{y\mapsto\int_{\mathcal{G}} \mathrm{e}^{-2\pi\mathrm{i}a\cdot y}F(a)\,\mathrm{d}a,\, y\in\mathbf{R}^q : F\in L^2(\mathcal{G})\right\} \tag{6.10}$$

的严格子集. 然而,这一观察结果并没有给我们带来太多麻烦,因为每一个 $F\in L^2(\mathcal{G})$ 都可以用一个函数 $F_\varepsilon\in C^{(0)}(\overline{\mathcal{G}})$(在 ε-精度下)来近似(在 $L^2(\mathcal{G})$-意义上),使得

$$\begin{aligned}
&\sup_{y\in\mathbf{R}^q} |F_{\hat{\mathcal{G}}}(y) - (F_\varepsilon)_{\hat{\mathcal{G}}}(y)| \\
\leqslant&\sup_{y\in\mathbf{R}^q}\left|\int_{\mathcal{G}} \mathrm{e}^{-2\pi\mathrm{i}a\cdot y}(F(a) - F_\varepsilon(a))\,\mathrm{d}a\right| \\
\leqslant&\left(\int_{\mathcal{G}} |F(a) - F_\varepsilon(a)|^2\mathrm{d}a\right)^{1/2}\left(\int_{\mathcal{G}} |\mathrm{e}^{-2\pi\mathrm{i}a\cdot y}|^2\mathrm{d}a\right)^{1/2} \\
\leqslant&\left(\int_{\mathbf{R}^q} |F_{\hat{\mathcal{G}}}(a) - (F_\varepsilon)_{\hat{\mathcal{G}}}(a)|^2\mathrm{d}a\right)^{1/2}\left(\int_{\mathcal{G}} |\mathrm{e}^{-2\pi\mathrm{i}a\cdot y}|^2\mathrm{d}a\right)^{1/2}
\end{aligned}$$

$$= \sqrt{\| \mathcal{G} \|}\,\varepsilon \tag{6.11}$$

总之,如果$\mathcal{G} \subset \mathbf{R}^q$是满足$\overline{\mathcal{G}} \subset \overline{\mathcal{F}_\Lambda}$的正则区域,那么 Paley-Wiener 空间 $B_{\overline{\mathcal{G}}}$ 是 $L^2(\mathbf{R}^q)$ 拓扑下空间 $C_{\mathcal{G}}^{(0)}$ 的完全化

$$B_{\overline{G}} = \overline{C_{\mathcal{G}}^{(0)}}^{\| \cdot \|_{L^2(\mathbf{R}^q)}} \tag{6.12}$$

在上述假设下,$\mathcal{G}$是满足$\overline{\mathcal{G}} \subset \overline{\mathcal{F}_\Lambda}$ 的正则区域,集合 $B_{\overline{\mathcal{G}}}$ 形成一个具有核

$$K_{\mathcal{G}}(x-y) = \int_{\mathcal{G}} \mathrm{e}^{2\pi i a \cdot (x-y)}\,\mathrm{d}a \tag{6.13}$$

的再生核空间. 事实上,根据(6.6),我们看到

$$\begin{aligned} | F_{\hat{\mathcal{G}}}(y) | &\leqslant \sqrt{\| \mathcal{G} \|} \sqrt{\int_{\mathcal{G}} | F(a) |^2 \mathrm{d}a} \\ &= \sqrt{\| \mathcal{G} \|} \sqrt{\int_{\mathbf{R}^q} | F_{\hat{\mathcal{G}}}(w) |^2 \mathrm{d}w} \end{aligned} \tag{6.14}$$

此外,标准傅里叶反演产生

$$\begin{aligned} F_{\hat{\mathcal{G}}}(y) &= \int_{\mathbf{R}^q} F_{\hat{\mathcal{G}}}(x) \left(\int_{\mathcal{G}} \mathrm{e}^{2\pi i a \cdot (x-y)}\,\mathrm{d}a \right) \mathrm{d}x \\ &= \int_{\mathbf{R}^q} F_{\hat{\mathcal{G}}}(x) K_{\mathcal{G}}(x-y)\,\mathrm{d}x \end{aligned} \tag{6.15}$$

对于所有 $y \in \mathbf{R}^q$,其中

$$\int_{\mathbf{R}^q} = \lim_{N \to \infty} \int_{\substack{|x| \leqslant N \\ x \in \mathbf{R}^q}} \cdots \tag{6.16}$$

因此,$B_{\overline{\mathcal{G}}}$ 是一个具有核(6.13)的再生核 Hilbert 空间.

备注 6.1 另外,在 Paley-Wiener 空间 $B_{\overline{\mathcal{F}_\Lambda}}$ 中,我们可以保证,再生核形成了一个具有离散正交性质的正交系.

7. 抽样展开的所有路径都通向再生核

Shannon 抽样定理的经典证明是基于傅里叶积分和复傅里叶级数的逆. 假设定义在实线上的信号 F 是平方可积且有限带宽的,比如$\mathcal{F}_{\mathbf{Z}}$,然后 F 可以表示为傅里叶变换 $F_{\hat{\mathbf{Z}}}$ 的逆,此时积分是在$\mathcal{F}_{\mathbf{Z}}$ 上的. 我们周期性地将 $F_{\hat{\mathbf{Z}}}$ 扩展到实线,并用复傅里叶级数展开得到的周期函数. 最后,我们交换积分和求和,抽样定理就出现了. 虽然这个证明很简单,但在两个方面并不是很有启发性.

(i) 人们无法从积分和求和的交换中得到任何见解,特别是,人们不会听到或感觉到证明的"心跳".

(ii) 信号 F 的充分条件,即有限带宽,是施加在傅里叶变换上的一个间接条件,而不是 F 本身.

在经典 Shannon 理论中,人们已经认识到,有限带宽函数的空间与对实线的限制为指数增长的整函数的 Paley-Weiner 空间相同. 而这个空间就是一个以 sinc 函数为再生核的再生核 Hilbert 空间. 然而,再生核性质并没有进入基数级数的各种方法的经典推导. Nashed 和 Nalter 介绍了一种新的方法来研究再生核 Hilbert 空间中函数的一般抽样定理,并说明了有多少抽样结果是其方法的特例. 在另一篇论文中,Nashed 和 Walter 展示了如何从允许抽样展开的函数空间构造再生核 Hilbert 空间.

鉴于再生核空间与抽样展开之间的这种密切关系,在过去的 25 年中,再生核 Hilbert 和 Banach 空间在信号分析以及逆问题和成像的应用中发挥了重要作用也就不足为奇了.

开创性的想法往往是由惊人的见解引发的. 许多这样的想法被彻底地推广到新的方向,并应用于不同的领域,远远超出了任何人的想象. 抽样理论的显著发展就是这样的一个例子.

总之,这一贡献的目标是表明今天 Shannon 抽样定理推广的重要根源是从 Gauss 的贡献开始的基本数论结果. 事实上,我们从 19 世纪初的 Gauss 圆问题出发,通过扩展 20 世纪末的 Hardy-Landau 格点恒等式,产生了具有高度实用性的新的多元 Shannon 抽样过程,从而提供了有关 Paley-Wiener 空间的一般再生核空间结构.

再比如第 3 章、第 20 章、第 23 章、第 24 章、第 25 章中所论及的超几何级数. 正如胡作玄先生所指出[①]:

三角函数、Bessel 函数、Legendre 函数展开成幂级数都是超几何级数的特殊情形. 一般的超几何级数

① 摘自《近代数学史》,胡作玄著,山东教育出版社,2006.

$$1+\frac{\alpha\cdot\beta}{1\cdot\gamma}x+\frac{\alpha(\alpha+1)\beta(\beta+1)}{1\cdot2\cdot\gamma(\gamma+1)}x^2+\frac{\alpha(\alpha+1)(\alpha+2)\beta(\beta+1)(\beta+2)}{1\cdot2\cdot3\cdot\gamma(\gamma+1)(\gamma+2)}x^3+\cdots \tag{1}$$

最早出现在Euler的《积分法导引》第二卷(1769)第11章中,其中他由积分

$$\int_0^1 u^{b-1}(1-u)^{c-b-1}(1-xu)^{-a}\mathrm{d}u$$

得出

$$B(b,c-b)\left[1+\frac{ab}{1\cdot c}x+\frac{a(a+1)b(b+1)}{1\cdot2\cdot c(c+1)}x^2+\cdots\right]$$

其中$B(b,c-b)$是第一类Euler积分,无穷级数满足微分方程

$$x(1-x)\mathrm{d}^2y+[c-(a+b+1)x]\mathrm{d}y\mathrm{d}x-aby\mathrm{d}x^2=0$$

这类方程后来被称为超几何微分方程,Euler还考虑了更一般的方程

$$x^2(a+bx^n)\mathrm{d}^2y+x(c+ex^n)\mathrm{d}y\mathrm{d}x+(f+gx^n)y\mathrm{d}x^2=0$$

这个方程后来被Gauss的老师Pfaff加以研究,并写进他的《分析研究》(*Disquisitiones analyticae*)(1797)中. Pfaff在书中专门有一章讲"超几何级数",正是他首次在上述意义下用"超几何"这个名称.

Gauss在1800年左右研究算术-几何平均以及椭圆积分时,多次遇到超几何级数的特殊情形,并知道它们满足超几何微分方程. 他在1805年9月3日写给Bessel的信中谈到,在天文计算中,这种级数收敛极快,这些无疑都促使Gauss系统地研究超几何级数,从而促使1812年论文的发表. 这篇论文以Gauss的完美风格写成,详尽而全面地讨论了超几何级数的性质. 首先,Gauss引进记号$F(\alpha,\beta,\gamma;x)$来表示超几何级数(1)表示的函数;其次,他十分明确地提出当$|x|<1$时级数收敛,当$|x|>1$时级数发散,当$|x|=0$时不定,这是最早的明确函数级数收敛的判据.

Gauss还引进了相关函数的概念,$F(\alpha,\beta,\gamma;x)$的相关函数是$F(\alpha\pm1,\beta\pm1,\gamma\pm1;x)$,加上对$\alpha,\beta,\gamma$等系统地置换,他得出$F(\alpha,\beta,\gamma;x)$以及15对相关函数所满足的15个等式,例如,

第15个等式为

$$0=\gamma(\gamma-1-(2\gamma-\alpha-\beta-1)x)F(\alpha,\beta,\gamma;x)+(\gamma-\alpha)(\gamma-\beta)xF(\alpha,\beta,\gamma+1;x)-\gamma(\gamma-1)(1-x)F(\alpha,\beta,\gamma-1;x)$$

由此可知,任何三个函数 $F(\alpha\pm m,\beta\pm n,\gamma\pm p;x)$(其中 m,n,p 是整数)之间存在线性关系,它们以有理函数为系数. Gauss 引进这些相关函数的目的是计算诸如

$$\frac{F(\alpha,\beta+1,\gamma+1;x)}{F(\alpha,\beta,\gamma;x)}$$

的商.

Gauss 还引入 Gauss 阶乘函数

$$\Pi(k,z)=\frac{1\cdot 2\cdot\cdots\cdot k\cdot k^{z}}{(z+1)(z+2)\cdots(z+k)}$$

并定义

$$\Pi(z)=\lim_{k\to\infty}\Pi(k,z)$$

借助于它们,可得

$$F(\alpha,\beta,\gamma;1)=\frac{\Pi(\gamma-1)\Pi(\gamma-\alpha-\beta-1)}{\Pi(\gamma-\alpha-1)\Pi(\gamma-\beta-1)}$$

特别是用 $\Pi(z)$ 可以求出一系列积分的值,例如 Gauss 极感兴趣的双纽线积分

$$A=\int_0^1\frac{\mathrm{d}x}{\sqrt{1-x^4}}=\frac{\Pi\left(\frac{1}{4}\right)\Pi\left(\frac{1}{2}\right)}{\Pi\left(-\frac{1}{2}\right)}$$

$$B=\int_0^1\frac{x^2\mathrm{d}x}{\sqrt{1-x^4}}=\frac{\Pi\left(\frac{3}{4}\right)\Pi\left(-\frac{1}{2}\right)}{3\Pi\left(\frac{1}{4}\right)}=\frac{\Pi\left(-\frac{1}{4}\right)\Pi\left(-\frac{1}{2}\right)}{4\Pi\left(\frac{1}{4}\right)}$$

因此推出

$$AB=\frac{\pi}{4}$$

Gauss 研究超几何级数的第二篇论文在他生前并没有发表（在他去世后收入《全集》第三卷(1876) 中），在这篇论文中，他明确地引入复变元，并引进了超几何微分方程的与 $F(\alpha,\beta,\gamma;x)$ 线性独立的第二个解. 他认识到，微分方程除了 0,1(及 ∞，他避免用无穷) 外，解处处存在，在这里他遇到了解析开拓问题以及单值性问题，他实际上已考虑到变元 x 过渡到复量 z 时级数的收敛，并提出经历不同道路的单值性问题，正如在其他领域一样，Gauss 的思想后来由其他数学家发表.

在 Gauss 之后，Kummer 首先发展了超几何微分方程. 1834 年他开始考虑 Gauss 的第一篇论文所遗留下来的问题，即不同变元的解之间的关系. 1836 年，他得到变元 x 变为 $\frac{1}{x}$, $1-x$, $\frac{1}{1-x}$, $\frac{x}{1-x}$, $\frac{1-x}{x}$ 后得出的另外24 个解，这是完全组，而且它们之间划分为6 个等阶组，这6 个组之间，其中有4 个3 组之间的线性关系. Kummer 在 1836 年的论文的结尾曾考虑到 α,β,γ 是实数而 x 是复数时会发生什么情况，不过他只考虑极为特殊的情形. 真正过渡到复值情形，无可争议应该归功于 Riemann，他是把微分方程引入复域的真正奠基者.

Riemann 在 1857 年的论文中把超几何级数称为 Gauss 级数. Riemann 的成就在于他把超几何级数及超几何微分方程由实域扩充到复域，并且避开了传统的那种研究函数必须用具体表达式来计算的方法，而是用“公理的”方法得出 Gauss，Kummer 等人通过 Riemann 计算才得出的关系式. 他最重要的成就在于把复变函数论引进常微分方程. 他的具体做法是引进复值函数 P 函数

$$P\begin{Bmatrix} a,b,c & \\ \alpha,\beta,\gamma;x \\ \alpha',\beta',\gamma' & \end{Bmatrix}$$

满足三个条件：

(1) 除了 a,b,c 之外，P 是 x 的有限单值函数.

(2) 任何三个 P 函数 P',P'',P''' 都存在线性关系

$$C'P' + C''P'' + C'''P''' = 0$$

其中,C',C'',C''' 是常数.

(3)P 函数可写成 $C_\alpha P^{(\alpha)} + C_{\alpha'}P^{(\alpha')}$,使

$$P^{(\alpha)}(x-Q)^{-\alpha}, P^{(\alpha')}(x-Q)^{-\alpha'}$$

在 $x = a$ 处仍是单值,且既不等于0,也不等于∞. 同样,P 函数可表示为

$$C_\beta P^{(\beta)} + C_{\beta'}P^{(\beta')}, C_\gamma P^{(\gamma)} + C_{\gamma'}P^{(\gamma')}$$

它们分别在 $x = b$,$x = c$ 处具有类似的性质,且6个量 $\alpha,\beta,\gamma,\alpha'$,$\beta',\gamma'$ 满足下面两个假定:

(1) 差 $\alpha-\alpha'$,$\beta-\beta'$,$\gamma-\gamma'$ 均非整数;

(2)$\alpha+\beta+\gamma+\alpha'+\beta'+\gamma' = 1$.

再比如一些我们在其他分支中也遇到过的概念,如正交性等,在本书中都有论及,比如第 8 章.

我们讨论有关本征函数正交性的一般问题[①].

假定一个定义在有界区间 $[0,L]$ 上的函数 $X(x)$ 满足本征值问题

$$\begin{cases} X''(x) + \lambda X(x) = 0 & (1) \\ f|_{x=0} = 0, g|_{x=L} = 0 & (2) \end{cases}$$

其中,$\lambda = \lambda_n$ 是第 n 个本征值,$X(x) = X_n(x)$ 是相应的本征函数($n = 1,2,3,\cdots$). 式(2) 表示 $X_n(x)$ 在 $x = 0$ 和 $x = L$ 的齐次边界条件(可以是各类边界条件). 现在讨论本征函数的正交性,即计算积分 $\int_0^L X_m(x)X_n(x)\mathrm{d}x\,(m \neq n)$.

方程(1) 给出

$$\frac{\mathrm{d}^2}{\mathrm{d}x^2}X_n = -\lambda_n X_n \tag{3}$$

因此我们有

$$\frac{\mathrm{d}}{\mathrm{d}x}(X_n') = -\lambda_n X_n, \frac{\mathrm{d}}{\mathrm{d}x}(X_m') = -\lambda_m X_m \quad (m \neq n) \tag{4}$$

① 摘自《数学物理方法》,顾樵编著,科学出版社,2017.

给以上两式两边分别乘以 X_m 和 X_n,得到

$$X_m \frac{\mathrm{d}}{\mathrm{d}x}(X_n') = -\lambda_n X_m X_n, X_n \frac{\mathrm{d}}{\mathrm{d}x}(X_m') = -\lambda_m X_m X_n \tag{5}$$

式(5) 中两式两边分别相减再积分,得到

$$\int_0^L \left[X_m \frac{\mathrm{d}}{\mathrm{d}x}(X_n') - X_n \frac{\mathrm{d}}{\mathrm{d}x}(X_m')\right]\mathrm{d}x = \int_0^L (\lambda_m X_m X_n - \lambda_n X_m X_n)\,\mathrm{d}x \tag{6}$$

计算式(6) 左边的积分

$$\begin{aligned}
&\int_0^L \left[X_m \frac{\mathrm{d}}{\mathrm{d}x}(X_n') - X_n \frac{\mathrm{d}}{\mathrm{d}x}(X_m')\right]\mathrm{d}x \\
&= \int_0^L X_m \mathrm{d}(X_n') - \int_0^L X_n \mathrm{d}(X_m') \\
&= [X_m X_n']_0^L - \int_0^L X_n' \mathrm{d}X_m - [X_n X_m']_0^L + \int_0^L X_m' \mathrm{d}X_n \\
&= X_m(L)X_n'(L) - X_m(0)X_n'(0) - \int_0^L X_n' \frac{\mathrm{d}X_m}{\mathrm{d}x}\mathrm{d}x - \\
&\quad X_n(L)X_m'(L) + X_n(0)X_m'(0) + \int_0^L X_m' \frac{\mathrm{d}X_n}{\mathrm{d}x}\mathrm{d}x \\
&= [X_n(0)X_m'(0) - X_m(0)X_n'(0)] - [X_n(L)X_m'(L) - \\
&\quad X_m(L)X_n'(L)] - \underbrace{\left(\int_0^L X_n' X_m' \mathrm{d}x - \int_0^L X_m' X_n' \mathrm{d}x\right)}_{=0} \\
&= [X_n(0)X_m'(0) - X_m(0)X_n'(0)] - \\
&\quad [X_n(L)X_m'(L) - X_m(L)X_n'(L)]
\end{aligned}$$

式(6) 右边的积分为

$$\int_0^L (\lambda_m X_m X_n - \lambda_n X_m X_n)\,\mathrm{d}x = (\lambda_m - \lambda_n)\int_0^L X_m X_n \mathrm{d}x \tag{7}$$

这样

$$\int_0^L X_m X_n \mathrm{d}x = \frac{Q}{\lambda_m - \lambda_n} \quad (m \neq n) \tag{8}$$

其中

$$\begin{aligned}
Q = {} & [X_n(0)X_m'(0) - X_m(0)X_n'(0)] - \\
& [X_n(L)X_m'(L) - X_m(L)X_n'(L)]
\end{aligned} \tag{9}$$

称为 Q 因子,它依赖于本征函数及其导数在两个端点 $x = 0$ 和 $x = L$ 的取值. 式(8) 是关于本征值问题(1)(2) 的一般性结果,

适用于该方程的任何形式的本征值与本征函数. 式(8) 表示,只要给定的边界条件满足 $Q = 0$,则相应的本征函数 $X_1, X_2, X_3, \cdots$ 在 $(0, L)$ 上是相互正交的.

再比如在泛函分析中有如下内容①.

设 X 是 Banach 空间,X^* 是其对偶空间.

定义1 元素系 $\{x_\alpha\} \subset X$ 称为在 X 内完备,倘若 $\{x_\alpha\}$ 的线性包的闭包是 X.

在 $L^2(0,1)$ 内的完备 O. N. S. $\Phi = \{\varphi_n(x)\}_{n=1}^\infty$ 称为完备正交规范系,记作 C. O. N. S.

定义2 称元素系 $\{x_\alpha^*\} \subset X^*$ 是整体的,倘若在 X 内不存在非零元 x 能使 $\langle x_\alpha^*, x\rangle = 0$ 对一切指标 α 成立.

下面的命题是空间 $L^p(0,1)$ 和 $L^q(0,1)$ 当 $1 < p < \infty$, $\frac{1}{p} + \frac{1}{q} = 1$ 时的对偶性质的直接推论.

命题1 为使 $\Phi = \{\varphi_\alpha\} \subset L^p(0,1)(1 < p < \infty)$ 在 $L^p(0,1)$ 内是完备的,必须且只需 Φ 是整体的.

证明 设 Φ 在 $L^p(0,1)$ 内完备,对某个函数 $f \in L^q(0,1)$,等式

$$\int_0^1 \varphi_\alpha(x) f(x)\,\mathrm{d}x = 0$$

对每一个 $\varphi_\alpha(x) \in \Phi$ 成立. 由于 Φ 的完备性,对任何 $g \in L^p(0,1)$ 都成立

$$\int_0^1 g(x) f(x)\,\mathrm{d}x = 0$$

这说明几乎处处有 $f(x) = 0$.

今设 Φ 是整体的,X_1 是 Φ 的线性包的闭包. X_1 是 $L^p(0,1)$ 的闭子空间,倘若 $X_1 \neq L^p(0,1)$,则由 Hahn-Banach 定理,存在

① 摘自《正交级数》,Б. С. Кашин, А. А. Саакян 著,孙永生,王昆扬译,北京师范大学出版社,2007.

一个$f \in L^q(0,1)$，$\|f\|_q > 0$，能使$\int_0^1 f(x)g(x)\mathrm{d}x = 0$对一切$g \in X_1$成立. 特别地，$\int_0^1 f(x)\varphi_\alpha(x)\mathrm{d}x = 0$对$\varphi_\alpha \in \Phi$成立，此事与$\Phi$的整体性矛盾.

设$\{x_n\}_{n=1}^\infty$是 Banach 空间X的元素序列，$X_k(k = 1,2,\cdots)$是集合$\{x_n : n = 1,2,\cdots, n \neq k\}$的线性包的闭包(依$X$范).

定义3　系$\{x_n\} \subset X$称为在X内最小，倘若对一切$k = 1,2,\cdots$，$x_k \notin X_k$.

定义4　系$\{x_n, y_n\}$，$x_n \in X$，$y_n \in X^*$，$n = 1,2,\cdots$称为双正交的，倘若当$n \neq m$时$\langle x_n, y_m\rangle = 0$；称为双正交规范的，倘若

$$\langle x_n, y_m\rangle = \begin{cases} 1, n = m \text{ 时} \\ 0, n \neq m \text{ 时} \end{cases} \quad (n, m = 1,2,\cdots) \tag{1}$$

若$\{x_n, y_m\}$是双正交规范系，则$\{y_n\}$是$\{x_n\}$的对偶系. 一般地说，对偶系不唯一. 不过易见有：

命题2　设$\{x_n\}_{n=1}^\infty$是Banach空间X内的完备系，则其对偶系(倘若存在)必唯一.

事实上，若$\{y_n'\}$，$\{y_n''\}$都满足式(1)(分别令$y_n = y_n'$及$y_n = y''_n$)，则对每一固定的$m = 1,2,\cdots$，将成立$\langle x_n, y'_m - y''_m\rangle = 0$，$n = 1,2,\cdots$，故由系$\{x_n\}$的完备性推出$y_m' = y_m''$.

定理　对一给定的序列$\{x_n\} \subset X$，为了存在序列$\{y_n\} \subset X^*$和$\{x_n\}$组成双正交规范系，当且仅当$\{x_n\}$在X内最小.

证明　若$\{x_n, y_n\}$是一双正交规范系，则$\{x_n\}$是最小的. 因为由(1)，对每个$x \in X_k$有$\langle x, y_k\rangle = 0$，而$\langle x_k, y_k\rangle = 1$，所以$x_k \notin X_k$，$k = 1,2,\cdots$.

反之，如果$\{x_n\}$最小，则由 Hahn-Banach 定理，考虑到$x_k \notin X_k$，而X_k在X内是闭子空间($k = 1,2,\cdots$)，我们可以找到泛函$y_k \in X^*$满足(1).

显然，当$X = L^p(0,1)$，$X^* = L^q(0,1)$，$1 \leqslant p < \infty$，$\frac{1}{p} + \frac{1}{q} = 1$时，每一 O. N. S. $\{\varphi_n\}_{n=1}^\infty$，$\varphi_n \in L^p(0,1) \cap L^q(0,1)$($n = 1,2,\cdots$)都满足式(1)，只需置$x_n = \varphi_n$，$y_n = \varphi_n$($n = 1,2,\cdots$)即可.

由定理可推出：

命题3 每一O. N. S. $\{\varphi_n(x)\}_{n=1}^{\infty}, x \in (0,1), \varphi_n \in L^p(0,1) \cap L^q(0,1)$ $\left(n = 1,2,\cdots, 1 \leqslant p < \infty, \frac{1}{p} + \frac{1}{q} = 1\right)$，在$L^p(0,1)$内最小，是其自身的对偶系.

读完本书您一定会有一种感觉，即在数学才能方面，人与人是有巨大差异的. 有些人看似不费力就能在广阔的领域上恣意奔跑，而我们仅仅是试图要理解其思想脉络都要拼尽全力，是我们不够努力吗？当然有这方面的因素，但更关键之处在于天分. 1960年，美国女作家哈珀·李发表了一篇名为《杀死一只知更鸟》的小说. 1961年，该书获当年度普利策奖，其中有一句金句是这样的："我们都知道，某些人灌输给我们的'人人生而平等'，实际上是个谬论——事实上，有些人就是比别人聪明睿智，有些人就是比别人享有更多的机会，因为他们生来如此，有些男人比别的男人挣钱多，有些女士做的蛋糕比别的女士更胜一筹——总而言之，有些人天生就比大多数普通人具有更高的天赋和才华."

在数学中我们格外地感受到了这种不平等！

刘培杰

2022年9月7日

于哈工大

刘培杰数学工作室
已出版(即将出版)图书目录——原版影印

书　　名	出版时间	定　价	编号
数学物理大百科全书.第1卷(英文)	2016－01	418.00	508
数学物理大百科全书.第2卷(英文)	2016－01	408.00	509
数学物理大百科全书.第3卷(英文)	2016－01	396.00	510
数学物理大百科全书.第4卷(英文)	2016－01	408.00	511
数学物理大百科全书.第5卷(英文)	2016－01	368.00	512
zeta函数,q-zeta函数,相伴级数与积分(英文)	2015－08	88.00	513
微分形式:理论与练习(英文)	2015－08	58.00	514
离散与微分包含的逼近和优化(英文)	2015－08	58.00	515
艾伦·图灵:他的工作与影响(英文)	2016－01	98.00	560
测度理论概率导论,第2版(英文)	2016－01	88.00	561
带有潜在故障恢复系统的半马尔柯夫模型控制(英文)	2016－01	98.00	562
数学分析原理(英文)	2016－01	88.00	563
随机偏微分方程的有效动力学(英文)	2016－01	88.00	564
图的谱半径(英文)	2016－01	58.00	565
量子机器学习中数据挖掘的量子计算方法(英文)	2016－01	98.00	566
量子物理的非常规方法(英文)	2016－01	118.00	567
运输过程的统一非局部理论:广义波尔兹曼物理动力学,第2版(英文)	2016－01	198.00	568
量子力学与经典力学之间的联系在原子、分子及电动力学系统建模中的应用(英文)	2016－01	58.00	569
算术域(英文)	2018－01	158.00	821
高等数学竞赛:1962—1991年的米洛克斯·史怀哲竞赛(英文)	2018－01	128.00	822
用数学奥林匹克精神解决数论问题(英文)	2018－01	108.00	823
代数几何(德文)	2018－04	68.00	824
丢番图逼近论(英文)	2018－01	78.00	825
代数几何学基础教程(英文)	2018－01	98.00	826
解析数论入门课程(英文)	2018－01	78.00	827
数论中的丢番图问题(英文)	2018－01	78.00	829
数论(梦幻之旅):第五届中日数论研讨会演讲集(英文)	2018－01	68.00	830
数论新应用(英文)	2018－01	68.00	831
数论(英文)	2018－01	78.00	832

刘培杰数学工作室
已出版(即将出版)图书目录——原版影印

书名	出版时间	定价	编号
湍流十讲(英文)	2018-04	108.00	886
无穷维李代数:第3版(英文)	2018-04	98.00	887
等值、不变量和对称性(英文)	2018-04	78.00	888
解析数论(英文)	2018-09	78.00	889
《数学原理》的演化:伯特兰·罗素撰写第二版时的手稿与笔记(英文)	2018-04	108.00	890
哈密尔顿数学论文集(第4卷):几何学、分析学、天文学、概率和有限差分等(英文)	2019-05	108.00	891
偏微分方程全局吸引子的特性(英文)	2018-09	108.00	979
整函数与下调和函数(英文)	2018-09	118.00	980
幂等分析(英文)	2018-09	118.00	981
李群,离散子群与不变量理论(英文)	2018-09	108.00	982
动力系统与统计力学(英文)	2018-09	118.00	983
表示论与动力系统(英文)	2018-09	118.00	984
分析学练习. 第1部分(英文)	2021-01	88.00	1247
分析学练习. 第2部分,非线性分析(英文)	2021-01	88.00	1248
初级统计学:循序渐进的方法:第10版(英文)	2019-05	68.00	1067
工程师与科学家微分方程用书:第4版(英文)	2019-07	58.00	1068
大学代数与三角学(英文)	2019-06	78.00	1069
培养数学能力的途径(英文)	2019-07	38.00	1070
工程师与科学家统计学:第4版(英文)	2019-06	58.00	1071
贸易与经济中的应用统计学:第6版(英文)	2019-06	58.00	1072
傅立叶级数和边值问题:第8版(英文)	2019-05	48.00	1073
通往天文学的途径:第5版(英文)	2019-05	58.00	1074
拉马努金笔记. 第1卷(英文)	2019-06	165.00	1078
拉马努金笔记. 第2卷(英文)	2019-06	165.00	1079
拉马努金笔记. 第3卷(英文)	2019-06	165.00	1080
拉马努金笔记. 第4卷(英文)	2019-06	165.00	1081
拉马努金笔记. 第5卷(英文)	2019-06	165.00	1082
拉马努金遗失笔记. 第1卷(英文)	2019-06	109.00	1083
拉马努金遗失笔记. 第2卷(英文)	2019-06	109.00	1084
拉马努金遗失笔记. 第3卷(英文)	2019-06	109.00	1085
拉马努金遗失笔记. 第4卷(英文)	2019-06	109.00	1086
数论:1976年纽约洛克菲勒大学数论会议记录(英文)	2020-06	68.00	1145
数论:卡本代尔1979:1979年在南伊利诺伊卡本代尔大学举行的数论会议记录(英文)	2020-06	78.00	1146
数论:诺德韦克豪特1983:1983年在诺德韦克豪特举行的Journees Arithmetiques数论大会会议记录(英文)	2020-06	68.00	1147
数论:1985-1988年在纽约城市大学研究生院和大学中心举办的研讨会(英文)	2020-06	68.00	1148

刘培杰数学工作室

已出版(即将出版)图书目录——原版影印

书 名	出版时间	定 价	编号
数论:1987 年在乌尔姆举行的 Journees Arithmetiques 数论大会会议记录(英文)	2020—06	68.00	1149
数论:马德拉斯 1987:1987 年在马德拉斯安娜大学举行的国际拉马努金百年纪念大会会议记录(英文)	2020—06	68.00	1150
解析数论:1988 年在东京举行的日法研讨会会议记录(英文)	2020—06	68.00	1151
解析数论:2002 年在意大利切特拉罗举行的 C. I. M. E. 暑期班演讲集(英文)	2020—06	68.00	1152
量子世界中的蝴蝶:最迷人的量子分形故事(英文)	2020—06	118.00	1157
走进量子力学(英文)	2020—06	118.00	1158
计算物理学概论(英文)	2020—06	48.00	1159
物质,空间和时间的理论:量子理论(英文)	2020—10	48.00	1160
物质,空间和时间的理论:经典理论(英文)	2020—10	48.00	1161
量子场理论:解释世界的神秘背景(英文)	2020—07	38.00	1162
计算物理学概论(英文)	2020—06	48.00	1163
行星状星云(英文)	2020—10	38.00	1164
基本宇宙学:从亚里士多德的宇宙到大爆炸(英文)	2020—08	58.00	1165
数学磁流体力学(英文)	2020—07	58.00	1166
计算科学:第 1 卷,计算的科学(日文)	2020—07	88.00	1167
计算科学:第 2 卷,计算与宇宙(日文)	2020—07	88.00	1168
计算科学:第 3 卷,计算与物质(日文)	2020—07	88.00	1169
计算科学:第 4 卷,计算与生命(日文)	2020—07	88.00	1170
计算科学:第 5 卷,计算与地球环境(日文)	2020—07	88.00	1171
计算科学:第 6 卷,计算与社会(日文)	2020—07	88.00	1172
计算科学. 别卷,超级计算机(日文)	2020—07	88.00	1173
多复变函数论(日文)	2022—06	78.00	1518
复变函数入门(日文)	2022—06	78.00	1523
代数与数论:综合方法(英文)	2020—10	78.00	1185
复分析:现代函数理论第一课(英文)	2020—07	58.00	1186
斐波那契数列和卡特兰数:导论(英文)	2020—10	68.00	1187
组合推理:计数艺术介绍(英文)	2020—07	88.00	1188
二次互反律的傅里叶分析证明(英文)	2020—07	48.00	1189
旋瓦兹分布的希尔伯特变换与应用(英文)	2020—07	58.00	1190
泛函分析:巴拿赫空间理论入门(英文)	2020—07	48.00	1191
卡塔兰数入门(英文)	2019—05	68.00	1060
测度与积分(英文)	2019—04	68.00	1059
组合学手册. 第一卷(英文)	2020—06	128.00	1153
* 一代数、局部紧群和巴拿赫 * 一代数丛的表示. 第一卷,群和代数的基本表示理论(英文)	2020—05	148.00	1154
电磁理论(英文)	2020—08	48.00	1193
连续介质力学中的非线性问题(英文)	2020—09	78.00	1195
多变量数学入门(英文)	2021—05	68.00	1317
偏微分方程入门(英文)	2021—05	88.00	1318
若尔当典范性:理论与实践(英文)	2021—07	68.00	1366
伽罗瓦理论. 第 4 版(英文)	2021—08	88.00	1408

刘培杰数学工作室
已出版(即将出版)图书目录——原版影印

书　名	出版时间	定　价	编号
典型群,错排与素数(英文)	2020－11	58.00	1204
李代数的表示:通过 gln 进行介绍(英文)	2020－10	38.00	1205
实分析演讲集(英文)	2020－10	38.00	1206
现代分析及其应用的课程(英文)	2020－10	58.00	1207
运动中的抛射物数学(英文)	2020－10	38.00	1208
2－纽结与它们的群(英文)	2020－10	38.00	1209
概率,策略和选择:博弈与选举中的数学(英文)	2020－11	58.00	1210
分析学引论(英文)	2020－11	58.00	1211
量子群:通往流代数的路径(英文)	2020－11	38.00	1212
集合论入门(英文)	2020－10	48.00	1213
酉反射群(英文)	2020－11	58.00	1214
探索数学:吸引人的证明方式(英文)	2020－11	58.00	1215
微分拓扑短期课程(英文)	2020－10	48.00	1216
抽象凸分析(英文)	2020－11	68.00	1222
费马大定理笔记(英文)	2021－03	48.00	1223
高斯与雅可比和(英文)	2021－03	78.00	1224
π与算术几何平均:关于解析数论和计算复杂性的研究(英文)	2021－01	58.00	1225
复分析入门(英文)	2021－03	48.00	1226
爱德华·卢卡斯与素性测定(英文)	2021－03	78.00	1227
通往凸分析及其应用的简单路径(英文)	2021－01	68.00	1229
微分几何的各个方面.第一卷(英文)	2021－01	58.00	1230
微分几何的各个方面.第二卷(英文)	2020－12	58.00	1231
微分几何的各个方面.第三卷(英文)	2020－12	58.00	1232
沃克流形几何学(英文)	2020－11	58.00	1233
仿射和韦尔几何应用(英文)	2020－12	58.00	1234
双曲几何学的旋转向量空间方法(英文)	2021－02	58.00	1235
积分:分析学的关键(英文)	2020－12	48.00	1236
为有天分的新生准备的分析学基础教材(英文)	2020－11	48.00	1237
数学不等式.第一卷.对称多项式不等式(英文)	2021－03	108.00	1273
数学不等式.第二卷.对称有理不等式与对称无理不等式(英文)	2021－03	108.00	1274
数学不等式.第三卷.循环不等式与非循环不等式(英文)	2021－03	108.00	1275
数学不等式.第四卷.Jensen 不等式的扩展与加细(英文)	2021－03	108.00	1276
数学不等式.第五卷.创建不等式与解不等式的其他方法(英文)	2021－04	108.00	1277

刘培杰数学工作室
已出版(即将出版)图书目录——原版影印

书　　名	出版时间	定　价	编号
冯·诺依曼代数中的谱位移函数:半有限冯·诺依曼代数中的谱位移函数与谱流(英文)	2021－06	98.00	1308
链接结构:关于嵌入完全图的直线中链接单形的组合结构(英文)	2021－05	58.00	1309
代数几何方法. 第 1 卷(英文)	2021－06	68.00	1310
代数几何方法. 第 2 卷(英文)	2021－06	68.00	1311
代数几何方法. 第 3 卷(英文)	2021－06	58.00	1312

书　　名	出版时间	定　价	编号
代数、生物信息和机器人技术的算法问题. 第四卷,独立恒等式系统(俄文)	2020－08	118.00	1199
代数、生物信息和机器人技术的算法问题. 第五卷,相对覆盖性和独立可拆分恒等式系统(俄文)	2020－08	118.00	1200
代数、生物信息和机器人技术的算法问题. 第六卷,恒等式和准恒等式的相等 问题、可推导性和可实现性(俄文)	2020－08	128.00	1201
分数阶微积分的应用:非局部动态过程,分数阶导热系数(俄文)	2021－01	68.00	1241
泛函分析问题与练习:第 2 版(俄文)	2021－01	98.00	1242
集合论、数学逻辑和算法论问题:第 5 版(俄文)	2021－01	98.00	1243
微分几何和拓扑短期课程(俄文)	2021－01	98.00	1244
素数规律(俄文)	2021－01	88.00	1245
无穷边值问题解的递减:无界域中的拟线性椭圆和抛物方程(俄文)	2021－01	48.00	1246
微分几何讲义(俄文)	2020－12	98.00	1253
二次型和矩阵(俄文)	2021－01	98.00	1255
积分和级数. 第 2 卷,特殊函数(俄文)	2021－01	168.00	1258
积分和级数. 第 3 卷,特殊函数补充:第 2 版(俄文)	2021－01	178.00	1264
几何图上的微分方程(俄文)	2021－01	138.00	1259
数论教程:第 2 版(俄文)	2021－01	98.00	1260
非阿基米德分析及其应用(俄文)	2021－03	98.00	1261
古典群和量子群的压缩(俄文)	2021－03	98.00	1263
数学分析习题集. 第 3 卷,多元函数:第 3 版(俄文)	2021－03	98.00	1266
数学习题:乌拉尔国立大学数学力学系大学生奥林匹克(俄文)	2021－03	98.00	1267
柯西定理和微分方程的特解(俄文)	2021－03	98.00	1268
组合极值问题及其应用:第 3 版(俄文)	2021－03	98.00	1269
数学词典(俄文)	2021－01	98.00	1271
确定性混沌分析模型(俄文)	2021－06	168.00	1307
精选初等数学习题和定理. 立体几何. 第 3 版(俄文)	2021－03	68.00	1316
微分几何习题:第 3 版(俄文)	2021－05	98.00	1336
精选初等数学习题和定理. 平面几何. 第 4 版(俄文)	2021－05	68.00	1335
曲面理论在欧氏空间 En 中的直接表示(俄文)	2022－01	68.00	1444
维纳－霍普夫离散算子和托普利兹算子:某些可数赋范空间中的诺特性和可逆性(俄文)	2022－03	108.00	1496
Maple 中的数论:数论中的计算机计算(俄文)	2022－03	88.00	1497
贝尔曼和克努特问题及其概括:加法运算的复杂性(俄文)	2022－03	138.00	1498

刘培杰数学工作室
已出版(即将出版)图书目录——原版影印

书　　名	出版时间	定　价	编号
复分析:共形映射(俄文)	2022—07	48.00	1542
微积分代数样条和多项式及其在数值方法中的应用(俄文)	2022—08	128.00	1543
蒙特卡罗方法中的随机过程和场模型:算法和应用(俄文)	2022—08	88.00	1544
线性椭圆型方程组:论二阶椭圆型方程的迪利克雷问题(俄文)	2022—08	98.00	1561
动态系统解的增长特性:估值、稳定性、应用(俄文)	2022—08	118.00	1565
群的自由积分解:建立和应用(俄文)	2022—08	78.00	1570

书　　名	出版时间	定　价	编号
狭义相对论与广义相对论:时空与引力导论(英文)	2021—07	88.00	1319
束流物理学和粒子加速器的实践介绍:第2版(英文)	2021—07	88.00	1320
凝聚态物理中的拓扑和微分几何简介(英文)	2021—05	88.00	1321
混沌映射:动力学、分形学和快速涨落(英文)	2021—05	128.00	1322
广义相对论:黑洞、引力波和宇宙学介绍(英文)	2021—06	68.00	1323
现代分析电磁均质化(英文)	2021—06	68.00	1324
为科学家提供的基本流体动力学(英文)	2021—06	88.00	1325
视觉天文学:理解夜空的指南(英文)	2021—06	68.00	1326
物理学中的计算方法(英文)	2021—06	68.00	1327
单星的结构与演化:导论(英文)	2021—06	108.00	1328
超越居里:1903 年至 1963 年物理界四位女性及其著名发现(英文)	2021—06	68.00	1329
范德瓦尔斯流体热力学的进展(英文)	2021—06	68.00	1330
先进的托卡马克稳定性理论(英文)	2021—06	88.00	1331
经典场论导论:基本相互作用的过程(英文)	2021—07	88.00	1332
光致电离量子动力学方法原理(英文)	2021—07	108.00	1333
经典域论和应力:能量张量(英文)	2021—05	88.00	1334
非线性太赫兹光谱的概念与应用(英文)	2021—06	68.00	1337
电磁学中的无穷空间并矢格林函数(英文)	2021—06	88.00	1338
物理科学基础数学. 第1卷,齐次边值问题、傅里叶方法和特殊函数(英文)	2021—07	108.00	1339
离散量子力学(英文)	2021—07	68.00	1340
核磁共振的物理学和数学(英文)	2021—07	108.00	1341
分子水平的静电学(英文)	2021—08	68.00	1342
非线性波:理论、计算机模拟、实验(英文)	2021—06	108.00	1343
石墨烯光学:经典问题的电解解决方案(英文)	2021—06	68.00	1344
超材料多元宇宙(英文)	2021—07	68.00	1345
银河系外的天体物理学(英文)	2021—07	68.00	1346
原子物理学(英文)	2021—07	68.00	1347
将光打结:将拓扑学应用于光学(英文)	2021—07	68.00	1348
电磁学:问题与解法(英文)	2021—07	88.00	1364
海浪的原理:介绍量子力学的技巧与应用(英文)	2021—07	108.00	1365
多孔介质中的流体:输运与相变(英文)	2021—07	68.00	1372
洛伦兹群的物理学(英文)	2021—08	68.00	1373
物理导论的数学方法和解决方法手册(英文)	2021—08	68.00	1374
非线性波数学物理学入门(英文)	2021—08	88.00	1376
波:基本原理和动力学(英文)	2021—07	68.00	1377
光电子量子计量学. 第1卷,基础(英文)	2021—07	88.00	1383
光电子量子计量学. 第2卷,应用与进展(英文)	2021—07	68.00	1384
复杂流的格子玻尔兹曼建模的工程应用(英文)	2021—08	68.00	1393

刘培杰数学工作室
已出版(即将出版)图书目录——原版影印

书　　名	出版时间	定　价	编号
电偶极矩挑战(英文)	2021—08	108.00	1394
电动力学:问题与解法(英文)	2021—09	68.00	1395
自由电子激光的经典理论(英文)	2021—08	68.00	1397
曼哈顿计划——核武器物理学简介(英文)	2021—09	68.00	1401
粒子物理学(英文)	2021—09	68.00	1402
引力场中的量子信息(英文)	2021—09	128.00	1403
器件物理学的基本经典力学(英文)	2021—09	68.00	1404
等离子体物理及其空间应用导论. 第1卷,基本原理和初步过程(英文)	2021—09	68.00	1405
拓扑与超弦理论焦点问题(英文)	2021—07	58.00	1349
应用数学:理论、方法与实践(英文)	2021—07	78.00	1350
非线性特征值问题:牛顿型方法与非线性瑞利函数(英文)	2021—07	58.00	1351
广义膨胀和齐性:利用齐性构造齐次系统的李雅普诺夫函数和控制律(英文)	2021—06	48.00	1352
解析数论焦点问题(英文)	2021—07	58.00	1353
随机微分方程:动态系统方法(英文)	2021—07	58.00	1354
经典力学与微分几何(英文)	2021—07	58.00	1355
负定相交形式流形上的瞬子模空间几何(英文)	2021—07	68.00	1356
广义卡塔兰轨道分析:广义卡塔兰轨道计算数字的方法(英文)	2021—07	48.00	1367
洛伦兹方法的变分:二维与三维洛伦兹方法(英文)	2021—08	38.00	1378
几何、分析和数论精编(英文)	2021—08	68.00	1380
从一个新角度看数论:通过遗传方法引入现实的概念(英文)	2021—07	58.00	1387
动力系统:短期课程(英文)	2021—08	68.00	1382
几何路径:理论与实践(英文)	2021—08	48.00	1385
论天体力学中某些问题的不可积性(英文)	2021—07	88.00	1396
广义斐波那契数列及其性质(英文)	2021—08	38.00	1386
对称函数和麦克唐纳多项式:余代数结构与 Kawanaka 恒等式(英文)	2021—09	38.00	1400
杰弗里·英格拉姆·泰勒科学论文集:第1卷. 固体力学(英文)	2021—05	78.00	1360
杰弗里·英格拉姆·泰勒科学论文集:第2卷. 气象学、海洋学和湍流(英文)	2021—05	68.00	1361
杰弗里·英格拉姆·泰勒科学论文集:第3卷. 空气动力学以及落弹数和爆炸的力学(英文)	2021—05	68.00	1362
杰弗里·英格拉姆·泰勒科学论文集:第4卷. 有关流体力学(英文)	2021—05	58.00	1363

刘培杰数学工作室
已出版(即将出版)图书目录——原版影印

书　　名	出版时间	定　价	编号
非局域泛函演化方程:积分与分数阶(英文)	2021—08	48.00	1390
理论工作者的高等微分几何:纤维丛、射流流形和拉格朗日理论(英文)	2021—08	68.00	1391
半线性退化椭圆微分方程:局部定理与整体定理(英文)	2021—07	48.00	1392
非交换几何、规范理论和重整化:一般简介与非交换量子场论的重整化(英文)	2021—09	78.00	1406
数论论文集:拉普拉斯变换和带有数论系数的幂级数(俄文)	2021—09	48.00	1407
挠理论专题:相对极大值,单射与扩充模(英文)	2021—09	88.00	1410
强正则图与欧几里得若尔当代数:非通常关系中的启示(英文)	2021—10	48.00	1411
拉格朗日几何和哈密顿几何:力学的应用(英文)	2021—10	48.00	1412
时滞微分方程与差分方程的振动理论:二阶与三阶(英文)	2021—10	98.00	1417
卷积结构与几何函数理论:用以研究特定几何函数理论方向的分数阶微积分算子与卷积结构(英文)	2021—10	48.00	1418
经典数学物理的历史发展(英文)	2021—10	78.00	1419
扩展线性丢番图问题(英文)	2021—10	38.00	1420
一类混沌动力系统的分歧分析与控制:分歧分析与控制(英文)	2021—11	38.00	1421
伽利略空间和伪伽利略空间中一些特殊曲线的几何性质(英文)	2022—01	68.00	1422
一阶偏微分方程:哈密尔顿—雅可比理论(英文)	2021—11	48.00	1424
各向异性黎曼多面体的反问题:分段光滑的各向异性黎曼多面体反边界谱问题:唯一性(英文)	2021—11	38.00	1425
项目反应理论手册.第一卷,模型(英文)	2021—11	138.00	1431
项目反应理论手册.第二卷,统计工具(英文)	2021—11	118.00	1432
项目反应理论手册.第三卷,应用(英文)	2021—11	138.00	1433
二次无理数:经典数论入门(英文)	2022—05	138.00	1434
数,形与对称性:数论,几何和群论导论(英文)	2022—05	128.00	1435
有限域手册(英文)	2021—11	178.00	1436
计算数论(英文)	2021—11	148.00	1437
拟群与其表示简介(英文)	2021—11	88.00	1438
数论与密码学导论:第二版(英文)	2022—01	148.00	1423

刘培杰数学工作室

已出版(即将出版)图书目录——原版影印

书　　名	出版时间	定　价	编号
几何分析中的柯西变换与黎兹变换:解析调和容量和李普希兹调和容量、变化和振荡以及一致可求长性(英文)	2021—12	38.00	1465
近似不动点定理及其应用(英文)	2022—05	28.00	1466
局部域的相关内容解析:对局部域的扩展及其伽罗瓦群的研究(英文)	2022—01	38.00	1467
反问题的二进制恢复方法(英文)	2022—03	28.00	1468
对几何函数中某些类的各个方面的研究:复变量理论(英文)	2022—01	38.00	1469
覆盖、对应和非交换几何(英文)	2022—01	28.00	1470
最优控制理论中的随机线性调节器问题:随机最优线性调节器问题(英文)	2022—01	38.00	1473
正交分解法:涡流流体动力学应用的正交分解法(英文)	2022—01	38.00	1475
芬斯勒几何的某些问题(英文)	2022—03	38.00	1476
受限三体问题(英文)	2022—05	38.00	1477
利用马利亚万微积分进行 Greeks 的计算:连续过程、跳跃过程中的马利亚万微积分和金融领域中的 Greeks(英文)	2022—05	48.00	1478
经典分析和泛函分析的应用:分析学的应用(英文)	2022—03	38.00	1479
特殊芬斯勒空间的探究(英文)	2022—03	48.00	1480
某些图形的施泰纳距离的细谷多项式:细谷多项式与图的维纳指数(英文)	2022—05	38.00	1481
图论问题的遗传算法:在新鲜与模糊的环境中(英文)	2022—05	48.00	1482
多项式映射的渐近簇(英文)	2022—05	38.00	1483
一维系统中的混沌:符号动力学,映射序列,一致收敛和沙可夫斯基定理(英文)	2022—05	38.00	1509
多维边界层流动与传热分析:粘性流体流动的数学建模与分析(英文)	2022—05	38.00	1510
演绎理论物理学的原理:一种基于量子力学波函数的逐次置信估计的一般理论的提议(英文)	2022—05	38.00	1511
R^2 和 R^3 中的仿射弹性曲线:概念和方法(英文)	2022—08	38.00	1512
算术数列中除数函数的分布:基本内容、调查、方法、第二矩、新结果(英文)	2022—05	28.00	1513
抛物型狄拉克算子和薛定谔方程:不定常薛定谔方程的抛物型狄拉克算子及其应用(英文)	2022—07	28.00	1514
黎曼-希尔伯特问题与量子场论:可积重正化、戴森-施温格方程(英文)	2022—08	38.00	1515
代数结构和几何结构的形变理论(英文)	2022—08	48.00	1516
概率结构和模糊结构上的不动点:概率结构和直觉模糊度量空间的不动点定理(英文)	2022—08	38.00	1517

刘培杰数学工作室
已出版(即将出版)图书目录——原版影印

书　名	出版时间	定　价	编号
反若尔当对:简单反若尔当对的自同构(英文)	2022－07	28.00	1533
对某些黎曼一芬斯勒空间变换的研究:芬斯勒几何中的某些变换(英文)	2022－07	38.00	1534
内诣零流形映射的尼尔森数的阿诺索夫关系(英文)	即将出版		1535
与广义积分变换有关的分数次演算:对分数次演算的研究(英文)	即将出版		1536
强子的芬斯勒几何和吕拉几何(宇宙学方面):强子结构的芬斯勒几何和吕拉几何(拓扑缺陷)(英文)	2022－08	38.00	1537
一种基于混沌的非线性最优化问题:作业调度问题(英文)	即将出版		1538
广义概率论发展前景:关于趣味数学与置信函数实际应用的一些原创观点(英文)	即将出版		1539
纽结与物理学:第二版(英文)	2022－09	118.00	1547
正交多项式和q－级数的前沿(英文)	2022－09	98.00	1548
算子理论问题集(英文)	即将出版		1549
抽象代数:群、环与域的应用导论:第二版(英文)	即将出版		1550
菲尔兹奖得主演讲集:第三版(英文)	即将出版		1551
多元实函数教程(英文)	2022－09	118.00	1552
球面空间形式群的几何学:第二版(英文)	2022－09	98.00	1566

联系地址:哈尔滨市南岗区复华四道街10号　哈尔滨工业大学出版社刘培杰数学工作室
网　　址:http://lpj.hit.edu.cn/
邮　　编:150006
联系电话:0451－86281378　　13904613167
E-mail:lpj1378@163.com